Handbook of Ecological Restoration

The two volumes of this handbook provide a comprehensive account of the rapidly emerging and vibrant science of the ecological restoration of both habitats and species.

Habitat restoration aims to achieve complete structural and functional, self-maintaining biological integrity following disturbance. In practice, any theoretical model is modified by a number of economic, social and ecological constraints. Consequently, material that might be considered as rehabilitation, enhancement, reconstruction, or re-creation is also included. Re-establishment and maintenance of viable, self-sustaining wild populations are the aims of species-centred restoration.

Restoration in Practice details the state of the science in a range of biomes within terrestrial and aquatic (marine, coastal and freshwater) ecosystems. Policy and legislative issues on all continents are also outlined and discussed. The accompanying volume, *Principles of Restoration*, defines the underlying principles of restoration ecology in relation to manipulations and management of the biological, geophysical and chemical framework.

The Handbook of Ecological Restoration will be an invaluable resource to anyone concerned with the restoration, rehabilitation, enhancement or creation of habitats in aquatic or terrestrial systems, throughout the world.

MARTIN PERROW is an ecological consultant at ECON, an organisation which he founded in 1990 to bridge the gap between consultancy and research. He specialises in the restoration and rehabilitation of aquatic habitats, and is a leading exponent of biomanipulation. In addition to presenting his work to the scientific community, Martin has endeavoured to communicate his findings to the general public, through appearances on television and radio and contributions to newspapers and magazines. Martin has travelled extensively, leading eco-tours, and is an award-winning natural history photographer.

ANTHONY (TONY) DAVY is Head of Population and Conservation Biology in the School of Biological Sciences at the University of East Anglia, where he has taught and researched a wide variety of topics in ecology and plant biology. His research interests include genetic variation and the evolutionary and physiological responses of plants to their environments.

Tony is the Executive Editor of the *Journal of Ecology*, Associate Editor of the *Biological Flora of the British Isles*, and was the Honorary Meetings Secretary of the British Ecological Society for a number of years.

Handbook of Ecological Restoration

Volume 2
Restoration in Practice

Edited by
Martin R. Perrow
ECON
University of East Anglia
and
Anthony J. Davy
University of East Anglia

CAMBRIDGE
UNIVERSITY PRESS

CAMBRIDGE UNIVERSITY PRESS
Cambridge, New York, Melbourne, Madrid, Cape Town, Singapore, São Paulo

Cambridge University Press
The Edinburgh Building, Cambridge CB2 8RU, UK

Published in the United States of America by Cambridge University Press, New York

www.cambridge.org
Information on this title: www.cambridge.org/9780521791298

First published 2002
Reprinted 2004
This digitally printed version (with corrections) 2008

A catalogue record for this publication is available from the British Library

Library of Congress Cataloguing in Publication data

Handbook of ecological restoration / [edited by] Martin R. Perrow, Anthony J. Davy.
    p.   cm.
Includes bibliographical references.
Contents: [1] Principles of restoration – [2] Restoration in practice.
ISBN 0 521 79129 4
1. Restoration ecology.   I. Perrow, Martin R. (Martin Richard), 1964–   II. Davy, A. J.
QH541.15 .R45 H36   2002
333.95´153 – dc21   2001043443

ISBN 978-0-521-79129-8 hardback
ISBN 978-0-521-04775-3 paperback

This book is dedicated to Morgan, Rowan and Bryony, in the hope that theirs and future generations are able to enjoy and live in harmony with all life on earth

Contents

Contributors

Paul Adam
School of Biological Sciences
University of New South Wales,
Australia

William M. Adams
Department of Geography
University of Cambridge
Cambridge CB2 3EN, UK

Janette R. Allen
Port Erin Marine Laboratory
University of Liverpool
Port Erin
Isle of Man IM9 6JA, UK

James Aronson
Centre d'Ecologie Fonctionelle et Evolutive
CNRS (UPR 9056)
34293 Montpellier, France

Robin F. Bay
Department of Biology
Colorado College
Colorado Springs Co 80903, USA

Anthony D. Bradshaw
School of Biological Sciences
University of Liverpool
Liverpool, UK

Peter Buckley
Imperial College at Wye
University of London
Wye
Ashford TN25 5AH, UK

M. K. Chakraborty
Central Mining Research Institute (CMRI),
Barwa Road,
Dhanbad 826001, India

Jeanne C. Chambers
Rocky Mountain Research Station,
USDA Forest Service,

University of Nevada,
920 Valley Road, Reno NV 89512, USA

Susan Clark
Department of Marine Sciences and
   Coastal Management
University of Newcastle
Newcastle upon Tyne NE1 7RU, UK

Alison M. Cochrane
10 Valentine Avenue
Parramatta
NSW 2150, Australia

David N. Cole
Aldo Leopold Wilderness Research Institute
Rocky Mountain Research Station
USDA Forest Service Missoula MT 59801, USA

David K. Conlin
Department of Biology
Colorado College
Colorado Springs CO 80903, USA

B. B. Dhar
Association of Indian Universities
New Delhi 110 002, India

Peter W. Downs
PWA Ltd
Corte Madera CA 94925, USA
and
School of Geography
University of Nottingham
Nottingham NG7 2RD, UK

James J. Ebersole
Department of Biology
Colorado College
Colorado Springs CO 80903, USA

Fatih Evrendilek
Department of Landscape Architecture
Mustafa Kemal University
31040 Anatakya
Hatay, Turkey

Stephanie Fluke
National Oceanic and Atmospheric
    Administration
Office of General Counsel for Natural Resources
St Petersburg FL 33702, USA

Mark S. Fonseca
National Oceanic and Atmospheric Administration
National Ocean Service
Center for Coastal Fisheries and Habitat Research
Beaufort NC 28516, USA

Bruce C. Forbes
Arctic Centre
University of Lapland
FIN-96101 Rovianemi, Finland

Martin J. Genner
Biology and Ecology Division
School of Biological Sciences
University of Southampton
Southampton SO16 7PX, UK

Sigurdur Greipsson
Department of Biological and Environmental
    Sciences
Troy State University
Troy 36082, USA

Ian D. Hannam
10 Valentine Avenue
Parramatta
NSW 2150, Australia

Stephen J. Hawkins
Marine Biological Association UK
Citadel Hill
Plymouth PL1 2PB, UK
and
Biology and Ecology Division
School of Biological Sciences
University of Southampton
Southampton SO16 7PX, UK

Karen D. Holl
Environmental Studies Department
University of California–Santa Cruz
Santa Cruz CA 95064, USA

Michael J. Hutchings
Department of Biology
University of Sussex
Brighton BN1 9QJ, UK

Satoshi Ito
Forest Science Division
University of Miyazaki
Miyazaki 889-2192, Japan

Daniel H. Janzen
Department of Biology
University of Pennsylvania
Philadelphia PA 19104, USA

Erik Jeppeson
National Environmental Research Institute
DK-8600 Silkeborg, Denmark

Tim A. Jones
DJEnvironmental
Harper's Mill
Berrynarbor
North Devon EX34 9TB, UK

Brian E. Julius
National Oceanic and Atmospheric Administration
National Ocean Service
Damage Assessment Center
Silver Spring MD 20910, USA

W. Judson Kenworthy
National Oceanic and Atmospheric Administration
National Ocean Service
Center for Coastal Fisheries and Habitat Research
Beaufort NC 28516, USA

G. Matt Kondolf
Department of Landscape Architecture and
Environmental Planning
University of California–Berkeley
Berkeley CA 94720, USA

Edouard Le Floc'h
Centre d'Ecologie Fonctionelle et Evolutive
CNRS (UPR 9056)
34293 Montpellier, France

John A. Ludwig
Division of Wildlife and Ecology
CSIRO
Winnellie
Darwin
NT 0822, Australia

F. Jane Madgwick
European Freshwater Programme
c/o WWF Denmark
DK-2200 Copenhagen N, Denmark

Jay D. McKendrick
PO Box 902
Palmer AK 99645, USA

Stéphane McLachlan
Environmental Science Program
University of Manitoba
Winnipeg R3T 2N2, Canada

Russ P. Money
Wetland Research Group
Department of Animal and Plant
Science
University of Sheffield
Sheffield S10 2TN, UK

Carlos Ovalle
CRI Quilamapu
Instituto de Investigaciones
   Agropecuarias (INIA)
Casilla 426
Chillán, Chile

Pauline M. Ross
Marine Biological Association UK
Citadel Hill
Plymouth PL1 2PB, UK
and
Faculty of Science and Technology
University of Western Sydney
Richmond
NSW 2753, Australia

Nirander M. Safaya
Reclamation Division
North Dakota Public Service Commission
Bismarck ND 58505, USA

Ilkka Sammalkorpi
Watercourse Planning and Restoration
Finnish Environment Institute
FIN-00251 Helsinki, Finland

Sue C. Shaw
Wetland Research Group
Department of Animal and Plant Science
University of Sheffield
Sheffield S10 2TN, UK

Sharon Shutler
National Oceanic and Atmospheric
   Administration
Office of General Counsel for Natural
   Resources
Silver Spring MD 20910, USA

Kevin S. Skinner
School of Geography
University of Nottingham
Nottingham NG7 2RD, UK

Alan J. A. Stewart
Department of Biology
University of Sussex
Brighton BN1 9QJ, UK

David J. Tongway
Gungahlin Homestead
CSIRO Sustainable Ecosystems
GPO Box 284
Canberra
ACT, Australia

Krystyna M. Urbanska
Geobotanisches Institut ETH
CH-8044 Zurich, Switzerland

Mohan K. Wali
OSU Environmental Science Graduate Program
and
School of Natural Resources
The Ohio State University
Columbus OH 43210, USA

Clive A. Walmsley
Countryside Council for Wales
Fford Penrhos
Bangor LL57 2LQ, UK

Nigel R. Webb
NERC Centre for Ecology and Hydrology
Winfrith Technology Centre
Dorchester DT2 8ZD, UK

Bryan D. Wheeler
Wetland Research Group
Department of Animal and Plant Science
University of Sheffield
Sheffield S10 2TN, UK

Scott D. Wilson
Department of Biology
University of Regina
Regina
Saskatchewan S4S 0A2, Canada

Ming H. Wong
Institute for Natural Resources and
    Environmental Management
and
Department of Biology
Hong Kong Baptist University
Hong Kong SAR
People's Republic of China

Joy B. Zedler
Botany Department and Arboretum
University of Wisconsin–Madison
Madison WI 53706, USA

Foreword

Ecosystem disturbance has been defined as an event or a series of events that changes the relationships between organisms and their habitats from their natural states, both spatially and temporally.[1] These changes may range from small to large, temporary to permanent, and slight to severe. The causative agents that lead to ecosystem degradation are many, but a recent United Nations report attributed the bulk of disturbance worldwide to five primary causes (with percentage of disturbance for each): overgrazing (35), deforestation (29), agricultural activity (28), overexploited vegetation (7), and industrial activity (1).[2] Acting in concert with extensive habitat loss (see below), the worldwide loss of biological diversity is under way.

The causative agents outlined function through at least seven major activities of disturbance: extensive clearing of natural vegetation for agriculture, abandonment of unproductive land, selective harvesting of desirable species and introduction of alien ones, mining, draining of wetlands, introduction of chemicals in the environment, and the impact of war.[3] Other activities, such as extensive urban expansion, can also be added to this list. The scales of these disturbances are often seen as local or regional, but, as Harlan Cleveland noted, 'Private decisions are partly public; domestic affairs are partly international; local issues are partly global.'[4]

Though estimates abound, the exact extent of land that is now disturbed or degraded by human activities is not known. Nonetheless, such estimates make it clear that ecosystem disturbance is a widespread phenomenon. Ecologists estimate that between one-third and one-half of the earth's land surface has been transformed by human action.[5] Recent studies calculate global ecosystem disturbance at well over 5000 million ha,[6] with nearly 550 000 ha degraded per day (an area larger than New York City), and approximately 200 million ha per year (an area larger than Poland).[7] Such staggering figures are corroborated by scientists separately studying agricultural, dryland and tropical ecosystems. Regardless of the exact figure for the area of disturbance, there is little doubt that many productive ecosystems are already seriously degraded.

The impact of human-induced global change looms large on the environmental horizon, which may make ecosystem restoration even more difficult than it is now. Most experts agree that climate change is occurring at a pace unprecedented in geological history. Changes, both in the physical nature of the earth (such as temperature regimes and melting of glaciers) and in its biota (such as changes in the phenology of animals and plants and the loss of coral reefs), provide compelling evidence that temperature changes afoot will have detrimental, unpredictable and unintended consequences. The effects of changing climate on the ecological amplitudes of biotic populations and species along environmental gradients will be crucial for restoration.

During the past three decades or so, a vast body of impressive scientific investigation has addressed the problems, practices and strategies for the rehabilitation and restoration of ecosystems from diverse regions of the world. Among these, ecological studies of drastic disturbances (such as mining) stand out. From diverse ecosystems, these have included such investigations as the nature of newly excavated geological materials and their weathering, lack of nutrients, mobilisation of toxic elements, patterns and processes of vegetation colonisation of new habitats, the diversity and stability of newly formed ecosystems, ecological succession, and ecosystem modelling. Our knowledge base has advanced enough, I believe, that we should be able to make predictions, with reasonable accuracy, on the course of establishing biological productivity for most ecosystems in the world.

However, our success in conveying the results of these investigations to policy makers, never so critical as now, has been much poorer than desirable. Granted that the scientists and decision-makers operate in disparate settings with vastly different traditions and *modus operandi*, both must work together more effectively. For while finding solutions to problems lies within the purview of scientists, mandating applications from the findings of science lies with the policy-makers and legislators; thus, science becomes a social enterprise. The art of true governing and leadership, in my view, rests on three basic attributes: communication, persuasion, and action. Active participation in all three on the part of the scientific community – ecologists in particular – is crucial in making such public policy.

Although gaps in ecological knowledge exist, these should not be viewed as a serious limitation to the development and implementation of appropriate restoration, rehabilitation, or reclamation strategies. Indeed, a new applied field of science – Restoration Ecology – has already emerged. It endeavors to examine the nature and extent of environmental disturbances and provide solutions for both preventing and mitigating the harmful effects of human activities on land and water resources.

This Handbook documents the state-of-the-science of restoration ecology. It presents the underlying *Principles of Restoration* (Volume 1) and in this Volume 2, *Restoration in Practice*, pointing to the direction of future efforts. Besides bearing testimony to the hard work and unswerving persistence of their editors, the volumes may prove to be a landmark with which future efforts are judged and, more importantly, may enhance the efficacy and speed of restoration.

References

1 Wali, M. K. (1987). The structure, dynamics, and rehabilitation of drastically disturbed ecosystems. In *Perspectives in Environmental Management*, ed. T. N. Khoshoo, pp. 163–183. New Delhi: Oxford and IBH Publishing.

2 United Nations (2000). *Population and Land Degradation*. United Nations Population Information Network (POPIN), Population Division, Department of Economic and Social Affairs. Available at http://www.undp.org/popin/fao/aland/land.html

3 Bazzaz, F. A. (1983). Characteristics of populations in relation to disturbance in natural and man-modified ecosystems. In *Disturbance and Ecosystems: Components of Response*, eds. H. A. Mooney and M. Godron, pp. 259–275. New York: Springer-Verlag.

4 Cleveland, H. (1981). Foreword: The extrapolation of metaphors. In *Ominous Trends and Valid Hopes*, ed. M. C. McHale, pp. i–iii. Minneapolis, MN: Hubert H. Humphrey Institute of Public Affairs, University of Minnesota.

5 Vitousek, P. M., Mooney, H. A., Lubchenco, J. & Mellilo, J. M. (1997). Human domination of earth's ecosystems. *Science*, **277**, 494–499.

6 Gibbs, H. K. (2001). Quantification of human-induced changes in global vegetation and associated climatic parameters. MS. thesis, The Ohio State University, Columbus, OH.

7 D'Aleo, J. (2000). Expanding deserts are the latest signs of a changing climate. *Weather Services International Corporation Bulletin*, 15 November 2000.

Mohan K. Wali
Columbus, Ohio

Preface

Few would deny the need for ecological restoration in a world that sees the continued impoverishment of our natural resources and habitats, coupled with the extinction of our fellow life-forms: earth's biodiversity. Inevitably, we are all driven by short-term goals in our daily lives, irrespective of whether these are for basic needs such as food, water and shelter or whether they are something rather more materialistic in nature. Particularly in more developed societies, however, we have little excuse for forgetting our responsibilities towards the planet and its other inhabitants. Despite the admirable efforts of the many environmental and ecological activists who seek to secure the future of the planet's habitats and biodiversity, and the advances made by recent international agreements, notably at the Earth Summit in Rio in 1992, future generations still face the prospect of ecological impoverishment.

Periodic crises, such as wars and other conflicts, and unwillingness to deviate from selfish short-term goals, perhaps for economic or political gain (for example the US avoidance of the Kyoto protocol) should not be allowed to obscure our vision. When crises have past and policies have been reversed, even though many things may have changed, the longer-term ecological problems and trends will remain to be faced. Scientists could fairly be accused of not doing enough. Thorough scientific investigation of complex issues rarely proceeds fast enough for the needs of practitioners faced with immediate problems. Nor have researchers always been as effective as they could have been at communicating the practical value of their findings. A number of recent events have highlighted the need for scientists to earn and justify the trust of society at large. However, restoration ecology has broken away from its early and ultimately sterile preoccupation with definitions of change and 'original' state. We recognise that change is continual and an integral part of the backdrop of our activities. The realities of global climate change and the increasing intervention in the genetic make-up of crops, amongst many other things, ensure that we cannot go back to some fixed, utopian state. Our efforts will be best directed forwards, armed with scientific knowledge of the interactions between taxa, trophic levels and the wider environment, a holistic understanding of what has happened in the past and a keen desire for improvement.

Perhaps for the first time an approach or philosophy has emerged that cuts across the disciplines, uniting peoples from a wide variety of backgrounds – a look at the delegate list of any Society of Ecological Restoration (SER) conference will verify this – under a common banner of trying to make things better. Restoration ecology embodies that movement and SER is doing a fine job in promoting the message.

This handbook, organised in two volumes, aims to produce a comprehensive account of the burgeoning and vibrant science of restoration ecology. Although it largely documents science, the book is intended to present the science in practical terms for those who may describe themselves primarily as practitioners, engineers or conservationists and includes a wealth of information on planning and legislative tools for planners or managers. Rather presumptuously, we have sought to inspire all concerned with restoration. Even the most hardened reductionists – the 'watchmakers' or more accurately 'watchdismantlers' – must have been inspired by Tony Bradshaw's words: 'the acid test of our understanding is not whether we can take ecosystems to bits on pieces of paper, however scientificially, but whether we can put them together in practice and make them work'.

Practice is the essence of this second volume. The first part, Part 6 in the work as a whole, outlines restoration policy and infrastructure across continents, using the Americas, Europe, Africa, Asia and Oceania as divisions. Cultural and political attitudes to restoration, the institutional and legal framework for restoration and application (successful or not) of such a framework to the restoration of particular biomes are the central themes discussed. The breadth of this subject matter could easily warrant a book on each region with contributions from many authors. Expecting a complete treatise on each aspect from a single or a few authors, given the constraints on space, is clearly inappropriate. Moreover, the resource base for applying policy and legislative instruments is highly variable in different parts of the world, as is the underpinning cultural attitude. At the very least, an increased understanding of how restoration may be applied in different regions has emerged. The illustration of the strengths and weaknesses of the different political and legislative frameworks in which restoration is undertaken has produced a fascinating mix of principles and potential models that could be applied in a wealth of circumstances.

The second Part of the volume (Part 7 of the work) outlines state-of-the science practice in the restoration of a broad range of biomes in 19 chapters. These inevitably focus on systems in which practical restoration and associated research have been most active. The first series of six chapters deals with systems in or associated with the sea ('Marine and coastal ecosystems'; 'Seagrasses'; 'Coral reefs'; 'Beaches'; 'Coastal dunes'; and 'Saltmarshes'). The aquatic motif continues with chapters on 'Rivers and streams', 'Lakes' and 'Freshwater wetlands'. The main contributions on lake restoration have been on shallow lakes and the chapter reflects this. Cold biomes are represented by chapters on 'Polar tundra' and 'High-elevation ecosystems'. Another series of chapters covers terrestrial biomes more or less limited by water supply and grazing ('Atlantic heathlands'; 'Calcareous grasslands'; 'Prairies'; 'Semi-arid woodlands and desert fringes'; and the particularly distinctive 'Australian semi-arid lands and savannas'). The final three chapters tackle the special challenges of restoring forest systems ('Temperate woodlands'; 'Tropical moist forest'; and 'Tropical dry forest').

The themes of each chapter are essentially the same with: (i) the distinctive features and processes of the biome concerned outlined in an introduction; (ii) the rationale for restoration determined by the interplay between the value (in ecological, sociological and economic terms) of the biome and the nature and extent of previous degradation; (iii) description and application of the principle strategies, techniques and tools for restoration, amply illustrated by case studies; and (iv) concluding remarks outlining what has worked or not worked and the direction of future efforts. The only notable deviation from this structure is the final chapter where the long-running and important restoration of Area de Conservación Guanacaste, a tropical dry forest in northwestern Costa Rica is documented. Differences in the nature and extent of the problem, the level of understanding of ecosystem functioning and the nature of the works required mean there are, inevitably, differences in the balance between the chapters. Constraining our authors in a structural strait-jacket would have been inappropriate.

What has also become clear is the difference in the depth, efficacy and predictability and style of restoration practice. In some habitats, restoration may be focused upon physical manipulation, removing constraints and encouraging natural recovery processes. Utilising flow and sediment dynamics in the restoration of rivers, vegetation succession in the restoration of forests and colonisation of invertebrates in the restoration of coral reefs all spring to mind. Other restorations require the prevention of similar processes, for example setting back or restriction of succession in the restoration of heathland. The manipulation of the biota within food webs may also be necessary. Nowhere is this more advanced than in lake ecosystems, where it is centred upon the biomanipulation of fish communities to initiate cascading trophic responses. The state-of-the-science is clearly far more advanced in some systems than others, with consequences for predictability of outcome that ultimately affect opportunities for restoration, because those holding the purse-strings like some guarantee of success. We hope that readers will take heart from experiences in other systems outside of their own and that there will be cross-fertilisation of ideas, principles and techniques.

Recording experiences in all possible habitats was clearly beyond the scope of even such a substantial handbook. We have striven to be as complete as possible and to make a statement about current knowledge and practice. We have also aimed to point the way forward, so that restoration can continue to learn from mistakes born of a lack of understanding. Consequently, we have tried to incorporate a range of habitats, concentrating on broad types, but including more specific, limited – in geographical terms – habitats where considerable effort has been made (e.g. calcareous grassland and Atlantic heathland). We hope that readers who do not find an account of the system they are concerned with will find relevant experiences in other systems, and be able to apply them.

Our contributors provide the strengths in the book. Any weaknesses in its structure and scope are ours. We are deeply indebted to all those that have contributed their time and efforts in what has proved to be a massive undertaking. It has been an honour to have Tony Bradshaw, John Cairns Jr. and Erik Jeppesen on the editorial board and we are grateful for their unfaltering support throughout the project's long history. We are grateful to CUP for taking up the challenge when we needed a new publisher, and particularly to Alan Crowden, who made the transition seamless. Shana Coates provided editorial assistance from the press. We are also grateful for the skilled copy-editing of Anna Hodson and for the essential support and hard work of Mark Tomlinson of ECON in the preparation of manuscripts. We thank the reviewers, many of whom were also authors, for generously sharing their insights and suggesting improvements to chapters This book was born out of what was a difficult time in both our lives; perhaps nothing worthwhile is ever easy. We hope it will inspire current and future restorers.

Martin Perrow and Tony Davy

Part 1 • Restoration policy and infrastructure

1 • The Americas: with special reference to the United States of America

MOHAN K. WALI, NIRANDER M. SAFAYA AND FATIH EVRENDILEK

INTRODUCTION

The burgeoning impacts of human activity on the local, regional and global environments have created a lot of public concern and debate, especially since the latter part of the last century. Environmental protection, conservation of natural resources and restoration of the affected ecosystems have become pressing issues of national and international significance. Scientists and policy-makers around the globe are currently grappling with these issues. In the Americas, and particularly in the United States of America, this call has become quite urgent because of the fast-paced, large-scale exploitation of natural resources, which often leaves behind a trail of environmental and ecological disturbance and degradation. Consequently, such terms as restoration, rehabilitation and reclamation are now commonly used in the scientific and non-scientific literature to describe those practices that help re-establish the structural and functional characteristics of a disturbed ecosystem to its natural or near-natural state.

Transformation of the natural ecosystems of the North and South American continents began with native Americans and has increased considerably since the days of European settlement. As the population of the Americas grew, housing and road construction followed; industrial and agricultural activities expanded and multiplied; mining for minerals, coal and other materials became essential; and the disposal of waste products became an equally inevitable necessity. The ecological impacts of such activities vary widely in their scope and significance. For example, large-scale deforestation, extensive sod breaking for cultivation of monoculture crops, and intensive grazing may result in unsustainable conversion of land use, extirpation of some plant and animal species, soil erosion, and weather modification. Mining scars the earth rendering it temporarily or permanently unproductive, and may cause serious surface and/or groundwater problems. Likewise, enormous networks of highways and urban sprawl take away land irreversibly from other natural uses and capabilities. In most case the impacts of such activities manifest themselves locally or regionally, but in some cases the impacts may be global in their significance, such as ozone depletion and global warming.

In the past three decades, especially since the 1972 United Nations Conference on Human Environment held in Stockholm, numerous scientific studies have focused on a wide array of problems and practices related to ecological restoration. These studies have produced a vast body of information on issues such as: loss of topsoil and organic matter; overgrazing of grasslands; deforestation; desertification; endangerment and loss of plant and animal species; and impacts of intensive agronomic practices, mining and other industrial activities on land and water resources. Most of these investigations employ a multidisciplinary approach in analysing as well as solving the problems. However, restoration of ecosystems disturbed by massive earth-moving operations, such as surface mining, calls for a high level of integration among a number of basic/applied sciences, engineering and economics. Also, because large-scale disturbances occur in a wide variety of ecosystems, ranging from forest to desert, the ecological and engineering challenges posed and the kind of technical expertise required are far greater than those required in most other cases of

Table 1.1. *Human population (actual and projected) in the Americas* ($\times 10^3$)

Region	1998	2025	2050	Annual growth rate 1995–2000 (%)	Population density per km² mid-1998
Northern America[a]	304,716	363,612	391,781	0.8	14
Central America[b]	130,457	188,504	222,502	1.9	53
Caribbean[c]	37,351	47,287	52,026	1.1	159
South America[d]	335,715	460,866	534,382	1.5	19
World	5,901,054	7,823,703	8,909,095	1.3	44

[a]Northern America includes Bermuda, Canada, Greenland, Saint Pierre and Miquelon, and United States of America.
[b]Central America includes Belize, Costa Rica, El Salvador, Guatemala, Honduras, Mexico, Nicaragua and Panama.
[c]Caribbean includes Anguilla, Antigua and Barbuda, Aruba, Bahamas, Barbados, British Virgin Islands, Cayman Islands, Cuba, Dominica, Dominican Republic, Grenada, Guadeloupe, Haiti, Jamaica, Martinique, Montserrat, Netherlands Antilles, Puerto Rico, Saint Kitts and Nevis, Saint Lucia, Saint Vincent and Grenadines, Trinidad and Tobago, Turks and Caicos Islands, and United States Virgin Islands.
[d]South America includes Argentina, Bolivia, Brazil, Chile, Colombia, Ecuador, Falkland Islands (Malvinas), French Guinea, Guyana, Paraguay, Peru, Suriname, Uruguay and Venezuela.
Source: United Nations (1998).

restoration. Indeed, an impressive body of knowledge has been accumulated for the restoration of drastically disturbed ecosystems.

The need to convey the results of restoration investigations to the law- and policy-makers has never been so paramount as it is now. For, while finding solutions to problems lies within the purview of scientists, mandating the applications of the findings of science lies with the policy-makers and legislators; in the latter context, science becomes a social enterprise. Active participation of the scientific community is crucial in making public policy. This has been occurring in many countries, and a new discipline, environmental law, has emerged with its roots in both ecological/environmental sciences and law. Some of the American nations have a strong infrastructure of laws, rules and policies that mandate or encourage protection of ecosystems from abuse. To appreciate fully the legal requirements and scientific procedures in the Americas that address the issues of ecosystem restoration, it is first necessary briefly to acquaint the reader with the types of natural ecological regions (biomes) that exist in these

two continents, and the type and extent of disturbances to which they are subject.

DIVERSITY OF BIOMES IN THE AMERICAS

For both ecological and economic systems, national boundaries have little meaning when one considers the interactions of regional and global interdependencies. Human populations depend on these systems and their linkages for survival. Consistent with the global trend, the human population of the Americas has been increasing (Table 1.1). Population growth, import and export linkages, and economic and ecological limits determine what is mined, grown and produced in a given region. Restoration and rehabilitation of degraded and destroyed ecosystems in the Americas, therefore, have significant implications for the well-being of communities and economies at both local and global scales. Ecological restoration strategies must be examined at a combination of different spatiotemporal scales: species, community, ecosystem and landscape (seascape). In order to succeed, these

Table 1.2. *Current land cover (km^2) of the Americas as depicted by DISCover*[a]

Land cover classes	Current vegetation	Potential vegetation[b]	Human-induced change[b]
Evergreen forest	10 036 268	14 200 623	−4 164 355
Deciduous forest	957 683	2 748 511	−1 790 828
Mixed forest	3 333 949	4 427 083	−1 093 134
Woody savanna	2 578 185	638 819	+1 939 366
Savanna	1 350 552	4 772 712	+3 422 160
Shrubland	6 724 171	6 286 924	−437 247
Grassland	2 795 509	2 968 184	−172 675
Desert	2 711 947	2 008 580	+703 367
Cropland	2 970 387	0	+2 970 387
Croplands mosaic	4 488 301	0	+4 488 301
Urban and built-up	104 484	0	+104 484
Wetlands	314 998	314 998	NA
Snow and ice	3 918 725	3 918 725	NA
Region total	42 285 159	42 285 159	42 285 159

[a]The Data and Information System data set (DISCover) was initiated by the International Geosphere Biosphere Programme and implemented through collaboration of many agencies (in particular, US Geological Survey Earth Resources Observation System Data Center) because of the need for global land cover data with known classification accuracy (Loveland & Belward, 1997). The 1-km resolution of DISCover captures the heterogeneity missed by the coarser resolution of past remote sensing estimates and is the first to utilize this resolution on a global scale (Loveland et al., 2000). Greenness classes were defined by monthly AVHRR NDVI composites from images taken between April 1992 and March 1993 (Loveland & Belward, 1997, slightly modified by H. Gibbs).
[b]Data on potential vegetation (Mathews, 1983) and human-induced change synthesized by H. Gibbs (pers. comm.).

plans must come from policies that account for socio-economic realities of the regions in which they are implemented.

North and South America are the third and fourth largest continents, together comprising an area of 42 million square kilometres. Extending from the Arctic to the sub-Antarctic latitudes, the Americas support every biome from the rainforests of the Amazon Basin to the arid steppes of Patagonia, with tremendous diversity within each system. Current land use patterns reveal the relative significance of croplands and urban systems in relation to natural systems (see Fig. 1.1; Table 1.2). The United States alone supports a tremendous diversity within its ecosystems, with more than 200 000 native plant and animal species (Stein et al., 2000). South America supports about 800 species of terrestrial mammals (19% of the world total) and an estimated 90 000 known species of flowering plants (more than one-third of the world's total), even though it only comprises 12% of the world's land area (Mares, 1986). The Caribbean supports the second richest region of marine biota in the world (Reid, 1992).

Tundra

The tundra is circumpolar, the northernmost biome, and typically receives less than 60 cm of annual precipitation. At its northern limits, it is a region of cold, lifeless desert, while at its southern limits, it is characterized by small-stature vegetation adapted to constant soil disturbance, strong abrasive winds, infertile soils, and a short growing season (usually not more than 60 days). Permafrost, a zone of subsurface soil that is continuously frozen, controls soil moisture. Low-lying, wet sites support grasses, sedges, dwarf shrubs and sphagnum moss; better-drained sites support dwarf trees. The dominant vertebrates are caribou (*Rangifer tarandus*), lemmings (species of *Lemmus*, *Dicrostonyx*, *Synaptomys*), arctic hares (*Lepus arcticus*) and musk oxen (*Ovibos moschatus*). Alpine tundra, the vegetated area above the tree lines of mountains, has similar features, with the exception of permafrost. One threat to this fragile ecosystem is the major oil development involving the 600 000 ha coastal plain of the 7.7 million ha Arctic National Wildlife Refuge in Alaska, characterized by some as the 'Serengeti of North America'.

Taiga

The taiga, or boreal forest, lies south of the arctic tundra, covering 11% of the earth's terrestrial surface. Winters are cold, but milder than polar regions; the average annual temperature is below 5 °C; rainfall is between 40 and 100 cm annually; soils are poor in nutrients. Boreal forests are dominated by relatively few tree genera, including the conifers species of pine (*Pinus*), spruce (*Picea*), larch (*Larix*) and fir (*Abies*), and the deciduous aspen (*Populus*) and birch (*Betula*). Droughts are frequent and the forests are adapted to fire. Caribou and moose (*Alces alces*) inhabit the taiga, often consuming as much as 50% of young plant growth. Humans, through logging, can also have a significant impact on this biome. Major logging operations are going on in the boreal forests of Canada, and Acharya (1995) reported that these were being 'destroyed' at a rate of more than 2 ha per minute. Schindler (1998) lamented that the North American and European ecologists have been preoccupied with the impacts of disturbance effects in South America while equal or greater impacts on the boreal ecosystems in North America and Europe were ignored.

Temperate rainforest

Large evergreen forests dominate the coastal areas of the Pacific Ocean, where abundant precipitation (200–380 cm per year) and ocean mists allow massive trees to develop. These forests are stratified and

support abundant epiphytes, although the soils are often leached of nutrients.

Temperate deciduous forest

The biome lies south of both the boreal forest and in many cases the subalpine areas. In this region the annual precipitation, in the form of rain or snow, exceeds evaporation. Once dominated by deciduous trees that are leafless during periods of frequent freezing temperatures, these forests have experienced major human disturbance. Depending on the latitude, the growing season ranges from three to nine months. The vertical structure of the forest is relatively simple compared to tropical rainforests, often with just one subcanopy of trees, shrubs and herbs. The forest floor is often covered with small, herbaceous plants, mosses and lichens. This biome has been heavily logged over the past 100 years, and has undergone numerous transformations as human habitation has increased. In addition to major urban development and construction of roadways, significant areas have been subjected to drastic disturbance. Many investigations on restoration have provided a rich body of knowledge on steps necessary for substrate stability and the establishment of a vegetation cover (see for example, Hutnik & Davis, 1973; Cairns *et al.*, 1977; Schaller & Sutton, 1978; Leopold & Wali, 1992; Keddy & Drummond, 1996). The cooperation of industry, private and government agencies and academic scientists is now apparent in a number of projects in this biome (see, for example, Boxes 1.1 and 1.2).

Temperate grasslands

South and west of the temperate deciduous forest biome are the temperate grasslands. Regions where rainfall is too low to support forest, but too high to allow deserts to form (25–80 cm per year) were historically covered by grasslands. Grasslands covered the centre of North America, and the southern tip of South America. In most years, evaporation exceeds precipitation, resulting in cycles of drought and fire. These systems are known in various parts of the world as prairies, steppes, pampas and velds. Large and small grazing animals such as

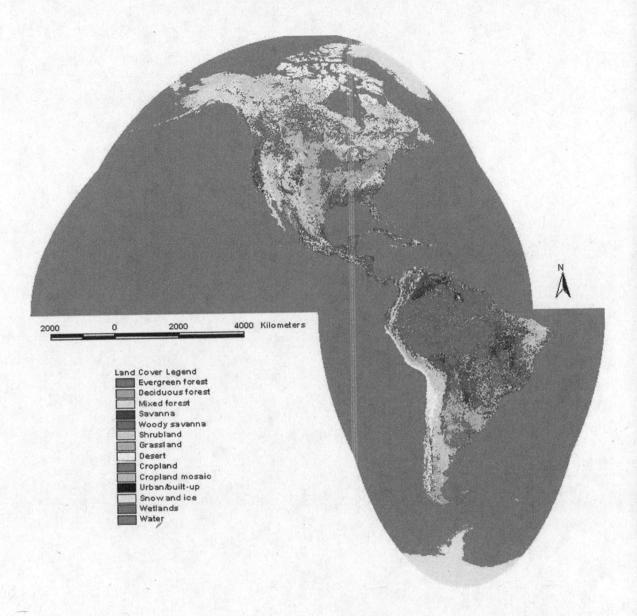

Land Cover Legend
- Evergreen forest
- Deciduous forest
- Mixed forest
- Savanna
- Woody savanna
- Shrubland
- Grassland
- Desert
- Cropland
- Cropland mosaic
- Urban/built-up
- Snow and ice
- Wetlands
- Water

2000   0   2000   4000   Kilometers

N

Fig. 1.1 Biomes of the Americas and their spatial hetero-geneity based on the 1-km resolution of DISCover remote sensing estimates, 1992–3. Greenness classes were defined by monthly AVHRR NDVI composites from images taken between April 1992 and March 1993. Adapted with permission from Loveland & Belward (1997), modified by H. Gibbs.

The original colour version of this map is available for download from www.cambridge.org/9780521047753

Box 1.1   Restoration of mined lands in the temperate deciduous forest biome

THE WILDS, OHIO

Over 3700 ha of abandoned surface mine land in southeastern Ohio was donated by American Electric Power company in the early 1970s for wildlife conservation, scientific studies and education programmes. The abandoned mine site had a rugged terrain with steep ravines and extensive erosion. Initially, seeding/planting was done directly into the spoil material, but later the area was recontoured and topsoil that could be salvaged was respread at the site. The reshaped site was seeded with a mixture of grasses, forbs and legumes. Currently, the vegetation at the site consists of alien species of European alder (*Alnus glutinosa*), black locust (*Robinia pseudoacacia*), autumn olive (*Elaeagnus angustifolia*), ailanthus (*Ailanthus altissima*), some species – probably invading and/or planted – of ailanthus, eastern cottonwood (*Populus deltoides*) and sycamore (*Platanus occidentalis*), and major species found in remnant stands are sugar maple (*Acer saccharum*), northern red oak (*Quercus rubra*), yellow (tulip) poplar (*Liriodendron tulipifera*), American elm (*Ulmus americana*), white ash (*Fraxinus americana*), big-tooth aspen (*Populus grandidentata*), red maple (*Acer rubrum*) and a dense understorey of grasses and other herbaceous species. Although the vegetation for the most part is indigenous or cultivated varieties of the temperate regions, a number of tropical animals (such as giraffes, elephants, rhinoceros and zebras) have been introduced in the area as a zoological park. Over 350 000 persons had visited The Wilds by 1998. Although the efforts in the 1970s and 1980s were on ensuring surface stability and rehabilitation, The Wilds now maintains an active programme of research in conservation biology, and in ecosystem restoration and management.

CARBON SEQUESTRATION, OHIO

Given the present concern about fossil fuel emissions and global climate change, attention is now being focused on the potential role of restoration in carbon sequestration. The total land area impacted in Ohio by mining in 1997 was 0.13 million ha of which 0.1 million ha has been reclaimed since the early 1970s. Native and introduced grass species are the predominant plants that have been used for reclamation, but some areas have been reclaimed using mixed and monoculture hardwood forest species. A chronosequence of 0, 5, 10, 15, 20 and 25-year-old reclaimed mine spoils in Ohio was studied to assess the rate of carbon sequestration by pasture and forest establishment in comparison with undisturbed pasture and forest (Akala & Lal, 2000). Over a period of 25 years, the soil organic carbon (SOC) pool of reclaimed pasture and undisturbed sites for 0–15 cm very nearly approximated each other but was lower in the reclaimed sites at 15–30 cm depth. For the reclaimed forest and undisturbed sites, SOC pool was approximately the same for both depths. The SOC pool of the 15–30 cm depth of the pasture site stabilised sooner than the forest site. Dynamic simulation modeling studies on carbon efflux rates that take into account soil erosion, plant production and biogeochemical cycles reveal the phenomenal potential of rehabilitation in sequestering carbon and lessening the impact of climate change (Wali *et al.*, 1999; West & Wali, in press).

bison (buffalo) (*Bison bison*), and burrowing mammals such as gophers (species of *Geomys, Pappogeomys, Orthogeomys, Thomomys, Zygogeomys*) and prairie dogs (*Cynomys* spp.) dominate the areas. Much of the productivity in grasslands is due to massive root growth; rainfall (ranging from 25 to 75 cm per year) and available soil water are major determinants of grass growth. Fire is an important ecological factor in grasslands, indeed, it maintains them.

In North America, as the gradient of decreasing precipitation occurs from east to west and south, it also causes an ecological gradient from the tall grass prairie (wetter), followed by mixed grass to short grass prairie (drier), and finally the desert grasslands. The prairies have historically been dominated by large herbivores. Because of rich soils that make farmlands productive, most grasslands have been converted to cropland or pastures for grazing. For

Box 1.2   Restoration of the longleaf pine (*Pinus palustris*) ecosystem

At one time, the longleaf pine ecosystem covered an estimated 36 million ha in the southeastern United States. Today, that area has been reduced to about 1 million ha, representing a 97% reduction; the exclusion of fires, preferred management given to commercial species, and urban development has caused this change. As a result, over 30 plant and animal species of these ecosystems are endangered or threatened, including the red-cockaded woodpecker (*Picoides borealis*) and gopher tortoise (*Gopherus polyphemus*). Through partnerships of private, government and academic agencies, the disturbance to longleaf pine ecosystem has abated, and now demonstrates a trend of areal increase. These partnerships have initiated longleaf pine habitat restoration projects on 20 different sites, totaling 525 ha, across the southeast. Several thousand additional hectares of potential restoration projects, involving over 20 private landowner partners, have been identified. The restoration plans include the reintroduction of fire and management techniques that favor longleaf pine. Widespread efforts are under way to produce large numbers of longleaf pine seedlings for plantings on sites once occupied by the pine. The Longleaf Alliance, an organisation of researchers, academicians, private groups and individuals, and public agencies, is devoted solely to the restoration of longleaf pine, and co-ordinates restoration and research activities.

example, Iverson (1988) notes that only 0.01% of the original unploughed prairie remains in Illinois. With a widespread conversion of prairies to agricultural land uses, and with increasing deforestation rates, numerous plant and animal species are being extirpated. Specific long-term studies on the restoration of both abandoned mine lands and those under the stipulations of new laws have provided good measures of ecosystem recovery in relation to time (Wali & Freeman, 1973; Iverson & Wali, 1992; Wali, 1999). There are several successful examples of rehabilitation and conservation from the American prairies and some are presented here (Boxes 1.3 and 1.4).

Tropical forest

Latitudes between 10° N and 10° S support tropical rainforest. There is little annual variation in rainfall (200–400 cm per year) and temperature (25 °C), although these features can vary widely on a daily basis. They do not form a continuous belt around the equator because of the influence of mountains, winds and oceans on precipitation patterns. This biome has the greatest diversity of flora and fauna in the world. Constant warm and wet conditions allow for evergreen nature of vegetation and high rates of primary productivity. Animal distributions are also stratified by height in the forest. Large herbivores include the tapir (*Tapirus* spp.); predators include jaguars (*Panthera onca*). The soils are highly leached. The rate of nutrient cycling is high because of rapid litter decomposition. A recent report (D'Aleo, 2000) provides some staggering data on deforestation. Originally in Brazil, forest cover constituted nearly one-third (7.3 million km^2) of Brazil's total area of 21.8 million km^2. The present extent of the forest cover is 4.61 million km^2. At the current 2.3% annual rate of deforestation, about 128 000 km^2 are disturbed each year. The rehabilitation processes after deforestation are ecologically documented in some examples from Brazil (see Box 1.5).

Tropical dry forests

These forests are found between 10° and 20° latitude, where the Intertropical Convergence Zone migrates seasonally. This produces a long dry season (three to six months) and distinct periods of productivity and dormancy. Soils in these forests are richer and not highly leached.

Tropical grasslands and savanna

Savannas are a mixture of grasses and scattered trees that occur where rainfall is between 50 and 200 cm per year. They are most extensive in tropical climates including South America. Variation in rainfall is usually extreme and ranges from 85 to

Box 1.3   Restoration of abandoned mine lands

THE UNITED STATES

From 1978, when Surface Mining Control and Reclamation Act (SMCRA)-based Abandoned Mine Lands (AML) programme was initiated, to the year 2000, 26 states and the native American tribes have restored about 13 521 ha of abandoned spoil and haul roads, 4737 ha of exposed pits, 230 453 km of highwalls, and 3102 ha of land degraded by gob piles, slurry and industrial and residential waste. Also under this programme, among many other hazardous conditions left by past mining, 734 km of clogged streams, 8601 ha of clogged stream lands and 2312 ha of subsidence features have been restored. (Based on data provided by Chuck Meyers, Office of Surface Mining, Washington, DC.)

NORTH DAKOTA

Since 1981, over 80 primary reclamation projects at abandoned underground and surface mine sites at a cost of $22 million have been completed by the Abandoned Mine Lands (AML) Division of the North Dakota Public Service Commission. In 1998, three reclaimed sites (Hazen-West, Noonan and Fritz) were evaluated for hazard abatement, soil development, erosion control, revegetation success, wildlife use, wetland and stockpond water quality, and improvements in land use capability. The Hazen-West site had been mined from 1952 to 1974, and covered over 388 ha, of which about 75 ha were reclaimed in 1991. The reclamation plan consisted of eliminating the hazardous pit and constructing two ponds and two

wetlands for wildlife enhancement. Good quality spoil salvaged from the site and mixed with coal fines was used to cover the backfilled and regraded pit area. The site was fertilised, seeded and mulched. The Noonan site had been mined from 1930 to 1963 and covered 566 ha of which 202 ha were reclaimed in 1994–5. A 3.2-km long, dangerous highwall was eliminated, and several large wetlands, diversions and a concrete weir were constructed for water management and wildlife use. A portion of water-filled pit was preserved as a fishing pond. A mixture of tall grasses and forbs was seeded to provide cover for nesting waterfowl and other wildlife. The Fritz site had been mined from late 1950s to 1967 for extraction of uraniferous lignite. The exposed lignite was burnt in the pit or in nearby kilns for concentrating uranium in its ash, which was sold to the Atomic Energy Commission. The 63-ha wasteland thus created contained acidic material and water and was contaminated with uranium, cadmium and molybdenum. Restoration of this site began in 1992. The contaminated material was identified and buried under 1 m of uncontaminated spoil, and the area was graded to a gently rolling topography. One stockpond, five wetland sumps and two terraces were constructed for surface water management, and to return the land to native grassland and grazing use. Within four to seven years of reclamation, all three sites established permanent vegetation of diverse grass and forb species. The soil quality, vegetative cover and productivity, and the quality of water in the constructed wetlands and ponds at all the three sites have improved significantly (Dodd & Ogaard, 1998).

---

150 cm per year but is seasonal and scattered. The vegetation is adapted to frequent fires.

Chaparral

These regions fall mostly between 32° and 40° N and S in western North America (California), and central Chile. They are characterised by Mediterranean climates that have hot, dry summers and cool, moist winters. About 65% of the annual precipitation falls during winter and there is at least one month when the temperature remains below 15 °C. The vegetation is adapted to drought, fire and

infertile soils. These systems are often considered to be shrublands rather than deserts with thickets of evergreen shrubs and small trees, often with sclerophyllous leaves. Fire is common in the chaparral region and the plants are adapted to it.

Desert

Deserts form where evaporation exceeds precipitation, often by a factor of 7 to 50, in every year. They cover 26% of Earth's land in two distinct belts centred on 20° N and S latitude (the Tropics of Cancer and Capricorn). Deserts form where dry air masses

Box 1.4 Conservation and reclamation in the grasslands biome under the SMCRA requirements

A case of well-planned and highly successful achievement of combining proactive reclamation with conservation was presented by Bellaire Corporation's Indian Head Mine in North Dakota. This is one of the oldest surface mines in the country, operated since 1922 under different ownerships. The pre-mine area was characterised by diverse topography, land uses and wildlife habitats. In the early 1980s, the then North American Coal Corporation began conducting extensive studies in preparation of mining a 1012-ha area that contained several wooded draws surrounded by native rangeland. The goal was to conserve the wooded draws, and mine around these precious wildlife habitats. Appropriate design plans for pit layout and surface water management, and creation of permanent ponds were adopted to achieve the goal. The wooded draws that were thus saved from destruction contained green ash (*Fraxinus pennsylvanica*), American elm (*Ulmus americana*), box elder (*Acer negundo*), silver buffaloberry (*Shepherdia argentea*), june berry (*Amelanchier alnifolia*), round-leaved hawthorn (*Crataegus rotundifolia*) and chokecherry (*Prunus virginiana*). The ground cover consisted of bluegrasses (*Poa* spp.), sedges (*Carex* spp.), dwarf wild indigo (*Amorpha nana*), buckbrush (*Symphoricarpos occidentalis*) and poison ivy (*Toxicodendron radicans*). Breeding bird censuses indicated 10 species in 1979, before mining around the wooded draws was started, 13 species in 1984 when mining was fully operational, and 15 species in 1992 when final reclamation and vegetation establishment was completed. Mourning dove (*Zenaida macroura*), least flycatcher (*Empidonax minimus*) and yellow warbler (*Dendroica petechia*) represented 64% of the breeding pairs. The wooded areas and the surrounding rangeland provided excellent habitat for mule deer (*Odocoileus hemionus*), white-tailed deer (*Odocoileus virginianus*), coyote (*Canis latrans*), sharp-tailed grouse (*Pedioecetes phasianellus*), ring-necked pheasant (*Phasianus colchicus*), great horned owl (*Bubo virginianus*) and over 40 species of other non-game birds. By skillfully avoiding deleterious impacts on the wooded draws, Bellaire was able to preserve the habitat. For this restoration project, Bellaire Corporation received the 1992 National Hall of Fame award from the US Department of the Interior Office of Surface Mining.

The agricultural land adjacent to the wooded areas that was actually mined and reclaimed under the full requirements of SMCRA covered 884 ha. The pre-mine land uses included cropland, hayland and native grassland. Most of the reclamation activities were completed by autumn 1993, and all temporary sedimentation ponds were reclaimed by spring 1995. The land was returned to all the pre-mining land uses, and the Bellaire Corporation started receiving partial bond releases as early as 1985. Monitoring of revegetation progress has been carried out for several years following the initial seeding/planting. Cropland yields on the reclaimed areas, including those of prime farmland, have exceeded the required yield standards on average by 26% in each year. The yields of reclaimed hayland and tame pastureland have been, on average, 37% higher than the required standard. The vegetative cover and productivity of the reclaimed native grasslands have also exceeded the required standards. On average, the reclaimed grasslands produced about 34% more than undisturbed grasslands of similar type. Many tracts of cropland, hayland/pasture and native grasslands have received final bond release. Grazing by livestock on reclaimed native grasslands has been ongoing without any adverse effects. Wildlife monitoring and assessment of groundwater resources have revealed no harmful effects either. For the project, Bellaire received the US Department of the Interior Office of Surface Mining's 'Best of Best' award for 1997's example of surface mine reclamation. (Source: the written records of Bellaire and North Dakota Public Service Commission.)

are produced by Hadley cell circulation, the rain shadow effect of mountains, or the presence of high-pressure systems. Drought resistant plants and animals inhabit all deserts. The density of plant cover varies from sparse to moderate, and the species composition can be grasses or shrubs. Cold deserts occur at high elevations where relatively low evaporation improves water availability. Such areas are

Box 1.5   Wetland Restoration and Conservation

THE UNITED STATES

The North American Waterfowl Management Plan (NAWMP) was initiated in 1986 by the United States and Canada, in recognition of the need for international co-operation to restore wetlands and associated grassland habitats for declining migratory bird populations. The plan became continent-wide with the addition of Mexico as a signatory in 1994. The three governments, with co-operation from non-federal partners, are implementing a strategy to restore waterfowl populations to levels of the 1970s by protecting, restoring and enhancing wetland and adjoining habitats. The plan focuses on regional 'joint venture' areas that are designed to conserve those wetland habitat complexes identified as critical to sustaining populations of breeding, migrating and wintering waterfowl. The 1998 update to the Plan calls for restoration and enhancement of 6.2 million ha and protection of an additional 4.9 million ha of wetlands. The North American Wetlands Conservation Act (NAWCA) has become an important mechanism for wetland conservation under NAWMP. Since 1991, NAWCA has provided $343 million for habitat restoration with an additional $782 million provided by over 1300 non-federal partners. Private conservation organisations like The Nature Conservancy, Ducks Unlimited, the Delta Waterfowl Association and the California Waterfowl Association play a major role in wetland restoration and conservation. To date, over 1.9 million ha of wetlands and surrounding uplands have been acquired, restored or enhanced in the United States and Canada. Nearly 4 million ha have been affected in biosphere reserves by education and management plan projects in Mexico. Waterfowl hunters also support wetland conservation from the sale of duck stamps that contributed an additional $43 million to habitat conservation in 1999. In addition to the ecological values that accrue from hydrological, biogeochemical and trophic processes within wetlands, the values of North American wetlands accrue to the United States economy. Waterfowl hunting and viewing generated $13.4 billion that supported an estimated 135 000 jobs in 1991. The NAWMP has proven effective in large-scale habitat conservation and ecological restoration, serving as a model for integration of similar continental migratory bird conservation programmes.

THE CACHE RIVER ECOSYSTEM

A partnership of federal, state and private interest groups formed the joint venture to establish the Cypress Creek National Wildlife Refuge in southern Illinois. A watershed management plan was developed to improve water quality by reducing erosion and sedimentation, and to preserve and restore the natural resources over 194 253 ha of the Cache River watershed in a manner that is compatible with a healthy economy and high quality of life. Situated in an abandoned channel of the Ohio River, the Cache River has been adversely impacted by channelisation and sedimentation associated with forest clearing, flood control and agricultural development over the last century. The Cache River gained international recognition in 1996 when the Cache River and associated Cypress Creek wetlands were added to UNESCO's list of 15 'Wetlands of International Importance'. Located at the crossroads of mid-continental climate zones, the Cache River basin's 20 natural community sites include the most notable cypress–tupelo swamps, southern flatwoods and bottomland forests, and upland forests and limestone/sandstone glades. The area supports 104 state and seven federally threatened or endangered species. Wetland restoration efforts in the region include revegetating forested bottomlands, establishment of riparian corridors and filter strips, and restoration of natural flow regimes to the Cache river and its tributaries.

THE TENSAS RIVER BASIN

The Tensas River Basin area is part of the Lower Mississippi Valley Joint Venture of NAWMP and includes 290 570 ha in northeastern Louisiana. Of the 92% of once-forested watershed, 85% of the area has been cleared for row-crop production since the 1950s. About 26 305 ha of bottomland swamp remain in the

Tensas River National Wildlife Refuge and Big Lake Wildlife Management Area. Loss of wetlands and riparian areas has degraded water quality, increased sedimentation and flooding, and caused loss of wildlife habitat and biodiversity. Federal, state and non-governmental agencies and local citizens formed collaborative partnerships to develop a Watershed Restoration Action Strategy. Best management practices, erosion control structures and reforestation measures have been implemented through the US Department of Agriculture's Environmental Quality Incentives and Wetland Reserves Programs (WRP). The US Department of the Interior Fish and Wildlife Service and state and federal partners work with private landowners to voluntarily protect and restore bottomland forests in the region. To date, an estimated 22 663 ha of farmland and 1619 ha and over 9 km of riparian area have been reforested, with another 19 425 ha enrolled in WRP.

dominated by shrubs as in the Great Basin of southwestern United States. Low annual rainfall (30 or less cm per year) and sparse plant cover characterise deserts. Perennial plants have a variety of adaptations to minimise water loss. Annual plants have adapted their life cycles to infrequent precipitation patterns: when rain comes, they develop quickly, flower, and die.

Wetlands and aquatic systems

The Americas support a tremendous diversity of lakes, rivers, estuaries and marine systems. In North America, Canada has over 750 000 km^2 of water bodies, the United States has over 470 000 km^2 and Mexico has about 50 000 km^2. In South America, Brazil contains over 55 000 km^2, and Colombia supports over 100 000 km^2 of water bodies. Wetlands, in particular, are vital as repositories of incredible biodiversity, critical for fish, bird and mammal species, and as sources of groundwater recharge. These systems are threatened, however, by agricultural and urban expansion. For example, the Everglades ecosystem in the southern coastal plain of Florida has been transformed by both draining and impounding water for agriculture and urban development since the early nineteenth century. 'Of the three traits that characterized the pre-drainage system in the Everglades: (i) habitat heterogeneity; (ii) large spatial extent; and (iii) a distinct hydrologic regime—the new water control works most directly affected the last, but the destruction of the system's hydrologic regime led, inevitably, to a reduction in the size and biotic diversity of the wetlands' (McCally, 1999). US President Bill Clinton signed into law the Comprehensive Everglades Restoration Plan, a 30-year, $7.8 billion project designed to eliminate dams and restore the original drainage patterns of the Everglades ecosystem. The implementation and effectiveness of this programme remains to be seen. It is worthwhile to record that of all habitat types, wetland restoration finds much support and financial subsidy from, besides the government, waterfowl hunters, sports fishery enthusiasts and other private agencies and supports jobs and generates revenue (Box 1.6).

NATURE AND EXTENT OF THE PROBLEM

Ecosystem disturbance may be defined as an event or a series of events that changes the relationship between organisms and their natural habitats/niches, both spatially and temporally (Wali, 1987). The changes may be small or large, temporary or permanent, with little to severe consequences. 'The most pervasive incidents of environmental degredation result from recurring and incremental impairments," notes Robinson (1992). Bazzaz (1983) has listed seven major activities as the causative agents of ecosystem disturbance: extensive clearing of natural vegetation for agriculture; selective harvesting of desirable species and introduction of alien ones; mining; draining of wetlands; introduction of chemicals in the environment; and the impact of war. Synthesising data from earlier studies, Houghton (1994) notes that deforestation rates in the tropics since the 1970s, and so in South America, have increased sharply. Ramakutty & Foley (1999) have specifically calculated the conversion of

Box 1.6   Restoration of deforested areas in tropical rainforest in Brazil

With 6 million ha of the Amazon Basin converted to pasture in 20 years, there is an urgent need to determine the potential for abandoned pastures to revert to forest. Near Paragominas, Para, Brazil, Buschbacher *et al.* (1992) examined succession on 15 different abandoned pastures of varying use intensities and ages since abandonment. They found significant potential for re-establishment on light-use areas that had been cleared and only lightly grazed. These areas developed closed canopies within 10 years and had high species diversity. Moderate-use areas, which had been repeatedly burned and heavily grazed before abandonment, recovered more slowly with a significant loss of diversity. Heavy-use areas, characterised by the use of machinery in the clearing process, showed very slow recovery. Pioneer species were slow to become established and species diversity was quite low. Their analysis led the authors to estimate a 500-year recovery period for heavy-use areas to return to mature forest. Clearly, lighter-use sites have much greater potential for rehabilitation to productive ecosystems, and use intensity should be minimised on current pastures to allow for reasonable recovery times. Loss of on-site regeneration capacity, slow seed dispersal, seedling predation and unfavorable microclimates all contributed to slow recovery of pasture areas.

After forest clearing, the use of forest plantations to re-establish forest cover can catalyse the recovery of secondary forest. These plantations should be contiguous with an undisturbed forest area as a seed source. Seed dispersers such as birds and bats were found to utilise plantations, accelerating the regeneration of small-seeded species. Large-seed dispersers, however, did not utilise plantations as extensively, creating a bias in regeneration. Even on abandoned strip mines, plantations can accelerate the recovery of forest diversity. Changes in physical and biological site conditions within the plantations facilitate succession by making soil surface conditions (light, temperature and moisture) favourable for seed germination. However, seed dispersal vectors are required to bring seeds to these favourable sites. Therefore, the plantation areas require an ecological connection to existing habitat for seed dispersers and seed sources. Plantations on a former bauxite mine showed significant recovery 10 years after planting, although seed dispersal vectors were found to be limiting recovery. Greater distance from undisturbed forest had a negative correlation with species diversity. Creating favourable habitat conditions for seed dispersers within forest plantations could further accelerate the rate of recovery (Parrota *et al.*, 1997*a*, *b*).

land to, and abandonment of, croplands between 1700 and 1992, noting the extent of accompanying land degradation in the Americas. The overall extent and magnitude of soil degradation in the Americas has been significant resulting from the combined activities of deforestation, resource exploitation, overgrazing and agricultural activities (Table 1.3).

Obviously, any restoration work needed will depend on the type of ecosystem disturbance encountered or expected. Also, like proactive or mitigative restoration, preventing ecosystems from getting damaged in the first place requires implementation of conservation policies and management plans. Many countries in the Americas have set aside some wilderness areas, which, because of their fragile ecosystems or aesthetic or scientific value, are left untouched in their pristine quality. Even such areas require constant management and care. However, the challenges involved in proactive and mitigative restoration vary widely depending upon the intensity and extent of the disturbance. For example, in the forest systems where logging operations are carried out, concomitant replanting of seedlings of same species, with appropriate management for minimising soil erosion and maximising survival rate of the planted seedlings, may suffice. Likewise, the impacts of agricultural practices on soil degradation may be minimised by using appropriate soil conservation practices, including minimum tillage, plant residue management, and crop species/cultivars that are genetically disease-pest-resistant and nutrient-uptake efficient. Overgrazing, which destroys native grasslands and exposes the

Table 1.3. *Soil degradation (area in Mha) of the Americas*

	South America	Central America	North America
Extent			
Water erosion	123	46	60
Wind erosion	42	5	35
Chemical	70	7	+
Physical	8	5	1
Total	243	63	96
Causative Factors			
Deforestation	100	14	4
Overexploitation	12	11	–
Overgrazing	68	9	29
Agricultural activities	64	28	63

Source: Based on Oldeman (1994).

land to erosion, is essentially a result of mismanagement and lack of proper scientific information. The restoration challenge posed by overgrazing includes dealing with the socio-economic reality of that region, in addition to applying proper scientific remedies.

Appropriate strategies for restoring drastically disturbed ecosystems can be developed based on a 'systems approach' (Wali, 1975). Revegetation of disturbed ecosystems is primarily a substrate/soil-driven process, and its success is dependent on the germ plasm (seeds, propagules) that can arrive naturally (or are seeded), and survive at these sites (e.g. Wali & Freeman, 1973; Bradshaw, 1983, 1997; Tilman, 1988; Gleeson & Tilman, 1990). Topography of the affected site may need to be modified to restore the area to a desired land use, to eliminate any potential hazards to the public, to re-establish proper surface drainage pattern, or a combination thereof. The presence or absence of organic matter (hence topsoil) is also a critical factor as it determines the rate at which the disturbed ecosystems recover toward long-term biological productivity (Stevenson, 1986; Logan, 1989, 1992). A schematic view shows factors and standards that must be considered in restoring/reclaiming a disturbed area (Fig. 1.2); it also shows how science and law work together in developing the appropriate standards and procedures for ecological restoration.

Cairns (1988a) calls restoration ecology the 'new frontier in both theoretical and applied ecology', no less in significance than biotechnology. For ecological restoration to be meaningful, understanding of ecological succession in a regional context is necessary (Whittaker, 1974). To comprehend the relative differences in stability of successional and self-maintaining communities, due recognition of regional climaxes is stressed. The successional aspects of restoration ecology have been discussed in many publications (e.g. Wali & Freeman, 1973; Woodwell, 1992; Marrs & Bradshaw, 1993; Wali, 1980, 1999; Keddy & Drummond, 1996; Walker, 1999), which provide ample guidance in these aspects.

Finally, consideration must also be given to the fact that revegetation of a site by the same type of species, genotypes or ecotypes that existed before disturbance may not be possible because of some edaphic stress factor that may have been unleashed as a result of the disturbance itself. The edaphic stress may be due to physical or chemical characteristics of the substrate material. Poor infiltration and/or water storage capacity resulting in droughty conditions; trace metal toxicities; nutrient deficiencies; saline, sodic or acidic soil conditions; low soil biological activity due to paucity of microflora, etc. constitute edaphic factors that can make revegetation of disturbed sites very difficult. In such situations, dependence on soil amelioration only may not prove as successful as the use of those plant species, cultivars or ecotypes that are naturally tolerant (adapted) to such edaphic stresses. This approach has worked well in production agriculture as well as in restoration ecology (Bradshaw *et al.*, 1965; Bradshaw, 1970; Brown *et al.*, 1972; Epstein & Norlyn, 1977; Safaya, 1979; Asay, 1979).

In this chapter, we have purposefully chosen to concentrate on restoration of drastically disturbed ecosystems (by mining) for several reasons. Mining of metals, minerals, aggregates and fossil fuels is widespread in the Americas (see Doan *et al.*, 1999), and will greatly expand in the future, given the current emphasis on economic development. The disturbances to ecosystems will increase not only by mining *per se*, but also because of the enormous

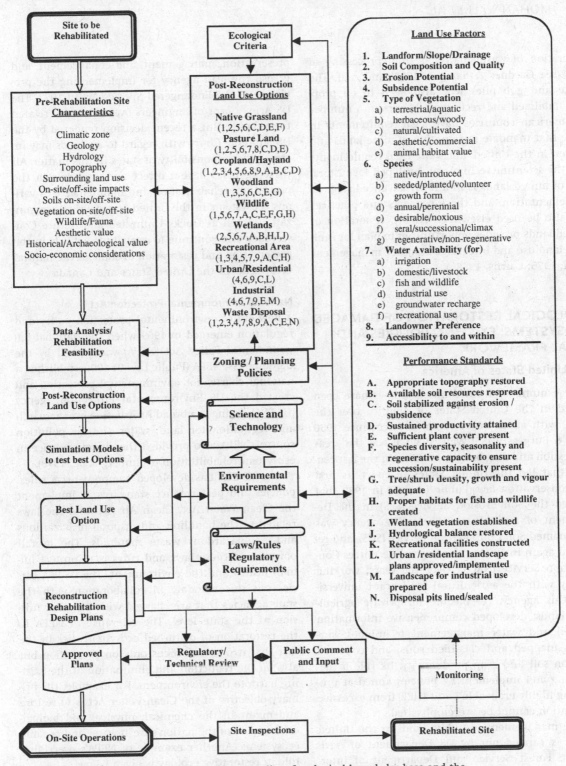

Fig. 1.2. A flow diagram showing the centrality of ecological knowledge base and the planning-and-decision process for the rehabilitation and management of degraded ecosystems. When fully implemented, the Surface Mining Control and Reclamation Act of the United States should closely match the expectations of restoration for multiple land use sustainability.

generation of waste materials from processing of ores (see Gardner & Sampat, 1998) and coal. The waste and gob piles will also need to be properly stabilized and reclaimed. Moreover, a number of American countries have legal requirements in place that mandate reclamation of mine lands and wastes. In the United States the legal requirements and the scientific technology developed for restoration of mined lands is in a fairly advanced stage of implementation, and these concepts and practices can also be used elsewhere. Finally, restoration of mined lands provides opportunities to plan for multiple land use and integrated resource management (Wali, 1975; Cairns, 1988*b*).

ECOLOGICAL RESTORATION OF DAMAGED ECOSYSTEMS: EXISTING POLICIES AND LEGAL FRAMEWORK

The United States of America

A large number of environmental laws have been passed in the United States of America over the years, with about 90 of them in force since 1950 (Wali & Burgess, 1985; Wali, 1987). It was the great depression and dust-bowl experience in the last century that led the United States to enact its first soil conservation law (Public Law 46) in 1933 that created the Soil Erosion Service (SES) in the Department of Agriculture (USDA). The agency was later named Soil Conservation Service (SCS), and renamed again in 1994 as the Natural Resources Conservation Service (NRCS). Over the years, working closely with the agricultural land grant universities, this agency, responsible for private agricultural lands, developed comprehensive information on soil and water management techniques; surveyed, mapped and classified soils; and collected data on soil and crop productivity. Its role in developing and implementing policies aimed at preventing highly erodeable lands (HEL) from excessive cultivation cannot be overemphasised.

The management of public lands in the United States is carried out by US Department of Agriculture Forest Service and Department of Interior's Bureau of Land Management. The US Fish and Wildlife Service is responsible for habitat preservation, management, and enhancement, and is also the lead agency for implementing the provisions of the Endangered Species Act of 1973. The US Army Corps of Engineers oversees issues related to wetlands. But a recent decision rendered by the US Supreme Court with regard to wetlands may increase the responsibility of states in this matter. All these agencies have a direct and vital role in the protection, rehabilitation and maintenance of various ecosystems in the United States. Private organizations such as Ducks Unlimited and Nature Conservancy also continue to contribute actively in the preservation and restoration of some of the natural ecosystems in the United States and Canada.

National Environmental Protection Act

A new era of environmental protection policy and regulation emerged in 1969 when the National Environmental Policy Act (NEPA) was passed by the US Congress. NEPA (Public Law 91-190) established a national charter for environmental protection and created the US Environmental Protection Agency (US-EPA), giving it a broad jurisdiction over research and regulation for land, water and air pollution control. NEPA also provided the citizens a voice in ecosystem rehabilitation (Holmberg *et al.*, 1978).

The US-EPA has developed comprehensive rules, policies and performance standards to implement the Clean Water Act, Clean Air Act, and the laws related to the handling and disposal of hazardous and non-hazardous waste materials. The overall objective of these laws and rules is to protect human health and the environment. The US-EPA carries out this mandate in collaboration with the state agencies that are charged with a similar mission at the state level. The relevance of NEPA to the restoration of disturbed ecosystems lies in the fact that it does not focus only on prevention but also on the reduction and elimination of the existing harm to the environment. For example, the primary objective of the Clean Water Act is to restore and maintain the chemical, physical and biological integrity of the nation's fresh water and oceanic ecosystems. Another example of NEPA's overriding role in restorative ecology is given below.

NEPA requires that all federal agencies prepare a concise public document, called an Environmental

Assessment (EA), prior to undertaking any action or project that may have a significant adverse effect on the human environment or an ecosystem. If the EA shows that adverse effect(s) is/are likely to occur, an Environmental Impact Statement (EIS) must be prepared. The EIS is a detailed written statement that provides a thorough analysis of the possible adverse effects of a proposed action or project. It is also required to include alternative plans that may have less adverse side-effects. EPA also recognises a category of actions called Categorical Exclusion, which are exempted from EIS preparation because individually or cumulatively they have no significant effect on the human environment. So, even before an EA is prepared the agency proposing an action has to make the categorical exclusion determination. If an EA shows that no significant impacts are likely to occur, the agency may prepare a Finding of No Significant Impacts (FONSI), instead of an EIS. But if an EIS is necessary, then a Notice of Intent (NOI) is published in the Federal Register of the US Government to inform the public that the agency intends to prepare an EIS for the proposed project. Thus, EIS procedure guarantees integration of scientific facts with the government decision-making process, and tries to ensure that only such actions are undertaken that have very little adverse effect on the environment. The EA/EIS procedures are required to be followed in the reclamation of abandoned coal-mined lands as well as the lands that are currently mined for federally leased coal.

Surface Mining Control and Reclamation Act

The legal and scientific framework for the restoration of drastically disturbed ecosystems is perhaps best exemplified by Public Law 95-87, Surface Mining Control and Reclamation Act of 1977 (SMCRA). Its historic importance was clear since its very inception (Imes & Wali, 1977, 1979). Although the mandates of SMCRA apply only to the coal-mined lands in the United States, the principles and technology developed to meet these mandates would be equally effective for restoration of ecosystems disturbed by most other means.

Prior to SMCRA, many coal-producing states in the United States had passed some form of legislation to minimise the adverse effects of surface

coal-mining (Beck, 1973; Bowling, 1978): West Virginia in 1939, Indiana in 1941, Illinois in 1943, Pennsylvania in 1945, Ohio in 1947 and North Dakota in 1969. Indeed the North Dakota law (Beck, 1973) provided the title for federal legislation. But two events triggered the need for a national policy on mining and reclamation. First was the publication of *Surface Mining and Our Environment: A Special Report to the Nation* from the US Department of the Interior (1967), in which the extent and magnitude of surface mining was recorded. Second, shortly after the 1973 energy crisis in the United States, it became clear that more domestic coal would have to be mined both to reduce the oil imports and meet the energy needs of the nation.

In its statement of findings, the SMCRA recognised the importance of mining for coal by both surface and underground methods for meeting the nation's energy needs. However, it also recognised that many surface mining operations result in disturbances of surface areas that destroy or diminish the utility of land for commercial, industrial, residential, recreational, agricultural and forestry purposes; cause erosion and landslides; contribute to floods and water pollution; destroy fish and wildlife habitats; impair natural beauty; and undermine efforts to conserve soil, water and other natural resources (SMCRA, Title I, Section 101). Thus, a nationwide programme to allow coal-mining under strict application of reclamation rules and standards was established to prevent the adverse effects of coal mining on the environment and restore the former uses and productivity of the lands disturbed by mining. The SMCRA created the Office of Surface Mining Reclamation and Enforcement (OSM) in the Department of the Interior, with powers to implement and enforce the provisions of this Act.

However, the SMCRA recognised that 'because of the diversity of terrain, climate, biological, chemical, and other physical conditions in areas subject to mining operations, the primary governmental responsibility for developing, authorizing, issuing, and enforcing regulations for surface mining and reclamation operations subject to this Act should rest with the States.' Thus, any state wishing to retain its primacy over coal-mining and reclamation within its borders was able to do so by developing a

State Regulatory Program (laws, rules, policies, etc.) that met the approval of the US Department of the Interior if the rules and standards contained therein were no less stringent than the federal counterparts. The states would receive federal funding to develop and implement their state programs to achieve the objectives of this Act.

The SMCRA also provided for the development and implementation of the Abandoned Mine Land (AML) programme that is totally dedicated to reclamation of lands that were degraded by mining operations prior to the enactment of this Act and were abandoned without any reclamation or restoration. The primacy states run the AML programme, which is funded by the Department of the Interior from a trust fund derived from reclamation fees (10 to 35 cents per ton of coal) levied on currently active and regulated coal-mining operations.

Prior to the SMCRA, mining for coal, metal ore, uranium, bentonite, sand, gravel, hard rock, etc., had left numerous sites throughout the United States in a state of utter degradation. The size and severity of disturbance varied according to the type of mining, the biogeoclimatic characteristics of the area, and the degree to which they may have been revegetated naturally or by human effort (Merrill & Safaya, 1984). The AML program required that all such sites be inventoried and classified into various categories, mainly (1) those that posed a direct threat to public health, safety and welfare (i.e. highwalls, surface instability, toxic radioactive substances), and (2) those that reflected environmental degradation due to increased potential for erosion, reduced biological productivity, chemical contamination, or any other condition. Priority for reclaiming the abandoned mine sites was also given in the same order as the categories listed above. Reclamation of the sites involved highwall reduction, landscape restoration, surface manipulation, use of ameliorative additives and fertilisers, and seeding/planting of adapted grasses, forbs, shrubs and trees. These efforts have paid off very well, as a large number of previously degraded ecosystems have been restored to a better level of stability and productivity (Boxes 1.3 and 1.4).

The SMCRA and its implementing regulations are very clear and specific about requirements and standards for post-1977 mining and reclamation

procedures. First of all, no one can engage in surface coal-mining without a valid permit from the surface mining regulatory agency. Permits or authorisations from several other federal, state and local agencies are also required. Furthermore, mining cannot begin without the mining company posting a bond as an assurance for the successful completion of all reclamation requirements. The SMCRA also provides for public and landowner participation and comment in the process. The permit application must contain accurate information about: (1) the company and its legal rights to mine a proposed area; (2) a detailed inventory of premining environmental conditions (geology, hydrology, land use, soils, vegetation, wildlife, threatened and endangered species, historical and archaeological resources, etc); (3) how the area will be mined and reclaimed; and (4) the methods that will be used to prove that successful reclamation has been achieved. The permit must also contain specific design plans for the construction of surface water management structures (sedimentation ponds, diversion channels, etc.), and plans for monitoring surface and groundwater, wildlife, etc.

All this information, which may require thousands of pages, is critically reviewed by the regulatory agency. After a mining permit is issued, the regulatory agency inspects and monitors all mining and reclamation activities at the mine regularly, and if there are any violations of law, rules or the permit condition noted, an enforcement action is taken that may result in a fine, cessation of mining, or revocation of the permit.

The core environmental requirements established by the SMCRA are as follows:

- Restore mined lands to former or better use
- Backfill and grade the mined areas to their approximate original contour
- Control erosion and attendant air and water pollution
- Minimise disturbance to the hydrological balance – surface and groundwater
- Remove, segregate and respread topsoil (plus subsoil in case of prime farmlands)
- Establish adequate vegetation on the mined lands.

In revegetating mined lands, the SMCRA, Section 515(b)(19), mandates that an operator shall:

establish on the regarded areas, and all other lands affected, a diverse, effective, and permanent vegetative cover of same seasonal variety native to the area of land to be affected and capable of self-regeneration and plant succession at least equal in extent of cover to the natural vegetation of the area; except, that introduced species may be used in the revegetation process where desirable and necessary to achieve the approved postmining landuse plan.

The reclaimed area must be seeded within three years of coal removal. After the last year of augmented seeding/planting, the permittee is required to assume responsibility for successful reclamation and revegetation for a period of five years in areas where average annual precipitation exceeds 65 cm, and for ten years in areas of 65 cm or less precipitation.

Obviously, the SMCRA expects that land disturbed by mining can and must be restored to the same landform and land use, with the same kind and amount of vegetation, as existed before mining. The surface drainage system and the underground resources of water, in both amount and quality, must also be restored to support the former uses of that land. The simple but sanguine mandate regarding revegetation of mined lands has spawned a comprehensive set of rules and standards for assessing its success.

Each land use category must meet its own set of standards; failing which, a final bond release cannot be granted. For a reclaimed native grassland to qualify for final bond release, its productivity, ground cover, diversity, seasonality and permanence must equal or exceed that of similar agricultural lands in the surrounding area under equivalent management, in the last two years of the responsibility period. Demonstration of equivalence for both productivity and cover must be made with 90% statistical confidence. For tame pasture lands only productivity and cover are required to meet the standard. For croplands and haylands, meeting the productivity standard is the only criterion, but for prime farmlands success must be evaluated over a period of three years. Vegetation data of the reclaimed areas are compared with those obtained from reference areas, or with an appropriately calculated technical standard, to determine the success of reclamation.

The success of post-mine woodlands, including the wooded riparian zones or floodplains alongside streams as wildlife and fish habitats is determined on the basis of tree and shrub numbers, vegetative ground cover, and an evaluation of species diversity, season variety and regenerative capacity of the vegetation so established. In the first place, the woody and herbaceous species required to be planted on the reclaimed lands must be approved by the regulatory agency in consultation with state counterparts of Fish and Wildlife, Forestry, and NRCS.

Any wetlands destroyed in the mining process must be replaced, and the total post-mine wetland area must equal the pre-mine area. For the initial inventory and identification of wetlands, consultation with NRCS is required to ensure compliance with the Wetland Conservation Provisions of the 1985 Food Security Act and the 1990 Food, Agricultural, Conservation, and Trade Act. The Army Corps of Engineers must also be notified. The seasonal, semi-permanent and permanent wetlands are reconstructed using appropriate engineering designs, suitably respread with previously saved wetland topsoil, and seeded with appropriate hydrophytic vegetation. The successful establishment of the wetlands is assessed on the basis of vegetation zones/communities, species composition, and the quality of water.

SMCRA's clarity of purpose, comprehensiveness, and on-the-ground achievements are unmatched. Once the SMCRA's legal stipulations are fully implemented, we believe, it should prove to be a model legislation bringing together effectively scientific research in the restoration plans for multiple land use as depicted (Fig. 1.2). A number of success stories as well as some that pose problems are displayed in the web site of US Department of the Interior's Office of Surface Mining (www.osm.gov). There are, however, some shortcomings in SMCRA and other areas that are discussed later.

Canada

Problems associated with drastic disturbance and ecosystem restoration in Canada are discussed by Thirgood (1978), Gunn (1995) and Ripley *et al.* (1978, 1996); the latter approach the problems in an ecosystematic perspective. In eastern Canada specifically,

Watkin (1979) identified three major problems that required reclamation urgently: sulphide-containing mines wastes which can produce acid tailings, highly alkaline asbestos tailings, and shoreline erosion on the Great Lakes.

One of the examples that caught national attention was the case of Sudbury, Ontario. Degradation, beginning with the logging of spruce, fir and pine forests in the late 1800s, intensified when the area became a major copper and nickel producing area. In extracting and processing ores, it polluted both the air and water in the region rendering thousands of hectares of land and water barren of life. Casting the story of the region in both a historical and ecological perspective *sans* effective governmental regulation, Gunn (1995) provides a comprehensive account of the restoration of Sudbury. Three lessons are clear from this story. First, had there been an effective government policy, many productive ecosystems would have been spared extreme degradation. Second, the tallest smokestacks of Sudbury were not only a hazard to the immediate area but also to a much larger area that received the emissions. Third, the dedication of many professionals, industry, and citizens can indeed be effective. Recent work from Canada on the restoration of alpine and subalpine areas (Macyk, 2000), western surface-mined coal lands (Fedkenheuer & Macyk, 2000), and oil shale mined lands reviews the successes and shortcomings of these efforts. However, lack of enforcement, scaled-down financial resources of governmental agencies responsible for environmental protection, and undue reliance on voluntary enforcement makes a strong case, according to the Environmental Law Institute (2000a), for a national framework.

Canada is a major producer of metals and minerals and exports nearly 80% of them. At the federal level, Canada now has a Minerals and Metals Policy which specifically addresses the current state of the natural environment, mine reclamation, and establishment of protected areas, and links restoration strongly to sustainable development (Government of Canada, 1996; Shinya, 1998). These pronouncements are further strengthened by the recent passage of the Canadian Environmental Protection Act (CEPA) (Government of Canada, 2000). CEPA 2000 includes some impressive fact sheets but none addresses ecosystem restoration directly. Hence, unlike in the United States, there is no federal agency that regulates or oversees the restoration or reclamation of lands disturbed by mining or other large-scale activities. This responsibility lies with the provincial governments of Canada. The provinces may differ in their legislative or administrative procedures for the restoration of disturbed ecosystems, but the overall goal is to minimise land degradation and air and water pollution. The concept of sustainability of the environment is underscored in all such efforts. Examples of two Canadian coal-producing provinces, Alberta and Saskatchewan, are given below.

In Alberta, reclamation of coal-mine lands is regulated under the Land Surface Conservation and Reclamation Act of 1973, which replaced the Act of 1963. The Act is implemented through a set of guidelines rather than rules. The guidelines are adjusted periodically as the technology evolves. There is a close co-ordination between the government and the industry in the area of reclamation research. Coal-mining is allowed through a stepwise process of approvals granted by the Energy Resources Conservation Board (which approves mining plans and issues mining licenses) and by the Department of Environment (which approves reclamation plans).

The permitting process involves sequentially: (1) preliminary disclosure; (2) environmental impact assessment; and (3) detailed review of licensing. Mined lands must be returned to the level of soil capability equivalent to that which existed before mining. There is no requirement for returning mined land to its approximate original contour or productivity *per se*. Topsoil and subsoil have to be salvaged for respreading only to the extent needed to return the land to its pre-mining level of soil capability. Erosion control is a major concern in the hilly areas, and considerable emphasis is placed on sediment control measures. The success of reclamation is assessed by ensuring that the desired soil capability has been attained, and there is no waiting or responsibility period for proving that fact.

In Saskatchewan, coal-mining is regulated under the provisions of the Environmental Management and Protection Act of 1983/4 and amendments thereto, and by the Mineral Industry Environmental Protection Regulations that went into effect in 1991. Neither the Act nor the regulations focus exclusively on coal-mining. The underlying philosophy of the Saskatchewan law and rules is to control pollution and return the land disturbed by mining or mineral exploration to a sustainable productive use. The Industrial Branch of Saskatchewan Environment and Resources Management is the primary regulatory agency for coal and all other mining operations.

Applications for reclamation approval require relatively very little information (legal land description, land use, topsoil/subsoil/overburden data, soil classification, schedule for various mining and reclamation activities, proposed cover soil depth, and the proposed land use). Plans for water and air pollution control and waste disposal are also included. Reclamation plans are submitted on an annual basis. Guidelines and mutually agreed-upon performance standards are the basic mechanisms for regulation. Reclamation should be concurrent with mining, and the disturbed lands should be returned to 'acceptable, predetermined land use' (preferably the pre-mining use, or that which is achievable) and 'ensure physical stabilization'. Regrading must be done within two years of mining disturbance, appropriate for the intended post-mining land use. Revegetation is required using 'good agricultural practices', but assessment of revegetation success *per se* is not required. Land is released, parcel by parcel, from the burden of reclamation, when the regulatory authority is satisfied, based on expert judgment.

Latin America

Since Mexico shares much more in common with South and Central American countries, many authors treat them together as Latin America. Most countries in this region now have overencompassing environmental legislation that addresses many issues of air and water pollution, and the protection of many newly established nature reserve or protected areas.

Latin America has been experiencing a major boom in mining investment since the early 1990s. We explored specific areas of legislation that would relate to ecosystem restoration. The environmental impact assessment process appears to be gaining ground. While such assessments are the 'linchpin' for preventing adverse environmental impacts in such countries as Bolivia and Chile, these seldom include specific measures that should be taken to prevent adverse effects (Environmental Law Institute, 2000b). Ecosystem rehabilitation or restoration do not seem to be a high priority.

Part of this is due to improved economic conditions and structural reforms that have opened up the economies of the region and made them more hospitable to foreign direct investment. In addition, several countries, such as Chile, Argentina, Bolivia and Peru, have introduced sectoral mining reforms and adjusted their legal, fiscal and environmental policies. This has led to a significant expansion of the mining sector in Latin America. Mining investment is new in several countries of the region, as in Argentina, and even for traditional mining countries, the magnitude of the projected investment has no precedent.

That companies developing resources all over the world have had an 'unprecedented access to a large portion of the earth's surface than ever before' during the 1990s (Otto, 1998) is especially true for South America. Almost three-quarters of these exploration activities were concentrated in five countries in 1996: Chile (18.2%), Mexico (16.6%), Peru (16%), Brazil (14.5%) and Argentina (8.8%). Productive investments in mining in Latin America over the next five years are projected to be the largest of any region in the world, accounting for 44% of the total world investment in the mining sector, or about US$17 billion. A compilation by the Economic Commission for Latin America and the Caribbean arrives at a forecast of $24 billion in mining investment by the year 2000 in only five countries (Argentina, Chile, Mexico, Peru and Venezuela). Much of this is for copper mining, which accounts for 66% of projected investments.

Mexico

Mexico's General Ecology Law on Ecological Balance and Environmental Balance, passed in 1988 and substantially amended in 1996, sets out principles for ecological planning and management, and the respective roles of state and municipal governments and the general public (Environmental Law Institute, 1998). With a total area of over 1 972 550 km^2, Mexico is the 14th largest country in the world but ranks fourth in its biological diversity because of a great diversity in climate and elevation (−10 m to 5700 m). More than 39% of the land is in permanent pasture, about 12% is arable land. The Ecology Law addresses the issues of degraded and desertified areas.

Forests and woodlands cover 26% (50 million ha) with an equal area of temperate and tropical forests. Despite forestry laws that protect valuable and sensitive forests, Mexico annually loses 1% of its temperate forests and 2.5% of tropical forests to farms and pastures. 'The laws are difficult to put into practice. . . . Most of the problems with Mexican forest laws are not evident from the laws as written. They are apparent from the laws as implemented' (Environmental Law Institute, 1998).

The environmental policy statement has been a part of Mexico's 1917 national constitution (Hernandez, 1997). Although Mexico's general environmental law recognises the 'preservation of the ecological equilibrium and the protection of the environment', environmental impact assessments are not required for all activities (Environmental Law Institute, 2000b). When an environmental impact statement is required, it must include a description of possible effects on the ecosystems, prevention and mitigation measures, and other measures to avoid or minimise negative effects on the environment. Mexico's Federal Mining Law of 1992, amended in 1996, regulates mining matters pursuant to Article 27 of the constitution. Its provisions are of public nature and to be observed all over the republic. Its application corresponds to the Federal Executive Order through the Ministry of Energy, Mines and Governmental Industry.

The population and economy are both expanding with the resultant need for resources. Poverty forces migration of workers to the north. An unprecedented continental treaty among Canada, the United States and Mexico called the North American Free Trade Agreement (NAFTA) was approved in 1992 and became effective in 1994. It is a most comprehensive and detailed treaty and, as Hernandez (1997) notes, may be as historic as the establishment of the European Union. The expectation is that Mexico will adopt laws such as the SMCRA although there is no apparent movement in that direction.

Argentina

Argentina, the second largest country (area 2 766 890 km^2) in South America, is mostly temperate, but arid in the southeast and sub-Antarctic in the southwest. Elevation ranges from −40 m to 6960 m and most of the land area is in permanent pasture (52%) and 19% comprises the forest and woodland area. Because the country possesses a wealth of extractable resources, legislation on mining goes as far back as 1813 (Albarracin, 1997). The laws were subsequently updated in 1886 and in 1919 came the Mining Code.

The legislative framework of Argentina has provisions for environmental impact assessment (EIA) for each of the exploration, discovery and mining phases which are reviewed individually (Environmental Law Institute, 2000b). The EIA requires such considerations as the impact and subsequent neutralisation of acid mine drainage but the law 'does not provide specific standards or best practices to address acid mine drainage' (Environmental Law Institute, 2000b). At the end of the mining activity, operators are required to submit an updated or amended EIA to include changes that may not have been contemplated in the original plan.

The protection of the environment, standards for air, land and water quality and the conservation of natural resources falls under the purview of the federal Mining Code. However, the actual enforcement is under the control of individual provinces. Rules exist that may penalise operators with fines and temporary or permanent closure of an operation. We were not able to trace specific examples of ecosystem restoration.

Bolivia

Bolivia, with an area of 1 098 580 km^2, and a rugged terrain ranging in elevation from 90 m to 6542 m, includes part of the Andes Mountains. About 53% of the area is forested and about 24% is in permanent pasture. Bolivia's Law on the Environment requires an EIA for all public and private projects (Environmental Law Institute, 2000b). However, the EIA enforcement varies on the type of activity; for example, prospecting and exploration activities do not need an EIA.

The Mining Law of 1997 addresses several issues that have a direct impact on ecosystem restoration. Once operations have ceased, the Environmental Regulation of Mining Activities 'stipulates that any area affected by an operator's mining activity (whether inside or outside the perimeter of his operations) must be rehabilitated when the mining activities have been abandoned for a period of more than three years' (Environmental Law Institute, 2000b). Referred to as closure plans, they address such specific environmental impacts as the physical and chemical stabilisation of waste, the control of pollution streams, surface draining and erosion control, and rehabilitation of the area. As this law was passed only recently, it is too early to evaluate the success of restoration.

Brazil

The largest country in South America, with an area of $8\,511\,965$ km^2, Brazil is mostly tropical. Brazil has enormous mineral wealth and at present produces over 60 minerals (Filho, 1997). The forests and woodlands area, representing 58% of the land, have been the subject of intense interest both from an ecological standpoint and because of possible impacts of widespread deforestation on erosion, soil quality and productivity, biological diversity and global climate change. About 22% of the land is in permanent pasture and more forested areas continue to be burned to create conditions for pasture and croplands. There have been numerous ecological studies in Brazil with some reference to rehabilitation (e.g. Uhl, 1988; Jordan, 1989; Buschbacher et al., 1992; Parrota et al., 1997a, b).

The legislative strategies of Brazil are in the 1981 National Environmental Policy Act, in which the environmental legislation has been modeled after the United States' NEPA. The powers for setting the national standards for environmental quality, including pollution and conservation of ecosystems, rest with the federal government. Granting of permits for activities that may pollute or harm ecosystems is also under federal jurisdiction. The state and local authorities have the regulatory autonomy over environmental protection and quality.

Brazil requires an EIA for some minerals, and it has some 'unique features' in that it is prepared by an interdisciplinary group of experts, when required (Environmental Law Institute, 2000b). But the experts are selected and paid by those proposing the project and the latter are responsible for accuracy of information contained therein. The Brazilian Institute for the Environment and Renewable Resources provides technical support to the Ministry of the Environment which has the ultimate responsibility and the National Council for the Environment which oversees the policy (Filho, 1997). The environmental policies still seem to be judged by economic development rather than by environmental disruption.

Chile

With an area of $756\,950$ km^2, Chile has about 22% forested land and 18% in pasture use. It has a diversity of climates ranging from temperate to Mediterranean to desert. Early large-scale disturbance (e.g. mining) was concentrated in the northern part of Chile where there is little agriculture and forestry, but recently projects are being developed in the south which has a high rainfall, possibly fragile ecosystems, and great tourism potential (Lagos, 1997).

Chile created a National Commission for the Environment in 1990 that became functional in 1994 with the ushering in of environmental laws (Lagos, 1997; Lagos & Velasco, 1998). The agency noted that in 1991–2 nearly 2200 environmental laws and regulations were on the books although these were either out of date or not applied at all. The environmental impact evaluation system responsible for environmental pollution and overseeing environmental management, however, has been effective since 1987. Only projects with significant environmental impacts are required to submit detailed environmental assessments.

Laws are in place that are intended to regulate such environmental impacts as the discharge of effluents into rivers, lakes, or the ocean, groundwater contamination, particulate and other (SO_2, CO_2, NO_2 and O_3) emissions from stationary sources, and other waste disposal. As the Environmental Law Institute (2000b) notes, the country's regulations do not state specifically the measures that will be implemented for mitigation of environmental impacts.

Colombia

Colombia, with an area of over 1.1 million km^2, has a tropical climate along the coast but is cooler in the highlands. Forests constitute about 48% of the land area and pastures about 46%. Mining legislation is contained in the new Mining Code of 1988. All mineral resources found on the surface or subsurface belong to the state which may carry out exploration and mining activities directly through its own decentralised entities or indirectly by granting mining titles to the private sector. The Code enacts in essence two types of mining activities: state mining and private-sector mining. Mining of coal, precious stones, salt and radioactive minerals may be performed under state mining only, while mining of the other mineral resources may be done under either system.

Peru

Peru has an area of over 1.28 million km^2 with tropical to cold climates in the Andes and an elevation ranging from 0 m to over 6700 m. Forests and woodlands constitute nearly 66% of the area, and 21% is in permanent pasture. In 1986, the national government issued a report which clearly emphasised the importance of mining activities in the degradation of soil, air and water resources (Nuñez-Barriga & Castaneda-Hurtado, 1998). The report also identified 16 critical areas where irreversible adverse changes may have taken or are taking effect.

Peru passed the Environmental and Natural Resources Code in 1990 which created the National Environmental System. In 1993, several legal measures were passed for mitigating pollution effects in the mining sector in 1993. EIAs are required for all new operations and those existing ones that need 50% or more modifications. But, as Environmental Law Institute (2000b) notes, many exploration activities are required to present only an 'Environmental Evaluation' rather than an 'Environmental Impact Study'. The former differs from the latter 'in the issues that must be addressed, the periods for approval and the requirement for conducting a public hearing. As a result, opportunities for promoting pollution prevention may be lost' (Environmental Law Institute, 2000b).

The Ministry of Energy and Mines is responsible for enforcing environmental regulations and is the authority in: (1) establishing the environmental protection policies for mining activities and issuing the corresponding rules; (2) approving the Environmental Impact Assessment (EIA) and the Program for Environmental Management and Adjustment (PAMA), and authorising their execution; (3) entering into administrative–environmental stability agreements with the holders of mining activities on the basis of the EIA or PAMA approved; and (4) controlling the environmental effects produced by mining activities on operational sites and influence areas, determining the holder's liability, in case of violations to the applicable environmental provisions, and imposing the sanctions provided for therein (Environmental Law Institute, 2000b). While mining has been an important component of Peruvian economy (45%–50% of minerals are exported), the overall enforcement of environmental regulation has been weak (Nuñez-Barriga & Castaneda-Hurtado, 1998).

Uruguay

A relatively small country (area 176 220 km^2) with nearly 77% of its land area in permanent pasture, Uruguay lacks any mountain barriers. The Mining Code in force in Uruguay was revised in 1982 and provides a suitable legal framework for the development of the mining industry. The National Mining and Geology section of the Ministry of Industry, Energy and Mining administers the regulations.

A three-step process is required in the extraction of minerals: (1) the prospecting permits allow companies to conduct geological surveys, geochemical or geophysical sampling and similar activities for one or more minerals in an area up to 200 000 ha; (2) after prospecting, permits are issued for identifying the characteristics, volume, quality and purity of minerals, and soil analyses; (3) 500-ha parcels for mineral acquisition are developed. Evaluation of impacts on biological resources, water and human habitation are required. It is not clear how ecosystem rehabilitation fits in the overall legislative scheme.

Venezuela

With an area of 912 050 km², Venezuela ranges in elevation from sea level to over 5000 m. The forested area constitutes about 34% and permanent pastures about 20% of the country. All land disturbance activities are controlled by the national government under the Mines Act of 1944. Venezuela is a major exporter of oil. The disturbance effects on savannas, their management and rehabilitation in Venezuela have been treated in several studies (see Solbrig *et al.*, 1992).

CONCLUDING REMARKS

The potential wealth of a nation lies in its land, its waters, its share of air and sunlight, and in the ability and vision of its people. Harnessing these resources without causing irreversible harm to the environment imposes a great challenge to human ingenuity, wisdom and sense of responsibility. Although economic growth and prosperity are the ultimate goal of all nations, these come at a price. As discussed above, expansion of agricultural, industrial and many other human activities has the potential to impact the environment adversely. Indeed, many ecosystems in the Americas are already affected and in danger of further degradation. However, the apparent polarity between economics and ecology is not necessarily an inevitable fact, because the very technology that fuels economic growth is also a live reservoir in which we find the technical solutions and economic means to protect our environment. Fortunately, most nations and their scientists and politicians are aware of the necessity of protecting their natural resources and environment in general. In the Americas, environmental law and restoration ecology are gaining both public support and the benefit of technical advancement.

In North America, development of natural resources and management of agricultural and non-agricultural lands and ecosystems have already become science and law based. The impetus came not only from individual scientists and the concerned public but also from prestigious scientific organisations (United States National Academy of Sciences, 1974; United States National Research Council, 1980, 1981, 1986; Shulze, 1996). Reclamation of coal-mine lands is required by law, which is quite comprehensive and exacting in the United States compared to Canada or any other nation in the two continents. There are a number of agencies that have an advisory, management or regulatory role in the use of land and water resources. However, there are also a number of other environment-damaging activities for which no appropriate laws or technical remedies exist. The legal and scientific philosophy embodied in the SMRCA regarding restoration of mine lands in the United States has a great potential of applicability to the restoration of lands disturbed by other activities both in the United States and other countries in the Americas (Safaya & Wali, 1992).

But several aspects of SMCRA could be strengthened, as was pointed out by McElfish & Beier (1990). These needed reinforcements to the law are given below:

- The law has resulted in the 'sophistication of state programs and those of the industry, federal oversight is deemed critical to keep up with the demands of regulation'.
- Permitting and bonding procedures need to be strengthened to ensure reclamation to SMCRA standards.
- Water quality concerns should be treated as critical. For example, a recent report notes that 'acid mine drainage from surface mines is estimated to last for 10–20 years, while estimates from acid drainage from underground mines vary from 10s–100s of years' (Demchak *et al.*, 2001).
- The SMCRA must address the problems of subsidence and impairment of water by underground mining. Although in 1997 the US Department of the Interior drafted new subsidence rules, the federal and state fragile lands were not addressed specifically. Thus, the new policy has been only partially successful (Watson *et al.*, 1999).
- To keep up with the original intent, more funds must be allocated to the reclamation of pre-1977 abandoned mine sites.
- The protection of areas unsuitable for mining should be firmly enforced.
- Many operators have followed the least expensive ways to rehabilitate, thus, a quick grass cover is used in areas that prior to mining were forested.

- Lastly, it is appropriate to point that even after the passage of the law 24 years ago, debates continue in some political circles on ways in which the provisions of the law could be weakened.

Further, unlike in respect of coal, no strong environmental law exists to protect the environment from mining of hard rock minerals such as silver, gold, lead, uranium and copper. Hard rock mining is regulated by 1872 Mining Law. Under this law, a person or company can lay claim to a 20-acre (8 ha) land parcel (or multiple parcels) at a nominal fee of $5 an acre (0.4 ha). Under this law, 270 million acres (109 million ha) of land may be eligible for mining (Wuerthner, 1998).

The devastating results from this 128-year-old law have been reported widely. An estimated 557 000 abandoned hard rock mine sites have been left unreclaimed and more than 19 000 km of rivers poisoned by mining wastes because they are exempt from federal regulation and rules that govern and regulate rehabilitation. The estimated minimum $32–72 billion in clean-up costs is borne by the taxpayers. Even within the National Parks, there are over 4000 abandoned mining sites. Although Congress passed a temporary moratorium in 1994 on the Mining Law of 1872, it has been ineffectual. A recent report on hard rock mining emphasises the importance of environmental restoration (United States National Research Council, 1999).

Such, however, is not the case for the mining of coal, oil and gas on federal lands. Other federal laws including the Mining Leasing Act of 1920 govern their extraction. Under the latter Act, lands are not sold, but leased, and companies pay a 12.5% royalty for these minerals to be removed. Government agencies can deny leasing if the environmental values are deemed to be diminished by mineral extraction. The above examples clearly illustrate the need for laws that are comprehensive and firmly science-based for the overall public good.

Most of the Central and South American countries emphasise economic development, but many important aspects of environmental rehabilitation are not on the forefront. Many professionals believe that while all countries have passed some environmental laws and regulations, few enforce them. However, the 1992 Earth Summit in Rio de Janeiro, the North American Free Trade Agreement, and a rising environmental consciousness reinvigorated a resolve to address serious environmental issues in the developing nations of Latin America. It is a significant development that nearly all of these nations are signatories to United Nations Agreements, Declarations and Protocols (see Table 1.4).

In order to bring about a practical, on-the-ground change, barriers to effective ecosystem restoration that remain must be addressed. Many of these are the same barriers that Mares (1986) noted for effective conservation of biological resources in Latin America. Some of these barriers are as follows:

- There is a lack of basic ecological studies in this region that limits the success of rehabilitation strategies. 'Too few scientists have gathered sufficient high-quality data on the status of South America's biota to determine exactly what is taking place. The data pertaining to widespread habitat destruction with concomitant loss of species are not strong' (Mares, 1986). An exchange of trained and experienced professionals would remedy this gap.
- There appears to be a serious disconnection between science and policy needs. Although there are laws and regulations, few of them seem to be enforced. This requires advancing the idea of interdisciplinary groups designed to address specific issues.
- A new cadre of environmentally literate and environmentally conscious young professionals needs to be trained, who can interact with each other with a degree of ease. This is a long-term agenda but one that no country can do without.
- There is a tendency to regulate environmental disturbances at the state or provincial level and to oppose national legislation that would bring about some uniformity (as in the United States). Revesz (1992) called this a 'race to the bottom' type of reasoning in environmental protection wherein economic competition among states results, most often, in lax environmental regulation and avoidance of environmental rehabilitation. Such thinking needs to be changed.
- The emphasis in Latin America appears to be on short-term advantages and hence a quick development of resources for export. Exports, foreign investments, and markets dominate the literature.

Table 1.4. *Nation–state signatories to international agreements which have a direct impact on environmental restoration*

Country	Biodiversity	Climate change	Climate Change – Kyoto Protocol	Desertification	Endangered species	Marine life conservation	Ozone layer protection
Argentina	•	•	•	•	•	•	•
Belize	•	•		•	•	•	•
Bolivia	•	•	•	•	•	•	•
Brazil	•	•		•	•		•
Canada	•	•	•	•	•		•
Chile	•	•	•	•	•	•	•
Colombia	•	•		•	•	•	•
Costa Rica	•	•	•	•	•		•
Cuba	•	•		•	•		•
Dominican Republic	•	•		•	•		•
Ecuador	•	•	•	•	•		•
El Salvador	•		•	•			•
Guatemala	•	•	•	•			•
Guyana	•	•		•			•
Honduras	•	•	•	•			•
Jamaica	•	•		•	•	•	•
Mexico	•	•	•	•	•		•
Nicaragua	•	•	•	•			•
Panama	•	•	•	•	•		•
Paraguay	•	•	•	•	•		•
Peru	•	•	•	•			•
Suriname	•			•	•		•
Trinidad and Tobago	•	•	•	•	•		•
United States	•	•	•	•	•	•	•
Uruguay	•	•	•	•	•		•
Venezuela	•	•		•	•	•	•

Without educating the general citizenry, no environmental rehabilitation policy or plan will succeed. An active participation of the public in the respective countries is crucial to any national environmental agenda.

As the Environmental Law Institute (1998) notes, laws enacted should be: (1) biologically sound; (2) socially adequate; and (3) institutionally practical. We agree. But, given the extent of ecosystem disturbances in Latin America, there is clearly a need in many of the countries of that region to create effective and enforceable environmental laws, rules and practices. This would establish appropriate mechanisms and standards for environmental restoration without much delay for the betterment of their ecosystems and their people.

ACKNOWLEDGMENTS

For invaluable help with the information in the Boxes, our sincere thanks are due to Drs Vasant

Akala, Evan Blumer, Robert Gates, Rattan Lal and
Roger Williams; to Ms Holly Gibbs for her assistance
with Fig. 1.1; and Dr T. R. Loveland for the use of
modified Fig. 1.1. Mr Jason Funk assisted at several
stages, our warmest thanks to him.

REFERENCES

Acharya, A. (1995). The fate of the boreal forests. *World Watch*, **8**, 20–29.

Akala, V. A. & Lal, R. (2000). Potential of land reclamation for soil organic carbon sequestration in Ohio. *Land Degradation and Development*, **11**, 289–297.

Albarracin, S. F. (1997). The new Argentine framework. *Resources Policy*, **23** (1–2), 33–43.

Asay, K. H. (1979). Breeding grasses for revegetation of surface mining spoils in western U.S. In *Ecology and Coal Resource Development*, vol. 2, ed. M. K. Wali, pp. 1007–1011. New York: Pergamon Press.

Bazzaz, F. A. (1983). Characteristics of populations in relation to disturbance in natural and man-modified ecosystems. In *Disturbance and Ecosystems: Components of Response*, eds. H. A. Mooney & M. Godron, pp. 259–275. New York: Springer-Verlag.

Beck, R. E. (1973). The North Dakota Surface Mining Control and Reclamation Law. In *Some Environmental Aspects of Strip Mining in North Dakota*, Education Series no. 5, ed. M. K. Wali, pp. 109–118. Grand Forks, ND: North Dakota Geological Survey.

Bowling, K. C. (1978). History of legislation in different states. In *Reclamation of Drastically Disturbed Lands*, eds. F. W. Schaller & P. Sutton, pp. 95–116. Madison, WI: American Society of Agronomy.

Bradshaw, A. D. (1970). Pollution and plant evolution. *New Scientist*, **48**, 497–500.

Bradshaw, A. D. (1983). The reconstruction of ecosystems. *Journal of Applied Ecology*, **20**, 1–17.

Bradshaw, A. D. (1997). The importance of soil ecology in restoration science. In *Restoration Ecology and Sustainable Development*, eds. K. M. Urbanska, N. R. Webb & P. J. Edwards, pp. 33–64. New York: Cambridge University Press.

Bradshaw, A. D., McNeilly, T. S. & Gregory, R. P. G. (1965). Industrialization, evolution and the development, of heavy metal tolerance in plants. *In Ecology and Industrial Society*, eds. G. T. Goodman, R. W. Edwards & J. M. Lambert, pp. 327–343. Oxford: Blackwell.

Brown, J. C., Ambler, J. E., Chaney, R. L. & Foy, C. D. (1972). Differential responses of plant genotypes to micronutrients. In *Micronutrients in Agriculture*, eds. J. J. Mordvedt, P. M. Giordano & W. L. Lindsay, pp. 389–418. Madison, WI: Soil Science Society of America.

Buschbacher, R., Uhl, C. & Serrao, E. A. S. (1992). Reforestation of degraded Amazon pasture lands. In *Ecosystem Rehabilitation: Preamble to Sustainable Development*, vol. 2, ed. M. K. Wali, pp. 257–274. The Hague: SPB Academic Publishing.

Cairns, J., Jr (1988a). Restoration ecology: the new frontier. In *Rehabilitating Damaged Ecosystems*, vol. 1, ed. J. Cairns, Jr, pp. 1–11. Boca Raton, FL: CRC Press.

Cairns, J., Jr (1988b). Increasing diversity by restoring damaged ecosystems. In *Biodiversity*, eds. E. O. Wilson & F. M. Peters, pp. 333–343. Washington, DC: National Academy Press.

Cairns, J., Jr, Dickson, K. L. & Herricks, E. E. (eds.) (1977). *Recovery and Restoration of Damaged Ecosystems*. Charlottesville, VA: University Press of Virginia.

D'Aleo, J. (2000). Expanding deserts are the latest signs of a changing climate. News release, 15 November 2000, Weather Services International, Billerica, MA.

Demchak, J., Skousen, P., Ziemkiewicz, P. & Bryant, G. (2001). Comparison of water quality from 12 underground coal mines in 1968 and 1999. *Green Lands*, **31**, 48–60.

Doan, D. B., Gurmendi, A. C., Torres, I. E. & Velasco, P. (1999). The mineral industries of Latin America and Canada. In *Area Reports International 1997, Latin America and Canada, United States Geological Survey Minerals Yearbook*, vol. 3, pp. A1–A9. Reston, VA: US Geological Survey.

Dodd, W. E. & Ogaard, L. A. (1998). Evaluating reclamation success at three AML sites in North Dakota. Presented at the *20th Annual Conference of the Association of Abandoned Mine Land Programs*, Albuquerque, NM, 20 September–1 October 1998.

Environmental Law Institute (1998). *Legal Aspects for Forest Management in Mexico*. Washington, DC: Environmental Law Institute.

Environmental Law Institute. (2000a). *Pollution Prevention and Mining: Research Guide for National Case Studies*. Washington, DC: Environmental Law Institute.

Environmental Law Institute. (2000b). *Pollution Prevention and Mining: A Proposed Framework for the Americas*. Washington, DC: Environmental Law Institute.

Epstein, E. & Norlyn, J. D. (1977). Sea-water-based crop production: a feasibility study. *Science*, 197, 249–251.

Fedkenheuer, A. W. & Macyk, T. M. (2000). Reclamation of surface mined coal lands in western Canada. In *Reclamation of Drastically Disturbed Lands*, eds. R. I. Barnhisel, R. G. Darmody & W. L. Daniels, pp. 567–594. Madison, WI: American Society of Agronomy.

Filho, C. A. V. (1997). Brazil's mineral policy. *Resources Policy*, 23, 45–50.

Gardner, G. T. & Sampat, P. (1998). *Mind over Matter: Recasting the Role of Materials in our Lives*, Worldwatch Paper no. 144. Washington, DC: World Watch Institute.

Gleeson, S. K. & Tilman, D. (1990). Allocation and the transient dynamics of succession on poor soils. *Ecology*, 71, 1144–1155.

Government of Canada (1996). *The Minerals and Metals Policy of the Government of Canada*. Ottawa, Ontario: Natural Resources Canada.

Government of Canada (2000). *The Canadian Environmental Protection Act*. Ottawa, Ontario: Natural Resources Canada.

Gunn, J. M. (ed.) (1995). *Restoration and Recovery of an Industrial Region*. New York: Springer-Verlag.

Hernandez, A. (1997). Mexico: policy and regulatory framework for mining. *Resources Policy*, 23, 71–77.

Holmberg, G. V., Horvath, W. J. & Lafevers, J. R. (1978). Citizens' role in land distrubance and reclamation. In *Reclamation of Drastically Disturbed Lands*, eds. F. W. Schaller & P. Sutton, pp. 69–94. Madison, WI: American Society of Agronomy.

Houghton, R. A. (1994). The worldwide extent of land-use change. *BioScience*, 44, 305–313.

Hutnik, R. J. & Davis, G. (eds.) (1973). *Ecology and Reclamation of Devastated Land*, 2 vols. New York: Gordon & Breach.

Imes, A. C. & Wali, M. K. (1977). An ecological–legal assessment of mined land reclamation laws. *North Dakota Law Review*, 53, 359–399.

Imes, A. C. & Wali, M. K. (1979). Governmental regulation of reclamation in the western United States. *Reclamation Review*, 1, 75–88.

Iverson, L. R. (1988). Land-use changes in Illinois, USA: the influence of landscape attributes on current and historical land use. *Landscape Ecology*, 2, 45–61.

Iverson, L. R., & Wali, M. K. (1992). Grassland rehabilitation after coal and mineral extraction in the western United States and Canada. In *Ecosystem Rehabilitation: Preamble to Sustainable Development*, vol. 2, ed. M. K. Wali, pp. 85–129. The Hague: SPB Academic Publishing.

Jordan, C. F. (ed.) (1989). *An Amazonian Rain Forest*. Paris: UNESCO.

Keddy, P. A. & Drummond, C. G. (1996). Ecological properties for the evaluation, management, and restoration of temperate deciduous forest ecosystems. *Ecological Applications*, 6, 748–762.

Lagos, G. (1997). Developing national mining policies in Chile: 1974–97. *Resources Policy*, 23, 51–69.

Lagos, G. & Velasco, P. (1998). Environmental policies and practices in Chilean mining. In *Mining and the Environment: Case Studies from the Americas*, ed. A. Warhurst, pp. 101–136. Ottawa, Ontario: International Development Research Centre.

Leopold, D. J. & Wali, M. K. (1992). The rehabilitation of forest ecosystems in the eastern United States and Canada. In *Ecosystem Rehabilitation: Preamble to Sustainable Development*, vol. 2, ed. M. K. Wali, pp. 187–231. The Hague: SPB Academic Publishing.

Logan, T. J. (1989). Chemical degradation of soil. *Advances in Soil Science*, 11, 187–222.

Logan, T. J. (1992). Reclamation of chemically degraded soils. *Advances in Soil Science*, 17, 13–35.

Loveland, T. R. & Belward, A. S. (1997). The International Geosphere-Biosphere Programme Data and Information System Global Land Cover Data Set (DISCover). *Acta Astronautica*, 41, 681–689.

Loveland, T. R., Reed, B. C., Brown, J. F., Ohlen, D. O., Zhu, Z., Yang, L. & Merchant, J. W. (2000). Development of a global land cover characteristics database and IGBP DISCover from 1 km AVHRR data. *International Journal of Remote Sensing*, 21, 1303–1330.

Macyk, T. M. (2000). Reclamation of alpine and subalpine lands. In *Reclamation of Drastically Disturbed Lands*, eds. R. I. Barnhisel, R. G. Darmody & W. L. Daniels, pp. 537–565. Madison, WI: American Society of Agronomy.

Mares, M. A. (1986). Conservation in South America: problems, consequences, and solutions. *Science*, 233, 734–739.

Marrs, R. H. & Bradshaw, A. D. (1993). Primary succession on man-made wastes: the importance of resource acquisition. In *Primary Succession on Land*, eds. J. Miles & D. W. H. Walton, pp. 221–248. Oxford: Blackwell.

Mathews, E. (1983). Global vegetation and land use: new high-resolution data bases for climate studies. *Journal of Climate and Applied Meteorology*, **22**, 474–487.

McCally, D. (1999). *The Everglades: An Environmental History.* Gainesville, FL: University of Florida Press.

McElfish, J. M., Jr & Beier, A. (1990). *Environmental Regulation of Coal Mining: SMCRA's Second Decade.* Washington, DC: Environmental Law Institute.

Merrill, S. & Safaya, N. M. (1984). Physical and chemical methods for enhancing productivity of abandoned mine lands in the Northern Great Plains. In *Proceedings of the National Symposium and Workshops on Abandoned Mine Land Reclamation*, eds. L. L. Schloesser, G. S. Anderson, N. M. Safaya & D. J. Thompson, pp. 562–600. Middlesex, UK: Science Reviews.

Nuñez-Barriga, A. & Castaneda-Hurtado, I. (1998). Environmental management in a heterogeneous mining industry: the case of Peru. In *Mining and the Environment: Case Studies from the Americas*, ed. A. Warhurst, pp. 137–180. Ottawa, Ontario: International Development Research Centre.

Oldeman, L. R. (1994). The global extent of soil degradation. In *Soil Resilience and Sustainable Land Use*, eds. D. J. Greenland & I. Szabolcs, pp. 99–118. Wallingford, UK: CAB International Publishers.

Otto, J. M. (1998). Global changes in mining laws, agreements and tax systems. *Resources Policy*, **24**, 79–86.

Parrota, J. A., Knowles, O. H. & Wunderle, J. M., Jr (1997*a*). Development of floristic diversity in 10-year-old restoration forests on a bauxite mined site in Amazonia. *Forest Ecology and Management*, **99**, 21–42.

Parrota, J. A., Turbull, J. W. & Jones, N. (1997*b*). Catalyzing native forest regeneration on degraded tropical lands. *Forest Ecology and Management*, **99**, 1–7.

Ramakutty, N. & J. A. Foley. (1999). Estimating historical changes in global land cover: croplands from 1700 to 1992. *Global Biogeochemical Cycles*, **13**, 997–1027.

Reid, W. V. (1992). How many species will there be? In *Tropical Deforestation and Species Extinction*, eds. T. Whitmore & J. Sayer, pp. 55–72. London: Chapman & Hall.

Revesz, R. L. (1992). Rehabilitating interstate competition: rethinking the 'Race-to-the-Bottom' rationale for federal environmental regulation. *New York University Law Review*, **1210**, 1233–1247.

Ripley, E. A., Redmann, R. E. & Maxwell, J. (1978). *Environmental Effects of Mining in Canada.* Kingston, Ontario: Centre for Resource Studies, Queens University.

Ripley, E. A., Redmann, R. E. & Crowder, A. A. (1996). *Environmental Effects of Mining.* Delray Beach, FL: St Lucie Press.

Robinson, N. (1992). Ecosystem rehabilitation: the environmental law framework. In *Ecosystem Rehabilitation: Preamble to Sustainable Development*, vol. 1, ed. M. K. Wali, pp. 129–142. The Hague: SPB Academic Publishing.

Safaya, N. M. (1979). Delineation of mineral stresses in mine spoils and screening for plants for adaptability. In *Ecology and Coal Resource Development*, vol. 2, ed. M. K. Wali, pp. 830–849. New York: Pergamon Press.

Safaya, N. M. & Wali, M. K. (1992). Applicability of US environmental laws in the developing countries: an analysis of ecological and regulatory concepts. In *Ecosystem Rehabilitation: Preamble to Sustainable Development*, vol. 1, ed. M. K. Wali, pp. 143–155. The Hague: SPB Academic Publishing.

Schaller, F. W. & Sutton, P. (eds.) (1978). *Reclamation of Drastically Disturbed Lands.* Madison, WI, American Society of Agronomy.

Schindler, D. (1998). A dim future for boreal waters and landscape. *BioScience*, **48**, 157–164.

Shinya, W. M. (1998). Canada's new minerals and metals policy. *Resources Policy*, **24**, 95–104.

Shulze, P. C. (ed.) (1996). *Engineering with Ecological Constraints.* Washington, DC: National Academy Press.

Solbrig, O., Goldstein, G., Medina, E., Sarmiento, G. & Silva, J. (1992). Responses of tropical savannas to stress and disturbance: a research approach. *Ecosystem Rehabilitation: Preamble to Sustainable Development*, vol. 2, ed. M. K. Wali, pp. 63–73. The Hague: SPB Academic Publishing.

Stein, B. A., Kutner, L. S. & Adams, J. S. (eds.) (2000). *Precious Heritage: The Status of Biodiversity in the United States,.* The Nature Conservancy and Association for Biodiversity Information. New York: Oxford University Press.

Stevenson, F. J. (1986). *Cycles of Soil: Carbon, Nitrogen, Phosphorus, Sulfur, Micronutrients.* New York: John Wiley.

Thirgood, J. V. (1978). Extent of disturbed land and major reclamation problems in Canada. In *Reclamation of Drastically Disturbed Lands*, eds. F. W. Schaller &

P. Sutton, pp. 45–68. Madison, WI: American Society of Agronomy.

Tilman, D. (1988). *Plant Strategies and the Dynamics and Structure of Communities*. Princeton, NJ: Princeton University Press.

Uhl, C. (1988). Restoration of degraded lands in the Amazon Basin. In *Biodiversity*, eds. E. O. Wilson & F. M. Peters, pp. 326–332. Washington, DC: National Academy Press.

United Nations (1998). *World Population 1998*, United Nations Publication ST/ESA/SER.A/176. Population Division, Department of Economic and Social Affairs. (www.undp.org/popin/wdtrends/p98/p98.htm)

United States Department of the Interior (1967). *Surface Mining and Our Environment: A Special Report to the Nation*. Washington, DC: Government Printing Office.

United States National Academy of Sciences (1974). *Rehabilitation Potential of Western Coal Lands*. Cambridge, MA: Ballinger Publishing Co.

United States National Research Council (1980). *Surface Mining: Soil, Coal, and Society*. Washington, DC: National Academy Press.

United States National Research Council (1981). *Coal Mining and Ground Water Resources in the United States*. Washington, DC: National Academy Press.

United States National Research Council (1986). *Abandoned Mined Lands: A Mid-Course Review of the National Reclamation Program for Coal*. Washington, DC: National Academy Press.

United States National Research Council (1999). *Hardrock Mining on Federal Lands*. Washington, DC: National Academy Press.

Wali, M. K. (1975). The problem of reclamation viewed in a systems context. In *Practices and Problems of Land Reclamation in Western North America*, ed. M. K. Wali, pp. 1–17. Grand Forks, ND: University of North Dakota Press.

Wali, M. K. (1980). Succession on mined lands. In *Adequate Reclamation of Mined Land? A Symposium*, pp. 23–46. Ankenny, IO: Soil Conservation Society of America.

Wali, M. K. (1987). The structure, dynamics, and rehabilitation of drastically disturbed ecosystems. In *Perspectives in Environmental Management*, ed. T. N. Khoshoo, pp. 163–183. New Delhi: Oxford University Press and IBH Publishing.

Wali, M. K. (1999). Ecological succession and the rehabilitation of disturbed terrestrial ecosystems. *Plant and Soil*, **213**, 195–230.

Wali, M. K. & Burgess, R. L. (1985). The interface of ecology and law: science, the legal obligation, and public policy. *Syracuse Journal of International Law and Commerce*, **12**, 221–253.

Wali, M. K. & Freeman, P. G. (1973). Ecology of some mined areas in North Dakota. In *Some Environmental Aspects of Strip Mining in North Dakota*, Education Series no. 5, ed. M. K. Wali, pp. 25–47. Grand Forks, ND: North Dakota Geological Survey.

Wali, M. K., Evrendilek, F., West, T. O., Watts, S. E., Pant, D., Gibbs, H. K. & McClead, B. E. (1999). Assessing terrestrial ecosystem sustainability: usefulness of carbon and nitrogen models. *Nature and Resources*, **35**, 21–33.

Walker, L. R. (1999). Patterns and process in primary succession. In *Ecosystems of Disturbed Ground*, ed. L. R Walker, pp. 585–610. Amsterdam: Elsevier Science Publishers.

Watkin, E. M. (1979). Reclamation practices and problems in eastern Canada. In *Ecology and Coal Resource Development*, vol. 1, ed. M. K. Wali, pp. 450–460. New York: Pergamon Press.

Watson, W. D., Lin, K. & Browne, T. (1999). US policy instruments to protect coal-bearing fragile lands. *Resources Policy*, **25**, 125–140.

West, T. O., & Wali, M. K. (in press). Modeling regional dynamics and soil erosion in disturbed and rehabilitated ecosystems as affected by land use and climate. *Water, Air, and Soil Pollution*.

Whittaker, R. H. (1974). Climax concepts and recognition. In *Vegetation Dynamics*, ed. R. Knapp, pp. 137–154. The Hague: Dr W. Junk.

Woodwell, G. M. (1992). When succession fails.... In *Ecosystem Rehabilitation: Preamble to Sustainable Development*, vol. 1, ed. M. K. Wali, pp. 27–35. The Hague: SPB Academic Publishing.

Wuerthner, G. (1998). High stakes: the legacy of mining. *National Parks*, **72**, 22–25.

2 • Europe

F. JANE MADGWICK AND TIM A. JONES

INTRODUCTION

The geography of Europe is such that it is not so much a continent, but more a peninsular append-age of Asia that is separated from Africa by the Mediterranean Sea. For the purposes of this chapter, and to reflect some ecological limits, the northern boundary lies beyond the Arctic circle at the edge of the permanent ice sheets, the southern bound-ary is the Mediterranean Sea and the eastern fron-tier is the eastern edge of the vast Ukrainian steppes. The expanding political boundaries of the European Union will soon include some of the eastern Baltic states and Hungary, most likely within a decade, reaching as far east as the Black Sea and as far south as Turkey.

The long history of human settlement in most parts of Europe means that landscapes have become highly modified, so that natural ecosystems have long since been replaced by systems that lie some-where on the spectrum between 'semi-natural' and 'artifical' or 'created' (Fig. 2.1).

Before considering the current attitudes to, con-straints on, and opportunities for ecological restora-tion in Europe, it is worth reflecting on the time-scale and process of landscape change that has resulted in the current gradient of human modification.

None of the remaining biomes, such as wetland or forest complexes, represents the variety of ecosys-tems that existed before the advent of cultivation. Two thousand years ago, the greater part of central lowland Europe was covered in dense forest. Only fragments of these primeval forests and swamps re-main to provide clues to our ancient natural history. The vast deltas, marshes and floodplains of rivers such as the Danube were quickly peopled by farmers and fishermen who used every local resource to cre-ate an economy. The mountains, moors and exten-sive flooded grasslands of the central plains were slower to change and were tamed just two cen-turies ago. Today it is only in the Arctic region that large blocks of natural ecosystems remain intact; the largest includes the Icelandic Highlands. Even here however, climate change accelerated by human activities is causing significant alterations.

Beginning 10 000 years ago and over the follow-ing centuries, complex 'cultural landscapes' devel-oped in Europe with the spread of settled agricul-ture. This began in the Mediterranean region. As a result of small-scale patterns of land use, these agri-cultural landscapes often supported a great variety of habitats in very small areas. Opening up of the forest also contributed to the extension of some, formerly rare habitats. Low-intensity grazing by do-mestic animals then maintained a great number of species-rich grassland communities. Indeed, many of the people-shaped landscapes are truly stun-ning, for example the flower-carpeted olive groves of Greece, the rolling vistas of cork oak dehesas in Spain, the terraced hillsides of Italy and the dra-matic moorlands of Scotland.

Human-made features have increasingly con-cealed the underlying pattern of distinct natural regions in Europe. Habitat types that we now recog-nise tend to have sharper boundaries than more nat-ural ecosystems so that woodlands or wetlands now end abruptly where cultivation and drainage begins, and complex mosaics of marshes, bogs and wood-lands are largely confined to parts of Scandinavia and Russia. The anthropogenic landscapes depend on continued human intervention and low-intensity

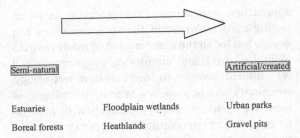

Semi-natural		Artificial/created
Estuaries	Floodplain wetlands	Urban parks
Boreal forests	Heathlands	Gravel pits

Fig. 2.1. Diagram to show the position of selected European habitats in relation to degree of human intervention.

land management practices for their structural and ecological diversity to be maintained. The dynamic influence of natural processes such as floods has been mostly lost due to the damming and taming of the great European rivers. In fact only one major European river, the Törnealven, on the border of Sweden and Finland, remains without a dam (Dynesius & Nilsson, 1994). Precious few stretches of river remain which retain their natural dynamics, with sand islands of varying ages, river cliffs, oxbow lakes and flooded forests. In addition, the large herbivores and carnivores that would have greatly influenced the dynamics of this European landscape have all but disappeared. For example, the last aurochs (the most primitive cattle breed) died in Jakarow Forest, just outside Warsaw, in 1627. This species was abundant after the last glaciation and occupied the richer grasslands and open woodlands of Europe before being hunted to extinction. The steppe marmot (*Marmota bobac*) and saiga (*Saiga tatarica*), Europe's only antelope, still survive within the remaining part of the herb-rich Ukrainian and Mongolian steppe, but they have largely been replaced by domestic herds and now depend on protected areas for their survival. Similarly, only around 3600 European bison (*Bison bonasus*) survive, about half in zoos, and half in small, isolated, free-living herds in Poland, Lithuania, Belarus, Russia and Ukraine. The brown bear (*Ursus arctos*), lynx (*Lynx lynx*), wolf (*Canis lupus*) and wolverine (*Gulo gulo*) are now confined to remote mountain areas, such as the Alps, Apennines, Carpathians, Pindus, Rhodope and Scandinavian mountains, where they struggle for survival alongside farming interests.

From the middle of the twentieth century onwards, agricultural intensification, rapid urban growth and the development of modern transport infrastructure led to the structural simplification and ecological impoverishment of large parts of Europe. This process has been most noticeable in northwest Europe, where regimented and 'sanitised' agricultural and urban landscapes now predominate. However, under communism, many of the countries of central and eastern Europe also experienced profound rural change during the twentieth century, with a drive for increased agricultural production and the implementation of engineering 'mega projects' such as the regulation of the Volga River in Russia. In southern Europe, over this same period, agricultural intensification (especially since EU membership for countries such as Greece, Portugal and Spain) and tourism development have also led to large-scale losses of semi-natural habitats. These changes have had a number of adverse ecological impacts, which have been the subject of considerable record and comment (e.g. Holdgate & Woodman, 1978; Hollis, 1992; Jones & Hughes, 1993; Tucker & Heath, 1994; EEA, 1995; Mannion, 1995; Møller, 1995; FAO, 1997; Benstead *et al.*, 1999; Finlayson & Davidson, 1999; IUCN, 1999; UNEP, 1999; Sih *et al.*, 2000):

- Reduction in habitat and species diversity and habitat heterogeneity
- Reduction in population and range of many species
- Fragmentation of habitats increasing the vulnerability of remaining isolated pockets to environmental or human-induced changes
- Reduction in the ability of naturally functioning ecosystems to provide economically important goods and services such as erosion protection, nutrient reduction or carbon retention.

These impacts have long been recognised by conservationists, but it is only in the last ten to fifteen years that their seriousness and scale have gained wider attention. Now, the need for ecological restoration is an increasingly accepted component of environmental management. In northern and western Europe at least, restoration measures are coupled with protection measures in most, if not all, nature conservation programmes.

THE CURRENT STATUS OF ECOLOGICAL RESTORATION

There are major differences within and between different regions of Europe in: (1) the relative 'intactness' of semi-natural habitats; (2) the relative technical, political and/or socio-economic priority that is being given to restoration; (3) the level and sources of funding and experience/expertise available.

For example, in northwest Europe, remaining areas of high natural value are so limited in extent and so highly fragmented, that a high and very costly degree of human intervention is required to arrest vegetation succession at a particular stage of development. Yet it is here that the majority of institutional and voluntary effort, research and funding is focused. A feature of northwest Europe's approach to nature conservation and the relatively high value placed by society on cultural landscapes is the myriad of environmental initiatives that operate at the local level. Nature restoration is often a central feature of environmental projects developed by municipalities, community groups, farmers and schools. For example, from 1989 to 1994, more than 1000 small-scale nature restoration projects have been carried out in Denmark to enhance nature, expand forest areas and promote recreational facilities. Many protected areas are tiny by international standards, but these fragments are all that remain and they are very highly valued. Visitor pressure stemming from the public's desire to enjoy such areas usually has to be tightly controlled to avoid irreversible damage.

In eastern Europe and some parts of southern Europe, there are still significant tracts of semi-natural habitats, including such renowned sites as the Danube Delta (Romania and Ukraine) and Bialowieza Forest (Poland and Belarus). Programmes for conservation and restoration in these areas rely largely on bilateral, EU and World Bank/Global Environment Facility (GEF) support and the engagement of western ecologists to work alongside the local experts. In the countries of central and eastern Europe, scientific specialists and long-term monitoring programmes abound, although there is little tradition of integrated, multidisciplinary management approaches. Furthermore, the economic crisis affecting most countries in the region over the last decade has led to the deterioration of many environmental research and monitoring programmes and very limited capacity to invest in new technology, techniques and approaches. Whilst the collapse of intensive agriculture and privatisation of land offer outstanding opportunities for innovative restoration projects, there can be a tendency both within the region and amongst external funding partners to think of restoration as an optional luxury. In fact, one only has to look at the ecological damage wrought by wholesale conversion of the Ukrainian steppe to intensive agriculture, or the massive regulation of major rivers, to recognise that restoration needs to be as much an integrated component of environmental management in eastern Europe as it is starting to become is some parts of western Europe.

In Mediterranean countries, people's activities have intricately shaped nearly every part of the landscape for thousands of years. Generally, the benefits of ecological restoration are poorly appreciated and economic objectives dominate the management of marginal lands, as illustrated by the continuing preference for plantations of exotic trees. Here, there is no contrast between wilderness and cultivated land. Ecological restoration is, for the most part, a relatively new and small-scale activity, with a large amount of variation between countries.

In the far north of Europe, much of the boreal forest and tundra zones appear virtually untouched, though they are suffering the effects of climate change and air-borne pollution. Ecological restoration (with the exception perhaps of treatment of lakes affected by acid rain and the improved management of dams) is seen as a low priority, while protection of the remaining intact ecosystems and control of pollution receive strong public and state support.

ECOLOGICAL AND SOCIO-ECONOMIC CONSTRAINTS

Although conservation in Europe initially focused on the establishment of protected areas, the failure of this approach in conserving biodiversity has naturally led to the development of a stronger focus on attempts to secure ecological functioning. This, in turn, has led to the acceptance of ecological restoration by most as an equal method of protection in a conservationist's 'toolbox'.

When considering ecological restoration, it is necessary to determine to what extent restoration is achievable – and desirable. In general, ecological targets for restoration are somewhere on a gradient between the treatment of some symptoms of ecological stress to the recovery of the intact, natural 'reference' condition including ecological functioning. The close links between biodiversity and the degree of natural dynamics within ecosystems is evident in comparative studies of most European habitat types. For example, in channelised rivers, where these edge habitats and floodplain features are sparse or absent, fish communities have tended to become dominated by relatively few species and the structure of the communities shifts from phytophilic (floodplain spawning species) to mainly lithophilic (main channel spawning) species. In contrast, large sections of rivers with a dynamic nature and intact floodplains support a wide range of fish species due to the occurrence of a large number of micro-environments. Two important relatively intact reference examples from central Europe are the Donau–Auen National Park just east of Vienna (60 species) and the Middle Vistula in Poland (44 species) (Zockler, 2000).

In practice, the conservationist must consider both ecological and socio-economic constraints in designing a restoration programme, since there are nearly always human-imposed limits as well as ecological limits on what is possible. In Europe, this means you can rarely 'turn back the clock' by putting the ecological processes into reverse. Instead, when considering the potential restoration targets, in addition to ascertaining the natural reference condition, the current constraints on ecological functioning and potential conflicts with economic objectives need to be understood and taken into account. For these reasons most conservationists in Europe recognise that ecological restoration in its true sense may only be possible to a limited extent, but that it still provides significant benefits. One example is the attempt to restore areas of the Rhône Delta in France which have been converted to rice fields (Box 2.1).

In severely damaged ecosystems, it can take many years of research and experimentation before sufficient understanding develops to describe the options open to decision-makers and to calculate the relative costs and benefits. This is true for example if you consider the attempts in the last decade at the restoration of eutrophic shallow lakes in northern and western Europe (Moss *et al.*, 1996; Madgwick,

Box 2.1 Restoration of rice fields, Rhône Delta

In the Rhône Delta, it is estimated that 40% of natural habitats have disappeared in the last 40 years, mostly due to rice cultivation. Less productive land was often abandoned and used for extensive grazing. However, the presence of salt in the groundwater and soil surface has resulted in the development of halophytic plant communities of low productivity and low diversity. The aim of restoration was to manage the water and grazing regimes to encourage the development of emergent plant communities typical of brackish Mediterranean marshes that were present before rice cultivation. Experimental treatment of 18 ha of former rice field plots over five years in the Camargue showed that artificial winter or summer flooding leads to the rapid decline of halophytes and an increase in emergent plant communities. It was also shown that grazing altered the nature of this succession. However, the lack of topographical variation due to successive levelling of these fields precluded depth and salinity gradients and hence the 'restoration' resulted in limited habitat heterogeneity, compared to natural marshes. The experiment nevertheless showed that artificially controlled seasonal flooding and grazing can be used to increase the forage value and provided improved feeding grounds for wintering wildfowl (Mesleard, 1994; Mesleard *et al.*, 1995, 1999).

1999; Jeppesen & Sammalkorpi, this volume). The public perception may be that enforcement of pollution reduction can be put in place to restore clear water and abundant, attractive aquatic life. It is true that in most cases nutrient reduction is a prerequisite for ecological restoration. Multiple lake studies have pointed to certain threshold levels of nutrient loading, above which restoration of the clear water condition is unlikely to be achieved (e.g. Jeppesen et al., 1999). But in reality, there are many places in northwest Europe where sufficient reduction of nitrogen and phosphorus loading to surface waters to enable shallow lake recovery may not be possible. For example, in some parts of The Netherlands, the phosphorus concentration in precipitation is high enough (90 μg l^{-1}) to make these limits impossible, even if agricultural pollution could be reduced substantially. In addition, due to internal buffering mechanisms that tend to keep the lake in alternative stable states, i.e. either an algal-dominated or plant-dominated condition, nutrient reduction is rarely sufficient to trigger ecological restoration and more drastic measures such as 'biomanipulation' through fish removal may be required (Jeppesen & Sammalkorpi, this volume). Even then, recovery may be uncertain due to human influences such as changed hydrology, recreational boating pressures or altered sediments or grazing bird populations (e.g. Madgwick, 1999).

Similarly, a major national effort has been made in Sweden to address the effects of acidification caused by 'acid rain', affecting around 20 000 lakes and vast areas of forest soils. The use of liming measures in Sweden might be more properly termed 'mitigation' than 'restoration'. As 85% of the acid load comes from other countries, in-lake measures are a holding mechanism, while the impact of international negotiations is awaited. The time-scale for the effects of major reductions to take effect (assuming they occur soon) is estimated to be around 50 years. Liming of surface waters is a controversial activity in Scandinavia, since some of the northern waters are naturally acidic and the liming of some wetland types leads to the deterioration of plant communities. Hence there is a careful balance to be struck between desirable and undesirable consequences and this has led to considerable debate between the various interest groups. The motivation for liming is to meet international and national commitments to maintain and restore biodiversity and to safeguard recreational fishing facilities. An intensive monitoring programme that has accompanied the liming of surface waters has shown that complete ecosystem restoration is not possible through liming, but that well-performed liming operations has resulted in improved conditions in around 8000 lakes so far (Henrikson & Brodin, 1995). The Swedish Environmental Protection Agency spends roughly 25 million Euros (€) each year on such rehabilitation measures involving the liming of water and adjacent land (T. Larsson, personal communication).

The scientific uncertainty, complexity of measures required and the need for a step by step approach to restoration projects of this kind can be a major hindrance to organisations who have limited access to resources for research and experimentation and who have increasing pressures to report on fixed targets of performance. Generally, a long-term institutional commitment is needed to bring about successful ecological restoration programmes, although sometimes positive results can be achieved rapidly and unexpectedly. Rarely, there is an adequate monitoring programme alongside the management measures, which helps to explain what happened (e.g. Zockler, 2000). Many projects start with a surge of enthusiasm and funding and then are abandoned when the results are unexpected and hard to explain. Hence, Europe is littered with short-term, failed ecological restoration projects. However, it is not only the unpredictability of results that can cause difficulties in trying to fund restoration projects. More significant is that ecological restoration, especially of habitats that are slow to respond or develop, takes time. Many current funding mechanisms will only cover projects for one to three years, sometimes five, but rarely more. Greater commitment to long-term investment is needed at the level of international institutions and national and regional governments, as well as by non-governmental organisations and the private sector.

DRIVING FORCES FOR ECOLOGICAL RESTORATION IN EUROPE

Establishment and enlargement of the European Union

There is a wide range of 'driving forces' that lie behind attempts to restore Europe's natural and semi-natural ecosystems. These range from legal and political obligations stemming from the Earth Summit in 1992, to the increased availability of land for nature, due to surplus agricultural capacity and the reduction of military use of land since the Second World War. Perhaps most of all, the growing political union and international co-operation in Europe also makes the twenty-first century a promising period for ecological restoration on a large, whole-landscape and long-term basis. Considerable opportunities are now appearing across the European Union for the promotion of ecological restoration through a range of financial instruments.

Each country in Europe has, of course, developed its own standards and approaches to nature conservation, including ecological restoration activities. But the development of the European Union has provided the possibility to develop coherent standards and to implement programmes at a spatial scale that reflects ecosystem boundaries and functions. For example, the establishment of the CORINE system of classification for European countries has enabled an assessment and planning for nature conservation within member states that takes into consideration the European significance of conservation measures.

Understanding of the ecological basis for restoration of key habitat types and increasing public interest in the state of the environment are widely recognised to be critical to underpin future restoration programmes. An increasing appreciation of the socio-economic benefits of restoration by decision-makers and stakeholders will assist in justifying such measures as part of development schemes. The establishment of the European Environment Agency (EEA) in 1994 (under the Regulation EEC/1210/90) has been a very significant contributor to this trend. Through the publishing of 'State of the environment' reports, it provides a regular, factual diagnosis of the state of the environment and anticipates future trends for the 15 EU member states and the 11 accession countries (see EEA 1995, 1998, 1999). The collation of the best available data has been achieved through the development of specialised European Topic Centres (e.g. on water, nature conservation and energy) who work with national data providers. The development of environmental scenarios is derived from 10-year economic scenarios to establish whether environmental targets will be met under existing and proposed policies. Due to this kind of analysis and information exchange, it is now more possible to assess the potential for ecological restoration of habitats on a European basis and to recognise the key driving forces and pressures that may inhibit success.

Nature conservation

Implementation of international agreements and national priorities

Scientists, naturalists and governments have been very active in pursuing protection of the most distinctive and rare habitats and species, and in the second half of the last century, the number of legal and political mechanisms to support nature conservation grew dramatically. There are now many global agreements and mechanisms that apply, for example, the 1992 Convention on Biological Diversity and the 1971 Convention on Wetlands. The Convention of Biological Diversity has required contracting parties to develop National Biodiversity Strategies and Action Plans. Most of these include restoration targets for both species and habitats. The Convention on Wetlands (the Ramsar Convention) has established specific guidelines on wetland restoration: 'Restoration as an element of national planning for wetland conservation and wise use'. These guidelines were approved by all member governments, including virtually every country in Europe, in 1999. The Ramsar Convention also provides for 'compensation' measures, which may include elements of restoration, to be taken in the event that a designated Ramsar site is adversely affected by activities considered to be in the 'urgent national interest'.

Some government agencies have initiated and funded large-scale ecological restoration projects

Box 2.2 Restoration of Lake Homborga, Sweden: one of Europe's most ambitious projects

The large wetland and lake area of Hornborgasjön (4124 ha) is typical of many post-glacial lakes in lowland Europe that have been successively drained by agriculture from the beginning of the nineteenth century to the mid twentieth century. Due to the invasion of reed (*Phragmites australis*) and willow (*Salix*) scrub, little open water remained by the 1960s. Restoration was entirely state funded and large areas of land were acquired by the Swedish Environment Protection Agency to facilitate restoration. Mechanical treatment, burning and raising water levels were used to restore the lake and large areas of seasonally flooded grasslands. The area characterised by a mosaic of wet grassland has increased from 50 ha to approximately 600 ha and numbers of breeding waders and wildfowl have dramatically increased. The area is lightly maintained by a variety of hardy grazing animals including traditional domestic breeds of sheep, cattle, ponies and pigs. Lakeside areas also provide a vital stopover for cranes (*Grus grus*) in spring and autumn. The project is seen as an excellent example of well-planned and executed ecological restoration. However, it is highly questionable whether this kind of project could be afforded elsewhere, even in Sweden. The restoration activities were implemented over a ten-year period, with intensive management from 1991 to 1996, and the total cost was €9 million, excluding maintenance and monitoring, which continue. One unexpected outcome was that the area became one of the most popular nature reserves in Sweden and so around €62 000 per year is derived from the visitors. For more information, see Hertzman & Larsson (1997, 1999).

within their remit for biodiversity conservation. The restoration of Hornborgasjön in Sweden is one example (Box 2.2).

Influential and active international non-governmental organisations (NGOs) (e.g. BirdLife International, the Worldwide Fund for Nature [WWF]) and national NGOs (e.g. the Royal Society for the Protection of Birds [RSPB] in the UK, Natuurmonumenten in The Netherlands and Naturreservaten in Flanders) are a hallmark of the European nature conservation movement. These organisations, which are particularly active in northwest Europe, take responsibility for the management of large areas of semi-natural habitat and have often been the pioneers of species recovery programmes and ecological restoration techniques that are required to meet international conservation targets (e.g. Box 2.5).

Within the European continent there are several relevant EU directives and regional mechanisms that require governments to take actions to further ecological restoration. Article 130 of the Treaty on the European Union (Maastricht, 1992) calls for prudent and rational utilisation of natural resources. The EU Habitats Directive (92/43/EEC) sets out priority habitats, gives a definition of each type and tries to protect these in EU countries and applicant states. Whilst based on scientific advice, these listings were ultimately determined by member states. Furthermore, it is likely that amendments will be needed if and when the EU is enlarged to accept new members from southern and eastern Europe. Together with the Birds Directive (79/409/EEC), the priority is to create the European ecological network (of special areas of conservation), called Natura 2000, and to integrate nature protection requirements into other EU policies such as agriculture, regional development and transport. The Habitats Directive requires member states to implement the measures necessary for restoring to a 'favourable conservation status' those species and habitats listed as priorities in the directive's Annexes. The measures for managing Natura 2000 sites are given in Article 6 of the Habitats Directive (European Communities, 2000). The Habitats and Birds directives require that 'compensatory measures' are taken in the case that a Natura 2000 site (designated under the Directive) is damaged, for reasons of overriding public interest. This was the case in southern Wales, where rehabilitation of a major freshwater wetland area was provided in compensation for the destruction of part of the Severn Estuary Natura 2000 site by construction of a tidal barrage.

Habitat-oriented approaches

The EU LIFE–Nature financial instrument has also been an important catalyst and source of funding for numerous nature restoration, rehabilitation and/or creation projects to benefit habitats that are considered priorities at the European level, notably those at Natura 2000 sites. In this way, experimental restoration efforts have been supported and the results magnified through the promotion of European-wide restoration efforts. The restoration of lowland heathland is a good example (Box 2.3).

Species-oriented approaches

Through the use of large, well-known or economically important species or 'flagship' species, programmes for large-scale ecological restoration have gained popular support, especially in northwest Europe. The concept is that by focusing efforts on the restoration of a 'flagship', 'indicator' or 'keystone species', benefits can be obtained for whole ecosystems. For example, declines in the bittern (*Botaurus stellaris*), otter (*Lutra lutra*) and Atlantic salmon (*Salmo salar*) in The Netherlands and the UK have acted as triggers for national habitat restoration programmes (e.g. Box 2.4). These programmes contribute to meeting the ecological requirements of the habitats in Annex I and species in Annex II of the Habitats Directive and Annex I of the Birds Directive. Whilst this approach is undoubtedly very successful in stimulating ecological restoration measures and public awareness, there is a danger that the need to reach population targets within a politically acceptable time-scale can result in the development of highly artificial habitat management programmes that have limited long-term success and a low cost–benefit ratio. There is also a lively debate about the overall conservation value (especially if compared with financial cost) of programmes to restore or reintroduce species that are rare in a given country or region, but common and not threatened on a European scale. On one hand, the desirability of maintaining or restoring natural ranges of distribution is a valid ecological argument. On the other hand, international assessment of conservation priorities may suggest that, in some cases, money could be better spent.

For example around €17 million will be invested by the Rhine states to assist the recovery of Atlantic salmon in the Rhine as part of the Action Programme launched by the International Commission for the Protection of the Rhine in 1987 (WWF, 2000). The stocking programme is quite intensive but the recovery is constrained by limited habitat availability, despite major improvements in water quality and the reconnection of many abandoned meanders (Christensen, 2001). However, it could be argued that these funds would be better invested in the restoration of salmon elsewhere, such as in the rivers feeding the Baltic Sea (e.g. the Vistula and Odra rivers in Poland). Here, remnant salmon populations survive together alongside other migratory fish species such as zähtre (*Vimba vimba*), shads (*Alosa* spp.) and brown trout (*Salmo trutta*). Since these river systems are relatively intact, fairly modest measures such as improving the functioning of fish-passes to address the needs of all migratory species and the reconnection of certain side channels to the main river could make a large contribution to the conservation of native fish populations.

Development of ecological networks

The European Ecological Network (EECONET) was a European initiative led by the Dutch government, which intended to promote a proactive approach to the enhancement of ecological status in Europe. The concept includes the use of core areas, buffer areas, corridors and stepping stones, together with the restoration of damaged habitats and the creation of new ones. The Principle for Restoration is one of the nine EECONET Principles. It states that 'where possible, parts of the biological and landscape diversity of Europe should be restored and/or (re)developed if it can be demonstrated by reference studies that the original state should be re-established'. It is intended that the network should be implemented through the development of the Natura 2000 network, together with the 'Emerald Network' of areas of special conservation interest (under Article 4 of the Bern Convention [The Convention on the Conservation of European Wildlife and Natural Habitats] and the Pan European Biological and Landscape

Box 2.3 Institutional and financial support for the restoration of Europe's lowland heathlands

Heathland was traditionally managed by grazing and burning and was once widespread on the sandy soils of northern Europe (e.g. Belgium, Germany, Netherlands and the UK) (see Webb, this volume). However, agricultural intensification during the twentieth century saw rapid loss and fragmentation of heathland through ploughing and application of inorganic fertilisers. Further areas were lost to urban development. Remaining heathland is vulnerable to scrub encroachment and reforestation due to abandonment of traditional management practices. In recent years, significant effort has been devoted to rehabilitation of degraded heathland and to the restoration of areas converted for agriculture and/or coniferous forestry plantations (Webb, this volume).

UK

Twenty percent of Europe's lowland heathland is found in the UK, making it the most important country in Europe in terms of area, even though the extent of UK lowland heathland has declined by 80% since 1800. 'Tomorrow's Heathland Heritage' is a major ten-year, €41 million project set up to reverse the decline of lowland heathland in the UK with the aim of reaching the national Biodiversity Action Plan target of 58 000 ha. The project is run by the statutory agency for nature conservation in England, English Nature, on behalf of a consortium of other bodies and has received major financial backing from the Heritage Lottery Fund, which was set up in the framework of the UK's national lottery. A significant amount of work has been done on the restoration of heathland on former arable land (Webb, this volume). For example, the RSPB (the BirdLife International partner in the UK) has restored 152 ha of heathland on farmland within its Minsmere nature reserve in eastern England.

A large part of the former cruise missile base at Greenham Common, southern England, decommissioned after the end of the cold war, is also being restored to lowland heathland. EU LIFE funding has supported important restoration work on the Dorset heathlands Natura 2000 sites.

THE NETHERLANDS

In The Netherlands restoration of wet heathland types that have been subject to partial drainage, nutrient enrichment and acidification processes involves hydrological restoration (e.g. ditch-filling), liming, sod-cutting, burning, mowing and grazing. These measures all require long-term planning and implementation. They are mostly funded and implemented by the state.

BELGIUM

Heathlands cover between 10 300 and 13 000 ha of Flanders and account for around one-third of the total surface of protected areas. With financial support from the EU LIFE–Nature programme, a project is being carried out in the Flanders region of Belgium between 1999 and 2003 with the aim of enhancing the conservation status of isolated fragments of heathland. The project builds on an Action Plan for heaths, mat-grasslands and associated habitats drawn up under an earlier LIFE project and which also involved significant elements of land purchase and tree clearance.

DENMARK

Denmark has suffered major losses and fragmentation of heathlands. All remaining areas are designated under the EU Birds Directive, but the habitat quality is highly degraded as shown by declines in populations of key indicator bird species such as golden plover (*Pluvialis apricaria*). A LIFE project was carried out in 1993–6 to experiment with habitat rehabilitation measures, including cost–benefit analysis.

Diversity Strategy (see http://www.ecnc.nl/). The experience of the Dutch Nature Policy Plan has been taken forward into the EECONET initiative which involves the creation of natural habitat features to link up 'core' nature areas such as forests, lakes and heaths.

Some of the most important ecological corridors in Europe have been identified by the World

Box 2.4   Bittern recovery project and reedbed restoration in the UK

The bittern (*Botaurus stellaris*) recovery project in the UK, led by the RSPB and English Nature, aimed at the creation and rehabilitation of beds of reed (*Phragmites australis*) in order to increase the number of booming males from their 1994 level of 16 to a target of 100 by 2020. In many countries (for example Sweden, Poland and some other Baltic states) the expansion of reedbed areas is considered to be undesirable as it is linked to the drying out of shallow lakes or abandonment of wet grasslands. There, bittern is relatively common and populations have expanded in recent decades. However, in the UK, the degree of loss of reedbed habitats due to agricultural drainage (10%–40% between 1945 and 1990: Bibby *et al.*, 1989) has led to intensive efforts for reedbed creation over the last 20 years. It was estimated in the UK that rehabilitation could support an additional 32 'boomers' but that 1040 ha additional habitat was needed for the remaining 52. Habitat creation was justified on the basis that, even in favourable conditions, natural reedbed expansion is too slow and reedbed can be readily established in two to five years (RSPB, 1994).

Since the dominance of *Phragmites* is desirable both for specialist reedbed birds e.g. marsh harrier (*Circus aeruginosus*), and for commercial harvest for reed thatching, the rural reed-growing and thatching industries and nature conservationists found a common cause in the re-establishment and continued management of reedbeds. A 10-ha reedbed could support one reed-cutter for the season, based on a crop yield of 750 bunches of reed per hectare. There is a huge potential for the production of commercially grown reed since it is estimated that around 1 million hectares in the UK are suitable and much of this is surplus arable production in pump-drained marshland. However, the EU agri-environment regulations currently provide too little incentive for the creation and management of reedbeds (Tyler, 1994; Hawke & José, 1996; Madgwick & Hawke, 1996).

Conservation Union (IUCN) and within these assessments, specific proposals for ecological restoration have been identified. Some other governments have since made steps towards the development of ecological networks, as part of their commitment to the Convention on Biological Diversity. For example a decree on nature conservation was adopted in Flanders in 1997, providing a legal basis for the establishment of core areas and stepping-stones and instruments to rehabilitate and restore degraded ecosystems, as well as to protect threatened species. Around €10 million are allocated for the implementation of the first Nature Development Plan. The funds are derived from a range of environmental taxes (e.g. levies for water pollution and waste management). In addition, co-operative efforts on wetland restoration have been triggered in Europe under the African–Eurasian Migratory Waterbird Agreement under the Bonn Convention (Convention on the Conservation of Migratory Species of Wild Animals). Other European-wide initiatives such as the IUCN 'Parks for Life' initiative, a major portfolio of activities related to the management of protected areas in Europe, have also sought to stimulate government action for ecological restoration. In this case, the extension or creation of protected areas is the focus, for the joint purposes of nature conservation and public enjoyment.

Large-scale, trans-boundary approaches

It is generally accepted that ecological restoration will only be truly successful if it is carried out at a large enough scale that reflects natural patterns of species organisation and ecological processes. Indeed, most international conservation organisations have recognised this need and called for concerted action. In Europe, there are relatively few large-scale examples of restoration programmes, partly due to the large number of individual countries and other ecological and institutional constraints described earlier.

The joint efforts of Denmark, Germany and the Netherlands over the protection and restoration of the Wadden Sea, one of the worlds' largest tidal wetlands, is one good example of international co-operation (see Box 2.5).

Box 2.5　Restoration of saltmarsh in the Wadden Sea

Every year this complex of islands, saltmarshes and mudflats forms a critical stopover for around 12 million individuals of 50 species of birds migrating along the East Atlantic flyway. It is also the spawning and feeding area for 102 species of North Sea fish, harbour seal (*Phoca vitulina*), grey seal (*Halichoerus grypus*) and bottlenose dolphin (*Tursiops truncatus*). This valuable area is fringed by one of the most industrialised areas of Europe. There have been substantial losses of habitats from port development and more than 33% of all saltmarshes have been lost to the development of dykes and flood defence. Other valuable habitat areas have been infilled, disturbed by tourism and hunting, or altered by water regime changes due to civil engineering. There is excessive algal growth and changes to plant and animal communities in coastal waters due to agricultural, industrial and urban pollution.

A Joint Declaration on the Protection of the Wadden Sea was signed in 1982 and a common secretariat for the Co-operation on the Protection of the Wadden Sea was put in place. In 1991, the trilateral policy was embedded in a common structure consisting of a Guiding Principle, a number of Management Principles and a set of common objectives for human use. The Guiding Principle of the trilateral Wadden Sea policy is 'to achieve, as far as possible, a natural and sustainable ecosystem in which natural processes proceed in an undisturbed way'. A Principle of Restoration has been established, which states that 'where possible, parts of the Wadden Sea should be restored if it can be demonstrated by reference studies that the actual situation is not optimal, and that the original state is likely to be re-established'. Saltmarshes have been embanked in the past, some only as a summer polder, but most of them as polders on the land side of a sea dyke. Most of the artificial saltmarshes on the Wadden Sea islands have an almost natural geomorphology whereas most mainland artificial saltmarshes have a geomorphology that is dominated by human structures: brushwood groynes, ditches and dams (Fig. B2.1). The trilateral policy takes as its starting-point the need to ensure that the present area of saltmarshes is not diminished and aims to

Fig. B2.1. Artificial saltmarshes on the Wadden Sea (top) compared to marshes on the mainland (bottom).

extend this. This will be achieved through the restoration of saltmarshes by opening summer dykes, provided that it is in line with the targets for the region, socio-economic conditions and coastal protection requirements. This will provide an increased natural morphology and dynamics, including natural drainage patterns of artificial saltmarshes. It will also provide an improved natural vegetation structure, including the pioneer zone, of artificial saltmarshes and favourable conditions for migrating and breeding birds.

Public opinion

Greater access to information and correspondingly greater awareness has ensured that there is a strong public lobby in support of environmental measures in many European countries. This is particularly true in western Europe where the NGO sector is more strongly developed and better resourced than in other regions.

Public outrage following environmental disasters has been a significant catalyst for restoration efforts in several cases (e.g. Doñana; see Box 2.6). However, at the local level, public opinion may work contrary to proposed restoration measures, as was the case at Chobham Common in the UK, where the British government disallowed a proposed heathland restoration project after protests from local residents concerned about possible restrictions of recreational access.

The growing water crisis

Increasing problems of inadequate water quantity and quality in Europe, and increasing incidents of damaging floods caused largely by overintensification of agriculture and loss of natural ecosystem

---

Box 2.6   Doñana: it took a disaster to stimulate effective restoration

Doñana in southern Spain is one of the largest and best-known wetlands in Europe. Here, major tourism and agricultural developments over the last 40 years have resulted in the dehydration of the internationally important wetland ecosystems through drainage, channelisation, construction of dykes and artificial drains and overabstraction of groundwater. While environmental NGOs in Spain had been campaigning for comprehensive measures to be put in place to restore these wetlands, it was not until the catastrophic toxic spill in 1998 from the Los Frailes mine near Aznalcóllar, that these plans were activated. The bursting of a tailings lagoon at the mine site resulted in 5 million m^3 of toxic sludge and wastewater flooding down the Guadiamar river and over farmland in the floodplain, including areas inside the Doñana National Park. Now two major projects have been put into operation: the Green Corridor of Guadiamar is a project aimed at remediating the area of the Guadiamar River and floodplain that was directly affected by the mining disaster. It further intends to carry out a range of measures to restore an eco-corridor along the Guadiamar, from the mountainous upper catchment, through Doñana to the coast (Junta de Andalucía, 2000). These works will be complemented by Doñana 2005, which aims to achieve a hydrological restoration of river stretches and water flows to the marshlands in the National Park (Ministerio de Medio Ambiente, 1999b). The regional and national authorities have been reluctant to work together on the restoration programmes. However, the Spanish response to the mining disaster was placed under an intense international spotlight and a number of international environmental bodies exerted pressure on the government (UNESCO, 1999). Partly as a result of this, a greater level of co-operation has been achieved and considerable efforts have been made to increase public awareness and participation in decision-making.

functioning, have led to the formulation of a major new piece of EU environmental legislation, namely the EU Water Framework Directive (WFD). The WFD requires EU member states to take a river basin approach to water management. It sets a clear environmental target of 'good water status' for all ground and surface waters in the EU. This will trigger investments in a range of measures within river basins. The Directive specifically recognises the important contribution that wetland restoration can make to achieving sustainable water use. Some governments have made significant progress in integrating land and water management, ahead of the establishment of the WFD. For example in Denmark, which has a high population density and intensive agricultural landscapes, a multiple approach to nutrient reduction is now being promoted. This includes addressing point source and diffuse loading, and enhancing the nutrient retention capacity of catchments (e.g. by re-establishing lost wetlands, re-meandering streams, creating buffer zones), as well as in-lake restoration measures, such as biological manipulation (see Jeppesen & Sammalkorpi, this volume). Other land use plans support the national effort to restore the ecological quality of lakes– for example, the afforestation plan, which aims to double the forest area in the next 60–80 years (5000 ha per year). Despite this laudable national commitment, there is no pretence that full ecological restoration will be achieved and the target is to restore clear water dominance over the summer months (Møller, 1995, 1998; Iversen et al., 1998; Jeppesen et al., 1999; Jeppesen & Sammalkorpi, this volume).

Many European countries have already begun implementing measures to restore regulated rivers and their floodplains as a means to improving water quality and/or quantity, absorbing excess floodwaters and regenerating wild fisheries. There are numerous projects ranging from small streams to some of Europe's largest rivers, for example the Danube (Box 2.7; see also Downs et al., this volume). A further example is the 1998 action plan of the International Rhine Commission, which by 2020 aims to restore 85% of the natural floodplains lost during the last 200 years. This will cost over €12 billion and will be funded by 12 countries as well

as the EU. It is justified by reducing the impacts of damaging floods, which have become more frequent and severe (e.g. the floods of 1993 and 1994/5) since the damming of the river from the mid nineteenth century. The floods of 1995 in the Rhine resulted in the deaths of 27 people, a further 200 000 people being evacuated, and an estimated €5.7 billion worth of damage (Hunt, 1997; RIZA, 2000)

The establishment of the European Centre for River Restoration in 1995 is a manifestation of the trend for wetland restoration being increasingly seen as a part of water management. In providing an international networking function, the Centre aims to improve technical exchange on river restoration across the European continent and to promote the benefits of river restoration more widely so that it becomes 'mainstreamed' into water management strategies. Whilst the initial impetus for the Centre came from UK, Dutch and Danish institutes and government agencies, it is significant that national networks of scientists and practitioners have now been spawned in Russia and southern Europe (see http://www.ecrr.org; Hansen et al., 1998).

The increasing trend towards privatisation of water supply and treatment, both in Europe and globally, may also provide a new driving force for protection and ecological restoration of water sources. There are some positive examples in the UK and France. For example, Essex and Suffolk Water (a subsiduary of Suez-Lyonnaise des Eaux) purchased the 168-ha lakes which form the Trinity Broads, part of the Norfolk and Suffolk Broads National Park, in the mid-1990s and forged partnerships with the national park authority, Environment Agency, NGOs and stakeholders to develop and implement a plan for restoration and maintenance of the water quality of these lakes, which form an important water supply. It was considered a better investment to restore this supply (to a clear water state, dominated by aquatic macrophytes), thus preventing the blue-green algal blooms that clogged the water filters and caused health concerns, than to redistribute and treat highly polluted water from the River Bure. Apart from the cost of land purchase and maintenance costs, a warden and a major biomanipulation programme (see Jeppesen & Sammalkorpi, this volume) for Ormesby Broad have been funded.

Box 2.7   Restoration of the Danube floodplains

The consequences of river alteration and floodplain loss along the Danube for flooding, water quality and biodiversity have been quite well documented (e.g. Ody & Sarano, 1993; EEA, 1995). In 1991, the Danube Environmental Programme was launched, involving six states bordering the Danube and funded by the United Nations, World Bank, European Union and other international agencies. The Task Force report focused on the threats to drinking water supplies for over 20 million people, as a result of the loss of the river's self-purification capacity and called for concerted action. Specific, acute toxic spills associated with the Kosovo War (spring 1999) and the failure of a mine storage lagoon at Baia Mare in Romania (January 2000) have led to additional needs for water quality and habitat restoration measures (Stephan et al., 1999; UNEP/OCHRA, 1999, 2000). The economic and social benefits of floodplain restoration are especially relevant to the downstream countries such as Bulgaria, Romania and Ukraine. A rough economic evaluation exercise estimated that the remaining Danube floodplains (1.7 million ha) provide goods and services worth €666 million, or €383 per hectare per year (Gren et al., 1994; WWF, 1993).

It was not until very recently that a comprehensive and systematic evaluation of the extent of remaining floodplains was carried out for the Danube and its five main tributaries, commissioned by UNDP/GEF as part of the Danube River Pollution Reduction Programme (Danube Pollution Reduction Programme, 1999). This study evaluated the current status of wetlands and the potential for wetland restoration. It was based on analysis and evaluation of the floodplain area in defined segments, using historical and current maps (e.g. topographical, geomorphological, land-cover maps), bio-indicators, and hydrological, economic and nutrient load data. It confirmed that the total loss of floodplain habitat was in excess of 80%, with only 7845 km^2 of floodplain wetlands remaining. The study recommended 17 floodplain sites for rehabilitation/restoration, considering their ecological importance, potential nutrient removal capacity and role in flood alleviation, as well as potential constraints such as proximity to industrial developments and settlements. In the spring of 2000, following years of careful discussion, the Ministers of Environment of Bulgaria, Moldova, Romania and Ukraine announced large-scale commitments for the creation of a green corridor in the lower Danube, which, when implemented, would include 500 000 ha of protected areas and 200 000 ha of priority sites for restoration (WWF, 2000). These countries will create an action plan and monitoring system for this programme and will share experiences in wetland restoration and pollution reduction. This European initiative is unique in scale and ambition. However, it remains to be seen how far this landmark political commitment will be underpinned by financial investment. Since resources are scarce in the region, it is hoped that international donors will play a significant role.

A programme for the reduction of nutrient inputs from the catchment through alteration of farming patterns is also under development. The success of the restoration programme so far (see Tomlinson et al., 2002) has brought considerable publicity and public awareness benefits for the company, which formerly had a very poor local image. Similarly, in northwest Germany (Weser–Ems district), a major drinking water company (serving about 850 000 people) began entering into voluntary agreements (based on compensatory payments) with farmers to apply organic farming methods in its water protection areas. This was partly in response to major NGO campaigns in the mid-1980s about the deteriorating state of water resources due to high nitrate and pesticide levels. Monitoring data has shown that nitrate levels underlying the fields fell from 125 mg l^{-1} in 1993 to 18 mg l^{-1} in 1997 (Lanz et al., 2000).

In the Baltic region, the Helsinki Convention (also known as HELCOM; see http://www.helcom.fi) focuses on the need to reduce nutrient and toxic inputs to the Baltic Sea, where the effects of eutrophication have become critical for marine life and fisheries. All the Baltic Sea states agreed to reduce their transport of nutrients from land to the Baltic Sea by approximately 50% in the period 1987 to 1995. Unfortunately this timetable was overoptimistic and proved impossible to achieve. Restoring wetlands is increasingly seen as complementary to other environmental measures for agriculture, forestry and wastewater treatment (e.g. Fleischer *et al.*, 1991). More than 220 000 km² of wetlands have been drained in the Baltic Sea basin. A recent study indicates that the existing wetlands of the Baltic Sea basin contribute to the nutrient reduction target by creating nitrogen sinks corresponding to roughly 15% of the total nitrogen load to the Baltic Sea. Thus, without these wetlands, costly measures would have to be implemented to achieve the same nitrogen reduction. A rough estimate showed that a further nitrogen reduction of about 10% would be achieved if drained wetlands could be restored (Folke & Jansson, 2000). Under the auspices of this convention, the MLW (Management Plans for Coastal Lagoons and Wetlands) project has given a significant focus to wetland restoration. Economic studies on the options for pollutant sinks in the Baltic Sea have further shown wetlands to be the most cost effective sink option in the majority of the Baltic countries, which include Poland, Germany, Denmark, Russia, Estonia, Latvia and Lithuania (Gren, 1999). The World Bank has recently initiated a large-scale Global Environment Facility financed project for the entire Baltic, which is expected to finance wetland restoration as a cost-effective way of providing natural wastewater treatment plants. In Denmark, ecological restoration has now become an integrated part of national water management plans. The 1997 Action Plan on the Environment (II) set a target of 49% reduction of nitrogen emissions to improve groundwater quality and reduce marine loading. The restoration of wetlands is one strategy for achieving this, alongside afforestation, groundwater protection and improved agricultural practice. It is estimated that an average nitrogen reduction of 350 kg ha⁻¹ per year can be obtained through denitrification processes in established wetlands (Iversen *et al.*, 1998). Approximately ECU 460 million was spent in the period 1989 to 1998 on rehabilitation and restoration in Denmark, mostly on wetlands, at an average cost of around ECU 8000 ha⁻¹ (Møller, 1998).

The EU pre-Accession funds (SAPARD, ISPA and Phare) that are available for the ten countries lining up to join the European Union offer some opportunities to invest in ecological restoration as a means of reaching Community environmental standards and integrating sustainable development principles into their sectoral policies. Funding is thus available for environmental measures that could include habitat restoration, for example to maximise water retention in catchments, to restore floodplains, or to absorb wastewater in rural areas with low populations. However, it seems likely that these governments will choose to prioritise large-scale infrastructure developments and conventional agricultural intensification schemes.

Disappearing subsidies and agricultural reform

In all EU member states, the Common Agricultural Policy (CAP) was the major post-war driving force behind the regulation of rivers, drainage and conversion of semi-natural land (such as forests or heathlands), simplification of landscape structure (e.g. removal of ponds, hedgerows and trees) and intensification of production through increased use of fertiliser and plant protection products. The CAP influences agriculture through a very wide range of subsidies and grants funded by an annual EU budget of around €50 million per year. The most significant of these are provided under 'product regimes' (e.g. for cereals, beef, etc.) in the form of price support, production subsidies and payments per hectare or per head of livestock. With very few exceptions (such as recent reforms in the beef sector), support regimes for different agricultural sectors are heavily biased in favour of intensive, high-yielding forms of production (e.g. for arable crops, olive oil production), which result in the more environmentally sensitive, low-input and low-output systems being marginalised. This has, not surprisingly, resulted in massive losses of habitats and species, as well as a reduction in rural employment, as recorded by many authors (e.g. Birdlife International, 1997; WWF, 1998; EEA, 1999).

Enlargement of the EU and discussions ongoing within the framework of the World Trade Organisation are expected to lead to significant changes in the CAP, with a move away from production subsidies towards more integrated rural development measures that will diversify rural economies. Although accounting for a small percentage of total CAP expenditure, reforms introduced in 2000 specifically promote rural development and sustainable farming and provide new legislation that gives member states the opportunity to take actions to reduce environmental degradation caused by agricultural activities supported by the CAP (Box 2.8).

It is increasingly obvious to conservationists everywhere that without the removal of 'perverse subsidies' (i.e. subsidies inducing behaviour that depletes biodiversity), nature conservation and restoration efforts will remain limited in scale, impact and sustainability, being highly vulnerable to ongoing impoverishment of the wider countryside. The reorientation of such subsidies to promote sustainable land use, nature conservation and restoration efforts is more likely to provide widespread, long-term benefits than seeking to mitigate the impacts of general subsidies with 'targeted' grants and stewardship schemes. The intricate balance between human use and biodiversity conservation in Europe and the rapid changes brought about through EU spending makes this an especially high priority and international conservation organisations that were founded in species conservation, such as WWF, have recently developed programmes that actually promote rural development. Increasingly, economic arguments for altering the spending patterns of the EU, and other key institutions, are used alongside arguments for nature conservation (Maltby, 1991; Dugan, 1993) (Box 2.9).

For semi-natural ecosystems, such as forests, where economic use and the pursuit of profit have traditionally dominated, ecological restoration is still new and relatively weak. There is great variation between countries and the potential benefits of restoration for providing multiple environmental benefits and rural development are rarely realised (Box 2.10). Historically, monoculture plantations have been the most common form of reforestation and there has been little consideration for

Box 2.8 Financial instruments in the EU Common Agricultural Policy that could be used to stimulate ecological restoration

The *Rural Development Regulation 1257/1999*, covering the period 2000 to 2006 accounts for about 10% of annual expenditure of the CAP and provides measures for the 'preservation and promotion of a high nature value and a sustainable agriculture respecting agricultural requirements' as well as for area-based compensation payments 'to maintain and promote sustainable farming systems which take particular account of environmental protection requirements' in 'naturally less favoured areas'. This could, for example, help reduce grazing pressure and hence trigger recovery of natural vegetation in certain regions. Agri-environment measures provide annual payments to farmers for extensifying farming practices for at least five years. The Regulation states that 'agri-environment commitments shall involve more than the application of good farming practice'. This could, for example, be used to encourage the reversion from arable production to permanent grassland. It could also influence cropping patterns, water consumption and fertiliser applications, potentially helping to provide a favourable context for restoration of a range of ecosystem types, including wetlands and herb-rich grassland.

Under the *European Agricultural Guidance and Guarantee Fund (EAGGF)*, Guarantee section, the options available include the use of agri-environment measures, mandatory requirements (EU and national law), or 'specific environmental requirements constituting a condition for direct payments', known as cross-compliance (Regulation 1259/1999). This provides the chance for governments to develop environmental standards in agriculture that would favour the restoration and maintenance of semi-natural habitats.

socio-economic or environmental effects, such as impoverished soil quality, reduced services to local people, poor resistance to storms, etc. EU agricultural policies have accentuated this trend. However, it is recognised that plantations (developed for economic purposes) may be a stepping-stone

Box 2.9   Options for reducing agricultural impact on Las Tablas de Daimiel National Park, Spain

This National Park contains the most significant wetlands in the Upper Guadiana Basin (Castilla–La Mancha) in Spain, forming part of a UNESCO Biosphere Reserve since 1980. The wetlands are fed by Aquifer no. 23 which covers around 5500 km^2. Over the past 30 years, assisted by national and European agricultural policies, grants and subsidies, irrigated agriculture has expanded in this region and now focuses on highly water-demanding crops such as sugar beet and maize. The Daimiel wetlands have been reduced greatly in extent from 22 000 ha to <9000 ha and are threatened with complete drying-out in drought years. Species losses and ecological degradation have resulted and this situation can only be stopped and reversed through actions to reduce water abstraction from the aquifer. Due to overabstraction, the Daimiel area was therefore supplied with water through the construction of a water transfer system from the Tagus Basin. Water abstraction restrictions according to the National Water Law (Ley de Aguas, 1985) were not properly enforced. The CAP provided compensation to farmers for reducing water consumption and a marked shift towards crops that use less water was observed by 1995, although this was insufficient to allow the

aquifers and wetlands to recover. The scheme was also very expensive, costing over €100 million in the first phase (1993–5). A recent study analysed the various options available for using different CAP scenarios to trigger a sustained reduction in water consumption in this area (WWF/Adena, 2000). This concluded that the application of environmental cross-compliance is an essential first step in promoting sustainable use of water resources in the area of Las Tablas de Daimiel, to ensure compliance with existing regulations. This, together with the Agenda 2000 changes to crop payments and support prices would result in a reduction in the consumption of water by 50% per hectare while having a minimal impact on farm incomes. It would cost €190 per hectare, i.e. less than half the cost of present-day policies, effectively saving €20 million, which could instead be invested in environmental protection and job creation. It is predicted that the elimination of all crop-related subsidies would result in a drop of 74% in water consumption while incomes would be barely affected, due to farmers switching to non-irrigated, low-input cropping which would greatly reduce farm costs. In any case, it was clear from the study that in order to enable a fuller recovery of the aquifer, a reduction of the relatively high rate of support for irrigated crops and for maize would be necessary.

towards restoration, providing canopy cover for the establishment of native shade-tolerant species (see Buckley *et al.*, this volume) and economic benefits which can be reinvested in restoration (see Howell, 1986) (Box 2.10). The imperative to 'clean up' and replant forest in France and Switzerland after the extensive impacts of Hurricane Lothar in December 1999 demonstrate clearly the priorities that continue to drive forest management. Even in Switzerland, which WWF rates as having some of the best-managed forests in Europe, structure and diversity is generally highly modified. In particular, dead and dying trees are rarely left in place outside specifically protected areas. Statistics released by the federal authorities in January 2001 showed that only 0.1% of Swiss forests are composed of dead trees.

In former communist countries there is now a general decline in agricultural production following the removal of the substantial subsidies provided by the centralised socialist economic planning regimes. In many areas, there is a need to reconsider the future use of agricultural lands, for example in Europe's major deltas, where drainage of polders to enable grazing or harvest of marsh hay or reeds for paper production is no longer, and perhaps never was, a viable option. Alternative land uses are being developed alongside ecological restoration programmes with a joint goal of sustaining and improving local incomes (Box 2.11).

Where applicable, accession to the EU will eventually bring greater financial incentives for the conservation and restoration of important natural habitats and features, for example through

Box 2.10   Government policies for forest restoration

There are very few natural forests left in Europe, with most of them occurring in Nordic countries. It is generally agreed that there is a need to expand the forest area in Europe via afforestation or reforestation (in the latter case the area has been forest in the last 50 years). The need is seen as especially great in southern, central–western and central–northern Europe. A comprehensive survey of the state of forestry in 19 countries in Europe (WWF, 2000) found though that although legal instruments to protect forests are adequate except in a few countries, such as Greece, both the policy goals and instruments concerning how much and what type of forests should be established are substandard. For example, with the exception of Estonia and Germany, there are only poorly developed policy goals regarding the conversion of plantations to semi-natural forests.

In Estonia, where forest currently occupies 47.5% of the national territory, there is a forest law policy goal to change the management of profitable forests towards more biodiverse and naturally structured stands. Also, most reafforestation work is done with seedlings with regional provenance. Here also, a substantial proportion of productive agricultural land is turning into forest naturally. This compares, for example, to the situation in France where forests occupy 27.6% of the national territory, but where there is no goal to convert plantations to natural forest types and government policy supports wood production and afforestation by artificial plantations, using fast-growing species and by the conversion of coppices to high forests.

In The Netherlands, there is a government target to reforest/afforest 75 000 ha of forest by 2018. However, this is proving very difficult to implement since land prices are high and there is intense competition for the limited space from agriculture, housing and other infrastructure. A plan for the restoration of around 30 000 ha of forest in floodplains as part of a 'Living Rivers' scenario has been promoted by NGOs (WWF, 1993). In Austria, which is 47% forest, the area increases around 2000 ha per year, around half of which is through natural regeneration.

In the UK, post-Second World War production policies mean that fast-growing conifers were used for blanket afforestation of the uplands to provide the maximum timber yield. National Parks and other high-value semi-natural landscapes and ecosystems were particularly affected. By the 1970s and 1980s upland afforestation (aided by government tax breaks) had become a subject of enormous controversy, with deep-ploughing of internationally important peatlands, acidification of upland water courses (leading to loss of aquatic invertebrates and declines in fish and waterbird populations) and loss of traditionally managed landscapes. Now, a more 'reasonable balance between forestry and the environment' is required. Although there is no firm policy goal for improving the environmental quality of plantations, some diversification of species and structure is taking place in second-generation plantations. Several large plantations that occur on ancient woodland sites owned by the private company Forest Enterprise, have been or are in the process of being restored to semi-natural types (see Buckley *et al.*, this volume).

In Portugal, afforestation has increased following a decline in governmental support for cereal farming. The most common species planted have been *Pinus pinaster* and *Eucalyptus globulus*. Following the entry of Portugal into the EU in 1986, most of the traditional, open, holm oak (*Quercus ilex*) 'montado' forest type had been cleared for agricultural use and the rest has been impacted by excessive cattle grazing. Now there are efforts aimed at conserving montado forests under the agri-environment scheme (EC Rule 2978/92) and for afforestation under EC Rule 2080/92, with *Pinus pinea, Quercus rotundifolia* and *Quercus suber* as main species. *Pinus pinea* can act as a pioneer for oaks, so there is potential for the development of oak forests in the future or the option of re-creating montado with concurrent production of pigs, grain, charcoal and cork, through selective clearance. The mean planted area under Regulation 2080/92 is the highest in the EU (31 ha in 1993–6) but 85% of forest properties are less than 3 ha in size.

For more information see Sollander (1999), Ledant (2000) and www.panda.org/europe/forests.

Box 2.11   Rural development linked to ecological restoration in central and eastern Europe

THE NEMUNAS DELTA (LITHUANIA/KALININGRAD)

In the Nemunas Delta area, which is shared by Lithuania and the Russian enclave of Kaliningrad, there are around 200 km^2 of drained polder grasslands. It is now necessary to consider the economic aspects of their use and the potential for environmental conservation, especially within the Lithuanian Regional Park, as a result of the declining economy, the removal of subsidies and declining agriculture, which has reduced forage production needs. The pumping and drainage costs of these peaty areas are prohibitively expensive so the disconnection of drains and use of sluices are being considered to save money and benefit wildlife through renaturalisation. However, conditions would be optimal for nature over most of this area if some extensive grazing management could be sustained. Innovative proposals for rural development based on bio-energy plantations and ecotourism are being investigated. However, extensive feasibility studies and local consultation are in progress, facilitated by NGOs, to determine the optimal mix of options for this area, since farmers are reluctant to give up their subsistence lifestyle for some unknown future (WWF, 1999; Ecotek, 2000).

THE DANUBE DELTA (ROMANIA/UKRAINE)

In the Danube Delta, the strategic breaching of the dykes in the Babina (2200 ha) and Cernovca (1580 ha) polder islands resulted, after only five years, in the nearly complete replacement of terrestrial vegetation, dominated by ruderals such as *Cirsium arvense* and salt-tolerant species (including *Sueda maritima*, *Artemisia santonicum*), by reedbeds and aquatic vegetation, including for example *Nymphoides peltata* and *Potamogeton* species. The diversity of birds in the Babina island increased from 34 to 72 species and the characteristic invertebrates, such as a subspecies of the large copper butterfly (*Lycaena dispar rutilans*) reappeared within two years of the recolonisation of the reedstands and aquatic habitats. Fifteen fishermen can now make their living as a result of the polder restoration (ICPDD, 1997).

agrienvironment schemes and the Natura 2000 network. However, environmental NGOs are currently playing a major role in working with landowners and institutions in such areas to develop management techniques and retrieve funds that will support ecological restoration alongside improved local incomes. For example, in the Morava floodplains (a major tributary of the River Danube), on the border of the Slovak Republic and Austria, reprivatisation of the land on the Slovak side is slowly proceeding, following the communist policies of 1945–89. The floodplain includes the largest complex of species-rich meadows in central Europe. Here, a local NGO has utilised bilateral support from the Dutch government and funds from the Global Environment Facility/World Bank and the EU Phare Programme to restore arable fields to floodplain meadows and to re-establish their importance for biodiversity, flood control and nutrient removal (Šeffer & Stanová, 1999).

Climate change and sea level rise

Climate change is likely to have significant effects on the distribution of European species and habitats as the twenty-first century progresses. One of the impacts already being felt is sea level rise, particularly around the southern North Sea, where climatic effects are compounded by tectonic lowering of the land relative to sea level. Important coastal habitats are threatened by increased erosion and permanent inundation, leading to moves in some countries to consider 'compensatory' habitat restoration or creation far inland. Changing rainfall patterns will lead to an increased frequency and duration of flooding in some areas but significantly decreased inundation elsewhere. Hence, floodplain restoration, to increase flood storage capacity, is one of the key strategies of adaptation for climate change currently being promoted through institutional linkages and joint activities under the UN Framework Convention on Climate Change (UNFCCC) and the Ramsar Convention on Wetlands. Articles 2.3 and 3.14 of the Kyoto Protocol call upon Annex I parties to implement the policies and measures in a manner that minimises the adverse effects of climate change and the impacts of response measures (Bergkamp & Orlando, 1999). In The Netherlands, for example, large-scale floodplain restoration measures are planned to accommodate the expected increase in flood volume in the Rhine, Meuse and Scheldt rivers (Hendrikesen, 1999; Pruijssen & Rheenen, 1999; de Jong, 2000; Garritsen et al., 2000). The Committee on Water Management in the 21st Century argued that safety can only be guaranteed by giving rivers more space. A policy decision was made in 1993 to create 7000 ha of 'new nature' along the Rhine and Meuse. More systematic involvement of lower tiers of government and international NGOs in the work of the International Commission for the Protection of the Rhine (ICPR) is helped in The Netherlands by a formal cooperation agreement between the Ministry of Water Management and the ministries dealing with nature conservation, called 'safety and wet nature'. The Delta Plan for the Major Rivers is to be completed at the end of 2001 and implemented by 2015 at a cost of around €1.05 billion. This will result in 3000–4000 ha of the previously mentioned target of 7000 ha of restored river floodplains. Unfortunately, there are relatively few incentives for floodplain restoration through European or national policies elsewhere in Europe and EU policies still threaten to cause further loss and deterioration (Jones, 2000).

The increased need for coastal flood defences also provides a major opportunity for the restoration of coastal marshes that were previously reclaimed from the sea. This is partly justified to take account of the estimated losses of intertidal habitats for conservation reasons. However, the economic argument is also strong and it is now widely recognised by coastal engineers that the presence of saltmarsh on the seaward side of sea walls dramatically reduces the cost of constructing and maintaining these walls (Zedler & Adam, this volume) (Box 2.12).

Mitigation of environmental impacts

Most European countries require Environmental Impact Analysis (EIA) of major infrastructure projects and developments, with EU member states being covered by a corresponding Directive. Partly as a result of the growing use of EIA approaches (including Strategic Environmental Analysis) a growing number of land use planning decisions require compensatory, 'mitigation' or restoration measures for the environmental impacts identified, in addition to 'avoidance' or 'minimisation of impact' measures. In many cases, such compensatory measures may involve habitat translocation, captive breeding programmes, restoration of degraded habitats, post-development restoration works or habitat creation. Despite the apparent success of some habitat translocation projects (Box 2.13), a review of a larger sample of such projects in the UK has concluded that neither habitat creation nor translocation provide compensation or acceptable mitigation for the loss of all or part of sites that are highly valued for nature conservation. The dividing line between acceptable and unacceptable use of habitat translocation for nature conservation is in fact a fine one. The official view in the UK is that it can be used for scheme enhancement, as a building-block for habitat creation, but it does not provide compensation for loss or damage to high value, non-replaceable sites (Anderson et al., 1993; Sheate, 1994). However, in the city of Peterborough, where the largest recorded great crested newt (Triturus cristatus) population occurs on land which is to be developed for Peterborough Southern Township, English Nature is advising on a large-scale translocation project. Part of this site is a candidate Special Area of Conservation under the EU Habitats Directive.

CONCLUDING REMARKS

Away from some coastal and arctic–alpine regions, natural and semi-natural habitats in Europe have become highly fragmented as the ecosystem processes which sustain them have been disrupted by intensive agriculture and urban–industrial development.

Box 2.12   Managed Retreat in East Anglia, UK

Encroachment from the sea is more pronounced along the East Anglian coast than elsewhere in the UK because of land subsidence at a rate of 1–3 mm per year due to isostatic readjustments under way since the last ice age (Boorman, 1992). It is estimated that if current trends continue, some 1500 ha of intertidal habitat will be lost from the Essex and Suffolk coasts over the next 25 years. Until recently, the principal flood defence strategy was to build massive, reinforced sea walls to protect low-lying coastal areas.

In the Blackwater estuary in Essex, horizontal saltmarsh erosion has been estimated to be proceeding at approximately 2 m per year (King & Lester, 1995). Hard sea defences prevent the landward movement of the estuary marshlands, resulting in 'coastal squeeze'. A retreat from maintaining existing sea defences is considered ecologically and economically prudent. Realignment of the sea defences is being implemented to create a natural 'soft' sea defence in the form of saltmarsh and to benefit nationally and internationally important bird populations. This has been developed through several experimental management projects established since 1991, that involved a partnership between English Nature (the national statutory conservation organisation), the National Rivers Authority (the organisation responsible for water management prior to the Environment Agency), the National Trust (landowner) and the MAFF (Ministry of Agriculture, Fisheries and Food, now DEFRA, Department of Environment, Food and Rural Affairs)

that are the main funders of flood defence. A 'Blackwater exchange' was set up to encourage multidisciplinary, technical exchange of experience between practitioners and stakeholders along the Essex coast and with agencies and stakeholders from America and Australia.

Although the Blackwater estuary managed retreat sites are relatively young, a flooding/drainage regime and sedimentation processes have been activated and characteristic saltmarsh geomorphic features have formed. Vegetation recolonisation has been rapid with the desired species and benthic and epiphytic invertebrate populations have become established on marsh surfaces and in tidal creeks. Continued monitoring will be necessary to determine to what extent the development of structural and functional characteristics of the wetlands occurs during maturation.

The institutional mechanisms for encouraging managed retreat as demonstrated in the experimental sites of the Blackwater estuary are complex. Coastal land falls under a wide range of regulatory agencies. In addition, agri-environmental scheme incentives are currently insufficient to attract farmers to convert their use of productive land to meet conservation goals. New, targeted payment schemes are therefore envisaged for managed retreat in future, prioritising areas on the basis of their importance for flood control and their value as habitat.

See Yozzo et al. (2000) for further explanation and references.

In western Europe, there are vast swathes of land used so intensively that nature reserves are the last refuges for species of native flora and fauna that were still common and widespread 40 or 50 years ago. Whilst nature conservationists and ecologists were among the first and most vociferous to raise the alarm and call for habitat restoration, rehabilitation and creation, it is only recently that the erosion of natural capital in Europe has started to have a wider impact.

More frequent severe flooding due to mismanagement of river basins, evidence for the accelerating impacts of climate change, contamination of

drinking water by agricultural chemicals, and a string of food scares (such as the BSE or 'mad cow disease' crisis) have served to focus the attention of decision-makers and the public at large on the consequences of environmental degradation. There is growing recognition that wise stewardship of the environment requires not only more appropriate management of those semi-natural areas and processes that remain, but also the restoration, where feasible, of what has been lost.

However, transforming this increasing awareness into large-scale practical action will require a paradigm shift amongst policy-makers at European

Box 2.13   Twyford Down, UK: restoration of valuable grassland habitats alongside motorway development

In 1990, there was a decision to construct the final section of the M3 motorway around the city of Winchester, southern England, and through an area which had national designations for its nature conservation and landscape value. This announcement created a national uproar of public opposition. Due to concerns over losses of habitat, it was decided to recreate downland on top of Twyford Down, an area of intensively farmed arable land, and on the restored route of the existing Winchester bypass. The Institute of Terrestrial Ecology worked with the consulting engineers designing the road on all ecological matters concerning the works. The aim of the habitat restoration measures was to create new areas of herb-rich downland with a characteristic, rich invertebrate fauna. This was carried out through a series of steps following the preliminary habitat surveys: site preparation (e.g. denutrification); introduction of plants (translocation, seeding and planting of pre-grown plants); and manipulation of management techniques to steer the development of plant and animal communities towards the desired grassland types (e.g. mowing or grazing). Approximately 0.5 ha was turfed and the remaining areas were seeded or planted with plug plants, all using materials of local provenance. An area of flood meadow that was directly in the line of the new road was also relocated through turf removal. The Highways Agency are funding botanical and invertebrate monitoring for a period of ten years. Preliminary results show that the translocations of turf and seeding have been very successful and that the invertebrate community is following expected trends of colonisation with characteristic downland species. Additional populations of the scarce chalkhill blue butterfly (*Lysandra coridon*) have developed due to the restoration sites and the small blue (*Cupido mimimus*) has appeared in the area for the first time. For more information see Snazell (1997).

and national levels. It is the so-called 'super tanker' policies and budgets driving agricultural and regional development policy that will have to be redirected if ecological restoration is to take its rightful place as a fully integrated component of environmental planning and management. There are some encouraging signs that radical change may at last be on the agenda. For example, at the time of writing (early 2001), the European Commission has just announced an initiative to tackle the crisis in Europe's beef industry by providing incentives for farmers to switch to organic production. Measures such as this, which have the potential to influence the way in which huge areas of land are managed, need to be integrated with the expertise of European restoration ecologists to reverse the wholesale loss of ecosystem processes and the habitats and species which depend on them.

As the European Union expands eastwards in the next decade, EU policies and expenditure will assume an ever-greater role in shaping the landscape of those countries currently queuing up to become member states. Expansion will in itself require some major policy and budgetary changes. It is to be hoped that these will take into account the lessons learned from the environmental degradation of the last 50 years in western Europe and provide for more sustainable rural development, incorporating support for ecological restoration where needed. Indeed, it is time to shatter the myth that eastern Europe has somehow been left untouched by the ravages of twentieth-century intensification and that ecological restoration is a low priority.

Though more can be done to further improve practical restoration methods and access to technical know-how, implementation, at least on a European scale, ultimately depends on political will. It is by no means certain that abundant scientific evidence and growing public concern will be sufficient to overcome powerful forces of inertia. It will take repeated demonstration of the long-term socio-economic advantages offered by ecological restoration, versus the long-term costs of 'business as usual'. Sadly, it will probably also take further environmental disasters before restoration ecology enters the mainstream of European planning and policy.

REFERENCES

Anderson, P. (1993). *Roads and Nature Conservation: Guidance on Impacts, Mitigation and Enhancement.* Peterborough, UK: English Nature.

Benstead, P. J., José, P. V., Joyce, C. B. & Wade, P. M. (1999). *European Wet Grassland: Guidelines for Management and Restoration.* Sandy, UK: RSPB.

Bergkamp, G. & Orlando, B. (1999). *Exploring Collaboration between the Convention on Wetlands (Ramsar, Iran, 1971) and the UN Framework Convention on Climate Change.* Background paper from IUCN. (www.ramsar.org)

Bibby, C., Housden, S., Porter, R. & Thomas, G. (1989). A conservation strategy for birds. Unpublished report. Sandy, UK: RSPB.

Birdlife International (1997). *A Future for Europe's Rural Environment: Reforming the Common Agricultural Policy.* Cambridge, UK: Birdlife International.

Boorman, L. A. (1992). The environmental consequences of climate change on British saltmarsh vegetation. *Wetlands Ecology and Management,* **2**, 11–21.

Christensen, H. (2001). The Atlantic salmon: restoration of freshwater habitats. Unpublished report. WWF.

Danube Pollution Reduction Programme (1999). *Evaluation of Wetlands and Floodplain Areas in the Danube River Basin.* Vienna: WWF.

de Jong, S. A., (2000). *Vergroting van de Afvoercapaciteit en berging in de Benedenloop van Rijn en Maas: Bestuurlijk advies Aangeboden aan de Staatssecretaris van Verkeer en Waterstaat door de Stuurgroep Integrale Verkenning Benedenrivieren.* Rotterdam, Netherlands: Dutch Ministry of Transport, Public Works and Water Management, Directorate, South Holland.

Dugan, P. J. (1993). *Wetlands in Danger.* London: Mitchel Beazley.

Dynesius, M. & Nilsson, C. (1994). Fragmentation and flow regulation of river systems in the northern third of the world. *Science,* **266**, 753–762.

Ecotek (2000). *Final Report: Pre-Feasibility Study of Nature Restoration in the Nemunas River Delta, Kaliningrad Oblast.* Denmark: WWF.

European Communities (2000). Managing Natura 2000 sites: provisions of Article 6 of the Habitats Directive 92/43/EEC. Brussels.

European Environment Agency (EEA) (1995). *Europe's Environment: The Dobris Assessment.* Copenhagen: European Environment Agency, and Luxembourg: Office for Official Publications of the European Communities.

European Environment Agency (EEA) (1998). *Europe's Environment: The Second Assessment.* Copenhagen: European Environment Agency, and Luxembourg: Office for Official Publications of the European Communities.

European Environment Agency (EEA) (1999). *Sustainable Water Use in Europe,* Part 1, *Sectoral Use of Water.* Copenhagen: European Environment Agency, and Luxembourg: Office for Official Publications of the European Communities.

FAO (1997). *State of the World's Forests.* Rome: FAO.

Finlayson, C. M. & Davidson, N. C. (collators) (1999). Global review of wetland resources and priorities for wetland inventory: summary report. In *Global Review of Wetland Resources and Priorities for Wetland Inventory,* eds. C. M. Finlayson & A. G. Spiers, Supervising Scientist Report no. 144, CD-ROM, Canberra, Australia: Environmental Research Institute of the Supervising Scientist.

Fleischer, S., Stibe, L. & Leonardson, L. (1991). Restoration of wetlands as a means of reducing nitrogen transport to coastal waters. *Ambio,* **20**, 271–272.

Folke, C. & Jansson, Å. (2000). Wetlands as nutrient sinks. In *Managing a Sea: The Ecological Economics of the Baltic,* eds. I. M. Gren, K. Turner & F. Wulff. London: Earthscan Publications.

Garritsen, T., Vonk, G. & de Vries, K. (2000). *Visions for the Rhine.* Lelystad, Netherlands: RIZA.

Gren, I. M. (1999). Value of land as a pollutant sink for international waters. *Ecological Economics,* **30**, 419–431.

Gren, I. M., Folke, C., Turner, K. & Bateman, I. (1994). Primary and secondary values of wetland ecosystems. *Environmental and Resource Economics,* **4**, 55–74.

Hansen, H. O., Boon, P. J., Madsen, B. L. & Iversen, T. M. (eds.) (1998). *River Restoration: The Physical Dimension,* a series of papers presented at the International Conference on River Restoration 1996, organised by the European Centre for River Restoration, Silkeborg, Denmark. New York: John Wiley.

Hawke, C. J. & José, P. V. (1996). *Reedbed Management for Commercial and Wildlife Interests.* Sandy, UK: RSPB.

Hendrikesen, G. (ed.) (1999). *State of Affairs: Room for Rhine Branches*. Lelystad, Netherlands: Rijkswaterstaat Directie Oost Nederland.

Henrikson, L. & Brodin, Y. (eds.) (1995). *Liming of Acidified Surface Waters*. Berlin: Springer-Verlag.

Hertzman, T. & Larsson, T. (1997). Hornborgasjön: from an ocean of reeds to a kingdom of birds. *Naturvårdsverket Report*, **4694**. (in Swedish)

Hertzman, T. & Larsson, T. (1999). *Lake Hornborga, Sweden: The Return of a Bird Lake*, Wetlands International Publication no. 50. Wageningen, Netherlands:

Holdgate, M. W. & Woodman, M. J. (eds.) (1978). The breakdown and restoration of ecosystems. *Nato Conference Series*, **1**. New York: Plenum Press.

Hollis, G. E. (1992). The causes of wetland loss and degradation in the Mediterranean. In *Managing Mediterranean Wetlands and Their Birds*, eds. C. M. Finlayson, G. E. Hollis & T. J. Davis.

Howell, E. A. (1986). Woodland restoration: an overview. *Restoration and Management Notes*, **4**, 13–17.

Hunt, C. (1997). *Mississippi/Rhine Exchange*, Report no. 1, *Reviving the Rhine*. Washington, DC: WWF.

ICPDD (1997). *Ecological Restoration in the Danube Delta Biosphere Reserve/Romania: Babina and Cernovca Islands*, 1st edn. Bucharest: Danube Delta Biosphere Reserve Authority. [Report can be obtained at WWF International]

IUCN (1999). *IUCN Report on the State of Conservation of Natural and Mixed Sites Inscribed on the World Heritage List and the List of World Heritage in Danger*. Gland, Switzerland: The World Heritage Convention IUCN.

Iversen, T. M., Grant, R. & Nielsen, K. (1998). Nitrogen enrichment of European inland and marine waters with special attention to Danish policy measures. *Environmental Pollution*, **102**, 771–780.

Jeppesen J., Søndergaard, M., Kronvang, B., Jensen, J. P., Svendsen, M. & Laurisden, T. L. (1999). Lake and catchment management in Denmark. *Hydrobiologia*, **395/396**, 419–432.

Jones, T. A. & Hughes, J. M. R. (1993). Wetland inventories and wetland loss studies: a European perspective. In *Waterfowl and Wetland Conservation in the 1990s: A Global Perspective*, IWRB Special Publication no. 26, eds. M. Moser, R. C. Prentice & J. van Vessem. Slimbridge, UK: IWRB.

Jones, T. J. (ed.) (2000). *Policy and Economic Analysis of Floodplain Restoration in Europe: Opportunities and Obstacles*, report prepared for WWF's European Freshwater Programme under the 'Wise Use of Floodplains' Life-Environment Project. Copenhagen: WWF Denmark. (www.panda.org/europe/freshwater)

Junta de Andalucía (2000). *The strategy for the Green Corridor of the Guadiamar River: Fundamentals of the Strategy*. Legal deposal: SE-1830–99. (octv.guadiamar@cma.junta-andalucia.es)

King, S. E. & Lester, J. N. (1995). The value of saltmarsh as a sea defence. *Marine Pollution Bulletin*, **30**, 180–189.

Lanz, K., Seul, H. & Peek, G. (2000). Three presentations on 'Voluntary agreements for water protection: the case of organic farming in Weser–Ems, Lower Saxony, Germany'. In *Implementing the EU water Framework Directive: A seminar series on water*, Proceedings of Seminar 1 on Water and Agriculture, pp. 113–126. Brussels, WWF European Policy Office. (www.panda.org/europe/freshwater)

Ledant, J. P. (2000) Forest restoration in the Mediterranean. Unpublished report. Gland, Switzerland: WWF International.

Madgwick, F. J. (1999). Restoring nutrient-enriched shallow lakes: integration of theory and practice in the Norfolk Broads, UK. *Hydrobiologia*, **408/409**, 1–12.

Madgwick, J. & Hawke, C. (1996). Reed: the forgotten crop? *Farming and Conservation*, **2**, 5–7.

Maltby, E. (1991). Wetlands and their values. In *Wetlands*, eds. M. Finlayson & M. Moser, pp. 8–26. Slimbridge, UK: International Waterfowl and Wetlands Research Bureau.

Mannion, A. M. (1995). *Agriculture and Environmental Change*. Chichester, UK: John Wiley.

Mesleard, F. (1994). Agricultural abandonment in a wetland area: abandoned ricefields in the Camargue, France: can they be a value for conservation? *Environmental Conservation*, **21**, 354–357.

Mesléard, F., Grillas, P. & Tan Ham, L. (1995). Restoration of seasonally flooded marshes in abandoned ricefields in the Camargue (southern France): preliminary results on vegetation and use by ducks. *Ecological Engineering*, **5**, 95–106.

Mesléard, F., Lepart, J., Grillas, P. & Mauchamp, A. (1999). Effects of seasonal flooding and grazing on the vegetation of former rice fields in the Rhône delta (Southern France). *Plant Ecology*, **145**, 101–114.

Ministerio de Medio Ambiente (1999a). *Summary of the Doñana 2005 Plan: Water Regeneration of the Basins and*

Streams Emptying into the Marshlands of Doñana National Park. Madrid.

Ministerio de Medio Ambiente (1999*b*). *Documento Marco para el Desarrollo de las Actuaciones del Proyecto Doñana 2005*, 2 vols. Madrid.

Møller, H. S. (ed.) (1995). *Nature Restoration in the European Union*, Proceedings of a seminar in Denmark, 29–31 May 1995. Copenhagen: National Forest and Nature Agency, Ministry of Environment and Energy.

Møller, H. S. (1998). Institutional criteria and guidelines for successful wetland restoration: wetlands returning in Denmark: policies for small-scale restoration projects, prepared for Convention on Wetlands 3rd pan-European Regional Meeting, Latvia, June 1998. Copenhagen: Ministry of Environment and Energy.

Moss, B., Madgwick, J. & Phillips, G. (1996). *A Guide to the Restoration of Nutrient-Enriched Shallow Lakes.* Norwich, UK: Broads Authority, Environment Agency and EU Life Programme.

Ody, D. & Sarano, F. (1993). *The Danube: For Whom and for What?* Paris: Equipe Cousteau.

Pruijssen, H. & Rheenen, J. van (1999). *Working Together with Nature in the River Regions.* Arnhem, Netherlands: Dienst Landelijk Gebeid.

RIZA (2000) *Visions for the Rhine: Regional Vision for the World Water Forum.* Lelystad, Netherlands: RIZA.

Royal Society for the Protection of Birds (RSPB) (1994). Reedbed Habitat Action Plan. Unpublished report. Sandy, UK: RSPB.

Šeffer, J. & Stanová V. (eds.) (1999). *Morava River Floodplain Meadows: Importance, Restoration and Management.* Bratislava: DAPHNE–Centre for Applied Ecology.

Sheate, W. (1994). *Making an Impact: A Guide to EIA Law and Policy.* London: Cameron May.

Sih, A., Jonsson, B. G. & Luikart, G. (2000). Habitat loss: ecological, evolutionary and genetic consequences. *Trends in Ecology and Evolution,* **15**, 132–34.

Snazell, R. (1997). *Ecology and Twyford Down.* Oxford: Institute of Terrestrial Ecology.

Sollander, E. (1999). *European Forest Scorecards.* Gland, Switzerland: WWF International. (www.panda.org/europe/forests).

Stephan, U., Stobel, U. & Klass, R. (1999). *Analysis of the Environmental Damage caused by the Bombing of Chemical and Petrochemical Industries in Pancevo and Novi Sad.* Halle, Germany: Dessau. [Report can be obtained at WWF]

Tomlinson, M. L., Perrow, M. R., Hoare, D., Pitt, J.-A., Johnson, S., Wilson, C. & Alborough, D. (2002). Restoration of Ormesby Broad through biomanipulation: ecological, technical and sociological issues. In *Management and Ecology of Lake Fisheries,* ed. I. G. Cow. Hull, UK: Hull International Fisheries Institute.

Tucker, G. M. & Heath, M. F. (1994). *Birds in Europe, their Conservation Status.* Cambridge, UK: BirdLife International.

Tyler, G. (1994). Management of reedbeds for bitterns and opportunities for reedbed creation. *RSPB Conservation Review,* **8**, 57–64.

UNEP (1999). *Global Environmental Outlook 2000.* London: Earthscan Publications. (www.unep.org)

UNEP, OCHRA (1999). The kosovo conflict: consequences for the environment. (www.unep.org)

UNEP, OCHRA (2000). Report on the cyanide spill at Baia Mare, Romania. (www.unep.org)

UNESCO (1999). *State of Conservation Report of Doñana National Park,* 23rd session of the World Heritage Committee, Marrakesh, Morocco, 29 November–4 December 1999. Paris: UNESCO.

World Wide Fund for Nature (WWF) (1993). *Living Rivers,* report of study commissioned by WWF Netherlands (a number of sub-reports in Dutch are cited). Zeist, The Netherlands: WWF.

World Wide Fund for Nature (WWF) (1998). *Agenda 2000: Reactions to the Proposed Regulations,* pack of policy papers. Brussels: WWF European Policy Office.

World Wide Fund for Nature (WWF) (1999). *Pre-Feasibility Study of Nature Restoration in the Nemunas Delta Regional Park.* Copenhagen: WWF Denmark.

World Wide Fund for Nature (WWF) (2000). *A Green Corridor for the Danube.* Vienna: WWF.

World Wide Fund for Nature (WWF)/Adena (2000). *Cross-Compliance: The CAP and Environmental Cross-Compliance in Spain: Practical Application in Las Tablas Daimiel.* Madrid: WWF/Adena.

Yozzo, D. J., Clark, R., Curwen, N., Graybill, M. R., Reid, P., Rogal, K., Scanes, S. & Tilbrook, C. (2000). Managed retreat: assessing the role of the human community in habitat restoration projects in the United Kingdom. *Ecological Restoration,* **18**, 234–244.

Zockler, C. (2000). *Wise Use of Floodplains – LIFE Environment Project: A Review of 12 WWF River Restoration Projects Across Europe.* Copenhagen: WWF European Freshwater Programme.

3 • Africa

WILLIAM M. ADAMS

INTRODUCTION

Africa is extremely diverse politically and ecologically. 'Sub-Saharan' Africa comprises 46 countries (from the huge, such as the Democratic Republic of Congo, Zambia or the Sudan to the tiny, e.g. Gabon, Burundi, or Gambia); it covers some 21 million square kilometres. To these may be added the countries of the Maghreb along the continent's Mediterranean coast (Libya, Tunisia, Algeria and Morocco), Egypt and South Africa.

The biogeographic description of Africa is far from simple (Meadows, 1996). White (1983) identified 20 vegetation units within eight formations. A simpler and cruder summary would identify five biomes (Meadows, 1996). The first is the tropical moist forest of central Africa (Gabon, Congo) and coastal West Africa (e.g. Nigeria, Cameroon or Sierra Leone: Grainger, 1996). The second is grassland and thorn-scrub savannas (e.g. in the Guinea and Sahel zones of West Africa, and through east and south–central Africa: e.g. Adams, 1996). The third is the semi-arid and arid environments around the deserts of the Namib and the Sahara (e.g. Seely, 1978; Cloudsley-Thompson, 1984). The fourth comprises the areas of Mediterranean shrub (in North Africa and in the highly biodiverse 'fynbos' biome of the southern part of South Africa: Allen, 1996). The fifth consists of montane environments (e.g. in the highlands of Ethiopia, East Africa, South Africa and in a few locations in West Africa (Taylor, 1996). There is a diversity of wetland environments (Hughes, 1996).

The dominant ecological factor in most African ecosystems is the amount and distribution in time of rainfall. Patterns of African rainfall are complex (Niewolt, 1982; Goudie, 1996). Except for small parts of the wettest zone of tropical moist forest, rainfall is seasonal. In West Africa, rainfall increases southwards from the Sahara through the Sahel (250–600 mm per year) and zones of increasingly wet and wooded savanna (the Sudan and Guinea savanna zones) towards the tropical forest near the coast. These areas have a single rainy season during the northern hemisphere summer. In south–central Africa, rainfall increases towards the northeast away from the Namib. Areas nearer the equator (for example East Africa) have a dual rainy season, complicated by the effect of the Indian Ocean trade winds.

African rainfall not only varies seasonally, but also within the rainy season. Much of it falls in the form of short, intense, convective storms, and rainfall is therefore also highly variable over short distances (Hulme, 1987). More critically, the semi-arid environments of Africa experience considerable interannual variability in rainfall. This came to scientific attention in the Sahel drought of 1972–4. The existence of major past climatic variations within the last 20 000 years was known by the 1950s (Grove, 1958), as was the fact of dry years earlier in the twentieth century, but the existence of climatic variations on an annual time-scale was little understood by climatologists at least until the 1970s (Grove, 1973; Lamb, 1979). With hindsight, it is now clear that the 1950s and 1960s were rather wet in the Sudan and Sahel zones of Africa (Hulme, 1996). Drought has recurred in recent decades (in the Sahel and the Horn of Africa in 1984, and in 1992 in southern Africa: Hulme, 1995), and palaeoenvironmental studies across the tropics have revealed the extent and scale of climatic variation over the Holocene (Roberts, 1998). 'Droughts' therefore need to be understood as an integral element of semi-arid ecosystems in Africa, to which it is reasonable to expect both ecology and human management systems to be adapted.

Africa is also remarkable for the extent to which its human population depends economically on rural resources. Africa is the least urbanised inhabited continent on earth, and only at the beginning of the twenty-first century are the majority of its inhabitants living in towns and cities. The settlement hierarchy in most countries is highly skewed, with a small number of very large cities, and high rates of rural–urban migration. This, combined with annual rates of population growth of up to 4%, creates distinctive and, in places, acute environmental problems associated with rapid, unplanned urban growth. Elsewhere, many millions of Africans remain dependent on semi-natural ecosystems, either as farmers, herders of domestic stock or fishers, hunters or gatherers. High population densities in savanna environments present particular challenges to human subsistence, particularly in times of low rainfall (Mortimore, 1989, 1998; Scoones, 1994). Economic and environmental management systems of considerable complexity have been developed by African resource users that allow farming and agriculture to provide sustainable livelihoods in areas of highly variable rainfall (e. g. Adams, 1992).

There has been considerable investment in ecological restoration projects in Africa, although relatively little work to restore land degraded by industrial processes and pollution, or to restore habitats on built-up land, because of the restricted scale of industrialisation and urbanisation. Restoration efforts have concentrated on rural land, and the perceived problem of human-induced degradation through renewable resource use, particularly in drylands (desert margins and savannas), and in rivers and their floodplains. There have also been attempts to remove human influence completely, and re-establish 'wilderness' for conservation purposes. These will be considered in the second half of this chapter. First, however, it is necessary to say something about the enterprise of environmental restoration as a whole in Africa, for it differs in a number of ways from experience elsewhere, and it faces two sets of critical problems. First, it faces a range of serious constraints, chief among which are a limited science base, poor economic performance and prospects, and weak governance. Second, much restoration activity has started from quite erroneous assumptions about the nature and scale of human impacts on nature. In particular, it has long been widely assumed that small farmers and pastoralists have had a recent and mostly highly destructive impact on soils, vegetation and wildlife. Recent research suggests that in fact this may not be the case, or at least may not always be the case. Before discussing what has been attempted in the way of restoration, it is necessary to consider both the institutional constraints on restoration, and the problematic concepts of naturalness that underpin many restoration projects.

CONSTRAINTS ON ECOLOGICAL RESTORATION

A number of factors have operated to limit the scale and effectiveness of ecological restoration in Africa. First, there is the constraint of limited scientific knowledge. Ecological restoration demands a high degree of knowledge of the dynamics of ecosystems, and an ability to be able to make realistic predictions about their response to stresses of various kinds. In Africa, long-term ecological monitoring is unusual, experienced ecologists are few in the Third World, and (with exceptions such as Deshmukh, 1980) there are few textbooks on ecology designed to be relevant to African readers. Scientific knowledge about African environments has predominantly been created quite recently, and by ecologists from outside Africa.

Formal scientific knowledge of Africa is not of long standing, remaining slight and conjectural until late in the nineteenth century, despite extensive trade across the Indian Ocean with the Middle East and Asia, and across the Sahara with the Arabic Mediterranean world, and from the Portuguese voyages of the sixteenth century that resulted in the establishment of coastal trading stations and the Atlantic slave trade (Iliffe, 1995: Reader, 1997). Even when the Victorian explorers had trekked and reported, little was known of African ecology when the race for colonial occupation was unleashed following the Berlin Conference in 1884 (Reader, 1997).

The European industrial powers that took military and administrative control of the various African states and political units, and vast tracts of land, knew remarkably little of the landscapes, ecosystems, people and societies they had grabbed.

The colonial annexation of Africa therefore took place at about the same moment that ecology was first being carved out as a discipline (McIntosh, 1985; Worster, 1985). The twenty-first century has therefore seen the science of ecology and scientific knowledge of the ecology of Africa advance hand in hand. A paper on the vegetation of South Africa was published in the new *Journal of Ecology* in 1916, and a monograph on African vegetation, soils and land classification, resulting from American work, in 1923 (Shantz & Marbut, 1923). The later 1920s saw British ecological expeditions to the East African lakes (Worthington, 1983) and to Nigeria (Richards, 1939). By that time ecological research on disease in Africa was well established, notably on the tsetse fly (Ford, 1971).

Ecological science was readily harnessed in the service of colonial development, and ecologists found a willing audience in the powerful but ignorant officers of the colonial state. In 1931, Phillips advocated a 'progressive scheme' of ecological investigation in East, Central and South Africa with the aim of helping to develop resources. Studies of vegetation and soils in Northern Rhodesia (Zambia) in the 1930s (Trapnell & Clothier, 1937; Trapnell, 1943) contributed to the development of concepts of carrying capacity and critical population density. The 'African Survey' (Hailey, 1938) and the contributory volume *Science in Africa* (Worthington, 1938) emphasised the importance of ecology to African development.

Following the end of the Second World War, institutions were put in place to promote and support science in Africa, for example the Conseil Scientifique pour L'Afrique (CSA)(Worthington, 1958), the British Colonial Research Council (under Lord Hailey), the French Office de Recherche Scientifique et Technique d'Outre Mer (ORSTOM), and the Belgian Institut pour la Recherche Scientifique en Afrique Central (IRSAC) (Worthington, 1983). Postwar colonial development drew even more heavily on a scientific approach to the environment and its management (e.g. Worthington, 1958). Historians describe the arrival of scientists and technicians and their engagement in development-related work in East Africa after the Second World War as a 'second colonial occupation' (Low & Lonsdale, 1976). The engagement with the development problems of rural Africans inevitably demanded involvement with the ecology of their production systems and the wider landscapes of which they were part. Scientific expertise in agriculture, grazing, forestry, disease, erosion and water supply won a place within what had been the predominantly legal, political and fiscal institutions of colonial government.

The response of policy-makers to the lack of formal scientific knowledge about African ecology was therefore to bring in colonial scientists to create it. In the last four decades of the twentieth century, in post-colonial Africa, it was undertaken by employing consultants, predominantly from Europe and North America. The importance of such outside influences on planning and development is considerable (Chambers, 1983). Foreign consultants and planners are ubiquitous in Africa, recruited and paid by First World aid donors determined to have timely and professional data and environmental assessments. Foreign experts can command a legitimacy, denied prophets in their own country, derived from their presentation of technology and experience, although this expertise may be less valuable than it at first appears. In the Awash Valley of Ethiopia, for example, French consultants planning water resource development lacked critical knowledge on political, ecological, social, cultural, scientific and technical aspects of the region (Winid, 1981); those in the field only had contact with the local elite, lacked supervision and failed to transfer technology. There are obviously pitfalls in a planning system dependent on foreign experts. Temperate-zone ecologists brought in as consultants may yet prove a poor (and costly) substitute for local expertise, perhaps glibly repeating dominant preconceptions and myths about tropical ecosystems (for example about overgrazing or desertification: see below and Leach & Mearns, 1996).

A second constraint on ecological restoration stems from the existence of significant institutional weaknesses in government, economy and civil society that restrict the extent to which science can be developed into policy. There are several dimensions to this. The first is the ongoing economic crisis. Economic growth in Africa has been uneven, both over time (with real economic decline in the 1980s, for example), and between countries. Far from enjoying a smooth economic 'development', involving industrialisation and wealth creation, from the 1980s deindustrialisation began, and Africa's share of export markets and world trade fell. Africa is still hugely dependent on the export of primary commodities, including oil, but the price of these has been depressed. Most African countries face serious economic and financial problems and are deeply in debt. Africa's economic ills include weak agricultural growth, declining industrial output, poor export performance, growing debt, declining institutions, and deteriorating socio-economic and environmental conditions. Infrastructure, which was created in the middle 40 years of the twentieth century at enormous cost, has deteriorated. Universities, set up with high hopes in the 1950s, have stagnated, starved of funding, their staff underpaid or even unpaid, their facilities falling further and further behind the sophistication of Western universities. Few good researchers trained overseas are willing to return to such institutions: they seek work in universities outside Africa or in international organisations, or go into business. Systematic ecological research by African institutions is rare in most of tropical Africa.

A second dimension of institutional weakness has been the human and environmental costs of conflict. Much of Africa has been ravaged by war in the second half of the twentieth century, with large areas devastated, and millions made homeless. This is no new phenomenon. The democratic republic of Congo (Zaïre) suffered a crippling civil war after independence, and is still experiencing war over four decades later; destabilisation and warfare cost Angola and Mozambique deeply in both human and economic terms, and the war in Angola persists. There has been conflict in most parts of the continent, most notably in Ethiopia, Eritrea, Somalia, Sudan, Rwanda, Uganda, Chad, Liberia and Sierra Leone. Even in countries nominally at peace, military spending exerts a heavy burden on government expenditure. Careful ecological research is an unimaginable luxury in a country ravaged by war.

A third dimension of institutional weakness relevant to ecological restoration in Africa is poor governance. This is a particular example of a general problem, a mixture of large and cumbersome government bureaucracies lacking skills and budgets, lack of democratic accountability, a high incidence of corruption, and weak institutions of civil society (Bayart, 1993).

These problems are widespread in Africa, and in places present critical constraints to effective conservation and development planning. However, they are not universal, and in a number of countries approaches to environmental planning have been developed that are innovative, creative and potentially effective. In Namibia, for example, the newly independent government from 1990 supported the innovative approach to conservation developed by the Namibia Wildlife Trust (a non-governmental organisation, NGO) based on the idea of 'conservancies' (Jones, 2000). This idea, influenced by experience with the 'CAMPFIRE' programme in Zimbabwe (Metcalfe, 1994; Murombedzi, 1999), involved the granting by the state of rights of ownership of game animals to the establishment of conservancies with a defined membership and boundary, a legal constitution and a representative committee. Conservancies could then sell hunting rights and raise money for development purposes (Jones, 2000). The development of this programme involved both the Ministry of Environment and an NGO, Integrated Rural Development and Nature Conservation (IRDNC). While the success of this programme in terms of development and conservation remain to be proven, this collaboration demonstrates the potential for promoting conservation when good government and strong institutions of civil society work effectively together.

The problem of naturalness

More fundamentally, however, ecological restoration in Africa has been influenced by a set of rather peculiar scientific and popular understandings of what is 'natural' in African ecology. This has had profound implications for the kinds of ecological restoration that have been attempted, and in as much as it has been based on mistaken assumptions about human impacts on ecosystems, it has contributed to the lack of success.

Colonial trade and conquest transformed European ideas of nature from the fifteenth century onwards. One of the ways in which newly discovered and astonishingly biodiverse worlds were conceived was in terms of the notion of paradise, or the Garden of Eden. The tropics (and especially tropical islands) were portrayed as idealised and 'natural' landscapes by Western thinkers (Grove, 1990). By the mid seventeenth century, it began to be recognised that imperial demands on paradise threatened to destroy their natural beauty and bounty, as for example on the Canary Islands and Madeira. In due course this led to an awareness of environmental limits and the need for conservation, as experience of ecological change on islands was translated into more general fears of environmental destruction on a global scale (Grove, 1990, 1995). In the nineteenth and twentieth centuries, European ideas about nature in Africa were influenced by very similar ideas. The variety and abundance of large mammals, the rurality of societies and low human population densities led observers to construe Africa as a particularly natural place, to a large extent unaffected by human action (Anderson & Grove, 1987).

In retrospect, this is a remarkably mistaken view. It fails to take account of the depth and scale of pre-colonial political entities and economies, or the transformations that they wrought in landscape and ecology (Connah, 1987). It ignores the scope and scale of the disastrously destructive dynamics of the trade in gold, ivory and enslaved people that preceded colonial annexation, or their impacts (and that of the firearms that made them possible) on ecology, economy and society (Iliffe, 1995; Reader, 1997). It also fails to take account of the destructiveness of the colonial annexation itself, and the forced connection of African societies to the world economy (Iliffe, 1981; Watts, 1983). Africa was indeed a 'Dark Continent' to arriving Europeans (Jarosz, 1992), but of the myths they wove about it, the notion that it was previously a scarcely inhabited place was particularly wide of the mark. There

were, of course, reasons for the mistake, apart from the sheer scale and diversity of the continent and the lack of any historical or environmental knowledge of it. In the savannas of East Africa, for example, rinderpest (introduced from India via northeast Africa), bovine pleuropneumonia and smallpox devastated the livestock economy in the early 1890s, and started a chain of ecological repercussions (chiefly the growth of scrub and trees on open grasslands, and changed incidence of fires). Livestock disease also triggered a sequence of social and economic changes, which were both influenced by, and in turn affected, ecological change. These were affected by the decisions about land annexation taken by German and British colonial governments, who found the plains of Maasailand emptied of cattle, and ripe for designation as a 'natural' place (Anderson & Johnson, 1988; Waller, 1988, 1990).

The implications of the view that Africa was (or should be) a 'natural' place, where extensive human impacts are recent, underpinned many areas of colonial policy, most notably in the area of conservation planning (MacKenzie, 1987, 1989). The designation of protected areas in Africa has very often involved the creation of wilderness, as is discussed further below (Neumann, 1998). The idea survived the end of colonial rule and the rise of modern states in independent Africa, and indeed lives on in the successive western myths about the lives of Hemingway or Karen Blixen to Walt Disney's *Lion King* or the television documentary about 'wild Africa' (Grzimek, 1960; Adams & McShane, 1992; Gavron, 1993).

The dependence of ideas of ecological restoration on ideas about naturalness is not of course confined to Africa. The many terms relating to ecological and environmental restoration (e.g. rehabilitation, reclamation, enhancement, creation, ecological recovery and mitigation) all address the issue of undoing or counteracting the effects of some kind of human action. One route through the semantic morass is to focus on the issue of specific human disturbance, as the US National Research Council does in defining restoration, habitat rehabilitation, habitat enhancement and habitat creation (Table 3.1).

The idea of restoration is therefore predicated on a particular idea of naturalness, or of what the ecosystem would look like in the absence of

Table 3.1. *Definitions of restoration*

Type of restoration	Definition
restoration	complete structural and functional return to a pre-disturbance state
rehabilitation	partial structural and functional return to a pre-disturbance state
enhancement	improvement of a structural or functional attribute
creation	birth of a new ecosystem that previously did not exist at the site

Source: Adams & Perrow (1999).

human intervention. This has presented particular problems for ecologists in Africa, for (unlike temperate regions, and especially Western Europe), there has, until comparatively recently, been relatively scant knowledge about when and what humans have done to the ecosystems of Africa.

Even in Europe, where recorded history is long, human occupation short, and there are plenty of peat bogs to preserve pollen, the concept of 'naturalness' has been a trap for the unwary ecologist. Early attempts to classify plant communities in North America had to be adapted in Europe to take account of the depth of human transformation of ecosystems, using concepts such as 'arrested' or 'deflected' succession and plagioclimax (McIntosh, 1985; Sheail, 1987). Nonetheless, it is the 'naturalness' of ecosystems on which arguments for conservation were built (Adams, 1996).

Peterken (1986) defines several kinds of naturalness that, while developed for the temperate woodlands of the UK, have some relevance to Africa: 'original naturalness' would be the state of an ecosystem before the first arrival of *Homo sapiens*; 'present naturalness' would be the state of an ecosystem now if humans had arrived but done nothing; if all human influences ended at once, ecosystems would be in a state of 'future naturalness'. In Africa, ecologists and policy-makers have long believed that anthropogenic ecological change is a serious problem. The problem is that they have been ignorant of environmental history, and have been unclear about the history of human occupation and transformation of landscape. They have mistaken Tansley's seminatural ecosystem (Tansley, 1939) for Peterken's 'original natural' ecosystem.

There are various reasons for this. The most important is the extent to which scientific ideas about African ecosystems have been linked to more popular concepts in the minds of policy-makers. Sociologists of science argue that no scientific idea is ever divorced from its social context, but certainly in the African case, the relative dearth of in-depth scientific research, the lack of strong indigenous research institutions and researchers, and the urgent nature of the problems that policy-makers and politicans have faced, have all led to excessively narrow and hasty links between field research, theory formation, and policy action.

Emory Roe (1991) argued that development planners tend to operate using self-referencing stories ('narratives') about the challenges that face them, including the environment. These provide reassurance that their understanding of problems is correct and their choice of solutions appropriate. Such narratives tend to be persistent in the face of contrary evidence, and cannot be overturned by simply showing that they are untrue in particular instances, but only by providing a better and more convincing story. Such narratives dominate thinking about the environment in Africa (Leach & Mearns, 1996), particularly in discussions of environmental degradation (e.g. in Ethiopia: Hoben, 1995) and desertification (Swift, 1996).

Sinclair & Fryxell (1985) offered a classic version of the standard desertification narrative. They argued that the 'normal' land use pattern in the Sahel was based on migratory grazing using seasonally available resources, and operated in a balanced and stable way for many centuries until the second half of the twentieth century, when aid funding, population growth, overgrazing, and profit-orientated agricultural practices led to extensive areas of the Sahel becoming bare of vegetation, a state exacerbated by feedback effects of soil on rainfall. This analysis of ecological crisis in the Sahel was a modern version of concerns about the 'spread of the Sahara' expressed (and refuted) in the 1930s (Stebbing, 1935; Jones, 1938), reawakened by the 'Sahel droughts' of the 1970s and 1980s (Warren, 1996). Desertification became one of the recurrent themes of environmentalist writing on the Third World in the 1980s. The United Nations Environment Programme (UNEP) claimed that desertification was 'a direct threat to 250 million people around the world and an indirect threat to a further 750 million people' (Williams & Balling, 1995, p. 8). The United Nations Conference on Desertification (UNCOD) was held in Nairobi in 1977, and the UN Convention to 'Combat Desertification in those Countries Experiencing Severe Drought and/or Desertification, especially in Africa' was adopted in 1994 and entered into force in 1996.

The most often quoted statistics on desertification relate to the rate at which deserts, and particularly the Sahara, are 'spreading'. Hellden (1988) questioned the classic work of Lamprey in 1975 (see Lamprey, 1988), who compared the boundary between desert and subdesert grassland and scrub in the southern Sudan. Lamprey (1988) suggested that the boundary had shifted 90–100 km between 1958 and 1975 (5.5 km per year). This statistic was taken up, repeated and exaggerated through the 1970s and 1980s. However, Hellden points out that the 1958 and 1975 vegetation boundaries were not comparable; the 1958 boundary was drawn from a vegetation map itself interpolated from a rainfall map, while the 1975 boundary was mapped from satellite imagery. Comparison of 1975 and 1979 satellite imagery showed no change in vegetation boundaries. Field surveys showed short-term impacts of drought between 1965 and 1974, but no systematic decline in crop production, no shift in dune field positions, and no major changes in vegetation cover (Thomas & Middleton, 1994).

In fact, there is considerable confusion about the nature, permanence and cause of ecosystem change in drylands, and a lack of authoritative ecological field research. A number of commentators have argued that desertification (and hence attempts to 'restore' degraded ecosystems) is an institutional as much as an ecological phenomenon (Thomas & Middleton, 1994). Swift (1996) argues that the desertification narrative serves the interests of three specific constituencies: national governments in Africa, international aid bureaucracies (especially UN agencies) and scientists. In the 1970s, recently independent African governments were restructuring their bureaucracies and strengthening central control over natural resources. The fact of drought, and the assumptions about human-induced environmental degradation linked to them, strengthened such claims and made centralised top–down environmental planning seem essential. Aid donors found in desertification a problem that seemed to transcend politics and legitimated large high-technology programmes (Swift, 1996). To scientists developing new fields such as remote sensing, desertification offered fertile terrain for expansion: satellite imagery seemed to offer answers without the need for lengthy and tedious fieldwork, and desertification became a source of funding and legitimacy for technicians and researchers.

The view that African ecosystems have been relatively little affected by human action until the twentieth century has had profound impacts for ecological restoration. This legacy, and the restoration adopted, will be described in the remaining sections of this chapter.

RESTORATION IN PRACTICE

Restoration of 'wilderness'

It has been argued above that conservation policy in Africa draws on notions of Africa as a wild and untransformed place. In the industrialised world, the dominant model for protected areas is that developed in the United States, where national parks were created in remote and sparsely populated areas from which indigenous people were evicted long ago. Such land has often been regarded as 'wilderness', and evidence of former occupation and ecosystem modification (by indigenous people) has been forgotten or downplayed. Neumann (1998) describes the ways in which the Anglo-American nature aesthetic was applied to the national parks of Africa, both at their inception (in the colonial period), and today. Unlike America, Africa was not emptied of people upon annexation and settlement, yet for the purposes of conservation, large tracts were adjudged to be empty, or empty enough to be treated conceptually as 'wilderness' (Turton, 1987).

In this conventional 'fortress conservation' model, national park landscapes were deemed to be 'natural', and humans were excluded. At least, certain humans were excluded, for while local farmers, hunters and other uneducated resource users were unacceptable, tourists with cameras, hotel proprietors and tour operators, scientists and sometimes big-game hunters were allowed. Most African government conservation departments have their roots in agencies established to prevent hunting (or 'poaching') by local people and to protect farmland from marauding wildlife. Land was set aside for 'game', and later in national parks. People were prevented from gaining a living in such areas, and were sometimes removed from them, for in a 'natural' area, people – particularly black people – were seen as an 'unnatural' presence, a disturbance to ecological equilibrium with their grazing animals, fires, spears and snares (MacKenzie, 1987, 1989).

The removal of people from protected areas in Africa is the most basic form of ecological restoration, and is obviously based on the notion that human occupation is 'unnatural'. In some instances this may be true, but in most it is not, and serious questions of human rights are raised by the imposition of 'wilderness' conditions through the eviction of farmers, hunter–gatherers or pastoralists. The economic impacts on people evicted from protected areas (for example the Mkomazi Game Reserve in northern Tanzania) can be considerable (Brockington & Homewood, 1996; Infield & Adams, 1999).

A good example of the creation of 'wilderness' is the Arusha National Park in Tanzania (Neumann, 1998). This consists of land taken from Meru people in 1896 following a punitive expedition by the German colonial regime. Following their defeat in battle, the Meru fled, and the land was either given to German settlers to farm, or gazetted as Forest or Game Reserve. The National Park was first established in 1960 by conversion of one of these reserves, and it grew through gradual purchase by international conservation bodies of farms no longer wanted by their owners in a newly independent Tanzania (Neumann, 1998). Conservation management of the park has sought to restore its 'natural' state, although there are no scientific data as to past ecological patterns, and no clear statement of what specific set of human disturbances are being disentangled. Local people are excluded, although legal claims by Meru people persist. Ironically, local people are employed to cut and burn the forest glades their ancestors once created and maintained through livestock grazing, in order to allow tourists to see wildlife.

Ecological restoration through land abandonment has been practised in other forest areas, for example in the Mgahinga Gorilla National Park in Uganda, again on land cleared of farmers (Infield & Adams, 1999). Institutional and legal problems abound in such circumstances, but are typically being met (sometimes with success) through various forms of community outreach programmes, and sometimes (as at Mgahinga) with revenue-sharing, targeted development aid and access for consumptive resource use (Infield & Adams, 1999). There can be significant ecological problems, such as the invasion of exotic plant species such as Australian wattles (*Eucalyptus* spp.).

Similar habitat restoration strategies are being tried in other parts of Africa, particularly in South Africa and Zimbabwe. The depressed state of commercial (white-owned) farms on poor land in the unsubisidised post-apartheid era has provided an opportunity for creative conservation. In a number of locations, businesses have been established to acquire and fence farmland and stock it with locally extinct large mammals as the basis for a safari tourist industry. These 'wildlife ranches' (often confusingly referred to as 'conservancies', but very different from those in Namibia, discussed above, in terms of their ownership) vary in the extent of the ecological restoration attempted, but some (for example that at Phinda Game Reserve in KwaZulu) can involve the creation or re-creation of extensive semi-natural ecosystems and associated large mammal communities (Varty & Buchanan, 2000) (Box 3.1). There is increasing interest in South Africa in partnerships between the state and the private sector to create conservation areas, for example through contractual parks and 'conservancies'. These developments often have an explicit restoration component. For example, the 20 000-ha Agulhas National Park will consist of a core of 12 000 ha of state-owned

Box 3.1   Species-centred restoration

Large private nature reserves and ranches provide an important opportunity for species-centred restoration of large mammals. Notable here is the black rhinoceros (*Diceros bicornis*), which has suffered catastropic population declines in the wild due to the value of its horn, both as an aphrodisiac in East Asia and for making dagger handles in the Middle East. Strategies for stabilising populations include capture and translocation to secure and closely patrolled national parks (for example Nairobi National Park in Kenya), or private reserves (e.g. Phinda Game Reserve in South Africa: Varty & Buchanan, 2000), and, more controversially, de-horning (Berger, 1992; Booth, 1992; Emslie & Brookes, 1999). The black rhinoceros is, in many African countries, effectively being managed in a kind of semi-captivity.

The strategy of captive breeding and release has been used relatively seldom in Africa, where the protection of species *in situ* and their habitats is a more cost-effective strategy. However, where species are approaching extinction in the wild, and a buoyant captive population exists, release may prove a useful strategy. One example here is the bald ibis (*Geronticus eremita*). There are 60 birds breeding in the wild breeding at a single site in Morocco, but 200 individuals in zoos. Despite the risks to the wild population due to disease and disruption of social structure, reintroductions may be a necessary and effective species restoration strategy (Anon, 2000).

land (some of it purchased by conservation NGOs in order to re-establish conservation management), surrounded by private land incorporated into the park on a contractual basis (Heydenrych *et al.*, 1999).

Ecological restoration and overgrazing

Many analyses of ecological change in sub-Saharan Africa have been based on problematic assumptions about livestock management or indigenous agriculture. The conventional view of rangeland management and mismanagement has been built around ideas of range conditions class and carrying capacity. The logic was that the environment is capable of supporting a certain fixed number of livestock (or biomass), and that for any given ecosystem this could be calculated primarily as a function of rainfall. There is a general relationship between rainfall and the productivity of herbivores (Coe *et al.*, 1976), and similar plots of livestock biomass against annual rainfall can be drawn (e.g. Bourn, 1978). If these regressions are taken to represent 'carrying capacity', it can be argued that at stocking levels lower than the line, pasture resources are being underused, and that at stocking levels above the line resources are being overused, such that there is likely to be adverse ecological change (e.g. extinction of palatable species and, eventually, loss of vegetation cover) and, ultimately, the death of excess stock.

Governments have tended to distrust people who are mobile and difficult to locate, tax, educate and provide with services, and conventional rangeland science has added a fear of mismanagement through overgrazing of seemingly fragile rangelands. Pastoral policy has been built on a perception of the need to maintain and restore rangeland condition. This has been done in two ways. First, governments have sought to 'combat desertification' directly through ecological interventions on the ground. A variety of methods have been used, including tree and shelter-belt planting and sand dune stabilisation (using mechanical fences or live planting).

A variety of techniques are used to control drifting sand on desert margins (Watson, 1990). Sand moves by creep or by saltation. Movement can be controlled by promoting deposition upwind of the problem area, enhancing the transport of sand within the problem area, reducing sand supply and deflecting moving sand (Table 3.2).

Techniques of this kind have been developed and applied in a number of parts of Africa, particularly in the northern Sahel, for example Mauritania (UNSO, 1991; Jensen, 1993), and in the Magreb. In 1975 the government of Algeria began to create the 'green dam', a physical barrier of bushes and shrubs (particularly *Eleagnus angustifolia*) and plastic wire mesh across the whole country, in an attempt to prevent Saharan sand blowing onto the northern

Table 3.2. *Approaches to sand movement control*

Action	Method
Promoting deposition	Ditches
	Barriers and fences
	Vegetation belts
Enhancing transportation (to clear an area of moving sand)	Aerodynamic streamlining of the land surface
	Surface treatment to enhance creep or saltation (e.g. with asphalt, latex, polyvinyl, sodium silicate, gelatine)
Reduction of sand supply	Surface treatment (a layer of gravel on the surface; chemical treatment; asphalt or oil; moisture; artificial salt crusts)
	Fences
	Vegetation
Deflection of moving sand	Fences and barriers
	Tree belts

Source: Watson (1990).

steppes (Ballais, 1994). In northern Sudan, the SOS Sahel Northern Province Community Forestry Project involved linked programmes that were organised to draw in women, men and schoolchildren to plant and maintain a variety of shelterbelts (Abdel & Ibrahim, 1997) (Table 3.2). The project concentrated on working with people, and enabling communities to engage in ecological restoration work (Table 3.3).

In a number of such projects, exotic plant species have been used, particularly in drier areas and those where supplementary watering or irrigation have not been possible. In the Sudanese project, experiments were carried out with 18 indigenous and exotic species for shelter-belts, and the exotic mesquite (*Prosopis juliflora*) was used for shelter-belts, while *Eucalyptus camaldulensis* was used for wind breaks. Other species were *Leptadenia pyrotechnica*, *Acacia mellifera*, *Pithecillobium dulce*, *Azardiracta indica* and *Albizia lebbekh*. *Prosopis* proved controversial since it is a potential weed of arable land, but its spread in the area was prevented because land-hunger was such that plants were prevented from invading cultivable land. In Mauritania, only *Prosopis juliflora* survived on dunes and around houses in areas with less than 150 mm of rainfall

per year, although a wider range of species could be grown on better soils or in wetter areas (UNSO, 1991).

The second approach to dryland ecological restoration in 'degraded' rangelands has been to develop policies aimed at the pastoralists themselves, and their institutions and economies (Swift, 1982; Horowitz & Little, 1987). This has typically involved a series of measures, including the control of livestock numbers, fencing to control grazing pressure, provision of watering points and sedentarisation (Table 3.4). None of these strategies fits with nomadic or semi-nomadic subsistence livestock production, so government pastoral policy has tended to emphasise sedentarisation, formal (i.e. freehold or leasehold) land tenure and capitalist production. In addition, the pursuit of higher productivity has involved the manipulation of range ecology through controlled burning, bush clearance and pasture reseeding.

There are wide gaps between pastoral policy prescriptions and the ways pastoralists actually manage their herds and rangelands. By implication, conventional pastoral policy has tended to emphasise the production of animal products from the slaughter of stock (i.e. meat and hides) rather

Table 3.3. *SOS Sahel Northern Province Community Forestry Project, Sudan*

Programme	Activity
Women's programme	Seed collection, storage and treatment
	Small private nurseries
	Tree planting around the home and village
	Raising awareness of the dangers of desertification
	Construction and use of improved stoves
	Marketing of seedlings for income generation
	Participation in project nurseries and planting sites
Men's programme	Seed collection, storage and treatment
	Small private nurseries
	Planting of trees in shelter-belts and windbreaks
	Raising awareness of the dangers of desertification
	Shelter-belt management
	Participation in project's main nurseries
	Motivation, organisation and management of communal shelter-belts
School programme	Classroom lectures, slide shows, puppet theatre
	Practical exercises
	Visits to places of interest
	Involvement in project nurseries and shelter-belt planting

Table 3.4. *Pastoral policy and arid land restoration*

The control of livestock numbers to match range
conditions, through destocking and the promotion
of commercial offtake, to improve stock health
and weight

Fencing and paddocking to allow grazing pressure on
particular pieces of land to be closely controlled

Provision of watering points to allow optimal
livestock dispersal

Disease control and stock breeding, both to improve
productivity and to improve stock health and weight

Sedentarisation and the imposition of formal
(i.e. freehold or leasehold) land tenure
and capitalist production

than products from live animals (milk or blood) that
typify many indigenous pastoral systems. Similarly,
pastoral development planning tends to focus
on cattle, whereas indigenous production systems
typically involve a mix of species. Indigenous
pastoral ecosystems seem well adapted to exploit
the spatial and temporal variability in production,
adapting herd composition and using movement to
maximise survival chances. The Turkana in Kenya,
for example, have mixed herds, with camels, which
use browse resources which are available even in
the dry season, and cattle, which are more productive in the wet season but have to move out of the
plains into the hills in the dry season (Coughenour
et al., 1985; Coppock *et al.*, 1986). Such systems offer
a relatively low output compared to modern capitalist systems such as ranching. However, they are remarkably robust in terms of providing a predictable,
if limited, livelihood.

Researchers have increasingly expressed reser-
vations about the universal applicability of the
concept of overgrazing, and with the unreflec-
tive links drawn between it and desertification
(Horowitz & Little, 1987; Mace, 1991). 'Overstocking'
or 'overgrazing' are rarely defined, and judgements
about carrying capacity are subjective, although

that subjectivity is rarely admitted. Homewood & Rodgers (1987, 1991) questioned the whole concept of carrying capacity. The data necessary to assess 'long-term degradation' of vegetation or desertification in most cases simply do not exist. Furthermore, there are real problems of obtaining adequate quantitative data either on the responses of vegetation to different stocking levels or on livestock numbers. Figures of 'carrying capacity' typically take no account of seasonal variations in fodder availability or annual variations, and rarely consider spatial mobility. They are therefore of little value in understanding either rangeland ecology or pastoral practice. Although they have become both entrenched and self-reinforcing, they are an unsatisfactory and sometimes dangerous basis for management.

Biologically based estimates of carrying capacity tend to be arbitrary, to be based on limited data and to derive from 'rule of thumb' and the experience of range ecologists rather than empirical scientific fieldwork (Mace, 1991). Arguably, the concept of carrying capacity is not appropriate in areas with great annual variation in primary productivity, and most estimates fail to take account of the variability and resilience of savanna ecosystems (Homewood & Rodgers, 1987). The productivity of semi-arid rangelands varies both seasonally and between years. The primary cause of this variation is rainfall. The high spatial and temporal variability of precipitation, particularly in sub-Saharan Africa, has increasingly been recognised. Other factors affecting productivity also vary over time and space, notably soil nitrogen and phosphate (Breman & de Wit, 1983) and the impact of fire (Norton-Griffiths, 1979). Ecological studies tend to be of short duration (often confined to the fieldwork period associated with a PhD thesis) and localised. Yet semi-arid grazing ecosystems undergo considerable and important fluctuations from year to year and place to place. Analyses of real or predicted degradation tend to be built on estimates of regional stocking rates. Such estimates are unreliable (livestock are difficult and expensive to count), they fail to take account of spatial and temporal variations the social processes of herd management and the political economy of livestock herding (e.g. the rise of absentee urban herdowners in the Sahel).

Homewood & Rodgers (1987) suggest that carrying capacity should be defined in terms of the density of animals and plants that allows the manager to get what they want out of the system: in other words that it is only possible to speak of a carrying capacity in the context of a particular management goal. What suits a nomadic pastoralist may not suit a rancher; many African systems have a subsistence stocking rate higher than commercial ranchers would adopt, giving low rates of production per animal but high output per unit area (Homewood & Rodgers, 1987). Actual stocking levels can and do exceed 'carrying capacity' for decades at a time (Behnke & Scoones, 1991). There is no one simple ecological succession towards an overgrazed state, but complex patterns of ecological change in response to exogenous conditions (especially rainfall), stock numbers and management. Ecological change can take many forms, not all of them serious, and it can proceed by diverse routes some of which can be reversed more easily than others, and some of which are more sensitive to particular management than others.

Scholes & Walker (1993) describe the evolution in understanding of savanna ecosystems over the duration of the South Africa Savanna Biome Programme at Nylsvley Nature Reserve north of Johannesberg. Whereas when this began in 1974, savanna ecosystems were seen as reasonably stable entities, by 1990 they were being understood as systems whose structure and function were largely driven by disturbances. The form and direction of savanna dynamics over long periods of time are determined by combinations of circumstances, each of which alone may be weak. Between these contingent events, savannas are 'in a state of continuous adjustment within defined limits' (Scholes & Walker, 1993, p. 256). Factors such as fire and elephant browsing are capable of shifting savanna ecosystems from one stable state to another (Dublin et al., 1990). Savanna management should not aim at establishing a fixed pattern in space and time, but allow 'a range of possibilities, based if possible on the historical pattern, with variation round it' (Scholes & Walker, 1993, p. 264). It is the temporal and spatial heterogeneity of savannas that enables the persistence of their high biodiversity.

Through the 1980s and 1990s, conventional thinking about carrying capacity and overgrazing began to be challenged by so-called 'new range ecology' (Behnke & Scoones, 1991). In drier rangelands, with greater rainfall variability, ecosystems exhibit non-equilibrial behaviour. Ecosystem state and productivity are largely driven by rainfall, and pastoral strategies are designed to track environmental variation (taking advantage of wet years, and coping with dry ones), rather than being conservative (seeking a steady-state equilibrial output). What appeared to be perversity or conservatism on the part of pastoralists is revealed to be highly adaptive (Behnke & Scoones, 1991; Prior, 1993; Scoones, 1994).

There is no single 'carrying capacity' of a pastoral ecosystem, represented by an equilibrium number of livestock, but a changing balance of livestock and range resources, with drought years first reducing the condition of stock and then (through disease, death and destitution-forced sales) reducing stock numbers. Good rain years then allow pastures to recover, allowing a lagged recovery of herd numbers as pastoralists track environmental conditions. Not only do herd managers need extensive knowledge of environmental conditions and opportunities in different areas open to them, and resilient multi-species herds, to survive under such conditions, but they also need institutions for the exchange and recovery of stock through kinship networks. Development strategies must support indigenous capacity to track rainfall and maintain social and economic networks, rather than demand a shift to a static, equilibrial capitalist form of production.

Ecological restoration of floodplains

One aspect of African ecology where there is no doubt about the nature, severity and cause of human impacts on the environment is the construction of dams. From the 1960s onwards, a series of major dams were constructed on African rivers. In the Zambezi Basin there are dams on the Kafue at Kafue Gorge and Itezhitezhi, and on the Zambezi itself at Kariba and Cabora Bassa. In West Africa, the Volta River is dammed at Akosombo, the Bandama (Ivory Coast) at Kossou, the Senegal at Manantali and Diama, and there are numerous dams in the Niger Basin (e.g. the Bakolori on the Sokoto and the Lagdo on the Benue, as well as the Sélingué, Sotuba, Markala, Karamsasso, Kainji and Jebba on the Niger itself). Both tributaries of the Nile are dammed, and the dam at Aswan controls flow of the combined river within Egypt.

Dams can have complex and extensive ecological impacts on in-stream and floodplain environments (Table 3.5), in particular by affecting the magnitude and timing of downstream flows, and reducing but extending flood peaks. The exact effect depends on the size of the dam and reservoir and the purpose of the project. Dams to store irrigation water for dry-season use are often filled early in a wet season before water is released downstream, and can change the regime of a river from one with a short flood season into a river with more moderate flows through the year. Changes in natural river hydrology affect the ecology of natural river and floodplain environments (and as a result, human economic use of resources such as grazing, agriculture and fishing).

The impacts of dam construction on running water ecosystems in tropical rivers is incomplete, but broadly clear (Welcomme, 1979; Lowe-McConnell, 1985; Payne, 1986; World Commission on Dams, 2000). Passage through a reservoir has a number of effects on water quality that can affect invertebrate drift, and river bed degradation downstream can lead to the loss of important in-stream spawning grounds. Many fish exhibit fairly short 'lateral' migrations or longitudinal migrations of greater length associated with breeding in response to seasonal fluctuations in the river (Welcomme, 1979; Lowe-McConnell, 1985). Reduction in flooding due to drought or dam construction can significantly cut recruitment, fish population numbers and the economic return to the fishing people, as for example in the Yaérés, the extensive floodplain of the Logone River systems above Lake Chad in Cameroon (Benech, 1992). Several fish species failed to spawn in the Phongolo floodplain in South Africa following river control and a reduction in the annual flood that isolated floodplain pools from the main channel (Jubb, 1972). On the Niger in West Africa, the Kainji Dam acts as a complete barrier to fish movement, and although catches at the foot of the dam are high, studies further downstream reported

Table 3.5. *Downstream environmental impacts of dams*

Environment	Environmental impact
Channel	Reduced water turbidity downstream of dams, increased clear water erosion, increased channel scour, changes in patterns of erosion/deposition and channel movement and associated channel evolution, leading to infrastructure damage; impacts on associated pools and wetlands
	Water chemistry (hydrogen sulphide; mercury) and aquatic biology
	Timing of flood flows and impact on lateral and longitudinal fish-spawning migrations and hence recruitment rate and fish ecology; risks of unpredictable high flows due to reservoir drawdown
Floodplains and inland deltas/swamps	Flood volumes and timing (extent and duration of aquifer response to flooding)
	Groundwater level and temporal and spatial dynamics of flooding
	Reduced silt deposition and floodplain ecological productivity; enhanced dry season water tables
	Changes in form and functioning of wetland ecosystems (especially riparian forests that depend on 50 yr+ flood events for regeneration)
	Changes in stable channel patterns, and implications for infrastructure (see under 'Channel' above)
	Flood control
	Insect or mollusc disease vector abundance
Downstream water bodies, coastal deltas and inshore/marine environments	Seasonal dynamics of saline/freshwater boundary breeding of crustacea, shellfish and finfish
	Mangrove ecology
	Fish reproduction

Source: Adams (2000).

significant reductions in fish catches between 1967 and 1969 (Lowe-McConnell, 1985), and associated reductions in fishing activity.

Dam construction also affects riparian and floodplain ecosystrems. Attwell (1970) and Guy (1981) described increased river bank erosion on the Zambezi downstream of the Kariba Dam in the Mana Pools Game Reserve due to unseasonal high flows, high flow levels, rapid fluctuations in flow levels, and the fact that water released by the dam is low in sediment for this increased erosion. The dam hastened normal river system dynamics and affected the regeneration of riparian *Acacia albida* woodland.

River floodplains frequently support woodland vegetation in areas of dry savanna bush (Hughes, 1994). These riparian or riverine woodlands are supported by high groundwater tables fed by river flows, and the outer edge of the forest is determined, in part at least, by the depth to water table. However, forest regeneration depends on periodic high flood flows. The maintenance of the forest therefore depends in a complex way on the annual and interannual pattern of river discharge. Alterations in that discharge pattern, for example from dam construction, can have significant impacts on the viability of the forest ecosystem. These impacts may only become clear decades after dam construction. Without the replenishment and maintenance of floodplain water tables, recruitment of riparian trees is unlikely to occur (Hughes, 1990, 1997).

In the floodplain of the Tana River in Kenya, groundwater recharge from the river, particularly at high flow stages, supports a narrow belt of forest 1–2 km wide in a region of only 300 mm of annual rainfall (Hughes, 1990). A series of dams has been built in the headwaters of the Tana. The forest, which exists in a series of discontinuous blocks, supports two endemic primates, the Tana River red colobus (*Colobus badius rufomitratus*) and the Tana mangabey (*Cercocebus galeritus galeritus*). The floodplain forest is diverse in structure and floristics, reflecting the complex geomorphology of the floodplain itself (Hughes, 1990). Evergreen species occur on heavier soils, trees such as figs on sandy levees and an endemic poplar (*Populus ilicifolia*) on point bars. There is a complex succession of low thorny scrub vegetation in oxbows and drier parts of the forest are dominated by species of *Acacia*. The Tana floods twice a year, with low flow periods between February and March and between September and October, and high flows, particularly in May, which inundate extensive areas of the floodplain. The dynamics of succession in the forest are complex, but the importance of river flooding patterns is clear. There is considerable interannual variation in flows, and studies of tree girths and growth rates suggest that past regeneration has been associated with high flood years that occur every few decades (e.g. 1961, three times the average annual maximum flow). Upstream dams reduce both the height and frequency of high flows in the lower Tana, and are likely to bring forest regeneration to an end (Hughes, 1990).

The physical removal of dams before their economic life is over is rarely feasible. Planning has therefore had to embrace management strategies to minimise impacts (World Commission on Dams, 2000). One approach considered for the restoration of floodplain ecosystems impacted by dams in Africa is the release of artificial floods to re-establish ecological function (Scudder, 1980, 1991; Horowitz & Salem-Murdock, 1991; Acreman & Howard, 1996). The simulation of the seasonal flood peak would make downstream production possible and also allow cultivation in the drawdown zone of the reservoir (Scudder, 1980).

Controlled flood releases to restore downstream ecological processes have been undertaken experimentally on several Africa rivers: for example on the Phongolo in Natal in South Africa (Scudder, 1991; Bruwer *et al.*, 1996), on the Senegal River between 1988 and 1990 (Horowitz & Salem-Murdock, 1991; Hollis, 1996) and on the Waza-Logone floodplain in Cameroon (Ngantou, 1994; Wesseling *et al.*, 1996). In each of these places, dam construction or other engineering works have created serious downstream ecological and economic impacts. On the River Senegal, about 0.5 million people practise flood-related cropping in floodplain 'waalo' land (Schmitz, 1986). The area cultivated has varied from about 150–200 000 ha in the 1960s (when rainfall was good) to about 20 000 ha in the drought years of the 1970s. Studies showed that the value of lost downstream production following construction of the Manantali Dam outweighed marginal benefits of HEP generation. Between 1904 and 1984 there was sufficient water both to generate 74 MW of power and to release an artificial flood large enough to inundate some 50 000 ha of land downstream in every year except the most severe drought years (1913, 1977 and 1979–84). An experimental release of an artificial flood has therefore been carried out.

There has been extensive study of the impacts of environmental change caused by dam construction in the Kafue Basin (Williams & Howard, 1977). The Kafue Gorge Dam was completed in 1972, and the second phase of the project, further generation capacity in the Kafue Gorge Dam and a second dam at Itezhitezhi above the Kafue Flats was completed in 1982, amidst considerable concern about environmental impacts on the ecology of the Kafue Flats. However, the Itezhitezhi Dam was built to allow releases of water for downstream wetland maintenance, and an annual release of an artificial flood of 300 m^3 s^{-1} over four weeks each March was agreed (Scudder & Acreman, 1996). Although releases have not mimicked natural flows very closely, and concern about environmental impacts continues, environmental impacts have not been as serious as some had feared.

The task of integrating the releases of water from upstream dams and the needs of people and ecosystems in downstream floodplains cannot simply be seen as a technical task. The diversity of downstream needs makes it effectively impossible to devise a single solution that automatically takes account of all interests. One approach to the complex planning required is to involve floodplain communities in the planning and management of releases. This has been done in the Phongolo floodplain in South Africa (Bruwer *et al.*, 1996). Here some 70 000 people depend on wetland resources sustained by the flooding of the river. The decision to build the Pongolapoort Dam was taken in the 1950s for political reasons, but it was only filled to 30% of capacity to avoid inundation of part of Swaziland. Surplus water was released from the dam to serve downstream communities, but the restructured floods up to 1984 were smaller and unpredictable in timing, and created a risky environment for floodplain resource use. In 1984 the reservoir was filled to capacity by floodwaters from Cyclone Dominoa, and larger releases began to be possible. This enabled the flooding patterns associated with the ecological conditions in the floodplain of the uncontrolled river to be re-established, although it took several years for an institutional framework to be established that allowed people in the floodplain to influence decisions about water releases (Bruwer *et al.*, 1996).

CONCLUDING REMARKS

Human impacts on African ecosystems are considerable, and growing, exacerbated by population growth, urbanisation, industrialisation and conflict. The extent and depth of mass poverty in Africa places a particular burden on planners to devise relations between people and ecosystems that are sustainable and deliver sustainable livelihoods (Carney, 1998). There will be growing need to understand the scale of anthropogenic impacts, where necessary to reduce their scale and scope, and to restore ecosystems they degrade. Inevitably, if Africans are to achieve acceptable levels of economic well-being (the 'sustainable development' that has proved so powerful a symbol of development thinking in the last decade: Adams, 2001), there will be further transformations of ecosystems across large parts of the continent.

The future therefore offers a considerable challenge to African ecologists, and ecologists outside the continent with knowledge and resources to undertake field science, to develop robust and sound understandings of African ecosystems. In the past ecological restoration has often been based on poor understanding of natural ecosystem dynamics and the causes and forms of human impacts. Much more research is needed to improve the science base for such interventions, and this will require significant strengthening of research institutions at all levels, and a serious attempt to re-educate non-scientific decision-makers, particularly in African governments and Western aid donor organisations. It will also require the development of an integrated understanding of not only the ecological but also the social and economic dimensions of environmental change, and an effective means of drawing African resource users into debates about ecosystem change and restoration.

REFERENCES

Abdel, M. & Ibrahim, M. (1997). Survival with moving sands: the Northern province Community Forestry Project Ed Debba–Sudan. *Desertification Control Bulletin*, 30, 65–73.

Acreman, M. & Howard, G. (1996). The use of artificial floods for floodplain restoration and management in sub-Saharan Africa. *IUCN Wetlands Programme Newsletter*, 12, 20–25.

Adams, J. S. & McShane, T. O. (1992). *The Myth of Wild Africa*. New York: Norton.

Adams, M. E. (1996). Savanna environments. In *The Physical Geography of Africa*, eds. W. M. Adams, A. S. Goudie & A. R. Orme, pp. 196–210. Oxford: Oxford University Press.

Adams, W. M. (1992). *Wasting the Rain: Rivers, People and Planning in Africa*. London: Earthscan Publications.

Adams, W. M. (2001). *Green Development: Environment and Sustainability in the South*, 2nd edn. London: Routledge.

Adams, W. M. & Perrow, M. R. (1999). Scientific and Institutional constraints on the restoration of European floodplains. In *Floodplains: Interdisciplinary Approaches*, eds. S. B. Marriott & J. Alexander, pp. 89–97. London: Geological Society.

Allen, H. A. (1996). Mediterranean environments. In *The Physical Geography of Africa*, eds. W. M. Adams, A. S. Goudie & A. R. Orme, pp. 307–325. Oxford: Oxford University Press.

Anderson, D. M. & Grove, R. H. (1987). The scramble for Eden: past, present and future in African conservation. In *Conservation in Africa: People, Policies and Practice*, eds. D. M. Anderson & R. H. Grove, pp. 1–12. Cambridge: Cambridge University Press.

Anderson, D. M. & Johnson, D. H. (1988). Ecology and society in northeast African history. In *The Ecology of Survival: Case Studies from Northeast African History*, eds. D. Johnson & D. M. Anderson, pp. 1–26. London: Lester Crook.

Anon (2000). *International Workshop on a Strategy for the Rehabilitation of Northern Bald Ibis* (Geronticus eremita), *Agadir, Morocco, March 1999*. Sandy, UK: RSPB.

Attwell, R. I. G. (1970). Some effects of Lake Kariba on the ecology of a floodplain of the mid-Zambezi Valley of Rhodesia. *Biological Conservation*, **2**, 189–196.

Ballais, J.-L. (1994). Aeolian activity, desertification and the 'green dam' in the Ziban region, Algeria. In *Environmental Change in Drylands: Biogeographical and Geomorphological Perspectives*, eds. A. C. Millington & K. Pye, pp. 177–198. Chichester, UK: John Wiley.

Bayart, J.-F. (1993). *The State in Africa: The Politics of the Belly*. Harlow, UK: Longman.

Behnke, R. H. & Scoones, I. (1991). *Rethinking Range Ecology: Implications for Range Management in Africa*. London: Overseas Developement Institute and International Institute for Environment and Development.

Benech, V. (1992). The northern Cameroon floodplain: influence of hydrology on fish production. In *Conservation and Development: The Sustainable Use of Wetland Resources*, eds. E. Maltby, P. J. Dugan & J. C. Lefeuve, pp. 155–164. Gland, Switzerland: IUCN.

Berger, J. (1992). Science, conservation and black rhinos. *Journal of Mammology*, **75**, 298–300.

Booth, M. (1992) *Rhino Road: The Black and White Rhinos of Africa*. London: Constable.

Bourn, D. (1978). Cattle, rainfall and tsetse in Africa. *Journal of Arid Environments*, **1**, 9–61.

Bremen, H. & de Wit, C. T. (1983). Rangeland productivity and exploitation in the Sahel. *Science*, **221**, 1341–1347.

Brockington, D. & Homewood, K. (1996). Wildlife, pastoralists and science: debates concerning Mkomazi Game Reserve, Tanzania. In *The Lie of the Land:*

Challenging Received Wisdom in African Environmental Change, eds. M. Leach & R. Mearns, pp. 91–104. Oxford: James Currey.

Bruwer, C., Poultney, C. & Nyathi, Z. (1996). Community-based hydrological management of the Phongolo floodplain. In *Water Management and Wetlands in Sub-Saharan Africa*, eds. M. C. Acreman & G. E. Hollis, pp. 199–211. Gland, Switzerland: IUCN.

Carney, D. (ed.) (1998). *Sustainable Rural Livelihoods: What Contribution Can We Make?* London: Department for International Development.

Chambers, R. (1983). *Rural Development: Putting the Last First*. Harlow, UK: Longman.

Cloudsley-Thompson, J. L. (ed.) (1984). *Sahara Desert*. Oxford: Pergamon Press.

Coe, M. J., Cummings, D. H. & Phillipson, J. (1976). Biomass and production of large African herbivores in relation to rainfall and primary production. *Oecologia*, **22**, 341–354.

Connah, G. (1987). *African Civilisations: Precolonial Cities and States in Tropical Africa: An Archaeological Perspective*. Cambridge: Cambridge University Press.

Coppock, D. L., Ellis, J. E. & Swift, D. M. (1986). Livestock feeding ecology and resource utilisation in a nomadic pastoral ecosystem. *Journal of Applied Ecology*, **23**, 573–589.

Coughenour, M. B., Ellis, J. E., Swift, D. M., Coppock, D. L., Galvin, K., McCabe, J. T. & Hart, T. C. (1985). Energy extraction and use in a nomadic pastoral ecosystem. *Science*, **230**, 619–625.

Deshmukh, I. (1980). *Ecology and Tropical Biology*. Oxford: Blackwell.

Dublin, H. T., Sinclair, A. R. E. & McGlade, J. (1990). Elephants and fire as causes of multiple stable states in the Serengeti–Mara woodlands. *Journal of Animal Ecology*, **59**, 1147–1164.

Emslie, R. & Brooks, M. (1999). *African Rhino: Status Survey and Conservation Action Plan*. Gland, Switzerland: IUCN.

Ford, J. (1971). *The Role of Trypanosomiasis in African Ecology: A Study of the Tsetse Fly Problem*. Oxford: Clarendon Press.

Gavron, J. (1993). *The Last Elephant: An African Quest*. London: HarperCollins.

Goudie, A. S. (1996). Climate: past and present. In *The Physical Geography of Africa*, eds. W. M. Adams, A. S. Goudie & A. R. Orme, pp. 34–59. Oxford: Oxford University Press.

Grainger, A. (1996). Forest environments. In *The Physical Geography of Africa*, eds. W. M. Adams, A. S. Goudie & A. R. Orme, pp. 173–195. Oxford: Oxford University Press.

Grove, A. T. (1958). The ancient erg of Hausaland, and similar formations on the South side of the Sahara. *Geographical Journal*, **124**, 526–533.

Grove, A. T. (1973). A note on the remarkably low rainfall of the Sudan Zone in 1913. *Savanna*, **2**, 133–138.

Grove, R. H. (1990). The origins of environmentalism. *Nature*, **345**, 11–14.

Grove, R. H. (1995). *Green Imperialism: Colonial Expansion, Tropical Island Edens and the Origins of Environmentalism, 1600–1800.* Cambridge: Cambridge University Press.

Grzimek, B. (1960). *Serengeti Shall Not Die.* London: Hamish Hamilton.

Guy, P. R. (1981). River bank erosion in the mid-Zambezi Valley downstream of Lake Kariba. *Biological Conservation*, **19**, 199–212.

Hailey, Lord (1938). *An African Survey.* London: Royal Institute for African Affairs and Oxford University Press.

Hellden, U. (1988). Desertification monitoring: is the desert encroaching? *Desertification Control Bulletin*, **17**, 8–12.

Heydenrych, B. J., Cowling, R. M. & Lombard, A. T. (1999). Strategic conservation interventions in a region of high biodiversity and a high vulnerability: a case study from the Agulhas Plain at the southern tip of Africa. *Oryx*, **33**, 256–269.

Hoben, A. (1995). Paradigms and politics: the cultural construction of environmental policy in Ethiopia. *World Development*, **23**, 1007–1021.

Hollis, G. E. (1996). Hydrological inputs to management policy for the Senegal River and its floodplain. In *Water Management and Wetlands in Sub-Saharan Africa*, eds. M. C. Acreman & G. E. Hollis, pp. 155–184. Gland, Switzerland: IUCN.

Homewood, K. & Rodgers, W. A. (1987). Pastoralism, conservation and the overgrazing controversy. In *Conservation in Africa: People, Policies and Practice*, eds. D. M. Anderson & R. H. Grove, pp. 111–128. Cambridge: Cambridge University Press.

Homewood, K. & Rodgers, W. A. (1991). *Maasailand Ecology.* Cambridge: Cambridge University Press.

Horowitz, M. M. & Little, P. D. (1987). African pastoralism and poverty: some implications for drought and famine. In *Drought and Hunger in Africa,* ed. M. H. Glantz, pp. 59–82. Cambridge: Cambridge University Press.

Horowitz, M. M. & Salem-Murdock, M. (1991). Management of an African floodplain: a contribution to the anthropology of public policy. *Landscape and Urban Planning*, **20**, 215–221.

Hughes, F. M. R. (1990). The influence of flooding regimes on forest distribution and composition in the Tana River floodplain, Kenya. *Journal of Applied Ecology*, **27**, 475–491.

Hughes, F. M. R. (1994). Environmental change, disturbance and regeneration in semi-arid floodplain rivers. In *Environmental Change in Drylands: Biogeographical and Geomorphological Perspectives*, eds. A. C. Millington & K. Pye, pp. 321–345. Chichester, UK: John Wiley.

Hughes, F. M. R. (1996). Wetlands. In *The Physical Geography of Africa*, eds. W. M. Adams, A. S. Goudie & A. R. Orme, pp. 267–286. Oxford: Oxford University Press.

Hughes, F. M. R. (1997). Floodplain biogeomorphology. *Progress in Physical Geography*, **21**, 501–529.

Hulme, M. (1987). Secular changes in wet season structure in central Sudan. *Journal of Arid Environments*, **13**, 31–46.

Hulme, M. (ed.) (1995). *Climate Change and Southern Africa: An Exploration of Some Potential Impacts and Implications in the SADC Region.* Norwich, UK: Climate Research Unit and Worldwide Fund For Nature.

Hulme, M. (1996). Climate change within the period of meteorological records. In *The Physical Geography of Africa*, eds. W. M Adams, A. S. Goudie & A. R. Orme, pp. 88–102. Oxford: Oxford University Press.

Iliffe, J. (1981). *A Modern History of Tanganyika.* Cambridge: Cambridge University Press.

Iliffe, J. (1995). *Africans: The History of a Continent.* Cambridge: Cambridge University Press.

Infield, M. & Adams, W. M. (1999). Institutional sustainability and community conservation: a case study from Uganda. *Journal of International Development*, **11**, 305–315.

Jarosz, L. (1992). Constructing the dark continent: metaphor as geographic representation of Africa. *Geografiska Annaler*, **74B**, 105–115.

Jensen, A. M. (1993). *Protection of Roads in Mauritania against Sand Encroachment.* New York: The United Nations Sudano-Sahelian Office.

Jones, B. (1938). Desiccation and the West African colonies. *Geographical Journal*, **41**, 401–423.

Jones, B. (2000). The evolution of a community-based approach to wildlife management in Kunene, Namibia. In *African Wildlife and Livelihoods: The Promise and Performance of Community Conservation*, eds. D. Hulme & M. Murphree, pp. 160–176. London: James Currey.

Jubb, R. A. (1972). The J. G. Strydon Dam, Phongolo River, northern Zululand: the importance of floodplain pans below it. *Piscator*, **86**, 104–109.

Lamb, P. J. (1979). Some perspectives on climate and climatic dynamics. *Progress in Physical Geography*, **3**, 215–235.

Lamprey, H. (1988). Report on the desert encroachment reconnaisance in northern Sudan, October 21–November 10, 1975. *Desertification Control Bulletin*, **17**, 1–7 (reprinted).

Leach, M. & Mearns, R. (1996). Challenging received wisdom in Africa. In *The Lie of the Land: Challenging Received Wisdom in African Environmental Change*, ed. M. Leach & R. Mearns, pp. 1–33. London: James Currey.

Low, D. A. & Lonsdale, J. A. (eds.) (1976). *History of East Africa*, vol. 3. Oxford: Clarendon Press.

Lowe-McConnell, R. H. (1975). *Fish Communities of Tropical Freshwaters.*, Harlow, UK: Longman.

Lowe-McConnell, R. H. (1985). The biology of the river systems with particular reference to the fish. In *The Niger and its Neighbours: Environment, History and Hydrobiology, Human Use and Health Hazards of the Major West African Rivers*, ed. A. T. Grove, pp. 101–140. Rotterdam: Balkema.

Mace, R. (1991). Overgrazing overstated. *Nature*, **349**, 280–281.

MacKenzie, J. M. (1987). Chivalry, social Darwinism and ritualised killing: the hunting ethos in central Africa up to 1914. In *Conservation in Africa: People, Policies and Practice*, eds. D. M. Anderson & R. H. Grove, pp. 41–62. Cambridge: Cambridge University Press.

MacKenzie, J. M. (1989). *The Empire of Nature: Hunting, Conservation and British Imperialism*. Manchester, UK: University of Manchester Press.

McIntosh, R. P. (1985). *The Background of Ecology: Concept and Theory*. Cambridge: Cambridge University Press.

Meadows, M. E. (1996). Biogeography. In *The Physical Geography of Africa*, eds. W. M. Adams, A. S. Goudie & A. R. Orme, pp. 161–172. Oxford: Oxford University Press.

Metcalfe, S. (1994). The Zimbabwe Communal Areas Management Programme for Indigenous Resources (CAMPFIRE). In *Natural Connections*, eds. D. Western, R. M. White & S. C. Strum, pp. 161–192. Washington, DC: Island Press.

Mortimore, M. J. (1989). *Adapting to Drought: Farmers, Famines and Desertification in West Africa*. Cambridge: Cambridge University Press.

Mortimore, M. J. (1998). *Roots in the African Dust: Sustaining the African Drylands*. Cambridge: Cambridge University Press.

Murombedzi, J. S. (1999). Devolution and stewardship in Zimbabwe's CAMPFIRE Programme. *Journal of International Development*, **11**, 287–293.

Neumann, R. P. (1998). *Imposing Wilderness: Struggles over Livelihood and Nature Preservation in Africa*. Berkeley, CA: University of California Press.

Ngantou, D. (1994). Rehabilitation of the Waza Logone floodplain, Cameroon. *IUCN Wetlands Programme Newsletter*, **10**, 9–10.

Niewolt, S. (1982). *Tropical Climatology*. Chichester, UK: John Wiley.

Norton-Griffiths, M. (1979). The influence of grazing, browsing and fire on the vegetation dynamics of the Serengeti. In *Serengeti: Dynamics of an Ecosystem*, eds. A. R. E. Sinclair & M. Norton-Griffiths, pp. 310–352. Chicago, IL: University of Chicago Press.

Payne, A. J. (1986). *The Ecology of Tropical Lakes and Rivers*. Chichester, UK: John Wiley.

Peterken, G. F. (1986). *Natural Woodland: Ecology and Conservation in Northern Temperate Regions*. Cambridge: Cambridge University Press.

Prior, J. (1993). *Pastoral Development Planning*. Oxford: Oxfam.

Reader, J. (1997). *Africa: A Biography of the Continent*. London: Hamish Hamilton.

Richards, P. W. (1939). Ecological studies on the rain forest of southern Nigeria. 1: The structure and floristic composition of the primary forest. *Journal of Ecology*, **26**, 1–61.

Roberts, N. (1998). *The Holocene: An Environmental History*, 2nd edn. Oxford: Blackwell.

Roe, E. (1991). Development narratives, or making the best of blueprint development. *World Development*, **19**, 287–300.

Schmitz, J. (1986). Agriculture de décrue, unités térritoriales et irrigation dans la vallée du

Sénégal. *Les Cahiers de Recherches Développement*, **12**, 65–77.

Scholes, R. J. & Walker, B. H. (1993). *An African Savanna: Synthesis of the Nylsvley Study*. Cambridge: Cambridge University Press.

Scoones, I. (1994). *Living with Uncertainty: New Directions in Pastoral Development in Africa*. Woburn, UK: IT Publications.

Scudder, T. (1980). River basin development and local initiative in African savanna environments. In *Human Ecology in Savanna Environments*, ed. D. R. Harris, pp. 383–405. London: Academic Press.

Scudder, T. (1991). The need and justification for maintaining transboundary flood regimes: the Africa case. *Natural Resources Journal*, **31**, 75–107.

Scudder, T. & Acreman, M. C. (1996). Water management for the conservation of the Kafue wetlands, Zambia and the practicalities of artificial flood releases. In *Water Management and Wetlands in Sub-Saharan Africa*, eds. M. C. Acreman & G. E. Hollis, pp. 101–106. Gland, Switzerland: IUCN.

Seely, M. K. (1978). The Namib dune desert: an unusual ecosystem. *Journal of Arid Environments*, **1**, 117–128.

Shantz, H. L. & Marbut, C. F. (1923). *The Vegetation and Soils of Africa*, American Geographical Society Research Series 13. New York: American Geographical Society and National Research Council.

Sheail, J. (1987). *Seventy-Five Years of Ecology: The British Ecological Society*. Oxford: Blackwell.

Sinclair, A. R. & Fryxell, J. M. (1985). The Sahel of Africa: ecology of a disaster. *Canadian Journal of Zoology*, **63**, 987–994.

Stebbing, E. P. (1935). The encroaching Sahara: the threat to the West African colonies. *Geographical Journal*, **85**, 506–524.

Swift, J. (1982). The future of African hunter–gatherer and pastoral people in Africa. *Development and Change*, **13**, 159–181.

Swift, J. (1996). Desertification: narratives, winners and losers. In *The Lie of the Land: Challenging Received Wisdom in African Environmental Change*, eds. M. Leach & R. Mearns, pp. 73–90. London: James Currey.

Taylor, D. (1996). Mountains. In *The Physical Geography of Africa*, eds. W. M. Adams, A. S. Goudie & A. R. Orme, pp. 287–306. Oxford: Oxford University Press.

Thomas, D. S. G. & Middleton. T. (1994). *Desertification: Exploding the Myth*. Chichester, UK: John Wiley.

Trapnell, C. G. (1943). *Soils, Vegetation and Agriculture of North-Eastern Rhodesia*. Lusaka: Government Printer.

Trapnell, C. G. & Clothier, J. N. (1937). *Soils, Vegetation and Agricultural Systems of North-Western Rhodesia*. Lusaka: Government Printer.

Turton, D. (1987). The Mursi and National Park development in the lower Omo Valley. In *Conservation in Africa: People, Policies and Practice*, eds. D. M. Anderson & R. H. Grove, pp. 169–186. Cambridge: Cambridge University Press.

UNSO (1991). *Lutte contre L'Ensablement en Mauritanie*. New York: The United Nations Sudano-Sahelian Office.

Varty, S. & Buchanan, M. (2000). *The Return: The Story of Phinda Game Reserve*. Johannesburg: Londolozi Publishers.

Waller, R. D. (1988). Emutai: crisis and response in Maasailand (1884–1904). In *The Ecology of Survival*, eds. D. Johnson & D. M. Anderson, pp. 73–112. London: Lester Crook.

Waller, R. D. (1990). Tsetse fly in Western Narok, Kenya. *Journal of African History*, **31**, 81–101.

Warren, A. (1996). Desertification. In *The Physical Geography of Africa*, eds. W. M. Adams, A. S. Goudie & A. R. Orme, pp. 342–355. Oxford: Oxford University Press.

Watson, A. (1990). The control of blowing sand and mobile desert dunes. In *Techniques for Desert Reclamation*, ed. A. S. Goudie, pp. 35–85. Chichester, UK: John Wiley.

Watts, M. J. (1983). *Silent Violence: Food, Famine and Peasantry in Northern Nigeria*. Berkeley, CA: University of California Press.

Welcomme, R. L. (1979). *The Fisheries Ecology of Floodplain Rivers*. Harlow, UK: Longman.

Wesseling, J. W., Naah, E., Drijver, C. A. & Ngantou, D. (1996). Rehabilitation of the Logone floodplain, Cameroon, through hydrological management. In *Water Management and Wetlands in Sub-Saharan Africa*, eds. M. C. Acreman & G. E. Hollis, pp. 185–198. Gland, Switzerland: IUCN.

White, F. (1983). *The Vegetation of Africa: A Descriptive Memoir*. Paris: UNESCO.

Williams, G. J. & Howard, G. W. (eds.) (1977). *Development and Ecology in the Lower Kafue Basin in the 1970s*. Lusaka: Kafue Basin Research Committee.

Williams, M. A. J. & Balling, R. C., Jr (1995). Interactions of desertification and climate: an overview. *Desertification Control Bulletin*, **26**, 8–16.

Winid, B. (1981). Comments on the development of the Awash Valley, Ethiopia. In *River Basin Planning: Theory and Practice*, eds. S. K. Saha & C. J. Barrow, pp. 147–165. Chichester, UK: John Wiley.

World Commission on Dams (2000). *Dams and Development: A New Framework for Decision-Making*. London: Earthscan Publications.

Worster, D. (1985). *Nature's Economy: A History of Ecological Ideas*. Cambridge: Cambridge University Press.

Worthington, E. B. (1938). *Science in Africa: A Review of Scientific Research relating to Tropical and Southern Africa*. London: Royal Institute of International Affairs.

Worthington, E. B. (1958). *Science in the Development of Africa: A Review of the Contribution of Physical and Biological Knowledge South of the Sahara*. Commission for Technical Cooperation in Africa South of the Sahara and the Scientific Council for Africa South of the Sahara.

Worthington, E. B. (1983). *The Ecological Century: A Personal Appraisal*. Cambridge: Cambridge University Press.

4 • Asia

4a • India

B. B. DHAR AND M. K. CHAKRABORTY

INTRODUCTION

Population growth and rapid industrialisation concurrent with an ever-increasing quest for better quality of life have resulted in a growing demand for energy and infrastructure in India. The significant impact on the region's environment and ecology has been exacerbated by the lack of concern and consideration for its degradation. According to the Indian Government's National Wasteland Development Board, there are some 12 million hectares of deserted wastelands. These are distributed throughout the territories and have resulted from large-scale exploitation and overexploitation of natural resources enhancing processes such as water erosion (Table 4.1). Moreover, exploitation of natural resources produces wastes and hazardous by-products that further degrade the environment and impede the quality of life of the very people that were supposed to benefit from the developmental activities. Disposal, management and reclamation of wastes is gaining greater importance in view of the consequent environmental hazards.

Mining activity to exploit natural mineral wealth to provide the supply of raw materials for development and industrialisation has been one of the major factors in the degradation of land. There are 3198 working mines in India, comprising 574 coal, 6 lignite (Table 4.2), 559 metallic and 2059 non-metallic mines. Of these, the coal fields cover by far the greatest area at 3.47 million hectares, with the non-coal sector covering only 0.69 million ha (Table 4.3). Mechanized open-cast coal mines removed 21.3 million cubic metres in 1975/6, reaching 309.6 million cubic metres in 1993.

Mining also generates large quantities of waste and hazardous by-products, which dramatically increase the environmental and ecological imbalance. The overburden (the consolidated and unconsolidated material overlying a coal seam or ore deposit disturbed and haphazardly mixed during mining activity) removal from coal and lignite mines alone was more than 1400 million cubic metres during 1997/8 (Table 4.2). This waste may occupy between two to six times the area of the pit itself. The quantity of waste generated per tonne of produce varies from mineral to mineral and industry to industry. According to Raju (1995), ore-to-waste ratios for iron, manganese, limestone and quartz are 1:4, 1:25, 1:2 and 1:1, respectively. In the steel industry this ratio varies from 1:0.5 to 1:1 (Pandey et al., 1996). For the multitude of smaller metallic mines distributed patchily throughout India, the resulting wasteland of over 141 000 ha, is typically thought to occupy 20% of lease area (Table 4.3).

Overburden and waste dumps present very rigorous conditions both for plants, fungi and microbial growth, because of the lack of essential plant nutrients, low organic matter content, unfavourable pH and either coarse texture or compacted structure.

The biological reclamation of mine wastes into a self-sustaining ecologically biodiverse stable land form that can prevent soil erosion and dust pollution and restore the aesthetic beauty of the environment has drawn the attention of several workers (Singh & Jha, 1984; Bradshaw, 1990; Tiwary, 1990). The field of biological reclamation of mine waste is of recent origin in India and the Asian region as a whole. Whilst the amount of work is inadequate at the moment, it is now receiving the attention of scientists and experimentation is now in progress. The basic premise is that the fundamentals of the ecosystem, including the processes of

Table 4.1. *Estimate of wastelands (Mha) in India*

State/union territory	Saline and alkaline lands	Wind-eroded area	Water-eroded area	Non-forest degraded area	Forest degraded area	Total
Andhra Pradesh	0.0240	—	0.7442	0.7682	0.3734	1.1416
Assam	—	—	0.0935	0.0935	0.0795	0.1730
Bihar	0.004	—	0.3892	0.3896	0.1562	0.5458
Gujarat	0.1214	0.0704	0.5235	0.7153	0.0683	0.07836
Haryana	0.0526	0.1599	0.0276	0.2404	0.0074	0.2478
Himachal Pradesh	—	—	0.1424	0.1424	0.0534	0.1958
Jammu and Kashmir	—	—	0.0531	0.0531	0.1034	0.1565
Karnataka	0.0404	—	0.6718	0.7122	0.2043	0.9165
Kerala	0.0016	—	0.1037	0.1053	0.0226	0.1279
Madhya Pradesh	0.0242	—	1.2705	1.2947	0.7195	2.0142
Maharashtra	0.0534	—	1.1026	1.1560	0.2841	1.4401
Manipur	—	—	0.0014	0.0014	0.1424	0.1438
Meghalaya	—	—	0.0815	0.0815	0.1103	0.1918
Nagaland	—	—	0.0508	0.0508	0.0878	0.1386
Orissa	0.0404	—	0.2753	0.3157	0.3227	0.6384
Punjab	0.0688	—	0.0463	0.1151	0.0079	0.1230
Rajasthan	0.0728	1.0623	0.6659	1.8010	0.1933	1.9934
Sikkim	—	—	0.0131	0.0131	0.0150	0.0281
Tamil Nadu	0.0004	—	0.3388	0.3392	0.1009	0.4401
Tripura	—	—	0.0108	0.0108	0.0865	0.0973
Uttar Pradesh	0.1295	—	0.5340	0.6635	0.1426	0.8061
West Bengal	0.0850	—	0.1327	0.2177	0.0359	0.2536
Union territories	0.0016	—	0.0873	0.0889	0.2715	0.3804
Total	0.7165	1.2926	7.3600	9.3694	3.5889	12.2721

Source: MOEF (1989).

primary and secondary succession, must be well understood to achieve ecological rehabilitation quickly and reliably. Any short-term solution sought without cognizance of these factors risks failure, both from an ecological as well as from an economic standpoint. Selection of species has been accorded the highest priority, considering their ability to colonise the degraded areas, fix atmospheric nitrogen, conserve soil, and produce fuel, fodder, etc. for the local populations of both humans and fauna (birds, butterflies etc.).

Reclamation of mine wastes thus represents the focus of habitat restoration in the Indian sub-continent. This paper aims to: (1) broadly highlight the legal framework and its application to restoration; (2) outline the institutional and political attitude to restoration of degraded mined out areas; and (3) illustrate the current state of knowledge.

LEGISLATIVE FRAMEWORK

The environmental management of industrial development projects, including the mining sector, appears to have only recently become standard practice in the region. In India, regulation is achieved

Table 4.2. *Number of coal and lignite mines and overburden removal in India*

State	Number of mines	Overburden removed (m³ × 1000)
Coal		
Andhra Pradesh	71	17 198
Assam	6	3950
Bihar	189	108 393
Jammu and Kashmir	2	–
Madhya Pradesh	129	397 679
Maharashtra	48	65 690
Orissa	21	35 534
Uttar Pradesh	3	25 972
West Bengal	105	471 913
Total coal-mines	574	1 126 329
Lignite		
Gujarat	3	19 019
Rajasthan	1	–
Tamil Nadu	2	55 109
Total lignite mines	6	74 128
All India	580	1200457

Source: Director General of Mine Safety (1997).

Table 4.3. *Summary of metalliferous mining lease in India by state (as on 31 March 1998)*

State	Number of leases	Lease area (ha)	Probable wasteland (ha)
Andhra Pradesh	1307	48 279	9 873
Assam	20	1 487	304
Bihar	437	48 365	9 890
Delhi	5	499	102
Goa	402	30 746	6 287
Gujarat	1391	29 984	6 131
Himachal Pradesh	51	3 630	742
Haryana	123	15 619	3 194
Jammu and Kashmir	15	1 536	314
Karnataka	1007	110 855	22 669
Kerala	171	5 397	1 103
Manipur	2	610	124
Meghalaya	20	4 062	830
Madhya Pradesh	1248	68 227	13 952
Maharashtra	225	15 005	3 068
Orissa	606	104 811	21 433
Rajasthan	1346	159 826	32 684
Sikkim	2	40	8
Tamil Nadu	625	19 718	4 032
Uttar Pradesh	131	9 332	1 908
West Bengal	110	14 454	2 955
Total	9244	692 491	141 613

Source: Indian Bureau of Mines (1999).

through various acts, such as the Environmental Protection Act of 1986, the Forest Conservation Act of 1980, the Mines and Minerals (Regulation & Development) Act of 1957 (amended in 1988 and 1994), as well as other provisions made by the government from time to time.

The Mines and Minerals (Regulation and Development) Act

The Mines and Minerals (Regulation and Development) Act ensures systematic development and conservation of mineral resources. Section 4A of this Act provides for the premature termination of the mining lease. The termination can be ordered only in the interest of the regulation of mines and mineral development. No premature termination is possible under the present Act to check dam-

age to environment and ecology, however. Moreover, under Section 18 of this Act, the Central (Federal) Government is empowered to frame rules to take steps to conserve and develop minerals as may be necessary. However, these rules also deal with the storage of overburden, waste rock, etc. which has environmental implications. Different types of waste and rejects generated during mining benefaction have to be stored in separate dumps. The dumps have to be properly secured to prevent escape of material in harmful quantities, which may cause the degradation of the environment. For example, the dumps should be sited on impervious ground to ensure minimum leaching. Wherever possible, waste rock should be backfilled into the mine excavation to restore the land to its original shape. If this is not possible, the dumps should be suitably terraced and stabilised through vegetation or otherwise. The tailings from benefaction plants should be disposed in specially prepared tailing disposal areas so that they cannot flow away and cause land degradation or damage to agricultural fields or pollute surface water bodies and groundwater.

Under the Act, the mining lessee is required to undertake phased restoration, reclamation and rehabilitation of lands affected by mining and complete this work before abandoning the mine.

Legislation surrounding hazardous industrial wastes

As regards industrial wastes, there is no clear-cut legislation, except the the Hazardous Wastes (Management and Handling) Rules of 1989 and the Manufacture, Storage and Import of Hazardous Chemical Rules of 1989/Amendment Rules, 1994. This is against a background of increasing production of solid wastes (fly ash, blast furnace slag, by-product gypsum, red-mud) in India: in 1985 around 33 million tonnes were generated but by 1995 this had increased to 74 million tonnes. This is due to large increases in production of fly ash (18 to 30 million tonnes) and especially blast furnace slag (7 to 34 million tonnes) (MOEF, 1995).

INSTITUTIONAL EFFORTS

Key institutions and their projects

In India there are a few well-established institutions that have been making constant efforts to improve the existing techniques of wasteland reclamation. Some technology packages have been developed for rehabilitating ecologically fragile rock areas in the Himalayas (Box 4.1), bauxite mine areas in plains (Box 4.2) as well as other areas of the country. The main institutions are: the Forest Research Institute, Dehradun (Box 4.1); State Forest Research and Education, Jabalpur (Box 4.2); Central Mining Research Institute, Dhanbad; Central Soil and

---

Box 4.1   Rehabilitation of Himalayan mines

The Forest Research Institute at Dehradun has specialised in the rehabilitation of fragile Himalayan areas and has developed several techniques involving plants for use in such hilly areas. Techniques include the construction of: mechanical structures, such as a series of stone walls and gabion check dams across the slope and channel flow; contour trenches on dumps (0.5 m deep with a vertical interval on the overburden of 2 m); and pits (30 × 30 × 30 cm). Planting seeds of pioneering shrubs, climbers and grass then follows. Shrubs are planted in contour trenches and trees, especially *Dalbergia sissoo* and *Acacia catechu*, are planted in pits.

These results have proven very satisfactory and there has been improvement in pH levels, increases (12%–14%) in the percentage of fine soil, significant increases in soil nitrogen levels to 24 tonnes/ha and over 300-fold increases in herb densities. At the end of eight years of experiments it was observed that the population of soil fauna was almost 20% greater than that found in adjacent natural forest areas, resulting in the attainment of a stable and diverse self-sustaining ecosystem.

Box 4.2   Establishing trees on bauxite and coal mine overburdens

The State Forest Research Institute, Jabalpur undertook planting trials with various tree monocultures on bauxite mine overburdens in 1979 and on coal-mine overburdens in 1982. Sites were surveyed to understand natural succession under planted tree cover and abandoned sites. Coal-mine overburdens appeared to provide more hospitable conditions for natural succession than bauxite mine overburdens, as 24 936 plants ha^{-1} survived on rehabilitated coal mine dumps compared to 14 580 plants ha^{-1} on equivalent bauxite mine sites.

Water Conservation Research and Training Institute, Dehradun (Box 4.3); and educational institutions,

Box 4.3   Reducing debris load in the Baldi River from an abandoned limestone mine

Scientists at the Central Soil and Water Conservation Research and Training Institute of the Indian Council of Agricultural Research in Dehradun, observed that nearly 35 210 tonnes of debris, originating from a 64-ha abandoned limestone mine, were rolling into the Baldi River (a tributary of the Ganga) annually. The state Public Works Department was reported to have been spending rupees one lakh (Rs. 100 000) (US$2000) per annum in clearing this debris, combined with the erection of gabion retaining walls along the road, since 1969. The Institute undertook the rehabilitation of the mine, developing a suitable, geotechnically sound, rehabilitation package for the area. The total debris retained over a three-year period was 53 300 tonnes (26 650 m^3). The comparative soil loss of various geotextiles (geojute, geonet, enviromat, etc.) varied from 14% to 130%. The cost of geotextile-based measures for rehabilitating the degraded sites was Rs. 8000 per 1000 m^2 (1 US$ = Rs. 46.00).

such as (1) Department of Botany, Banaras Hindu University, Varanasi (Box 4.4) and (2) Center for Mine Environment, Indian School of Mines, Dhanbad. Besides these, several large-scale mining companies like Coal India, National Mineral Development Corporation, Ambuja Cements Ltd and Tata Iron and Steel Company Ltd have developed techniques to reclaim mined areas using scientific inputs and their own local methodology.

Through the initiatives of such organisations, methodologies for achieving biological reclamation have gained a sound footing. Various approaches to tackle sewage sludge and power-station fly ash, and the use of soil amendments and biofertilisers has now been attempted in India.

Research initiatives

Several research programmes have been launched at different locations in recent years to further the development of suitable techniques and plant species for the establishment of vegetation on mine spoils (Jha & Singh, 1994; Noronha, 1996; J. S. Singh *et al.*, 1996; R. S. Singh *et al.*, 1996) (Box 4.4).

The Central Mining Research Institute, a premier institute in the field of mining and mine environment, was given a challenging scientific assignment of determining methods to stabilise and restore coal mine waste dumps abandoned for over 10 to 30 years. Funds were provided by the government's Ministry of Coal from 1993 to 1996. This experiment was successfully completed and several old mined areas were put on a path of restoration using local plant species and techniques developed in pot experiments (J. S. Singh *et al.*, 1996).

Together with these experimental efforts, research and development initiatives are also being applied to develop appropriate low-cost technologies suitable for land restoration (Mathur, 1993). In the Western world, research has resulted in reduced costs for recontouring mined land, and machines have been developed into high-volume low-cost tools (Goris, 1980).

Development of reclamation protocols

Primarily as a result of the various initiatives, working reclamation procedures for mine wastelands are now in existence (Fig. 4.1). This includes techniques for physical, nutrient and biological reclamation (Box 4.5).

CONCLUDING REMARKS

In India, vast areas of land have been degraded, particularly by the exploitation of natural mineral wealth, in association with rapid industrialisation. There is thus a great need for the restoration or reclamation of mines and mine wastes and the control and management of hazardous substances. Whereas developed countries have formulated rules and guidelines for land restoration, until recently, there were no such regulations regarding restoration of wasteland in India. There was thus little to fulfil the cultural and political aspirations of the people.

Box 4.4   Revegetation of mine spoils in the Northern Coalfields

The Department of Botany at Banaras Hindu University, Varanasi did an intensive ecological study on the revegetation of mine spoils at Jayant and Bina projects in the Northern Coalfields Ltd, Singrauli (1993–5). Field experiments trialled revegetation of tree monocultures seeded with: (1) grasses and legumes; (2) crop plants; and (3) ground seeding and mixed tree cultures seeded with grasses and legumes, in combination with fertiliser applications. Pot culture experiments were also undertaken to screen suitable tree species for revegetation of mine spoils (Fig. B4.1), and to test the feasibility of direct seeding of tree species on flat and sloping lands. Of the 40 species tested, 34 species were found to be suitable for the revegetation of mine spoil.

The tree monoculture experiment at Jayant consisted of collecting data on standing biomass and net production levels, fine root biomass, rate of litter fall and leaf litter decomposition, N_2-fixation efficiency and photosynthetic and transpiration rates. Physicochemical properties of mine spoils such as pH, bulk density, water-holding capacity, total organic carbon, total soil N, available nutrients ($NaHCO_3$-P, nitrate N and ammonium-N) were estimated for plots planted to different species. Soil microbial biomass, C, N and P, nitrification, and total N mineralization rates were also quantified. These findings suggest that native legumes are more efficient in utilising different soil properties than exotic legumes, at least in the short term (J. S. Singh *et al.*, 1996).

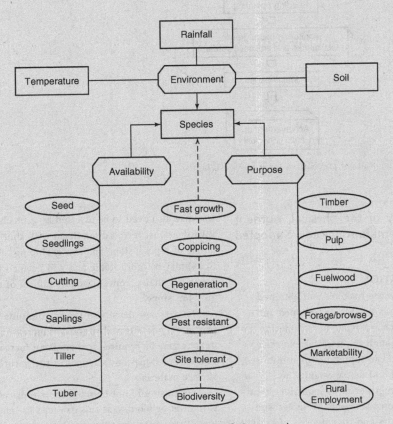

Fig. B4.1. Factors determining screening of plant species.

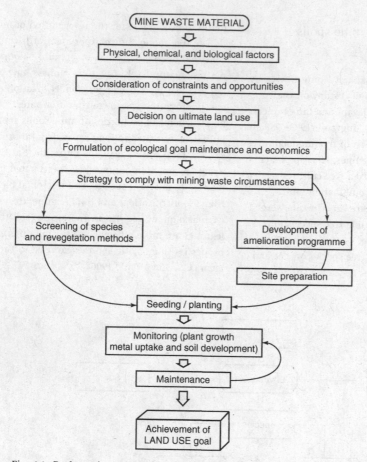

Fig. 4.1. Reclamation procedures for mine wastelands.

Box 4.5 Summary of the physical, nutrient and biological reclamation procedure adopted in Indian mines

PHYSICAL RECLAMATION

Sites with high slope, waste rock and soil obtained from stripping operations need to be reshaped and graded to obtain a shape amenable to reclamation (Singh *et al.*, 1994), although the need for extensive reshaping of mine spoil can sometimes be minimised by good mine planning and management. The maximum angle and length of mine spoil slope that will be stable depends on site-specific variables, soil and topsoil characteristics and the intensity of rainfall. The erosion potential of the different materials at the site may also need to be assessed by geotechnical investigations. Terraced landform with short, steep (angle of repose) slope and gently sloping terraces (<5%) may be more stable and will have a higher land capability than a conventional landform of around 15%–18% slope.

The drainage density of the surrounding area provides a guide to soil runoff. An increase in drainage density may be required if there is an increase in the gradient of slopes and changes in the nature of the surface materials.

Topsoil should not be stripped or replaced when it is too wet or too dry, as this can lead to compaction, loss of structure and loss of viability of seeds and mycorrhizal inoculum. Topsoil should be replaced along the contours wherever possible. This will help in erosion control by reducing water flow down slope and increasing water storage.

Until an adequate cover of vegetation has been established, it is imperative that provision be made to control erosion from disturbed areas.

NUTRIENT RECLAMATION

Nutrient reclamation involves the use of an appropriate reagent for stabilisation and/or fertilisation of the mine wasteland. Application of Ca, NH_4 or sodium lignin-sulphonates is recommended.

Sodic and acidic spoils may require chemical amendments. Liming and fertilisation may accelerate decomposition of litter, breakdown of soil organic matter and release of nitrogen in mine land. Low-fertility areas need application of inorganic fertiliser and/or organic matter and planting N_2-fixing plants. On acidic waste dumps, it is necessary to control further oxidation and weathering of the pyretic waste material and soil erosion. Various organic wastes (e.g. animal excreta, sewage sludge, bone) may be used.

For revegetation of steep, unstable and inaccessible topography, hydro-seeding technology can be applied. Some of the chiefly available materials which can be used are:

- Sawdust to make the soil more porous but also to hold soil moisture. It also improves soil structure and acts as a plant protection material for controlling the development of nematode populations.
- Fly ash with a high silica content (50%–55%) encourages the conversion of nutrients into an easily available form for plant uptake.
- Press-mud, a waste from sugar factories, is a good source of NPK and calcium.
- Gypsum both buffers soil pH and helps loosen the soil.

BIOLOGICAL RECLAMATION

The initial revegetation effort must establish the building blocks for a self-sustaining system, so that successional processes lead to the desired vegetation complex. The best time to establish vegetation is determined by the seasonal distribution and reliability of rainfall. All the preparatory works must be completed beforehand (Srivastava *et al.*, 1989).

The technology developed can be summarised step by step as follows:

- During the first year all preparatory and prerequisite protection activities should be carried out. These include: (1) fencing the entire area by 4–6-strand barbed wire of 1.2 m height to control feral and native animal populations; (2) broadcasting seeds of appropriate shrubs and grasses; (3) developing nurseries to raise appropriate numbers of seedlings of shrubs and trees and slips of grasses; (4) planning soil and moisture conservation measures; (5) beginning soil workings, i.e. digging pits (45 × 45 × 45 cm); and (6) applying biofertilisers.
- During the second year, to accelerate the process of revegetation, cuttings and saplings of a number of shrubs and slips of grasses should be planted or seeds sown. Tree saplings of appropriate types should be planted in the pits. Soil conservation structures should also be constructed. Watering plants in drier areas, especially in the establishment phase, may be required.
- During the third year, tree species should be planted in all areas and any casualties from the first plantings replaced. Some of the soil and water conservation measures, such as check dams, gully control and related works should be completed. Over subsequent years, appropriate interculture and tending works should be carried out, including replacement wherever necessary. Maintenance and repairs of soil and water conservation structures should also be undertaken.

The species selected for establishment will depend on the future land use of the area, soil conditions and climate (Fig. B4.1). If the objective is to restore the native vegetation and fauna then the species are predetermined. In other cases laboratory or pot experiments help select species on the basis of their vigour, vitality and adaptability.

Restoration of natural fertility is a very long-drawn process, whereas establishing the vegetation with slow-release biofertilisers can quickly restore the fertility. Biofertilisers with efficient strains of micro-organisms, like blue-green algae, *Azolla rhizobium*, *azospyrillum*, *acetobactor* or *mycorhischi*, fix atmospheric nitrogen into a form easily utilised by the crop species.

The use of algae, lichens and mosses to stabilise tailings and other overburdens accelerates the establishment of higher species. Inoculation with microflora including mycorrhiza, which are not sensitive to heavy-metal concentrations, is an important technique. Introducing earthworms, termites and other soil-living insects has also been suggested to aid building up the topsoil.

Of late, the subject has drawn the attention of government agencies and regulators, and recent amendments (1988 and 1994) to the Mines and Minerals (Regulation and Development) Act of 1957 now mean that provision should be made for reclamation after degradation of land caused by excavation, dumping and wash-off from waste dumps. To the best of our knowledge, this type of legislation has not been made in respect of any other industry.

It is heartening to note that most of the mining companies have now started to include land reclamation programmes in their developmental plans. The Ministry of Environment and Forests has also started carrying out the environmental appraisal of new mining projects. A few well-established institutions have also been making constant efforts to improve the existing techniques of wasteland reclamation. Coupled with recent research initiatives, progress has been made and the processes involved in undertaking successful reclamation are now better understood and project protocols are available. However, only a few successful projects have been undertaken and with a large number of failures, good practices do not get reflected.

Consequently, in spite of all the recent developments, progress, in terms of actual land reclaimed, has been negligible (only 7800 ha of mined land reclaimed by 1998/9) (Table 4.4). Concerted effort needs to be made to achieve success in reclamation of the open-cast mined-out areas in particular.

Future approach

As a result of continuous research in rehabilitating mine waste areas, we would like to suggest the following as a way forward for those who really want to achieve restoration of degraded land in the region:

- To ensure continued and sustained progress of the mining industry, the environmental dimension clearly needs to be integrated into the planning, design, development, operation and working of mines.
- To improve reclamation technology, continuous and extensive research work is necessary to establish methodologies of land reclamation suited to different geo-mining conditions, along with selection of plant species specifically suited to particular mine spoils, always with an eye to conserving native biodiversity.
- Pilot-scale projects should be launched to demonstrate the efficacy of the different land reclamation techniques for different mine/industrial wastes. Work must also continue in experiments with humic acid, sewage sludge and fly ash as amendment for mine spoil.

We must realise that mining is only an interim use of land and hence mine management must plan the post-mining reclamation of the land to be disturbed prior to starting the operations. Properly designed and executed, eco-friendly developmental work need not be a threat to the environment. It may not be possible to use the same methodology for reclamation of waste or wasteland in all the mines or industry in the country. It is necessary to look at each mining and industrial project as a unique entity and develop site-specific reclamation and land use planning practices. Overburden and waste-dump management must be an integral part of any project implementation plan.

Table 4.4. *Summary of leasehold, wasteland and reclaimed area in different mines in India*

Minerals	Leasehold area (ha)	Wasteland area (ha)	Reclaimed area (ha)
Coal	45 579	15 791	2 674
Bauxite	12 502	1 834	516
Iron	10 349	4 374	2056
Soapstone	1 689	240	8
Building stone	462	45	1
Barytes	160	103	6
Magnesite	531	531	20
Manganese	4 053	972	209
Silica sand	105	10	00
Feldspar	29	5	00
Silimanite	77	30	00
Dolomite	451	64	34
Kyanite	63	5	00
Granite	17	8	5
Lignite	51 800	7 368	840
Fluorite	32	2	00
Clay	932	584	129
Rare earths	2 877	15	1 11
Copper	7 604	1 166	235
Uranium	1 000	222	87
Diamond	381	31	3
Calcite	28	13	11
Limestone	27 306	7 395	866
Total (Non-coal)	122 460	25 026	5 141
Total	168 039	40 818	7 816

Environmental risks posed by any such waste(s) need to be taken care of at all costs, not only to stop further degradation but also (if possible) to attenuate earlier faults. It is much better to anticipate and assess the problem and to take up preventive or ameliorative measures at the very beginning of the project so that no adverse environmental impact takes place.

In conclusion, the successful restoration of degraded land will depend on social and political change and continued motivation and awareness of the people, coupled with a strict institutional and legal framework for restoration. Voluntary organisations, NGOs and the government can all play a vital role. Future programmes of restoration should reflect the linkage between environment and society. It has been observed that restoration of damaged ecosystems and cultural attitudes need a longer time-scale than the political and national planning cycle. This needs to be kept in mind when framing policies and programmes.

REFERENCES

Bradshaw, A. D. (1990). Ecological approaches to handling of mine waste. *Mine and Mineral World*, 22–23.

Director General of Mine Safety (1997). *Statistics of Coal*. Dhanbad, India: Government of India.

Goris, J. M. (1980). *Reducing costs for recontouring mined land*. US Bureau of Mines Information Circular no. 8823. Washington, DC: US Bureau of Mines.

Indian Bureau of Mines (1999). *Indian Mineral Year Book*. Nagpur, India: Government of India.

Jha, A. K. & Singh, J. S. (1994). Rehabilitation of mine spoils with particular reference to multipurpose trees. In *Agroforestry Systems for Sustainable Land Use*, eds. P. Singh, P. S. Pathak & M. M. Roy, pp. 237–249, New Delhi: Oxford University Press and IBH Publishing Co.

Mathur, R. B. (1993). Problems and prospects of rehabilitation of mining areas: Coal India scenario. *Minetech*, **14**, 16–20.

MOEF (1989). *1985–1989 Developing India's Wastelands*. New Delhi: National Wastelands Development Board.

MOEF (1995) *Industries' Solid Wastes in India: Report*. New Delhi: Ministry of Environments and Forests.

Noronha, L. (1996). Ecological management in Goa. *Mining Environmental Management*, June 1996, 17–20.

Pandey, H. D., Bhattacharya, S., Maheshwari, G. D., Prakash, O. M. & Mediratta, S. R. (1996). Research needs in environmental and waste management in iron and steel industries. In *Proceedings of Environmental and Waste Management in Metallurgical Industries*, pp. 1–21. Bihar, India: NML.

Raju, M. S. (1995). *Solid Waste Management in Mines*. Souvenir, Mines Environment and Mineral Conservation Week. Hyderabad: Indian Bureau of Mines.

Singh, J. S. & Jha, A. K. (1984). Ecological aspects of reclamation and revegetation of coal mine spoils. In *Proceedings of the National Workshop on Environmental*

Management of Mining Operations in India: A Status Paper, ed. B. B. Dhar, pp. 74–86. Varanasi, India: Banaras Hindu University.

Singh, J. S., Singh K. P. & Jha A. K. (1996). *An Integrated Ecological Study on Revegetation of Mine Spoil: Concepts and Research Highlights*. Varanasi, India: Banaras Hindu University. Report to Central Mine Planning and Design Institute Ltd, Ranchi.

Singh, R. S., Chaulya, S. K., Tewary, B. K. & Chakraborty, M. K. (1994). *Reclamation Techniques for Fresh Overburden Dumps Constructed by Dragline*, Souvenir, Mines Environment and Mineral Conservation Week. Hyderabad: Indian Bureau of Mines.

Singh, R. S., Chaulya, S. K., Tewary, B. K. & Dhar, B. B. (1996). Restoration of a coal-mine overburden dump: a case study. *Coal International*, March 1996, 83–88.

Srivastava, S. C., Jha, A. K. & Singh, J. S. (1989). Changes with time in soil biomass C, N and P of mine spoils is a dry tropical environment. *Canadian Journal of Soil Science*, **69**, 849–855.

Tiwary, S. N. (1990). Waste control in mines. *Indian Mining and Engineering Journal*, June 1990, 16–19.

4b · China: progress in the reclamation of degraded land

MING H. WONG AND ANTHONY D. BRADSHAW

INTRODUCTION

General land use and mineral production in China

Careful husbandry of the land resource has been a conspicuous feature of Chinese land use since the earliest times (Wang *et al.*, 1994). Since the period of the Warring States (720–221 BC) the Chinese have investigated the harmonious relationship among Tian (heaven or universe), Di (earth or resource) and Ren (people or society), advocating man and nature to be one. From this a systematic set of principles for managing the relationships between man and the environment was developed. In particlar, principles of holism, symbiosis, circulation and self-reliance were emphasised. The harmony between Tian and Ren, and between this generation and the next, is the final objective of all human activities.

There is little doubt that this attitude came out of, and was made imperative by, the intensity of the population settlement in the central lowlands plains of China. There was little room or facility to migrate into new undeveloped lands. To survive, people had to work with, rather than against, natural forces, by resorting to integrative production, and develop a technology based on self-reliance. As a result there were no wastes in ancient China, either in town or country, everything was valued, and used as fuel, forage or fertiliser within the local human/natural ecosystem, leading to remarkable systems of husbandry sustainable over millennia, the details of which can now be unravelled (e.g. Guo & Bradshaw, 1993).

As a result the land surface itself was cared for. Although it became used intensively and became heavily modified in many areas by terracing etc., this was carried out with a care which is visible today. However, in the remarkable industrial development of China since the 1950s, especially in the Great Leap Forward (1958–63), exploitation of mineral resources took place with urgency, with little concern for conservation and restoration, and considerable damage occurred, which was compounded with the previous wholesale destruction of the land surface and its ecosystems by the Sino-Japanese war and occupation. As a result, systematic land reclamation was only initiated in the late 1950s and not carried out on any scale until the advent of specific legislation in the late 1960s. The total amount of land reclamation now achieved is over 10 000 km^2, over one-third originating from mining and industrial enterprises (Pan, 2000).

Out of the total land of about 9.6 million km^2 in China, only 14.2% is cultivated, 0.5% is under fruit trees, 34.8% is under grazing, 17.2% is under forests, 2.6% is used for industry, communication and urban purposes, and 3.5% is covered by water. Much of the remaining 27.2% is high and barren desert, or under permanent snow cover (Commission of Integrated Survey of Natural Resources, 1990).

For the reasons given, mineral exploitation has been a major cause of damage. It can be classified in four main groups: metals, industrial minerals (such as lime or soda ash), construction materials, and energy minerals (i.e. coal, natural gas, oil, etc.). China is a significant producer of coal on an international scale, and has become the world's largest producer, with an annual production of $136\,000 \times 10^4$ t in 1995 (Table 4.5).

In addition, the reserves of tungsten, tin, stibium, zinc, titanium, tantalum and thalium in China rank first: reserves of lead, nickel, mercury, molybdenum and niobium rank second: and the reserve of aluminium ranks fifth in the world. The gross output of

Table 4.5. *Coal production (10^4 t) in China between 1990 and 1995*

Year	1990	1991	1992	1993	1994	1995
	107 988	108 741	111 600	115 100	124 000	136 000

Source: State Statistical Bureau (1998).

10 main non-ferrous metals increased from 13 000 t in 1949, to 952 400 t in 1978, 4.25 million t in 1995, and 6.16 million t in 1998 (Commission of Chinese Environmental Yearbook, 1997, 1999). The rapid increase in production has been mainly due to improvement of equipment and technology from the end of the 1970s to the beginning of the 1980s. There was a steady increase of production of copper, aluminium, lead, zinc and tin in China from 1991 to 1996 (Table 4.6).

Underground mining is the principal mining method, accounting for 70% of the total ore production. In addition, extensive small-scale or artisanal mining plays an important economic role and provides a livelihood for a large number of people. It has been estimated that there were 5 million artisanal miners working some 280 000 mines in 1994 in China (UNEP, 1997). This further aggravates problems of vegetation destruction, hydrological disruption, noise and air pollution, and severe contamination of surface and underground water. According to the National Land Management Bureau, the land area disturbed by mining in China is about 3 million ha, including 1 million ha caused by village and township enterprises and 2 million ha by national industrial and mining enterprises. It has been estimated that the rate of production of derelict land by mining in China is 20 000 ha per year, and will exceed 33 000 ha per year by the end of the twentieth century (Gao *et al.*, 1996).

However a large amount of derelict land has been caused by other factors. The northern and western regions of China are naturally very arid and many of their soils composed of loose sandy material, mostly loess in origin. The most arid areas, especially in Inner Mongolia, are effectively desert, but these grade into arid grassland, traditionally exploited by an agro-pastoral regime. Excessive exploitation, particularly in recent times, has resulted in desertification of these grasslands, so that as much as 10% has been seriously damaged by wind erosion.

Loss of plant cover has also left these areas with their deep loess deposits open to serious erosion by water. This has been exacerbated by arable cultivation which has intensified with the growth and spread of population in these regions, so that as much as 1.796 million km^2 have been affected (Yang, 1999). The eroded material has been carried down into the major rivers, especially the Huang Ho (the Yellow River). Although once the origin of the fertility of the lowland plains, in its present excess it now causes enormous problems.

A third factor has been the removal of tree cover for development and, more importantly, for fuel wood which has become seriously deficient in many areas as populations have grown substantially. This particularly lays open steep slopes to new and serious erosion (Zhu *et al.*, 1996).

The land of China is therefore a land of considerable contrasts. Many areas remain carefully tended by methods developed over millennia. But where resources are being exploited there can be havoc. This applies not only in the specific areas where there are major resources such as coal, but also on the

Table 4.6. *Production (10^4 t) of copper, aluminium, lead, zinc and tin in China from 1991 to 1996*

	1991	1992	1993	1994	1995	1996
Copper	56	66	75	74	108	112
Aluminium	96	110	126	150	187	190
Lead	32	37	41	47	61	72
Zinc	61	72	89	108	108	116
Tin	4	4	5	7	7	7

Source: State Statistical Bureau (1998).

outskirts of developing towns where there is quarrying for building materials and disposal of domestic and industrial wastes, no longer carefully husbanded as in the past. The latter is obvious in the Special Administrative Region of Hong Kong.

The intensity of agriculture has meant that wild areas containing the country's biodiversity are mainly in the hilly regions. Very little biodiversity remains in the agricultural lowlands. This situation was exacerbated during the Cultural Revolution when the local people were encouraged to rid the land of every bird and animal, to protect the food of the people. This has meant that few nature reserves have been established in the lowlands, although the need for them is now understood. A number have been established taking advantage of the land that has traditionally never been exploited around Buddhist temples, such as Dinghushan (Guandong Province) and Tian Mu (Zhejiang Province). Others

are in upland forested areas; one of the best known is the Wolong Preserve (Sichuan) established particularly for the giant panda (*Ailuropoda melanoleuca*). Endangered species are an important focus (Enderton, 1985).

The position is therefore changing. In the early days of the People's Republic, China was relatively isolated from international thinking about the environment. But in 1972, during the later stages of the Cultural Revolution, China attended the United Nations Conference on the Human Environment in Stockholm. This lead to an increased awareness of environmental matters and the formation of a small Environmental Protection Office, which has subsequently evolved into the National, and now the State, Environmental Protection Administration (SEPA), with a wide range of legislation and supporting organisations, so that China now manifests levels of environmental concern equivalent to that of many more advanced countries (Fig. 4.2) (Ross, 1998).

POLICY FOR LAND RECLAMATION

Reclamation of mined land

Because of the great size of China, the focus in this chapter will be on mining. But mining itself illustrates the restoration problems facing the country. In general, land reclamation after mining in China has developed at a slow pace. By the end of 1990, only 2% of the total 2 million ha wasteland (produced by nation-run industrial and mining enterprises) had been reclaimed, and is of poor quality. However, the reclamation ratio has increased throughout the years; only 0.7%–1% of derelict land was reclaimed at the beginning of 1980, about 2% by end of 1980, 6.7% at the beginning of 1990, reaching 13.3% by 1994. In 1994, 133 300 ha of land destroyed by mining had been brought back to use, an area equalling the total areas reclaimed in 1992 and 1993 (Liu, 1995). A total of 3.5 million ha of land was reclaimed for cultivation between 1987 and 1995 throughout the country (Yang, 1999).

However, there are major problems related to reclamation of mined land. These include insufficient enforcement of environmental law, lack of

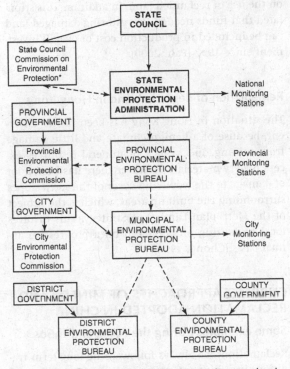

Fig. 4.2. The organisation of environmental protection in China. Continuous arrows, management relationship; dashed arrows, advisory relationship. Redrawn from Jahiel (1998).

initiatives of the mining companies, use of inappropriate reclamation techniques, and lack of funding for reclamation and overall planning (Lan & Rong, 1992).

The challenge posed by existing damaged land is considerable. If half the total amount of damaged land originating from all causes was reclaimed to agriculture (6.7 million ha), the annual grain output could be increased by 40 billion kg (Pan, 2000).

Reclamation of eroded and desertified land

Despite the extent of the damage, rehabilitation of eroded land has so far been concentrated in small catchments, easily handled by local people. Up to the present, 9 000 catchments totalling 400,000 km^2 have been treated. The desertified areas are being tackled by planting 1 million ha of shelter-belts and the rehabilitation of 44 million ha of seriously degraded grassland. Afforestation is a major activity, notably in the 'Three North' areas (north, northeast and northwest) which make up 1.85 billion ha (Yang, 1999). The work has been assisted by the setting up of nine research stations. The effectiveness of the reafforestation has been shown (Liu & Wu, 1985), but more recent evidence suggests that it is not altogether effective.

Policy in land reclamation

In earlier times reclamation of damaged land was a local responsibility. Although a number of excellent projects were carried out, such as the East Lake in Shaoxing City (Zhejiang) (an old quarrying area) and Dingshan Park in Bejing (the refuse disposal site for the Imperial Palace), there was little systematic effort. Following the beginnings in the early 1970s, the Ministry of Land and Resources, under the State Council, became responsible for the unified administration, supervision and inspection of the nationwide work on land reclamation, with the aid of the local governments. Regulations on Land Rehabilitation, the first comprehensive measures for China, were issued in October 1988, in order to minimise the reduction of land resources and to improve the environmental quality. Twenty-five provincial governments have drawn up the Implementation Measures for Regulations on Land Reclamation, and a number of cities and counties have also drafted the Administrative Rules of Land Reclamation, which are suited for local situations.

Basically, the Regulations mean that stipulations have been added to the revised Land Administration Law of China clearly stating that: (1) whoever destroys the land is responsible for reclamation; (2) administrative departments of mining companies are in charge of land reclamation; (3) land reclamation should be planned at the very beginning, and should be part of the overall plans of mining companies; (4) mining companies should be responsible for any expenses incurred for land reclamation and any reimbursement made due to land loss; (5) the operation procedures of land reclamation are defined; and (6) the ownership and use right of the rehabilitated land are rationally stipulated. Policies related to the promotion of land reclamation have been formulated. These included exemption of tax on the use of reclaimed land. In addition, it is stipulated that funds needed for restoring damaged land can be included in production cost or overall investment since 1997 (Pan, 2000).

Remediation of degraded land in Hong Kong

The situation in Hong Kong has been rather different, because of a benign climate and little damage from mining due to lack of mineral resources except quarry materials. However there has been a lot of damage to the woodland cover of the steep slopes surrounding the built-up areas, which is the subject of the Metroplan Landscape Strategy. This involves local replanting of woodland on a progressive annual basis (Chong, 1999).

CURRENT APPROACHES OF MINED LAND RECLAMATION ADOPTED IN CHINA
Some examples during the 1950s and 1960s

Reclamation has always followed the long-term tradition in China of transforming degraded land into farmland. During the 1950s and 1960s the following mined lands were reclaimed: Huauren lead and zinc mine (decommissioned tailings pond),

Xiaoguan aluminium mine of the Zhengzhou Aluminium Company (Henan) (mine wastes), Bantan tin mine (Guangdong) (scraped topsoil to rehabilitate mined area), and Magezhuang iron mine near Tangshan City (Hebei) (backfilling mined pits with waste rocks or replacing tailings on flood beaches).

Land rehabilitation in the 1970s and 1980s

During the 1970s and 1980s, efforts began to be made to promote a more systematic basis to land rehabilitation, paying particular attention to the basic problems to be overcome. For example, methods for prediction of subsidence caused by coal-mining, theories of anti-deformation for construction, and comprehensive technology for rehabilitating subsidence land caused by coal mining, have been developed by the Tangshan (Hebei) Branch of China Academy of Coal Sciences (Yan *et al.*, 2000). The following techniques have been developed: (1) trial sites designed on subsidence basins to study soil quality and crop yield, in order to search for the impact of subsidence caused by coal-mining on the physical, chemical and biological characteristics of the affected soil; (2) land destruction evaluation and classification models established to reveal the destruction characteristics and patterns of landscape change caused by mining, and also the relationship between subsidence and deformation; (3) in-house subsidence simulation trials conducted to obtain the development and distribution pattern of surface fracture; and (4) new land resources allocation and rehabilitation management models derived using technical and economic evaluations on the common rehabilitation practice, procedure and technology. In addition, quantitative analytical methods such as comprehensive fuzzy evaluation and grey cluster analysis have been recently developed to conduct rehabilitation investigations, planning and evaluation activities related to land rehabilitation projects. Equivalent work is in progress at Pingdingshan (Henan).

Ecological restoration

Restoration should mean returning the land to its original state. But for China the concept has been broadened to making the rehabilitated land fit for farming or other productive purposes, depending on the local situations, for which the better word is reclamation (Bradshaw, 1997). This is widely used in the Chinese literature although, significantly, it is now termed 'recultivation' in the English version of the Land Management Act. More emphasis is being put on the ecological aspects of rehabilitation including: (1) screening appropriate topsoil, plant species and fertiliser; (2) identifying the structure of root distribution and growth pattern of the pioneer species; (3) investigating the migration pattern of heavy metals; (4) testing the characteristics of waste refilling, e.g. the use of coal refuse or fly ash for backfilling and for plant growth; and (5) integration of landscape optimisation, sustainable development, harmony of man with nature, etc. into ecological restoration (Peng, 2000; Tang & Gao, 2000). This approach is capable of being further developed particularly by careful analysis and experimental management of soil fertility (Miao & Marrs, 2000). Ecological thinking is therefore being applied, but there has been little emphasis so far on restoring original ecosystems, which would be considered true ecological restoration by Western scientists.

Land reclamation testing sites and demonstration sites

More than 20 testing sites dealing with reclamation work related to coal, ferrous and non-ferrous minerals, gold mining, etc., and 22 demonstration sites for developing reclaimed land for eco-agriculture purpose have been set up nationwide since 1989 and 1995 respectively. Three comprehensive sites have been established in Tongshan (Jiangsu), Huaibei (Anhui) and Tangshan (Hebei) by the State Land Administration in 1994. All these sites have summed up experience on different aspects of land reclamation which included legislation, policy-making, technology, funds collection, standardised administration, and use of reclaimed land (Pan, 2000; Peng, 2000). Two sites are being treated under a co-operation initiative between the Australian and Chinese governments for non-ferrous metals at Zhong Tiao Shan (Shanxi) and Tong Ling (Anhui)

Table 4.7. *Main technical requirements for reclamation of derelict land for subsequent use in different ways*

Aspect	Usage	Technical requirements
Agriculture	Farmland	For grain farming, the land is levelled and covered with topsoil. The topsoil should have a thickness not less than 0.5 m with a humus layer of 0.2–0.3 m. The filler materials should be free of harmful elements. Otherwise an insulation layer with a thickness of not less than 0.4 m should be provided and compacted. Required soil properties: good hydraulic condition; mass density >1.5 g cm^{-3}; clay:sand ratio 1:3 or 1:2; porosity >40%–45%; contents of soluble sodium sulphate and magnesium sulphate <5%; content of sodium chloride <0.01%; pH 6–8.
Forestry	Planting trees or for use as an orchard	The land should be appropriately sloped and covered with topsoil. For planting trees, the topsoil thickness should not be less than 0.3m. The hole for tree planting should be deepened and filled with a layer of soil >1m. If the fill material contains harmful elements, an 0.4m insulation layer should be spread and compacted via ramming.
Fishery	Fishpond or water storage pond	The bank slope should not be too steep and the water area should not be too large. The water quality should be up to the standard for a fishery.
Construction	Civil or industrial buildings	The backfilled foundation should be compacted through ramming, and the buildings should be reinforced.
Recreation	Stadium, park and swimming pool, sanatorium, hospital	Same as above.

Source: Li (2000).

mines. From these a set of national guidelines is being developed (Freak, 1998, 1999).

USAGE AND REQUIREMENT OF RECLAIMED LANDS

Different usage of reclaimed lands

Since reclaimed mined lands are mainly used for agriculture, forestry, fishery, construction sites, and recreational grounds, each type of usage has technical requirements (Table 4.7). These are essentially based on ecological principles, involving the establishment of integrated farming systems, comprising agriculture, fishery, poultry and livestock breeding. The crops and fodders are used as feeds for fowl and animals while the excrement of the animals becomes pond or farmland fertiliser. The spent compost after fungal cultivation can also be used as livestock feeds, as well as pond and farmland fertiliser. Different models of reclamation and ecological construction have been established in the coal-mining subsidence areas in Tangshan (Yan *et al.*, 2000).

Restoration of soil fertility and choice of species

It is essential to restore soil fertility and raise production capacity of the reclaimed land for agricultural purposes. This can be achieved by soil amelioration and selection of suitable crop variety. For soil amelioration, the following methods are commonly adopted (Li, 2000): (1) green manuring – planting of perennial or annual leguminous crops to improve the physicochemical properties of the reclaimed soil; (2) manure application – extensive application of organic manure to increase the organic matter and also improve the physicochemical properties of the reclaimed soil; (3) improvement of soil structure – addition of clay or sand to improve the clay:sand ratio of the tillable layer; and (4) chemical treatment – addition of lime to acid soil, and gypsum, calcium chloride and sulphuric acid to alkaline soil.

Selection of appropriate plant species or varieties is achieved through greenhouse and field trials, as well as empirical analogy. The selected plant species or varieties should be adapted to poor soil conditions, possess a high growth rate and a high yield. Indigenous species should be chosen wherever possible.

Mine wastes containing toxic levels of heavy metals

Mine wastes containing elevated concentrations of heavy metals such as in Dexing copper mine (Jiangxi) (Chen et al., 2000; Tang & Gao, 2000), and lead and zinc mines at Shaoguan and Lechang (Guangdong) (Lan et al., 1998; Ye et al., 1999, 2000) impose metal toxicity problems to plant growth. Various experiments conducted in different mine sites generally indicate that in situ immobilisation together with vegetation is the most effective and economic method for the reclamation of metal-contaminated areas. This involves the use of proper inorganic and organic amendments and choice of suitable plant species. In addition, continuous monitoring and aftercare is also important for ensuring durability.

The two demonstration sites established under the China Australia Mine Waste Management Project at Wu Gong Li tailings dam at Tong Ling (Anhui) and Mao Jia Wan tailings dam at Zhong Tiao Shan (Shanxi) copper mines have been rehabilitated to housing and parkland, and to agriculture respec-

tively (Freak, 1999). For the former site, a staged rehabilitation programme was undertaken, consisting of the implementation of an interim surface drainage, stabilisation of the tailings dam surface with suitable vegetation and final rehabilitation with the construction of the preferred land use. For the latter site, a staged programme was also launched in order to test the establishment of healthy stands of crops which can be implemented on a large scale. The dust problem was also suppressed using a variety of experimental methods including water irrigation and vegetation.

It has been claimed that the infertile, dust-generating tailings with residual heavy metals can be made to produce crops suitable for human consumption (Zhou et al., 1999). However the use of reclaimed land for agricultural production is a major concern, as there may be health hazards, due to uptake of potentially toxic metals by crops. Risk assessment is therefore necessary to reveal the amount of toxic metals taken up by crops, which may subsequently enter the food chain.

CONCLUDING REMARKS

In the past, rehabilitation of damaged land was carried out as a local responsibility in response to local needs. With the pressures arising out of the Great Leap Forward and the subsequent rapid industrialisation in progress at present, rehabilitation had not kept pace with damage. But the situation is appreciated by the state government and administrative mechanisms have been set up to remedy this and to reduce the inherited damage. Reclamation will surely forge ahead (Pan, 2000). There is an increasing awareness of the need to make it an operation that must be carried out universally. Because of the pressures from food production and development for the practical use of all land, this 'restoration' will mostly be in the form of reclamation and rehabilitation rather in its more perfectionist form. There is, so far, only slight interest in the more complete ecological restoration being advocated in the West, although some investigations have begun (Zhuang & Yau, 1999).

Nevertheless a holistic approach is widely recommended, in which government officials, mine operators and different scientists should co-operate to tackle the problems. Reclamation must be an integral and important part of mining operations, which should be planned at the very beginning and then applied without delay. However, at the moment many of the users and destroyers of land do not have the ability to carry out reclamation (Luo et al., 2000). Deadlines need to be imposed for reclaiming existing derelict land caused by mining activities. Source of funding and its usage should be explicitly stipulated. Clear guidelines should be provided as to the quality and maintenance of the reclaimed land.

Perhaps the most serious problem is to ensure that the necessary work is actually carried out. Despite the extensive bureaucracy, it has not been easy to ensure that the legislation is adhered to, especially among the smaller enterprises, now so numerous. At the same time there is insufficient awareness of the best technology. The following points in particular, not universally accepted at the moment, must become so:

- Land reclamation is a long-term process, and revegetation should be effective and the biodiversity of reclaimed land should be maintained.
- Productivity of planted species should achieve a permitted level, and their survival rate should exceed 80%–90%.
- The reclaimed land should achieve a better standard or at least maintain the same standard as that before mining.
- Toxic mine wastes should be remediated using inert and organic covers or amendments and planted with appropriate plant species.
- Special attention and special reclamation techniques should be developed, if the reclaimed land is to be used for agricultural production.
- Original land morphology should be restored, and if possible original cover soil should be used for reclamation.
- The imbalance of material flows between reclaimed land and adjacent areas should be minimised.

- Toxic waste materials, including potentially acid materials, should be properly treated.
- Special attention should be given to minimising disturbance of wildlife communities, including birds, mammals and fish.

ACKNOWLEDGMENTS

The authors thank Professor H. M. Chen, C. Y. Lan and L. Gao for providing information, and the Croucher Foundation and the Research Grants Council of the University Grants Committee, Hong Kong for financial support.

REFERENCES

Bradshaw, A. D. (1997). What do we mean by restoration? In *Restoration Ecology and Sustainable Development*, eds. K. M. Urbanska, N. R. Webb & P. J. Edwards, pp. 8–16. Cambridge: Cambridge University Press.

Chen, H., Zheng, C. C. & Tu, C. (2000). The current status of nutrition and heavy metals in soil–plant systems of copper mine tailings. In *Mined Land Reclamation and Ecological Restoration for the 21st Century, Proceedings of Beijing International Symposium on Land Reclamation*, May 2000, pp. 303–306. Beijing.

Chong, S. L. (1999). Restoration of degraded lands in Hong Kong. In *Remediation and Management of Degraded Lands*, eds. M. H. Wong, J. W. C. Wong & A. J. M. Baker, pp. 185–193. Boca Raton, FL: CRC Press.

Commission of China Environment Yearbook (1997). *China Environment Yearbook*. Beijing: Chinese Environment Science Press.

Commission of China Environment Yearbook (1999). *China Environment Yearbook*. Beijing: Chinese Environment Science Press.

Commission of Integrated Survey of Natural Resources (1990). *Database of Chinese Natural Resources*, vol. 1, pp. 1–2. Scientific Press.

Enderton, C. (1985). Nature preserves and protected wildlife in the People's Republic of China. In *China Geographer*, vol. 12, *Environment*, eds. C. W. Pannell & C. L. Salter, pp. 117–140. Boulder, CO: Westview Press.

Freak, G. (1998). Rehabilitation guidelines: a systematic approach to mine waste management in China. In *Land*

Reclamation: Achieving Sustainable Benefits, eds. H. R. Fox, H. M. Moore & A. D. McIntosh, pp. 425–436. Rotterdam: Balkema.

Freak, G. (1999). Chinese rehabilitation trials. *Mining Environmental Management*, May 1999, 7–9.

Gao, L., Wang, Y. & Ge, F. (1996). Environmental management and pollution control in mine. In *Proceedings: Restoration and Management of Mined Lands: Principles and Practice*, December 1996, Guangzhou.

Guo J. Y. & Bradshaw A. D. (1993). The flow of nutrients and energy through a Chinese farming system. *Journal of Applied Ecology*, **30**, 86–94.

Jahiel, A. R. (1998). The organisation of environmental protection in China. In *Managing the Chinese Environment*, ed. R. L. Edmonds, pp. 33–63. Oxford: Oxford University Press.

Lan, C. Y., Shu, W. S. & Wong, M. H. (1998). Reclamation of Pb/Zn mine tailings at Shaoguan, Guangdong Province, PRC: the role of river sediment and domestic refuse. *Bioresource Technology*, **65**, 117–124.

Lan, Y. & Rong, D. (1992). The situation of mineral resources of China and the counter measures. *Journal of Natural Resources*, **7**, 304–311.

Li, S. (2000). China's mining-affected land reclamation and utilization techniques. In *Mined Land Reclamation and Ecological Restoration for the 21st Century, Proceedings Beijing International Symposium on Land Reclamation*, May 2000, pp. 139–146. Beijing.

Liu, C. & Wu, K. (1985). The effects of forest on water loss and soil conservation in the loess plateau of China. In *China Geographer*, vol. 12, *Environment*, eds. C. W. Pannell & C. L. Salter, pp. 25–38. Boulder, CO: Westview Press.

Liu, R. (1995). Review of reclamation of wasteland in China. *Reclamation and Greening of Mined Wasteland*, 1995, 1–6.

Luo, M., Jiang, Y., Li, G., Ju, Z. & Zhong, M. (2000). Research on wasteland reclamation and its implementation mechanism in China. In *Mined Land Reclamation and Ecological Restoration for the 21st Century, Proceedings Beijing International Symposium on Land Reclamation*, May 2000, pp. 54–59. Beijing.

Miao, Z. & Marrs, R. (2000). Ecological restoration and land reclamation in open-cast mines in Shanxi Province, China. *Journal of Environmental Management*, **59**, 205–215.

Pan, M. (2000). Land reclamation in China: review, trend

and strategy. In *Mined Land Reclamation and Ecological Restoration for the 21st Century, Proceedings Beijing International Symposium on Land Reclamation*, May 2000, pp. 1–6. Beijing.

Peng, D. (2000). Review and prospects of land reclamation and ecological restoration in China. In *Mined Land Reclamation and Ecological Restoration for the 21st Century, Proceedings Beijing International Symposium on Land Reclamation*, May 2000, pp. 33–37. Beijing.

Ross, L. (1998). China: environmental protection, domestic policy trends, patterns of participation in regimes and compliance with international norms. In *Managing the Chinese Environment*, ed. R. L. Edmonds, pp. 85–111. Oxford: Oxford University Press.

State Statistical Bureau (1998). *China Energy Statistical Yearbook, 1991–1996*. Beijing: Statistical Publishing House.

Tang, W. & Gao, L. (2000). Ecological reconstruction for sustainability of mineral regions. In *Mined Land Reclamation and Ecological Restoration for the 21st Century, Proceedings Beijing International Symposium on Land Reclamation*, May 2000, pp. 112–117. Beijing.

UNEP (1997). Industry and environment. *Mining Facts and Figures*, **20**, 1–91.

Wang, R., Ouyang, Z. & Yang, Z. (1994). Sustainability indicator in China: a theoretical and practical concern. In *Wealth, Health and Faith: Sustainability Study in China*, eds. R. Wang, J. Zhao, Z. Ouyang & T. Niu, pp. 1–15. Beijing: Environmental Science Press.

Yan, Y., Lu, J., Chen, D., Zhang, G. & Yang, Z. (2000). Study on the models of land reclamation and ecological reconstruction of the coal mining subsidence areas in Tangshan. In *Mined Land Reclamation and Ecological Restoration for the 21st Century, Proceedings Beijing International Symposium on Land Reclamation*, May 2000, pp. 156–165. Beijing.

Yang, C. F. (1999). Land degradation and its control strategies in China. In *Remediation and Management of Degraded Lands*, eds. M. H. Wong, J. W. C. Wong & A. J. M. Baker, pp. 175–184. Boca Raton, FL: CRC Press.

Ye, Z. H., Wong, J. W. C., Wong, M. H., Baker, A. J. M., Shu, W. S. & Lan, C. Y. (1999). Revegetation of Pb/Zn mine tailings, Guangdong Province, China. *Restoration Ecology*, **8**, 87–92.

Ye, Z. H., Wong, J. W. C. & Wong, M. H. (2000). Vegetation response to lime and manure compost amendments on acid Pb/Zn mine tailings: a greenhouse study. *Restoration Ecology*, **8**, 289–295.

Zhou, L. B., van de Graff, R., Dai, H. W., Wu, Y. L. & Wall, L. N. (1999). Rehabilitation of copper mine tailings at Zhong Tiao Shan and Tong Ling, China. In *Remediation and Management of Degraded Lands*, eds. M. H. Wong, J. W. C. Wong & A. J. M. Baker, pp. 111–122. Boca Raton, FL: CRC Press.

Zhu, Z. (1996). Features of distribution and assessment for control measures of desertification in China. *China Environmental Science*, **16**, 328–334. (in Chinese)

Zhuang, X. Y. & Yau, M. L. (1999). The role of plantations in restoring degraded lands in Hong Kong. In *Remediation and Management of Degraded Lands*, eds. M. H. Wong, J. W. C. Wong & A. J. M. Baker, pp. 201–208. Boca Raton, FL: CRC Press.

5 • Oceania

IAN D. HANNAM AND ALISON M. COCHRANE

INTRODUCTION

Oceania comprises the geographic region of Australia, New Zealand, Norfolk Island, Fiji, Solomon Islands and Tonga. The islands of the Oceania region cover 30 million square kilometres and number more than 10 000. More than 90% of this area under sovereign jurisdiction is ocean. The history of each of these countries is quite different, because of the American, Australian, British, French or New Zealand colonial heritage (Boer *et al.*, 1998). As recently as 33 000 years ago, no human beings lived in Oceania, apart from Australia and New Zealand. With the imposition of European culture in Oceania, these societies continue to draw on a wide array of ecosystems, species and genetic variants to meet their ever-changing needs. The diversity of nature in the region (Box 5.1) is the source of its biological wealth, supplying food, raw materials and a wide range of goods, services and genetic materials for agriculture, medicine and industry. It is essential that this biological diversity be preserved as a matter of principle (United Nations, 1992*c*; Flannery, 1994). Long-term sustainability in Oceania is reliant on the retention of biological diversity. Degradation of the natural ecosystems has occurred in many of the islands as a result of a number of terrestrial ecological factors. Population growth, particularly in the atoll countries, has resulted in expansion of urban areas, and has placed stress on the limited areas of arable land (Boer *et al.*, 1998). Natural ecosystems and natural diversity are now more threatened than at any time in the past as more habitats are converted to exclusively human use (Saunders *et al.*, 1993). Many species are losing a considerable part of their genetic variation, making them increasingly vulnerable to pests, disease and climate change. For these reasons, there has been considerable effort in the region over the past decade to reform and improve the environmental law and policy to manage ecological systems (de Klemm & Shine, 1993).

Many environmental law regimes have evolved for the management and protection of the Oceanic environment. Specific environmental law and policy issues include the preparation of powers to conserve, restore and re-establish ecosystems, and the conservation of biological diversity and the use of biological resources in a sustainable manner (de Klemm & Shrine, 1993; Preston, 1995). In this chapter, the term 'restoration ecology', in the context of policy and legislation, means 'the basic characteristics and capabilities of environmental law and policy to manage ecosystem values, and to re-establish or re-instate ecosystem values.'

This aim of this chapter is to identify the main ecological issues of the Oceania region and outline how the region has responded with environmental policy and law to conserve and restore natural ecosystems. Restoration is implicit in the legislation and policy for biodiversity, native vegetation conservation environmental planning assessment. A number of the basic ecological terms used in the chapter, as defined in their legislative and policy context are set out in Box 5.2.

INTERNATIONAL ENVIRONMENTAL LAW

Since the early 1920s, over 200 multilateral environmental treaties, agreements, conventions and protocols have been established worldwide, for flora and fauna conservation, fisheries protection, pollution management, regional conservation

Box 5.1   Biodiversity in Oceania

AUSTRALIA

Australia is mainly an old, weathered, eroded landscape, flat and generally dry, with a variable climate, especially rainfall. Significant areas are environmentally fragile due to the thin or poor quality of soil and the arid or semi-arid nature of the climate (McTainsh & Boughton, 1993). Australia's unique plants and animals reflect its long isolation from other landmasses and their wildlife (Australia, 1996a). Australia is one of the world's 12 most biologically diverse countries that together account for 75% of total global diversity (Environment Protection Authority NSW, 1997). The continent has a high percentage of endemic species: 6 families of mammals, 4 families of birds and 14 families of flowering plants are endemic to Australia (Australia, 1996b). At a more detailed species level, 82% of mammals, 45% of land birds, 89% of reptiles and approximately 93% of frogs are unique to the Australian continent (Salt, 1998). Loss of biological diversity from over-clearing native vegetation (Australia, 1996b) may account for the 20 mammal, 20 bird and 76 plant species that have become extinct since 1788 (Department of the Environment, 1998; Krockenberger, 1998).

NEW ZEALAND

Much of New Zealand's flora is endemic. Giant gum-producing forests, rainforest, ferns and flax, alpine and subalpine herb fields and scrub and tussock have survived but their range has been dramatically reduced. About 10% to 15% of the land area is covered with native flora and is protected in national parks and reserves. Native fauna is limited with the only indigenous mammals being bats. Bird life is abundant; the most famous birds are the flightless kiwis (Apteryx spp.), a forest-dwelling bird. New Zealand has suffered a loss of biodiversity, including about 44 species of endemic birds, since human settlement. Introduced species such as pigs, goats, dogs, cats and sheep have had a deleterious effect on the environment. Over 400 species of plants and animals are threatened. Indigenous forests have been reduced from 78% to about 28% of the total land area. Lowland forests have been reduced to 15% of the pre-Maori extent and only 10% of the tussock grasslands that existed in 1840 remain (McBride, 1997).

SMALLER ISLANDS
Lord Howe Island

There are 241 native species of plants on the island, including 105 endemic species. Sixteen of these are considered rare, endangered or vulnerable. There are at least 129 native and introduced bird species, mostly vagrants, and 27 breed regularly on the island. Seven species of birds breed within the islands, in greater concentrations than anywhere else in the world. Approximately 10% of the island's vegetation has been cleared for agriculture, and another 10% has been subject to physical disturbance. In the lower-lying areas, destruction of native vegetation has been virtually complete where clearings have been made for settlement, grazing and agriculture. There are adequate samples of intact lowland vegetation in the less accessible parts of the island, some of them in special flora reserves.

Norfolk Island

Norfolk Island National Park is administered under the Commonwealth *Environment Protection and Biodiversity Conservation Act 1999*. Norfolk Island has 174 species of native plants, of which 50 are endemic. Many of the endemic plants are rare, but the famous Norfolk Island pine (*Araucaria heterophylla*) is still abundant. Introduced plants are now common on the island and some, African olive (*Olea europaea* ssp. *cuspidata*), lantana (*Lantana* spp.) and Hawaiian holly (*Ilex anomala*), have become pests and weeds. The fauna is a mixture of native and introduced species. There are no mammals or amphibians and the only reptiles are seen exclusively on the small rocks and islands surrounding Norfolk Island itself. Bird life is diverse. Forty-three species are native, and include white (*Gygis alba candida*) and sooty (*Sterna fuscata*) terns, Norfolk Island kingfisher (*Todiramphus sanctus norfolkiensis*), Norfolk Island green parrot (*Cyanomorphus cookii*) and the Norfolk Island morepork or boobook owl (*Ninox novaeseelandiae undulata*) which is probably the rarest bird in the world. Clearing and depleting the forest by logging caused extensive loss of habitat. By 1987 it was confirmed that only one female owl remained. Loss of nesting hollows, competition for hollows with introduced species, predation by introduced rats and cats, and the impact of the use of pesticides for rat extermination severely reduced owl numbers.

Fiji

Around 3000 plant species have been identified on the Fijian islands, and about one-third are endemic. Almost half of Fiji's total area is forest and woodland. Severe drought and floods have caused widespread damage in the western and northern divisions. Mangrove and soil degradation in the coastal zones, pollution of coastal waters and the rapidly depleting fisheries resources in the coastal waters and loss of biological diversity are other issues of environmental concern. There is currently no legislation to protect Fiji's endangered species such as the crested iguanas (*Brachylophus vitiensis*) and peregrine falcons (*Falco peregrinus*) (Pulea, 1999).

Solomon Islands

The Solomon Islands have a high level of speciation and population variation. Of the 163 species of land birds that breed in the Solomon Islands, 72 are unique and 62 occur elsewhere in the world but are represented in the Solomon Islands by unique races or subspecies (Robinson, 1992). The major threats to species are logging, fishing, and plantation monoculture and increasing slash-and-burn agriculture, which is a traditional practice but a major cause of forest depletion.

Tonga

Tonga supports surprisingly few species of birds. The main land attractions are the flowering plants, such as frangipani (*Plumeria rubra*) and several species of native hibiscus (*Hibiscus* spp.). More than 100 species of tropical fish live in the reefs around the islands, where hard and soft corals and black coral grow in abundance. Unique to Tonga are the huge fruit bats (*Pteropus tonganus*) with a wing span approaching 2 metres. Deforestation of rainforests is continuing with land being cleared for agriculture and settlement. Permanent crops account for 43% of the land use and arable land accounts for 24% of the land use. Large areas of rainforest and bushland remain on outer volcanic islands.

protection, Antarctic conservation, settlement of disputes, civil liberties in relation to environmental damage, protection of world cultural and natural heritage, endangered species, and landscape protection (United Nations, 1996). Principles and standards in many of these instruments have been applied in the Oceanic region to form national environmental laws and policies to prevent ecosystem degradation (de Klemm & Shine, 1993; Boer, 1995).

Key environmental declarations

The *Stockholm Declaration on the Human Environment* (1972 Stockholm Declaration) and the *Action Plan for the Human Environment* have influenced international and national environmental action for three decades. They sought common effort to preserve and improve the human environment, particularly through the enhancement of environmental policies, environmental planning and legal measures. Ten years later, the *Nairobi Declaration* urged the intensification in effort to protect and improve the environment from deforestation, soil and water degradation and desertification. The *World Charter for Nature* (United Nations, 1982) proposed that ecosystems be managed to achieve and maintain optimal sustainable productivity, and the method to achieve international cooperation on such an objective was spelt out in the *Charter of the United Nations* (United Nations, 1996). Another influential instrument in the Oceanic region was the *World Conservation Strategy*, with its three main objectives: to maintain essential ecological processes and

Box 5.2   Ecological definitions

BIODIVERSITY

The variability among living organisms from all sources (including terrestrial, marine and other aquatic ecosystems and the ecological complexes of which they are part). This includes: (a) diversity within species and between species; and (b) diversity of ecosystems (Commonwealth of Australia *Environment Protection and Biodiversity Conservation Act 1999*, Section 528) (EPBCA).

ENVIRONMENT

The same definition of 'environment' is used in the more prominent environmental law of Oceania, in particular the EPBCA (Section 528) and the New Zealand *Resource Management Act 1991* (Section 2) (RMA). In these Acts 'environment' includes '(a) all ecosystems and their constituent parts, including people and communities; (b) natural and physical resources; (c) the qualities and characteristics of locations, places and

areas; and the social, economic and cultural aspects of anything mentioned in (a–c).

ECOSYSTEM

A dynamic complex of plant, animal and micro-organism communities and their non-living environment interacting as a functional unit (Section 528 EPBCA).

ECOLOGICAL COMMUNITY

An assemblage of native species that inhabits a particular area in nature (Section 528 EPBCA).

HABITAT

The biophysical medium or media (a) occupied (continuously, periodically or occasionally) by an organism or group of organisms; (b) once occupied (continuously, periodically or occasionally) by an organism, or group of organisms, and into which organisms of that kind have the potential to be reintroduced (Section 528 EPBCA).

life-support systems; preserve genetic diversity; and the sustainable use of species and ecosystems. (IUCN, 1980; Australia, 1983). However, it was through the *World Commission on Environment and Development* (WCED, 1987) and *Caring for the Earth* (IUCN, 1991) strategies, that the concept of 'ecologically sustainable development' became prominent in the region.

The Rio Declaration and Agenda 21

These two instruments (United Nations, 1992*a, b*) have had major influence in environmental law and policy development in Oceania (Boer, 1995; Williams, 1997). The *Rio Declaration on Environment and Development* sought to increase co-operation among states, key sectors of society, individuals; to work towards international agreements to protect the integrity of the global environmental and developmental system; and to promote the enactment of effective environmental laws and regulation. The role of 'sustainability' (Chapter 39) gives special attention to the delicate balance between

environmental law and developmental concerns (Malanczuk, 1995). Chapter 17.G of *Agenda 21* defined 'small island developing states', such as those in Oceania, as a special case for both environment and development, presenting special challenges to the planning for sustainable development. *Agenda 21* sets out clear objectives to be pursued by states in conserving biodiversity, the development of relevant national strategies, undertaking studies in biodiversity conservation and the sustainable use of biological resources.

Commission on Sustainable Development

The *Commission on Sustainable Development* (CSD), the major institutional outcome of the 1992 UNCED Conference, facilitates, monitors and evaluates the implementation of the UNCED commitments to the integration of the environmental management objectives embodied in the *Rio Declaration* and the *Statement of Forest Principles of 1992*. The CSD monitors international environment negotiations, and identifies land resources, desertification, forests and biodiversity as major sectoral issues.

Convention on Climate Change

The *Framework Convention on Climate Change* 1992 (FCCC) (United Nations, 1995) recognises the role and importance of terrestrial ecosystems as a sink and reservoir for potential greenhouse gases and that human activities have substantially increased atmospheric concentrations (Rawson & Murphy, 2000). Two of the principal contributors to greenhouse gases in Oceania are changes in land use cover and land use through deforestation, biomass burning, cultivation, use of organic manure and application of nitrogenous fertilisers and livestock. Control and prevention of vegetation clearance, a principal cause of ecosystem loss and fragmentation, is one of the key concerns of the FCCC (Article 1). Article 4.2 is an implicit agreement for developed countries to adopt policies and legislation with the aim of returning greenhouse gas emissions to their 1990 levels.

Kyoto Protocol

The FCCC responds to changes in scientific understanding and political will in the Kyoto Protocol of 1997 (Kyoto, 1997). Article 2 of the Protocol promotes sustainable development and calls for each party to implement policies and measures to protect and enhance sinks and reservoirs of greenhouse gases, taking into account their commitments under various international environmental agreements, sustainable forest management practices, afforestation and reforestation.

Convention on Biological Diversity

The objectives of the *Convention on Biological Diversity* 1992 (CBD) (United Nations, 1996) have been wisely applied in Oceania (Australia, 1996b; Williams, 1997), particularly to: conserve biological diversity, and to encourage the sustainable use of its components, and the fair and equitable sharing of the benefits arising out of the utilisation of genetic resources, including by appropriate access to genetic resources, and by appropriate transfer of relevant technologies (see Miller & Lanou, 1995; Swanson, 1997). Biological diversity is being significantly reduced by human activities (see Boxes 5.1 and 5.5).

Convention to Combat Desertification

Dregne & Chou (1994) estimate that around 55% of land in Australia and the Pacific is degraded. Under the *Convention to Combat Desertification* 1992 (CCD) (United Nations, 1996) 'desertification' means land degradation in arid, semi-arid and dry sub-humid areas resulting from various factors, including climatic variations and human activities (Article 1). There has been a good response to the objective to prevent and or reduce land degradation, rehabilitate partly degraded land and reclaim desertified land. Long-term integrated strategies for rehabilitation, conservation and sustainable management of land and water resources are recognised as necessary for ecological restoration in the region (e.g. New Zealand Ministry for the Environment, 1996).

Forest Principles

The 1992 *Forest Principles* are a non-legally-binding statement for a global consensus on the management, conservation and sustainable development of forests (Tarasofsky, 1995). They are relevant to the management of natural and planted forests of Oceania, especially the austral, boreal, subtemperate, temperate, subtropical and tropical forests. The *Principles* acknowledge that forests embody complex and unique ecological processes, which are the basis for their present and potential capacity to provide resources to satisfy human needs as well as environmental values.

Convention concerning the Protection of the World Cultural and Natural Heritage

The 1972 *Convention concerning the Protection of the World Cultural and Natural Heritage* (WHC) was adopted by the United Nations Economic, Social and Cultural Organisation (UNESCO) (Boer *et al.*, 1998). The World Heritage Committee (Article 8), is responsible for the 'World Heritage List' – the properties that form part of the cultural and natural heritage (Articles 1 and 2) (Boer & Fowler, 1996). Numerous sites in Oceania have been listed. Australia has 13 sites on the World Heritage List, including the Great Barrier Reef, Kakadu National Park, Macquarie Island and the Lord Howe Island Group. New Zealand has three sites listed: Tongariro National Park, Te Wahipounamu – South West New Zealand and the New Zealand sub-Antarctic Islands. The Solomon Islands are represented on this list by East Rennell, which was inscribed in 1998.

South Pacific Regional Environment Program

The *South Pacific Regional Environment Program* (SPREP) was established to monitor the environment of the region. SPREP has developed and implemented regional programmes for biodiversity and natural resource conservation, climate change and integrated coastal management, waste management, pollution prevention, environmental management, planning and institutional strengthening, environmental education, information and training.

The Barbados Declaration

This 1994 Declaration originated from the United Nations *Global Conference on Sustainable Development of Small Island Developing States* and is a comprehensive framework to implement strategies and measures to enhance sustainable development as defined in *Agenda 21*. It has identified 15 key areas of environmental concern including climate change and sea level rise, natural and environmental disasters, coastal and marine resources, and land and biodiversity resources. It encourages the small island nations to ratify the FCCC and to adopt legislation to support sustainable development.

THE NATIONS OF THE REGION

Australia

Over 75% of Australia's population is along the coastline which places environmental stresses on the unique coastal environments, including streams and rivers, lakes, wetlands, estuarine areas, lagoons, beaches and inshore waters. Land use demands include mining, recreation, forestry, agriculture, housing and industry (Farrier, 1999). Australia is a federation of six self-governing states and two self-governing territories. There are also a number of external territories that comprise islands such as Norfolk Island and Macquarie Island, together with a substantial part of Antarctica. The federal government's powers and responsibilities are defined in the Australian Constitution and the state and territory governments are responsible for all other matters (Crawford, 1991). Environmental powers are not specifically dealt with in the Australian Constitution and are not the sole province of any one sphere of government. The federal government instead makes use of other constitutional powers available to it, in order to enact legislation of environmental significance (Ramsay, 1997) (see Box 5.3). The states and territories make their own environmental laws (Bates, 1992; Boer, 1995). Further, various Commonwealth, state and territory co-operative and community-based programmes make a substantial contribution to achieving Commonwealth environmental management goals, including the National Heritage Trust (see *Journal of the National Heritage Trust*, Australia).

New Zealand

New Zealand is located in the South Pacific Ocean, 1600 km southeast of Australia. It stretches 1600 km from north to south and consists of two large islands and a scattering of smaller ones. The Maori settlement of New Zealand began over 1000 years ago. In 1840, the *Treaty of Waitangi* was signed between several Maori chiefs and representatives of the British Crown giving European settlers rights of settlement and governance in return for recognition of Maori authority of Maori rights, possessions and interests. Many more Europeans followed after the Treaty was signed and a Westminster system of government was eventually established (McBride, 1997). New Zealand is a unitary state, with no formal comprehensive written constitution, upper house or entrenched bill of rights (Richardson & Palmer, 2000).

Box 5.3 National environment protection and biodiversity conservation legislation and policy in Australia

ENVIRONMENT PROTECTION AND BIODIVERSITY CONSERVATION ACT 1999

The most prominent national environmental law in Australia to maintain and restore its ecology is the *Environment Protection and Biodiversity Conservation Act 1999* (Commonwealth of Australia) (EPBCA). This legislation aims to achieve national consistency in the protection of the environment, improve Australia's biodiversity, and protect matters of national environmental significance (Section 3). The EPBCA takes particular interest in the protection of World Heritage Property, wetlands of international importance, listed threatened species and communities, migratory species, and protecting the marine environment. It strengthens intergovernmental co-operation, minimises duplication under bilateral agreements and accredits environmental assessment and approval processes. These are objectives of Australia's *Intergovernmental Agreement on the Environment* (Australia, 1992). The objects of the EPBCA promote ecologically sustainable development through the conservation and ecologically sustainable use of natural resources, promote conservation of biodiversity, and promote a co-operative approach to the protection and management of the environment with community input (Glanznig & Kennedy, 2001). The main principles of ecologically sustainable development, as it is applied in the EPBCA, include:

- The use of integrated decision-making processes (economic, environment and social considerations).
- Applying the 'precautionary principle' which states that where there are threats of serious or irreversible environmental damage, lack of full

scientific certainty should not be used as a reason for postponing measures to prevent environmental degradation.
- The responsibility of the present generation to preserve the environment for future generations.
- The conservation of biological diversity and ecological integrity (Australia, 1992).

Bilateral agreements

Bilateral agreements ensure consistency in national environmental assessment and decision-making. They are made between the Commonwealth, state or self-governing territory (Section 44) to protect the environment, promote the conservation and ecologically sustainable use of natural resources, establish rules for environmental assessment, and minimise duplication in environmental assessment. They establish the environmental protection standards and the criteria for management plans (Section 45(2)). The agreements will cover Australia's obligations under declared global conservation values, including: the World Heritage Properties; the declared Ramsar wetlands; the listed threatened species and ecological communities; and for migratory species under the *Convention on the Conservation of Migratory Species of Wild Animals*, Bonn, June 1979; the *Agreement between the Government of Australia and the Government of the People's Republic of China for the Protection of Migratory Birds and their Environment*, October 1986 (CAMBA); and the *Agreement between the Government of Japan and the Government of Australia for the Protection of Migratory Birds and Birds in Danger of Extinction and their Environment*, Tokyo February 1974 (JAMBA). Agreements will promote the management of these areas in accordance with their respective conservation management principles and conservation status, and will seek to improve the conservation status of each species or community.

Smaller islands

Lord Howe Island

Lord Howe Island is located in the South Pacific Ocean, 700 km northeast of Sydney, and is part of the state of New South Wales. It is the eroded remnant of a large shield volcano that erupted from

the sea floor intermittently for about 500 000 years, some 6.5 to 7 million years ago. The Lord Howe Island group was inscribed on the World Heritage List in 1982. The *Lord Howe Island (Amendment) Act 1991* provides for the preparation of a management plan, under the Lord Howe Island Regional

Environmental Plan (REP) 1986. The Lord Howe Island REP aims to conserve the World Heritage values of Lord Howe Island; to ensure that appropriate planning controls are implemented and to conserve and enhance its environmental heritage. Part 2 of the REP zones the land for rural, permanent park reserve, marine and environment protection, and sets out the objectives for each zone type and states the type of activity permissible within the zone (Department of Urban Affairs and Planning, 1997).

Norfolk Island

Norfolk Island is a small volcanic outcrop in the South Pacific Ocean and is the highest point in a huge chain of mountains running 2850 km from New Zealand to New Caledonia. Erosion of the island's coast by the surrounding sea has reduced it to about one-third its original size, forming the high cliffs of its rugged coastline. The island is administered under the *Norfolk Island Act 1979*. All legislation passed by Norfolk Island's Legislative Assembly must be approved by the Administrator (appointed by the Governor-General of Australia to administer the island as a territory under the Commonwealth of Australia), before they become laws (Bates, 1992).

Solomon Islands

The Solomon Islands are located due east of Papua New Guinea, and have an island population of approximately 367 800. The total land area of the islands is 28 370 km², and is divided into a capital territory and seven provinces. The Constitution of the Solomon Islands gives land ownership rights to the indigenous islanders and currently 90% of land is owned by customary tribal or village groups. The seven provinces have legislative and administrative powers under the *Provincial Government Act 1981*. The *Environment Protection and Preservation Culture Ordinances 1989* of Temotu Province allow customary activities in protected places for the taking of protected species. The *Wild Birds Protection Ordinance 1914* prohibits the taking of certain birds, but is not enforced. The *Fisheries Regulations* of the national government set minimum size limits for catches of turtles and crocodiles.

Fiji

Fiji is a maritime, archipelagic nation comprising of approximately 300 islands of which 97 are inhabited. About 94% of Fiji's population of 822 000 is located on two main islands, whilst 6% is distributed amongst the remaining scattered islands. The territorial limits enclose an area of 1.3 million km², but only 1.5% of this is dry land. Around 54 Acts in Fiji have relevant provisions for environmental management, but poor enforcement results in severe environmental degradation (Richardson, 1996). In January 1995, the Fijian Cabinet approved the drafting of a new Sustainable Development Bill for environmental impact assessment, sustainable forestry management, resource management, biodiversity conservation and formation of national parks.

Tonga

Tonga is located in the South Pacific and consists of an estimated 170 islands and islets in three main groups: the Tongatapu in the south, the Ha'apai in the centre, and the Vava'u in the north. The capital, Nuku'alofa, is on Tongatapu. Tonga's population is estimated at 98 200. Fewer than 40 of Tonga's islands are inhabited, and it is the only nation in Oceania never to have been colonised by a foreign power. Tonga was the first small Oceanic island to establish national parks and reserves to protect its unique flora and fauna. Tonga has seven officially protected areas, including five national marine parks and reserves, one national historic park, and the Eua National Park.

ENVIRONMENTAL LAW AND POLICY REGIMES FOR ECOLOGICAL RESTORATION

There are a number of environmental law and policy regimes in Oceania which are important for the successful long-term management and conservation of its terrestrial ecosystems. The structure of the regimes varies, having been shaped by the political and legal history of the particular country (Bates, 1992; Bonyhady, 1992; Williams, 1997).

Box 5.4   New Zealand Resource Management Act 1991 and Biosecurity Act 1993

RESOURCE MANAGEMENT ACT 1991

After a decade of social, economic and legislative reform, the *Resource Management Act* was introduced in 1991. Founded on the concepts of integrated resource management and sustainability, it is well structured to achieve ecological restoration. In the Act, 'integrated resource management' means 'the sharing and co-ordination of the values and inputs of a broad range of agencies, public, and other interests when conceiving, designing and implementing policies, programs or projects'. The Act promotes the sustainable management of land and physical resources (Sections 5 and 6), which is: managing the use, development, and protection of natural and physical resources in a way, or at a rate, which enables people and communities to provide for their social, economic, and cultural well-being, and for their health and safety. This is achieved through provisions aimed at sustaining the potential of natural and physical resources for future generations; safeguarding the life supporting capacity of air, water, soil, and ecosystems; and avoiding, remedying, or, mitigating any adverse effects of activities on the environment. Factors to be considered include preservation of the marine environment, protection of outstanding natural features and landscapes, protection of areas of significant habitat and fauna, and the relationship between the indigenous people and their culture and traditions (Crengle, 1993). A system of National Science Strategies is used to co-ordinate the strategic planning of environmental and scientific issues. One of the first strategies to emerge was the *National Science Strategy for Sustainable Land Management* (New Zealand Ministry for the Environment, 1996).

BIOSECURITY ACT 1993

The *Biosecurity Act 1993* was implemented to exclude, eradicate and effectively manage pests and unwanted organisms in New Zealand. The National and Regional Pest Management Strategies (Part 5) considers the economic well-being, biodiversity, soil resources or water quality, human health or the relationship of Maori and their culture and traditions with their ancestral land. Responsibilities of landowners, regional councils and others and the cost-sharing for the management of the pest are allocated according to the nature and extent of threats posed by an organism.

A common thread that now exists in much of the environmental law in Oceania is the promotion of sustainable management of natural and physical resources, having filtered down from the various international environmental law instruments (Australia, 1992; Williams, 1997).

Many environmental laws in Oceania have a direct role in the restoration of ecological values, specifically those that manage aspects of the environment that directly, or indirectly, manage the essential processes for ecological restoration. In this chapter, a 'regime of legislation' means a group of laws that deal with a specific area of environmental management. In the case of Australia and some other small island countries, a number of individual laws can be in a regime (see Bates, 1992), whereas for New Zealand, many different areas of environmental law reside in its comprehensive *Resource Management Act 1991* (Williams, 1997) (see Box 5.4).

Soil conservation

Soil conservation law has had a major role in landscape rehabilitation (e.g. New South Wales *Soil Conservation Act 1938* and the South Australian *Soil Conservation and Landcare Act 1989* [and the former *Water and Soil Conservation Act 1967* of New Zealand]), particularly the control of soil degradation and conservation of soil resources (Gardner, 1994; Hannam, 2001a). Their provisions directly aim at restoring or improving landscape management through: property planning; soil erosion control techniques; community land use advisory groups; natural resource planning (catchment management schemes); and use of compliance and enforcement tools. The effectiveness of this regime of law for ecological restoration has been improved by linking it to the native vegetation law, e.g. New South Wales *Native Vegetation Conservation Act 1997* (see Box 5.5) and South Australian *Native Vegetation Act 1991*, where the vegetation law controls the land use activities that contribute directly to ecosystem destruction and soil degradation, particularly the activities that deplete natural ecosystems, e.g. overgrazing and clearing land.

Forestry

The original role of forest law which was to manage forest on public land (e.g. New South Wales *Forestry Act 1916*) has now expanded to include public and private forest land management, encouraging farm agroforestry and plantations. Two recent developments are the Tasmanian *Private Forests Act 1994* and the New South Wales *Plantations and Reafforestation Act 1999* which were introduced with

---

Box 5.5 The native vegetation conservation legislation, New South Wales, Australia

INTRODUCTION

Many indigenous ecosystems of New South Wales have been severely modified and degraded by fire, frequent cutting, partial and total removal, and insect and chemical attack (Benson, 1991; Benson & Redpath, 1997). Only isolated remnants remain of some ecosystems, underrepresenting a particular region's former vegetation types with serious consequences in terms of biodiversity, genetic deterioration, loss of evolutionary potential, and resultant non-viable populations (Decker, 1993). There are many reasons to conserve native vegetation in southeastern Australia, including biodiversity and natural heritage values; economic factors; and as part of the nation's duty to future generations (Australia, 1992). For these reasons, over time, a variety of legislation and legal instruments were used to protect, conserve and restore native vegetation, the principal Acts being the *Western Lands Act 1901*, *Forestry Act 1916*, *Soil Conservation Act 1938* and the *Crown Lands Act 1989*. Under the *Environmental Planning and Assessment Act 1979*, environmental planning instruments are used to control the disturbance and destruction of native vegetation on fragile lands, including littoral rainforest, urban bushland, coastal wetlands and threatened species habitat (Mossop, 1992).

The *Native Vegetation Conservation Act* was proclaimed in December 1997 (Lee *et al.*, 1998) to replace the former laws. It is a comprehensive piece of environmental law with a wide range of modern ecological concepts, land use planning and legal controls (Department of Land and Water Conservation, 1997). Under this Act, native vegetation means 'indigenous vegetation, and includes trees, understorey plants; groundcover; and wetland ecosystem plants'. The argument supporting the need to review and reform the environmental law of native vegetation built up over many years (Farrier, 1991; Stratford, 1993), but many resisted based on private property rights attitudes, which prolonged the argument (Bonyhady, 1992; Gillespie, 1995). The earlier vegetation controls were mainly a licensing mechanism for proposed clearing, and were introduced into respective legislation to supplement existing conservation measures and control soil erosion and land degradation. The controls were limited to specific parts of the landscape and to some types of environmentally sensitive land.

ECOLOGICAL PHILOSOPHY

The principle of 'ecologically sustainable development' is embedded in the *Native Vegetation Conservation Act 1997*. Its Objects (Section 3(1)) specify that there will be: a regional approach to conserving native vegetation; consideration of social, economic and environmental interests of native vegetation; an emphasis on protecting native vegetation communities with high conservation values; an improvement in the condition of native vegetation; revegetation and rehabilitation will be a priority; land clearing will be controlled; and the ecological significance of native vegetation will be promoted.

The principal features for ecological restoration

The Act has a variety of provisions to ensure that native vegetation is maintained in community structure, as an ecosystem, or in a landscape.

- *Public participation procedures.* The Native Vegetation Advisory Council advises on the status of vegetation and prepares the vegetation conservation strategy, similar to the state *Biodiversity Strategy* (New South Wales National Parks and Wildlife Service, 1999). Regional Vegetation Committees have representatives from the land use and conservation sectors and they prepare regional vegetation management plans. The public makes submissions on the draft plans and codes of land use practice. The regional plans are a statutory instrument and include: regulatory standards for native vegetation conservation; native species and their habitats; soil and water conservation; archaeologically, geologically or anthropologically sensitive areas; and social and economic aspects of land management.
- *Licence control.* A landholder can make an application to clear native vegetation, which is subject to a rigorous assessment.
- *Native vegetation codes of practice.* These are designed to regulate the management of native vegetation for specific land use activities and they set the environmental assessment and restoration standards for vegetation communities.
- *Property agreements.* This is a mechanism for landholders to develop strategic plans to manage vegetation on their land. The plans identify areas of land and vegetation to be set aside for conservation or rehabilitation, and vegetation management practices. Financial assistance is provided under the Native Vegetation Management Fund (Part 7 of the Act) The restoration of degraded native vegetation and the protection of high conservation value vegetation are priorities.
- *Conservation and remedial measures.* These provisions allow immediate remedial action if a person is contravening, or about to contravene the Act. Legal conditions apply to repair environmental damage, and rehabilitate land and vegetation.
- *Prosecution provisions.* These allow offences under the Act to be dealt with before a Local Court or the New South Wales Land and Environment Court.
- *Restraint provision.* This allows any person to bring proceedings in the Land and Environment Court for an order to remedy or restrain a breach of any regional vegetation management plan, code of practice, or a restoration notice.

SELF-REGULATION

The Act has self-regulation opportunities for the rehabilitation and restoration of vegetation. Any person can participate in the drafting and preparation of the regional vegetation management plans, and the native vegetation codes of practice (Parts 3 and 4), and rural property owners can voluntarily enter into a property agreement (Part 5). To assist with the social acceptance of the legislation, there are a number of 'exempt situations' to allow 'routine' land use practices to take place without development consent, e.g. clearing native vegetation for fire emergency, noxious weed control, and operations undertaken to enable building of stock yards, farm buildings, dams, and fences.

FORENSIC ECOLOGY: CONTRAVENTION OF NATIVE VEGETATION LEGISLATION

Historical records show that illegal clearing has occurred in most major vegetation associations with a high number of instances in the dry woodland and grassland communities. These communities had been extensively used for pastoral agriculture, but, due to favourable topographic characteristics and good-quality soils, they are targeted for dryland cropping. Under arable agriculture these lands generally have a higher financial return than for sheep and cattle grazing. The main forms of illegal clearing in the coastal zone have been selective logging of wet sclerophyll forest and piecemeal clearing of already disturbed areas (initially for grazing) for rural residential development.

Criminal offence

There has been reasonably good success with criminal convictions for contravention of native vegetation law in New South Wales. Illegal vegetation clearing cases are heard before the New South Wales Land and Environment Court, a specialist Supreme Court. Of 23 major cases brought before the Court since 1996, 19 resulted in convictions, with a total monetary penalty of A$500 000. Three thousand hectares of land set are set aside under court order for long-term ecological restoration. All cases involved some form of settlement, to identify 'set-aside' lands for permanent conservation and restoration.

Scientific evidence

To establish a prima facie case of illegal clearing and bring it before the court involves rigorous and substantial scientific evaluation and documentation of physical and botanical characteristics. Other forms of testimony are also required to satisfy the legal test of 'beyond reasonable doubt'. This aspect of criminal law is now referred to as 'forensic ecology', where the main forms of evidence include quantitative field sampling data, oblique photography, remotely sensed data (aerial photographs, satellite imagery), and evidence on environmental harm and ecological loss. There is also the oral evidence of admissions and witness statements. Scientists who assemble the evidence are trained in criminal investigation techniques.

Field investigation

Legally and ecologically reliable methods to quantify and describe severely disturbed areas of vegetation have been developed (Dewar & Jones, 1997). Hundreds of field sites have been investigated, including areas where native vegetation has been destroyed or severely modified by bulldozing, pushing and stacking, ploughing, and burning the patches of vegetation. The range of communities and structural formations investigated include: dense chenopod shrubland, open coolabah woodland, tableland dry forest, coastal wet forest, depauperate rainforest, and coastal heathland. Remotely sensed imagery is interrogated to determine the pattern and sequence of clearing, density of vegetation removed and spatial area of removal. Given the high standard of proof required, a high degree of accuracy is needed to determine exact dates of clearing, determination of age and stage of vegetation and the biological effects of the methods used to clear, e.g. the application of herbicide. A number of the court cases have involved substantial argument on the validity of scientific evidence, indicating the ongoing uncertainty of the court's interpretation of certain aspects of the ecological sciences (see *Director-General Department of Land and Water Conservation* v. *Bungle Gully Pty Ltd* (1998) NSWLEC 6 (6 February (1998); *Evidence Act 1995*, Part 2.2 – Documents).

Expert evidence

A variety of expert scientists and social scientists provide evidence to prove illegal clearing 'beyond reasonable doubt' to minimise any possible defences, and to provide the court with a comprehensive description of the environmental harm caused by illegal land clearing (Lee *et al.*, 1998). Often, assistance is sought from experts from the National Parks and Wildlife Service and the Royal Botanic Gardens, and from ecological consultants employed by the state Crown Solicitor's Office. The expert evidence involves:

- The use of scientific method to establish the nature and extent of native vegetation loss, e.g. botanical surveys and descriptions.
- Establishing the nature and extent of environmental harm and ecological loss, particularly the impact on threatened flora and fauna populations and habitat loss.
- Proving the timing and extent of illegal clearing using satellite imagery.
- Economic analysis to establish financial gain from illegal clearing.

ENVIRONMENTAL HARM AND ECOLOGICAL LOSS

A key environmental outcome of illegal clearing investigations is the valuable scientific and ecological data on the harm caused to the environment from disturbance or removal of native vegetation, and the loss of environmental integrity from the damage to flora and fauna, particularly loss of habitat of rare, threatened and endangered species. Field sampling data is used to reconstruct profiles of the original ecosystem. It is verified through sampling and characterisation of adjacent, similar vegetation communities and examining habitat profiles and fauna records from species identified in the local area (Saunders *et al.*, 1993). Prediction models have been developed based on principal habitat components, and when applied to the reconstructed vegetation profile they indicate which threatened or endangered species may have occurred in the cleared vegetation (*Threatened Species Conservation Act 1995 Data Base*, 1998). Removal of patches of vegetation affects connectivity values and contributes to local and regional fragmentation and extinctions (Decker, 1993; see: Smith, in (unreported) *Director-General Department of Land and Water Conservation v. Prime Grain Pty Ltd and Limthono Pty Ltd* 1998 New South Wales Land and Environment Court 40, 4 March 1998, Talbot J; *(unreported) Director-General Department of Land and Water Conservation v. Orlando Farms*, New South Wales Land and Environment Court Nos. 50040, 50045, 50048 and 50051 of 1997, 19 May 1998, Lloyd J.) The cumulative effect of clearing increases soil erosion hazard, flooding, and dryland salinity by the raised local and regional water table (Benson, 1991).

LAND RESTORATION

A feature of native vegetation litigation has been the court's action to require the preparation of a vegetation rehabilitation management plan in accordance with the biodiversity principles specified by the court. The objective is to rehabilitate land, or permanently set land aside for long-term flora and fauna conservation. Orders associated with these cases contain legal conditions to:

- Exclude stock, or limit stock numbers at sustainable grazing levels.
- Establish and maintain wildlife conservation reserves.

- Establish and maintain fences to exclude grazing animals from the rehabilitation area.
- Control noxious weeds and feral animals.
- Control soil erosion.
- Replant native vegetation.
- Monitor the progress of rehabilitation periodically.

A benefit of the court-order process is that, through negotiation, valuable portions of a greater range of ecotypes than those affected by the illegal clearing are set aside. Set-aside lands have included ecologically valuable areas of chenopod shrubland; semi-arid wetland; riverine forest; open woodland; and brigalow community. This process makes a substantial contribution to the conservation and protection of flora and fauna at a local and regional level. It has also added valuable areas to conservation generally.

OTHER REMEDIAL MEASURES

There are civil actions in the *Native Vegetation Conservation Act 1997* to achieve ecological restoration. They may be used in conjunction with, or as an alternative to, criminal litigation. The civil actions include a legal order which can be invoked for the immediate cessation of an activity that is causing environmental harm, and a provision to invoke a remedial notice to restore the ecological values of the affected area (Sections 46–47 of the Act). These actions are applied to control soil degradation, revegetate land, replant, or encourage natural regeneration to restore threatened species habitat. Remedial tools have been used with good success on riverine and terrestrial environments. Other provisions in the Act which support the use of the criminal or civil actions for ecological restoration include:

- The specific power to obtain information about a possible offence.
- The authorisation of scientific and non-scientific officers lawfully to enter land to investigate native vegetation contraventions. A person who obstructs an authorised officer is guilty of an offence.

- The power to serve a remedial notice to a landholder without any prior notice.
- The provision to enable any person to bring proceedings in the Land and Environment Court for an order to remedy or restrain a breach of the Act.
- The provision to bring proceedings against a director of a corporation.

COMMENT

The introduction of the native vegetation conservation legislation in New South Wales has been significant for the conservation and restoration of native vegetation ecosystems. Conceptually, the *Native Vegetation Conservation Act 1997* is quite an advanced piece of environmental law, based on key global ecological principles, including the principle of ecological sustainability. The legislation allows a precautionary approach to be taken, and to provide for the long-term conservation of native vegetation ecology in southeastern Australia. The community participation provisions enable all parts of society to have input in biodiversity conservation. There are also mechanisms for the accountability of all decision-making (the advisory council, restraints provision, register of

applications and appeal provisions) to the community. After five years the Act will be reviewed (by law) and changes made to improve its administrative, technical and ecological effectiveness. The Act has a comprehensive range of compliance and enforcement provisions, enabling flexible, strategic management of illegal clearing when self-regulation fails. Despite many achievements, in the period since January 1998 it has been found that at least the following aspects of the Act could be amended to improve the overall performance of the native vegetation conservation law:

- Including measurable and enforceable scientific criteria as part of the ecological definitions.
- Rewriting the legal exemptions to be quantitatively measurable, able to be clearly understood by rural landholders and enforced by the agencies.
- Expanding the voluntary conservation provisions to include unilateral commitments, contracts, easements or servitudes, and management agreements, to increase the flexibility in deciding set-aside lands.
- Reforming the environmental assessment process to be more fully ecologically accountable.

specific financial incentives to encourage ecologically sustainable forest management on private land. In response to concerted community action (Watson, 1990; see e.g. *Jurasius v. Forestry Commission of New South Wales and Ors* (1990) 71 LGRA 79-107), there is now a greater emphasis on forest conservation, establishing reserves, undertaking scientific studies, biodiversity management and education. Legislative reform has given the principal environment management organisations (those with environmental assessment, environment protection, and land and water conservation responsibilities) a major legislative role in forest management decision-making. National forest policies now cover the role of forests in national ecological restoration, retaining more forested land under forest, and expanding natural forest on former agricultural land. These policies support the principal global policy to reduce greenhouse gases and improve biodiversity conservation (Rawson & Murphy, 2000). In Australia, Regional

Forest Agreements are agreements between the Commonwealth and state governments for a period of 20 years to establish ecologically sustainable management of whole forest estates. This reserve system safeguards biodiversity, old growth, wilderness and other natural and cultural values of forests. Forests outside the reserves will be available for wood production, subject to codes of practice that ensure long-term sustainability and contribute to the conservation of these natural and cultural values.

Biodiversity

There are no specific 'biodiversity acts' in Oceania but many laws have provisions which enable the 'frequency and variety of life in all its forms, levels and combinations' to be considered (de Klemm & Shine, 1993; Cunningham, 1996). These are the laws that promote the concept of ecological sustainability (e.g. New South Wales *Environmental Planning* and *Assessment Act 1979*, New Zealand *Resource Management Act 1991* – various Parts, but especially Part V, New South Wales *Native Vegetation Conservation Act 1997*), and apply conservation and ecologically sustainable land use standards in the environmental planning and assessment processes. The introduction of national biodiversity strategies (New Zealand Ministry for the Environment, 1997; New South Wales National Parks and Wildlife Service, 1999) has improved the recognition of the role of legislation in biodiversity conservation. Importantly, this legislation applies over agricultural landscapes, where the greatest biodiversity losses have occurred in the past, as well as to the conservation and management of protected and reserved areas. The provisions in the 'biodiversity' law to conserve and restore natural ecosystems include the environmental assessment processes; bilateral conservation agreement process; identifying and monitoring biodiversity; making bioregional plans; protecting threatened species and ecological communities; protecting habitat; establishing protected areas; and conducting environmental audits. The development of biological legal and institutional profiles has led to a greater use of the biodiversity concept in environmental law and policy reform (Preston, 1995; Glowka, 1998) (see Boxes 5.3 and 5.4).

Environmental planning and assessment

Environmental planning and assessment legislation protects the human and natural environment by prescribing environmental assessment standards for specific land use activities, and determining significant environmental impact from changes in land use. Most Australian states have specific environmental planning and assessment laws and there are similar provisions in the New Zealand law (Bates, 1992; Williams, 1997). The legislation protects the natural environment and has a very significant environmental policy development role, especially the development of regional and local environmental plans. Regional plans set broad rules, criteria and policies for land use allocation, conservation and environmental protection, and how they are applied in regional and local decision-making. The provision to make State Environmental Planning Policies (Part 2)

under the New South Wales *Environmental Planning and Assessment Act 1979* has been a very successful legal protection tool for many ecotypes, including wetlands, coastal littoral rainforest, and the habitat of some threatened fauna (see Mossop, 1992).

Protection of the environment

This regime of law contributes to ecological restoration by managing 'pollution' activities: protecting, restoring and enhancing environmental quality by reducing human health risks; and preventing environmental degradation by better waste management, reducing discharge of harmful substances, and reducing point source pollution, particularly air and water pollution (see New Zealand *Resource Management Act 1991*, New South Wales *Protection of the Environment Administration Act 1991* and the *Protection of the Environment Operations Act 1997*). Environment protection law produces strategies for environment protection standards, guidelines and protocols and puts in place the controls over the activities that impact on the environment, including technological or chemical processes that produce substances that impact on the environment (New Zealand *Biosecurity Act 1993*). Environment protection laws are managed by regulatory authorities with powers to conduct environmental audit, make inquiries, authorise restoration, and to remedy and restrain harm to the environment. They also undertake criminal prosecution. There is a growing trend to reformulate environment protection legislation under the concept of ecologically sustainable development, by constructing provisions which embellish the notions of the precautionary principle, intergenerational equity, and the conservation of biological diversity (see the New South Wales *Protection of the Environment Operations Act 1997*).

Native vegetation conservation

The specific laws and policies to conserve native vegetation were introduced in recognition of the decline and fragmentation of native vegetation ecosystems from land clearing and overgrazing (Dendy & Murray, 1996; Department of Land and Water

Conservation, 1997). The objective of legislation like the South Australian *Native Vegetation Act 1991*, New South Wales *Native Vegetation Conservation Act 1997* and Queensland *Vegetation Management Bill 1999*, is to protect vegetation with a high conservation value and ensure the ecological sustainability of vegetation communities, and encourage revegetation and rehabilitation of vegetation (Native Vegetation Advisory Council, 1999). These laws set detailed ecological standards to assess the biodiversity, habitat values, flora and fauna values, map vegetation, and assess threatened species. There are provisions to form vegetation scientific advisory bodies, prepare bioregional plans of management, enable public consultation, and to develop property plans for the voluntary protection of nature on private property (New Zealand Department of Conservation, 1997). Native vegetation laws have comprehensive powers for criminal enforcement and to impose civil notices to restore the land and native vegetation ecosystems (Bates & Franklin, 1999). A comprehensive outline of the New South Wales *Native Vegetation Conservation Act 1997* appears in Box 5.5. It is upheld as a reasonable model of biodiversity protection legislation because it contains many specific provisions to manage and restore ecosystems.

Reafforestation

This regime of legislation and policy encourages voluntary reafforestation of land and the establishment of native and non-native forest plantation for a wide range of environmental management purposes e.g. the Tasmanian *Private Forests Act 1994*. The New South Wales *Plantations and Reafforestation Act 1999* encourages plantation establishment as a land restoration measure, to control dryland salinity, soil degradation, and improve biodiversity management. The main operational aspect of this Act is its *Plantations and Reafforestation (Code) Regulation 2001*, an environmental standard to establish and manage plantation forest. This Act has a significant ecological restoration role by regulating replanting of trees or shrubs, soil erosion control, conservation of native animals and plants and their habitat, conserving threatened species, protection of indigenous sites, and fire management.

Pastoral land management

In Australia and New Zealand the pastoral land law regime presides over a vast area of rural land under leasehold tenancy. It controls the land management activities that deplete vegetation and cause land degradation. Legal conditions are applied to leases to limit stock grazing levels, outline vegetation management and soil management conditions, and define standards for monitoring the condition of the pastoral land and to prevent land degradation (see the South Australian *Pastoral Land Management Act 1989*) (Hannam, 2001*b*).

Carbon rights legislation and the environment

This new area of environmental law is based on the important ecological link between environmental restoration and control of greenhouse gases (Rawson & Murphy, 2000; Hannam, 2001*a*). The *Carbon Rights Legislation Amendment Act 1998* of New South Wales, believed to be the first 'carbon sequestration' law in the world, amends the *Conveyancing Act 1919* and *Forestry Act 1916*, enabling the acquisition and trading in the legal right of carbon sequestrated by trees and forests. It allows land management agencies to acquire and trade in such rights, with a fundamental ecological restoration role. A covenant that gives access to, or the maintenance of trees or forests of any carbon sequestration right manages these rights. A legal agreement provides for vegetation to be owned on land the subject of a right, and to be vested in that person. This law separates ownership of stored carbon from ownership of land or trees, allowing forest plantation owners to sell 'carbon rights' separate from 'timber rights', opening the way for a market in stored carbon, and ultimately, the future creation of carbon credit schemes.

CONCLUDING REMARKS AND FUTURE CONSIDERATIONS

A feature of the development of environmental law and policy in Oceania is its continual improvement to adapt to the land use pressures which humans place on the environment, and the outcome of these pressures. This chapter has shown that there are many environmental law regimes in the region to manage and restore ecological harmony. However, it is prudent for nations in the region to review periodically their legislative structures and take the necessary steps to improve these laws and supportive policies to more capably manage ecological restoration. Some of the improvements that can be made include using provisions to:

- Enable governments, landowners and the community to work together to achieve ecological restoration by sharing responsibility between these groups.
- Enable the development of policy, guidelines and ecological standards, land use codes of practice and indicators of the ecological limits of land use.
- Enable ecological restoration to be achieved through a mix of regulatory, part regulatory, and non-regulatory means, incentive programmes, and specialist advisory groups.
- Control unsustainable land use practices and take enforcement action.
- Enable conservation and management of natural resources in an integrated way.
- Protect biodiversity and ecological values.
- Apply a geographic perspective to ecological restoration including development of national, regional and local biogeographic plans.
- Encourage voluntary conservation through incentives which are based on ecologically sustainable land management (Australia, 1996b).

Bioregional plans

Regional plans and strategies can be prepared for one, or a number of ecosystems and set out administrative, policy, ecological, land capability, land use and sociological information (Farrier, 1999). The statutory plan should divide a region into a series of land units according to their ecological capabilities. Land management policies, guidelines and acceptable standards of land use may be developed for each land unit, including the circumstances where an environmental impact assessment may be required (Bosch & Booysen, 1992). A regional plan may specify when particular land use activities (e.g. use of fire) are prohibited altogether, or prohibited beyond a specified level.

Land management codes

A land management code is a self-regulation system that defines the ecological capabilities of land use in the form of specific land management conditions and guidelines. They are applicable for ongoing routine land use. A code also denotes land use circumstances where an approval or a licence based on an environmental assessment may be required.

Voluntary agreements

This is a voluntary system initiated by a landholder to enter into a formal agreement with a local agency for a prescribed course of management over a specified time period. As a 'two-way' system, if the landholder agrees to specific conservation and land management practices, including cessation of agricultural use in some areas, the landholder could be eligible for financial and technical support (e.g. training for land managers in land assessment and biodiversity surveys). Property agreements offer one of the best mechanisms for improving land management. It brings parties together with a common commitment and a common community goal to rehabilitate land by reducing land degradation and protecting biodiversity.

Environmental planning and assessment

The legislation may include an approval system for activities with a potentially high environmental impact, involving a major change or an intensification in land use, e.g. clearing extensive areas of native vegetation, or changing the direction of surface water flow. These controls would be defined in a bioregional plan or under a separate development control system (Farrier, 1999). In some circumstances, an environmental assessment, an environmental impact statement or threatened species impact statement (e.g. threatened species legislation) may be required.

Conservation areas and ecological reserves

Conservation easements provide a flexible means of achieving biodiversity protection from lessees or landholders (Bates, 1992). Regional analysis can pinpoint significant biodiversity conservation areas, and land with high conservation values could be targeted. General ecological criteria for selecting conservation reserves can be established in a bioregional plan. The aim of establishing conservation areas and ecological reserves is to protect an adequate representation of each ecological community (West, 1993). Availability of incentives, such as land rate relief and financial and land management support under a property agreement, will increase the attractiveness of elective conservation.

The nations of Oceania have identified the ecological issues facing the region and responded with environmental laws and policies to meet these various problems. The numerous international conventions and national environmental laws that are in place give recognition of the impact that such ecological issues can have on the biodiversity of the region. The *Native Vegetation Conservation Act 1997* model (Box 5.5) is one example of legislation which brings in a variety of provisions which enable the conservation and restoration of ecosystems to be approached through public participation and enforcement. The existing laws and policies need to be reviewed in light of their practical implementation and effectiveness. Provisions that build on ecologically sustainable development and provide incentives are encouraging. At a national level, the smaller nations of Oceania need to adopt and incorporate environmental laws to more capably manage ecological restoration. Such laws need to be enforceable and protect biodiversity and ecological values. Nations need to be aware of the continuing pressure humans place on the environment, and respond to these pressures by adapting laws and policies to incorporate ecological restoration across the region. Nations have a basic obligation to ensure the diversity of nature in the region is retained for future generations.

REFERENCES

Australia (1983). *National Conservation Strategy for Australia: Living Resource Conservation for Sustainable Development.* Canberra: Australian Government Publishing Service.

Australia (1992). *National Strategy for Ecologically Sustainable Development.* Canberra: Australian Government Publishing Service.

Australia (1996a). *State of the Environment Australia.* Collingwood: CSIRO Publishing.

Australia (1996b). *The National Strategy for the Conservation of Australia's Biological Diversity.* Canberra: Department of Environment, Sport and Territories.

Bates, G. M. (1992). *Environmental Law in Australia.* Sydney: Butterworth.

Bates, G. M. & Franklin, N. (1999). *Compliance with the Native Vegetation Conservation Act 1997.* Sydney: Australian Centre for Environmental Law, University of Sydney.

Benson, J. (1991). The effect of 200 years of European settlement on the vegetation and flora of New South Wales. *Cunninghamia,* **29**, 343–501.

Benson, J. & Redpath, P. (1997). The nature of pre-European native vegetation in south-eastern Australia: a critique of Ryan, D. G., Ryan, J. R. & Starr, B. J. (1995) *The Early Australian Landscape: Observations of Explorers and Early Settlers. Cunninghamia,* **5**, 285–328.

Boer, B. W. (1995). Institutionalising ecologically sustainable development: the roles of national, state, and local governments in translating grand strategy into action. *Willamette Law Review,* **2**, 307–358.

Boer, B. W. & Fowler, R. J. (1996). *The Management of World Heritage in Australia,* Part II. Canberra: Department of Environment, Sport and Territories.

Boer, B. W, Ramsay, R. & Rothwell, D. R. (1998). *International Law in the Asia Pacific.* London: Kluwer Law International.

Bonyhady, T. (ed.) (1992). *Environmental Protection and Legal Change.* Sydney: The Federation Press.

Bosch, O. J. & Booysen, J. (1992). An integrative approach to rangeland condition and capability assessment. *Journal of Rangeland Management,* **45**, 116–122.

Crawford, J. (1991). The constitution and the environment. *Sydney Law Review,* **13**, 11.

Crengle, D. (1993). *Taking into Account the Principle of the Treaty of Waitangi: Ideas for the Implementation of Section 8 of the Resource Management Act 1991.* Wellington, New Zealand: Ministry for the Environment.

Cunningham, N. (1996). Biodiversity: economic incentives and legal instruments. In *Environmental Outlook, no 2, Law and Policy,* eds. B. W. Boer, R. Fowler & N. Cunningham. Sydney: The Federation Press.

de Klemm, C. & Shine, C. (1993). *Biological Diversity Conservation and the Law: Legal Mechanisms for Conserving Species and Ecosystems.* Environmental Policy and Law

Paper no 29. Gland, Switzerland: IUCN and World Conservation Union.

Decker, W. (1993). Fragmented landscapes: is our wildlife in danger? *ECOS*, **77**, 30–34.

Dendy, T. & Murray, J. (eds.) (1996). *From Conflict to Conservation: Native Vegetation Management in Australia: A Focus on the South Australian Program and Other Initiatives*. Adelaide: South Australian Department of Environment and Natural Resources.

Department of the Environment (1998). *Reform of Commonwealth Environment Legislation: Consultation Paper*. Canberra: Department of the Environment, Commonwealth of Australia.

Department of Land and Water Conservation (1997). *A Proposed Model for Native Vegetation Conservation in New South Wales: A White Paper*. Sydney: Department of Land and Water Conservation.

Department of Urban Affairs and Planning (1997). *Lord Howe Island Development Control Plan: Setbacks, Site Coverage and Landscaping*. Sydney: Department of Urban Affairs and Planning.

Dewar, R. & Jones, H. (1997). *Sampling Design for Compliance Officers*. Compliance workshop 1997. Sydney: Department of Land and Water Conservation.

Dregne, H. E. & Chou, N. T. (1994). Global desertification dimensions and cost. In *Degradation and Restoration of Arid Lands*, ed. H. E. Dregne, pp. 249–282. Lubbock, TX: Texas Technical University.

Environment Protection Authority (New South Wales) (1997). *New South Wales State of the Environment*. Sydney: Environment Protection Authority.

Farrier, D. (1991). Vegetation conservation: the planning system as a vehicle for the regulation of broadacre agricultural land clearing. *Melbourne University Law Review*, **8**, 26–59.

Farrier, D. (1999). *The Environmental Law Handbook*. Melbourne: Redfern Legal Centre Publishing.

Flannery, T. (1994). *The Future Eaters: An Ecological History of the Australasian Lands and People*. Melbourne: Reed Books.

Gardner, A. (1994). Development of norms of land management in Australia. *Australasian Journal of Natural Resources Law and Policy*, **1**, 127–166.

Gillespie, A. (1995). Common property, private property and equity: clash of values and the quest to preserve biodiversity. *Environmental Planning and Law Journal*, **12**, 388.

Glanznig, A. & Kennedy, M. (2001). Land degradation and native vegetation clearance in the 1990s: Addressing biodiversity in Australia. In *Response to Land Degradation*, eds. E. M. Bridges, I. D. Hannam, L. R. Oldeman, F. Penning deVries, S. J. Scherr & S. Sombatpanit, pp. 395–403, Enfield, NH: Science Publishers Inc.

Glowka, L. (in collaboration with Shine, S., Sanroa, O., Farce, M. & Gunning, L.) (1998). *A Guide to Undertaking Biodiversity Legal and Institutional Profiles*. Environmental Law and Policy Paper no. 35. Gland Switzerland: IUCN (in collaboration with IUCN Environmental Law Centre, Bonn, Germany).

Hannam, I. D., (2001a). Global view of the law and policy to manage land degradation. In *Response to Land Degradation*, eds. E. M. Bridges, I. D. Hannam, L. R. Oldeman, F. Penning deVries, S. J. Scherr & S. Sombatpanit, pp. 385–394, Enfield, NH: Science Publishers Inc.

Hannam, I. D. (2001b). Policy and law for rangeland conservation. In *Rangeland Desertification*, eds. O. Arnalds & S. Archer, pp. 165–179. Dordrecht: Kluwer Academic Publishers.

IUCN (1980). *World Conservation Strategy: Living Resource Conservation for Sustainable Development*. Gland, Switzerland: IUCN.

IUCN (1991). *Caring for the Earth: A Strategy for Sustainable Living*. Gland, Switzerland: IUCN, United Nations Environment Program and World Wildlife Fund.

Krockenberger, M. (1998). Falling down: land clearing in Australia. *Habitat*, **26**, 13–20.

Kyoto (1997). *The Kyoto Protocol to the Convention on Climate Change*. Bonn, Germany: Climate Change Secretariat, UNEP.

Lee, E., Baird, M. & Lloyd, I. (1998). Commentary: State Environmental Planning Policy no. 46 – Protection and management of native vegetation. *Environmental Planning and Law Journal*, **15**, 127–135.

Malanczuk, P. (1995). Sustainable development: some critical thoughts in the light of the Rio Conference. In *Sustainable Development and Good Governance*, eds. K. Ginther, E. Denter & P. J. I. M. de Waart, pp. 23–52. Dordrecht: Martinus Nijhoff.

McBride, T. (1997). Country report: New Zealand. *Asia Pacific Journal of Environmental Law*, **2**, 159–182.

McTainsh, G. H. & Boughton, W. C. (eds.) (1993). *Land Degradation Processes in Australia*. Melbourne: Longman.

Miller, K. R. & Lanou, S. M. (1995). *National Biodiversity Planning: Guidelines Based on Early Experiences around the World*. Baltimore, MD: IUCN and World Resources Institute.

Mossop, D. (1992). Coastal wetland protection. *Environmental Planning and Law Journal*, **9**, 331–359.

Native Vegetation Advisory Council (1999). *Towards a Native Vegetation Conservation Strategy*. Sydney: Department of Land and Water Conservation.

New South Wales National Parks and Wildlife Service (1999). *New South Wales Biodiversity Strategy*. Sydney: New South Wales National Parks and Wildlife Service.

New Zealand Department of Conservation (1997). *Voluntary Protection of Nature on Private Property*. Wellington, New Zealand: New Zealand Ministry for the Environment.

New Zealand Ministry for the Environment (1996). *Sustainable Land Management: A Strategy for New Zealand*. Wellington, New Zealand: New Zealand Ministry for the Environment.

New Zealand Ministry for the Environment (1997). *The State of New Zealand's Environment 1997: Our Biodiversity*. Wellington, New Zealand: New Zealand Ministry for the Environment.

Preston, B. (1995). The role of law in the protection of biological diversity in the Asia–Pacific region. *Environmental Planning and Law Journal*, **12**, 264–277.

Pulea, M. (1999). Country report: Fiji. *Asia Pacific Journal of Environmental Law*, **4**, 61–68.

Ramsay, R. (1997). Country reports: Australia. *Asia Pacific Journal of Environmental Law*, **2**, 153–158.

Rawson, A. & Murphy, B. (2000). *The Greenhouse Effect, Climate Change and Native Vegetation*. Background Paper of the Native Vegetation Advisory Council of New South Wales no 7. Sydney: Department of Land and Water Conservation.

Richardson, B. (1996). Conservation of wetlands in Fiji: reconciling customary resource use with government regulation. *Asia Pacific Journal of Environmental Law*, **1**, 47–77.

Richardson, B. & Palmer, K. (2000). The emerging 'citizenship' discourse in environmental law: a New Zealand perspective. *Environmental Planning and Law Journal*, **17**, 99–117.

Robinson, D. (1992). Endangered species protection and environmental management in the Solomon Islands. *Environmental Planning and Law Journal*, **9**, 51–59.

Salt, D. (1998). The lucky country's luck is running out. *Australasian Science*, **19**, 21–22.

Saunders, D. A., Hobbs, R. J. & Ehrlich, P. R. (eds.) (1993). *Nature Conservation* vol. 3, *The Reconstruction of Fragmented Ecosystems: Global and Regional Perspectives*. Chipping Norton, NSW: Surrey Beatty.

Stratford, E. (1993). Ideology, environment and legislation: South Australian attitudes to vegetation. *Australian Geographical Studies*, **31**, 14–25.

Swanson, T. (1997). *Global Action for Biodiversity: An International Framework for Implementing the Convention on Biological Diversity*. London: Earthscan Publications in association with IUCN.

Tarasofsky, R. (1995). *The International Forest Regime: Legal and Policy Issues*. Gland, Switzerland: IUCN, World Conservation Union and World Wide Fund for Nature.

United Nations (1982). *World Charter for Nature*. Nairobi, Kenya: United Nations Environment Program.

United Nations (1992*a*). *The Rio Declaration on Environment and Development*. New York: United Nations.

United Nations (1992*b*). *Agenda 21*. New York: United Nations.

United Nations (1992*c*). *Convention on Biological Diversity*. Nairobi, Kenya: United Nations Environment Program.

United Nations (1995). *United Nations Framework Convention on Climate Change*. Nairobi, Kenya: United Nations Environment Program.

United Nations (1996). *Handbook of Environmental Law*. Nairobi, Kenya: United Nations Environment Program.

World Commission on Environment and Development (WCED) (1987). *Our Common Future*. Oxford: Oxford University Press.

Watson, I. (1990). *Fighting Over the Forests*. Sydney: Allen & Unwin.

West, N. E. (1993). Biodiversity of rangelands. *Journal of Rangeland Management*, **46**, 2–13.

Williams, D. D. (1997). *Environmental and Resource Management Law in New Zealand*. Wellington, New Zealand: Butterworth.

Part 2 • The biomes

6 • Marine and coastal ecosystems

STEPHEN J. HAWKINS, JANETTE R. ALLEN, PAULINE M. ROSS AND MARTIN J. GENNER

INTRODUCTION

Few marine ecosystems can be considered pristine. Even the open ocean has been subject to a wholesale reduction of animals at the top of the food web such as sharks, tuna, syrenians, seals and whales. Mineral extraction (e.g. hydrocarbons and manganese nodules) is creeping into deeper and deeper water as demand for resources in short supply increases and technology is developed to supply that demand. Impacts markedly increase in shelf and coastal waters where human activity is greatest, leading to considerable degradation and modification of habitats as well as exploitation of many species. Clearly, intervention is required if marine and coastal ecosystems are to continue to function with some semblance of natural processes.

Restoration of marine ecosystems has lagged behind that of terrestrial and freshwater ecosystems. In part this is because of their extensive nature and common ownership which often precludes effective intervention. In recent years, however, an impressive body of work has developed, mainly on coastal systems, particularly those dominated by particular habitat-providing biotas such as coral reefs, seagrass beds, mangroves and saltmarshes. Less work has been done on other marine habitats. Key publications on recovery, restoration or rehabilitation (see Frid & Clark, 1999; Hawkins et al., 1999a for definitions) of marine and coastal ecosystems include various papers in Thayer (1992) and Frid & Clark (1999).

We begin this chapter by summarising the major impacts on marine ecosystems. Then we examine the key features of marine ecosystems influencing their potential for recovery and restoration. We focus upon case studies from rocky shores and the disused docks of industrialised estuaries of which

we have direct experience, and review experiences in the rehabilitation/restoration of mangrove systems (see also Cintron-Molero, 1992; Ellison, 2000; Imbert et al., 2000; Lewis, 2000). Other marine systems, not dealt with elsewhere in this volume, are briefly covered in order to provide the context for the chapters elsewhere in this volume on seagrasses, saltmarshes and coral reefs. We compare open coastal, more enclosed estuarine systems and the special case of mangroves. This chapter draws largely on previous reviews (Hawkins & Southward, 1992; Hawkins et al., 1992; Hawkins et al., 1999a) with some updating and broadening of material. We aim to indicate some successful examples of restoration, but also to be realistic and indicate where rehabilitation or recovery is the most cost-effective approach or the only viable option. Some speculations on what may be done in the future are also given.

IMPACTS AND DEGRADATION OF MARINE ECOSYSTEMS

The largest scale of degradation is that of system-wide disruption. Offshore examples include eutrophication of regional seas such as the Baltic, Adriatic, North and Irish Seas (Ryther & Dunstan, 1971; Bonsorff et al., 1997; Allen et al., 1998; Schiewer, 1998) and disruption by the direct and indirect effects of fishing activities (Jennings & Kaiser, 1998; Hall, 1999). Figure 6.1 shows how fishing activities extended throughout the North Sea in the nineteenth century (from Jennings & Kaiser, 1998). On the coast more localised impacts generally occur. These can include: heavy exploitation of near-shore fish, shellfish and bait collection (e.g. Meehan, 1982;

121

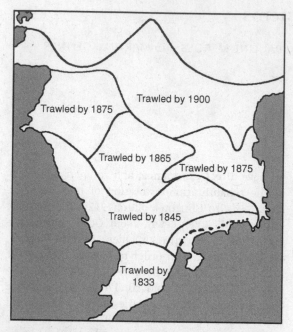

Fig. 6.1. Spread of fishing activities in the North Sea in the nineteenth century. From Jennings & Kaiser (1998).

Newton *et al.*, 1993); localised pollution of estuaries and lagoons including eutrophication (e.g. Castel *et al.*, 1996; Morand & Briand, 1996); intensive aquaculture (e.g. Kaspar *et al.*, 1985; Talbot & Hole, 1994); extraction of building materials from beaches, offshore sediments and reefs (e.g. Kenny & Rees, 1994); coastal defence schemes (e.g. Wolff, 1997); land reclamation and infilling (e.g. McLusky *et al.*, 1991); dumping of mine and dredge spoil (e.g. Somerfield *et al.*, 1995); heavy recreational usage (e.g. Schiel & Taylor, 1999); and introduced (e.g. *Sargassum:* Farnham *et al.*, 1973) and artificial species (e.g. *Spartina anglica*: Gray *et al.*, 1991) (general reviews in Raffaelli & Hawkins, 1996; R. B. Clark, 1997; Jennings & Kaiser, 1998; Hall, 1999; Hawkins *et al.*, 1999*b*; Crowe *et al.*, 2000).

On many densely populated coastlines very little of the coast is natural, and localised piecemeal degradation can scale up to coast-wide disruption. The essentially artificial coastline of dykes and polders of most of the Netherlands is perhaps the most extreme example (Wolff, 1997). Large areas of coastline elsewhere in the world are heavily modified by engineering works to various degrees – especially depositing shores and estuaries. These works include promenades, piers, port installations (breakwaters, wharves, quays, jetties and docks) and various sea defences such as sea walls, groynes and offshore breakwaters. Human predation of invertebrates (e.g. mussels, sea urchins, winkles, limpets, stalked barnacles, crabs) and collection of seaweed is very common worldwide, with shores throughout Asia, Africa and South America being subject to intense collecting, much at a subsistence level (see Castilla, 1999 for review). The same considerations apply to much of the coast of France, Spain and Portugal and the European Mediterranean and the Atlantic Islands (Azores, Canaries, Madeira). Additionally, coastal systems are impacted by acute incidents such as oil spills (e.g. Southward & Southward, 1978) and toxic algal blooms (Southgate *et al.*, 1984). The latter are thought in many cases to be prompted by eutrophication (e.g. Smayda, 1990) but can also be a strictly natural phenomenon (Cloern, 1996).

Chronic pollution is also ubiquitous, including localised sewage and industrial discharges plus inputs from diffuse sources including the atmosphere and road runoff (Castel *et al.*, 1996). Persistent organic compounds can find their way up the food web and bio-accumulation has been widely shown for heavy metals (Bryan & Langston, 1992). Although effects of these pollutants have been shown at the molecular, cellular and individual levels, few good examples exist of population or community level effects (but see Millward & Grant, 1995, for nematodes). Tributyl tin (TBT) leached from antifouling paints has been shown to have widespread effects on many species (Bryan & Gibbs, 1991). *Nucella lapillus* and other muricid neogastropods have been particularly susceptible with disruption of hormonal systems (Spooner *et al.*, 1991) leading to masculinisation and sterility of females (Gibbs & Bryan, 1986; Ellis & Pattisina, 1990) and widespread reduction in numbers (Hawkins *et al.*, 1994; Evans *et al.*, 1996).

RECOVERY, RESTORATION AND REHABILITATION IN PRACTICE

Definitions and relevant features

It is worth briefly considering terms and definitions in the light of the special features of marine ecosystems. In the rest of this volume natural recovery is seen as the most appropriate mechanism for restoration. Rehabilitation is usually considered a pragmatic option that enables partial return to a specified state rather than a complete return to pre-impact natural conditions. This approach is useful where baseline natural conditions are not known or are highly dynamic (M. J. Clark, 1997).

Marine ecosystems, including many of those of the coastal zone, are largely open. Most of the species recruit from remote sources via widely dispersed propagules or larvae. Additionally, succession is usually rapid. Thus there is considerable potential for natural recovery by recolonisation from un-impacted populations. In many cases there is little physical alteration of the habitat during succession, particularly on rocky substrates and in coarser more mobile sediments. Thus while recovery is rapid, opportunities for active restoration are limited. The openness of most marine systems usually prevents manipulation to improve water quality. The only actions that can be taken are via globally enacted international conventions limiting harmful inputs and activities.

Even restoration of open systems can be attempted, however, where biologically generated habitat modification occurs. For example, increased biologically generated structure can be created by large macrophytes in shallow water. Rooted macrophytes such as seagrasses, saltmarsh vegetation and mangrove trees can markedly alter the hydrodynamic and sedimentary characteristics of whole systems (Furukawa et al., 1997). They also provide surfaces for colonisation and refuges for more mobile species. Dispersal of these rooted macrophytes can be very limited (Orth et al., 1994), hence the need to encourage colonisation by active planting. Kelp beds can also modify the environment but to a lesser extent (Renaud et al., 1997). Kelps do, however, provide additional habitat for a diverse range of epiphytic species (Lambert et al., 1992).

Coral reefs are the ultimate example of biogenic carbonate structures created entirely by the corals and associated crustose algae (Sammarco, 1996). More modest biogenic structures include mussel beds and oyster reefs, reefs built by worms (e.g. Sabellaria), and dense sheets of barnacles. Many

of these 'ecosystem engineers' are exploited (e.g. kelps, mussels) or impacted by recreational use of the foreshore. In sediments, bioturbation can influence both the structure and the chemistry of the sediments (Jones & Jago, 1993).

Enclosed waters such as estuaries, bays and lagoons can be amenable to restoration by manipulation of water quality. The physical environment can be manipulated by altering flushing regimes and by artificial mixing. Thus, 'bottom–up' forcing processes can be altered. The benthos can exert considerable influence on the water column in enclosed or semi-enclosed bays, estuaries or lagoons. Macrophytes can sequester nutrients. The phytoplankton standing crop may be influenced by filter feeders in a 'top–down' manner (Officer et al., 1982).

Another feature of many coastal ecosystems is 'keystone species', especially on rocky shores (Paine, 1994). In recent years, however, this concept and its ubiquity have been questioned (Menge et al., 1994; Raffaelli & Hawkins, 1996). In the northeast Atlantic, limpets of the genus Patella are probably a keystone grazing species on the mid levels of all but sheltered shores (Southward & Southward, 1978; Hawkins et al., 1992). Their removal leads to a massive proliferation of algae. Clearly restoration should concentrate on keystone species – if and when they occur and have been identified. This is particularly the case if they are the subject of targeted fisheries (e.g. whelks in Chile: Castilla & Duran, 1985).

Open marine and coastal ecosystems

Rocky coasts

Although not as severely impacted as some coastal habitats, rocky shores are subject to a variety of impacts (reviewed in Hawkins & Southward, 1992; Crowe et al., 2000). Rocky shores are also very open systems with large imports and exports of both material (largely detritus) and propagules (larvae, spores, etc.).

Kelp beds

In the immediate subtidal zone of rocky shores considerable effort has been given to restoring kelp beds in areas where they are harvested for commercial purposes or to enhance fisheries. Much of this work has been carried out in California (see Schiel & Foster, 1992 for review). Schiel & Foster point out the variable and often uncertain level of success of many restoration schemes. This is in part due to complex interaction of broadscale natural events (e.g. El Niño: Dayton & Tegner, 1984) and unnatural processes (e.g. widespread reduction in sea otters (*Enhydra lutris*): Tegner, 1989, and extensive fishing: Tegner & Levin, 1983). Broadscale processes can be coupled with more localised impacts such as changes in sedimentation regimes due to coastal construction, urban runoff and pollution inputs, particularly sewage, all of which may influence changes in kelp distribution and abundance in time and space.

Two main approaches to kelp restoration have been used. There have been attempts to increase or restore hard substrate suitable for kelp growth. Measures have also been taken to reduce the main grazer (sea urchins), which can proliferate leading to extensive barren areas. Both approaches have been coupled with attempts to reseed with adults or juvenile kelp. A novel approach has been to use plastic imitation kelp plants that sweep the rock surface and remove urchins, thereby allowing juvenile plants to establish (Vasquez & McPeak, 1998). In addition, other components of the kelp forest system have been restocked, ranging from suspected top predators (sea otters, e.g. Estes & Duggins, 1995) to species of commercial interest (abalones *Haliotis rufescens*).

On a localised scale, reduction of grazers has proved effective but labour intensive. Up to 50 hectares was restored by Kelco at Point Loma in 1981 (Schiel & Foster, 1992). Artificial reefs have been successfully used as habitat for kelp in localities close to existing stands which can supply propagules, but not when isolated. Anchor-like devices bearing attached kelp plants have been introduced into areas where hard substrate nuclei had been lost.

The success of kelp restoration schemes has been difficult to judge because of the design of the trials which, as Schiel & Foster (1992) point out, are often unreplicated and do not have control areas. Therefore, it is difficult to show whether restorative techniques have contributed to improvement or whether broadscale natural recovery has happened. Even where successful it has been very costly. Box 6.1 highlights some of the problems caused by natural fluctuations in determining baselines against which to set restoration targets and judge success.

Box 6.1   Problems in restoring fluctuating marine systems: kelp bed examples

A major problem in defining objectives for the restoration of marine ecosystems is their tendency to fluctuate. Kelp beds are a fine example: temporal variability can be forced by variation in the physical environment, or by biological factors such as changes in populations of important consumers.

Work in California has shown how kelp beds shrink and grow in relation to large-scale fluctuations in seawater temperature. These are caused by changes in the regional upwelling regime, in turn driven by hemispheric El Niño events (see Fig. B6.1a–c). Decreases in kelp beds could be attributed to a variety of factors including pollution, increased coastal sediment, or decrease in keystone predators such as sea otters leading to increases of grazing sea urchins. Incorrect restorative action could be taken if the fluctuations are caused by extrinsic bottom–up forcing factors such as temperature and nutrient regimes.

Cyclical changes in population density of grazing sea urchins can lead to extensive kelp-free barren grounds. This has been reported in Canada (Breen & Mann, 1976a, b; see Vadas & Elner, 1992 for critique) and more recently in Norway (Hagen, 1995; Stein *et al.*, 1998). Population explosions have been attributed to lack of predators but are more likely to be caused by massive stochastic recruitment events. Eventually the urchin population crashes due to starvation, disease or some combination of both and kelps reappear. These fluctuations are driven by variation in top–down grazing control. Both examples illustrate the problem in marine and coastal restoration: what is the baseline or target condition to aim for in restoration? Without long-term data sets and in-parallel experiments to understand system function, it would be easy to put in place incorrect restorative actions. Manipulation is possible with top–down controlled systems, but little can be done about El Niño-scale events.

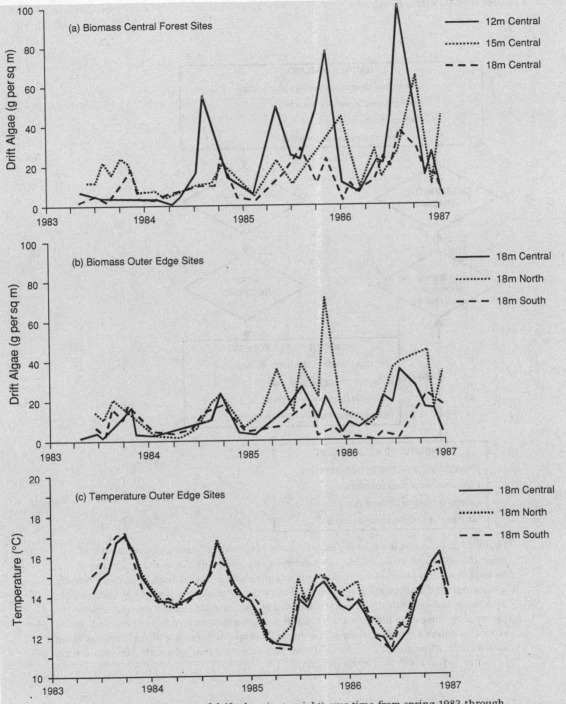

Figure B6.1. (a, b) Biomass of drift algae (wet weight) over time from spring 1983 through to the end of 1986. Samples were collected from 400 m² per site at approximately monthly intervals. (a) Central forest sites, (b) outer edge sites, (c) bottom temperatures from the three outer edge sites of the kelp forest, 18 m S, 18 m central and 18 m N, 1983 to 1987. 18 m N is significantly warmer than the other two sites; the former two sites were not significantly different. From Tegner & Dayton (1991).

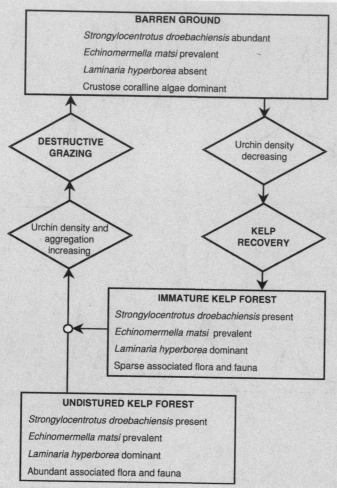

Figure B6.2. Processes involved in fluctuations in abundance of urchin (*Strongylocentrotus droebachiensis*), parasitic nematode (*Echinomermella matsi*) and kelp (*Laminaria hyperborea*). Scenario based on observed sea urchin outbreak dynamics at Værøy Island, northern Norway: a barren ground configuration is established after the initial destruction of the undisturbed kelp forest. Subsequent kelp recovery is followed by recurrent destructive grazing which prematurely interrupts the succession towards a mature, undisturbed kelp forest, and the ecosystem enters a cylical domain which precludes proper termination of the outbreak. From Hagen (1995). More recently Stein *et al.* (1998) have suggested that nematode infection is likely to be less important than initially thought.

Collection of key consumers

Collection of food and bait can severely disrupt rocky shore communities. Abundance of target species is reduced and population structure is distorted to smaller size classes. Non-target species are also affected: species growing in or amongst structuring species such as mussels are also removed. Removal of grazers or predators will also have knock-on effects on other species as their prey proliferate. Hockey & Bosman (1986) showed, using multivariate methods, that exploited shores converged towards a common degraded state whilst unexploited shores in reserves tended to show site-specific differences. At the degraded sites many ephemeral species were present. The dramatic influence of human impact on rocky shore communities has been further demonstrated in Chile (Moreno *et al.*, 1984; Duran & Castilla, 1989; reviewed in Paine, 1994), where changes in reserves following complete bans on collecting were charted (see Box 6.2).

Recovery from pollution

Rocky coastlines are impacted by acute pollution incidents and more chronic contamination from diffuse sources. Here we outline catastrophic oil spills, using one of the longest studied, the *Torrey Canyon* spill, as a case study (see Box 6.3) and the insidious effects of leachates from antifouling paints.

The *Torrey Canyon* spill provides an example of recovery rather than restoration – although recovery

Box 6.2   Heavily exploited rocky shores

Excellent work on the effects of human predation on shore communities has been done in Chile using large-scale experiments (see reviews by Paine, 1994; Castilla, 1999). Moreno *et al.* (1984) excluded shoreline harvesters (*Mariscadores*) from a 6000-m² reserve in Chile. There, people traditionally harvested algae and marine invertebrates for both local consumption and export. Limpets (*Fissurella picta*) rapidly recovered in numbers (see Fig. B6.3); in time this led to a reduction in algae. The causal link between grazers and algal

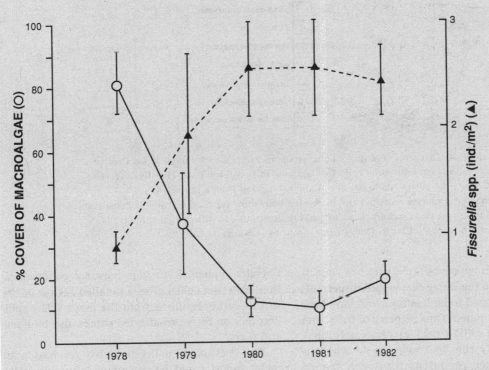

Figure B6.3. Changes in the percent cover of macroalgae (mean ± S.D., *n* = 3) following human exclusion beginning in May 1978 at Mehiun, Chile. Modified from Paine (1994) after Moreno *et al.* (1984), from Hawkins *et al.* (1999a).

abundance was confirmed by parallel experimentation (Jara & Moreno, 1984). This work was taken further in 1982 with 500 m of waterfront being designated as a human exclusion zone, with control areas (Duran & Castilla, 1989). Once people were excluded, the predator *Concholepas concholepas* and two grazing limpets increased in abundance. Mussels (*Perumytilus purpuratus*) declined, allowing algae to increase.

This was a transient intermediate state as algae were subsequently eliminated by grazers, leading to barnacle domination (see Fig. B6.4). Paine (1994) points out that the effect of human exploitation occurs on a landscape scale. The pattern of zonation, both pre-exclusion and in the controls, is very widespread (20° S to 44° S) along the whole coast of Chile.

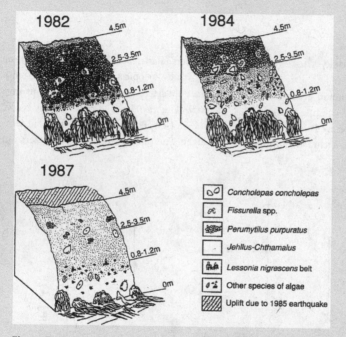

Figure B6.4. Diagrams of ecological change on an exposed rocky shore at Las Cruces, Chile, following exclusion of collectors initiated in December 1982. The 1982 overview represents the before condition; the gradual transition from a mussel- to a barnacle-dominated shoreline and increasing body dimensions of the mobile consumers (*Fissurella* spp. and *Concholepas concholepas*) are illustrated. Modified from Paine (1994), based on Duran & Castilla (1989), from review by Hawkins *et al.* (1999a).

was hindered by inappropriate clean-up (see Box 6.3). Fortunately, since the *Torrey Canyon* spill dispersants have been developed to be less toxic and have been more sparingly applied. Thus recovery of the shores after the *Braer* oil spill in 1993 was reasonably rapid, although possibly this had more to do with the prevailing weather conditions than with the application of chemical dispersants (Newey & Seed, 1995). Recovery from the *Sea Empress* spill (1996) was also reasonably quick, although *Fucus* proliferation did occur on one shore where dispersants were

liberally applied (R. Crump, personal communication). Petersen (2001) gives a detailed review of the vast literature resulting from the *Exxon Valdez* spill. Recovery on the worst-affected shores also took several years.

The question remains if active restoration of rocky shores could be attempted after oil spills. In contrast to saltmarshes, we think probably not. However, minimal intervention following oil spills reduces the risk of slowing recovery by inappropriate treatment. We also await with interest the

Box 6.3   *Torrey Canyon* oil

Dramatic changes followed the 1967 *Torrey Canyon* oil spill in Cornwall, UK (Smith, 1968). Massive use of dispersants killed large numbers of limpets and other grazers, totally distorting rocky-shore community structure over many kilometres of coastline. Long-term recovery has been recorded for the worst affected shores (Southward & Southward, 1978; Hawkins *et al.*, 1983; Hawkins & Southward, 1992). The early generation dispersants used were more toxic than the oil and on shores where they were not applied recovery was reasonably rapid (e.g. Godrevy Island, two to three years). In contrast on shores where dispersants were applied recovery took at least 10 years (Southward & Southward, 1978) and possibly as long as 15 years (Hawkins *et al.*, 1983, Fig. B6.5; Hawkins & Southward, 1992). The widespread kills of grazers led to a massive explosion of algae: ephemerals at first followed by fucoids that persisted for four to five years. Barnacles which survived the initial oiling and dispersant application reasonably well, were reduced to very low levels two to three years after the spill, having succumbed to the dense cover of algae and the increase of the predatory dog-whelk *Nucella* which occurred under the canopy. The dense algal canopy formed an ideal nursery for juvenile limpets and very heavy recruitment under the canopies in 1968 and 1969 led to a population explosion. They restricted further *Fucus* recruitment and probably hastened the demise of the *Fucus* stands by feeding on the holdfasts. As food decreased they abandoned their normal homing habits and migrated along the shore in fronts, before dying from starvation. This reduction in grazing pressure allowed a subsequent pulse of *Fucus* recruitment in the early 1980s (Hawkins & Southward, 1992), confirming that recovery was occurring through a series of damped oscillations as proposed by

Southward & Southward (1978). Eventually the shore returned to the level of small-scale temporal fluctuations and spatial patchiness typical of these shores (Fig. B6.5). Clearly large-scale and inappropriate use of dispersants made the impact of the oil spill much worse and slowed recovery.

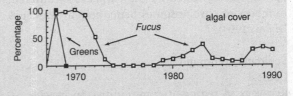

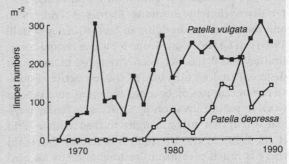

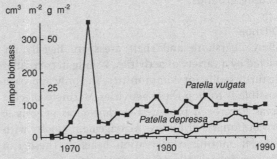

Figure B6.5. Changes at Porthleven following recovery from the dispersant treatment after the *Torrey Canyon* oil spill. From Hawkins & Southward (1992).

results of various bioremediation trials in which microbial activity was enhanced to speed the breakdown of oil (Prince, 1997).

A good example of chronic pollutants are tributyl tin (TBT) leachates from antifouling paints that have severely impacted the dog-whelk *Nucella lapillus* (see Bryan & Gibbs, 1991 for review). Since restrictions on TBT paints were introduced in the mid-1980s some recovery of populations has occurred, but this has been slow (Evans *et al.*, 1996). Dog-whelks do not have a planktonic stage, laying eggs in cases that emerge as juvenile crawlaways. These can raft and probably have some limited buoyancy enabling populations to be re-established in areas where dog-whelks were essentially extinct due to TBT pollution.

To speed recovery, transplants could be used from uncontaminated sites such as the exposed north coast of Cornwall where dog-whelks could be rapidly collected. There would, however, be problems due to small-scale spatial variation in the genetics of *Nucella* (e.g. Kirby *et al.*, 1997) reflecting direct development, complicated by the existence of different numbers of chromosomes being present (Pascoe & Dixon, 1994).

Sediment communities
Coastal
Recovery of sediment communities on beaches after disruption by dumping (Barnes & Frid, 1999) and maintenance dredging in ports (Quigley & Hall, 1999) can be very slow. Scope for active restoration is limited and it has not been attempted to any great degree. Paradoxically, one of the few active restoration programmes on beaches and mudflats has been work to reduce invasion by *Spartina anglica* by physical disruption using a light tracked vehicle (Frid *et al.*, 1999). In the UK and elsewhere in the world there is much pressure to restore mudflats for bird feeding grounds.

Offshore
Many nearshore and shelf areas are heavily impacted by a variety of activities. Fishing in particular disturbs sediment communities. The only approach possible is to stop these activities. Marine protected areas have been proposed worldwide as pragmatic precautionary fisheries management tools with broader marine conservation benefits. These can have various levels of protection (see Jennings & Kaiser, 1998; Hall, 1999 for excellent overviews) from absolute exclusion ('no take zones') to less strict regions, where particular types of gear are excluded ('zoning') or fishing is banned in some seasons (Box 6.4). There is strong evidence for the efficacy of this approach in tropical reef fisheries, perhaps because of the territoriality of many of the larger and more valuable species. Towed gear in general (Jennings & Kaiser, 1998; Hall, 1999; Kaiser & De Groot, 2000) and scallop dredging in particular are acknowledged to have major impacts on non-target

species and the sea bed itself (Bradshaw *et al.*, 2000, in press; Kaiser & De Groot, 2000; Veale *et al.*, 2000).

Mangrove ecosystems

Mangrove forests have integral ecological roles and values (Lugo & Snedaker, 1974; Lewis, 1982; Streever, 1997). Damage to, and loss of, mangrove forests has been extensive (Crooks & Turner, 1999). At the same time as mangrove forests were being degraded and removed for the construction of marinas (Kaly *et al.*, 1997), shrimp farming (Stevenson *et al.*, 1999) and other urban needs, research began into the issues related to their restoration and rehabilitation (Teas, 1977; Lewis, 1982, 2000).

A chronology of mangrove restoration
Initially, restoration and rehabilitation projects were focused on reversing altered hydrology (Lewis, 1982; Kusler & Kentula, 1990; Streever, 1997; Turner & Lewis, 1997) and establishing persistent vegetation cover (Thorhaug, 1990; Saenger & Siddiqi, 1993; Field 1996*a b*; Saenger, 1996; Lewis, 2000; Streever, 1999). Draining, dredging and filling are modifications of hydrology which have either dried out or caused wetland destruction inadvertently (Kusler & Kentula, 1990; Lewis, 1990*b*).

In general, the first critical issue is to remove the degrading forces such as cattle grazing on the landward edge (Box 6.5) and to restore the hydrological regime (i.e. tidal inundation and flushing). The removal of structures such as culverts (e.g. Streever *et al.*, 1996; Turner & Lewis, 1997; Streever, 1998) and enhancing drainage by re-excavating areas which have become filled (Lewis, 1990*a b*; Turner & Lewis, 1997) are typically undertaken. The consequent impact on the vegetation and fauna is poorly understood however, with a few exceptions (Box 6.5).

An extensive literature on the most cost-effective way to restore vegetation structure has grown up (Teas, 1977; Rabinowitz, 1978; Lewis, 1982; Thorhaug, 1990; Field, 1996*a*). Approaches to establishing vegetation, include allowing natural regeneration and planting of propagules and seedlings where the natural regeneration has been insufficient (Field, 1996*a*). Alternatives include raising propagules and transplanting them as small trees into the field. It is clearly cheaper to collect propagules and plant them as it has been widely documented that propagules have good survival and growth rates where adult trees are absent and where they are planted in an appropriate substratum and elevation (Rabinowitz, 1978; Lewis, 1982, 1990b; Thorhaug, 1990; Kaly & Jones, 1998).

Thus, the initial period in mangrove restoration and rehabilitation focused on restoring the structure of the forest. Once the successful planting of a mangrove forest to provide an equivalent structure was routine, it was natural that 'functional equivalence' became the driving force of restoration or rehabilitation programmes (Lewis, 2000). The pressure to do so stemmed largely from the United States and Australia where legislation and the concept of 'no net loss' wetland mitigation started to drive the process (Kusler & Kentula, 1990; Ambrose, 2000; Huggett, 2000). Pivotal to this approach was the concept that habitat losses must be offset by habitat gains, and that restoring or rehabilitating a habitat can substitute those destroyed by development and maintaining ecosystems assets and functions (Crooks & Turner, 1999).

Box 6.4   Recovery of sediment communities from scallop dredging

A closed area was established in 1989, off Port Erin on the Isle of Man, as part of a programme of scallop fisheries management (Fig. B6.6). The prime aim was to investigate the utility of rotational closure of inshore fishing grounds. Despite buoying and policing, the exclusion zone has been the target of occasional illegal fishing incursions. It has proved extremely valuable, however, in assessing the effects of dredging on sea-bed epibenthos. The account below is largely based on recent reviews by Bradshaw et al. (2000, in press).

The closed area has enabled scallop numbers to increase above the background of heavily fished grounds. Experiments of dredging at different intensities within the closed area and comparison with adjacent commercially dredged areas have been run. These have shown that the experimentally disturbed plots became more like the commercially dredged ones and less like adjacent undisturbed plots. In particular dredged areas became more like each other whilst undredged plots show greater spatial heterogeneity (Fig. B6.7) – an interesting echo of Hockey & Bosman's (1986) work on rocky shores.

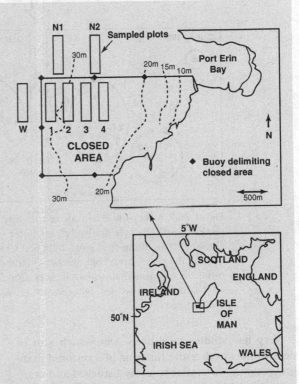

Figure B6.6. Map showing layout and location of the sampling plots in the closed area. Plots 2 and 4 have been undredged since 1989, plots 1 and 3 experimentally dredged every two months since 1995, and plots N1, N2 and W are exposed to commercial scallop dredging. Inset shows location of the Isle of Man in the Irish Sea. From Bradshaw et al. (2000).

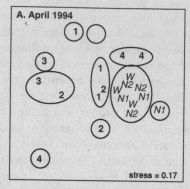
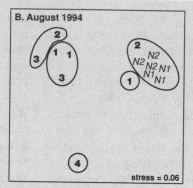

From Jan 1995, plots 1 and 3 experimentally dredged every 2 months

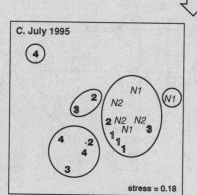
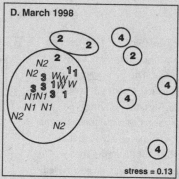

Figure B6.7. MDS plots of benthic community data before and after experimental dredging. For all dates, dredged plots cluster separately from, and are more tightly clustered than, undredged plots. Italicised text represents commercially dredged plots outside the closed area; bold numbers, undredged plots inside the closed area; open numbers, experimentally dredged plots inside the closed area (after January 1995). Plots are based on standardised, reduced, 4√ transformed, pooled count data. From Bradshaw *et al.* (in press).

Very few studies have been done which aim to determine whether the function of a restored mangrove habitat is similar to a less disturbed and more 'natural' habitat (Kaly & Jones, 1998). Rehabilitation projects are often planned to support biodiversity, yet the establishment of faunal assemblages, particularly invertebrates, has been left to the immigration of fauna from adjacent wetlands or in the water column (Lewis, 1990b; Levin et al., 1996; Kaly & Jones, 1998). This may lead to species with good dispersal capacities dominating (Zedler, 1999). There is no evidence that the active introduction of fauna would accelerate the colonisation process (Lewis, 1990b), but there is some evidence that the process of recovery of fauna takes more time than the flora (Levin et al., 1996). It has been suggested that rehabilitation outcomes should include a range of taxa, not just plants, so that key components of the entire food web can be assessed (Zedler, 1999; Zedler & Callaway, 2000).

The third most recent phase in the development of ecological restoration of mangroves is in ecological goal-setting (Lewis, 2000; Zedler & Callaway, 2000). Proactive approaches have been developed

Box 6.5   Rehabilitation of a degraded temperate mangrove forest, Kooragang Island, New South Wales, Australia

Kooragang Island (32° 51′ S 151° 43′ E) has been a site for mangrove and saltmarsh rehabilitation since 1991. It is a temperate mangrove and saltmarsh area, which has been subject to industrialisation and grazing by cattle in the past. Saltmarsh bordering four creeks was sampled in 1995 before any culverts were removed to determine eggshell density of the common saltmarsh mosquito (*Aedes vigilax*) and percentage cover of saltmarsh and mangrove flora. Samples were then taken two years later after culverts were removed from two creeks, the two other creeks being left intact. The aim of this was to determine the effect of culvert removal on the rehabilitation of the saltmarsh flora and fauna. Seedlings of *Avicennia marina*, *Triglochin striata* and *Sarcocornia quinqueflora* increased significantly in cover at the creeks where the culverts were removed relative to the creeks which did not have culverts removed. Overall, the mean number of eggshells decreased significantly at the creeks where culverts were removed compared to the creeks where culverts were left intact (Turner & Streever, 1999). At the same time benthic core samples were taken to assess the number of invertebrates greater than 500 μm, but these were resampled only six months later. Twenty-five taxa were found in samples – but there was no evidence to suggest that abundance of individual species in areas directly affected by removal of culverts changed more dramatically than in areas not affected by the removal of culverts (Streever, 1998).

such as 'conservation banking' whereby the habitat is restored in advance of any proposed development. Coupled with this have been suggestions on logic and experimental designs to provide a better understanding of coastal wetland functioning (Chapman & Underwood, 2000; Zedler & Callaway, 2000). Given the broad range of regional settings, however, it is difficult to predict how a specific rehabilitation will progress (Zedler, 1999). It is similarly difficult to predict what effect the alternative outcomes of human efforts will have to return the habitat to some desired prior condition (Zedler, 1999). Most researchers agree that restoring a mangrove forest to the original functional condition is impossible and may even be undesirable (Kusler & Kentula, 1990; Kaly & Jones, 1998; Chapman & Underwood, 2000; Zedler & Callaway, 2000).

Self-repair of disturbed mangrove habitats

It has been extensively reported that mangrove forests can self-repair if the hydrology and the propagules of mangroves from adjacent stands are not disrupted or blocked (Stevenson *et al.*, 1999). Mangrove propagules are thought only to be dispersed short distances (Clarke, 1993; McGuinness, 1997), so an assessment of a viable stand of mature mangrove trees is required before planting occurs. Given these issues and the historical nature of the debate in restoring mangrove habitats, there are, however, some readily identifiable characteristics of disturbed or degraded mangrove forests. Apart from altered hydrology, one of the most common features of disturbed mangrove habitats is that the canopy has been removed. The removal of canopy increases the amount of light to the floor of the forest. In these disturbed areas, survival and growth of seedlings is greater than in areas of the mangrove forests still shaded by the canopy (Minchinton, 2001). Characteristics of the soil such as moisture and nutrients (low levels of phosphorus and nitrogen) are also found in disturbed compared to undisturbed mangrove forests (Kaly *et al.*, 1997). Coupled with this there are differences in the densities of macrofauna (McGuinness, 1990; Skilleter, 1996). Studies on macroinvertebrates and other measures of physical structure in mangrove forests have found decreasing abundance and species richness along a gradient of increasing damage (Skilleter, 1996). Not all organisms are, however, equally sensitive. The abundance of crabs has been found in several studies to be greater in disturbed than undisturbed mangrove and saltmarsh habitats, increasing the amount of organic material in the soil and aerating the soil through the activity of burrowing (McGuinness, 1990; Smith *et al.*, 1991; Kelaher *et al.*, 1998*a*). An inverse relationship, however, between damage and the density of crabs was found by Kaly *et al.* (1997) (fewer crabs with increasing damage). This covaried with greater compaction of soils in damaged mangrove areas. In a manipulative experiment, Kelaher *et al.* (1998*a*), investigating the impact of boardwalks in temperate mangrove forests, found a greater abundance of crabs in disturbed areas because of the absence of root material in the soil (Kelaher *et al.*, 1998*b*). It is now believed that crabs are one of the first macroinvertebrates to colonise disturbed mangrove forests and may act as keystone species (Robertson, 1986; Smith *et al.*, 1991).

Issues of scale in mangrove restoration

The main critical gaps in our understanding of restoring mangrove habitats relate to scale. We have little idea of the time involved to replace the structure of a mature mangrove forest and less idea of when this will function similarly to less disturbed mangrove forests. Similarly, there is debate on how to compensate and what is the spatial extent of the compensatory forest required (Zedler, 1999; Zedler & Callaway, 2000). These issues will continue to be relevant for tidal mangrove forests as they have been discussed for other wetland types.

Enclosed and semi-enclosed coastal systems

Throughout the developed and developing world major cities and conurbations are situated on estuaries and coastal lagoons, usually with considerable industrialisation and installations of major ports. Estuaries can also be modified in rural areas by levees to reclaim and protect agricultural land. The hydraulics of estuaries and lagoons can be much modifed by dredged channels, bar openings, entrainment of sediments, embankments and channelisation. All the inputs into the catchment influence the estuary and they often become sinks for pollutants from large tracts of the hinterland. Thus there are few pristine estuaries in densely populated areas and lagoons are threatened by a variety of developments. The fragility of coastal lagoons has

been recognised in Europe by the granting of priority habitat status by the European Union.

Estuarine ecosytems are driven by the exchange between river and sea modulated by tidal movements. Various coastal engineering schemes, including reclamation, interfere with the natural hydrological processes in estuaries and lagoons. A simple but drastic form of large-scale restoration is to simply allow or reintroduce tidal inundation to reclaimed or freshwater marshes and then let nature take its course. This can also be effective in returning freshwater systems to brackish ones with additional benefits such as mosquito control (Lin & Beal, 1995). Measures are often simple: sluices can be opened or removed and saline water is allowed to ingress into freshwater-dominated marsh. Halophobic vegetation dies off and is replaced by saltmarsh and mudflat (see Zedler & Adam, this volume). Here, we concentrate on a rehabilitation of former commercial docks in highly urbanised estuaries.

Rehabilitation of disused docks

Disused docks became a common feature of many of the traditional ports in the UK in the 1970s and 1980s (Fig. 6.2). This is a common pattern worldwide in macrotidal estuaries, where small enclosed basins have been rendered obsolete by increases in ship size and the move to containerisation. Perhaps the two biggest sets of docks that ceased commercial operation in the UK were those in London and Liverpool. We focus on the Mersey Docks, where much work has been undertaken. The Mersey has the reputation of being one of the most polluted estuaries in Europe (e.g. R. B. Clark, 1997). Over the last 20 years it has improved considerably as a result of de-industrialisation of the region and pollution abatement via the Mersey Basin Campaign (Hawkins *et al.*, 1999*b*). For many years the docks on the Mersey were some of the most polluted habitats. Once shipping ceased in the upper docks they presented a considerable challenge to various urban renewal schemes wishing to capitalise on a potentially attractive waterside location.

Water quality problems resulted essentially from the nutrient-rich and often sewage-contaminated water from the Mersey. There were three main types: (1) phytoplankton blooms, resulting in turbid or

Fig. 6.2. Former commercial ports in the UK with dock basins no longer used for commercial shipping (●) and sites of major redevelopment (■).

brightly coloured water; (2) periods of low oxygen concentration in deeper water, with associated release of foul-smelling hydrogen sulphide gas and mortality of fauna; and (3) contamination of water with faecal indicator organisms and hence possibly pathogens (Allen & Hawkins, 1993; Wilkinson et al., 1996). These water quality problems are typical of many redeveloped docks throughout the UK (Box 6.6).

Comparing lakes and docks

The sudden switch from turbid to clear-water conditions in the South Docks, Liverpool could also indicate two alternative stable states: a phytoplankton-rich turbid state associated with eutrophication, and a clear-water state with low nutrient

loading (Scheffer et al., 1993). This has frequently been reported for lakes (see Jeppesen & Sammalkarpi, this volume). When the main components of each state, along with their positive (enforcing) or negative (repelling) interactions, are considered, it is apparent that a large number of possible interactions tend to enforce the existence of the alternative states and buffer transfer from one to the other (Fig. 6.3). Unlike the fish–zooplankton–macrophyte-based control of freshwaters (e.g. Scheffer, 1990; Scheffer et al., 1993) the role of benthic filter feeders is crucial in higher salinity systems.

The main enforcing interaction within the turbid state is the reduction of hypolimnetic dissolved oxygen concentrations, which prevents colonisation by benthic filter feeders. Dense phytoplankton populations may cause oxygen depletion during the decay of blooms. Increasing absorption of solar radiation in the surface layers might also enhance thermal stratification (Sathendranath et al., 1991). In the clear-water state the filter feeders reduce algal blooms and benthic oxygen concentrations remain high, ensuring the continued survival of benthic fauna. In the presence of a larval supply of filter feeders, mixing alone may be sufficient action, as natural recruitment of filter feeders should occur when conditions are made favourable. Biomanipulation by the transplant or culture of a filter feeding population will speed the progression, or enable it to happen where a suitable larval supply is unreliable or non-existent. In some systems, such as the South Docks, top–down control by filter feeders was sufficient to lead to a healthy sustainable ecosystem without the need for additional mixing. Mixing would, however, also increase the rate of supply of phytoplankton to the filter feeders.

Observations from natural systems show that in shallow marine environments control of phytoplankton by benthic suspension feeders can be very effective and that the filter feeders can be resistant to short-term change in water quality (Officer et al., 1982). Artificial mixing would help to ensure survival of filter feeding benthos at all depths in enclosed systems, so that occasional phytoplankton blooms or prolonged hot weather would not

Box 6.6   Rehabilitation of disused docks

At its peak Liverpool had over 100 docks stretching from the river mouth to 10 km upstream (Ritchie-Noakes, 1984), constituting a major modification of the shoreline. Intertidal areas of mudflat, saltmarsh and creeks were replaced by the hard substrate of retaining walls, permanently submerged sediment and extensive areas of standing brackish water (15–30 salinity).

The South Docks (Fig. B6.8) were closed to commercial shipping by 1972 and abandoned with the gates of the docks left open leading to rapid silting up. Dredging of the docks by the Merseyside Development Corporation began in 1981 and finished in 1985, with water gradually being replaced as dredging progressed (Hawkins *et al.*, 1992).

Severe water-quality problems were recorded in the South Dock in the years following redevelopment, despite limited exchange of water with the Mersey estuary. Detailed monitoring began in 1988 and severe water-quality problems were evident (Allen, 1992;

Wilkinson *et al.*, 1996). Persistent dinoflagellate blooms caused poor water clarity, colouring the water orange–brown. The dominant species was the dinoflagellate *Prorocentrum minimum*, that has been linked to paralytic shellfish poisoning outbreaks in other areas. Dissolved oxygen levels were often very low at the bottom of the deeper docks. Oxygen depletion was attributed to decay of phytoplankton cells and a high demand from the organically rich sediments, coupled with a tendency for thermal stratification (Hawkins *et al.*, 1992). Deoxygenation also led to increased release of inorganic phosphates from the sediments, bad smells and death of fish and benthos.

Water quality improvements in the Albert Dock (no mixing) were the result of a decline in the frequency and persistence of phytoplankton blooms. It is likely that the observed improvements in water quality were largely due to control of the phytoplankton populations by mussel filtration; it was calculated that the entire volume of the Albert Dock passed through the mussels in one to three days.

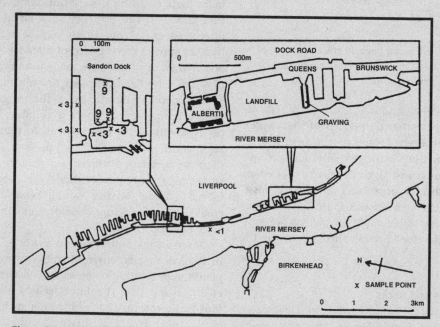

Figure B6.8. Map of the docks under study in Liverpool showing major basins investigated. South Docks expanded on right insert. Secchi disc extinction depths are given for Sandon Dock. From Hawkins *et al.* (1992).

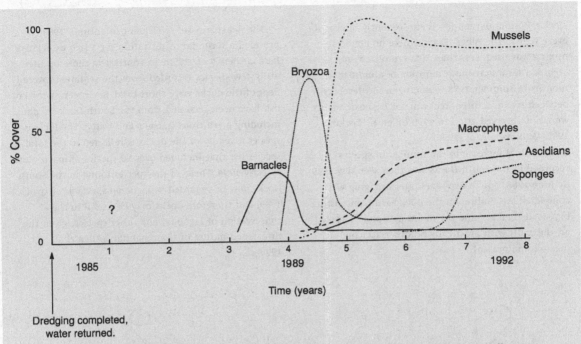

Figure B6.9. Schematic diagram of sequence of colonisation of major groups in the Albert Dock after dredging and refilling with water. Modified from Allen (1992).

Continued monitoring of water quality has shown that improvements were sustained over a period of seven years (1990–7). Between 1990 and 1995, however, new recruitment of mussels on the walls was poor (Wilkinson *et al.*, 1996). In 1996, a new recruitment of mussels was observed throughout the South Docks which probably occurred in late 1995 (Fielding, 1997). Therefore the system appears to be sustainable over the longer term.

Improvements in water quality have allowed colonisation by a relatively diverse estuarine/marine fauna. Initially, the walls that had been covered in sediment and the newly dredged bottom were devoid of benthic organisms.

Remedial measures to improve water quality were attempted based on previous experience at Sandon Dock used between 1978 and 1982 as an experimental fish and shellfish farm. Artificial mixing was used successfully to eliminate anoxic conditions, and improvements in water clarity in Sandon were attributed to the filtering effect of the large naturally settled mussel (*Mytilus edulis*) population growing on

ropes and walls in the dock (Russell *et al.*, 1983; Hawkins *et al.*, 1992). In the South Docks, the Graving Dock was selected as a pilot site due to its semi-isolated nature. A helical type air-lift mixer was deployed in this dock along with a large population of mussels bought from a fish farm in mesh tubing suspended from buoyed long-lines (Allen & Hawkins, 1993).

Preliminary investigations of the walls and sediment throughout the South Docks in 1988 indicated that benthic communities were impoverished. However, in late summer and autumn 1988, a large natural settlement of mussels occurred in the South Docks, being particularly dense in the Albert Dock, with mussels almost completely covering the dock wall. In the summers immediately following the experimental introduction or natural settlement of mussels, marked improvements in water clarity were observed in both the Graving and Albert Docks: median Secchi disc extinction depths over the summer algal bloom season improved from about 1 m to 3 m or more, with the disc even being visible on occasion sitting on the bottom in 6 m of water (Hawkins *et al.*,

1999*a*). Bottom oxygen levels subsequently improved every year in the Albert Dock, where no artificial mixing was used. Less than 20% saturation was recorded for a maximum duration of almost three months in summer 1988, reducing to low levels only between two and three weeks in 1989. Anoxic events were not observed at all in 1990 (Allen & Hawkins, 1993; Wilkinson *et al.*, 1996).

A gradual increase in the number of species recorded in the South Docks was then seen (Fig. B6.9). By June 1988, one to five years after dredging was completed (depending on the dock basin), encrusting bryozoans had become the major occupier of space on the walls with practically no other attached macrofauna present.

The dense mussel settlement in autumn 1988 provided a secondary substratum for a rich associated fauna (Allen *et al.*, 1995). In contrast to those on the walls, few species recorded from the sediment-covered floor of the docks were short lived. Seventeen species of fish have been recorded from the South Docks to date, including a sea trout (*Salmo trutta trutta*). The increased species diversity of the docks is reflected in the total species list that included over 90 species of macrobiota by early 1994. Three of the species found in the South Docks may be regarded as lagoonal specialists (Barnes, 1989) and therefore docks may be useful in the conservation of lagoonal and other species, given the threatened nature of these habitats (Allen *et al.*, 1995).

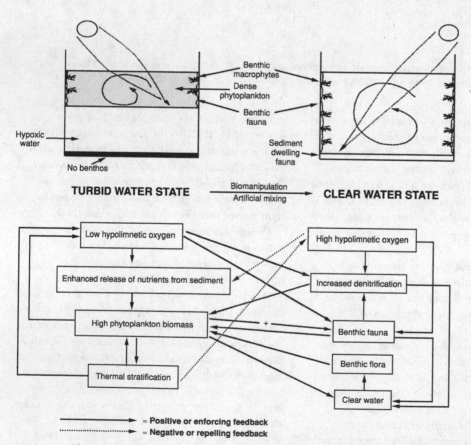

Fig. 6.3. Qualitative model of possible alternative stable states in the South Dock and associated feedback mechanisms. Modified from Allen (1992).

cause reversal to the turbid water state. The longevity of bivalves when compared to zooplankton enhances their ability to control phytoplankton from the initial stages of the bloom avoiding the lag in zooplankton-controlled systems.

Recently similar approaches have been advocated on larger scales such as whole estuaries, especially using oyster reefs (e.g. in Chesapeake Bay: Ulanowicz & Tuttle, 1992; Breitberg et al., 2000; Mann, 2000).

FUTURE CHALLENGES FOR MARINE RESTORATION

Scientific rigour

Restoration, although science-based, has more in common with agriculture or engineering. Inevitably short cuts are taken and restoration schemes often tend to be exploitation of opportunities or rapid responses to threats rather than planned scientific studies. Many authors have urged the adherence to the basic scientific method when undertaking restoration schemes (see comments by Underwood, 1996; Grayson et al., 1999; Zedler & Callaway, 2000). With untouched control areas and restored treatments the success of any manipulations applied can be judged in the context of natural change. However, although this is a laudable and achievable aim, it can sometimes be very difficult in practice. Often the topography of the area prevents this from happening; for example in Liverpool it was impossible to have treatment and control docks as they were all very different sizes and dredged of silt at different times.

Scaling up: restoring coastlines

Most coastal restoration work has focused particularly on habitats, biotopes, assemblages or species, usually on a localised scale. Coastal ecosystems are strongly interconnected with different habitats forming feeding areas or refuges for different life-history stages of mobile species or the larvae of more sedentary or sessile adults. This provides a pool of species for recolonisation, once basic functional attributes (water quality, biogenic

structure) have been put in place. It also provides a challenge as restoring one habitat may positively or negatively influence other adjacent habitats. Thus, seagrass restoration may in turn aid saltmarsh recovery or restoration. Conversely, attempts to restore shellfish populations associated with oyster reefs may be less successful next to saltmarsh and seagrass beds as these provide corridors for large mobile predators (Micheli & Peterson, 1999).

The other approach is to focus on the physical regime based on the assumption (hope?) that the chemistry and biology of the system will follow. This is best exemplified by schemes that reintroduce a tidal element to reclaimed land by altering the hydrological regime (e.g. Falk et al., 1994).

Overall, the challenge is to reverse piecemeal degradation of coastlines by a landscape-scale approach to restoration. Given the size of coastal ecosystems with coastal sedimentary cells and recruitment pools operating over at least 100 km scales, this will be extremely difficult. However, encouraging progress is being made at the whole catchment/estuary/coastline scales; good examples being work on Chesapeake Bay (Hennessey, 1994; Swanson, 1994), the Wadden Sea (Essink, 1992; Wolff, 1997) and the Mersey estuary (reviewed in Hawkins et al., 1999b). These approaches also link degradation within a freshwater catchment with problems downstream in estuaries. Only this holistic scaling-up will reverse loss of habitat and ecosystem services on the scale at which marine and coastal ecosystems function. Sadly it will only be possible in certain areas and will be constrained by the patchwork-quilt nature of ownership and regulatory responsibility of most coastal systems – including the large commons of open water.

Open seas and oceans

On the shelf and in large enclosed seas action is much more difficult. Eutrophication can be tackled by reducing inputs from rivers and the atmosphere that in turn will reduce the likelihood of incidence of harmful algal blooms ('red tides'). Steps are under way throughout the world to reduce the fishing effort to stabilise individual fish stocks:

such measures will also benefit non-target species and whole sea-bed communities (Jennings & Kaiser, 1998; Hall, 1999). Proposed closure of sea-bed areas (no-take zones) will greatly help in this regard. Climate change and redistribution of current patterns are closely intertwined. Whilst emissions of greenhouse gases can be regulated, nothing can be done for the seas alone. Nor does anything need to be done – such changes have happened throughout the history of the oceans and the open nature of the ocean will result in the redistribution of organisms to regions where they can survive and function. The only real measures that can be implemented for the pelagic realm of the open ocean are species-centred conservation initiatives for large animals. Examples include sea otters (see Estes *et al.*, 1998, for review), cetaceans (e.g. Hay, 1985), turtles (e.g. Bjorndal, 1995) and large fish (e.g. tuna and large game fish: Speer, 1996; shark: Manire & Hueter, 1995; Vas, 1995).

CONCLUDING REMARKS

Restoration must be geared towards a defined end point. At the most ambitious level this may be return to pre-impact state and functioning of the whole system. This can often be a difficult objective as the pre-impact state is often not known or may not be stable. At the most limited extent it may be reintroduction and re-establishment of a single component such as an exploited species. In many cases restoration can be at best focused on a few elements. To date much effort has been spent on restoring biogenic structure (open systems) or water quality (semi-enclosed systems).

Recovery and restoration of marine systems must occur as a series of stages. First impacts must be identified. This is essential as much effort may be spent on reducing the wrong suspected impact. Second steps must be taken to stop or reduce the risk of further impacts (Fig. 6.4). Once impacts have ceased, or at least been reduced, then natural recovery can occur. Usually this is rapid given the extensive dispersal of most marine plants and animals.

In most cases the best and most cost-effective course of action for an open marine ecosystem is to stop the damaging activity and let rapid natural recolonisation occur. If there is experimental evidence that a keystone species occurs in a particular community then stopping exploitation of that species can lead to recovery of the community as a whole (e.g. in Chile, conservation of the predatory whelk *Concholepas*: Castilla & Duran, 1985). Similarly, during clean-up operations following acute pollution events such as oil spills, steps should be taken to prevent further damage by inappropriate action that might kill functionally important species. To enable rapid recovery of such ecosystems, being informed and brave enough to do nothing is probably the best approach.

Active restoration may be used to speed up natural recovery and may be attempted even in open systems when biogenic structure is conferred by a species or an assemblage – so-called 'ecosystem engineers' (Lawton, 1994). Coral reefs, oyster reefs, mussel beds, seagrass beds, saltmarsh and mangroves are all examples where a strong structural element is conferred by the dominant biota, whether invertebrates or macrophytes. Active steps can be taken to restore the physical structure of coral reefs and other biogenic structures damaged by both chronic and acute impacts. Recruitment can be encouraged by providing suitable settlement substrata, for example transplanting chunks of seed coral to encourage vegetative proliferation (see Clark, this volume). However, we should also consider that many marine invertebrate larvae follow specific cues to ensure that they only settle next to conspecifics or congenerics; thus in order to stimulate settlement in recovered environments, these cues need to be provided. This also works well for oyster reefs and mussel beds. Similarly rooted macrophytes (e.g. seagrasses) can be planted out to form the nuclei of new beds and this approach has met with some success (see Fonseca *et al.*, this volume). Mangroves have also been restored by out-planting but there are difficulties because of the size and longevity of the plants and the complex interactions of vegetation with the hydrology of the wetland. Direct planting techniques of saltmarsh vegetation, seagrasses and corals all depend on the ability to

WHAT CAN AND SHOULD BE DONE?

1. Stop impacts (chronic)
or
Prevent impacts (acute)

2. Initial clean-up (if appropriate)

OPEN SYSTEMS

CLOSED SYSTEMS

3. Wait ?
Actions
(i) Do nothing
(ii) stop unnecessary interventions
(iii) need to inform managers of
     likely timescale of recovery

Advantages: a) cheap
               b) natural

Disadvantages: a) slow
               b) perception of
                 political inertia
                 ("doing nothing")

4. Speed succession ? ☑
Actions
(i) modify physical environment
(ii) restore structure artificially
(iii) enhance settlement/recruitment
(iv) transplants
(v) biomanipulate

Disadvantages: a) often untried
                technology
            b) could fail
            c) can cause
               unforeseen problems
            d) costly even if
               successful

Advantages: a) restore to a defined
                state
            b) can speed natural
               processes
            c) in case of "ecosystem
               engineers" giving
               biogenic structure
               recovery might not
               occur to the original
               state
            d) politically "doing
               something"

Fig. 6.4. Summary of decision-making involved in restoration/rehabilitation of marine ecosystems. ✓ indicates whether action should be taken.

vegetatively proliferate. This often compensates for limited dispersal or recruitment once stands are established.

In contrast to open coastlines, there is much more scope to restore enclosed or semi-enclosed systems such as estuaries, lagoons, fjords, or rias. Our own work on restoration of disused docks in Liverpool has shown analogies to freshwater systems and the efficacy of using bottom–up manipulation by artificial mixing and top–down biofiltration by benthic suspension feeders to improve water quality. These docks have been managed to support healthy and interesting ecosystems, but these do not represent in any way the original conditions of a network of creeks and marshes that would have typified the Mersey, pre-development. This may be an example of what Frid & Clark (1999) have called rehabilitation rather than restoration. Such small-scale rehabilitation via biomanipulation works best in the wider context of catchment-wide amelioration schemes (e.g. the Mersey Basin Campaign).

Overall, compared to less open freshwater and terrestrial systems the measures that can be employed – even in the most tractable marine systems – are at best attempts to nudge nature (Hawkins et al., 1999a).

ACKNOWLEDGMENTS

This work was supported by a variety of sources over the years including NERC, NCC and the Merseyside Development Corporation. Thanks are due to colleagues at Liverpool and Manchester Universities including Keith White, Hugh Jones, George Russell, Keith Hendry, Vanessa Wanstall, Nicola Fielding, Steve Wilkinson and Wei Zhong Zheng. During assembly of the manuscript Lydia Mathgauer, Toby Collins and Dorothea Summefeldt all helped greatly. Martin Perrow deserves thanks for very positive and tolerant editing.

REFERENCES

Allen, J. R. (1992). Hydrography, ecology and water quality management of the South Docks, Liverpool. PhD dissertation, University of Liverpool.

Allen, J. R. & Hawkins, S. J. (1993). Can biological filtration improve water quality? In Urban Waterside Regeneration, Problems and Prospects, eds. K. N. White, E. G. Bellinger, A. J. Saul, M. Symes & K. Hendry, pp. 377–385. Hemel Hempstead, UK: Ellis Horwood.

Allen, J. R., Wilkinson, S. B. & Hawkins, S. J. (1995). Redeveloped docks as artificial lagoons: the development of brackish-water communities and potential for conservation of lagoonal species. Aquatic Conservation: Marine and Freshwater Ecosystems, 5, 299–309.

Allen, J. R., Slinn, D. J., Shammon, T. M., Hartnoll, R. G. & Hawkins, S. J. (1998). Evidence for eutrophication of the Irish Sea over four decades. Limnology and Oceanography, 43, 1970–1974.

Ambrose, R. F. (2000). Wetland mitigation in the United States: assessing the success of mitigation policies Wetlands (Australia), 19, 1–27.

Barnes, N. & Frid, C. L. J. (1999). Restoring shores impacted by colliery spoil dumping. Aquatic Conservation: Marine and Freshwater Ecosystems, 9, 75–82.

Barnes, R. S. K. (1989). The coastal lagoons of Britain: an overview and conservation appraisal. Aquatic Conservation: Marine and Freshwater Ecosystems, 2, 65–94.

Bjorndal, K. (ed.) (1995). A Global Strategy for the Conservation of Marine Turtles. Gland, Switzerland: IUCN.

Bonsdorff, E., Blomqvist, E. M., Mattila, J. & Norkko, A. (1997). Coastal eutrophication: causes, consequences and perspectives in the archipelago areas of the northern Baltic Sea. Estuarine Coastal and Shelf Science, 44 (Suppl.), 63–72.

Bradshaw, C., Veale, L. O., Hill, A. S. & Brand, A. R. (2000). The effects of scallop dredging on gravelly seabed communities. In Effects of Fishing on Non-Target Species and Habitats, eds. M. J. Kaiser & S. J. de Groot, pp. 83–104. Oxford: Blackwell.

Bradshaw, C., Veale, L. O., Hill, A. S. & Brand, A. R. (in press). The effect of scallop dredging on Irish Sea benthos: experiments using a closed area. Hydrobiologia.

Breen, P. A. & Mann, K. H. (1976a). Destructive grazing of kelp by sea-urchins in eastern Canada. Journal of the Fisheries Research Board of Canada, 33, 1278–1283.

Breen, P. A. & Mann, K. H. (1976b). Changing lobster abundance and the destruction of kelp beds by sea-urchins. Marine Biology, 34, 137–142.

Breitberg, D. L., Coen, L. D., Luckenbach, M. W., Mann, R., Posey, M. & Wesson, J. A. (2000). Oyster reef restoration: convergence of harvest and conservation strategies. Journal of Shellfish Research, 19, 371–377.

Bryan, G. W. & Gibbs, P. E. (1991). Impacts of low concentrations of tributyl tin (TBT) on marine organisms: a review. In Metal Ecotoxicology: Concepts and Applications, eds. M. C. Newman & A. W. McIntosh, pp. 323–361. Boston, MA: Lewis Publishers.

Bryan, G. W. & Langston, W. J. (1992). Bioavailability, accumulation and effect of heavy metals in sediments with special reference to UK estuaries: a review. Environmental Pollution, 76, 89–131.

Castel, J., Caumette, P. & Herbert, R. (1996). Eutrophication gradients in coastal lagoons as exemplified by the Bassin Darcachon and the Etang du Prévost. Hydrobiologia, 329, 9–28.

Castilla, J. C. (1999). Coastal marine communities: trends and perspectives from human-exclusion experiments. Trends in Ecology and Evolution, 14, 280–283.

Castilla, J. C. & Duran, L. R. (1985). Human exclusion from rocky intertidal zone of central Chile: the effects on *Concholepas concholepas* (Gastropoda). *Oikos*, **45**, 391–399.

Chapman, M. G. & Underwood, A. J. (2000). The need for a practical scientific protocol to measure successful restoration. *Wetlands (Australia)*, **19**, 28–49.

Cintron-Molero, G. (1992). Restoring mangrove systems. In *Restoring the Nation's Marine Environment*, ed. G. W. Thayer, pp. 223–278. College Park, MD: Maryland Sea Grant College.

Clark, M. J. (1997). Ecological restoration: the magnitude of the challenge: an outsider's view. In *Restoration Ecology and Sustainable Development*, eds. M. Urbanska, N. R. Webb & P. J. Edwards, pp. 353–377. Cambridge: Cambridge University Press.

Clark, R. B. (ed.) (1997). *Marine Pollution*, 4th edn. Oxford: Clarendon Press.

Clark, S. & Edwards, A. J. (1999). An evaluation of artificial reef structure as tools for marine habitat rehabilitation in the Maldives. *Aquatic Conservation: Marine and Freshwater Ecosystems*, **9**, 5–21.

Clarke, P. J. (1993). Dispersal of grey mangrove (*Avicennia marina*) propagules in southeastern Australia. *Aquatic Botany*, **45**, 195–204.

Cloern, J. E. (1996). Phytoplankton bloom dynamics in coastal ecosystems: a review with some general lessons from sustained investigation of San Francisco Bay, California. *Reviews of Geophysics*, **34**, 127–168.

Crooks, S. & Turner, R. K. (1999). Integrated coastal management: sustaining estuarine natural resources. *Advances in Ecological Research*, **29**, 241–289.

Crowe, T. P., Thompson, R. C., Bray, S. & Hawkins S. J. (2000). Impact of anthropogenic stress on rocky intertidal communities. *Journal of Aquatic Ecosystem Stress and Recovery*, **7**, 273–297.

Dayton, P. K. & Tegner, M. J. (1984). Catastrophic storms, El Niño and patch stability in a southern Californian kelp community. *Science*, **224**, 283–285.

Duran, L. R. & Castilla, J. C. (1989). Variation and persistence of the middle rocky intertidal community of central Chile, with and without human harvesting. *Marine Biology*, **103**, 555–562.

Ellis, D. V. & Pattisina, L. A. (1990). Widespread neogastropod imposex, a biological indicator of global TBT pollution. *Marine Pollution Bulletin*, **21**, 248–253.

Ellison, A. M. (2000). Mangrove restoration: do we know enough? *Restoration Ecology*, **8**, 219–229.

Estes, J. A. & Duggins, D. O. (1995) Sea otters and kelp forests in Alaska: generality and variation in a community ecology paradigm. *Ecological Monographs*, **65**, 75–100.

Estes, J. A., Tinker, M. T., Williams, T. M. & Doak, D. F. (1998). Killer whale predation on sea otters linking oceanic and nearshore ecosystems. *Science*, **282**, 473–476.

Essink, K. (1992). Misfunctional use of the Wadden Sea aiming for high natural value. *Publication Series Netherlands Institute for Sea Research*, **20**, 35–43.

Evans, S. M., Evans, P. M. & Leksona, T. (1996). Widespread recovery of dog-whelks, *Nucella lapillus* (L), from tributyl tin contamination in the North Sea and Clyde Sea. *Marine Pollution Bulletin*, **32**, 263–269.

Falk, K., Noehr, H. & Rasmussen, L. M. (1994). Margrethe-Kog and the artificial saltwater lagoon: evaluation of a habit restoration project in the Danish Wadden Sea. *Environmental Conservation*, **21**, 133–144.

Farnham, W. F., Fletcher, R. L. & Irvine, L. M. (1973). Attached *Sargassum* found in Britain. *Nature*, **243**, 231–232.

Field, C. (1996a). General guidelines for the restoration of mangrove ecosystems. In *Restoration of Mangrove Ecosystems*, ed. C. Field, pp. 233–250. Okinawa, Japan: International Society for Mangrove Ecosystems.

Field, C. (1996b). Rationale for restoration of mangrove ecosystems. In *Restoration of Mangrove Ecosystems*, ed. C. Field, pp. 28–35. International Society for Mangrove Ecosystems.

Fielding, N. J. (1997). Fish and benthos communities in regenerated dock systems on Merseyside. PhD dissertation, University of Liverpool.

Frid, C. L. J. & Clark, S. (1999). Restoring aquatic ecosystems: an overview. *Aquatic Conservation: Marine and Freshwater Ecosystems*, **9**, 1–4.

Frid, C. L. J., Chandrasekara, W. U. & Davey, P. (1999). The restoration of mud flats invaded by common cord-grass (*Spartina anglica* CE Hubbard) using mechanical disturbance and its effects on the macrobenthic fauna. *Aquatic Conservation: Marine and Freshwater Ecosystems*, **9**, 47–61.

Furukawa, K., Wolanski, E. & Mueller, H. (1997). Currents and sediment transport in mangrove forests. *Estuarine Coastal and Shelf Science*, **44**, 301–310.

Gibbs, P. E. & Bryan, G. W. (1986). Reproductive failure of the dog-whelk, *Nucella lapillus*, caused by imposex induced by tributyl tin from antifouling paints. *Journal of the Marine Biological Association of the UK*, **66**, 767–777.

Gray, A. J., Marshall, D. F. & Raybould, A. F. (1991). A century of evolution in *Spartina anglica*. *Advances in Ecological Research*, **21**, 1–61.

Grayson, J. E., Chapman, M. G. & Underwood, A. J. (1999). The assessment of restoration of habitat in urban wetlands. *Landscape and Urban Planning*, **43**, 227–236.

Hagen, N. T. (1995). Recurrent destructive grazing of successively immature kelp forests by green sea urchins in Vestfjorden, Northern Norway. *Marine Ecology Progress Series*, **123**, 95–106.

Hall, S. (ed.) (1999). *The Effects of Fishing on Marine Ecosystems and Communities*. Oxford: Blackwell.

Hawkins, S. J. & Southward, A. J. (1992). Lessons from the *Torrey Canyon* oil spill, recovery and stability of rocky shore communities. In *Restoring the Nation's Marine Environment*, ed. G. W. Thayer, pp. 583–688. College Park, MD: Maryland Sea Grant College.

Hawkins, S. J., Southward, A. J. & Barrett, R. L. (1983). Population structure of *Patella vulgata* during succession on rocky shores in south-west England. *Oceanologica Acta*, Special Issue, 103–107.

Hawkins, S. J., Allen, J. R., Russell, G., White, K. N., Conlan, K., Hendry, K. & Jones, H. D. (1992). Restoring and managing disused sites in inner city areas. In *Restoring the Nation's Marine Environment*, ed. G. W. Thayer, pp. 473–542. College Park, MD: Maryland Sea Grant College.

Hawkins, S. J., Proud, S. V., Spence, S. K. & Southward, A. J. (1994). From the individual to the community and beyond: water quality, stress indicators and key species in costal ecosystems. In *Water Quality and Stress Indicators in Marine and Freshwater Systems: Linking Levels of Organisations*, ed. D. W. Sutcliffe, pp. 35–62. Ambleside, UK: Freshwater Biological Association.

Hawkins, S. J., Allen, J. R. & Bray, S. (1999*a*). Restoration of temperate marine and coastal ecosystems: nudging nature. *Aquatic Conservation: Marine and Freshwater Ecosystems*, **9**, 23–46.

Hawkins, S. J., Allen, J. R., Fielding, N. J., Wilkinson, S. B. & Wallace, I. D. (1999*b*). Liverpool Bay and its estuaries: human impact, recent recovery and restoration. In *Ecology and Landscape Development: A History of the Mersey Basin*, ed. E. F. Greenwood, pp. 155–165. Liverpool, UK: Liverpool University Press and National Museums and Art Galleries on the Merseyside.

Hay, K. A. (1985). Status of the right whale, *Eubalaena glacialis*. *Canadian Field Naturalist*, **99**, 433–437.

Hennessey, T. M. (1994). Governance and adaptive management for estuarine ecosystems: the case for Chesapeake Bay. *Coastal Management*, **22**, 119–145.

Hockey, P. A. R. & Bosman, A. L. (1986). Man as an intertidal predator in Transkei: disturbance, community convergence and management of a natural food resource. *Oikos*, **46**, 3–14.

Huggett, D. (2000). Designing and building dynamic coasts and wetlands: developing a no net loss approach. *Wetlands (Australia)*, **19**, 50–59.

Imbert, D., Rousteau, A. & Scherrer, P. (2000). Ecology of mangrove growth and recovery in the Lesser Antilles: state of knowledge and basis for restoration projects. *Restoration Ecology*, **8**, 230–236.

Jara, H. F. & Moreno, C. A. (1984). Herbivory and structure in a midlittoral rocky community: a case in southern Chile. *Ecology*, **65**, 28–38.

Jennings, S. & Kaiser, M. J. (1998). The effects of fishing on marine ecosystems. *Advances in Marine Biology*, **34**, 201–352.

Jones, N. S. & Jago, C. F. (1993). In situ assessment of modification of sediment properties by burrowing invertebrates. *Marine Biology*, **115**, 133–142.

Kaly, U. L. & Jones, G. P. (1998). Mangrove restoration: a potential tool for coastal management in tropical developing countries. *Ambio*, **27**, 656–661.

Kaly, U. L., Eugelink, G. & Robertson, A. I. (1997). Soil conditions in damaged North Queensland mangroves. *Estuaries*, **20**, 291–300.

Kaiser, M. J. & De Groot, S. J. (2000) *Effects of Fishing on Non-Target Species and Habitats*. Tunbridge Wells, UK: Gray Publishing.

Kaspar, H. F., Gillespie, P. A., Bayer, I. C. & MacKenzie, A. L. (1985). Effects of mussel aquaculture on the nitrogen cycle and benthic communities in Kenepuru Sound, Marlborough Sounds, New Zealand. *Marine Biology*, **85**, 127–136.

Kelaher, B. P., Chapman, M. G. & Underwood, A. J. (1998*a*). Changes in benthic assemblages near boardwalks in temperate urban mangrove forests. *Journal of Experimental Marine Biology and Ecology*, **228**, 291–307.

Kelaher, B. P., Underwood, A. J. & Chapman, M. G. (1998b). Effect of boardwalks on the semaphore crab *Heloecius cordiformis* in temperate urban mangrove forests. *Journal of Experimental Marine Biology and Ecology*, **227**, 281–300.

Kenny, A. J. & Rees, H. L. (1994). The effects of marine gravel extractions on the macrobenthos: early post-dredging recolonization. *Marine Pollution Bulletin*, **28**, 442–447.

Kirby, R. R., Berry, R. J. & Powers, D. A. (1997). Variation in mitochondrial DNA in a cline of allele frequencies and shell phenotype in the dog-whelk, *Nucella lapillus* (L.). *Biological Journal of the Linnean Society*, **62**, 299–312.

Kusler, J. A. & Kentula, M. E. (1990). *Wetland Creation and Restoration: The Status of the Science*. Washington, DC: Island Press.

Lambert, W. J., Levin, P. S. & Berman, J. (1992). Changes in the structure of a New England (USA) kelp bed: the effects of an introduced species. *Marine Ecology Progress Series*, **88**, 303–307.

Lawton, J. H. (1994). What do species do in ecosystems? *Oikos*, **71**, 367–374.

Levin, L. A., Talley, D. & Thayer, G. (1996). Succession of macrobenthos in a created saltmarsh. *Marine Ecology Progress Series*, **141**, 67–82.

Lewis, R. R. III (1982). Mangrove Forests. In *Creation and Restoration of Coastal Plant Communities*, ed. R. R. Lewis III, pp. 153–171. Boca Raton, FL: CRC Press.

Lewis, R.R. III (1990a). Creation and restoration of coastal plain wetlands in Florida. In *Wetland Creation and Restoration: The Status of the Science*, eds. J. A. Kusler & M. E. Kentula, pp. 73–101. Washington, DC: Island Press.

Lewis, R. R. III (1990b). Creation and restoration of coastal plain wetlands in Puerto Rico and the US Virgin Islands In *Wetland Creation and Restoration: The Status of the Science*, eds. J. A. Kusler & M. E. Kentula, pp. 103–122. Washington, DC: Island Press.

Lewis, R. R. III (2000). Ecologically based goal-setting in mangrove forest and tidal marsh restoration. *Ecological Engineering*, **15**, 191–198.

Lin, J. & Beal, J. L. (1995). Effects of mangrove marsh management on fish and decapod communities. *Bulletin of Marine Science*, **57**, 193–201.

Lugo, A. E. & Snedaker, S. C. (1974). The ecology of mangroves. *Annual Review Ecology and Systematics*, **5**, 39–64.

Manire, C. A. & Hueter, R. E. (1995). Human impact on the shark nursery grounds of Tampa Bay. *Florida Acadamy of Science*, **58**, 107.

Mann, R. (2000). Restoring the oyster reef communities in the Chesapeake Bay: a commentary. *Journal of Shellfish Research*, **19**, 335–339.

McGuinness, K. A. (1990). Effects of oil spills on macro-invertebrates of saltmarshes and mangrove forests in Botany Bay, New South Wales, Australia. *Journal of Experimental Marine Biology and Ecology*, **142**, 121–135.

McGuinness, K. A. (1997). Dispersal, establishment and survival of *Ceriops tagal* propagules in a north Australian mangrove forest. *Oecologia*, **109**, 80–87.

McLusky, D. S., Bryant, D. M. & Elliott, M. (1992). Impact of land claim on macrobenthos, fish and shorebirds on the Forth Estuary, eastern Scotland. *Aquatic Conservation: Marine and Freshwater Ecosystems*, **2**, 211–222.

Meehan, B. (ed.) (1982). *Shell Bed to Shell Midden*. Melbourne: Australian Institute of Aboriginal Studies and Globe Press.

Menge, B. A., Burlow, E. L., Blanchette, C. A., Navarrete, S. A. & Yamada, S. B. (1994). The keystone species concept: variation in interaction strength in a rocky intertidal habitat. *Ecological Monographs*, **64**, 249–286.

Micheli, F. & Peterson, C. H. (1999). Estuarine vegetated habitats as corridors for predator movements. *Conservation Biology*, **13**, 869–881.

Millward, R. N. & Grant, A. (1995). Assessing the impact of copper on nematode communities from a chronically metal-enriched estuary using pollution-induced community tolerance. *Marine Pollution Bulletin*, **30**, 701–706.

Minchinton, T. E. (2001). Demographic consequences of canopy and substratum heterogeneity to the grey mangrove *Avicennia marina*. *Journal of Ecology*, 2001, **89**, 888–902.

Mitsch, W. J. & Wilson, R. F. (1996). Improving the success of wetland creation and restoration with know-how, time and self-design. *Ecological Applications*, **61**, 77–83.

Morand, P. & Briand, X. (1996). Excessive growth of macroalgae: a symptom of environmental disturbance. *Botanica Marina*, **39**, 491–516.

Moreno, C. A., Sutherland, J. P. & Jara, F. H. (1984). Man as a predator in the intertidal zone of southern Chile. *Oikos*, **46**, 359–364.

Newey, S. & Seed, R. (1995). The effects of the *Braer* oil spill on rocky intertidal communities in south Shetland, Scotland. *Marine Pollution Bulletin*, **30**, 274–280.

Newton, L. C., Parkes, E. V. H. & Thompson, R. C. (1993). The effects of shell collection on the abundance of gastropods on Tanzanian shores. *Biological Conservation*, **63**, 241–245.

Officer, C. B., Smayda, T. J. & Mann, R. (1982). Benthic filter feeding: a natural eutrophication control. *Marine Ecology Progress Series*, **9**, 203–210.

Orth, R. J., Luckenbach, M. & Moore, K. A. (1994). Seed dispersal in a marine macrophyte: implications for colonisation and restoration. *Ecology*, **75**, 1927–1939.

Paine, R. T. (1994). *Marine Rocky Shores and Community Ecology: An Experimentalist's Perspective*. Oldendorf/Luhe, Germany: Ecology Institute.

Pascoe, P. L. & Dixon, D. R. (1994). Structural chromosomal polymorphism in the dog-whelk *Nucella lapillus* (Mollusca, Neogastropoda). *Marine Biology*, **118**, 247–253.

Petersen, C. H. (2001). The *Exxon Valdez* oil spill in Alaska: acute, indirect and chronic effects on the ecosystem. In *Advances in Marine Biology*, vol. 39, eds. A. J. Southward, P. A. Tyler, C. M. Young & L. A. Fuiman, pp. 1–103. London: Academic Press.

Prince, R. C. (1997). Bioremediation of marine oil spills. *Trends in Biotechnology*, **15**, 158–160.

Quigley, M. P. & Hall, J. A. (1999). Recovery of macrobenthic communities after maintenance dredging in the Blyth Estuary, north-east England. *Aquatic Conservation: Marine and Freshwater Ecosystems*, **9**, 63–73.

Rabinowitz, D. (1978). Early growth of mangrove seedlings in Panama, and an hypothesis concerning the relationship of dispersal and zonation. *Journal of Biogeography*, **5**, 113–133.

Raffaelli, D. & Hawkins, S. J. (eds.) (1996). *Intertidal Ecology*, 2nd edn. Dordrecht: Kluwer Academic Publishers.

Renaud, P. E., Riggs, S. R., Ambrose, W. G., Schmid, K. & Snyder, S. W. (1997) Biological–geological interactions: storm effects on macroalgal communities mediated by sediment characteristics and distribution. *Continental Shelf Research*, **17**, 37–56.

Ritchie-Noakes, N. (ed.) (1984). Liverpool's historic waterfront: the world's first mercantile dock system. London: HMSO.

Robertson, A. I. (1986). Leaf-burying crabs: their influence on energy flow and export from mixed mangrove forests (*Rhizophora* spp.) in northeastern Australia. *Journal of Experimental Marine Biology and Ecology*, **102**, 237–248.

Russell, G., Hawkins, S. J., Evans, L. C., Jones, H. D. & Holmes, G. D. (1983). Restoration of a disused dock basin as a habitat for marine benthos and fish. *Journal of Applied Ecology*, **20**, 43–58.

Ryther, J. H. & Dunstan, W. M. (1971). Nitrogen, phosphorus, and eutrophication in the coastal marine environment. *Science*, **171**, 1008–1013.

Saenger, P. (1996). Mangrove restoration in Australia: a case study of Brisbane international airport. In *Restoration of Mangrove Ecosystems*, ed. C. Field, pp. 36–51. Okinawa, Japan: International Society for Mangrove Ecosystems.

Saenger, P. & Siddiqi, N. A. (1993). Land from the sea: the mangrove afforestation program of Bangladesh. *Ocean and Coastal Management*, **20**, 23–39.

Sammarco, P. W. (1996). Comments on coral reef regeneration, bioerosion, biogeography and chemical ecology: future directions. *Journal of Experimental Marine Biology and Ecology*, **200**, 135–168.

Sathendranath, S., Gouveia, A. D., Shetye, S. R., Ravindran, P. & Platt, T. (1991). Biological control of surface temperature in the Arabian Sea. *Nature*, **349**, 54–56.

Scheffer, M. (1990). Multiplicity of stable states in fresh-water systems. *Hydrobiologia*, **200**, 475–486.

Scheffer, M., Hosper, S. H., Meijer, M. L., Moss, B. & Jeppesen, E. (1993). Alternative equilibria in shallow lakes. *Trends in Ecology and Evolution*, **8**, 275–279.

Schiel, D. R. & Foster, M. (1992). Restoring kelp forests. In *Restoring the Nation's Marine Environment*, ed. G. W. Thayer, pp. 279–342. College Park, MD: Maryland Sea Grant College.

Schiel, D. R. & Taylor, D. I. (1999). Effects of trampling on a rocky intertidal algal assemblage in southern New Zealand. *Journal of Experimental Marine Biology and Ecology*, **235**, 213–235.

Schiewer, U. (1998). 30 years' eutrophication in shallow brackish waters: lessons to be learned. *Hydrobiologia*, **363**, 73–79.

Skilleter, G. A. (1996). Validation of rapid assessment of damage in urban mangrove forests and relationships with molluscan assemblages. *Journal of Marine Biological Association of the UK*, **76**, 701–716.

Smayda, T. S. (1990). Novel and nuisance phytoplankton blooms in the sea: evidence for a global epidemic. In

Toxic Marine Phytoplankton, eds. E. Graneli, B. Sundstroem, L. Edler & D. M. Anderson, pp. 29–40. Amsterdam: Elsevier.

Smith, J. E. (ed.) (1968). Torrey Canyon *Pollution and Marine Life*. Cambridge: Cambridge University Press.

Smith, T. J. III, Boto, K. G., Frusher, S. D. & Giddins, R. L. (1991). Keystone species and mangrove forest dynamics: the influence of burrowing by crabs on soil nutrient status and forest productivity. *Estuarine Coastal and Shelf Science*, **33**, 419–432.

Somerfield, P. J., Rees, H. L. & Warwick, R. M. (1995). Interrelationships in community structure between shallow-water marine meiofauna and macrofauna in relation to dredgings disposal. *Marine Ecology Progress Series*, **127**, 103–112.

Southgate, T., Wilson, L., Cross, T. F. & Myers, A. A. (1984). Recolonisation of a rocky shore in southwest Ireland following a toxic bloom of the dinoflagellate *Gyrodinium aureolum*. *Journal of the Marine Biological Association of the UK*, **64**, 485–492.

Southward, A. J. & Southward, E. C. (1978). Recolonisation of rocky shores in Cornwall after the use of toxic dispersants to clean up the *Torrey Canyon* oil spill. *Journal of the Fish Research Board of Canada*, **35**, 682–706.

Speer, L. (1996). Conserving and managing straddeling, highly migratory fish stocks: United Nations takes a major step forward for marine conservation. *Fisheries*, **21**, 4–5, 45.

Spooner, N., Gibbs, P. E., Bryan, G. W. & Goad, L. J. (1991). The effect of tributyl tin upon steroid titers in the female dogwhelk *Nucella lapillus*, and the development of imposex. *Marine Environmental Research*, **32**, 37–49.

Stein, A., Leinaas, H. P., Halvorsen, O. & Christie, H. (1998). Population dynamics of the *Echinomerlla matsi* (Nematoda) – *Stongylocentrotus droebachienssis* (Echinoida) system: Effects on host fecundity. *Marine Ecology Progress Series*, **163**, 193–201.

Stevenson, N. J., Lewis, R. R. & Burbridge, P. R. (1999). Disused shrimp ponds and mangrove rehabilitation. In *An International Perspective on Wetland Rehabilitation*, ed. W. J. Streever, pp. 277–297. Dordrecht: Kluwer Academic Publishers.

Streever, W. J. (1997). Trends in Australian wetland rehabilitation. *Wetlands Ecology and Management*, **5**, 5–18.

Streever, W. J. (1998). Preliminary example of a sampling design assessment method for biomonitoring studies that rely on ordination. In *Wetlands for the Future*, eds.

A. J. McComb & J. A. Davis, pp. 539–552. Adelaide: Gleneagles Publishing.

Streever, W. J. (1999). *An International Perspective on Wetland Rehabilitation*. Dordrecht: Kluwer.

Streever, W. J., Wiseman, L., Turner, P. & Nelson, P. (1996). Short-term changes in flushing of tidal creeks following culvert removal. *Wetlands (Australia)*, **15**, 22–30.

Swanson, A. P. (1994) Watershed restoration on the Chesapeake Bay. *Journal of Forestry*, **92**, 37–38.

Talbot, C. & Hole, R. (1994). Fish diets and the control of eutrophication resulting from aquaculture. *Journal of Applied Ichthyology – Zeitschrift für angewandte Ichthyologie*, **10**, 258–270.

Teas, H. J. (1977). Ecology and restoration of mangrove shorelines in Florida. *Environmental Conservation*, **4**, 51–58.

Tegner, M. J. (1989). The feasibility of enhancing red sea-urchin (*Strongylocentrotus franciscanus*) stocks in California: an analysis of the options. *Marine Fisheries Review*, **51**, 1–22.

Tegner, M. J. & Levin, L. A. (1983). Spiny lobsters and sea-urchins: analysis of a predator–prey interaction. *Journal of Experimental Marine Biology and Ecology*, **73**, 125–150.

Tegner, M. J., Dayton, P. K., Edwards, P. B. & Riser, K. L. (1997). Large-scale, low-frequency oceanographic effects on kelp forest succession: a tale of two cohorts. *Marine Ecology Progress Series*, **146**, 117–134.

Thayer, G. W. (ed.) (1992). *Restoring the Nation's Marine Environment*. College Park, MD: Maryland Sea Grant College.

Thorhaug, A. (1990). Restoration of mangroves and seagrasses: economic benefits for fisheries and mariculture. In *Environmental Restoration Science and Strategies for Restoring the Earth*, ed. J. J. Berger, pp. 265–281. Washington, DC: Island Press.

Turner, P.A. & Streever, W. J. (1999). Changes in productivity of the saltmarsh mosquito *Aedes vigilax* (Diptera: Culicidae), and vegetation cover following culvert removal. *Australian Journal of Ecology*, **24**, 240–248.

Turner, R E. & Lewis, R. R. III (1997). Hydrologic restoration of coastal wetlands. *Wetlands Ecology and Management*, **4**, 65–72.

Ulanowicz, R. E. & Tuttle, J. H. (1992). The trophic consequence of oyster stock rehabilitation in Chesapeake Bay. *Estuaries*, **15**, 298–306.

Underwood, A. J. (1996). Detection, interpretation, prediction and management of environmental disturbances: some roles for experimental marine ecology. *Journal of Experimental Marine Biology and Ecology*, **200**, 1–27.

Vadas, R. L., Sr & Elner, R. W. (1992). Plant–animal interactions in the northwest Atlantic. In *Plant–Animal Interactions in the Marine Benthos*, eds. D. M. John, S. J. Hawkins & J. H. Price, pp. 33–60. Oxford: Clarendon Press.

Vas, P. (1995). The status and conservation of sharks in Britain. *Aquatic Conservation: Marine and Freshwater Ecosystems*, **5**, 67–79.

Vasquez, J. A. & McPeak, R. H. (1998). A new tool for kelp restoration. *California Fish and Game*, **84**, 149–158.

Veale, L. O., Hill, A. S. & Brand, A. R. (2000). Effects of long-term physical disturbance by commercial scallop fishing on subtidal epifaunal assemblages and habitats. *Marine Biology*, **137**, 325–337.

Wilkinson, S. B., Zheng, W., Allen, J. R., Fielding, N. J., Wanstall, V. C., Russell, G. & Hawkins, S. J. (1996). Water quality improvements in Liverpool Docks: the role of filter feeders in algal and nutrient dynamics. *Marine Ecology, Naples*, **17**, 197–211.

Wolff, W. J. (1997). Development of the conservation of Dutch coastal waters. *Aquatic Conservation Marine and Freshwater Ecosystems*, **7**, 165–177.

Zedler, J. B. (1999). The ecological restoration spectrum. In *An International Perspective on Wetland Rehabilitation*, ed. W. J. Streever, pp. 301–318. Dordrecht: Kluwer Academic Publishers.

Zedler, J. B. & Callaway, J. C. (2000). Evaluating the progress of engineered tidal wetlands. *Ecological Engineering*, **15**, 211–225.

7 • Seagrasses

MARK S. FONSECA, W. JUDSON KENWORTHY, BRIAN E. JULIUS,
SHARON SHUTLER AND STEPHANIE FLUKE

INTRODUCTION

Seagrasses are marine flowering plants consisting of 12 genera and approximately 60 species growing in all of the world's oceans with the exception of the most polar regions (den Hartog, 1970; Phillips & Meñez, 1988). Nearly all seagrasses grow in unconsolidated sediments in water depths ranging from the intertidal zone to as deep as 35–50 m. They are vascular plants anchored to soft sediments by a functional and complex rhizome and root system, with the exception of the genus *Phyllospadix* which grows on solid substrates along the Pacific coast of the United States. The seagrass leaf canopy baffles the flow of water, and together with their rhizome and root mat seagrasses stabilise sediments, cleanse the water column of fine particles, and recycle nutrients between the sediments and overlying waters (Fonseca, 1996). Numerous species of invertebrates and large vertebrates consume seagrasses as a portion of their diet, and the complex structure and physical stability provided by seagrasses form the basis for productive ecosystems consisting of plant and animal epiphytes, benthic macroalgae, invertebrates, mobile vertebrates and numerous other organisms (Thayer *et al.*, 1984). Many of the animal species that utilise seagrasses rely on their structural complexity to provide shelter and sources of food for their juvenile stages. This is one of the most important biological functions of the seagrass ecosystem.

The lack of taxonomic biodiversity in seagrasses is compensated by a wide diversity of size and morphological growth forms. The size and biomass of seagrass varies over an order of magnitude (Kenworthy *et al.*, 2000), resulting partly from genotypic differences as well as from phenotypic plasticity within individual species. For example, the canopy height of the smallest species known, *Halophila decipiens*, usually never exceeds 10 cm while species of *Zostera* can have canopies exceeding 5–7 m in height. The diversity of clonal growth forms and sexual reproductive strategies is accompanied by phenotypic variation that allows the limited number of seagrass species to occupy a wide range of environmental conditions from wave-swept shorelines to relatively deeper regions of continental shelves. Only a few other macroscopic plants growing in the ocean are capable of filling the niche type that seagrasses occupy.

Even though taxonomic biodiversity is limited in seagrasses, the diversity of size and morphological forms is accompanied by different growth and survival strategies uniquely adapted to the environments where the plants thrive. The range of growth strategies is also responsible for the patterns of seagrass bed development seen throughout the world. This is especially evident in multi-species tropical seagrass communities where distinctive successional processes are evident in the formation of stable climax communities and in their response to disturbance (Zieman, 1982). In tropical seagrass communities, colonising and climax species can be readily distinguished from one another and the unique attributes of these species can be utilised to enhance their protection and restoration (Fonseca *et al.*, 1987).

RATIONALE FOR RESTORATION

Fortunately, in many countries, the battle to recognise seagrasses as critical coastal ecosystems worthy of conservation and restoration has been won.

This recognition can be credited to the publication of thousands of papers from dozens of countries around the world representing years of research. To the best of our knowledge, research has yet to record a seagrass bed which is anything but a faunal-rich, highly productive ecosystem, that stabilises the sea floor, limits coastal erosion and filters the water column (Wood *et al.*, 1969). Thus, the ecological and sociological value of seagrasses has been broadly established (Wyllie-Echeverria *et al.*, 2000). Where these values are not recognised, it often appears to be the result of local political and development interests overriding conservation values (personal observation).

Threats to seagrass ecosystems and causes of degradation arise from a wide variety of sources. Eutrophication, coastal construction, motor vessel operation, fishing practices and many other activities have led to both local and regional losses of seagrasses (Short & Wyllie-Echeverria, 1996; Fonseca *et al.*, 1998a). Losses of seagrass also occur through natural processes such as disease (Muelstein, 1989; Robblee *et al.*, 1991), tropical cyclones (Preen *et al.*, 1995) and overgrazing by invertebrates (Rose *et al.*, 1999). Where the species composition and life-history strategies promote recolonisation, seagrasses can recover naturally from perturbations (Preen *et al.*, 1995). However, in many instances either the severity of the environmental modification responsible for the declines or the extremely slow rate of natural recovery leads to long-term losses. For example, in climax tropical communities dominated by *Thalassia testudinum* the time to full recovery in severely damaged vessel grounding sites can be more than a decade (Kenworthy *et al.*, 2000; Whitfield *et al.*, in press). In these instances, loss of seagrasses leads to numerous undesirable and difficult-to-reverse conditions, most importantly the elimination of habitat structure and the sediment stabilisation properties of the canopy and rhizome mat. A negative feedback on the ecosystem results; once the seagrass cover is lost and with it the self-sustaining properties of the system provided by the seagrasses, modification of the sediments and degradation of the water column may proceed without interruption. Seagrass restoration then becomes a much more difficult task, because it is nearly impossible to replace the attributes seagrasses provide, and a way to correct the physicochemical properties of the system must be found before reintroduction of the seagrasses can begin.

We posit that the issues regarding seagrass restoration are not the technology of planting and raising seagrass beds, but the failure to apply basic ecological principles in implementing restoration actions. Seagrasses can be readily transplanted and when sites are appropriately selected (see below and discussion in Fonseca *et al.*, 1998a), significant restoration successes have emerged. In fact, new technologies are continually being developed in both the deepwater (Perth, Western Australia: Fonseca *et al.*, 1998b; E. Paling, personal communication) and shallow water (Tampa Bay, Florida: J. Anderson, personal communication) approaches. Also, improvements in large-scale seeding techniques are being advanced which have promise with some seagrass species (Granger *et al.*, 2000; Orth *et al.*, 2000). We are only just beginning to recognise the many situations in which opportunities for substantial restoration have either been squandered or serious mistakes in site selection have been made, largely because those involved did not understand the habitat requirements and/or the life history of the plants with which they were working.

The ecological value of seagrasses translates into enormous commercial and social benefits. For example, in the Indian River Lagoon, Florida seagrass meadows have been described as the marine equivalent of tropical rainforests providing the ecological basis for fisheries worth about US$25 000 per hectare or a total of approximately 1 billion dollars a year (Virnstein & Morris, 1996). Seagrass-dependent fisheries and wildlife communities are the economic foundation for commercial and recreational fishermen as well as for a variety of industries and people that utilise the coastal zone for commerce and personal enjoyment. Socially, these values are transferred to the health and well-being of the families of these user groups and the regional economies of nations worldwide. The many physical, biological, economic and social attributes combine to make seagrasses an essential and ecologically important habitat in coastal marine ecosystems (Wyllie-Echeverria *et al.*, 2000); consequently they are in need of restoration where they have been anthropogenically injured or lost (Fonseca *et al.*, 1996; Sheridan, 1999).

PRINCIPLES OF RESTORATION

We base our assessment of the status of seagrass restoration on a perspective from within the United States legal framework. Seagrass beds in United States coastal waters are generally viewed as public trust resources, and such injuries to these resources are considered losses suffered by the public. A number of federal and state laws include liability provisions which allow the public to be compensated for injuries to seagrasses (for example, the United States's National Marine Sanctuary Act of 1972, 16 USC 1431 *et seq.*). To evaluate this loss in a fair and reasonable manner, we must consider not only the static loss in area and/or degree of the injury, but also the loss of resource services provided by the seagrass bed between the time it is injured and the time it recovers to 100% of pre-injury conditions (Fonseca *et al.*, 2000a). This approach is consistent with the 'no-net-loss' of wetlands policy that has become a benchmark of restoration strategies in the United States. Our more recent approach substitutes for the 'mitigation' or 'replacement ratio' used to identify the amount of habitat to be generated to offset the amount lost. In the past, use of replacement ratios has frequently led to undercompensation of lost resources because lost interim ecological services were not addressed.

Effecting no-net-loss and achieving recovery of interim resource services requires that the injured site be fully rehabilitated (on-site restoration), alternative compensatory restoration sites be found (off-site) or some degree of both. To limit the scope of discussion, we are focusing on in-kind restoration (i.e. seagrass service loss replaced by seagrass service gains). On-site restoration can often be achieved, but may require engineering interventions to 'fix' the site, such as filling excavation holes caused by vessel groundings or altering water flow. Off-site selections, in our experience, have had higher probabilities of restoration failure because inexperienced resource managers choose inappropriate sites. They are frequently under the impression that open habitat areas are prime sites for restoring seagrass when, in reality, the sites selected either cannot support seagrass, or currently support only low levels of seagrass. This fallacy has been addressed in detail in several publications (Fredette *et al.*, 1985; Fonseca *et al.*, 1987, 1998a; Fonseca, 1992, 1994). Suffice it to say, Fredette *et al.*'s (1985) condition 'If seagrass does not grow there now, what makes you think it can be established?' best sums the problem. Recently, Calumpong *et al.* (in press) listed the criteria for off-site selection that can be used to avoid off-site selection problems. By giving attention to these details of site selection, the probability of successful restoration can be greatly enhanced.

RESTORATION IN PRACTICE

We have dealt previously with what we consider to be the status of this aspect of restoration (Fonseca *et al.*, 1998a). However, there are at least four major deficiencies in the process of seagrass restoration. First, the choice of an appropriate metric for evaluating restoration has been elusive. We present here for the first time findings from a panel of United States seagrass experts that considered what are the appropriate metrics for tracking the performance of a seagrass restoration project. Second, setting fair, reasonable and consistent ratios for replacement of damaged seagrasses has also been at issue. We review the methodology used by the National Oceanic and Atmospheric Administration (NOAA) for defining the interim loss of resource services accrued by damage to seagrass beds and the process of computing compensatory restoration. Third, we feel that the weakest part of seagrass restoration has been the selection of the restoration site. We delve into the pitfalls of site-selection strategy – the point in the process where most plans go awry. For completeness, we briefly review the extant methodologies for restoring seagrass beds. Fourth and finally, finding a realistic basis for computing cost of these projects has been a vexing issue for years. Here we provide an evaluation of cost for the planning, implementation and monitoring of a seagrass restoration project based on a United States federal court case successfully prosecuted by NOAA.

Definition of injury and evaluation of lost interim services

Defining lost resource services

Computation of lost resource services requires three assessments: (1) area of habitat lost; (2) the length of time needed for the functions associated with that area (and lost to the ecosystem at large during the period of the injury) to recover to their pre-impact levels; and (3) the shape of that recovery function (Fonseca *et al.*, 2000a). Using seagrass ecosystems as an example, if 1 hectare of seagrass were destroyed today and replanted tomorrow and, for argument's sake, reached standards of equivalency (e.g. shoot density, biomass, coverage) in two years, the interim loss of ecological services over this two-year period would be relatively low. However, if the restoration of this site were not undertaken immediately and if the site required seven years to reach its pre-impact state, the level of compensation due the public for the interim losses from this same 1-hectare injury would be substantially higher. This highlights the weakness of fixed compensation ratios.

Actual projects rarely enjoy tight temporal coupling between either the injury and on-site repair work, or between the injury and the additional restoration required to compensate for the ecological services lost from the time of the injury until full recovery. Among other issues, it is very difficult to consistently locate and successfully create new seagrass habitat that meets ecologically responsible site-selection criteria, especially those criteria which preclude simply substituting naturally unvegetated bottom for vegetated bottom (Fonseca *et al.*, 1998a). Finding large areas of suitable substrate for restoration in close proximity to the impacted area is rare, and often results in restoration at sites physically

removed from the impact area. Thus, any functions affected by spatial elements of ecosystem linkages are lost (i.e. geographic setting). Second, the lost production was removed from a specific point in time. Therefore, in some instances it cannot be returned in a way to avoid disruption of ecosystem functions, such as the loss of last year's spawn of herring or set of bay scallops that might occur as a result of injury to a seagrass bed. Moreover, if there were a longer period of time between the injury and full recovery from the injury, then one could argue that replanting conducted a long time after an impact has less value than ones conducted sooner. This realisation is the basis for NOAA's more recent approach to objectively and quantitatively standardise the problem of computing interim lost services by habitat equivalency analysis (HEA). This approach provides a basis for setting replacement ratios and arriving at a quantity of persistent area of given quality that has been defined as an appropriate metric of success (Fonseca, 1989, 1992, 1994; Fonseca *et al.*, 1998a, 2000a).

Determination of interim loss and its implementation into the restoration process is tightly integrated with the establishment of a restoration plan. While such a plan must identify the mechanics of the physical restoration itself, the plan must also have a clear definition of injury, site selection, monitoring protocols and success. As mentioned earlier, those guidelines have been established (Fonseca, 1989, 1992, 1994), but have not yet been quantitatively coupled with the issue of interim loss to determine replacement ratios.

Recently, NOAA developed and implemented HEA using basic biological data to quantify interim lost resource services (NOAA, Damage Assessment and Restoration Program, 1997a). While sharing many of the same principles as other methods incorporating interim losses into replacement ratio calculations for wetlands (Unsworth & Bishop, 1994; King *et al.*, 1993), HEA focuses on the selection of a specific resource-based metric(s) as a proxy for the affected services (e.g. seagrass short-shoot density in the example discussed below), rather than basing its calculations on a broad aggregation of injured resources. Determination of this metric was one of the conclusions from the expert panel as discussed in Box 7.5 (biomass, as opposed to shoot density, has not yet been adopted because of a lack of empirical data on the recovery rate of belowground biomass, whereas recovery rate of shoots is a robust data set; this choice is an extremely generous concession to the responsible parties). This approach has the advantage of making HEA applicable not only to a wide range of different habitats, but to injuries to individual species as well (see Chapman *et al.* [1998] for a discussion of HEA applied to the calculation of compensation for historic salmon losses). Additionally, the selection of a resource-based metric allows for differences in the quality of services provided by the injured and replacement resources to be captured and incorporated into the replacement ratio (NOAA, Damage Assessment and Restoration Program, 1997*b*). Without specification of a quantifiable resource metric, analysis of the recovery of the resource following injury and/or the success of the restoration project may be difficult to evaluate precisely. For example, in the wetlands context, alternative metric specifications may lead to significantly different maturity horizons (Broome *et al.*, 1986) as well as the level of functional equivalence ultimately achieved by the restoration project (Zedler & Langis, 1991).

Box 7.1   Application of the habitat equivalency analysis (HEA)

An example of applying HEA to habitat restoration occurred in a recent federal court case to provide compensation for the loss of 1.63 acres of seagrasses (turtlegrass, *Thalassia testudinum*) within the Florida Keys National Marine Sanctuary (*United States of America* v. *Melvin A. Fisher et al.*, 1997). Extremely energetic hydrodynamic conditions at the injury site, together with intense grazing of the seagrass by nocturnal herbivores prevented successful establishment of seagrass plantings. Therefore, off-site restoration was chosen in the form of in-kind (same species) repair of *T. testudinum* beds previously damaged by boat propeller scars (prop scars). This approach focused initially on planting a native pioneering seagrass species, *Halodule wrightii*, to facilitate the eventual recovery of the slower-growing *T. testudinum*. This sequence, termed 'compressed succession' (M. Moffler, personal communication) promotes more suitable conditions for the slower-growing *T. testudinum* to encroach naturally upon the prop scar while temporarily stabilizing the site and preventing additional erosion with a more rapidly growing species (see Box 7.3). Project success was to be quantified by four parameters: (1) at planting, a minimum average of one horizontal *H. wrightii* rhizome apical per planting unit must be installed; (2) 75% survival of planting units at the end of year 1; (3) seagrass shoot density (as compared to nearby natural beds); and (4) achieve the target acreage of bottom coverage within a three-year monitoring period. Additionally, if monitoring indicated that performance standards were not being met or were not projected to be met, remedial plantings of those affected areas were designed into the plan. However, all remedial plantings reset the monitoring clock for that portion of the project. The ultimate success criterion was unassisted persistence of target bottom coverage by the seagrass plantings for three years, using photo-documentation to provide a common basis of assessment, perception, and historical reference.

Key factors in the National Oceanic and Atmospheric Administration (NOAA)'s development of restoration plans have been issues of pre-project planning, particularly regarding site suitability. Here, sites were reviewed for suitability using the following criteria: (1) they were adjacent to natural seagrass beds at similar depths; (2) they were anthropogenically disturbed; (3) they existed in areas that were not subject to chronic storm disruption; (4) they were not undergoing rapid and extensive natural recolonisation by seagrasses; (5) seagrass restoration had been successful at similar sites; (6) there was sufficient area to conduct the project; and (7) similar quality habitat would be restored as was lost. The restoration of seagrass prop scars created by vessel impacts represented NOAA's preferred approach to seagrass restoration off-site. In order to select a planting site that could accommodate the project's size, the amount of restored area was computed using the HEA.

Description of the compensatory restoration scaling approach

Accurate determination of the appropriate target scale of compensatory restoration[1] projects is necessary to ensure that the public and the environment are adequately compensated for the interim service losses. For injuries to seagrass resources, NOAA has employed HEA as the primary methodology for scaling compensatory restoration projects. The principal concept underlying HEA is that the public and the environment can be made whole for injuries to natural resources through the implementation of restoration projects that provide resources and services of the same type, quality and comparable value. HEA has been applied in cases centered on seagrass injuries because those incidents typically meet the three criteria defined by NOAA: (1) the primary category of lost on-site services pertains to the biological function of an area (as opposed to direct human uses, such as recreational services); (2) feasible restoration projects are available that provide services of the same type and quality and are comparable in value to those lost; and (3) sufficient data on the required HEA input parameters exist or are cost-effective to collect. If these criteria are not met for a particular injury, other valid, reliable approaches and methodologies are available for scaling the chosen compensatory restoration projects (NOAA, 1997b). These criteria for the use of HEA were upheld by the US District Court (*United States of America* v. *Melvin A. Fisher et. al.* 1997 92-10027-CIV-DAVIS). Of equal importance to the Mel Fisher decision was the decision by the US District Court in *United States of America* v. *Great Lakes Dredge & Dock Co.* 1999 97-2510-CIV-DAVIS to uphold the use of the HEA as a proper method by which to scale compensatory restoration.

At its most basic level, HEA determines the appropriate scale of a compensatory restoration project by adjusting the project scale such that the present value of the compensatory project is equal to the present value of interim losses due to the injury of that action (e.g., freshwater diversion projects intended to create wetland acreage).[2] This 'balancing' of gains and losses is accomplished through a four-step process (NOAA, 1997a). First (step 1), the extent, severity, and duration of the injury (from the time of the injury until the resource reaches its point of maximum recovery), and functional form of the recovery curve must be determined, in order to calculate the total interim resource service losses. Next (step 2), the resource services provided by the compensatory project over the full life of the project must be estimated to quantify the benefits attributable to the restoration. This step is analogous to the previous one and requires estimation of both the time required for the compensatory restoration project to reach its maximum level of service provision and the functional form of the maturity curve. After these resource service losses and gains have been quantified, the scale of the compensatory project is adjusted until the projected future resource service gains are equal to the interim losses associated with the injury (step 3). This process is depicted graphically in Fig. 7.1, where the scale of the compensatory restoration project is adjusted until the area under the maturity curve (the total resource service gains, represented by area B) is equal to the interim lost resource services (represented by area A). Because these services are occurring at different points in time, they must be translated into comparable present value terms through the use of a discount rate.

---

[1]Compensatory restoration refers to any action taken to compensate for interim losses of natural resources and services that occur from the point of the injury until recovery of those resources/services to baseline. Conversely, primary restoration refers to actions that return the injured natural resources and services to baseline.

[2]In some instances, it may be beneficial to all parties involved to implement a project where the total discounted gains from the compensatory project exceed the total discounted losses. This situation occurs when the scale of the preferred project can only be adjusted according to a binary or stepwise function rather than a continuous function, or when the resulting amount of natural resources/services generated by a restoration action cannot be tightly controlled following implementation.

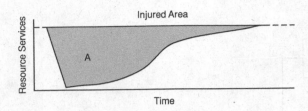

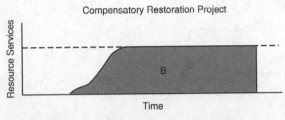

Fig. 7.1. Graphical depiction of how habitat equivalency analysis (HEA) sets the compensatory restoration to equal interim loss of resource services. This is achieved by setting the total services (hectare–years) lost until complete recovery back to pre-injury conditions (area A) is equalled by the services rendered under the compensatory project (area B).

Discounting is a standard economic procedure that adjusts for the public's preferences for having resources available in the present period relative to a specified time in the future. Because of discounting, plantings that occur longer after an impact are worth less in present-value terms than plantings conducted shortly after an impact, and therefore more planting must be done as time elapses. Finally (step 4), appropriate performance standards associated with the compensatory restoration must be developed to ensure that the project provides the anticipated level of services. Well-defined and measurable standards are essential to the success of the project regardless of whether the restoration will be implemented by the parties responsible for the original resource injury or by the management agency (trustees) using monetary damages which are recovered.

In Box 7.5, we present the outcome of a national workshop that set the stage for NOAA to provide reasonable and fair assessments of injuries to seagrasses and the effort needed to recover the lost resources which must be assumed by the responsible party.

The importance of site selection

Clearly one of the largest problems with seagrass transplanting is finding an appropriate place to conduct the restoration and install the plantings. It is not advised to plant seagrasses in areas with no history of seagrass growth, or where the aforementioned disturbances have not ceased. Planting should not be done under those circumstances because of the low probability of success. Planting may be done in open, unvegetated areas among patches of seagrass, but only for the goal of experimental manipulations and/or the evaluation of planting techniques (keeping in mind that these among-patch locations are not a strong test of the efficacy of a technique as they are embedded within viable seagrass territory). Seagrass patches migrate, alternately colonising currently unvegetated sea floor and dying out where seagrass is located presently (Marba *et al.*, 1994; Marba & Duarte, 1995; Fonseca *et al.*, 1998*a*, 2000*b*). Thus, the spaces between the patches today may be naturally colonised by seagrasses in the future.

Campbell *et al.* (2000) provide a decision strategy for assessing the selection planting sites that include measures of light, epiphytisation, nutrient loading, water motion, depth, proximity of donor site and alternative actions (Fig. 7.2). Similarly, Fonseca *et al.* (2000*a*) and Calumpong *et al.* (in press) give the following criteria for the selection of a restoration site away from the original injury site:

- It is at depths similar to nearby seagrass beds
- It was anthropogenically disturbed
- It exists in areas that are not subject to chronic storm damage
- It is not undergoing rapid and extensive natural re-colonisation by seagrasses
- Seagrass restoration has been successful at similar sites
- There is sufficient area to conduct the project
- Similar quality habitat would be restored as was lost.

These selection criteria have been used successfully in the US Federal Court as the basis for seagrass restoration projects (*United States of America v. Melvin A. Fisher et al.* 1997). By considering these

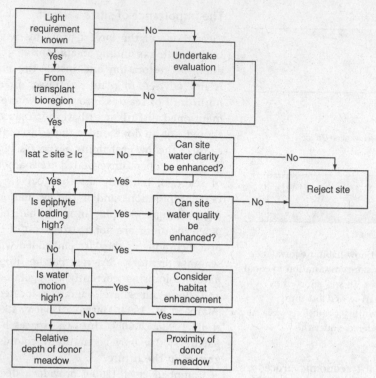

Fig. 7.2. Decision flow diagram regarding site selection for restoration. From Campbell *et al.* (2000). Isat = saturation irradiance, Ic = compensation irradiance.

criteria, it is apparent that transplantation should probably not be undertaken for the purposes of enhancing recovery from natural disturbance events as these events have both an ecological and evolutionary function in determining the survival and fitness of the seagrass ecosystem. When possible, rehabilitation of the primary injury site should be performed to restore or accelerate the recovery of baseline service flows, with compensatory restoration used to compensate the public for interim service losses that accrue while the site reaches its pre-injury levels of service provision.

Critical factors influencing transplant success

Numerous factors have been determined to affect transplanting success. Some are of the crop-risk type, are extrinsic and cannot be controlled. Others involve issues of protocol. In a survey of North American seagrass planting projects, Fonseca *et al.* (1998*a*) listed the following as factors that had the

potential of being controlled by those conducting transplants:

- Similarity of environmental conditions of donor and recipient beds.
- Choice of species: preferably same as that lost, but pioneering species may be substituted to initiate a project.
- Presence of grazers or sediment burrowers: these bioturbating organisms may need to be excluded or plantings may have to be conducted in large patches to dissuade them from their activities.
- Source of planting material: similar depths and environmental conditions and from over as broad a geographic area as possible to ensure genetic diversity.
- Time of year: seagrass should be planted at a time to ensure the longest period before seasonal stressors.
- Cost: many variations of cost have been given but standardised costs are elusive; based on recent cases in the US federal court, a contracted project that

Box 7.2 The need for physical stabilisation after boat grounding: *United States of America v. Great Lakes Dredge & Dock Co.*

Another example of the difficulties faced in primary restoration of seagrass habitat was recently illustrated in another Federal Court case (*United States of America v. Great Lakes Dredge & Dock Co. and Coastal Marine Towing Inc.* 1999) where a large tugboat grounded on a shallow seagrass–*Porites* coral bank in 1993 and destroyed 7200 m² of habitat. The grounding site was located in an exposed, high-energy environment where seagrass transplantation was deemed inappropriate. The expert case team assembled by NOAA and the US Department of Justice recommended that the primary restoration plan should include filling and regrading the trench made by the tugboat to physically stabilise the site. It was assumed that once the site was physically stable the seagrasses would recolonise naturally, but slowly. Interim losses of seagrass would be compensated off-site in a plan similar to the plan for the *United States of America* v. *Melvin Fisher et al.* described previously. Attempts to negotiate a settlement with the responsible parties proceeded for five years without a resolution. In September 1998 the site was impacted by Hurricane Georges which severely damaged the partially recovered portions of the injury and effectively set back the recovery clock (Whitfield *et al.*, in press). Seagrass beds adjacent to the injured area were unaffected. The impact of the hurricane confirmed initial concerns that the grounding site was physically unstable and vulnerable to further injury. Clearly, there are cases in high-energy environments where physical restoration is needed to stabilise injured sites to promote the recovery of seagrasses. In the primary restoration plan for this case, the amount of sediments excavated by the grounding called for substantial *in situ* engineering to recreate the bank structure. This is not surprising, since it took nature between 500 and 1000 years to form the bank that was destroyed. Recent studies have shown that large vessel groundings are becoming more common on seagrass banks in south Florida and elsewhere in the Caribbean (Whitfield *et al.*, in press). Whitfield *et al.* (in press) have documented the instability of these injuries and recommend that physical regrading is necessary to prevent further damage during severe storms.

includes site surveys, planting, monitoring and reporting will cost (in 1996 US dollars) ~US$630 000 per hectare.

It is essential to study the substrate-energy (exposure) regime, and optical water quality (clarity or light availability) of the area that will be transplanted so that suitable source materials can be identified. Areas exposed at low tides should be carefully mapped so as to place plantings with minimal exposure to air, unless the plants are regularly occurring in the intertidal zone (e.g. in the Pacific Northwest of the United States). Planting in high wave energy or tidal current areas will require planting in larger groups to avoid disruption (Fonseca *et al.*, 1998a, b). Planting in larger groups also appears to be an effective method of deterring physical disruption of the planting by marine organisms. However, as suggested by Addy (1947), matching water depths, temperature, salinity, water clarity and plant size remain some of the best general guidelines for matching donor and recipient beds.

The characteristics of the species, such as fast growing vs. slow growing; pioneering vs. climax, annual vs. perennial growth, etc. must be considered before conducting transplantation work. For example, *Halodule* spp. and *Halophila* spp. are fast-growing pioneering species while *Thalassia* spp. and *Enhalus* spp. are slow-growing climax species. *Halophila* spp. rapidly colonise disturbed areas like those with moving sand bars and are under-canopy species, requiring low light. Although a climax species may have been disturbed, it is often advisable to first install a faster-growing species to stabilise the environment.

Another important factor in the selection of seagrass for transplanting, besides their intrinsic recovery rate, is their growth habit (Short & Short, 2000). Transplanting can be rendered almost wholly ineffective if meristematic regions of these plants are

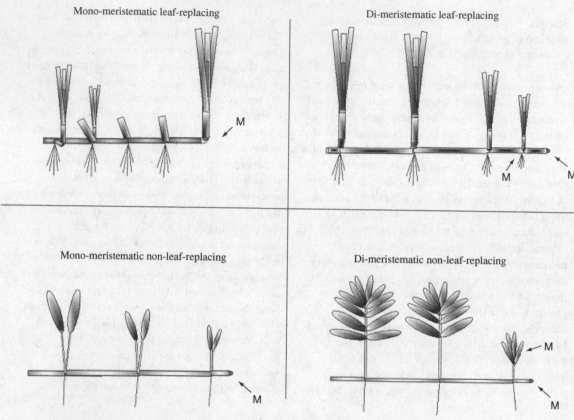

Mono-meristematic leaf-replacing

Di-meristematic leaf-replacing

Mono-meristematic non-leaf-replacing

Di-meristematic non-leaf-replacing

Fig. 7.3. The four basic growth forms of seagrass. In the case of the mono-meristematic, leaf-replacing form, each terminal shoot on the runner is a viable planting unit – comparatively low modular integration is present meaning that most often each shoot has high potential for contributing to spatial colonisation. The other three forms require at least three to four short shoots be maintained on the runner for a complete planting unit, as well as an intact rhizome apical meristem. From Short & Short (2000).

damaged or not incorporated in sufficient quantity in a planting unit to initiate recolonisation. Short & Short (2000) summarise the morphotypes of seagrass (Fig. 7.3).

Seagrass grazers can have disastrous effects on plantings. Seagrass grazers include sea-urchins, gastropods and herbivorous fishes. Some migratory waterfowl such as geese and ducks have been observed to decimate seagrass plantings (personal observation). Significant grazing of natural *Syringodium filiforme* beds points out the general susceptibility of seagrasses to grazing (Rose *et al.*, 1999). Fonseca *et al.* (1994 and references therein) found significant disturbance by rays in Tampa Bay,

Florida, indicating that it would be necessary to use exclosure cages to ensure the survival of transplanted seagrasses in some areas. Recently, we have seen that planting in clumps of at least 20–50 cm on a side deters many animals from disturbing the plantings (authors' unpublished data).

Minimisation of disturbance to the source bed is paramount in seagrass transplanting so as not to exacerbate injury to local populations. With present techniques focusing on the use of wild, vegetative stocks, this may be achieved by conducting the transplantation in phases, or dispersing the collection effort, thus allowing the source bed to recover. Harvesting of donor stock should also be done from

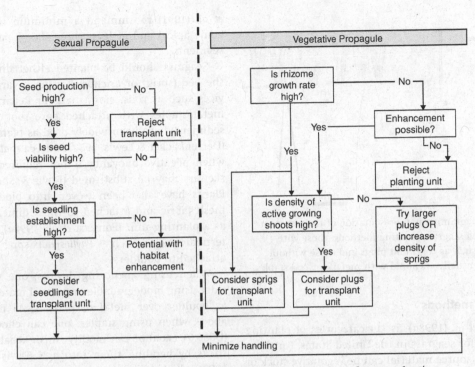

Fig. 7.4. Decision tree for choosing seedlings or whole, mature plants for transplanting. We caution that the technology for seed establishment is not as developed as for the use of sprigs or cores. From Campbell *et al.* (2000).

beds over as broad a geographical area as possible. This may help avoid loss of genetic diversity in the planted bed (*sensu* Williams & Orth, 1998) and may actually incorporate the full local range of genetic diversity into the planting.

Fortunately, many pioneering species can be harvested with minimal disturbance to the beds (Fonseca *et al.*, 1994). However, for climax species, harvesting of donor beds may cause long-lasting damage and harvesting from these beds should only occur when the beds are under some anthropogenic source of physiological stress that does not seem likely to abate or if they are in imminent danger of physical removal (e.g. dredging).

The size of the source or donor bed should first be assessed to determine if recovery will proceed after removal of the sods, cores or sprigs. This is especially true when transplanting vegetative stock, as a large amount of material is needed. Spacing harvesting at ~0.25 m for small cores or sods (<0.15 × 0.15 m) is often sufficient to avoid long-lasting damage. More-

over, Fonseca *et al.* (1998*a*) suggested that *Ruppia, Halophila, Halodule* and *Zostera* spp. can recover in small patches (<0.25 m²) within a year with shoot density returning to normal. Furthermore, Fonseca *et al.* (1998*a*) cautioned that patches >~30 m² in high-current areas may never recover. Campbell *et al.* (2000) also provide a decision tree for selection of planting stock for both sexual and asexual propagules that focuses on intrinsic propagation rates (Fig. 7.4).

Choosing the time-frame for planting is an obvious concern, and as with all crops, the appropriate time for seagrass varies with geographical region. In general, the best strategy is to plant at a time just after the period of highest seasonal stress, when natural populations are experiencing recovery. For example, eelgrass (*Zostera marina*) should be planted in the autumn in North Carolina, and other mid-Atlantic regions in the United States, because summer is the period of maximum physiological stress at that location (Moore *et al.*, 1997).

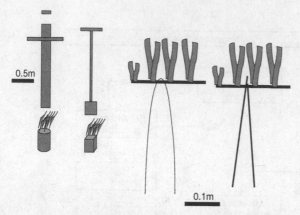

Fig. 7.5. Diagrammatic representation of the two most widely used seagrass planting methods; those with sediment such as cores and plugs and those without sediment, usually anchored with metal or wood staples.

Planting methods

Fonseca *et al.* (1998a) list 14 categories of planting methods for seagrass in the United States. For these methods, source material can be vegetative stock or seeds. Transplantation using vegetative stock typically requires available wild stock as a source and is labour-intensive and invariably expensive. However, it often gives faster, more reliable coverage than seed methods (but see review by Orth *et al.*, 2000). Most projects today are carried out using either small sods or sprigging of sediment-free units (Fig. 7.5).

Sediment-free methods

For most sediment-free methods, plants are dug up using shovels, the sediment is shaken from the roots and rhizomes and the plants are placed in flowing seawater tanks or floating pens. For vegetative stocks, Fonseca *et al.* (1998a) recommend a minimum of one apical shoot per planting unit. The number of short shoots on a long shoot should be maximised whenever possible, so as to derive benefits from the clonal nature of the plant. Also, the plants should be collected and planted on the same day, kept in water with the same ambient temperature and salinity, and kept as moist as possible when out of the water. When using vegetative stocks of *Thalassia testudinum*, Tomasko

et al. (1991) recommend a minimum of one rhizome apical and at least three shoots per rhizome segment.

Seagrass should be planted either directly into the bed (sprig) or anchored using a variety of devices such as rods, rings, nails or Rebar. U-shaped metal staples with attached bare root sprigs (no sediment) have been widely used as planting units (Derrenbacker & Lewis, 1982; Fonseca *et al.*, 1982) or, when negative buoyancy is not required, bamboo skewers may be substituted (Davis & Short, 1997). Plants have also been woven into biodegradable mesh fabric and attached to the sediment surface as a planting unit (Fonseca *et al.*, 1998a). Rocky intertidal species, such as *Phyllospadix* spp., have been attached to boulders.

When using anchoring devices, one must consider using biodegradable or natural materials such as boulder over metal or plastics. As mentioned above, when using staples, one can choose metal (US$0.01 each) or can modify 'shish kebab' bamboo sticks by bending them into a V (Davis & Short, 1997) which when purchased in bulk could cost only US$0.006 a piece. In tropical areas where bamboo is plentiful, this could be a more economical medium to use. Bamboo is also biodegradable. Using either kind of staple, planting units are made by grouping plants and attaching the root–rhizome portion under the bridge of a staple and securing the plants with a paper-coated metal twist-tie. This can either be prepared beforehand or the planting unit can be pinned directly to the substrate during planting.

When using nails, boulders or Popsicle sticks (Merkel, 1988), the technique is more or less similar and the planting unit is tied to the anchoring instrument. Frames, such as Short's TERF device (F. T. Short, in Fonseca *et al.*, 1998a), have great promise for rapid and non-diver-assisted planting at depth. A cage deployment system that has shoots attached to the bottom is lowered onto the sea floor and retrieved after the shoots have rooted and their paper ties have decomposed. This eliminates the need for divers in deeper water, can be used in chemically polluted areas, and provides initial protection of the plantings from biological disturbance.

Seagrass with sediment methods

The sod or turf method consists of planting a shovel-full of seagrass with sediment and rhizomes intact. This is the easiest method, and is most applicable for hard, compact substrates and deep-rooted and large species such as *Enhalus acoroides*. The only equipment needed are shovels and large basins for the sods. However, if the donor site is far away, transporting the sods may present a problem as the weight of the material is a physical burden. Some species, such as *E. acoroides*, *Posidonia* spp. and *Thalassia* spp. may have very deep root–rhizome systems requiring removal of a tremendous amount of sediment to harvest the belowground plant structures all intact (Fig. 7.6). To our knowledge, this has only been accomplished in Western Australia (by E. Paling, of Murdoch University; see review in Fonseca *et al.*, 1998*b*). Furthermore, harvesting an entire sod may constitute one of the most severe perturbations in a seagrass meadow, inhibiting recovery in the donor bed.

The plug method utilises tubes as coring devices to extract the plants with the sediment and rhizomes intact. The plugs are planted directly into the seagrass bed after creation of a hole to receive the contents of the tube. The core tubes are usually made of 4–6-cm diameter PVC plastic pipe with caps for both ends to initially create a vacuum and keep sediments from washing out the bottom. The tube is inserted into the sediment, capped (which creates a vacuum), pulled from the sediment and capped at the other end to avoid losing the plug. This can only be done with soft but cohesive sediments and generally only for small species to avoid excessive leaf shearing (unless extreme care is taken to avoid the shearing, which adds measurably to the cost of the process). When the donor bed is far away from the planting site, many tubes are needed which also adds to the cost.

Sod pluggers extract a plug out of the donor bed which is then extruded into a peat pot; the method was first used by Robilliard & Porter (1976)

Fig. 7.6. Harvest of *Posidonia* sod near Perth, Western Australia. Photo courtesy E. Paling.

and modified by Fonseca *et al.* (1994). Because these pots are typically only a few cm across, they may be inserted into the bottom by liquefying the sediment with a hand tool. After the peat pot is planted, its side walls must be ripped off or torn down and the pot pushed into the sediment to allow the rhizomes to spread out.

Sowing of seed

Seed planting holds promise for large-scale restoration but is currently more applicable only in low-energy areas where the seeds can settle and germinate and where there are few seed predators. This method was first introduced by Thorhaug (1974) with *Thalassia testudinum*. A seedling grow-out method for *T. testudinum* has been registered by Lewis (1987). The availability of seeds must also be considered. Large areas in the Chesapeake Bay have been established by sowing seeds from a small boat (R. J. Orth, personal communication). Work continues in this highly promising area (Orth *et al.*, 2000). Experiments using seeds pelletised to increase their density to facilitate sinking and seeds embedded in biodegradable mesh are presently being carried out by Granger and his colleagues (Granger *et al.*, 2000). Experiments on planting depth also indicate that at least for *Zostera marina*, seeds should be within the top 2 cm of the sediment for best germination and that sowing densities should be 400–1000 seeds per square metre (Granger *et al.*, 2000).

Laboratory cultured stocks

This approach uses plants reared and grown in the laboratory from plant fragments. It may become especially applicable for large-scale plantings where a large amount of planting units is needed. This technique also holds promise for reducing or eliminating donor bed damage and this has been shown to be minimal for pioneering species, such as *Halodule wrightii* and *Syringodium filiforme* (Fonseca *et al.*, 1994). This approach also has the potential to maintain donor stocks for unscheduled plantings and could theoretically supply genetically variable and disease-resistant plants.

Several aspects of this approach remain controversial. So far, three species have been successfully propagated in the laboratory, *Ruppia maritima*, *Halophila decipiens* and *H. engelmannii* (M. Durako,

personal communication). *Ruppia maritima* has been successfully transplanted from laboratory culture stock (Bird *et al.*, 1994), but all these species are naturally fast growing (i.e. pioneering) and it is unclear whether laboratory culture is a cost-effective means of restoring naturally prolific species. Moreover, questions regarding the ability to maintain genetic structure of the population have not been solved. Given the growing emphasis on mechanised plantings using wild stock, laboratory culture will probably only be cost-effective when techniques are developed for slow-growing species, hence avoiding long-term donor bed impacts.

Monitoring the restoration

Monitoring of the restoration project is necessary to provide data required to evaluate the viability of the project based on the performance standards (defined below). This permits timely identification of problems or conditions that may require corrective action to ensure the success of the project.

Monitoring schedule and activities

Field collection of data for performance monitoring should occur for four years after planting. Original plantings should be monitored for three years and potential remedial plantings in year 2 should be monitored for three years for a total monitoring period of four years. Under this schedule the monitoring would be conducted as follows:

year 1 – day 60, 180, 365
year 2 – day 180, 365
year 3 – day 180, 365
year 4 – day 180, 365

The precise dates are weather-dependent. In carbonate sediments, each surviving planting unit should receive an additional spike of constant-release phosphorous fertiliser (0–39–0, nitrogen–phosphorus–potassium) at day 60 of year 1. Alternatively, bird roosting stakes could be installed about every 5–10 m along scars (see Box 7.3).

Data collection

Monitoring should focus on documenting the numbers of apicals at planting time, planting unit survival, shoot density and areal coverage under the following schedule and definitions. This monitoring

Box 7.3 Transplanting strategies

RESTORING SLOW-GROWING SPECIES

Durako et al. (1992) developed a strategy for restoring slow-growing species such as Thalassia testudinum by mimicking natural succession. They initially plant another, faster-spreading congener, such as Halodule wrightii to achieve 'compressed succession' (see Box 7.1). This temporary substitution of a faster-growing species for a slow-growing one promotes a more suitable condition for the slow-growing one to establish itself while the faster one stabilises the sediment and provides a functional seagrass habitat.

Based on a series of ecological field experiments (Fourqurean et al., 1995) and transplanting studies (Kenworthy et al., 2000), we have determined that 'compressed succession' of H. wrightii can be enhanced by fertilising transplants with bird roosting stakes. The procedure works by placing bird roosting stakes over the transplants. Seabirds roosting on the stages defecate nutrients into the water, stimulating rapid growth of H. wrightii transplants. In cases where there is a mixed meadow of T. testudinum and H. wrightii recovery of an injured area can be accelerated by placement of stakes alone.

STEPS IN A GENERALIZED PROTOCOL

- Study the site to be restored and determine at least the following parameters; (a) seagrass bed history (species composition, cause of loss); (b) exposure to air and waves and currents; (c) substrate type – avoid clays and high organic sediments; (d) rate of siltation – plants often cannot withstand much more than 25% burial and vertical growth is possible only for some species; and (e) presence of animal disturbance.
- Determine time-frame and budget by evaluating the typical staffing requirements. Merkel (1992) estimated a minimum of seven persons were required for intertidal, bare-root (e.g., staple technique) and nine persons for subtidal bare-root planting. The use of scuba incurs higher costs that need to be factored in. In the Philippines, Calumpong et al. (1992) accomplished planting of 32 0.125 m^2 sods in the inter- and subtidal with nine persons in one day. Fonseca et al. (1994) provide a comparison among several methods using timed trials. They found that collection + fabrication + planting costs ranged from ~1.2 to 3.5 minutes per planting unit with peat pot plugs being the most rapid method and cores being the slowest method.

It is imperative that one recognises what is not timed in that report, such as mobilisation and demobilisation, both daily and on a project-scale basis, travel, reporting and monitoring. The timed trials also did not accurately measure the effect of boredom on the speed of the process.

- Locate a donor bed that matches the conditions at the planting site. For vegetative methods, this should be near enough so the shoots can be planted the same day. Overnight storage of material, particularly bare-root material, should be placed in moving seawater at ambient temperatures and salinities. To our knowledge, there has not been a sufficiently controlled experiment to determine the storage capability for seagrasses.
- Be prepared to manage the workforce with regard to the tedium of tasks. To avoid boredom, varying tasks among individuals can be a useful strategy.
- Carefully delineate plots to facilitate monitoring.
- Consider all the aforementioned potential costs. These include site delineation, reports, mobilisation and demobilisation, insurance, overhead, benefits, mapping, planting operations, monitoring, remedial planting and a 10% profit margin for contractors.
- Conduct thorough monitoring (see below) and be prepared to conduct remedial plantings.

protocol applies to original plantings for three years (years 1–3) and to remedial plantings for three years (years 2–4).

1. Apical counts. Prior to planting, one planting unit out of every 100 collected should be examined for the number of rhizome apicals.

2. Survival. Each site should be examined for survival of all planting units during each survey in year 1 (days 60, 180 and 365) or until coalescence. Survival of each species should be expressed as a percentage of the original number, but the actual whole number should also be reported.

3. Shoot density. A separate (from survival) random

selection of three planting units per 100 planted should be assessed for number of shoots per planting unit at each survey time until coalescence begins. After some planting units begin to coalesce, three randomly selected locations per 100 m² (100 planting units) should be surveyed for shoot density over a 1 m² area at 0.0625 m² (25 cm×25 cm) resolution. Shoot density should be monitored for three years.

4. Areal coverage. The randomly selected planting units (may be the same as shoot density selection) should be surveyed for coverage at each survey time starting at day 180 of year 1. Measurements should be taken at a 0.0025 m² (5 cm × 5 cm) resolution prior to coalescence and over a 1 m² area at 0.0625 m² (25 cm × 25 cm) resolution after coalescence for each seagrass species present at each survey time. Areal coverage should also be monitored for three years.

5. Video tape transects. Five 100-m transects along randomly selected portions of the planted area should be video tape recorded to establish permanent visual documentation of the progression of areal coverage of seagrass through time. A tape measure should be laid along the central (long) axis of the scar and should be included in the video tape to allow physical reference of locations within the scar. Video recordings should be taken at each survey time during the monitoring period of three years. Observation-based assessment of success may be substituted if quadrats are used in accordance with a Braun–Blanquet survey method (Fonseca *et al.*, 1998*a*) or if the data are obtained from the video tape (making the observational data base available for cross-checking). The same number of sample points must be obtained with the same spatial extent (i.e. survey each scar). Similarly, Braun–Blanquet observations of cover at every metre along each scar may also be obtained from the video tape to obtain estimates of planting performance.

Reporting requirements

Monitoring reports should include copies of raw data gathered in each survey, an analysis of the data, and a discussion of the analysis. Originals of all video tapes recorded since the previous report should be provided with each new report. Originals of all video tapes and other photography should be turned over to the permitting agency following project completion by the party conducting the monitoring.

Remedial plantings and/or project modifications

If data from a monitoring report establishes that the performance standards are not being met or are projected not to be met, remedial plantings of those

---

Box 7.4   Performance standards

Although it is the overall objective to restore the species that was injured, performance criteria may also be based on the success of planting of pioneering seagrass species, as found when *Halodule* is installed to expedite the recovery of *Thalassia* (Fonseca *et al.*, 1998*a*).

APICALS
A minimum average of one horizontal rhizome apical per unit should be maintained in all original planting and remedial planting.

SURVIVAL
The survival rate shall be considered successful if a minimum of 75% of the planting units have established themselves by the end of year 1. If it is determined that less than 75% survival has occurred by the end of year 1, then remedial planting should occur during the next available planting period to bring the percentage survival rate to the minimum standard by the next monitoring survey.

GROWTH
The third success criterion should be the measured growth rate of bottom coverage. The growth rate should be considered successful if, starting after one year, the planting is projected to achieve the total desired acreage of bottom coverage, with 95% statistical confidence, within the three-year monitoring period for original plantings. If this criterion is not met, then remedial planting should occur during the next available planting period.

affected seagrass species should occur. If there is a recurring problem with survival of plantings or replantings in a particular area, remedial planting should occur in another suitable area in as close proximity as possible, subject to the approval of permitting agencies.

Based on past experience in seagrass restoration efforts, it is assumed that 30% of the planted area should require remedial planting in year 2. All original plantings should be monitored for three years. Remedial plantings should also be monitored for three years.

Box 7.5 National workshop for defining the metrics of assessment for seagrass restoration projects

National Oceanic and Atmospheric Administration (NOAA) is designated as a trustee for natural resources under several United States laws,[1] and, in that capacity, is authorised to act on behalf of the public to seek compensation for injuries to its trust resources. Under each of these statutes, natural resource damage claims are composed of three basic components: (1) the cost of restoring the injured resource to baseline; (2) compensation for interim lost resource services[2] from the time of the injury until restoration occurs; and (3) the cost of performing the damage assessment. Habitat equivalency analysis (HEA) is one of the more frequently used methodologies available to natural resource trustees for calculating the appropriate scale of restoration projects necessary to compensate for interim resource service losses (Fonseca et al., 2000a). The basic approach underlying HEA is to determine the amount of compensatory habitat to be restored, enhanced and/or created, such that the total services provided by the compensatory project over its functional lifespan are equal to the total services lost due to the resource injury.[3]

While HEA is conceptually and computationally straightforward, proper application of this approach requires a detailed understanding of the biological and ecological processes that affect the recovery and productivity of injured and restored habitats. In order to gain a better understanding of these processes, NOAA is undertaking a systematic, expert review of the ecological assumptions made within the HEA framework for a number of habitats for which HEA is most frequently applied. Seagrass habitats were selected as the first habitat to review for two primary reasons: (1) NOAA expects that due to the frequency of injuries to seagrasses (more than 70 000 hectares of injured seagrass in Florida alone), HEA will be commonly applied in cases involving injuries to seagrass habitats; and (2) the relatively small number of species of seagrasses present within areas under NOAA's trusteeship made this habitat a logical starting-point from which to develop and refine the review process for more diverse, complex habitats.

WORKSHOP PARTICIPANTS

Under the direction of NOAA's Southeast Fisheries Science Center, several academic experts in addition to NOAA staff were selected to discuss the underlying assumptions of HEA for seagrass environments. The workgroup was assembled to be geographically diverse, as well as to reflect a range of specialties within the field of seagrass biology/ecology. The workgroup participants were: Susan Bell, University of South Florida; Kenneth Moore, Virginia Institute of Marine Science; Mary Ruckelshaus, Florida State University; Frederick Short, University of New Hampshire; Charles Simenstad, University of Washington, Mark Fonseca,

---

[1] Notably, the Comprehensive Environmental Response, Compensation, Liability Act, the Clean Water Act, the National Marine Sanctuaries Act, and the Oil Pollution Act of 1990.

[2] Services here refer to the functions that one resource performs for another or for humans.

[3] For a detailed discussion of HEA, see National Oceanic and Atmospheric Administration, Damage Assessment and Restoration Program (March 1995, revised October 2000), Habitat Equivalency Analysis: An Overview, unpublished report.

NMFS Southeast Fisheries Science Center (Moderator); John Cubit, NOAA Damage Assessment Center; Brian Julius, NOAA Damage Assessment Center; Arthur Schwarzschild, NMFS Southeast Fisheries Science Center; Erik Zobrist, NOAA Restoration Center.

WORKSHOP CONCLUSIONS

Prior to the workshop, all participants were provided with background materials on the theory and application of HEA, as well as a series of null hypotheses developed by NOAA to capture the major issues relative to implementation of HEA for seagrass environments. Each of the null hypotheses discussed is presented below, accompanied by a summary of the conclusions reached by the workgroup.

Null hypothesis 1: Recovery of functional attributes can be forecast based on seagrass biomass and density alone[4]

The general conclusion of the workgroup was that seagrass biomass represents a more comprehensive metric of habitat function than seagrass shoot density. In general, biomass was cited as preferable to density measures because density measures do not capture the belowground component of seagrass systems, which may be important in determining the long-term persistence of recovering systems (particularly for *Thalassia testudinum*); and seagrass shoot density develops much more rapidly than function, and may overshoot baseline shoot density before achieving equilibrium.

Canopy volume (shoot density times the height of the seagrass canopy) was proposed as a preferable measure to biomass. While in most cases canopy volume would be expected to be highly correlated with total (above- and belowground) biomass, canopy volume has the added advantage of being easily measured in a non-destructive manner. Despite the apparent advantages of the canopy volume measure for capturing within-patch functions, the workgroup cited among-patch attributes of landscape structure, scale and setting that would also be expected to significantly influence functional performance and recovery rates.

Cover and connectedness among vegetated patches were among the measures cited as important in capturing the among-patch aspects of seagrass habitats.

Null hypothesis 2: Forecasting seagrass bed recovery is independent of scale

The general conclusion of the workgroup was that seagrass bed recovery likely will be influenced by the spatial scale of the injury. One rationale provided for rejecting this hypothesis was that current data suggest seagrass beds modify their environment, and thus the size, severity and shape of an injury will affect the recovery process. Other participants suggested that as the scale of the injury increases, landscape features will be increasingly important in determining and restoring functional attributes of seagrass beds. Stated differently, participants expected differences in the recovery of a continuous seagrass bed versus a bed of the same total density or biomass, but distributed in discrete patches.

Null hypothesis 3: Overcompensation responses by seagrass in injured areas (e.g. generation of shoot densities higher than un-impacted controls) does not constitute enhanced ecological/biological functions

The general conclusion of the workgroup was that overcompensation responses by seagrass in injured areas do not constitute enhanced functions. Higher seagrass density does not necessarily indicate higher production. In addition, external controls, such as light availability, will serve as limiting factors on the function of a particular habitat. The workgroup also concluded that the presence of an observed overcompensation response may be an artifact of density-based measures of recovery, while biomass-based measures for the same area may not exhibit the same response.

The outcome of this workshop set the stage for NOAA to provide a reasonable and fair technical basis to assess seagrass injuries and support the recovery of the lost resource services for which the responsible party is liable.

---

[4]Functional attributes refer to the range of ecological services provided by seagrass habitats including, but not limited to, primary production, faunal use, nutrient filtration, and sediment stabilization.

Table 7.1. *Top: General distribution (%) of costs by task (United States of America v. Salvors Inc.); bottom: summary of costs by specific actions (Fisher Natural Resource Damage Assessment claim)*

Task	Percentage of total costs
Map and ground-truth	5.5
Planting	18.5
Monitoring	58.7
Contractor	8.3
Government oversight	9.1

Type of cost	US$ (1996 values)
Damage assessment costs	
Federal assessment costs (up to 26 October 1996)	211 130
Interest on federal assessment costs at judgment	26 553
Subtotal	237 683
Restoration costs	
Primary restoration costs (vessel-generated holes in sea floor – restoration deemed not feasible with current technology)	0
Restoration site selection analysis	5 465
National Environmental Policy Act compliance/permitting costs	14 695
Preparation of map/ground-truthing sites	14 314
Collection, preparation and installation of planting units	64 846
NOAA restoration oversight/supervision costs	17 650
Subtotal	116 970
Monitoring costs	
Monitoring of compensatory prop scar areas	205 650
Contractor profit on restoration/monitoring work	29 028
Grand total for claim	589 331

Costs of restoration

From our experience, there is a general set of factors that drive up the cost of seagrass transplanting, particularly inappropriate site selection, inexperience, and disturbance events (requiring remedial planting). Consistent estimates of planting costs in dollars remain elusive, but recent restoration plans in the United States that have been litigated in the federal courts have shown the full cost of a restoration distributed among the various tasks (Table 7.1) at ~US$590 000 for a 1.55 acre area or ~US$940 000 per hectare (1996 dollars). Two important points here are that: (1) the actual costs of collecting and installing planting units is less than 20% of the actual cost of the entire project; (2) while monitoring costs at first glance may appear high relative to planting costs, it is important to note that monitoring represents a labour-intensive, multi-year effort to ensure that performance standards are met and necessary mid-course corrections are undertaken. The majority of planting costs on the other hand are incurred at a single point in time. This cost pattern is not unique to seagrass projects, but is commonly observed in natural-resource restoration projects across different types of habitats. We consider these data to be much more indicative of the real costs of executing a restoration project than previously presented (e.g. Fonseca et al., 1982).

CONCLUDING REMARKS

In this chapter we have dealt with what we consider to be some of the critical issues that must be addressed in the implementation of effective restoration projects. These issues include: (1) choice of an appropriate metric, representative of the array of services provided by a resource, by which to measure success; (2) evaluation of lost interim resources; (3) appropriate selection criteria for off-site restoration projects; and (4) accurate project cost estimation. A fifth issue presented itself as we edited the paper – the role of disturbance. Disturbance is a fundamental ecological process and we noticed that it repeatedly worked its way into our discussions, signalling its obvious but subtle role in influencing the outcome of restoration projects. Finally we review methods, but we do not view these to be a weak link in the process, *per se*. The weakness in methods arises when workers do not study past efforts. Rather, failure of restoration arises in general from not considering the broader context of ecological injuries, particularly issue (3). When restoration plans are sent to us for consideration, the first aspect of the plan that we look at is the choice of a restoration site. Almost without fail for those with little restoration experience, a site is selected that is not damaged and does not need repair (e.g. planting in spaces among naturally patchy seagrass).

In the United States as elsewhere around the world, we have largely won the battle to recognise the value of seagrasses as a national resource. However, the acceptance by US federal courts of our metrics for assessing success, the concept of interim resource service losses and the methods for quantifying them, and the logic for selecting planting sites has given us an unprecedented ability to foster effective restoration of these habitats. More importantly, perhaps, is the signal that this has sent to the development community and responsible parties: that this resource is of vital national importance and its destruction cannot be tolerated by the public. While transplanting seagrass is not technically complex, in order to meet the goal of maintaining or increasing seagrass area, careful attention to detail must be paid to the entire process of planning, planting and monitoring – a process that does not lend itself to oversimplification. As with all terrestrial crops, there are inherent risks with seagrass and failures will inevitably occur. Given that despite collective millennia of human experience we trade stock futures on the probability of successful cultivation of food crops, restoration of seagrass ecosystems will suffer from at least this kind of risk.

REFERENCES

Addy, C. E. (1947). Eel grass planting guide. *Maryland Conservationist*, **24**, 16–17.

Bird, K. T., Jewett-Smith, J. & Fonseca, M. S. (1994). Use of *in vitro* propagated *Ruppia maritima* for seagrass meadow restoration. *Journal of Coastal Research*, **10**, 732–737.

Broome, S. W., Seneca, E. D. & Woodhouse, W. W., Jr (1986). Long-term growth and development of transplants of the salt marsh grass *Spartina alterniflora*. *Estuaries*, **9**, 63–74.

Calumpong, H. P., Phillips, R. C., Moppets, E. G., Estacion, J. S., de Leon, R. O. & Alava, M. N. (1992). Performance of seagrass transplants in Negros Island, central Philippines and its implications in mitigating degraded shallow coastal areas. In *Proceedings of the 2nd RP-USA Phycology Symposium/Workshop*, eds. H. P. Calumpong & E. G. Meñez, pp. 295–313. Laguna, Philippines: Philippine Council for Aquatic and Marine Research and Development.

Calumpong, H. P., Fonseca, M. S. & Kenworthy, W. J. (in press). Seagrass transplantation. In *Global Seagrass Research Methods*, eds. F. T. Short & R. G. Coles Amsterdam, The Netherlands: Elsevier.

Campbell, M. L., Bastyan, G. R. & Walker, D. I. (2000). A decision-based framework to increase seagrass transplantation success. In *Proceedings of the 4th International Seagrass Biology Workshop*, eds. G. C. Pergent-Martini, M. C. Buia & M. C. Gambi, pp. 332–335. Corsica, France: Società Italiana di Biologia Marina, Biologia Marina Mediterranea.

Chapman, D., Iadanza, N. & Penn, T. (1998). Calculating resource compensation, an application of the service-to-service approach to the Blackbird Mine, hazardous waste site. NOAA Technical Paper no. 97–1. *Contributions in Marine Science*, **32**, 41–48.

Davis, R. & Short, F. T. (1997). An improved method for transplanting eelgrass, *Zostera marina* L. *Aquatic Botany*, **59**, 1–16.

den Hartog, C. (1970). *The Seagrasses of the World*. Amsterdam: North-Holland.

Derrenbacker, J., Jr & Lewis, R. R. (1982). *Seagrass Habitat Restoration, Lake Surprise, Florida Keys*. Tampa, FL: Mangrove Systems Inc.

Durako, M. J., Hall, M. O., Sargent, F. & Peck, S. (1992). Propeller scars in seagrass beds: an assessment and experimental study of recolonisation in Weedon Island State Preserve, Florida. In *Proceedings of the 19th Annual Conference on Wetlands Creation and Restoration*, ed. F. J. Webb, Jr, pp. 42–53. Tampa, FL: Hillsborough Community College.

Fonseca, M. S. (1989). Regional analysis of the creation and restoration of sea-grass systems. In *Wetland Creation and Restoration: The Status of the Science*, vol. 1, *Regional Reviews*, eds. J. A. Kusler & M. E. Kentula, pp. 175–198. Corvallis, OR: Environmental Research Laboratory.

Fonseca, M. S. (1992). Restoring seagrass systems in the United States. In *Restoring the Nation's Marine Environment*, ed. G. W. Thayer, pp. 79–110. College Park, MD: Maryland Sea Grant College.

Fonseca, M. S. (1994). *A Guide to Planting Seagrasses in the Gulf of Mexico.* Galveston, TX: Texas A&M University.

Fonseca, M. S. (1996). The role of seagrasses in nearshore sedimentary processes: a review. In *Estuarine Shores: Hydrological, Geomorphological and Ecological Interactions,* eds. C. Roman & K. Nordstrom, pp. 261–286. New York: John Wiley.

Fonseca, M. S., Kenworthy, W. J. & Phillips, R. C. (1982). A cost-evaluation technique for restoration of seagrass and other plant communities. *Environmental Conservation,* **9**, 237–241.

Fonseca, M. S., Thayer, G. W. & Kenworthy, W. J. (1987). The use of ecological data in the implementation and management of seagrass restorations. *Florida Marine Research Publication,* **42**, 175–187.

Fonseca, M. S., Kenworthy, W. J., Courtney, F. X. & Hall, M. O. (1994). Seagrass planting in the southeastern United States: methods for accelerating habitat development. *Restoration Ecology,* **2**, 198–212.

Fonseca, M. S., Kenworthy, W. J. & Courtney, F. X. (1996). Development of planted seagrass beds in Tampa Bay, Florida, U.S.A. 1: Plant components. *Marine Ecology Progress Series,* **132**, 127–139.

Fonseca, M. S., Kenworthy, W. J. & Thayer, G. W. (1998a). *Guidelines for the Conservation and Restoration of Seagrasses in the United States and Adjacent Waters.* Silver Spring, MD: U.S. Department of Commerce, National Oceanic and Atmospheric Administration. Also retrievable in .pdf format from: http://shrimp.bea.nmfs.gov/library/digital.html/ and http://www.cop.noaa.gov

Fonseca, M. S., Kenworthy, W. J. & Paling, E. (1998b). Restoring seagrass ecosystems in high disturbance environments. In *Ocean Community Conference,* 16–19 November 1998. Washington, DC: Marine Technology Society.

Fonseca, M. S., Julius, B. E. & Kenworthy, W. J. (2000a). Integrating biology and economics in seagrass restoration: how much is enough and why? *Ecological Engineer,* **15**, 227–237.

Fonseca, M. S., Kenworthy, W. J. & Whitfield, P. E. (2000b). Temporal dynamics of seagrass landscapes: a preliminary comparison of chronic and extreme disturbance events. In *Proceedings of the 4th International Seagrass Biology Workshop,* eds. G. Pergent, G. C. Pergent-Martini, M. C. Buia & M. C. Gambi, pp. 373–376. Corsica, France:

Fourqurean, J. W., Powell, G. V. N., Kenworthy, W. J. & Zieman, J. C. (1995). The effects of long-term manipulation of nutrient supply on competition between the seagrasses *Thalassia testudinum* and *Halodule wrightii* in Florida Bay. *Oikos,* **72**, 349–358.

Fredette, T. J., Fonseca, M. S., Kenworthy, J. W. & Thayer, G. W. (1985). Seagrass transplanting: 10 years of U.S. Army Corps of Engineers research. In *Proceedings of the 12th Annual Conference on Wetland Restoration and Creation,* ed. F. J. Webb, pp. 121–134. Tampa, FL: Hillsborough Community College.

Granger, S. L., Traber, M. S. & Nixon, S. W. (2000). The influence of planting depth and density on germination and development of *Zostera marina* seeds. In *Proceedings of the 4th International Seagrass Biology Workshop,* eds. C. Pergent-Martini, M. C. Buia & M.C. Gambi, pp. 55–58. Corsica, France:

Kenworthy, W. J., Fonseca, M. S., Whitfield, P. E. & Hammerstrom, K. (2000). Experimental manipulation and analysis of recovery dynamics in physically disturbed tropical seagrass communities of North America; implications for restoration and management. In *Proceedings of the 4th International Seagrass Biology Workshop,* eds. G. Pergent, C. Pergent-Martini, M. C. Buia & M. C. Gambi, pp. 385–388. Corsica, France:

King, D. M., Bohlen, C. C. & Adler, K. J. (1993). *Watershed Management and Wetland Mitigation: A Framework for Determining Compensation Ratios.* University of Maryland System Draft Report no. UMCEES-CBL-93-098. Washington, DC: Office of Policy, Planning and Evaluation.

Lewis, F. G. III (1987). Crustacean macrofauna of seagrass and macroalgae in Apalachee Bay, Florida, USA. *Marine Biology,* **94**, 219–229.

Marba, N. & Duarte, C. M. (1995). Coupling of seagrass (*Cymodocea nodosa*) patch dynamics to subaqueous dune migration. *Journal of Ecology,* **83**, 381–389.

Marba, N., Cebrian, J., Enriquez, S. & Duarte, C. M. (1994). Migration of large-scale subaqueous bedforms measured using seagrass (*Cymodocea nodosum*) as tracers. *Limnology and Oceanography,* **39**, 126–133.

Merkel, K. W. (1988). Eelgrass transplanting in South San Diego Bay, California. In *Proceedings of the California Eelgrass Symposium,* eds. K. W. Merkel & R. S. Hoffman, pp. 28–42. National City, CA: Sweetwater River Press.

Merkel, K. W. (1992). *A Field Manual for Transplantation Techniques for the Restoration of Pacific Coast Eelgrass Meadows.* National City, CA: Pacific Southwest Biological Services Inc.

Moore, K. H., Wetzel, R. J. & Orth, R. J. (1997). Seasonal pulses of turbidity and their relations to eelgrass

(*Zostera marina*) survival in an estuary. *Journal of Experimental Marine Biology and Ecology*, **215**, 115–134.

Muelstein, L. K. (1989). Perspectives on the wasting disease of eelgrass *Zostera marina*. *Diseases of Aquatic Organisms*, **7**, 211–221.

National Oceanic and Atmospheric Administration (NOAA), Damage Assessment and Restoration Program (1997a). *Habitat Equivalency Analysis: An Overview*. Policy and Technical Paper no. 95–1. Silver Spring, MD: NOAA Damage Assessment Center.

National Oceanic and Atmospheric Administration (NOAA), Damage Assessment and Restoration Program (1997b). *Scaling Compensatory Restoration Actions, Guidance Document for Natural Resource Damage Assessment Under the Oil Pollution Act of 1990*. Silver Spring, MD: NOAA Damage Assessment Center.

Orth, R. J., Harwell, M. C., Bailey, E. M., Bartholomew, A., Jawad, J. T., Lombana, A. V., Moore, K. A., Rhode, J. M. & Woods, H. E. (2000). A review of issues in seagrass seed dormancy and germination: implications for conservation and restoration. *Marine Ecology Progress Series*, **200**, 277–288.

Phillips, R. C. & Meñez, E. G. (1988). *Seagrasses*. Smithsonian Contribution to the Marine Sciences no. 34. Washington, DC: Smithsonian Institution Press.

Preen, A., Lee-Long, W. J. & Coles, R. G. (1995). Flood and cyclone related loss, and partial recovery of more than 1000 km^2 of seagrass in Hervey Bay, Queensland Australia. *Aquatic Botany*, **52**, 3–17.

Robilliard, G. A. & Porter, P. E. (1976). *Transplantation of eelgrass* (Zostera marina) *in San Diego Bay*. San Diego, CA: Undersea Sciences Department, Naval Undersea Center.

Robblee, M. B., Barber, T. R., Carlson, P. R., Durako, M. J., Fourqurean, J. W., Muehlstein, L. K., Porter, D., Yarbro, L. A., Zieman, R. T. & Zieman, J. C. (1991). Mass mortality of the tropical seagrass *Thalassia testudinum* in Florida Bay (USA). *Marine Ecology Progress Series*, **71**, 297–299.

Rose, C. D., Sharp, W. C., Kenworthy, W. J., Hunt, J. H., Lyons, W. G., Prager, E. J., Valentine, J. F., Hall, M. O., Whitfield, P. & Fourqurean, J. W. (1999). Sea urchin overgrazing of a large seagrass bed in outer Florida Bay. *Marine Ecology Progress Series*, **190**, 211–222.

Sheridan, P. (1999). Trajectory for structural equivalence of restored and natural *Halodule wrightii* beds in Texas. *Gulf Research Reports*, **10**, 81–82.

Short, F. T. & Short, C. A. (2000). Identifying seagrass growth forms for leaf and rhizome marking applications. In *Proceedings of the 4th International Seagrass Biology Workshop*, eds. G. Pergent, C. Pergent-Martini, M. C. Buia & M. C. Gambi, pp. 131–134. Corsica, France:

Short, F. T. & Wyllie-Echeverria, S. (1996). Natural and human-induced diturbance of seagrasses. *Environmental Conservation*, **23**, 17–27.

Thayer, G. W., Kenworthy, W. J. & Fonseca, M. S. (1984). The ecology of eelgrass meadows of the Atlantic coast: a community profile. U.S. Fish and Wildlife Service, FWS/OBS-84/02. Washington, DC: US Department of the Interior.

Thorhaug, A. (1974). Transplantation of the seagrass *Thalassia testudinum* König. *Aquaculture*, **4**, 177–183.

Tomasko, D. A., Dawes, C. J. & Hall, M. O. (1991). Effects of the number of short shoots and presence of the rhizome apical meristem on the survival and growth of transplanted seagrass *Thalassia testudinum*. *Contributions in Marine Science*, **32**, 41–48.

Unsworth, R. E. & Bishop, R. C. (1994). Assessing natural resource damages using environmental annuities. *Ecological Economics*, **11**, 35–41.

Virnstein, R. W. & Morris, L. J. (1996). *Seagrass Preservation and Restoration: A Diagnostic Plan for the Indian River Lagoon*. Technical Memorandum no. 14. Palatka, FL: St Johns River Water Management District.

Whitfield, P. E., Kenworthy, W. J., Fonseca, M. S. & Hammerstrom, K. (in press). Role of storms in the expansion and propagation of disturbances initiated by motor vessels in subtropical seagrass beds. *Journal of Coastal Research*.

Williams, S. L. & Orth, R. J. (1998). Genetic diversity and structure of natural and transplanted eelgrass populations in the Chesapeake and Chincoteague Bays. *Estuaries*, **21**, 118–128.

Wood, R. J. F., Odum, W. E. & Zieman, J. C. (1969). Influence of seagrasses on the productivity of coastal lagoons. In *Memorias de Symposium Internacional Lagunas Costeras*, eds. A. Castanares, A. Ayala & F. B. Phleger, pp. 495–502. Mexico, DF: UNAM-UNESCO.

Wyllie-Echeverria, S., Arzel, P. & Cox, P. A. (2000). Seagrass conservation: lessons from ethnobotany. *Pacific Conservation Biolology*, **5**, 329–335.

Zedler, J. B. & Langis, R. (1991). Comparisons of constructed and natural salt marshes of San Diego Bay. *Restoration and Management Notes*, **9**, 21–25.

Zieman, J. C. (1982). *The ecology of seagrasses of south Florida: a community profile*. US Fisheries and Wildlife Service. Washington, DC: US Department of the Interior.

8 • Coral reefs

SUSAN CLARK

INTRODUCTION

Coral reefs rank amongst the most diverse and productive natural ecosystems, their high productivity stemming from efficient biological recycling, high retention of nutrients and structural complexity, which provides habitat and food for vast numbers of organisms. The world conservation strategy (IUCN/UNEP, 1985) declared the coral reef as 'One of the essential life-support systems' necessary for food production, health and other aspects of human survival and sustainable development.

There is no universally accepted definition of a coral reef, and here it is defined as 'a diverse assemblage of animals, plants and minerals that is based on calcium carbonate platforms formed over millions of years through successive growth, deposition and consolidation of skeletal remains of hermatypic (reef-building corals) and coralline-secreting algae'. Although the primary reef building organisms are the hermatypic corals, calcareous algae are almost as important in reef-building in certain environments, and a wide range of invertebrates make an important contribution to reef growth and development. Reef-building corals comprise a symbiotic association with tiny, unicellular algae (generally termed zooxanthellae) which reside within the endoderm tissue of corals. These algae trap light energy during photosynthesis and transfer the products to the coral host. In return, the coral provides the algae with nitrogen, carbon dioxide and shelter. The symbiotic plant–animal relationship is the key factor explaining the restricted environmental requirements of scleractinian corals. It allows corals to flourish in nutrient poor waters through recycling of nutrients. The coral benefits from use of photosynthetically produced carbon as its major food source and the enhancement of the calcification process, which is essential for skeletal growth.

After Darwin's 1836 classification, coral reefs can be divided into four major structural types:

- *Fringing reefs* are the most common type of reef structure. They develop adjacent to shorelines along rocky coasts of uplifted islands or shores of exposed limestone islands where hard substrates are available, but can, in some cases, develop over soft substrate.
- *Barrier reefs* develop parallel to coastlines in areas where suitable hard foundations exist on submerged offshore shelf platforms. They are thus separated from land by a deep lagoon.
- *Atolls* are horseshoe- to circular-shaped, emergent coral reef systems that arise from deep-sea platforms, such as submerged volcanic sea-mounts or shelf platforms, towards the surface. They are usually found in deep water with a central lagoon and no landmass.
- *Patch reefs* develop in shallow water in depositional environments where favourable environmental conditions occur. They can be described as isolated or discontinued patches of fringing reefs with seaward accretion.

Fringing reefs and patch reefs are particularly susceptible to coastal activities, eutrophication and pollution, especially back-reefs due to their close proximity to land. Being further from land, barrier reefs are less susceptible to pollution than other reef types, while atoll reefs are particularly vulnerable to climate change, sea level rise and resource exploitation. Reef-building corals have existed for approximately 450–500 million years and

Table 8.1. *Summary of the major human uses of reef resources*

Extractive uses	Non-extractive uses
Food – reef fish, molluscs and other invertebrates	Recreation (scuba diving, snorkelling, boating)
Aquarium fish and corals	Transportation
Building and industrial materials	Use as natural harbours and breakwaters
Extractive recreation (collecting corals/shells)	Waste and sewage disposal (e.g. deep water with good flushing)
Land reclamation (foundations for construction and development)	Education (trails and excursions – without collections)
Education (collection of material for museums, important bio-chemicals)	Research (measurements and observations – without collections)
Research (collection of material for lab studies, extraction of chemicals)	Physical protection from storm surges and wave action
Ornaments and jewellery (curio trade)	Marine reserves and parks
Mariculture – pearl oyster, edible seaweeds	

reef communities are generally considered to be re-silient over geological time-scales. However, the individual plants and animals that form the living coral reef are considered to be fragile and sensitive to human disturbances (Brown, 1996).

Coral reefs are the major repository of biodiversity in tropical marine ecosystems. The diverse habitat of coral reefs provides shelter and a source of food as well as breeding grounds to support a large number of fish species and invertebrates. Coral reefs also provide a wide variety of services to coastal communities. For example, they provide major coastal defence in the form of a natural barrier against wave action and erosion, they support important reef fisheries, provide a variety of livelihoods for coastal communities and attract tourism and recreational activities in many countries. The socio-economic benefits of coral reefs are widely recognised and have recently been calculated to be in the order of US375\times 10^9$ per year (Costanza *et al.*, 1997).

This chapter provides an introduction to coral reef habitats and ecological processes, then evaluates various techniques used for restoration and rehabilitation of reef systems and presents existing data in terms of the potential advantages, disadvantages and lessons learnt. Given the limited literature

on this subject and the importance of corals as the main reef-builders the focus of this chapter will be on hermatypic corals.

NATURE OF THE PROBLEM

Principal uses of coral reefs

Due to the great diversity of resources provided by coral reefs a wide variety of extractive and non-extractive uses have evolved (see Table 8.1). Over the last two decades rapid population growth combined with increased use of coastal areas has led to increased pressures on coral reef resources. Globally, non-sustainable fisheries practices, such as blast fishing, poison fishing and the live fish trade, have increased dramatically in response to demographic changes (Öhman *et al.*, 1993). Furthermore, large quantities of coral are harvested for the ornamental trade and the live aquarium industry (Green & Shirley, 1999).

Although there are a number of positive benefits as a result of extractive and non-extractive use including food production and security, formal and non-formal employment and non-market products for home use, competition for reef resources can lead to conflicts between different user groups, overexploitation and destruction of critical habitats.

Major threats to reef systems

Generally, threats to reefs are discussed under two broad categories: natural and anthropogenic, although the distinction is not always clear. Natural threats include storms, hurricanes, red tides, outbreaks of coral predators (e.g. crown-of-thorns starfish *Acanthaster plancii*) and mass mortalities of certain herbivores (Brown, 1996). A wide range of human disturbances has been shown to have an impact on reefs including sewage discharge, chemical pollution, oil spills, coastal engineering, ship groundings, tourist damage and runoff of sediment, fertilisers and pesticides as a result of changing land-use (Salvat, 1987; Hatcher *et al.*, 1989). Although the sources of disturbance may be distinctive the actual responses of the reef community may be very similar. In the context of reef restoration it is more informative to distinguish between those impacts that result in structural damage (e.g. storms, ship groundings and coral mining) versus those that result in biological damage (e.g. predation, and disease). Although the latter can result in severe mortality they generally leave behind a stable reef framework that can be recolonised by corals. In contrast, physical damage can result in mortality of the organisms, loss of the reef framework and production of large volumes of sediment and rubble that may inhibit coral settlement.

Disturbances rarely occur in isolation and the response of corals to any given disturbance will depend on the timing and history of events prior to the disturbance. For example, in Jamaica, reef degradation over the last two to three decades has been brought about by the synergistic effects of natural and anthropogenic disturbances. A series of hurricanes during the 1980s was exacerbated by overfishing and a natural mass mortality of sea-urchins, which, in turn, resulted in sustained algal growth that has continued to prevent recovery of corals (Hughes, 1994).

Wilkinson (1999) discussed a third category of threats: natural impacts accentuated by anthropogenic activities, in particular, extreme weather events exacerbated by global climate change. Recently, several strong El Niño Southern Oscillation events have been correlated with widespread warm-water anomalies and associated 'bleaching' and mortality of corals (Glynn, 1984; Wilkinson *et al.*, 1999).

RATIONALE FOR REEF RESTORATION

Coral reef ecosystems may take from one to several decades to achieve full ecological recovery, depending on the type and extent of damage, duration of impact and the nature of the reef affected. Lack of reef recovery at many sites has been shown to be directly related to recurrent events. For example, storms at Heron Island in the Great Barrier Reef (Connell, 1997) and oil spills at Eilat, Red Sea (Loya & Rinkevich, 1980). In contrast a number of studies have shown that the presence of surviving colonies accelerates coral regeneration and reef recovery (Shinn, 1976; Pearson, 1981).

General indices used to measure reef recovery have included species diversity, percentage cover of coral, and survivorship and growth rates (Pearson, 1981). One of the major problems encountered is that coral communities can exhibit high cover but low diversity. For example in Florida, reef recovery in terms of live coral cover was achieved within five years although the community was dominated by just one species (Shinn, 1976). This illustrates the importance of defining what is meant by reef recovery. Many studies refer to 'recovery of the reef to its pre-impact state'. There are two problems with this approach. First, information on the ecological state of the habitat and communities prior to the disturbance is rarely available. Second, coral reefs are dynamic systems that can undergo change due to a wide range of environmental and physical factors at both temporal and spatial scales. Thus, the more appropriate goal for recovery is 'recovery to a state indistinguishable from adjacent, undamaged areas'. Reef recovery is then gauged by comparing quantitative measurements of parameters on a recovering reef with those on an adjacent 'model' reef, chosen because it is subject to similar environmental and physical conditions.

The global decline in reef health has led to concern over the ability of reef systems to withstand both repeated and an increasing number of perturbations (Sammarco, 1996; Wilkinson, 1996). The main reason behind the interest in reef restoration is the fact that natural reef recovery takes a long time in terms of human time-scales and there is a perceived need to reduce recovery periods. Generally the term 'restoration' encompasses several types of human intervention or manipulation that are designed to return (to a close approximation) a degraded system to the condition it was prior to the disturbance. This includes rehabilitation which is defined as 'any activity which aims to change a degraded system to a stable functional state' (Pratt, 1994) and pre-emptive restoration which refers to activities to offset expected or irretrievable damage to an otherwise unaffected part of the ecosystem (Hueckel *et al.*, 1989). Reef restoration is considered here as 'a proactive program designed to speed natural recovery to an end point that has aesthetic value and is functional as a coral reef system' (after Miller *et al.*, 1993).

Widespread reef degradation in the early 1990s led to the formation of new initiatives (i.e. International Coral Reef Initiative; Global Coral Reef Monitoring Network) to address the apparent decline in reef status. At the same time, there has been an increased recognition of the need for research to develop practical methods to restore degraded reefs (Yap, 2000).

Despite the successful application of restoration activities to improve resource management in agriculture, forestry, lakes, wetlands (Cairns, 2000) seagrasses (Fonseca, 1994) and mangroves (Field, 1996) such interventions have rarely been applied to coral reef systems. Woodley & Clark (1989) found the primary reef rehabilitation techniques to be passive, that is mitigation of the impact to allow natural recovery processes. Active rehabilitation was limited to small-scale studies involving transplanting adult colonies, repairing injured colonies and removing predators.

The case for restoration

Perhaps the most difficult question facing coastal managers is when to recommend reef restoration. The obvious answer would be where a reef system has been substantially damaged to the point where it can no longer provide the services and functions it formerly provided. In terrestrial systems the case for restoration has largely depended on the value of the degraded resource or habitat and the recognised benefit (ecological and social) of restoring its function, service or use (Cairns, 2000). Economic considerations associated with damaged reef ecosystems arise from production and revenue losses associated with reef fisheries and coastal tourism, to non-use values such as costs for alternative coastal protection as a result of coral reef degradation (Cesar, 2000). Although economic considerations are important, it is vital that any feasibility study for ecological restoration is based on robust scientific criteria as well as the perceived or actual monetary value.

PLANNING REEF RESTORATION
Decision tools, appraisal techniques and adaptive management

Prior to any restoration project it is important to assess whether the degradation is permanent or a transient natural condition (Done, 1995). This can be a major problem as normal cycles of recovery are slow and cannot be detected without long-term monitoring programmes that are usually expensive to undertake. Furthermore, it should not be assumed that removal of the stress or a pollutant will lead to recovery of corals, other essential conditions may include restoration of 'on-site ecological and biogenic conditions' and availability of coral propagules.

Recently, it has been suggested that various decision tools and appraisal techniques such as cost–benefit analysis framework (Spurgeon, 1998) and habitat equivalency analysis (Milon & Dodge, 2001) could be involved in the planning process. The benefits of using these tools include assessing the justification for coral reefs, improving the overall effectiveness of reef restoration programmes and assessing the appropriate extent to which damaged habitats should be restored or compensated.

Feasibility studies must evaluate the site-specific conditions at the restoration site because these are difficult to correct and will have a direct effect on reef community succession and development. Such

Table 8.2. *Relationship between the level of damage and the consequences of this on the reef system*

Level of damage	Measures of damage	Outcome on the reef ecosystem
Organism	Physiological properties, e.g. reduced growth and reproduction	Individual organisms but not structure and function affected
Species	Changes in species richness, coral cover and species diversity	Structure but not function is affected
Community	Indicators of reef malfunction (e.g. shift to an algal-dominated community)	Structure and function affected
Reef framework	Loss of habitat, structural reef framework and topographical complexity	Structure and function severely affected

conditions include substrate stability, suitability of settlement surfaces, status of the remaining reef community, sources of coral recruits, biotic factors (i.e. grazing, predation and succession), water quality and hydrodynamics.

Reef restoration is largely limited by incomplete knowledge on the ecosystem processes. The use of adaptive management practices as advocated by Thom (1997) helps address the uncertainty associated with planning restoration projects. It also provides a framework whereby knowledge gained during the monitoring can be transferred back into project management to help make informed decisions on performance and guide actions, such as rectifying problems. The frequency and duration of the monitoring will depend on the technology being used and the goals of the project. Pratt (1994) suggested that monitoring should be diagnostic in that it should be able to tell managers what is working and what is not. Currently there is no evidence that structural measures (e.g. species richness and community composition) can be reliably used to infer ecosystem function in reef systems. If restoration created reef systems that function the same as natural reef system then managers would be more certain that investments in restoration are worthwhile.

Damage assessment

It is vital to have some knowledge of what reef characteristics have been lost or altered by a disturbance. Biological damage can occur on three levels:

organism, species and community, which manifest themselves in different ways (see Table 8.2). In addition, physical disturbances (e.g. storms, hurricanes, dredging and dynamite damage) can result in destruction of the three-dimensional biogenic reef framework. It is therefore important that investigations of damage assessment incorporate methods to address all these issues.

Restoration goals

For most degraded coral reefs, recovery will be a slow process; therefore it is important that both short- and long-term goals are set. From a conceptual viewpoint the long-term goal of reef restoration could be defined as the return of the full range of services and functions provided by the original reef. However, corals are long-lived species and the development of a coral reef ecosystem from the first settlement of coral larvae will take decades. Even under the most optimistic scenario, one would not expect to see the first indicators of reef recovery in less than five years following restoration.

Evaluation criteria

At present, our knowledge of the most appropriate restoration techniques to use is limited by a lack of criteria to judge their effectiveness. The criteria adopted for monitoring effectiveness and success need to be clearly stated and relevant to the original goal and objectives of the project. However, given the different processes under investigation the indicators used to judge success will depend partly on the type of restoration adopted. For example, the general indices used to assess effectiveness of coral transplantation include an increase in coral cover and diversity, coral growth and survival rates; whereas artificial reef structures will address recruitment rates and survival of juveniles, at least in the short-term. Another major constraint in developing standard criteria is that the time-frames of research projects do not match the long-term (>10 years) periods necessary to understand temporal variability in coral reef communities.

To judge effectiveness of any large-scale practical restoration project will require the integration of both environmental and socio-economic criteria. To date, coral reef restoration success has largely been considered in terms of ecological criteria alone and a rigorous cost–benefit analysis has never been published. Few reef restoration projects have involved local communities, largely due to the fact that most studies are at the experimental stage. However, Bowden-Kerby (1999) and Heeger et al. (1999) reported that involving local communities in reef restoration projects raised the awareness of the value of what is being restored.

Measuring the ecological success of reef restoration

Most studies have involved measurement of particular criteria (e.g. coral cover) after a fixed period to confirm recovery. Ecological indicators that suggest reef restoration is progressing successfully include: (1) increasing coral cover, (2) attainment of large size and old age in coral colonies, (3) accumulation of reef framework, (4) increasing biomass of fish and other reef resources, and (5) accumulation of diverse populations.

An alternative approach would be to adopt defined stages of recovery but this implies knowledge about natural successional stages, which are still only partly understood. The complexity and dynamics of coral reefs make the evaluation of restoration programmes very difficult, and it is therefore important to regard restoration as an ongoing process and not as a final product. To this end both the reference and restored systems should be monitored and evaluated with greater awareness of their dynamics.

REEF RESTORATION TECHNIQUES

The primary aim of this section is to describe the most commonly used reef restoration techniques, evaluate their appropriateness and provide guidelines on their application. Generally, reef restoration techniques are reported in the literature in three categories: (1) small discrete research projects, (2) practical pilot studies or applied research to meet a management objective, and (3) opportunistic studies which have attempted to save corals threatened by some type of disturbance or threat.

Coral transplantation

Reasons for transplantation

Coral transplantation involves the collection of either whole colonies or fragments of colonies from a healthy reef environment (usually referred to as the donor site) and their relocation to another site, usually a degraded reef area with similar environmental conditions. The transplantation of healthy adult coral colonies to damaged areas is considered to accelerate natural recovery through bypassing the early stages of the life cycle of juvenile corals which are subject to high mortality and introducing a source of larvae not available locally (Harriot & Fisk, 1988a). Coral transplantation studies generally fall into two categories; those that use whole coral colonies and those that use coral fragments.

Coral transplantation has been conducted for a variety of purposes including: (1) to accelerate reef recovery after ship groundings (Gittings et al., 1988; Hudson & Diaz, 1988); (2) to replace corals killed by sewage, thermal effluents or other pollutants (Maragos, 1974; Birkeland et al., 1979); (3) to save coral communities or locally rare species threatened by pollution, land reclamation or pier construction (Plucer-Rosario & Randall, 1987; Newman & Chuan, 1994; Muñoz-Chagin, 1997); (4) to accelerate recovery of reefs after damage by crown-of-thorns starfish (*Acanthaster plancii*) or red tides and El Niño warming events (Harriot & Fisk, 1988b; Guzmán, 1991); (5) to aid recovery of reefs following dynamite fishing or coral mining (Auberson, 1982; Yap et al., 1990, 1992; Clark & Edwards, 1995); (6) to mitigate damage caused by recreational activities (Rinkevich,

1995; Oren & Benayahu, 1997); and (7) to enhance the attractiveness of underwater habitat in tourism areas (Bouchon et al., 1981).

Transplantation using coral fragments

Fragmentation is a common mode of asexual reproduction, which refers to the pieces that are naturally broken off source colonies and become reestablished as a new colony (Highsmith, 1982). Artificially produced fragments have been used in a large number of coral transplantation studies (Harriot & Fisk, 1988b; Kaly, 1995; Bowden-Kerby, 1997, 1999; Franklin et al., 1998; Lindahl, 1998; Heeger et al., 1999; Nagelkerken et al., 2000). Guzman (1991) had considerable success in transplanting small (4–7 cm) fragments of Pocillopora damicornis and P. elegans (see Box 8.1). Comparable transplantation studies in other regions (Bowden-Kerby, 1997; Franklin et al., 1998) have had less success, emphasising that generalisations for reef management should be made cautiously due to the variability in biological and physical conditions within different localities.

Transplantation with whole colonies

Studies that have used whole coral colonies as transplants have tended to focus on a practical management objective or have tested the feasibility of transplantation under different environmental conditions (Maragos, 1974; Birkeland et al., 1979; Auberson, 1982; Clark & Edwards, 1995; Yap et al., 1998). There are no general trends as survival rates have depended upon site specific conditions and the species used.

Potential weaknesses of transplantation methodologies

Impact on the donor site

The damage at the donor site depends on how much material is removed and whether collection involves whole colonies or fragments. For example, Lindahl (1998) reported a lack of recovery in donor areas two years after collection of coral colonies for fragments. Clark & Edwards (1995) reported a small number of new coral recruits in their donor areas after 16 months but that there was no significant recovery measured by changes in coral cover. To minimise impacts at the donor site Harriot & Fisk (1988b) recommended that at least 50% of the donor colonies

should be left intact when being used for fragments. Furthermore, Edwards & Clark (1998) suggested that donor areas should be sufficiently large and rich in coral communities so that they are not significantly impacted by collection of whole colonies. Transplantation of gravid adult colonies might result in the greatest return for a given effort (cost) and loss of material from the donor area, although this has not been tested (Richmond & Hunter, 1990).

Recently, a number of workers have proposed strategies to minimise impacts at the donor site. Rinkevich (1995) first used the term 'gardening coral reefs' to describe the use of small colonies or fragments which have been cultured in nursery areas or spawned gametes and shed planula larvae which have been reared and grown in the laboratory, prior to transplantation into degraded areas. These approaches have been attempted for Acropora species (see Box 8.2), Stylophora pistillata planulae and autotomised fragments of the soft coral Dendronephthya hemprichi (Oren & Benayahu, 1997) and for the culture of small coral fragments from four species, Acropora nasuta, Pocillopora verrucosa and Porites cylindrica and P. nigrescens (Franklin et al., 1998), with varying levels of success.

Attachment failure

Various methods have been used to attach transplants to the substrate, including cement, epoxy resin, cable ties, nylon strings, plastic-coated wires or large staples, with different levels of success. Several workers have found significant losses (for fragments and colonies) where transplantation sites (e.g. shallow reef-flat sites) are exposed to storms or strong wave action even when attached (Birkeland et al., 1979; Alcala et al., 1982; Clark & Edwards, 1995; Kaly, 1995). The two major consequences of attachment failure is risk of damage to adjacent colonies at the transplant site due to movement of the loose transplants and the collateral damage to the donor site with no positive effects at the transplantation site.

Mortality rates and survivorship of transplants

There is considerable evidence that transplanted colonies tend to have higher mortality rates than

Box 8.1   Large-scale coral transplantation using coral fragments

SITE
Cano Island Biological Reserve, Costa Rica, East Pacific.

PERIOD OF STUDY
Three years.

PURPOSE OF RESTORATION
To demonstrate the feasibility and present need for reef restoration on the Pacific coast of Costa Rica and the eastern Pacific region in general, within the context of management of marine protected areas.

NATURE OF THE PROBLEM
Several species of *Pocillopora* were decimated by the 1982–3 El Niño and a subsequent dinoflagellate bloom. The need for restoration was based on a lack of sexual reproduction, severe bioerosion and lack of recovery over eight years.

METHODOLOGY
Large numbers of *Pocillopora damicornis* and *P. elegans* fragments ($n = 110$) were transplanted to two shallow reefs (0–5 m depth) comprising dead pocilloporid reef framework. Between one and five fragments (4–7 cm) from surviving colonies less than 5 km away were removed and transported in buckets to the site. Fragments were attached to 30 cm stakes using wire.

RESULTS
Survivorship of fragments was between 79% and 83% after three years. In addition, fragmentation caused an increase of 41%–115% in new colonies. Fragments attached to the reef framework within five months and some colonies measured 21 cm in diameter after three years.

CONCLUSIONS
High survivorship and growth over three years suggested that coral transplantation for reef management was a feasible tool in the eastern Pacific region. Guzman (1991) suggested that biological reserves could play an important role in restoration programmes, by providing a protected habitat where clones of surviving organisms can be restored prior to export to restock exploited or damaged reefs.

LESSONS FOR MANAGEMENT
In 1991, Guzman presented a strong case for the transplantation of *Pocillopora* spp., based on available scientific information. However, after 14 years of monitoring, Guzman (1999) found that sexual reproduction could naturally take 10–12 years and that the restoration efforts had perhaps been premature. These findings highlight the importance of site-specific information but demonstrate that due to the long time-scales for life-history processes, even when the biology of the target species is understood, good advice at the time may turn out to be flawed. However, the transplantation study provided information on survivorship and growth of *Pocillopora* fragments which may have wider application for restoration of Eastern Pacific reefs.
Source: Guzman (1991, 1999).

undisturbed colonies (Plucer-Rosario & Randall, 1987; Yap *et al.*, 1992;). Clark & Edwards (1995) found a trade-off between growth rates of transplants and survivorship in some *Acropora* species. Other studies have demonstrated size-dependent survivorship (i.e. greater survival of larger fragments) of transplanted *Acropora* fragments (Harriot & Fisk, 1988b; Bowden-Kerby, 1997; Smith & Hughes, 1999). Moreover, Yap *et al.* (1998) found no difference in mortality between large and small fragments in two species of *Porites*. These studies demonstrate that mortality of

fragments and whole colonies is highly site- and species-specific and that it may not be appropriate to depend on generalisations.

Reduced fecundity in transplanted colonies
Rinkevich & Loya (1989) reported reduced fecundity in *Stylophora pistillata* as a result of the stress of transplantation. Smith & Hughes (1999) found substantially reduced fecundities in fragments of three *Acropora* species relative to intact control colonies. However, in the Maldives *Acropora* colonies surviving

Box 8.2   Feasibility studies to test coral transplantation in sheltered habitats using unattached fragments and cultured colonies

SITES
Pohnpei, Micronesia, Pacific Ocean and southwest Puerto Rico, Atlantic Ocean.

PERIOD OF STUDY
A series of experiments were set up over different time periods ranging from three months for size-specific mortality to two years for feasibility studies.

PURPOSE OF RESTORATION
Series of feasibility studies to determine size-specific mortality and growth of *Acropora* fragments in two habitats, reef-flat rubble and back-reef sand. Whole colony transplants using colonies cultured from fragments over a two-year period were also conducted.

METHODOLOGY
Puerto Rico

Branches of *Acropora cervicornis* were transplanted into a staked area (25 m^2) of reef-flat rubble (1 m depth). Three size classes of *Acropora* fragments (3–5 cm, 8–12 cm and 15–22 cm) were tested for size-specific mortality in reef-flat rubble and back-reef sand. Coral colonies that had been cultured from fragments were transplanted into 15 lagoon sites and colony mortality and fish recruits monitored. The response of fragments (8–12 cm) to different degrees of sand contact (i.e. direct on sand, firmly supported in contact with sand and firmly supported above sand) was tested over two to three months.

Pohnpei

Single and double branches of *Acropora* spp. were transplanted into sand at 1.5 and 3 m depths.

RESULTS
Puerto Rico

Coral cover increased from 2.2% to 24.5% at the transplanted site after one year and there was an increase in fish abundance compared to the control area. Survival of fragments was found to be strongly size-dependent in the sand habitat, although securing small fragments to wire in sand increased survival significantly. Mortality was reduced for all fragments suspended above the substratum. Survivorship was high for unattached fragments in all size categories in reef-flat rubble.

Pohnpei

Survivorship was significantly greater at the deeper site and in double branches.

CONCLUSIONS
A two-step 'coral gardening' methodology for back-reef and reef-flat rehabilitation was advocated. This involves the use of reef-flat rubble areas as a nursery to culture unattached fragments into larger colonies (i.e. 25–50 cm^3 colonies) over two to three years. These cultured colonies can then be used as a source of further fragments for transplantation or transplanted intact to create patches in back-reef lagoons.

LESSONS FOR MANAGEMENT
These 'gardening' methods may have application for the creation of new patch reefs to enhance fisheries habitat, particularly in rural island communities. Additional benefits include raising community awareness amongst local communities and reducing destructive fishing practices.
Source: Bowden-Kerby (1997, 1999).

six months after transplantation had similar gametogenic development to undisturbed colonies (S. Clark & A. J. Edwards, unpublished data). Thus, for some coral species both donor colonies and fragments taken from them might have significantly reduced sexual reproductive capacity, for at least several months after transplantation.

Potential strengths of transplantation methodologies

Increase in coral cover and diversity

Transplantation using fragments can provide a small increase in live coral cover but it does not provide the faunal diversity associated with a natural healthy reef system. This applies particularly to those studies that have focused on a few species (e.g. fast-growing *Acropora* species). Despite laudable efforts the existing studies have only demonstrated that coral fragmentation may be appropriate for a small number of 'weedy' species.

Transplantation using whole colonies can provide an instant increase in live coral cover and faunal diversity. However, collection of whole colonies results in negative impacts (e.g. decline in coral cover) at the donor sites and a potential reduction in fecundity depending on the timing of transplantation relevant to the reproductive period. The main application for the use of whole colonies or coral aggregations is in cases where reefs are threatened by coastal alteration. There is however, a lack of information on whether transplanted corals play any role in increasing local recruitment. Clark & Edwards (1995) found that transplantation did not enhance natural recruitment compared to the provision of stable surfaces in the form of artificial reefs.

Pre-emptive restoration case studies

A few studies have used transplantation techniques to rescue and relocate corals threatened by human activities. For example, coral reef fauna endangered by the construction of a pier at Cozumel Island, Mexico was relocated using transplantation techniques to a similar reef habitat within 600 m of the original location (Muñoz-Chagin, 1997). Several complete coral aggregations and 23 796 animals were attached to specially constructed concrete structures to promote a firm base for the transplants away from the sea floor. Initial monitoring, after one month, showed that mortality was low even after Hurricane Roxanne had impacted Cozumel just after the completion of transplantation. In Singapore, a community-based coral transplantation project was conducted to relocate corals threatened by dredging for land reclamation (Newman & Chuan, 1994). The project, which received funding from the Hong Kong Bank 'Care for Nature', used volunteer divers to enhance public awareness of coral reefs. Unfortunately, as in so many cases, there is a lack of information on the survival of the transplanted corals.

Transplantation of massive frame-building species

Despite the importance of massive frame-building species there have been few attempts to use them in fragmentation studies. Vago & Turak (1995) advocated a simple and rapid technique for the renovation of damaged coral reef communities based on the transplantation of coral cores to natural substrata. Small cores (3 × 5 cm) of *Porites lutea* were removed, using an underwater pneumatic drill, from donor colonies and placed into similar-sized cuts on the target reef. Preliminary studies over six months showed no mortality and two-dimensional growth over the adjacent substrate. However, given the small amount of transplant material used it is only likely to be effective for small-scale restoration projects.

Applications of transplantation

The desire to preserve coral colonies threatened by human activities is an entirely laudable goal but scientists should beware of setting a dangerous precedent. Edwards & Clark (1998) pointed out that if decision-makers think that corals can always be transplanted if they get in the way of development, then there is little incentive to resolve the management issues leading to the coral reefs being threatened in the first place. Also, the widespread emphasis on transplanting corals *per se* as 'flagship' species is misleading because they are but one component of a complex ecosystem whose other components thus tend to be ignored (but not always, e.g. Newman & Chuan, 1994; Muñoz-Chagin, 1997).

Use of artificial reef structures to increase habitat complexity

Basis of artificial reefs

The slow rates of natural reef recovery have prompted interest in the potential of artificial reef structures (ARS) to rehabilitate damaged reefs and coastal habitats. Although artificial reefs have been used as a tool to enhance fisheries production for many years (Seaman & Sprague, 1991), there has been little substantive research on their application in reef rehabilitation programmes. One of the reasons for the high productivity of coral reefs is the amount of surface area and textural variety provided by the reef framework. Hence, for artificial reef structures to be successful they must mimic and perform the same functions as natural reefs (e.g. provide stability, topographical relief, shelter and refuge).

Types of artificial reefs

Artificial reefs have been used for a variety of purposes and their design and planning reflects their use and economic viability. To date, a wide range of materials have been used including scrap metal, automobiles, rubber tyres, bamboo, PVC pipes and concrete (White et al., 1990). Results from ARS constructed from tyres are highly variable. In the Philippines, Gomez et al. (1982) demonstrated successful colonisation by 30 species of coral on automobile tyre reefs at depths of 16–23 m and estimated it would take 15 years to develop a community with 50% coral cover. Munro & Balgos, (1995) on the other hand, found that tyre reefs were unsuccessful due to breakage. Gilliam et al. (1995)

evaluated the potential of a tyre–concrete aggregate, which used tyre shreds mixed into concrete, as a suitable reef building material. Monitoring over 28 months showed that the reefs acquired a diverse assemblage of reef fish and invertebrates and that there was no significant difference in community composition when compared to gravel–concrete reefs of the same design.

A number of artificial reef studies (Schumacher, 1977; Fitzhardinge & Bailey-Brock, 1989; Thongtham & Changsang, 1999) have demonstrated that concrete provides a suitable material for settlement of corals. Clark & Edwards (1994) showed that concrete ARS can satisfy the critical biological requirements (i.e. provide substrate for the settlement and growth of corals and habitat to restore reef fish populations) associated with reef rehabilitation (Box 8.3). Moreover, Hudson et al. (1990) used cementation of precast concrete domes and other shapes resembling corals that eventually served as colonisation sites for hard and soft corals.

Box 8.3 Artificial reef structures as habitat replacement

SITE
Malé Atoll, Maldives, Indian Ocean.

PERIOD OF STUDY
Four years (1990–4) and sporadic ongoing monitoring

PURPOSE OF RESTORATION
An experimental artificial reef programme was initiated in 1990 to discover whether it is feasible to use a bio-engineering approach to 'kick-start' natural reef recovery.

PROJECT GOALS AND OBJECTIVES
The main goals of the project were to determine: (1) whether provision of stable surfaces in the form of concrete artificial reef structures (ARS) would be sufficient to attract reef fish, allow coral settlement and growth and (2) whether transplantation of corals

to such surfaces was justified by significantly accelerating recovery.

STATEMENT OF THE PROBLEM
Traditional mining of corals for the construction industry has resulted in severe physical degradation on shallow reef-flat areas. Reefs mined for corals have shown virtually no natural recovery over a 20-year period.

METHODOLOGY
Four sets of artificial reef structures (360 t weight) of varying topographic diversity, stabilising effect and cost were deployed on the reef flat (0.8–1.5m). These included 1-m³ hollow concrete blocks (SHEDS) with various infills, flexible concrete mattresses (Armorflex) and chainlink fencing anchored by paving slabs. Coral recruitment and reef fish colonisation were monitored at the ARS, on three sites on the mined reef flat and on three healthy unmined reef flats which served as controls. Survivorship and growth of coral transplants on one set of Armorflex was followed simultaneously.

RESULTS

Restoration of fish

Within one year of deployment the ARS, with the exception of the chainlink fencing, had similar or greater species richness and densities of reef fish than control pristine reef flats. However, the community structure of the fish populations on the ARS was significantly different to that of unmined reef flats. Design features (e.g. shaded crevices) of the ARS created habitats that attracted certain fish such as *Myripristis vittata*, *Pempheris vanicolensis* and *Apogon apogonoides* that are not commonly found on the natural reef flat.

Restoration of corals

Substantial natural coral recruitment occurred on the larger reef structures (SHED areas), which were each supporting *c.* 500 colonies, some of which were approaching 25 cm in diameter after 3.5 years (Fig. B8.1). The coral community that developed on the SHED areas was dominated by acroporids (50%) and pocilloporids (45%), with very few massive colonies (5%). This compares favourably with the unmined control areas that were also dominated by acroporids (77%), followed by pocilloporids (9%), with massive species making up the remaining 14%. Lack of coral recruitment on the chainlink fencing and the mined controls was attributed to the absence of stable, sediment-free surfaces.

Evaluation of coral transplantation

Twenty-nine months after transplantation 41%–59% of corals still survived on the concrete mattresses (Fig. B8.2). Overall, 19%–37% of the transplanted colonies were swept off the mats by wave action and a further 18%–29% of colonies became loose (i.e. unattached but still present within transplant area) during the early part of the study. Mortality rates of corals, which remained attached, were between 23% and 31% over the study period. The presence of transplants did not enhance natural coral recruitment.

CONCLUSIONS

The project has achieved:

- Increased understanding of the physical and biological conditions conducive to coral recruitment and fish colonisation.

Fig. B8.1. Concrete SHED (Sheppard Hill Energy Dissipator) blocks showing branching coral colonies 3.5 years after deployment. The largest colonies had diameters approaching 25 cm.

Fig. B8.2. Transplanted coral colonies which have survived on the Armorflex concrete mattresses 29 months after transplantation.

- Preliminary cost–benefit analysis of different rehabilitation strategies.
- Initiation of coral recovery but on a limited spatial scale.

However, the ARS used did not promote abiotic conditions associated with unmined sites (i.e. stability of substrate, topographical relief) and ecosystem functions such as capacity for sea defence and contribution to reef framework within the 3.5-year monitoring period.
Sources: Edwards & Clark (1992, 1998) and Clark & Edwards (1994, 1995, 1999).

Reef balls: artificial reef modules

A new technology 'Reef Balls' shows promise for restoring degraded coral reefs, creating new fishing and scuba diving sites (Box 8.4). However, the only information available comes from the manufacturers (Reef Ball Ltd, 2000) and an independent evaluation is not available.

Applications of artificial reefs

ARS can help to restore coral reefs in two principal ways: (1) by providing shelter and refuge for reef fish and mobile reef invertebrates and (2) by providing suitable substrates for the settlement and colonisation by corals and other sessile reef fauna. However, benefits are usually proportional to the amount of new habitat created and eventually colonised.

Despite the potential advantages of ARS for reef restoration (Table 8.3) there are some concerns; for example, ARS serve as aggregating devices and attract reef fish and other organisms from adjacent reef systems. Furthermore, given the large areas of degraded reefs in need of rehabilitation it would be impractical to recommend the use of ARS due to the expense and effort required to ensure their success. However, if the degraded area has a high value (e.g. for shore protection or recreational tourism) the costs/investment could be justified. One of the main advantages of ARS over transplantation is the wide range of surface and topography provided for the colonisation of a wide range of organisms, something that transplantation of a single species cannot achieve.

Box 8.4   New artificial reef technology: 'Reef Balls'

Recently, a company in the United States has created a specific product to mimic both the appearance and function (i.e. food, shelter and protection) of a natural reef. Their patented product known as the 'Reef Ball' system is composed of concrete and has a textured swiss-cheese-like exterior and internal surface. Reef Balls are made by pouring concrete into a fibreglass mould containing a central polyform buoy surrounded by various sized inflatable balls to make holes (Fig. B8.3). These balls are inflated to different pressures to vary hole sizes and provide a textured appearance. One of the main advantages of Reef Balls is that they can be floated and towed behind even small boats for deployment.

Fig. B8.3. Schematic diagram of an individual Reef Ball mould showing design features.

The casting technique allows great flexibility in determining the weight and surface texture and various mould sizes are available. Any type of concrete can be used, including end-of-day waste, but the manufacturers recommend the use of a 'marine-friendly' concrete to make the concrete suitable for growth of marine life.

Table 8.3. *The potential advantages and disadvantages of the use of artificial reef structures for reef rehabilitation.*

Advantages	Disadvantages
Provide stability in high-energy environments	May break up, leading to pollution
Provide a variety of surfaces for settlement of reef fauna	Aesthetics (i.e. unlikely to simulate natural reefs for decades)
Provide shelter, refuge and cryptic habitats for fish	Site selection and placement may result in navigation problems
Rapid colonisation of juvenile corals on certain reef substrates (e.g. concrete)	May be costly and time-consuming (construction and deployment)
Can enhance the biomass of harvestable resources	Act as fish-attracting devices – thus transfer organisms from one site to another

Mineral accretion technology

Hilbertz *et al.* (1977) introduced the concept of mineral accretion technology (MAT), that is, using electrolysis of seawater to precipitate calcium and magnesium minerals to 'grow' a crystalline coating over artificial structures to make construction materials. The mineral accretion material is largely comprised of aragonite ($CaCO_3$) and brucite ($Mg(OH)_2$) at the cathode, and has very similar chemical and physical properties to reef limestones (Fig. 8.1). Higher current densities result in faster growth but weaker material dominated by brucite, while lower current densities produce slower deposition dominated by harder aragonite (Hilbertz, 1992). Cathodes and anodes can be made in any size and shape, with current flow dependent on their spacing and surface area. Typically, the cathode is built out of expanded steel mesh constructed as simple geometric forms such as cylinders, sheets, triangular prisms or pyramids. The anode can be lead, graphite, steel or coated titanium.

MAT has been used to develop artificial reefs in Jamaica, Maldives and the Seychelles. It has been suggested that corals grow at very rapid rates on these structures due to the change in the environment produced by electrical currents, which speed up formation and growth of both chemical limestone rock and the skeletons of corals (Global Coral Reef Alliance, 1999). However, this has not been

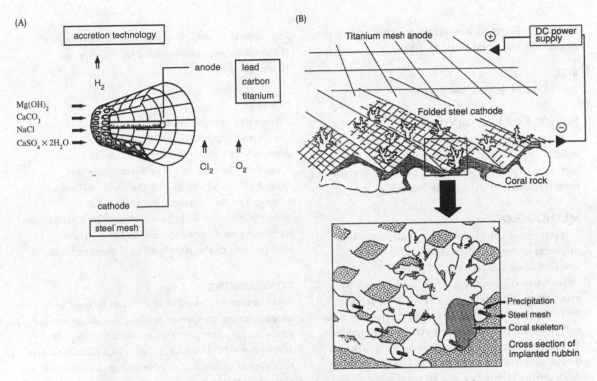

Fig. 8.1. (A) Diagrammatic representation of the artificial reef substrate created by mineral accretion in seawater through electrolysis. (B) Diagram showing the transplantation of coral nubbins onto test squares of mat. After van Treeck & Schumacher (1997).

substantiated by data on coral growth rates in natural reefs compared to those on mineral accretion reefs. A group of independent scientists is currently carrying out applied research to evaluate the application of MAT for reef restoration (Box 8.5).

Applications of MAT

Although preliminary results indicate that the technology can be used both for transplantation of fragments and natural colonisation, it is labour-intensive and requires specialised equipment and skills (Table 8.4). The concept that MAT may have a role in meeting the requirements of recreational scuba diving has not been subject to market research and it is difficult to see how divers would pre-fer MAT structures to natural reefs and traditional wrecks.

Reef stabilisation and cementation

Ship grounding incidents can result in severe, but localised, reef framework damage depending on the size of the vessel, the type of incident and whether the vessel remains on the reef (Precht, 1998). In addition to major framework damage there is a risk of secondary damage to adjacent undamaged reefs due to the production of large volumes of unconsolidated rubble and crushed sediment. An evaluation of reef restoration capabilities and methodologies to restore reefs damaged by large ship groundings in Florida (Miller et al., 1993) recommended:

Box 8.5   Mineral accretion technology: a new approach for coral transplantation

SITE
Gulf of Aqaba, Jordan, Red Sea.

PURPOSE OF RESTORATION
To test the feasibility of combining *in situ* creation of natural artificial reef structures using mineral accretion technology (MAT) with the transplantation of coral fragments (referred to as 'nubbins').

METHODOLOGY
Test squares of chicken wire (1 × 3 m) with wire thickness of 1 mm and mesh size of 10 mm were installed at depths between 1 and 18 m. The test squares were tied to the natural reef horizontally using a zigzag folding design to reduce grazing by fish. Six of the most common species on the natural reefs were selected: *Pocillopora damicornis, Acropora variabilis, A. squarrosa, Montipora danae, Stylophora pistillata* and *Pavona varians*. Coral nubbins were inserted into the mesh during electrolysis and all became naturally cemented to the artificial substrate within eight weeks.

The electrodes were disconnected and survivorship of the nubbins monitored after three, seven and 12 months.

RESULTS
Survivorship was highly variable depending on the species and depth. After 12 months *Pocillopora damicornis* had the lowest survival rate (36%) and *A. variabilis* the highest (72%). Survivorship was dependent upon how quickly the coral nubbins were cemented to the substrate, the presence of the corallivorous snail *Drupella*, which showed a preference for feeding on *Acropora* species, and the variable susceptibility to the treatment by different species.

CONCLUSIONS
Initial results suggested that MAT can be used with transplantation to shorten initial colonisation rates of semi-natural substrates. This led to the concept of developing MAT into modules which can accommodate recreational diving and thereby reduce pressures on natural reefs.
Source: Van Treeck & Schumacher (1997, 1998, 1999).

cementing the fractured reef framework, removing loose rubble and fine sand, repairing excavations, coral transplantation, increasing habitat complexity and manipulating key species groups which influence community structure.

Removal of sediment (mechanical or hydraulic)
Large quantities of sediment and rubble can be removed by suction dredges or pipeline devices floated over the reef to collect and pump off sediments. The disposal of the sediment must be considered and the usual options are land disposal or disposal offshore. Hydraulic removal involves the mixing of sediment with water to form a slurry which is easier to transport off the site. Removal by mechanical means (e.g. by clamshell dredge) may result in sediment damage to the adjacent reef area. Measures to protect reefs from sedimentation include the use

of silt nets in shallow reef environments to contain the sediment, and the construction of solid berms or settlement ponds on land to receive and hold the sediment.

Emergency triage
One of the main goals is to rescue damaged resources by placing them in a safe location until the damage assessment and salvage work is completed. This may involve righting overturned colonies, lifting boulders, collecting broken fragments and cementing structures.

Structural restoration
Various engineering approaches have been recently attempted to restore the physical structure and habitat complexity of the reef surface to enhance rates of recovery within ship grounding sites in the United

Table 8.4. *Potential advantages and disadvantages associated with mineral accretion technology (MAT)*

Advantages	Disadvantages
Produces a substrate with a limestone character	Stability in high-energy environments is largely unknown
Can be built in a variety of shapes and sizes and integrate with the natural reef	Long-term survival rates of corals are unknown
Suitable for attachment of coral nubbins or transplants	Time-consuming and initially requires skilled labour
Provides a suitable surface for natural coral recruitment	Application to different types of physical environments have not been tested
Specialised equipment can be reused	Initial set-up costs are expensive
May provide recreational reefs	No evaluation of diver responses to MAT

Table 8.5. *Potential advantages and disadvantages associated with structural restoration including clearing the reef framework and stabilisation of damaged corals*

Advantages	Disadvantages
Reduces secondary damage of corals due to increased sedimentation	Requires feasibility studies to assess potential for secondary damage
Provides sediment-free surface/substrates for settlement	Technical problems ensuring the area is kept free of sediment and rubble
Reduces mortality of damaged corals by securing loose colonies	Does not provide habitat complexity (i.e. three-dimensional relief)
Minimises abrasion and smothering of newly settled larvae and juvenile corals	Expensive in terms of equipment required (depends on scale of damage)
Appropriate for areas with small rubble	Time-consuming and labour-intensive

States (NOAA, 1997; Jaap, 2000). Structural restoration design is site-specific and depends upon the settlement claim between the responsible parties (i.e. ship owners, insurance companies and the trustee). Methods used include the use of quarried blocks to fill blow holes and cavities in the reef structure, and use of concrete blocks and mats to stabilise unconsolidated sediment.

Application of stabilisation and cementation

To date, attempts at structural reef restoration have been limited to ship grounding incidents in the United States which are driven by the legal framework and funded by large settlement claims. The transfer of this technology to other areas will depend on whether expensive structural restoration is the best use of available resources. Clearly, the high costs and technology requirements will limit its application in many developing countries (Table 8.5).

Coral mariculture

Recent work in the field of coral mariculture demonstrates that a few species can be successfully reared from larvae (Yates & Carlson, 1992; Szmant, 1999) (Box 8.6). The seasonal timing of coral spawning is predictable for some locations, which facilitates the opportunity to utilise coral gametes for coral culturing. Heyward *et al.* (1999) used larval rearing ponds with mesh enclosures to assess settlement success and larval survival following mass spawning in Western Australia. Initial results indicated high recruitment success on terracotta settlement tiles and further work is being conducted with flow rates and equipment modifications to develop coral seeding practices. Sammarco *et al.* (1999) proposed an application for coral mariculture whereby larvae are cultured in laboratory tanks until competent to settle, then seeded into the centre of eddies to promote enhanced local settlement within target reefs. This has not been fully tested however.

Box 8.6   Coral mariculture and transplantation

SITE
Bantayan reef, Central Philippines.

PURPOSE OF RESTORATION
To develop and test coral seeding as a method to reintroduce corals to a degraded reef, thereby reducing the need to damage intact coral colonies.

METHODS
Coral planulae were collected in laboratory aquaria from wild adult colonies of *Pocillopora damicornis*, allowed to settle and then reared for up to six months (to at least 10 mm diameter) to seed degraded shallow reefs. Transplants from size cohorts (<3 mm, 3–6 mm, 6.1–10 mm and >10 mm) were used to evaluate the relationships between colony size, growth rates and mortality.

RESULTS
Mean settlement success rate was significantly greater for laboratory-reared planulae (52.9%) compared to 14.3% for planulae released directly to the natural reef. Similarly, first-week mortality rates were 30% and 81% respectively. Survivorship of transplants >10 mm diameter over six weeks was 95%.

CONCLUSIONS
Preliminary studies suggest that extended laboratory rearing of *P. damicornis* to a size of 10 mm (about six months) prior to transplantation will ensure high survivorship of transplants.

LESSONS FOR MANAGEMENT
Certain species can be reintroduced to degraded sites by settling planulae in laboratory aquaria, rearing them to a minimum size and transplanting them onto appropriate substrate. *Pocillopora damicornis* appears to be well suited to this technique as it has been well studied, is easily maintained under laboratory conditions and is opportunistic and ubiquitous. However, other species may not prove so easy to rear under laboratory conditions.
Source: Raymundo *et al.* (1999)

Application of coral mariculture
The feasibility of coral mariculture on a large scale and relative costs of the various approaches, which may be considerable, need to be evaluated. Further information is also required on the optimal size for release of juveniles, habitat requirements of juveniles and the methods of attaching the juveniles *in situ* on the reef. A major benefit to reef restoration would be the supply of large numbers of target species at adequate sizes to restock damaged reefs and to provide sustainable, commercial harvesting for the curio trade or local fisherfolks (Table 8.6).

There are however, some concerns over the laboratory culturing and release of organisms. Namely, the potential to alter the genetics of natural populations, spread diseases and possibly introducing organisms with inappropriate behaviour for survival. To minimise such problems brood stock would need to be collected from reefs and individuals of appropriate provenance.

CONSTRAINTS OF REEF RESTORATION
Lack of fundamental research

The principal scientific constraint is the lack of understanding of reef systems. Coral reefs are highly dynamic systems, their condition (growth or decline) reflecting the change in environmental conditions in which they are found. Although reef ecologists have made considerable advances in understanding biological processes (i.e. life histories of key species, biological interactions, recruitment rates and trophic interactions) there is a lack of information on natural recovery processes, particularly the time-scales for recreating physical structure in reef systems.

Scientists have a responsibility to ensure that reef restoration techniques are based on sound scientific principles before widespread implementation. Any future scientific research should be orientated towards applied research that will generate information to provide the guidelines needed for coastal managers. Another major challenge to effective reef restoration lies in the need to integrate

Table 8.6. *Potential advantages and disadvantages associated with coral mariculture for laboratory-reared juveniles*

Advantages	Disadvantages
May reduce high natural mortality rates of early juvenile stages	Problems with attachment of laboratory-reared juveniles in the field
Settlement on ARS can be induced in the laboratory and transferred directly to the field	Risk of high losses and predation during early life stages
Can select fast-growing species	Specific facilities for mariculture required
Reduces damage to reef through collection of donor colonies for transplantation	Only a few species have been shown to be suitable
Can release larvae after fertilisation to specific reef locations	Costs unknown but likely to be high due to the level of equipment and husbandry required
Adult colonies can be returned to field after spawning	Application on a large scale is unknown
	Time-consuming

scientific knowledge on the structure, function and dynamics of the ecosystems with economic and social pressures that have a direct impact on the system. There is a clear need for a systems approach to ecosystem restoration involving changes in human behaviour and attitude as well as scientific technology (Cairns, 2000; Hackney, 2000). This will involve: (1) more scientific information on reef recovery processes (coral recruitment, larval dispersal and settlement patterns, succession, predation and coral–algae–fish interactions), (2) integration of scientific knowledge with economics and social pressures which impact directly or indirectly on the system, (3) improving dissemination of generic information to the policy-makers and decision-makers,

and (4) incorporating long-term monitoring in all projects.

Lack of available material

The data reviewed here is not exhaustive, as one of the major problems is that information on reef restoration is not readily available. For example, many reef restoration projects in Southeast Asia are reported only in technical reports and grey literature that is difficult to source. Field (1998) reported similar problems in attempts to review literature on mangrove rehabilitation that led him to propose the need for an archival system by an international agency. As a consequence, many reef restoration projects will be carried out without reference to lessons learnt from other projects, leading to unnecessary duplication of effort.

Costs

The principal constraint of reef restoration is cost, and this varies considerably between the different techniques available (Table 8.7). It is important to emphasise that in most cases full costs of restoration activities are not available. Total costs should include damage assessment, appraisal and feasibility studies, as well as capital, operational and monitoring costs. Other factors to be considered include the potential damage to a donor site during transplantation and the opportunity cost (i.e. potential or next best use of the reef) of a site. Ultimately, a wide range a factors will affect the magnitude of the restoration costs, such as site location, complexity of restoration programme, availability of materials and funds, skilled labour and period of monitoring. These make a thorough evaluation difficult. For example, Spurgeon & Lindahl (2000) found that costs for three reef restoration projects varied enormously, depending upon the type of restoration technique adopted. At the lower end of the scale costs were approximately US$10 000 per hectare for low-cost transplantation and at the higher end of the scale, US$5 million for structural restoration in ship grounding incidents.

Table 8.7. *Comparison of costs involved in different reef restoration techniques*

Restoration technique	Costs	Reference
Transplantation of whole colonies onto concrete structures for rehabilitation of degraded reef flats	500 colonies required about 250 person–hours on site (i.e. excluding travel time to the site and preparation). Consumable costs were about US$400 and 140 hours of boat time were needed	Edwards & Clark (1998)
Different concrete artificial reef structures to rehabilitate degraded reef flats	US$310 000 for chainlink fencing to US$990 000 for Armorflex concrete mats and US$3.15 million for concrete blocks per hectare (costs included equipment, deployment, labour and transport but excluded monitoring)	Clark & Edwards (1999)
Transplantation using different methods of attachment and handling techniques for fragments	For 1 ha of reef with a density of 24 500 fragments costs were US$58 000 (costs excluded access to site)	Kaly (1995)
Transplantation using *Acropora* fragments in low to moderately exposed shallow reefs without use of scuba	Costs were US$7000 per hectare for 2.5 kg of corals per m^2	Spurgeon & Lindahl (2000)
Community-based coral farming using fragments reared in nursery areas before transplantation to degraded reefs	Costs were US$3600 per hectare for 20 000 fragments at 12.5% cover (costs included equipment, local labour, transport and monitoring)	Heeger *et al.* (1999)
Structural restoration using quarried boulders embedded in cement to restore reef structure at the MV *Columbus Islinan* grounding site	US$3.76 million in natural resource damages claims to cover damage assessment, removal of rubble and ship debris, structural restoration and monitoring	NOAA (1997)

A number of studies refer to the costs of transplantation (Vago & Turak, 1995) or advocate low-cost approaches (Bowden-Kerby, 1997; Lindahl, 1998). However, actual costs are never stated. Due to the variation in costs between different regions some authors provide details of labour and travel in terms of hours so that this can be adapted to the local pay rates and the type of transport used (Birkeland *et al.*, 1979; Muñoz-Chagin, 1997). Few studies have provided details of ARS costs although are they generally considered to be expensive due to capital and operational costs. Clark & Edwards (1999) found the success of the methods was inversely related to their complexity, design features and monetary costs.

CONCLUDING REMARKS

The world-wide decline in the health and status of coral reefs and the long time-scales for reef recovery processes have prompted interest in techniques to either restore or rehabilitate degraded coral reefs. Given the current interest in reef restoration there is an urgent need for reliable information on the effectiveness of biological and structural reef restoration technology as claims for environmental damages to coral reefs escalate and restoration projects are promoted as a 'cure-all' for habitat destruction. Virtually all reef restoration projects have only been recently established and few have reached the point where they can be critically evaluated with respect to their ultimate success in restoring reef structure and function. Two key points must be made. First, broad generalisations about which techniques are appropriate for different types of reef degradation must be avoided due to the site-specific nature of the results from restoration studies. Second, the importance of selecting criteria to assess reef recovery at the appropriate scale and relevant to the aims cannot be overemphasised.

Coral transplantation is the most widely used restoration technique. However, many studies have focused on a research aspect without reference to a management problem making a rigorous evaluation difficult. A key issue in assessing whether transplantation is an appropriate option is the importance of establishing whether natural recovery is limited by larval supply or post-settlement mortality. Edwards & Clark (1998) suggested that in general, unless receiving areas are failing to recruit juvenile corals, natural recovery processes are likely to be sufficient in the medium to long term and that transplantation should be viewed as a tool of last resort. This view is supported by recent findings following long-term monitoring of transplantation and natural recovery in the Eastern Pacific (Guzman, 1999). Therefore, in many cases it is more appropriate to reinstate ecological or structural conditions known to be conducive to natural recovery or regeneration. However, small-scale transplantation can in certain circumstances be justified to save colonies threatened by destruction of their primary habitat.

There is good evidence that artificial reef structures can provide habitat complexity and offer some degree of habitat or structural restoration depending on the materials used and design of the ARS (Clark & Edwards, 1999; Jaap, 2000). However, the potential advantages must be weighed against the substantial monetary costs, which limit ARS application on a large scale. Similarly, mineral accretion technology is unlikely to have widespread application due to the time-scales and the mone-

tary costs involved. Perhaps a more appropriate use of this technology would be to test its application in stabilising unconsolidated rubble and sediment, thereby providing a semi-natural substrate for natural colonisation without the need of transplantation. Although structural restoration efforts to repair reef framework can be justified on scientific merit the costs and technology involved have restricted its application to ship grounding sites in the United States, which are financed by recovered damages. Furthermore, these projects have been recently established and it will be a few years before they can be evaluated with reference to their original aims.

The field of coral mariculture has been slow to develop although there is growing interest in its potential application for reef restoration, particularly within reef areas that are recruitment-limited. Various low-cost applications have been suggested and it is now timely to evaluate different types of collection (e.g. from natural spawning slicks or from adult colonies reared in aquaria) and seeding practices, including the use of competent larvae released directly in the field or within larval ponds. Alternative techniques that involve rearing larvae to small juveniles before transplanting to the field may increase survival rates, but will encounter problems in attaching the juveniles to the natural reef.

One of the main goals of reef managers is to minimise further ecological losses and prevent loss of biodiversity. Whilst this must remain the primary goal, it is just as important to plan for the restoration or repair of ecosystems that have been severely damaged in the past. The major constraint in evaluating reef restoration is that, apart from a few exceptions, there has been inadequate provision for monitoring over time-scales from five to ten years. It would therefore seem more appropriate to gather detailed information on the performance of existing reef restoration projects than conduct more research on the development of techniques. This would require a multidisciplinary team to conduct a detailed scientific and cost–benefit analysis, one of the outputs of which would be selection of existing reef restoration projects for more detailed monitoring and evaluation.

To date, most of the information available on reef restoration is derived from coral transplantation studies. Furthermore, transplantation and other restoration techniques have generally been carried out at small spatial scales (e.g. less than 100 m^2) and it is difficult to see how such studies could be used in large-scale restoration programmes akin to reforestation. There are no guidelines to suggest the best technique to use in a given situation or at what level of damage does reef restoration become necessary. Thus, at this time all reef restoration projects can still be regarded as pioneer studies from which varied lessons can be learnt and mixed levels of success expected (Yap, 2000). Finally, the major constraint to widespread application of any reef restoration technique is that those countries with the most serious reef degradation and largest areas to restore can least afford restoration programmes.

REFERENCES

Alcala, A. C., Gomez, E. D. & Alcala, L. C. (1982). Survival and growth of coral transplants in Central Philippines. *Kalikasan, Philippines Journal of Biology*, 11, 136–147.

Auberson, B. (1982). Coral transplantation: an approach to the re-establishment of damaged reefs. *Kalikasan, Philippines Journal of Biology*, 11, 158–172.

Birkeland, C., Randall, R. H. & Grim, G. (1979). *Three Methods of Coral Transplantation for the Purpose of Re-establishing a Coral Community in the Thermal Effluent Area of the Tanguisson Power Plant*. University of Guam Marine Laboratory Technical Report no. 60. Guam: University of Guam.

Bouchon, C., Jaubert, J. & Bouchon-Navaro, Y. (1981). Evolution of a semi-artificial reef built by transplanting coral heads. *Tethys*, 10, 173–176.

Bowden-Kerby, A. (1997). Coral transplantation in sheltered habitats using unattached fragments and cultured colonies. *Proceedings of the 8th International Coral Reef Symposium*, 2, 2063–2068. Miami, FL.

Bowden-Kerby, A. (1999). A community coral reef initiative: coral reef restoration in rural Pacific settings. In *Program and Abstracts, International Conference on Scientific Aspects of Coral Reef Assessment, Monitoring, and Restoration*, pp. 188–189. National Coral Reef Institute, Nova Southeastern University.

Brown, B. E. (1996). Disturbances to reefs in recent times. In *Life and Death of Coral Reefs*, ed. C. Birkeland, pp. 354–379. New York: Chapman & Hall.

Cairns, J., Jr (1988). Restoration ecology: the new frontier. In *Rehabilitating Damaged Ecosystems*, vol. 1, ed. J. Cairns Jr, pp. 1–11. Boca Raton, FL: CRC Press.

Cairns, J., Jr (2000). Setting ecological restoration goals for technical feasibility and scientific validity. *Ecological Engineering*, 15, 171–180.

Cesar, H. S. J. (ed.) (2000). *Collected Essays on the Economics of Coral Reefs*. Kalmar, Sweden: CORDIO.

Clark, S. & Edwards, A. J. (1994). The use of artificial reef structures to rehabilitate reef flats degraded by coral mining in the Maldives. *Bulletin of Marine Science*, 55, 726–746.

Clark, S. & Edwards, A. J. (1995). Coral transplantation as an aid to reef rehabilitation: evaluation of a case study in the Maldives. *Coral Reefs*, 14, 201–213.

Clark, S. & Edwards, A. J. (1999). An evaluation of artificial reef structures as tools for marine habitat rehabilitation in the Maldives. *Aquatic Conservation of Marine and Freshwater Ecosystems*, 9, 5–21.

Connell, J. H. (1997). Disturbance and recovery of coral assemblages. *Coral Reefs*, 16 (Suppl.), S101–S113.

Costanza, R., d'Arge, R., de Goot, R., Farber, S., Grasso, M., Hannon, B., Limburg, K., Naeem, S., O'Neil, R., Paruelo, J., Raskin, R., Sutton, P. & van den Belt, M. (1997). The value of the world's ecosystem services and natural capital. *Nature*, 387, 253–260.

Done, T. (1995). Remediation of degraded coral reefs: the need for broad focus. *Marine Pollution Bulletin*, 30, 686–688.

Edwards, A. J. & Clark, S. (1992). Re-establishment of reef fish populations on a reef flat degraded by coral quarrying in the Maldives. *Proceedings of the 7th International Coral Reef Symposium*, 1, 593–600.

Edwards, A. J. & Clark, S. (1998). Coral transplantation: a useful management tool or misguided meddling? *Marine Pollution Bulletin*, 37, 474–487.

Field, C. D. (ed.) (1996). *Restoration of Mangrove Ecosystems*. Okinawa, Japan: International Society for Mangrove Ecosystems.

Field, C. D. (1998) Rehabilitation of mangrove ecosystems: an overview. *Marine Pollution Bulletin*, **37**, 383–392.

Fitzhardinge, R. C. & Bailey-Brock, J. H. (1989). Colonisation of artificial reef materials by corals and other sessile organisms. *Bulletin of Marine Science*, **44**, 567–579.

Fonseca, M. S. (1994). *A Guide to Planting Seagrasses in the Gulf of Mexico*. Texas A& M University.

Franklin, H., Muhando, C. A. & Lindahl, U. (1998). Coral culturing and temporal recruitment patterns in Zanzibar, Tanzania. *Ambio*, **27**, 651–655.

Gilliam, D. S., Banks, K. & Spieler, R. E. (1995). Evaluation of a novel material for artificial reef construction. In *Proceedings of the 6th International Conference on Aquatic Habitat Enhancement*, Japan, pp. 345–350.

Gittings, S. R., Bright, T. J., Choi, A. & Barnett, R. R. (1988). The recovery processes in a mechanically damaged coral reef community: recruitment and growth. *Proceedings of the 6th International Coral Reef Symposium*, **2**, 225–230.

Global Coral Reef Alliance (1999). http://fas.harvard.edu/~goreau/restr.accret.html

Glynn, P. W. (1984). Widespread coral mortality and the 1982/3 El Niño warming event. *Environmental Conservation*, **11**, 133–146.

Gomez, E. D., Alcala, A. C. & Alcala, L. C. (1982). Growth of some corals in an artificial reef off Dumaguete, Central Visayas, Philippines. *Kalikasan, Philippines Journal of Biology*, **11**, 148–157.

Green, E. P. & Shirley, F. (1999). *The Global Trade in Coral*. Cambridge: World Conservation Press.

Guzman, H. M. (1991). Restoration of coral reefs in Pacific Costa Rica. *Conservation Biology*, **5**, 189–195.

Guzman, H. M. (1999). Large-scale restoration of Eastern Pacific reefs: the need for understanding regional biological processes. In *Program and Abstract, International Conference on Scientific Aspects of Coral Reef Assessment, Monitoring, and Restoration*, p. 97. National Coral Reef Institute.

Hackney, C. T. (2000). Restoration of coastal habitats: expectation and reality. *Ecological Engineering*, **15**, 165–170.

Harriott, V. J. & Fisk, D. A. (1988a). Coral transplantation as a reef management option. *Proceedings of the 6th International Coral Reef Symposium*, **2**, 375–379. Townsville, Australia: Great Barrier Reef Marine Park Authority.

Harriott, V. J. & Fisk, D. A. (1988b). *Accelerated Regeneration of Hard Corals: A Manual for Coral Reef Users and Managers*. Great Barrier Reef Marine Park Authority Technical Memorandum 16.

Hatcher, B. G., Johannes, R. E. & Robertson, A. I. (1989). Review of research relevant to the conservation of shallow tropical marine ecosystems. *Oceanography and Marine Biology Annual Review*, **27**, 337–414.

Heeger, T., Cashman, M. & Sotts, F. (1999). Coral farming as alternative livelihood for sustainable natural resource management and coral reef rehabilitation. In *Proceedings of the Oceanography International 99, Pacific Rim*, Singapore, pp. 171–186.

Heyward, A. J., Rees, M. & Smith, L. D. (1999). Coral spawning slicks harnessed for large-scale coral culture. In *Program and Abstracts, International Conference on Scientific Aspects of Coral Reef Assessment, Monitoring, and Restoration*, pp. 188–189. National Coral Reef Institute.

Hilbertz, W. H. (1992). Solar-generated building material from seawater as a sink for carbon. *Ambio*, **21**, 126–129.

Hilbertz, W. H., Fletcher, D. & Krausse, C. (1977). Mineral accretion technology: applications for architecture and aquaculture. *Industrial Forum*, **8**, 75–84.

Highsmith, R. C. (1982). Reproduction by fragmentation in corals. *Marine Ecology Progress Series*, **7**, 207–226.

Hudson, J. H. & Diaz, R. (1988). Damage survey and restoration of MV *Wellwood* grounding site, Molasses Reef, Key Largo National Marine Sanctuary, Florida. *Proceedings of the 6th International Coral Reef Symposium*, **2**, 231–237.

Hudson, J. H., Robin, D. M., Tilmant, J. T. & Wheaton, J. W. (1990). Building a coral reef in SE Florida: combining technology with aesthetics. *Bulletin of Marine Biology*, **44**, 1067.

Hughes, T. P. (1994). Catastrophes, phase shifts and large-scale degradation in a Caribbean coral reef. *Science*, **265**, 1547–1551.

Hueckel, G. J., Ruckley, R. M. & Benson, B. L. (1989). Mitigating rocky habitat loss using artificial reefs. *Bulletin of Marine Science*, **44**, 913–922.

IUCN/UNEP (1985). *The Conservation and Management of Renewable Marine Resources in the Red Sea and Gulf of Aden Region.* UNEP Regional Seas Reports and Studies no. 64. Nairobi: United Nations Environment Programme.

Jaap, W. C. (2000) Coral reef restoration. *Ecological Engineering*, **15**, 345–364.

Kaly, U. L. (1995). *Experimental Test of the Effects of Methods of Attachment and Handling on the Rapid Transplantation of Corals.* CRC Reef Research Centre Technical Report no. 1. Townsville, Australia: Cooperative Research Centre for Ecologically Sustainable Development of the Great Barrier Reef.

Loya, Y. & Rinkevich, B. (1980). Effects of oil pollution on coral reef communities. *Marine Ecology Progress Series*, **3**, 167–180.

Lindahl, U. (1998). Low-tech rehabilitation of degraded coral reefs through transplantation of staghorn corals. *Ambio*, **27**, 645–650.

Maragos, J. E. (1974). *Coral Transplantation: A Method to Create, Preserve and Manage Coral Reefs.* University of Hawaii Sea Grant Programme AR- 74-03. Honolulu, HI: University of Hawaii.

Miller, S. L., McFall, G. B. & Hulbert, A. W. (1993). *Guidelines and Recommendations for Coral Reef Restoration in the Florida Keys National Marine Sanctuary.* Report to the National Undersea Research Center, Wilmington, NC: University of North Carolina.

Milon, J. W. & Dodge, R. E. (2001). Applying habitat equivalency analysis for coral reef damage assessment and restoration. *Bulletin of Marine Science*, **69**, 975–988.

Muñoz-Chagin, R. F. (1997). Coral transplantation program in the Paraiso coral reef, Cozumel Island, Mexico. *Proceedings of the 8th International Coral Reef Symposium*, **2**, 2075–2078.

Munro, J. L. & Balgos, M. C. (eds.) (1995). *Artificial Reefs in the Philippines.* ICLARM Conference Proceedings no. 49. Manila, Philippines: International Centre for Living Aquatic Resource Management.

Nagelkerken, I., Bouma, S., van den Akker, S. & Bak, R. P. M. (2000). Growth and survival of unattached *Madracos mirabilis* fragments transplanted to different reef sites, and the implication for reef rehabilitation. *Bulletin of Marine Science*, **66**, 497–505.

Newman, H. & Chuan, C. S. (1994). Transplanting a coral reef: a Singapore community project. *Coastal Management in Tropical Asia*, **3**, 11–14.

NOAA (1997). http://www.sanctuaries.nos.noaa.gov/special/columbus.html

Öhman, M. C., Rajasuriya, A. & Lindén, O. (1993). Human disturbances on coral reefs in Sri Lanka: a case study. *Ambio*, **22**, 474–480.

Oren, U. & Benayahu, Y. (1997). Transplantation of juvenile corals: a new approach for enhancing colonisation of artificial reefs. *Marine Biology*, **127**, 499–505.

Pearson, R. G. (1981). Recovery and recolonisation of coral reefs. *Marine Ecology Progress Series*, **4**, 105–122.

Plucer-Rosario, G. P. & Randall, R. H. (1987). Preservation of rare coral species by transplantation: an examination of their recruitment and growth. *Bulletin of Marine Science*, **41**, 585–593.

Pratt, J. R. (1994). Artificial habitats and ecosystem restoration: managing for the future. *Bulletin of Marine Science*, **55**, 268–275.

Precht, W. F. (1998). The art and science of reef restoration. *Geotimes*, **43**, 16–20.

Raymundo, L. J. H., Maypa, A. P. & Luchavez, M. M. (1999). Coral seeding as a technology for recovering degraded coral reefs in the Philippines. *Phuket Marine Biology Centre Special Publication*, **19**, 81–90.

Reef Ball (2000). http://www.reefball.com

Richmond, R. H. & Hunter, C. L. (1990). Reproduction and recruitment of corals: comparisons among the Caribbean, the tropical Pacific and the Red Sea. *Marine Ecology Progress Series*, **60**, 185–203.

Rinkevich, B. (1995). Restoration strategies for coral reefs damaged by recreational activities: the use of sexual and asexual recruits. *Restoration Ecology*, **3**, 241–251.

Rinkevich, B. & Loya, Y. (1989). Reproduction in regenerating colonies of the coral *Stylophora pistillata*. In *Environmental Quality and Ecosystem Stability*, vol. 4B, *Environmental Quality*, eds. E. Spanier, Y. Steinberger & M. Luria, pp. 259–265. Jerusalem: Israel Society for Environmental Quality Sciences.

Salvat, B. (1987). *Human Impacts on Coral Reefs: Facts and Recommendations.* Antenne, French Polynesia: Muséum Ecole Pratique des Hautes Etudes.

Sammarco, P. W. (1996). Comments on reef regeneration, bioerosion, biogeography and chemical ecology: future directions. *Journal of Experimental Marine Biology and Ecology*, **200**, 135–168.

Sammarco, P. W., Brazeau, D. A. & Lee, T. N. (1999). Enhancement of reef regeneration processes: supplementing coral recruitment processes through larval seeding. In *Program and Abstracts, International Conference on Scientific Aspects of Coral Reef Assessment, Monitoring, and Restoration*, p. 169. National Coral Reef Institute.

Schumacher, H. (1977). Initial phases of reef development, studied at artificial reef types of Eilat (Red Sea). *Helgolander wss Meersunters*, **30**, 400–411.

Seaman, W., Jr & Sprague, L. M. (1991). *Artificial Habitats for Marine and Freshwater Fisheries*. London: Academic Press.

Shinn, E. A. (1976). Coral reef recovery in Florida and the Persian Gulf. *Environmental Geology*, **1**, 241–254.

Smith, L. D. & Hughes, T. P. (1999). An experimental assessment of survival, re-attachment and fecundity of coral fragments. *Journal of Experimental Marine Biology and Ecology*, **235**, 147–164.

Spurgeon, J. P. G. (1998). The socio-economic costs and benefits of coastal habitat rehabilitation and creation. *Marine Pollution Bulletin*, **37**, 373–382.

Spurgeon, J. P. G. & Lindahl, U. (2000). Economics of reef restoration. In *Collected Essays on the Economics of Coral Reef*, ed. H. S. J. Cesar, Kalmar, Sweden: CORDIO.

Szmant, A. M. (1999). Coral restoration and water quality monitoring with cultured larvae of *Montastraea 'annularis'* and *Acropora palmata*. In *Program and Abstracts, International Conference on Scientific Aspects of Coral Reef Assessment, Monitoring, and Restoration*, pp. 188–189. National Coral Reef Institute.

Thom, R. M. (1997). System-development matrix for adaptive management of coastal ecosystem restoration projects. *Ecological Engineering*, **8**, 219–232.

Thongtham, N. & Chansang, H. (1999). Influence of surface complexity on coral recruitment at Maiton Island, Phuket, Thailand. In *Proceedings of the International Workshop on the Rehabilitation of Degraded Coastal Systems*, pp. 93–100.

Vago, R. & Turak, E. (1995). Renovation of coral reefs: a recolonisation technique. In *Ecological System Enhancement Technology for the Aquatic Environment*,

International Conference, October 1995, Tokyo, pp. 879–884.

van Treeck, P. & Schumacher, H. (1997). Initial survival of coral nubbins transplanted by a new coral transplantation technology: options for reef rehabilitation. *Marine Ecology Progress Series*, **150**, 287–292.

van Treeck, P. & Schumacher, H. (1998). Mass diving tourism: a new dimension calls for new management approaches. *Marine Pollution Bulletin*, **37**, 499–504.

van Treeck, P. & Schumacher, H. (1999). Artificial reefs created by electrolysis and coral transplantation: an approach ensuring the compatibility of environmental protection and diving tourism. *Estuarine Coastal and Shelf Science*, **49**, 75–81.

White, A. T., Chou, L. M., De Silva, M. W. R. N. & Guarin, F. Y. (1990). *Artificial Reefs for Marine Habitat Enhancement in Southeast Asia*. ICLARM Education Series no. 11. Manila, Philippines: International Centre for Living Aquatic Resources Management.

Wilkinson, C. R. (1996). Global change and coral reefs: impacts on reefs, economies and human culture. *Global Change Biology*, **2**, 547–558.

Wilkinson, C. R. (1999) Global and local threats to coral reef functioning and existence: review and predictions. *Marine and Freshwater Research*, **50**, 867–878.

Wilkinson, C., Lindén, O., Cesar, H., Hodgson, G., Rubens, J. & Strong, A. E. (1999). Ecological and socio-economic impacts of 1998 coral mortality in the Indian Ocean: an ENSO impact and a warning of future change? *Ambio*, **28**, 188–196.

Woodley, J. D. & Clark, J. R. (1989). Rehabilitation of degraded coral reefs, Coastal Zone '89. *Proceedings of the 6th Symposium on Coastal and Ocean Management*, **4**, 3059–3075.

Yap, H. T. (2000). The case for restoration of tropical coastal ecosystems. *Oceanography and Coastal Management*, **43**, 841–851.

Yap, H. T., Licuanan, W. Y. & Gomez, E. D. (1990). Studies on coral reef recovery and coral transplantation in the northern Philippines: aspects relevant to management and conservation. In *UNEP Regional Seas Reports and Studies no. 116*, ed. H. T. Yap, pp. 117–127. Nairobi: United Nations Environment Program.

Yap, H. T., Aliño, P. M. & Gomez, E. D. (1992). Trends in growth and mortality of three coral species (Anthozoa: Scleractinia), including effects of transplantation. *Marine Ecology Progress Series*, **83**, 91–101.

Yap, H. T., Alvarez, R. M., Custodio, H. M. III & Dizon, R. M. (1998). Physiological and ecological aspects of coral transplantation. *Journal of Experimental Marine Biology and Ecology*, **229**, 69–84.

Yates, K. R. & Carlson, B. A. (1992). Corals in aquariums: how to use selective collecting and innovative husbandry to promote reef conservation. *Proceedings of the 7th International Coral Reef Symposium*, **2**, 1091–1095.

9 • Beaches

CLIVE A. WALMSLEY

INTRODUCTION

Beaches are unconsolidated deposits of sand, pebbles, cobbles and boulders that extend from mean low water to the top of the foredune or the limit of storm tide influence. They occur along an estimated 40% of the world's coastline (Bird, 1996). Most beaches are predominantly composed of sand; that is particles between 0.062 and 4 mm in diameter. Shingle beaches are uncommon worldwide, with most found in northwest Europe, New Zealand, Japan and Namibia. Shingle is generally defined as sediment dominated by particles between 4 and 200 mm diameter. Wave energy is a key factor determining the sediment composition and form of beaches. Large waves breaking on a beach create a high wave energy environment with a steep beach profile generally consisting of coarse well-sorted sediments, while low wave energy beaches mostly have gentle slopes and finer less well-sorted sediments. However, other factors, such as sediment availability, can determine beach composition with many sandy beaches found on high wave energy coasts. There are enormous variations in substrate origin, particle size composition, tidal range, wave exposure, currents and winds along the world's beaches influencing beach morphology, profile and stability. Bird (1996) and Nordstrom (2000) review the geomorphological processes that shape beaches and the beach management techniques used to control erosion or accretion. On many sections of coast, silt, clay or rock outcrops are mixed with sand and shingle, particularly at the interface between beach and saltmarsh, rocky shore or cliff habitats. This chapter focuses specifically on the restoration of sand and shingle beaches.

Almost all beaches possess native infauna but vegetation is absent or ephemeral on many beaches, especially shingle or cobble beaches, because of their instability and exposure. The meiofauna, including nematodes and copepods, occupies the interstitial spaces between particles on beaches while a macrofauna of molluscs, polychaetes and amphipods burrows within the beach. On temperate sandy beaches the strandline is generally sparsely vegetated by summer annuals, such as sea rocket (*Cakile maritima*), saltwort (*Salsola kali*) and orache (*Atriplex* spp.), often backed by fore dunes dominated by perennial dune-building grasses, such as marram grass (*Ammophila arenaria*), American beachgrass (*A. breviligulata*) and lyme grass (*Leymus arenarius*). On tropical beaches, grasses, forbs and woody species are all generally important elements of the beach community (Doing, 1985). Buckley (1988) identified two main components in tropical Australian strandline communities, perennials established at the landward edge of the beach, including creepers like railroad vine (*Ipomoea pes-caprae*), and small short-lived herbaceous species, such as *Salsola kali* and spurges (*Euphorbia* spp.) that establish from seed. Unlike temperate beach communities, there are characteristic strandline-fringing trees on tropical beaches, including coconut (*Cocos nucifera*), taccada (*Scaevola sericea*) and whistling pine (*Casuarina equisetifolia*), which germinate in the landward areas of the beach but rarely survive to maturity except at or beyond the limit of tidal disturbance.

Seedling establishment and plant survival are threatened by many environmental stress factors on beaches, including erosion or accretion, poor nutrient status, large diurnal temperature fluctuations, drought, salt spray, tidal inundation and predation (Davy & Figueroa, 1993; Maun, 1994; Davy et al., 2001). The extent of tidal influence and beach sediment composition are crucial factors determining the stability of the beach and the ability of vegetation to establish on it (Scott, 1963; Walmsley & Davy, 1997a). An assessment of vegetation on North American beaches has shown that annuals, geophytes and hemicryptophytes are overrepresented in beach floras compared to Raunkiaer's world norm while phanerophytes are underrepresented (Barbour et al., 1985). These trends may reflect the selective advantage of life forms that have subterranean perennating organs, such as hemicryptophytes, in exposed and unstable beach environments, especially in the temperate regions. Coarse, highly porous beach substrates are rapidly leached of nutrients and water soon percolates through the surface. Hence, large-seeded plants that have sufficient reserves to rapidly establish and plants with deep or diffuse root systems also have a selective advantage (Davy et al., 2001).

CAUSES OF HABITAT LOSS AND DEGRADATION

The prevalence of beach erosion along so many of the world's coasts is probably the most serious threat of habitat loss. The present widespread erosion of beaches has a multitude of causes, including isostatic land subsidence, sea level rise, increased storminess, loss of terrestrial or marine sources of sediment, artificial structures or other human interference (Bird, 1996). Certainly the most ubiquitous threat is the global rise in mean sea level. Mean global temperature will rise by a 'best estimate' of 2 °C by 2100 and cause a substantial rise in global sea level as a result of ocean expansion and the melting of polar ice sheets and glaciers. The Intergovernmental Panel on Climate Change (1995) has produced scenarios ranging from a 15 to 95 cm rise in global sea level by 2100 with a 'best estimate' of 50 cm. That this rise will result in more extensive and severe erosion of beaches is not in doubt, only the extent of the impact is uncertain. Unfortunately, our ability to predict the frequency and severity of extreme events is very low, and it is these that are responsible for the majority of changes in beach geomorphology (Lee, 1993). However, further breaching of coastal defences, loss of entire beaches and the extermination of coastal vegetation communities and fauna through erosion and inundation is likely. These changes will often require an increase in coastal management, whether defending the current coastline or creating a new one ('coastal realignment'), that will threaten natural beach habitats. Property developments, especially coastal resorts, will continue to destroy some beaches, particularly in developing countries, but the recognition of their recreational importance may reduce future losses. Recreational activity itself sometimes causes damage to vegetation through trampling and vehicle use or disturbance to nesting birds or turtles. Oil contamination as a result of serious oil spills and beach litter often require mechanical removal resulting in damage to beach biota. Other human uses, including aggregate mining, grazing, groundwater extraction, military use and waste disposal, may cause localised habitat loss or damage.

RATIONALE FOR RESTORATION

Economic and social importance

Beaches have been exploited for coastal developments ranging from ports and industrial plants to tourism resorts and housing. They are locally important sources of aggregates and heavy minerals. Beaches play a crucial role as sea defences preventing inundation of property, agricultural land and natural habitats. Their value for coastal protection is illustrated by the growth in beach nourishment rather than hard structure construction along the coasts of North America and Europe since the 1960s. Since the mid nineteenth century coastal tourism has developed extensively in Europe and North America. In Spain, some 3500 km or 50% of the coastline has been developed for tourism over the last 40 years (Nordstrom, 2000). It is only during the last two decades that the growth in international travel has resulted in coastal tourism becoming a worldwide phenomenon. For example, Cancun, Mexico has developed from a small fishing village to become a world-class resort with more than 20 000 hotel rooms alongside a city with a population of 400 000, over the last 20 years. There will inevitably be further resort developments in coastal areas because of the predicted growth in global tourism. Consequently, the economic and social importance of beaches as a recreational resource will continue to increase rapidly.

Environmental importance

Although both the fauna and flora generally include many cosmopolitan species, the linear zonation of communities across beaches and the wide variation in abiotic conditions have led to the development of many specialist species confined to beaches. Some rare and threatened specialists are totally reliant on beach habitat, including those plants and invertebrates confined to shingle, and sea turtles. Beaches are vitally important nesting grounds for sea turtles, seals and sea-lions. They are also crucial nesting sites for birds, particularly terns, gulls and plovers, and the beach infauna is an important feeding resource for shore-birds. Their conservation is considered particularly important because beach habitat is very limited in its extent when compared with other habitats.

Coastal vegetated shingle and sand dunes (including pioneer beach communities) have been identified as priority habitats for conservation under the European Community Habitats Directive and the UK's Biodiversity Action Plan, because of their limited extent and declining distribution as well as the threatened status of some specialist species reliant on them (UK Biodiversity Group, 1999). Targets to recreate or restore 200 ha of shingle, 240 ha of sand dune and 10 000 ha of intertidal sand and mud flats in England are based on a commitment to maintain the extent of coastal habitats despite predicted losses due to both human development and erosion, primarily related to sea level rise, over 20 years (Pye & French, 1993). The UK Biodiversity Action Plan has reaffirmed the objective to create and restore coastal habitats to offset predicted losses (UK Biodiversity Group, 1999).

CONSTRAINTS

Emphasis on beach nourishment for sea defence

Among the thousands of beach nourishment projects undertaken world-wide relatively few have monitored their impact on the ecology of the beach.

Very few have sought to enhance beach habitats and monitor the ecological outcome of restoration work. Many of those that have produced ecological benefits have merely created or extended the beach, thereby providing nesting habitat for sea turtles or birds (see Boxes 9.1 and 9.2). Other beach restoration projects have largely relied on natural recolonisation to vegetate the new beach (see Box 9.3) or planted a monoculture of dune grass to stabilise it (see Box 9.4). Very few projects have sought to introduce a diversity of species comparable to that present in the original beach vegetation (see Box 9.5). The limited number of ecological restoration projects results from an emphasis on sea defence and recreational objectives.

Inadequate specification and monitoring

Crucial phases for a successful nourishment and habitat restoration project include the agreement of objectives or targets for restoration, the collection of baseline data, careful planning and design, successful implementation through good supervision and project management and monitoring. The engineering performance of beach nourishment projects is generally monitored in terms of shoreline and beach volume changes derived from detailed beach profile measurements made on a regular basis (e.g. Davis et al., 2000). Generally, far less rigorous monitoring of ecological outcomes has been carried out as an afterthought. There is no reason why ecological restoration should not apply standards similar to those used for the engineering specification. Standard guidelines for both physical engineering and biological monitoring have been developed for beach nourishment projects in Florida (Nelson, 1985; Stauble & Hoel, 1986). There is a need to develop guidelines for ecological restoration projects covering the specification of fill selection, beach profile design, the timing of nourishment work, the use of appropriate nourishment methods, the source and method of revegetation and the detailed monitoring of ecological outcomes.

RESTORATION IN PRACTICE

The restoration of beach communities requires, first, the reinstatement of suitable beach sediments

Box 9.1 Consequences of beach nourishment for sea turtle breeding in the United States of America

Thirty percent of the world population of the loggerhead turtle (*Caretta caretta*) and populations of the endangered hawksbill (*Eretmochelys imbricata*), leatherback (*Dermochelys imbricata*) and green (*Chelonia mydas*) turtles breed on the subtropical beaches of the United States. Beach nourishment projects designed to restore beaches can have negative impacts on the reproductive success of sea turtles (Crain *et al.*, 1995). The ability of adults to build nests and the survival of both eggs and hatchlings may be reduced by: (1) artificial beach profile and structure; (2) sand compaction; (3) changes to the gaseous, thermal and moisture environment; (4) importation of contaminated sediments; (5) changes to nutrient availability; and (6) osmotic conditions.

Nourishment during reproductive activities can reduce both nesting and hatching success as a result of disturbance, including light pollution (Wolf, 1988). Nourishment operations should be carried out outside of the turtle nesting season because heavy machinery or pipelines can prevent adult turtles from reaching beaches to lay eggs; and noise, light and human disturbance may discourage nesting or result in aborted nesting attempts. On beaches nourished during the nesting season, the relocation of nests has been successful, and is preferable to deep burial

(Flynn, 1992). Concern that the relocation of sea turtle eggs might decrease hatching, hatchling survivorship or hatchling sex ratios has been shown to be unfounded (Spadoni & Cummings, 1992). Hatching success was similar regardless of whether nests were relocated on natural or nourished beaches in Florida.

Despite the many potential impacts of nourishment on turtle nesting and nests, many monitoring programmes have shown no significant difference in hatching or emergence rates on natural and nourished beaches, and on a nourished beach at Boca Raton, Florida hatchling emergence and hatchling weights were greater on the nourished beach than in adjacent areas (Nelson *et al.*, 1987; National Research Council, 1995). Moreover, nourishment projects can create or maintain nesting habitat that would not otherwise be available. The numbers of nests and turtle crawls have been shown to increase significantly after beach nourishment improved or protected nesting habitat in Florida (Flynn, 1992; Spadoni & Cummings, 1992). Beach nourishment has been responsible for improving turtle nesting habitat on at least seven beaches in Florida (National Research Council, 1995). The careful specification of the fill material, creation of a natural beach profile, nourishment outside of nesting seasons and the avoidance of compaction should enable many more nourishment projects to enhance rather than damage turtle nesting habitat.

by nourishment and, second, the restoration of beach vegetation. There have been no attempts to reintroduce beach infauna but assessments of the recovery of macroinvertebrate populations on nourished beaches suggest that natural recovery should occur within two years as a result of immigration accelerated by sea dispersal. Most studies of the intertidal fauna have recorded only short-term changes in species composition and abundance over a few weeks or months (National Research Council, 1995). However, almost all studies have been carried out on subtropical North American coasts and the rate of recovery on cooler temperate coasts may be appreciably slower.

Beach nourishment

Beach nourishment, replenishment, recharge or restoration is the addition of imported sediments onto a beach to prevent further erosion and recreate beach habitat for sea defence, recreational, or more rarely, environmental purposes. In an essentially erosive environment, beach nourishment needs to be a cyclical process for the foreseeable future but in other localities a single operation may be sufficient. Fill can be sourced from offshore areas, accreting beaches or inland and may be deposited at various locations along the shore profile (see Fig. 9.1). Generally, damage to biota is least likely if fill is deposited

Box 9.2   Nourished beaches as bird nesting habitat

Beach habitat has been specifically created for nesting California least terns (*Sterna antillarum browni*) at a number of coastal sites in California using dredged material. These sites have contributed to the rapid growth in the least tern population since the 1980s, along with the protection of breeding areas from disturbance and predation (Powell, 1998). Monitoring of sandy beach and lagoon habitat created at Batiquitos Lagoon, California in 1994–5 showed that the threatened western snowy plover (*Charadrius alexandrinus nivosus*) also benefited, and nested on four out of five created areas (Powell & Collier, 2000). The number of plover nesting attempts increased dramatically from five in 1994 to 38 in 1997 on the new beaches. Another beach site created in the mid-1970s showed similar nest productivity

rates (fledglings per nest) to the natural site suggesting that, after initial fluctuations, artificially created beaches provide suitable nesting habitat.

Management of nourished sites can improve habitat for the threatened piping plover (*Charadrius melodus*), which nests on beaches and foredunes. Colonisation was successfully encouraged by the erection of signs and fences that restricted human access, and prevented trampling and beach cleaning on a nourished beach at Ocean City, New Jersey (Nordstrom, 2000). The fencing allowed the formation of pioneer dunes colonised by eight native plant species. A natural community on a nourished beach was thus established with minimal intervention where it would have otherwise had almost no ecological value. In 1989, some 12% of breeding pairs of piping plovers on Long Island, New York were found on beaches composed mainly of fill material (Downer & Liebelt, 1990).

in the near shore and beach areas. Beach nourishment has become an increasingly popular method of maintaining coastal defence because of its limited visible environmental impact and relatively low capital costs. It is a viable alternative to engineered hard structures and the principal technique for beach restoration (National Research Council, 1995). It can also provide an opportunity to restore or create threatened beach habitats. The US Committee on Beach Nourishment and Protection concluded that, where feasible, 'projects should incorporate design features that would enhance biological resources of concern' (National Research Council, 1995).

Since early attempts to nourish beaches in California in 1919 and New York in 1922, some 640 km of beaches along the United States coast have been nourished (Davison *et al.*, 1992). There have been nourishment projects around the coasts of Europe from the Baltic coast of Poland to Italy, with many projects in the UK and Holland (Bird, 1996). Elsewhere, there have been nourishment projects in Egypt, Nigeria, South Africa, Malaysia, Singapore, Hong Kong, Japan and Australia. The very extensive literature concerned with the planning,

methods, economics and impacts of beach nourishment has been frequently reviewed (e.g. Davison *et al.*, 1992; National Research Council, 1995; Bird, 1996; Nordstrom, 2000). Most beach nourishment around the world has been on sandy beaches. However, there have been major shingle beach nourishment projects in Sussex, Dorset, Hampshire and Kent on the south coast of England (McFarland *et al.*, 1994; Bird, 1996). Nourishment is likely to remain an important method of coastline management. It has been predicted that between 23 and 47 million m^3 of shingle and 36 to 83 million m^3 of sand will be needed for beach nourishment in the UK between 1996 and 2016 (Humphreys *et al.*, 1996).

Most ecological studies of beach nourishment have focused on negative impacts either on the borrow site (the source location for fill) or the nourished beach. The detrimental effects of nourishment on the beach biota are likely to be inversely related to the need, because those areas requiring urgent nourishment are also lacking in beach and dune habitat (Nordstrom, 2000). Hence, it is probable that nourishment would provide net ecological benefits on many severely eroded beaches, unlike the construction of hard sea defences. Nevertheless, it is

Box 9.3   Vegetation colonisation after beach nourishment at Perdido Key, Florida

The impact of beach nourishment on a previously undisturbed barrier island at Perdido Key, Florida was monitored over four years (Looney & Gibson, 1993; Gibson & Looney, 1994; Gibson et al., 1997). Although the aggregate was dredged to maintain an adjacent navigable inlet and deposited to mitigate coastal erosion, the control of adverse impacts on existing habitat and the enhancement of beach habitat were important goals owing to the area being a National Park. Initially the possibility of depositing dredged spoil along 6.4 km of coast was considered but this would have raised the maximum height of the beach significantly above the maximum dune height on the barrier island and affected the supply of salt spray reaching the dunes. Salt spray has been identified as an important ecological factor influencing germination and vegetation zonation, and providing a source of nutrients. The final design created a lower profile beach along 8 km of coastline so that despite a 150-m increase in beach width, the overwash could reach the original strandline during storm events. This was advantageous for the development of a 'natural' profile and primary succession across the beach. In the first season after nourishment the vegetation was dominated by the strandline annual sea rocket (Cakile constricta), while the dune grass sea oats (Uniola paniculata) subsequently became dominant as a result of seedling and vegetative tiller recruitment. All of the species permanently established on the new beach were important within the fore dune and strandline communities before nourishment. Sand fences along the strand trapped sand and propagules resulting in the establishment of large Uniola paniculata and seaside primrose (Oenothera humifusa) populations. After four years' recolonisation similar species diversity to the original beach had been attained on the nourished area, but overall vegetation cover was much lower, with less pioneer dune vegetation and more strand vegetation. Although these trends in cover and biodiversity indicated that the vegetation structure of the nourished beach would eventually resemble the original community, it was predicted that it would be many years before the vegetation was comparable to the original foredunes (Looney & Gibson, 1993; Gibson et al., 1997). Proactive restoration could have accelerated this process.

important to be aware of these impacts in order to minimise any damage to existing infauna or vegetation, and maximise ecological benefits. Although most beaches are nourished to meet sea defence or recreational objectives, understanding the ecological problems associated with these projects and the methods of mitigating the impacts can help inform the development of practical recommendations for ecological restoration projects in future.

Profile

Artificially constructed beach profiles are rarely the same as the natural slopes or even the designed profile because of the limited ability of machinery to contour it suitably (Stauble & Hoel, 1986). The development of eroded scarps and steep profiles can undermine and destroy newly established vegetation and discourage the nesting of birds and sea turtles. It is important that every effort is made to create a facsimile of the natural profile and thereby avoid subsequent erosion or accretion (see Box 9.3). A natural profile is vital for the establishment of shingle-beach vegetation because it reduces the probability of severe erosion or accretion destroying newly established vegetation that is least able to endure disturbance (Walmsley, 1995). Nourishment schemes should be planned as several small fill sections interspersed with undisturbed beach rather than one continuous large nourishment area to accelerate the process of recolonization by infauna (Reilly et al., 1980).

Burial

Burial is the most immediate impact of nourishment. Some temporary loss of buried infauna from both intertidal and subtidal nourished areas is almost certain and unavoidable, but the rate of recovery after burial can vary greatly.

Box 9.4 Creating an artificial barrier island
system at Køge Bugt, Denmark

An 8-km-long artificial barrier island system was
constructed at Køge Bugt on the Baltic coast of Sjlland,
Denmark in 1977–8. The profile of the 300-m-wide
island was modelled on a nearby system. It was built
using offshore marine sand pumped to the site and
planted with *Ammophila arenaria*. The natural
development of vegetation and changes in
geomorphology were monitored between 1978 and
1990 (Vestergaard & Hansen, 1992). The creation of a
beach profile based on neighbouring beaches ensured
that there was little change in profile during the five
years after construction. The natural colonisation of
the beach by *Leymus arenarius* and *Elytrigia juncea* and
the spread of *A. arenaria* from plantings on the dunes
led to the accumulation of wind-blown sand and the
formation of 1.5-m-high pioneer dunes on the upper
beach after 12 years. The beach was naturally colonised
by strandline species, including *Cakile maritima*,
Honckenya peploides and *Salsola kali* during the first
season. There were fluctuating numbers of species on
the beach during the five years after construction with
most species exterminated after three years
(Fig. B9.1). However, after 12 years, the diversity of
species had risen substantially, including more
perennials, like *Lathyrus japonicus*, red fescue
(*Festuca rubra*) and creeping thistle (*Cirsium arvense*).

The project successfully provided coastal protection
and established dune vegetation at the landward edge
of the beach aided by the planting of *A. arenaria* on the
dunes and the substantial inputs of marine sand to
the beach system.

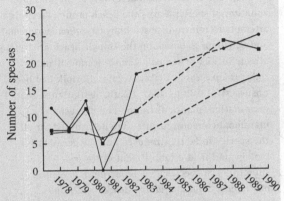

Fig. B9.1. The number of vascular plant species
recolonising an artificial barrier island beach created
at Køge Bugt, Denmark in 1977–8. The variation in
species number occurring on the outer beach (▲), inner
beach (■) and fore dune (●) between 1978 and
1990 are shown. The rapid colonisation of the beach
by around ten pioneer species in the first season after
nourishment was followed by extirpation and
recolonisation during the next five years. Adapted from
Vestergaard & Hansen (1992) (© Swets & Zeitlinger).

Experimental studies show that dominant swash-
zone species on American beaches, including mole
crab (*Emerita talpoida*) and coquina clams (*Donax*
spp.), are able to survive immediate burial by up
to 10 cm of sediment (Nelson, 1989). A study of
the capacity of seven common near-shore subtidal
species to emerge after burial showed that some
were able to emerge through up to 40 cm of fill, although survival was dependent on sediment composition and temperature (Maurer *et al.*, 1986). Several
macrofauna species were shown to recolonise the
nourished section of Folly Beach, South Carolina by
migrating through 0.6 to 1 m of fill (Lynch, 1994).
Although it is common to deposit more than a metre of fill during nourishment, no studies have inves-

tigated the capacity of fauna to emerge from such
depths. Offshore nourishment allows the gradual
transfer of material to the beach thereby reducing
impacts on beach fauna and vegetation, but it is
only suitable where tidal currents would result in
sea-dumped sediments accumulating on the beach.
However, offshore dumping does bury the original benthos causing the death of bottom-dwelling
sedentary species and may increase water turbidity.

Particle size

Nelson (1989) reviewed studies examining the effects of beach nourishment on faunal communities and concluded that significant mortality tended

Box 9.5   Restoration of shingle-beach vegetation at Sizewell, Suffolk, UK

In 1994 shingle-beach vegetation was restored along 1 km of coast to mitigate the damage caused by construction of a power station at Sizewell, Suffolk, UK (Walmsley, 1995). Some 10 800 container-grown plants of 11 species were planted according to the natural zonation of species across the beach profile. Previous attempts to reintroduce sea holly (*Eryngium maritimum*) and *Calystegia soldanella* on the shingle beach at Shingle Street, Suffolk or sea kale (*Crambe maritima*) and *Lathyrus japonicus* at Blakeney Point, Norfolk had been unsuccessful (White, 1967; Heath, 1981). Hence, seed germination, plant cultivation and plant-stock maintenance requirements were determined for all the species to be reintroduced and the germination biology and field establishment of six key shingle-beach species (*Crambe maritima, Eryngium maritimum, Glaucium flavum, Honckenya peploides, Lathyrus japonicus, Rumex crispus*) were investigated to identify suitable restoration methods.

Techniques based on seedling emergence showed that there was only a very small germinable seed bank within 13-cm-deep cores taken from the beach (Walmsley, 1995). The naturally occurring seed bank was an unsuitable resource for restoration because of its small size and the difficulty of stripping the surface layer without rapidly dispersing the seeds deep within the substrate. Seed representative of the local genetic stock was collected from the locality before disturbance because the extent of genetic variation in beach species is generally unknown. The six-year construction project necessitated the prolonged storage of the seed. The seed was stored over silica gel (to maintain low humidity) and in the dark at 5 °C in order to maintain seed viability. Innate seed dormancy that required either scarification or stratification to enable germination was identified in five of the six key species (Walmsley & Davy, 1997b). Seed stored for seven years retained high viability in all six species but the ageing resulted in a reduced range of environmental conditions permitting germination. The sowing of stored seed on the beach would have resulted in slow

Table B5.1. *Seed germination under controlled laboratory conditions compared with percentage emergence when sown directly on a shingle beach at Sizewell, Suffolk, UK*

Species	Seed germination[a] (% ± S.E.)	Total emergence (% of viable seed sown)
Crambe maritima	59 ± 6.3	18
Eryngium maritimum	22 ± 4.0	4.7
Glaucium flavum	88 ± 2.2	34
Lathyrus japonicus	91 ± 1.8	1.3
Rumex crispus	100 ± 0	9.5

[a]The incubated *C. maritima* seeds were excised from their pericarp, *L. japonicus* seeds were acid scarified (96% H_2SO_4 for 45 min) and *E. maritimum* and *G. flavum* seeds were stratified at 2 °C for 15 weeks before incubation.
Source: Adapted from Walmsley & Davy (1997b).

and erratic germination because of a combination of innate dormancy and restrictive germination requirements, while the use of pre-treatments and controlled environmental conditions could ensure high germination rates for the production of container-grown plants. Directly sowing seeds of the perennial species onto the shingle beach proved to be unsuccessful because innate and induced seed dormancy caused poor and erratic germination (see Table B5.1) and seedling survival was low because of drought and high-temperature stress during the summer and salt-spray damage and tidal inundation during the winter (Walmsley & Davy, 1997c). Only the short-lived monocarpic species *Glaucium flavum* (yellow horned-poppy) germinated readily and developed to maturity during the two seasons before the entire site was inundated by 15–25 cm of sediment, extirpating all of the vegetation. *Glaucium flavum* seedlings showed higher mortality and slower growth on sandy areas of the beach compared to the sparse seedlings in coarser substrates. A glasshouse experiment showed that the higher mortality in sandy areas probably resulted from density-independent substrate effects, emphasising the importance of creating a substrate texture suitable for plant establishment.

In contrast, container-grown plants of all six species established rapidly and showed very low mortality during the season after planting (Walmsley & Davy, 1997a). Storm tides inundated the container-grown plant field trial causing up to 35 cm erosion or up to 49 cm accretion of sediment about seven months after planting. Over the next eight months a high percentage of plants that had been buried or eroded by less than 10 cm recovered, and some regenerated from much greater depths (see Table B5.2). Only *G. flavum* showed almost no recovery even amongst those plants buried or eroded by less than 10 cm. The large-scale restoration of the beach was carried out using container-grown plant stock largely because of the ability of the perennial plants to establish extensive root or rhizome systems necessary for subterranean growth after burial or regrowth from totipotent stock following erosion. The use of semi-mature plants minimised the risk of catastrophic failure. However, the predominance of innate dormancy among the beach species and the low probability of seedlings growing to maturity were also key factors. It was recognised that strandline

Table B5.2. *The absolute maximum depth (cm) of accretion and erosion from which container-grown plants of six shingle-beach species emerged and re-established within a field experiment on the shingle beach at Sizewell, Suffolk, UK. The plants were 31 weeks old when planted on the beach and they were buried or eroded approximately 33 weeks later*

Species	Absolute maximum depth of erosion tolerated (cm)	Absolute maximum depth of accretion tolerated (cm)
Crambe maritima	22	16
Eryngium maritimum	19	18
Glaucium flavum	4	20
Honckenya peploides	26	41
Lathyrus japonicus	8	37
Rumex crispus	7	15

annuals were able to recolonize naturally while monocarpic species like *Glaucium flavum* were suitable for establishment by sowing directly onto the beach where sufficient fine fraction was present.

to occur only when sediments with particle size distributions significantly different from the existing substrate were used, especially fill containing high amounts of organic matter, shelly material or fine sediments. Similarly, inappropriate fill material may have a negative impact on birds by deterring nesting. It has been recommended that particle size distribution data should be collected at all borrow areas so that material can be matched (Stauble & Hoel, 1986). It is also important that the particle

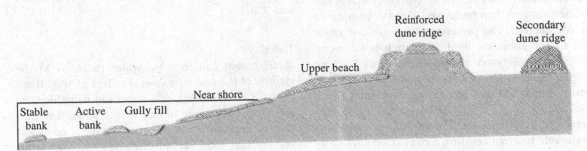

Fig. 9.1. Potential locations for beach nourishment along the shore profile. The deposition of fill near shore or on the beach is likely to have least ecological impact. Adapted from van der Wal (1998).

size distribution along the natural beach is properly assessed prior to nourishment to determine the fill specification. However, the use of fill containing fine sand or silt (<0.15 mm diameter) should be avoided even where the indigenous beach substrate has a similar particle size distribution because it increases the likelihood of beach compaction and threatens the ability of infauna to emerge (Stauble & Hoel, 1986). Most fill has been dredged from offshore marine sources but the use of terrestrially sourced sediments has been successful. Sand from inland quarries used for beach nourishment at Myrtle Beach, South Carolina initially caused reductions in biota but there was rapid recovery with some areas subsequently exhibiting species enrichment (Baca & Lankford, 1988).

Compaction

Nourishment can produce a beach surface that may remain compacted for between one and seven years, depending on substrate type, weather and wave conditions (Moulding & Nelson, 1988). The monitoring of beach nourishment schemes on Sand Key, Florida for between four and eight years revealed that the method of construction was a key factor determining compaction (Davis et al., 2000). Suction dredging of offshore sediment and its transfer to the site as slurry produced a more compact beach substrate than the use of a dragline to extract sediment and a conveyor-belt to deposit it. Although a compacted beach surface may extend the life of the beach nourishment scheme the loose-packed sediment produced by the conveyor-belt method may be ecologically more desirable, especially where turtle nesting or other fauna are likely to be harmed by compaction. The presence of fine-grained material in the substrate used for shingle-beach nourishment projects on the south coast of England caused the creation of a compacted beach with less permeability and mobility than the original beach (McFarland et al., 1994). Although tilling a compressed beach surface will improve its porosity it is preferable to avoid creating a compacted surface by specifying the appropriate particle size distribution for the fill and a nourishment method that avoids compaction.

Salinity

High salinities in fill material and the increased deposition of salt spray in foredunes during the deposition of the substrate as slurry are potential threats to biota on nourished beaches (Nelson, 1993; National Research Council, 1995). It is difficult to avoid the high salt concentrations in fill but transferring slurry to the nourishment site during periods of light or offshore winds can minimise salt-spray deposition.

Timing

The life cycle of species and the timing of works are major factors determining the impact of nourishment on fauna (Reilly et al., 1980). For example, the coquina clam Donax variabilis migrates offshore during the winter and returns to the intertidal zone in spring so work during the spring is likely to prevent its return, resulting in the absence of adult clams throughout the season. Species recruited from pelagic larvae are likely to recover rapidly if work is carried out before spring recruitment while most species that spend their entire life history within the beach, including many amphipods, will be affected for considerably longer, regardless of timing (Reilly et al., 1980). Beaches should be nourished during winter months with works completed before pelagic larval recruitment or the return of offshore overwintering adults in the spring. Similarly, nourishment work should be avoided during the nesting season of birds or turtles as disturbance can reduce reproductive success, but nourishment with appropriate fill outside of the nesting season can create or enhance nesting habitat (see Boxes 9.1 and 9.2).

Turbidity

Nourishment can increase water turbidity in the vicinity of the beach by several orders of magnitude (Reilly et al., 1980). Yet, this effect can be reduced by selecting a sediment source with a small fine fraction or depositing the substrate using a conveyor-belt rather than as pipeline slurry, the latter method potentially creating copious amounts of sediment-rich effluent. Suspended solids have been implicated as a cause of pelagic larval mortality during dredging operations (Mileikovsky, 1970), thereby reducing the ability of species with pelagic larvae to recolonize nourished beaches.

Sand transport

If relatively fine sand is used for beach nourishment it is likely that aeolian transport may result in secondary effects on adjacent foredunes. However, nourishment sediments have been shown to be less prone to wind-borne transport than natural beach sand. Nourishment had no significant effects on biodiversity of Dutch foredunes, although the vigour of *Ammophila arenaria* was significantly greater for two years after nourishment due to additional sand accumulation (Loffler & Coosen, 1995). Subsequently both sand accumulation and vigour declined. After extensive beach nourishment at Perdido Key, Florida an analysis of vegetational succession using Markovian models indicated that the successional processes in existing vegetation were unaffected by nourishment (Gibson *et al.*, 1997). The effects of nourishment on beach infauna are certainly far greater than any secondary impacts on adjacent vegetation. The effects of sand accretion are discussed further by Greipsson (this volume).

Establishing vegetation

Source of propagules

There is evidence that seeds of beach plants have been transported long distances by sea (Guppy, 1906; Ridley, 1930). Of more importance is the ability of either seed or vegetative fragments to migrate to restored beaches from adjacent undisturbed areas. On the disturbed beach at Sizewell, Suffolk, U.K. (see Box 9.5), only strandline species established readily from local seed and the germinable seed bank within the shingle beach was very small, although this was partly a consequence of innate seed dormancy rather than a lack of propagules (Walmsley & Davy, 2001). Monitoring of vegetation establishment showed that there was sufficient migration of propagules onto the nourished beach at Kòge Bugt for the natural development of a diverse vegetation community (see Box 9.4). The ability of adjacent areas to provide a suitable source of propagules will be determined by site-specific factors, such as suitable vegetation, prevailing wind direction and tidal currents.

Breaking seed dormancy

The prevalence of both innate and secondary seed dormancy in many beach species has been recognised as an important factor preventing rapid colonisation (Walmsley & Davy, 1997b; Davy *et al.*, 2001). An array of dormancy mechanisms, including hard seed or fruit coats requiring scarification, a physiological need for stratification and sensitivity of germination response to variations in temperature and salinity, may provide selective advantage by inhibiting germination under conditions unfavourable for survival. Consequently, restoration of beach vegetation requires an understanding of germination requirements and methods to release seed dormancy. Some species are unsuitable for sowing directly in the field and require cultivation from seed under controlled conditions or vegetative propagation. Beach species producing seed that requires scarification, such as sea bindweed (*Calystegia soldanella*) and sea pea (*Lathyrus japonicus*), have been scarified and cultivated for introduction as container-grown plants (Guinon & Allen, 1990; Walmsley & Davy, 1997b).

Overcoming environmental constraints
Disturbance

Occasional storm tides cause catastrophic erosion or accretion on shingle beaches that destroys whole cohorts of seedlings and many plants (Walmsley & Davy, 1997c), and large-scale seedling mortality often results from accretion or erosion in foredunes (Maun, 1994). Smooth cordgrass (*Spartina alterniflora*) beds at the seaward edge of shingle beaches in New England facilitate the establishment and persistence of a sparse cobble-beach plant community by reducing water velocities and greatly increasing substrate stability. Experimental studies confirmed that the distribution of four cobble-beach plant species was confined to sections of stabilised beach behind *S. alterniflora* swards because seedlings were unable to emerge and survive elsewhere (Bruno, 2000).

Clearly, catastrophic failure of restoration projects is quite likely. The use of seeding techniques to restore shingle-beach vegetation is generally too risky, so a matrix of vegetation has been established using container-grown plants at Sizewell (Walmsley & Davy, 2001) (see Box 9.5). However, the stochastic nature of these failures mean that repeated attempts at restoration may be justified, with some dune sites successfully restored after four, five or six attempts (Woodhouse, 1982). On cobble beaches fringed by saltmarsh, the restoration of the *Spartina* sward to create beach stability is probably a prerequisite to the restoration of the cobble-beach community.

Particle size

Scott (1963) described the correlation between the presence of a fine particle fraction and vegetation distribution and recognised the importance of this factor in controlling plant distribution on British shingle beaches. Field experiments have shown that substrate composition was the main factor determining seed germination, seedling survivorship and mortality, growth and fecundity of container-grown plants on a shingle beach at Sizewell (Walmsley & Davy, 1997b, c). In very coarse substrates, seeds are buried too deeply for successful emergence, the moisture and nutrient retention capacity of the substrate is too low for high rates of germination and survivorship of both seedlings and adult plants is very low. On sandy beaches the particle size composition is much less important for plant establishment.

The appropriate particle size composition is crucial for the establishment of shingle-beach vegetation (Walmsley & Davy, 2001). In practice, only 10% to 20% of fine fraction (<2 mm particle size) greatly improves germination, seedling survival and plant establishment (Walmsley & Davy, 1997a, c) (see Box 9.5). Similarly, at inland aggregate mines in the northeast United States most species could not be established where the fine fraction was less than 15% (Gaffney & Dickerson, 1987). On shingle beaches a matrix of areas with contrasting compositions may be appropriate so that some areas are predominantly unvegetated while others may favour species of sandy environments (Walmsley, 1995).

Organic matter and nutrients

Organic matter content is limited on most beaches, with the accumulation of tidal litter along the strandline the main source, so it has been identified as an important limiting factor possibly controlling vegetation distribution on beaches (Tansley, 1949; Randall, 1988). Organic matter improves nutrient and water retention and prevents seeds from being buried deep within shingle (Fuller, 1987; Randall, 1988). Beaches are also normally nutrient-limited, with rainwater, decomposition products from drift and litter, and sea spray the most important sources of plant nutrition (Scott, 1960). However, it has been demonstrated that the foreshore and beach act as a reservoir of sand able to provide sufficient readily mineralised nitrogen for plant growth (Fay & Jeffrey, 1992).

On shingle beaches organic-matter and fertiliser treatments produced no significant improvement in plant growth, and organic matter only slightly reduced the mortality of container-grown plants (Walmsley & Davy, 2001). Overall, the minor benefits to plantings of fertiliser and organic matter were outweighed by the additional costs and time required to apply the treatments, especially as the addition of small amounts of sand was shown to have similar effects (see Box 9.5). The absence of beach cleaning and recontouring works, which frequently occur on nearby unvegetated beaches, aided the natural development of a diverse shingle strandline community on Black Rock Beach, Brighton, UK (Packham et al., 1995). Preventing bulldozing and beach cleaning may be a cost-effective method of promoting revegetation by allowing the accumulation of debris thereby improving the organic matter and nutrient status.

Salinity and drought

Beach plants are not generally true halophytes and, although they are able to tolerate moderate salinity, the prolonged saline conditions experienced by saltmarsh species cannot be endured (Barbour et al., 1985). Nevertheless, many strandline and beach plants exhibit greater tolerance of salinity compared to dune species. The beach and fore dune environment has been historically considered xeric, but xylem water potentials, appreciable stomatal densities and moderately low stomatal indices indicate that plants do not generally experience significant water stress (de Jong, 1979; Davy et al., 2001). However, occasional drought conditions may explain the prevalence of deep-rooted species.

The sowing or planting of beach vegetation should be timed to maximise the period of establishment before the vegetation is exposed to storm tides and salt spray. On temperate beaches the planting of shingle-beach vegetation immediately after the threat of equinoctial spring tides has been recommended (Walmsley & Davy, 2001). But on more stable beaches autumn plantings may be favoured if the probability of damage due to summer drought is greater than the chance of inundation by storm tides or salt-spray damage.

Vegetation structure and composition

Strandline vegetation encourages sand and organic-matter accumulation on beaches and is often a precursor to pioneer dune vegetation so there is considerable merit to its restoration. However, ubiquitous strandline annuals, such as *Cakile maritima* and *Salsola kali*, are readily dispersed by sea and have been observed to naturally colonise recontoured beaches within one season (Vestergaard & Hansen, 1992; Gibson & Looney, 1994). The ability of strandline annuals to colonise rapidly from adjacent beaches led to their restoration being considered unnecessary at Sizewell (Walmsley & Davy, 2001) (see Box 9.5). *Cakile maritima* has been used as a primary colonist, despite being an alien species, in restoration of beach vegetation at Spanish Bay, California, because it rapidly establishes and stabilises sand without becoming invasive or competitive (Guinon & Allen, 1990).

Perennial dune grasses are widely utilised for beach as well as dune planting because they are easily multiplied vegetatively, harvested, transported, stored and planted. The techniques for creating pioneer dune using vegetative offsets of dune-building grasses have been widely reported (see Greipsson, this volume). Nourished beaches are likely to have limited sand resources so the use of sand fences or brushwood thatching is an important method of trapping sand to establish pioneer dune grasses

(Ranwell & Boar, 1986) (see Box 9.3). Owing to their greater ability to endure saltwater inundation and establish on strandlines *Leymus arenarius* and sand couch (*Elytrigia juncea*) have been recommended in preference to *Ammophila arenaria* for planting on European beaches (Ranwell & Boar, 1986).

Unfortunately, most plantings on beaches and dunes have been monocultures of dune grasses. Many other species of temperate and tropical fore-dunes and beaches, such as *Ipomoea pescaprae*, beach-bur (*Ambrosia chamissonis*) and sea sandwort (*Honckenya peploides*), could be replanted but they have been widely ignored because of the pre-eminent dune-building ability of the perennial beach grasses (see Box 9.4). On the Pacific coast of North America, sea figs (*Carpobrotus* spp.) have been used very successfully to stabilise beaches, dunes and cliffs. The native *C. aequilaterus*, although less vigorous than the introduced *C. edulis* (Hottentot fig) and hybrid plants, may be successfully established on beaches using cuttings, which root readily (Woodhouse, 1982). The Hottentot fig and hybrid plants should be avoided because they are very competitive and can dominate vegetation. Beach sagewort (*Artemisia pyenocephala*) is a pioneer coloniser of Pacific beaches that can be transplanted or seeded and propagated by cuttings (Cowan, 1975). Seashore elder (*Iva imbricata*) is a widespread forb on beaches along the southern coasts of North America (Colosi & McCormick, 1978). It has been identified as useful for dune stabilisation and successfully established from bare-rooted cuttings. It has been demonstrated that a range of species can be successfully introduced as container-grown plants to restore shingle-beach vegetation at Sizewell (see Box 9.5). It is recommended that the diversity of beach species capable of establishment on beaches be utilised more widely in future restoration projects.

CONCLUDING REMARKS

The restoration of beach ecosystems requires either the reinstatement of suitable beach sediments by nourishment or the recontouring of the beach profile after disturbance and, subsequently, the restoration of beach vegetation. Nourishment projects can protect turtle or bird nesting sites from erosion and create new nesting habitat, but the nourishment operations must be conducted in an environmentally sensitive manner. Beach nourishment has been shown to reduce beach fauna biodiversity and species density but once nourishment work has been completed, infauna should recover within one or two years. Although most nourishment work has been done for coastal defence or recreation purposes, it has provided invaluable information that provides guidance on methods for habitat restoration. Nourishment projects should be carried out over winter and outside relevant breeding seasons to avoid harm to infauna, turtles and birds. It is important to match the nourishment fill to the original beach material, particularly the particle size distribution. Careful contouring to create a stable beach profile and avoiding substrate compaction are also essential requirements. Strandline annuals rapidly recolonise nourished beaches and should not generally require restoration, but disturbance often prevents rapid revegetation by perennials. The use of vegetative offsets to establish dune grasses is a well-established practice and it has been shown that monocarpic species can be established from seed. The use of container-grown stock for restoration of perennial species is recommended because of the need to rapidly establish mature plants capable of enduring disturbance and the prevalence of seed dormancy mechanisms.

Future growth in coastal tourism and the predicted rise in global sea level and increased storminess may increase the need for effective, environmentally sensitive beach nourishment and revegetation programmes. While engineers set clear targets for the effectiveness of beach nourishment schemes, in terms of coastal defence, this is rarely the case for minimising the ecological impacts of nourishment or creating habitats. Those projects that do monitor impacts of nourishment or habitat restoration generally do so in the short term; there is a need for monitoring over many years rather than a few months. Successful projects are likely to be carefully planned and designed with people expert in engineering, coastal geomorphology and ecology working as a multidisciplinary team. The belief that beach vegetation will rapidly recolonise restored beaches naturally can only be affirmed for strandline species. In contrast, perennial beach species may take many years to establish vegetation able to tolerate natural disturbance and, in some localities, aesthetically unacceptable vegetation may establish where natural beach and dune establishment is expected. Ecological restoration is likely to be the only method of rapidly re-establishing vegetation that replicates the diversity and resilience of natural vegetation.

REFERENCES

Baca, B. J. & Lankford, T. E. (1988). *Myrtle Beach Nourishment Project: Biological Monitoring Report to City of Myrtle Beach*. Myrtle Beach, SC: Coastal Science and Engineering Inc.

Barbour, M. G., de Jong, T. M. & Pavlik, B. M. (1985). Marine beach and dune plant communities. In *Physiological Ecology of North American Plant Communities*, eds. B. F. Chabot & H. A. Mooney, pp. 296–322. New York: Chapman & Hall.

Bird, E. C. F. (1996). *Beach Management*. Chichester, UK: John Wiley.

Bruno, J. F. (2000). Facilitation of cobble beach plant communities through habitat modification by *Spartina alterniflora*. Ecology, **81**, 1179–1192.

Buckley, R. (1988). Plant succession under repeated disturbance: establishment and growth of strandline forbs and creepers on unstable beaches in the wet tropics. In *The Ecology of Australia's Wet Tropics*, ed. R. L. Kitching, pp. 307–311. Chipping Norton, NSW: Surrey Beatty.

Colosi, J. C. & McCormick, J. F. (1978). Population structure of *Iva imbricata* in five coastal dune habitats. *Bulletin of the Torrey Botanical Club*, **105**, 175–186.

Cowan, B. (1975). Protecting and restoring native dune plants. *Fremontia*, **2**, 3–10.

Crain, D. A., Bolten, A. B. & Bjorndal, K. A. (1995). Effects of beach nourishment on sea turtles: review and research initiatives. *Restoration Ecology*, **3**, 95–104.

Davis, R. A., Wang, P. & Silverman, B. R. (2000). Comparison of the performance of three adjacent and differently constructed beach nourishment projects on the Gulf Peninsula of Florida. *Journal of Coastal Research*, **16**, 396–407.

Davison, A. T., Nicholls, R. J. & Leatherman, S. P. (1992). Beach nourishment as a coastal management tool: an annotated bibliography on developments associated with artificial nourishment of beaches. *Journal of Coastal Research*, **8**, 984–1022.

Davy, A. J. & Figueroa, M. E. (1993). The colonisation of strandlines. In *Primary Succession on Land*, ed. J. Miles & D. H. W. Walton, pp. 113–131. Oxford: Blackwell.

Davy, A. J., Willis, A. J. & Beerling, D. J. (2001). The plant environment: aspects of the ecophysiology of shingle species. In *Ecology and Geomorphology of Coastal Shingle*, eds. J. R. Packham, R. E. Randall, R. S. K. Barnes & A. Neal, pp. 191–201. Otley, UK: Westbury Publishing.

De Jong, T. M. (1979). Water and salinity relations of Californian beach species. *Journal of Ecology*, **67**, 647–663.

Doing, H. (1985). Coastal fore dune zonation and succession in various parts of the world. *Vegetatio*, **61**, 65–75.

Downer, R. H. & Liebelt, C. E. (1990). *1989 Long Island Colonial Waterbird and Piping Plover Survey*. Stony Brook, NY: New York Department of Conservation.

Fay, P. J. & Jeffrey, D. W. (1992). The foreshore as a nitrogen source for marram grass. In *Coastal Dunes*, eds. R. W. G. Carter, T. G. F. Curtis & M. J. Sheehy-Skeffington, pp. 177–188. Rotterdam: Balkema.

Flynn, B. (1992). Beach nourishment, sea turtle nesting, and nest relocation in Dade County, Florida. In *New Directions in Beach Management, Proceedings of the 5th Annual National Conference on Beach Preservation Technology*, ed. L. S. Tait, pp. 381–394. Tallahassee, FL: Florida Shore and Beach Preservation Association.

Fuller, R. M. (1987). Vegetation establishment on shingle beaches. *Journal of Ecology*, **75**, 1077–1089.

Gaffney, F. B. & Dickerson, J. A. (1987). Species selection for revegetating sand and gravel mines in the Northeast. *Journal of Soil and Water Conservation*, **42**, 358–361.

Gibson, D. J. & Looney, P. B. (1994). Vegetation colonisation of dredge spoil on Perdido Key, Florida. *Journal of Coastal Research*, **10**, 133–143.

Gibson, D. J., Ely, J. S. & Looney, P. B. (1997). A Markovian approach to modeling succession on a coastal barrier island following beach nourishment. *Journal of Coastal Research*, **13**, 831–841.

Guinon, M. & Allen, D. (1990). Restoration of dune habitat at Spanish Bay. In *Environmental Restoration: Science and Strategies for Restoring the Earth*, ed. J. J. Berger, pp. 70–80. Washington, DC: Island Press.

Guppy, H. B. (1906). *Observations of a Naturalist in the Pacific between 1896 and 1899*, vol. 2, *Plant Dispersal*. London: MacMillan.

Heath, S. E. (1981). A brief survey of the botany at Shingle Street, Suffolk. *Transactions of the Suffolk Naturalists' Society*, **18**, 249–253.

Humphreys, B., Coates, T., Watkiss, M. & Harrison, D. (1996). *Beach Recharge Materials: Demand and Resources*. CIRIA Report no. 154. London: Construction Industry Research and Information Association.

Intergovernmental Panel on Climate Change (1995). *IPCC Second Assessment: Climate Change 1995*. Geneva, Switzerland: IPCC.

Lee, E. M. (1993). The political ecology of coastal planning and management in England and Wales: policy responses to the implications of sea-level rise. *The Geographical Journal*, **159**, 169–178.

Loffler, M. & Coosen, J. (1995). Ecological impact of sand replenishment. In *Directions in European Coastal Management*, eds. M. G. Healy & J. P. Doody, pp. 291–299. Cardigan, UK: Samara Publishing.

Looney, P. B. & Gibson, D. J. (1993). Vegetation monitoring of beach nourishment. In *Beach Nourishment Engineering and Management Considerations*, eds. D. K. Stauble & N. C. Kraus, pp. 226–241. New York: American Society of Civil Engineers.

Lynch, A. E. (1994). Macrofaunal recolonisation of Folly Beach, South Carolina, after beach nourishment. MSc dissertation, University of Charleston, South Carolina.

Maun, M. A. (1994). Adaptations enhancing survival and establishment of seedlings on coastal dune systems. *Vegetatio*, **111**, 59–70.

Maurer, D., Keck, R. T., Tinsman, H. C., Leathem, W. A., Wethe, C., Lord, C. & Church, T. M. (1986). Vertical migration and mortality of marine benthos in dredged material: a synthesis. *International Review of Hydrobiology*, **71**, 49–63.

McFarland, S., Whitcombe, L. & Collins, M. (1994). Recent shingle beach renourishment schemes in the UK: some preliminary observations. *Ocean and Coastal Management*, **25**, 143–149.

Mileikovsky, S. A. (1970). The influence of pollution on pelagic larvae of bottom invertebrates in marine nearshore and estuarine waters. *Marine Biology*, **6**, 350–356.

Moulding, J. D. & Nelson, D. A. (1988). Beach nourishment issues related to sea turtle nesting. In *Proceedings of the Symposium on Coastal Water Resources*, Technical Publication Series no. TPS-88-1, eds. W. L. Lyke & T. J. Hoban, pp. 87–93. Bethesda, MD: American Water Resources Association.

National Research Council (1995). *Beach Nourishment and Protection*. Washington, DC: National Academy Press.

Nelson, W. G. (1985). *Physical and Biological Guidelines for Beach Restoration Projects*, Part 1, *Biological Guidelines*. Report no. 76. Gainesville, FL: Florida Sea Grant College.

Nelson, W. G. (1989). An overview of the effects of beach nourishment on the sand beach fauna. In *Problems and Advances in Beach Nourishment, Proceedings of the 2nd Annual Conference on Beach Preservation Technology*, ed. L. S. Tait., pp. 295–310. Tallahassee, FL: Florida Shore and Beach Preservation Association.

Nelson, W. G. (1993). Beach restoration in the Southeastern US: environmental effects and biological monitoring. *Ocean and Coastal Management*, **19**, 157–182.

Nelson, W. G., Mauck, K. A. & Fletemeyer, J. (1987). *Physical Effects of Beach Nourishment on Sea Turtle Nesting on Delray Beach, Florida*. Technical Report no. TR-87-15. Vicksburg, MS: US Army Corps of Engineers.

Nordstrom, K. F. (2000). *Beaches and Dunes of Developed Coasts*. Cambridge: Cambridge University Press.

Packham, J. R., Harmes, P. A. & Spiers, A. (1995). Development of a shingle community related to a specific sea defence structure. In *Directions in European Coastal Management*, eds. M. G. Healy & J. P. Doody, pp. 369–371. Cardigan, UK: Samara Publishing.

Powell, A. N. (1998). Western snowy plover (*Charadrius alexandrinus nivosus*) and California least tern (*Sterna antillarum browni*). In *Status and Trends of the Nation's Biological Resources*, eds. M. J. Mac, P. A. Opler, C. E. Packett-Haecker & P. D. Doran, pp. 626–631. Reston, VA: US Geological Survey.

Powell, A. N. & Collier, C. L. (2000). Habitat use and reproductive success of Western Snowy Plovers at new nesting areas created for California Least Terns. *Journal of Wildlife Management*, **64**, 24–33.

Pye, K. & French, P. W. (1993). *Targets for Coastal Habitat Re-Creation*. English Nature Science Report no. 13. Peterborough, UK: English Nature.

Randall, R. E. (1988). The vegetation of Shingle Street,

Suffolk in relation to its environment. *Transactions of the Suffolk Naturalists' Society*, **24**, 41–58.

Ranwell, D. S. & Boar, R. (1986). *Coast Dune Management Guide*. Huntingdon, UK: Institute of Terrestrial Ecology.

Reilly, F. J., Cobb, D. M. & Bellis, V. J. (1980). Biological implications of beach replenishment. In *Utilisation of Science in the Decision-Making Process, Proceedings of 6th Annual Conference*, eds. N. P. Psuty & D. McArthur, pp. 269–280. Arlington, VA: The Coastal Society.

Ridley, H. N. (1930). *The Dispersal of Plants throughout the World*. Ashford, UK: L. Reeve & Co.

Scott, G. A. M. (1960). The biology of shingle beach plants with special reference to the ecology of selected species. PhD dissertation, University of Wales, Bangor, UK.

Scott, G. A. M. (1963). The ecology of shingle beach plants. *Journal of Ecology*, **51**, 517–527.

Spadoni, R. H. & Cummings, S. L. (1992). A common sense approach to the protection of marine turtles. In *New Directions in Beach Management, Proceedings of the 5th Annual National Conference on Beach Preservation Technology*, ed. L. S. Tait, pp. 1–19. Tallahassee, FL: Florida Shore and Beach Preservation Association.

Stauble, D. K. & Hoel, J. (1986). *Physical and Biological Guidelines for Beach Restoration Projects*, Part 2, *Physical Engineering Guidelines*. Report no. 77. Gainesville, FL: Florida Sea Grant College.

Tansley, A. G. (1949). *The British Islands and their Vegetation*, vol. 2. Cambridge: Cambridge University Press.

UK Biodiversity Group (1999). *Tranche 2 Action Plans*, vol. 5, *Maritime Species and Habitats*. Peterborough, UK: English Nature.

van der Wal, D. (1998). The impact of the grain-size distribution of nourishment sand on aeolian sand transport. *Journal of Coastal Research*, **14**, 620–631.

Vestergaard, P. & Hansen, K. (1992). Changes in morphology and vegetation of a man-made beach-dune system by natural processes. In *Coastal Dunes*, eds. R. W. G. Carter, T. G. F. Curtis & M. J. Sheehy-Skeffington, pp. 165–176. Rotterdam: Balkema.

Walmsley, C. A. (1995). The ecology of shingle-beach vegetation in relation to its restoration. PhD dissertation, University of East Anglia, Norwich, UK.

Walmsley, C. A. & Davy, A. J. (1997a). The restoration of coastal shingle vegetation: effects of substrate composition on the establishment of container-grown plants. *Journal of Applied Ecology*, **34**, 154–165.

Walmsley, C. A. & Davy, A. J. (1997b). Germination characteristics of shingle beach species, effects of seed ageing and their implications for vegetation restoration. *Journal of Applied Ecology*, **34**, 131–142.

Walmsley, C. A. & Davy, A. J. (1997c). The restoration of coastal shingle vegetation: effects of substrate composition on the establishment of seedlings. *Journal of Applied Ecology*, **34**, 143–153.

Walmsley, C. A. & Davy, A. J. (2001). Habitat creation and restoration of damaged shingle communities. In *Ecology and Geomorphology of Coastal Shingle*, eds. J. R. Packham, R. E. Randall, R. S. K. Barnes & A. Neal, pp. 409–420. Otley, UK: Westbury Publishing.

White, D. J. B. (1967). *An Annotated List of the Flowering Plants and Ferns on Blakeney Point, Norfolk*. Norfolk, UK: The National Trust.

Wolf, R. E. (1988). Sea turtle protection and nest monitoring summary: Boca Raton South Beach nourishment project. In *Problems and Advancements in Beach Nourishment, Proceedings of the 1st National Conference on Beach Preservation Technology*, ed. L. S. Tait, pp. 273–283. Tallahassee, FL: Florida Shore and Beach Preservation Association.

Woodhouse, W. W. (1982). Coastal sand dunes of the U.S. In *Creation and Restoration of Coastal Plant Communities*, ed. R. R. Lewis, pp. 1–44. Boca Raton, FL: CRC Press.

10 • Coastal dunes

SIGURDUR GREIPSSON

INTRODUCTION

Coastal sand dunes have a world-wide distribution and are found in all continents except Antarctica (van der Maarel, 1993*a*, *b*). Consequently, coastal dune ecosystems show great biogeographical variation, with communities typical of tundra, Mediterranean, temperate and tropical environments (Bliss, 1993; de Lacerda *et al.*, 1993; Olsson, 1993; Randall, 1993). Dunes are found on a relatively narrow hinterland behind the shore. They may form continuous ecosystems along hundreds of kilometres along coastlines and, in some cases, extend inland; in Australia coastal sand dunes stretch 10 km inland from the shore of Brisbane (Walker *et al.*, 1981).

Formation of coastal sand dunes

Sand dunes are generally formed where sand deposition exceeds erosion. The sand usually originates from terrestrial erosion or animal skeletons and is transported via rivers to the coast. Coastal sands are mainly composed of material derived from quartz, silica, basalt (volcanic) and feldspar of terrestrial origin and carbonate sands of marine origin (Brown & MacLachlan, 1990). Sand dunes can prograde, generally as a result of continuous sand deposition on the shore (Greipsson & El-Mayas, 1994), but isostatic uplift of a landmass can also lead to the formation of new dunes. Sand dunes form complex ecosystems (Fig. 10.1), which show a variety of landscapes, including the shore, fore dunes, main dunes with windward and leeward slopes, wet dune slacks, and back dunes, with plateaus and hollows that support grasslands, shrub lands and forests (Ranwell, 1972;

Hesp, 1991). Woodhouse (1982) recognised a simpler subdivision with only three zones: pioneer zone, the intermediate or scrub zone, and the back dune or forest zone.

Sand dune formation is affected by the proximity to rivers with high sediment load. For instance, the sand dunes at Tentsmuir, Scotland next to the mouth of the River Tay are one of the most rapidly prograding (up to 14 m per year) coastal areas of the UK (Crawford & Wishart, 1966). The net sediment transportation away from river mouths along the shore affects sand deposition on the shores. When the sand is washed ashore its movement is affected by dune elevation, beach slope, beach width, coastline orientation and local topography (Goldsmith, 1989). Sand dune formation is also influenced by the embayment size and the prevailing winds (Hesp, 1991).

Vegetation on the dunes plays an important role in the formation and stabilisation of the coastal sand dunes. Dune grasses accumulate sand around their foliage and their ability to grow upward through accreting sand layers also affects dune formation (Wagner, 1964; Maun, 1998). Foliage of dune grasses in turn slows erosive activity of the wind on the dunes and this in turn increases sand accretion on the lee side of the dune (Olson, 1958; Ranwell, 1972; Chapman, 1976). In general, dunes grow initially rapidly upward, and after reaching an elevation of 5–10 m, their growth rate is usually decreased. The height of dunes varies according to sand supply, climate and exposure related to local topography (Ranwell & Boar, 1986). However, exceptionally high dunes can form such as the 90-m dunes at Doñana, Spain (Garcia Novo, 1979).

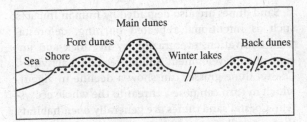

Fig. 10.1. Schematic drawing of different ecozones found on coastal dunes.

Environmental constraints for plants

The environment on the dunes is characterised by strong winds, sand movements (accretion and erosion), high evaporation, salinity and limited availability of macronutrients, which all influence ecological processes on the dunes. A gradient is usually found such that these environmental extremes diminish away from the shore (Table 10.1). Sand dunes are therefore stressful habitats and plants

have evolved special strategies for survival in this extreme environment.

Sand accretion on fore dunes may fluctuate to a great extent with wind or tide and sand movement is considered among the most important factors that affect the distribution of plant communities on sand dunes (Moreno-Casasola, 1986).

Seedlings of dune plants show plastic responses in biomass allocation in response to sand accretion (Harris & Davy, 1987). For instance, seedlings of lyme grass (*Leymus arenarius*) and sand couch (*Elytrigia juncea*) respond to burial by allocating biomass to shoots (with lower root mass ratios) and leaves (longer leaves) growth rather than to the roots (Harris & Davy, 1987; Greipsson & Davy, 1996a). Dune grasses do not only respond to sand accretion by different growth behaviour but also by different physiological responses. For instance, rates of photosynthetic CO_2 uptake were enhanced after emergence from burial in American beach grass

Table 10.1. *Various abiotic factors that affect distribution of plant communities on the coastal sand dune ecosystem.*

Stress factors[a]	Zones of coastal sand dune ecosystems				
	Shore	Fore dunes	Main dunes	Dune slacks	Back dunes
Wind exposure	++	++	++	−	+
Sand accretion	+	++	+	+−	−
Sand erosion	++	++	+−	−	−
Soil salinity	++	+	−	+−	−
Salt spray	++	++	+	+−	−
Watertable	−	−	−	++	+
Light intensity	++	++	+	+	−
Evaporation	++	++	+−	+−	−
Desiccation	+−	++	+	−	−
Heat stress	+−	++	++	+−	−
Soil moisture	−	++	++	−	−
Soil N	−	−	+−	+	++
Soil P	−	−	+−	++	++
Fire	−	+−	+	−	++

[a]Stress level is indicated by: ++ = intensive stress, + = moderate stress, +− = low stress, − = no stress.

(*Ammophila breviligulata*) and sand reed (*Calamovilfa longifolia*) (Yuan *et al.*, 1993).

Exposure to salt (mainly NaCl) derived from the sea can limit the distribution of plants on coastal sand dunes. Long periods of inundation in seawater are usually rare on the sand dunes because of the rapid drainage of sands (Rozema *et al.*, 1985). Exposure to salinity on the dunes is therefore episodic rather than continuous in duration (Barbour *et al.*, 1985). Only the most salt-tolerant plants grow on the shore and fore dunes. For instance, of all sand dune plants in the UK, *Leymus arenarius* grows closest to the sea (Moore, 1971; Garcia-Novo, 1976). *Leymus arenarius* can survive salt concentration higher than 12% whereas other dune grasses such as *Elytrigia juncea* can tolerate 6% and *Ammophila arenaria* (marram grass) only 2% (Benecke, 1930). Airborne salt spray is greatest on the shore and fore dunes but it declines inland. Salt spray is a limiting factor for plant growth on coastal dunes (Rozema *et al.*, 1985). Dune plants that tolerate airborne salt spray have evolved a protective wax in the composition of their epicuticule (Ahmad & Wainwright, 1977).

Sand dunes contain generally low amounts of macronutrients, especially nitrogen and phosphorus (e.g. Ranwell, 1972; Koske *et al.*, 1975; Houle, 1997*a*, *b*), which limits vegetation growth (Willis, 1985; Olff *et al.*, 1993).

NATURE OF THE PROBLEM

Sand dunes are generally fragile ecosystems that are prone to degradation and consequent destruction. Degradation of sand dunes can be caused by natural disturbances such as high tides, storm surges or hurricanes. Hurricanes can have a great impact and even deflate coastal dunes. In the Gulf of Mexico, it usually takes about five to ten years for coastal dunes to recover after such drastic disturbance (Barbour *et al.*, 1987). Furthermore, the distribution of plants along the south Brazilian coast has been linked to natural disturbance (Costa *et al.*, 1996). In addition to natural disturbances the current rise in sea level due to global climate changes (see below) poses a threat to coastal sand dunes around the world.

Sand dunes are also degraded by human impacts such as intentional repeated burning, deforestation, cultivation, overgrazing by livestock and uncontrolled development and recreational activities. Coastal dune grasses can show a decline in vigour, which in turn can pose a threat to the whole ecosystem. Coastal sand dunes are generally open habitats that are easily invaded by alien species. Alien species can alter the functioning of the ecosystem and even the formation of sand dunes.

Impact of global climate change

Concentrations of atmospheric CO_2 are steadily rising, mainly due to burning of fossil fuel, with resulting changes in the global climate (e.g. Karl *et al.*, 1997). The most serious threat to coastal ecosystems in the near future is the expected rise in sea level due to thermal expansion of seawater and addition of meltwater from glaciers and the polar regions. Pertinent questions deal with the capability of coastal vegetation to cope with this rise in sea level. Coastlines that are retreating due to isostatic movements of land can give us some indications on how coastal vegetation reacts to increasing sea levels. A coastline retreat ranging from 1 to 50 m per year has been experienced on the Louisiana Delta of the Mississippi River. In this case the sand dune grass sea oat (*Uniola paniculata*) has not been able to keep up with the pace of this rapid coastal retreat. Part of the unsuccessful colonisation rate of *U. paniculata* was related to its low seed production (Hester & Mendelssohn, 1987). If sea level rises rapidly the existence of coastal dunes is threatened unless there is an available escape route for vegetation on back dunes or even further landward.

Sea level rise is expected to result in changes in groundwater level, which in turn will influence slack vegetation (Vestergaard, 1997). Global climate change is predicted to result in increased intensity and frequency of hurricanes on coastal areas (see above) and precipitation patterns that could affect coastal vegetation. For instance, vegetation on dune slacks on the tropical coast of Veracruz, Mexico showed dynamic relations with precipitation patterns: years of exceptionally high precipitation resulted in high groundwater tables in dune slacks which in turn killed many trees and shrubs (Martinez *et al.*, 1997). Long-term fluctuations in precipitation pattern affect plant composition by increasing plant cover on mobile dunes and decreasing it in the dune slacks during years of high precipitation (Martinez *et al.*, 1997). Changes in precipitation patterns will therefore affect coastal vegetation and such information needs to be incorporated into conservation and restoration programmes.

Elevated ambient temperatures and atmospheric CO_2 levels may have direct or indirect effects on coastal vegetation. The productivity of the annual grass *Vulpia fasciculata* that grows on maritime sands in the UK was strongly influenced by elevated temperatures but not by elevated CO_2 levels (Firbank *et al.*, 1995). Elevated CO_2 levels could, however, have potential impact on dynamics of plant populations where competition of species could be altered (Bazzaz, 1990). Elevated ambient temperatures could affect distribution pattern of coastal plants, although the migration of coastal plants is also influenced by other factors, such as soil fertility and changes in precipitation patterns (Firbank *et al.*, 1995). It is predicted that coastal species may migrate northward in the northern hemisphere as a result of global warming. Future restoration and conservation of coastal sand dunes must incorporate models of global climate changes.

Development and recreational uses

The development of roads, tracks, car parks, buildings, housing and other structures near coastal sand dunes generally disrupts dune processes, particularly sand nourishment of dunes and development of pioneer dunes (Woodhouse, 1982; Sanjaume & Pardo, 1992). Developments on the fore dunes restrict the amount of sand available to the back dunes. As a result there may be a general loss of habitat (Sanjaume & Pardo, 1992; Sidaway, 1995). Sand extraction from the dunes for building material adds to this problem (Mitchell, 1974; Kirkpatrick & Hassall, 1981). Service pipelines including stormwater and sewage pipes that cross sand dunes also create disturbance (Ritchie & Gimingham, 1989).

The major impact that recreational activities have on coastal dunes is direct trampling of vegetation, which can result in dune erosion (Olsauskas, 1995; Sidaway, 1995). Other recreational impacts include disturbance of fauna especially birds during their breeding season and pollution (litter, oil etc.) that can contaminate soils or surface water (Sidaway, 1995). Public pressure on sand dunes can be assessed by using image processing of aerial digital photography recording indicators such as path length, vegetation cover and extent of bare sand (Curr *et al.*, 2000). This information should allow coastal managers to monitor sand dune ecosystems and prevent damage by introducing voluntary zoning or seasonal restrictions to certain parts of dune ecosystems (Sidaway, 1995), building formal public access-ways or simply fencing off sensitive areas. The location of access-ways and protective fences needs to consider sand dune dynamics and human use of the dunes and beach. Use of motor vehicles on dunes should always be prohibited. Education through signposts providing information and directions to the access-ways may successfully influence the behaviour of people on the dunes (Sidaway, 1995).

Overgrazing

Sand dune vegetation is vulnerable to damage from grazing and trampling by livestock (Woodhouse, 1982). Bitter panicum (*Panicum amarum*) disappeared from dunes in South Atlantic and Gulf coasts of the United States as a result of overgrazing (Woodhouse, 1982). Grazing reduces rooting depth, which may be critical in draught conditions (Ranwell & Boar, 1986). Trampling by sheep creates paths, which can initiate blowouts and eventually mobile dunes (Woodhouse, 1982). Rabbits have also influenced vegetation of coastal sand dunes by their burrowing as well as grazing (Ranwell & Boar, 1986). Uncontrolled grazing of livestock on coastal sand dune vegetation has resulted in destabilisation and eventual erosion of dunes resulting in sand drifts that, in a few cases, have threatened nearby communities (Woodhouse, 1982; Ranwell & Boar, 1986; van Dijk, 1992).

Restoration sites need to be protected from livestock and rabbit grazing especially after the initial planting when plants are small and vulnerable. Protective fencing is often required to enclose restoration sites to prevent grazing by rabbits and livestock and even to prevent public access (Mitchell, 1974; Ranwell & Boar, 1986; Ritchie & Gimingham, 1989).

On the other hand, grazing can be used as a management tool. For instance, in The Netherlands where eutrophication on the dunes due to excess nitrogen deposition (derived from air pollution) has resulted in encroachment of grasses, grazing by livestock has been suggested to improve biodiversity and to maintain open landscapes on coastal sand dunes (van Dijk, 1992; Kooijman & de Haan, 1995). Mowing has also been used to increase biodiversity (Anderson & Romeril, 1992).

Decline of vigour

Loss of vigour in sand dune grasses has been termed 'the *Ammophila* problem' and it manifests itself as a decrease in height and density of foliage, with reduced number of tillers and seed yield (Wallen, 1980); it can be seen in transplanted dune grasses within a few years (Seliskar & Huttel, 1993). Decline of vigour and consecutive die-out of dune grasses are usually followed by colonisation and eventual replacement by other plants on the dunes (van der Putten *et al.*, 1988, 1989; Seliskar & Huttel, 1993; Greipsson & Davy, 1994*a*). Various factors have been associated with the decline of vigour, such as competition between plants, plant age, soil pH, soil microorganisms, soil aeration and reduced root growth (Eldred & Maun, 1982; Barbour *et al.*, 1985; Seliskar, 1995). Competition for soil nutrients is likely to be responsible for the decline of even the most vigorous dune plants (Watkinson *et al.*, 1979). Sometimes, decline in vigour occurs without interspecific competition (Greipsson & Davy, 1994*a*). Factors responsible for decline of vigour may vary in importance between sites (Eldred & Maun, 1982).

Decline of vigour of dune grasses can result in erosion and regression of the whole ecosystem. This in turn represents an economic problem where dune grasses are used in restoration (van der Putten *et al.*, 1989; Seliskar, 1995; Greipsson & El-Mayas, 1996). Studies on decline of vigour should be taken into consideration in the aftercare and long-term management of sand dune ecosystems.

Invaders

Coastal sand dunes are usually characterised by open habitat that can be frequently disturbed and are therefore prone to invasion by alien plant species. Alien plants can alter ecosystem functioning by changing the groundwater table, increasing nutrient status of the soil or by outcompeting native plants. In turn, this can change vegetation composition or even the biodiversity of the dunes. Disturbed coastal sites are more prone to invasion of alien species than naturally undisturbed sites. Sand dune communities in the UK and Denmark had an average of 13% and 18% of alien species respectively (Andersen, 1995*a*; Crawley, 1987).

Sand-stabilising plants have been introduced into different parts of the world. *Ammophila arenaria*, introduced from Europe to the west coast of the United States in the late 1800s in order to stabilise coastal sands (Wiedermann *et al.*, 1974), now dominates most of the dunes north of San Francisco Bay (Aptekar & Rejmanek, 2000), where it has altered the coastal landscape by creating new fore dunes (Wiedermann *et al.*, 1974). Sea-dispersed rhizome fragments colonise new coastal dunes and it has been estimated that viable rhizomes of *A. arenaria* can be transported up to 505 km along the coast in 13 days (Aptekar & Rejmanek, 2000). Conifers have been planted to a great extent in Europe in order to stabilise back dunes. In the UK about 4000 ha of coastal dunes were planted with conifers (*Pinus maritima*) from 1922 to 1952 (Ranwell & Boar, 1986). These conifer plantations changed the physical environment of the dunes, and generally resulted in lower groundwater tables and loss of flora and fauna associated with the original sand dune ecosystem (Sturgess, 1992; Packham & Willis, 1997). Some of these plantations have since been eliminated in an attempt to restore the dunes to their original state (Sturgess, 1992).

Invasive N_2-fixing species may have a competitive advantage over indigenous vegetation in N-limited sand dunes. Nitrogen-fixing *Acacia cyclops* and *A. longifolia* (coastal wattle) were introduced from Australia to the Cape, South Africa in order to stabilise the sand. They have invaded the coast of South Africa widely and have increased the N content of the sand, which has resulted in a shift in species composition (Avis, 1995). Another example is invasion by the N_2-fixing coastal shrub sea buckthorn (*Hippophaë rhamnoides*), a widespread problem in Europe, where this species forms very dense populations (Binggeli *et al.*, 1992; Packham & Willis, 1997).

Other invasive aliens such as *Acer pseudoplatanus*, *Rhododendron ponticum* and *Rubus spectabilis* have been introduced on coastal sand dunes in Europe for amenity purposes (Binggeli *et al.*, 1992).

RATIONALE FOR RESTORATION

Coastal sand dunes serve generally as a buffer against the impacts of wind and sea waves (Ranwell & Boar, 1986), stabilise sand before it inundates nearby agricultural lands and they also provide a sand source for the beach. Dunes are considered as a valuable natural habitat, with high amenity and recreational value.

Varied landscapes within sand dune ecosystems support a high biodiversity, including ground-nesting birds (Woodhouse, 1982). Frequent disturbance associated with salt spray or sand instability is considered to be an important factor in maintaining high biodiversity (Espejel, 1987), although major disturbances such as hurricanes may be detrimental (Barbour *et al.*, 1987). Coastal sand dune ecosystems often harbour a significant proportion of a particular flora of a region. For instance, the 800–ha sand dune system at Braunton Burrows, UK contains about 350 vascular plant species, or about 17% of the UK flora (Willis, 1985). Ranwell & Boar (1986) consider that about 50% of the vascular plants in the UK can be found on coastal ecosystems. Similarly, the Mediterranean coast of Egypt at Burg El-Arab contains about 1000 plant species, which is about 50% of the vascular plants in Egypt (Ayyad, 1973).

Large numbers of endemic plants are found on coastal sand dunes. For instance, about 8% of the vascular plant species found on coastal sand communities in Wales were considered to be endemic to Europe (Rhind & Jones, 1999) and coastal dune vegetation on Baja California contains about 30% endemic species (Johnston, 1977).

UNDERLYING THEMES OF RESTORATION

Plant succession on dunes

Succession has been demonstrated along chronosequences of sand dunes (Cowles, 1899; Warming, 1909; Olson, 1958). Ecosystem functioning also changes along succession and is manifested as increased net ecosystem production, nutrient pools (nitrogen and phosphorus) and nutrient cycling (Lichter, 1998). Succession occurs generally because, for each species, the probability of establishment changes with time, along with changes in soil conditions, and competition with other plants (Watkinson *et al.*, 1979; Lichter, 2000). It has been suggested that soil microorganisms may influence the plant succession on coastal sand dunes (Webley *et al.*, 1952; Brown, 1958). A model was proposed whereby differences in plants' susceptibility to soil pathogens determine the outcome of an interspecific competition and are therefore the underlying mechanism in plant succession on the dunes (van der Putten *et al.*, 1993). On prograding dunes on Lake Michigan, replacement of species was demonstrated from fore-dune grasses to evergreen shrubs and bunchgrass (after 100 years) to a mixed pine tree forest (after 345 years) (Lichter, 1998).

The facilitation model of succession, where pioneer species improve environmental conditions for later arriving species, applies particularly to coastal sands where the substrate has not been influenced by organisms beforehand (Connell & Slatyer, 1977). Although facilitation is often the driving force behind succession, examples to the contrary can also be found. For instance, the facilitation model was not found to apply to a primary succession on fore dunes in the subarctic Hudson Bay, Canada (Houle, 1997*a, b*) or on fore dunes in southern Brazil (Cordazzo & Seeliger, 1993). Furthermore, sand dunes have been described as being in a phase of constant succession due to continuous extrinsic disturbances (Watkinson *et al.*, 1979). Plant succession has not yet been described in a single model (Pickett *et al.*, 1987) and models incorporating mechanistic richness are needed to describe succession on sand dunes. Studies on plant succession can give valuable information for restoration of damaged ecosystems. Models of plant succession are particularly useful in long-term management of sand dune ecosystems in order to predict the patterns of vegetational changes following restoration.

Role of arbuscular mycorrhizal fungi

Arbuscular mycorrhizal fungi (AMF) (order Zygomycota) form symbiotic relationships with roots of vascular plants, including the majority of those inhabiting coastal sand dunes (Brundrett, 1991). In tropical dunes of the Gulf of Mexico 97% of the vascular plants formed AMF symbioses (Corkidi & Rincon, 1997 *a*, *b*). Exceptions to this high occurrence of AMF symbiosis can be found, particularly in arctic and antarctic regions, where a low proportion of plant species form AMF symbioses (Bledsoe *et al.*, 1990; Christie & Nicolson, 1983). In addition, species that belong to Brassicaceae and Chenopodiaceae generally do not form AMF symbioses, including their coastal representatives.

The benefits of AMF to sand dune plants are well documented as enhancement of growth, mainly by improving nutrient (especially phosphorus) and water uptake (Sutton & Sheppard, 1976; Forster & Nicolson, 1981; Koske & Polson, 1984; Read, 1989). In field plantings of dune grasses the benefits of AMF were related to improved establishment, more vigorous growth and inflorescences, and generally increased survival (Nicolson & Johnston, 1979; Gemma & Koske, 1989, 1997). In addition to bringing direct nutrient improvements, AMF protects sand dune grasses against harmful soil nematodes (Little & Maun, 1996) and fungi (Newsham *et al.*, 1994). The AMF protection of plants against pathogens appears to depend on the synergistic effects of soil pathogens. As a result of accumulation of species of pathogens in the rhizosphere the balance between AMF and soil pathogens may be compromised, causing decline of vigour (Greipsson & El-Mayas, 2000). The protective role of AMF could be related to induction of certain biochemical processes by the roots,

such as increased lignification of cell walls, chitinase activity or the production of phytoalexins or endoproteolytic activity (Sharma *et al.*, 1992; Filion *et al.*, 1999).

Total AMF root colonisation is usually high on sand dune grasses: 90% in *Ammophila arenaria* on dunes in the UK (Nicolson & Johnston, 1979); 80% in *A. breviligulata* on dunes in Rhode Island (Koske & Polson, 1984). Several factors appear to influence AMF colonisation on roots. First, it is seasonal, typically with highest values in the summer (Nicolson & Johnston, 1979; Siguenza *et al.*, 1996; Greipsson & El-Mayas, 2000). An exception was however found in northern Greece, where the highest AMF root colonisation occurred in the spring (Vardavakis, 1992). Second, AMF root colonisation generally increases with dune stability (Koske & Polson, 1984; Vardavakis, 1992; Siguenza *et al.*, 1996; Corkidi & Rincon, 1997*a*, *b*). In Baja California, AMF root colonisation of the pioneer sand dune plants was low (<1%) but plants in the fixed dunes had up to 80% AMF root colonisation (Siguenza *et al.*, 1996). There are exceptions however (Logan *et al.*, 1989).

Root colonisation by AMF is generally not species-specific and various groups of AMF are found to be associated with sand dune plants around the world (Jehne & Thompson, 1981; Koske & Halvorson, 1981; Giovannetti & Nicolson, 1983; Bergen & Koske, 1984; Giovannetti, 1985; Sylvia, 1986; Puppi & Riess, 1987; Rose, 1988; Gemma *et al.*, 1989; Vardavakis, 1992; Nicolson & Johnston, 1979; Blaszkowski, 1994*a*, *b*).

The density of AMF propagules in the soil influences root colonisation. The density of AMF spores may be seasonal, with highest values in the autumn (Puppi & Riess, 1987; Gemma & Koske, 1989; Greipsson & El-Mayas, 2000), or exceptionally in northern Greece, where the highest density occurred during the summer (Vardavakis, 1992). Seasonal differences in spore density were not observed in coastal dunes in North California and Florida (Sylvia, 1986; Rose, 1988). Density of AMF spores generally increases with dune stability (Koske & Polson, 1984; Sylvia, 1986; Rose, 1988; Vardavakis, 1992; Siguenza *et al.*, 1996; Corkidi & Rincon, 1997*b*; Greipsson & El-Mayas, 2000). There is also considerable variation in spore densities from sand dunes in different locations world-wide (Table 10.2).

Table 10.2. *Spore densities of arbuscular mycorrhizal fungi based on 100 g dry soil from sand dunes in different locations worldwide*

Location	Spore density	Reference
Slowinski National Park, Poland	34.7[a]	Blaszkowski (1994a, b)
Cape Cod, Massachusetts	23.5[a]	Bergen & Koske (1984)
Lake Huron, USA	0–680[b]	Koske et al. (1975)
Rhode Island, USA	140–351[b]	Koske & Halvorson (1981)
Italy	max 250	Puppi & Riess (1987)
Northeast Florida	0–677[b]	Sylvia (1986)
Baja California	max < 400	Siguenza et al. (1996)
New South Wales, Australia	0–1100[b]	Koske (1975)

[a] Average numbers.
[b] Range.

Nutrient cycling

Nutrient input into coastal sand dune ecosystems is mainly dependent on the rate of atmospheric deposition, abundance of free-living N_2-fixing microorganisms in soils, abundance of symbiotic N_2-fixing plant species, input by seawater (flooding and airborne spray) and organic debris (Walker et al., 1981). Atmospheric N input from dry and wet deposition on the East Frisian Islands, Germany has been estimated at about 1.5 g N m^{-2} per year (Gerlach et al., 1994). Nitrogen fixation by free-living soil bacteria has been estimated to be about 6 g N m^{-2} per year in dune slacks (Stewart, 1965). Nitrogen-fixing plants such as *Hippophaë rhamnoides* can add significant amounts of N to the ecosystem (17.9 g N m^{-2} per year) (Becking, 1970). Seawater was found to add little N to the ecosystem (0.26 g N m^{-2} per year)

(Olff et al., 1993). Nitrogen input in coastal sand dunes has also been attributed to decomposition of organic debris (algae) washed ashore (Fay & Jeffrey, 1992). Nitrogen losses from sand dune ecosystems were attributed to denitrification, leaching and litter removal (grazing) (Olff et al., 1993). Phosphorus may also be a limiting nutrient on coastal sand dunes, especially at low soil pH where P is complexed with iron and/or aluminium (Kooijman et al., 1998). In addition to the poor nutrient pool of sand dunes, the nutrient retention is very poor (Kellman & Roulet, 1990).

Biomass and litter production and nutrient cycling generally increase with succession (Kellman & Roulet, 1990). Primary succession is concomitant with the increase of soil organic matter (SOM) (Olson, 1958; Olff et al., 1993; Kooijman et al., 1998), especially with the increasing N in the soil (Marrs et al., 1983; Olff et al., 1993; Kooijman et al., 1998). The availability of N in soil is to a large extent controlled by SOM through decomposition and mineralisation (Kooijman, et al., 1998). In addition, the SOM may affect the cation exchange capacity of soils (Kooijman et al., 1998) and water holding capacity (Olff et al., 1993; Lichter, 1998).

Increase in soil N content often occurs rapidly with succession. For instance, the total soil N content increased from 1 to 15 g N m^{-2} during 16 years of succession on coastal dunes in the Netherlands (Olff et al., 1993). These changes are in agreement with the resource ratio hypothesis of primary succession were competition for nutrients is the driving factor in the earliest phase of the succession (Tilman, 1985).

RESTORATION PLANNING

A restoration site should be carefully investigated to determine the appropriate restoration techniques and to develop a long-term management plan (Box 10.1). Reconstruction is the first step in restoration but consideration must be given to the landscape of previous dunes, availability of sands and priorities of end users. Cost analysis is essential prior to any restoration project. For example, van der Putten (1990) ranked the cost of different planting methods of *Ammophila arenaria* on the coast of

Box 10.1 Factors to consider in restoration of damaged coastal sand dunes

Identify degradation processes

    Natural: high storm surge, rapid or reduced sand accretion, frequent sea flooding

    Anthropogenic: overgrazing by livestock, mining, trampling (Fig. B10.1)

Is dune reconstruction needed ?

    Use of dune-building fences

    Use of bulldozers

Is sand stabilisation needed ?

    Selection of native plants in the restoration process

Are propagules of native sand dune plants adapted to this locality available ?

Collecting propagules of native sand dune plants

Transplanting techniques: seed or clonal offsets

Propagation of clonal offsets of sand dune plants in nurseries

    Layout design of transplanted material

Use of soil stabilisers

Application of fertilisers

    Rate, timing and type of fertilisers

Transplanting secondary species (N_2-fixers)

Transplanting shrubs and trees

Protect the restoration site from grazing and/or trampling

Long-term management of the site

    Control alien invaders

    Monitor vegetation cover and plant succession

Fig. B10.1. Destruction of coastal sand dunes by mining.

the Netherlands in the order: culms with rhizomes > bundles of culms > rhizomes > seeds. Comparisons should be made of available restoration techniques. Key factors relating to planning the restoration work are: (1) when can the restoration work begin? (2) what labour, machinery and materials are required? (3) how much funding is available? and (4) is the restoration compatible with the desired land use? The availability of seed of native plants and/or clonal offsets is also critical. Appropriate inoculation of effective soil microorganisms should be evaluated for transplanted materials. Restored coastal sand dunes require long-term monitoring of vegetation cover, and management to avoid deterioration of vegetation and erosion of the dunes. Long-term management of sand dunes may involve: (1) replanting sites where previous establishment failed or die-back is experienced; (2) applying fertiliser when required to enhance growth; and (3) transplanting shrubs and trees on the dunes to facilitate succession and control of invasive alien plants. Management should be incorporated into the framework of integrated coastal zone management, which integrates remote sensing data, and a decision-support system (Doody, 1995; van Zuidam, et al., 1998). Detailed restoration designs or management plans tend to be site-specific and are not provided here.

Site preparation

Reconstruction of fore dunes may be required where sand dunes have eroded. Factors to consider during dune reconstruction are: (1) the availability of sand (whether there is enough sand on the site, or transportation is required); (2) the type of sand required; (3) the location and shape of the previous dune system; (4) the location of the fore dune; (5) the characteristics of any remnant dune; and (6) available funds. The shape of the reconstructed dunes should conform to existing nearby dunes. This can be achieved by using earth-moving equipment or by constructing dune-building fences, which trap sand. Heavy machinery can, however, leave behind destructive tracks on the dunes and should be used with caution. Using dune-building fences to reconstruct dunes is cost-effective compared to using earth-moving equipment, especially in remote areas, but the rate of dune formation using dune-building fences depends upon the amount of sand blown from the beach. Moreover, dune-building fences must be supported by new fences when they become buried by sand and, eventually, by establishing dune vegetation.

A good understanding of dune soils is required before restoring the dunes as these control the type of vegetation that can establish. Beach nourishment of dredged sand is one method of long-term maintenance of shores experiencing erosion (Hillen & Roelse, 1995; de Ruig, 1998; see Walmsley, this volume). Sand that has been added via beach nourishment (i.e. adding dredged materials) needs to be treated prior to restoration however. Treatments of the dune soils may include desalinisation, additions of chemicals to alter the acidity levels and addition of nutrients.

Use of seed or clonal offsets

The advantages and availability of plant propagules (seed or clonal offset) need to be determined prior to implementation of a restoration program, as the lack of propagules of native locally adapted plants is often a limiting factor. Furthermore, the scale of use dictates whether planting must be from seed rather than transplantation of clonal offsets. Planning ahead is essential to the successful use of clonal offsets, since it is difficult to hold planting stock in the nursery for more than one or two years (Woodhouse, 1982).

Availability of seed is often a limiting factor as sand-dune grasses produce generally sparse amounts of seeds and populations are maintained principally by clonal growth (Wagner, 1964; Huiskes, 1979; Harris & Davy, 1986; Hester & Mendelssohn, 1987; Maun & Baye, 1989; Cordazzo & Davy, 1994a, b). Low seed yield of dune grasses is mainly caused by pollen self-incompatibility, ovule abortion (Wagner, 1964; Cordazzo & Davy, 1994a; Wright, 1994) and low density of flowering spikes (Greipsson & Davy, 1997). Fertiliser application can increase the density of flowering spikes. For instance, a moderate application of a nitrogen fertiliser (54 kg ha^{-1}) on sand dunes in Oregon, USA resulted in a three-fold increase in the density of flowering spikes of A. arenaria (Brown & Hafenrichter, 1948). Similarly, seed yield of L. arenarius was increased dramatically by additions of N at 50–100 kg ha^{-1} on coastal dunes in Iceland (Greipsson & Davy, 1997). Seed yield of L. arenarius can be expected to be 334 kg ha^{-1} on semi-cultivated fields in Palmer, Alaska (Wright, 1994).

Seed collection can be accomplished by hand or by special harvesting machines (Fig B10.2). The latter can, however, have a detrimental impact on the sand dune environment and must be used with great care. Seed collection by hand has much less impact on the sand dunes and is an ideal method for collecting seed of small populations. Seeds are dried, threshed and cleaned before storage in a dry place. It is important during restoration to maintain genetic variability within the propagated species and seed of locally adapted dune species should be used whenever possible.

The advantage of using clonal offsets is their availability and also that they can be used on sites where sand is accreting rapidly or flooding is expected, especially on fore dunes (Woodhouse, 1982). The disadvantage of using clonal offsets is mainly associated with high manpower requirement and subsequent high cost of the operation (Wright, 1994). Clonal offsets of dune grasses can be dug up mechanically or by hand from nearby dunes but great care should be taken in minimising the environmental impact of such an operation (Dieckhoff, 1992). The supply site should be as close as possible to the restoration site in order to decrease cost of transportation and the whole site should be fertilised to ensure regrowth of the material. Clonal offsets that are dug up can be used directly on the restoration site or grown for two years in a nursery in order to increase their size before transplantation. The disadvantages of relying on clonal offsets dug directly from coastal dunes are summarised in Box 10.2. Establishing a local nursery ensures constant supply of material when needed.

Use of arbuscular mycorrhizal fungi

Colonisation of roots of sand dune plants by AMF depends on the presence of AMF propagules in the soil. Propagules are usually lacking in unvegetated coastal sands (Sylvia, 1986; Greipsson & El-Mayas, 2000) but they can spread to unvegetated sites by passive means (i.e. via wind, water, animals and insects) (Sylvia & Will, 1988). This process, however, can be very slow and erratic. An important factor in the inoculation process is the number of species of AMF to be used. In general, several species of AMF are found in any particular sand dune ecosystem (see section 'Underlying themes of restoration' above). Therefore, isolates containing a variety of AMF species are recommended for introduction on restoration sites (Sylvia & Burks, 1988). The effectiveness of AMF isolates must be carefully pre-examined. Different AMF isolates were found to vary in their ability to enhance growth of sand dune grasses and should, therefore, be screened before

they are considered for inoculum development (Sylvia & Burks, 1988; Greipsson & El-Mayas, 2000). Use of exotic AMF inocula in restoration should be treated with caution.

Nutrient amendments and aftercare

It is important to evaluate the responses of particular dune plant species to combinations and rates of fertiliser application prior to any restoration programme, for economic reasons. The type of fertiliser used depends on the species. Sand dune grasses generally require a high proportion of nitrogen but lupins and shrubs or trees require a higher proportion of phosphorus. Appropriate application rates vary according to location and season. Timing of fertiliser application is important since nutrient retention is generally low on sand dunes and rapid-release fertilisers can leach away quickly. Slow-release fertilisers have the advantage of releasing the chemicals over a relatively long period of time. Application of a slow-release fertiliser was found to increase seedling size and emergence of *Ammophila breviligulata* and *Calamovilfa longifolia* (Maun & Krajnyk, 1989; van der Putten, 1990). However, they are usually less cost-effective than the ordinary rapid-release formulations. Fertilisers can also enhance growth of vegetation on partly established dunes. However, fertilisers should be used with caution, as extensive application generally results in greater biomass productivity of the grasses, reduction in biodiversity and enhanced establishment of alien species (Woodhouse, 1982; Willis, 1985). Therefore, long-term management of dune ecosystems needs to consider the effects of fertilisers on the interactions between species and hence on the course of succession (DiTommaso & Aarssen, 1989).

Monitoring vegetational cover

Vegetation cover on coastal sand dunes has been estimated using false-colour imagery by remote sensing (van der Putten, 1990). To standardise these images, vegetation cover was measured manually in fixed plots and compared with colour density ratios of the corresponding plots on the false-colour images. The false-colour images could then be converted into values of vegetation cover according to colour density ratios (van der Putten, 1990). False-colour imagery can even differentiate between cover of different plant species (Everitt et al., 1999). Changes with time in vegetation cover have been studied by integrating remote sensing and geographical information systems (GIS) (Shao et al., 1998). Furthermore, GIS can be used to assess vegetation succession on the coastal dunes (van der Veen et al., 1997).

Box 10.2 Ecological approaches to restoration of coastal sand dunes in Iceland by using a dune-building grass: *Leymus arenarius*

In a large-scale sand dune restoration a reliable and economical source of seed from adapted native populations is needed. Seed of *Leymus arenarius* are harvested in the autumn from coastal sand dunes around the country. Coastal sand dunes that are harvested are carefully monitored and fertilisers are added if needed in order to increase seed production. However, excessive use of fertilisers can increase cover of competing vegetation. Seed harvesting is mechanised (Fig. B10.2) but small populations are harvested by hand. Consequently, seed are threshed and coated with diatomous earth and stored in cold hermetic containers during the winter. Seed germination is often a critical factor in the establishment. Therefore, seed germination is tested in each population. Subsequently, seeding rate can be calculated for each population. Seeds are sowed on restoration sites using precision seed-drillers and fertilisers are applied at the same time (Fig. B10.3). Seedling establishment and survival depends on judicious fertiliser application. Aftercare on restored sand dunes involves fertiliser application in order to avoid regression and secondary erosion and to accelerate succession of the sand dunes. Container-grown plants (derived from seed) are used where sands are accreting at fast rates. Dune-building fences can be used to reconstruct fore dunes and to avoid inundation of plants with seawater.

Fig. B10.2. Mechanised collection of native seed of *Leymus arenarius*.

Fig. B10.3. Mechanised seeding of *Leymus arenarius*.

RESTORATION IN PRACTICE

Sand stabilisation is an ancient practice that involves erecting dune-building fences and transplanting dune grasses on damaged or deflated
dunes. Dune restoration should include the use of
indigenous species and avoid establishment of invasive alien species, which have potential to alter
the functioning of the ecosystem and can prevent
establishment of native plants (Avis, 1995). Use of
fast-growing cover crops to stabilise dunes can be
justified but it may provide competition for native plants or enhance the establishment of native
plants (Box 10.3). Use of native plants in restoration practices requires a comprehensive ecological
understanding of the plants involved. Questions
relating to seed formation, seed germination,
growth of seedlings and mature plants need to be
answered.

The major emphasis in sand dune restoration has
been the establishment of the aboveground component of dune grasses (i.e. tillers and leaves). However,
this approach has overlooked soil dynamics especially the role of soil micro-organisms such as AMF.

An ecologically sound approach to dune restoration may be encouraged by using AMF, especially
where the aim is to create a self-sustaining ecosystem (Gemma *et al.*, 1989; Greipsson, 1998).

Long-term management on restored sand dunes
should follow initial restoration and this involves
maintenance of vigorous growth of the dune grasses
by judicious use of fertilisers and control of invasive aliens. Introduction of species in order to
sustain succession is needed in restoration, since
colonisation by native plants can take a considerably long time. For instance, coastal fore dunes that
were stabilised using *A. arenaria* did not contain
the same assemblages of plants or arthropods as
undisturbed fore dunes after 12 years (Webb *et al.*,
2000).

Site preparation

The initial step in restoration of deflated sand dunes
is to use structures that can accumulate sand. Dune-
building fences can be erected in order to trap sand
and reconstruct new dunes. Large dune ridges can

Box 10.3 Restoration and conservation of the coastal sand dunes at Asilomar, Pacific Grove, California

The establishment of the sand dune ecosystem at Asilomar (peace at sea) is a good example how dunes can be restored and conserved in spite of high recreation pressure on the beach. The degenerated drifting sand dunes were initially stabilised using hydromulching including fertiliser and seed of cover crops such as annual and perennial ryegrasses (*Lolium* spp.). A sprinkler irrigation system was installed on the dunes and the irrigation system was used for one year. This enhanced germination of seed, resulting in continuous cover of the *Lolium* spp. which gradually declined in the second year and native sea fig (*Carpobrotus aequilaterus*) and beach sagewort (*Artemisia pyenocephala*) were transplanted into the dunes. At the same time, seed of other native plants such as seaside paintedcup (*Castilleja latifolia*), sea rocket (*Cakile maritima*), mock heather (*Haplopappus ericoides*), sand verbena (*Abrona latifolia*), beach bur (*Ambrosia chamissonis*) and *A. pyenocphala* were broadcasted over the dunes. Within five years a good protective cover of native plants grew on the dunes (Woodhouse, 1982). Today, other plant species have colonised the dunes and growth of some plants such as tree lupin (*Lupinus arboreus*) need to be carefully monitored and controlled. A local nursery growing native plants is situated close to the dune system. Accessways for pedestrians have been built across the dunes, which allow easy access for people to walk from the hotel buildings to the beach (Fig. B10.4). The dunes are fenced off to avoid vehicular traffic. Dune-building grasses (mainly *Leymus mollis*) are protected by fences on the fore dunes.

Fig. B10.4. Wooden footpath on coastal sand dunes at Asilomar, California.

be built up by a programme of sequential fencing carried out over several years. Construction of dune-building fences depends on the type of dunes and fences may have to be designed by trial and error. The material for the fence should preferably be cost efficient, disposable and biodegradable, since fences will become buried with sand. Sand accumulation is facilitated by constructing fences with a material of optimum porosity of 50% and such fences can accumulate as much as 3 m of sand in three months (Ranwell & Boar, 1986). However, it is essential to follow up the dune process by transplanting dune grasses on to the newly reconstructed dune to stabilise the surface and to allow the dune to grow further. Dune grasses could also provide further shelter for other plant species to colonize the dunes. A successful reconstruction of a coastal sand dune ecosystem is described in Box 10.4.

Box 10.4   Reconstruction of a sand dune ecosystem: example from Køge Bay Seaside Park, Copenhagen, Denmark

Sand dune ecosystems have been reconstructed for various purposes. Dunes have been reconstructed to reinforce coastal protection (Ritchie & Gimingham, 1989) or to extend beaches for recreational purposes (Sylvia, 1988). Sand can simply be dredged and pumped ashore or accumulated by using dune-building fences. One example of a reconstructed coastal sand dune ecosystem is from the Køge Bay Seaside Park that is situated 15 km south of the city of Copenhagen, Denmark. It was reconstructed by artificial dunes of 500 ha for recreational purposes and coastal protection (Andersen, 1995b). Sand and clay material were pumped from the bottom of the Baltic Sea to reconstruct sand dunes ashore (Andersen, 1995b). The newly created dunes were seeded with a mixture of commercial grass species (Festuca rubra, F. arundinacea, Lolium multiflorum and L. perenne). Management of the dunes for the first few years was mainly by fertilisation and control of invasive aliens such as Hippophaë rhamnoides and Rosa rugosa. Also, the back-dune grasslands were mowed and human trampling was restricted (Andersen, 1995b). Native plants including dune grasses colonised the dunes rapidly and the total number of vascular plants was 26 in the first year and increased to 91 species after 13 years (Andersen, 1995b). However, the proportion of annual plants was more than 50% of the vegetation in the first year. The biodiversity of the dunes was similar to other Danish sand dune ecosystems with the one exception that the back dunes have less biodiversity than natural dunes (Andersen, 1995b).

Mulches can be used to stabilise the dune surface temporarily. They can keep the surface of the sand moist and increase the soil organic matter of the sand as the mulch breaks down. Mulches that can be used on sand dunes include chopped straw, peat, topsoil, leaf litter and wood pulp. Mulching is particularly useful in large-scale restoration since it can be applied mechanically using a mulch spreader and it is usually incorporated in the sand by disc harrowing. This method has proved to be beneficial for the establishment of A. arenaria on dunes in the Netherlands (van der Putten, 1990).

Use of seed

The use of seed of native dune grasses is effective in restoration of large areas, especially where seeding can be accomplished mechanically and where sand accretion is not rapid. However, using seed could be disadvantageous when germination is erratic and initial growth of seedlings is slow.

Dune grasses should be sown just below the sand surface where seed can germinate and the seedlings can emerge to the surface. The optimal position for germination is at a shallow depth where a buried seed can imbibe water readily and sense high diurnal temperature fluctuation (Harty & McDonald, 1972; Huiskes, 1977; Greipsson & Davy, 1994a, b, 1996a, b). Emergence potential varies between species of dune grasses. Optimal germination of seed of Ammophila arenaria was found at depths of 2.5 to 3.75 cm (Harty & McDonald, 1972). Drilling seed of Leymus arenarius at 5–10 cm depth allowed good emergence of seedlings and yet avoided high temperature fluctuations and the severe desiccation risk of the surface sand layers (Greipsson & Davy, 1996a). Mechanised seeding can be accomplished by using an ordinary seed drill (Fig. B10.3). Seeding can be accomplished in the spring or autumn. Seeding rates of dune grasses should be based on germination tests (Greipson & Davy, 1994b). After seeding is accomplished, the restoration site should be firmed with a crawler tractor or equivalent machinery. Surface seeding is not likely to be successful for dune grasses (Harty & McDonald, 1972; Mitchell, 1974; Greipsson & Davy, 1994b).

Rapid germination is advantageous in restoration (Greipsson & Davy, 1994b, 1997). Seed of sand dune grasses generally shows dormancy (Greipsson & Davy, 1994b). A cold period (stratification) pretreatment of seed can alleviate seed dormancy (Wagner, 1964; Seneca & Cooper, 1971; Greipsson & Davy, 1994b). Other methods of improving germination of dormant seed involve treatments with plant hormones such as GA₃ (Greipsson, 2001).

Seedling establishment can usually be improved by adding fertilisers and soil stabilisers to the surface (Mitchell, 1974). Seedling establishment on sand dunes is generally erratic and this has been related to adverse water relations, lack of nutrients and sand movements (Huiskes, 1977; Maun, 1985; Maun & Baye, 1989; Greipsson & Davy, 1997). On coastal sand dunes in Iceland, overall establishment of *L. arenarius* was characteristically low (less than 2%) but it was significantly improved by fertiliser application (Greipsson & Davy, 1997).

Burial with sand is a major hazard for plants on the dunes. Seeds can be buried beyond their capacity to emerge to the surface and seedlings may be unable to elongate as fast as the sand accretes. Emergence from burial in sand involves both the ability to survive in darkness (Sykes & Wilson, 1990) and the potential to elongate and penetrate a relatively thick sand layer. The potential of etiolation depends partly on seed mass (Maun & Lapierre, 1986; Davy & Figueroa, 1993) and partly on inherent ability. Therefore, it would be beneficial to select relatively large-seeded populations for restoration of areas with potentially high sand accretion (Greipsson & Davy, 1995).

Use of clonal offsets

Stabilisation of drifting coastal sands by transplanting clonal offsets of dune-building grasses is an old practice in Europe and dates back to the fourteenth or fifteenth century in Scotland (Hobbs *et al.*, 1983).

A clonal offset is derived from the smallest division of a rhizome that can grow into a new individual. Problems associated with collection of clonal offsets include: (1) lack of ready availability during the spring, when plant material is most needed;

(2) unreliable quality of plant material; (3) disturbance to the source dunes; and (4) transplanting one or two species is not suitable for all sites on the dunes (Dieckhoff, 1992). A viable clonal offset should contain foliage with at least 15 to 30 cm of rhizome attached. It is important to avoid damaging the foliage of clonal offsets during transplantation (Hobbs *et al.*, 1983). Care should be taken in storing clonal offsets in damp sand during transportation. The restoration site should be prepared by mechanically digging shallow trenches (20–30 cm deep) to receive the clonal offsets. After planting trenches are filled with firmly packed sand. Transplanting should be carried out in the season of relative dormancy of the plants (Ranwell & Boar, 1986). Planting can be mechanised by using a tree-planting machine, but transplanting by hand is preferred on steep hills or on small restoration sites. Offsets should be transplanted by spacing them 0.3 m to 0.9 m apart in staggered rows or triangular arrays to increase efficiency of sand stabilisation. The layout design of the transplanted offsets can influence dune formation: even planting on deflated sands promotes uniform sand deposition, whereas clustered planting can result in dune formation. Transplants should be fertilised in order to increase tillering rate.

Pieces of rhizomes can also been used as offsets and species-dependent orientation of the planted rhizomes can affect subsequent growth behaviour. Rhizomes of *A. arenaria* that were planted horizontally showed new horizontal rhizome development and had more foliage than those planted vertically. *Leymus arenarius* did not show this difference in behaviour (Ritchie & Gimingham, 1989). The planting depth of rhizomes is also species-dependent; for *A. arenaria* it should not exceed 20 cm whereas for *L. arenarius* maximum depth is 10 cm (Hobbs *et al.*, 1983).

Use of arbuscular mycorrhizal fungi

One of the oldest reports of the occurrence of AMF in sand dune plants is from Stahl (1900). However, it was not until 1974 that the importance of AMF in restoration was emphasised (Daft & Nicolson, 1974). Despite this, even today AMF inoculation is usually omitted when plants are raised in nurseries for restoration purposes and it becomes important to inoculate plants on site. This can be achieved by pre-inoculating nursery grown offsets with effective AMF isolates and transplanting them at regular intervals on site.

Nutrient amendments and aftercare

Fertilisers are usually applied at the same time or immediately after clonal offsets are transplanted or seeds are sown, to ensure high survival and vigorous growth of plants. Addition of N fertilisers is essential for establishment of dune grasses (Brown & Hafenrichter, 1948; Boorman, 1977; Wright, 1994). The response to addition of P fertilisers varies however, and in some cases an increase in growth has not been observed (Boorman, 1977). High rates of fertilisers are usually added. Application of 560–680 kg ha^{-1} of fertilizer (20% N, 20% P and 10% K) enhanced growth of newly transplanted clonal offsets or seedlings in Alaska (Wright, 1994).

Chemical soil stabilisers can be applied to restoration sites in order to stabilise the sand surface temporarily, decrease evaporation, and reduce extreme temperature fluctuation in the sand. Soil stabilisers are not an alternative to a restoration programme but are usually applied on restoration sites after seeds have been sown or offsets transplanted. Soil stabilisers include starch, cement, bitumen, oil, rubber, synthetic latexes, resins, plastics and oil–latex emulsions (Mitchell, 1974) as well as commercial products such as 'Soil Seal' and 'Coherex'. However, great care should be taken in using soil stabilisers that could be polluting or harmful for the environment. The disadvantages of using soil stabilisers include high cost, difficulties of application, increased runoff during rainfall, tendency to crack and lift off in high wind and possible leaching of undesirable chemicals (Ranwell & Boar, 1986).

Sequential introduction of species

The restoration of coastal sands dunes usually involves programmes where species are sequentially introduced on the dunes in order to promote succession. Establishment of sand-stabilising grasses is only the first step. For instance, in Oregon an initial planting of *Ammophila arenaria* was followed after one year by transplantation of *Cytisus* sp. (broom) and, a year later, by transplanting native *Pinus contorta* (coast pine) onto the dunes (Wiedermann *et al.*, 1974). Similarly, in southern Australia the native dune grasses *Spinifex hirsutus* and *Ehrharta villosa* (veldt grass) have been used to stabilise dunes. This was followed by transplanting *Lupinus digitatus*, *Lupinus arboreus* (tree lupin), *Carpobrotus cyanophyulla* (pigface), *Chrysanthemoides monilifera* and *Acacia sophorae* (Mitchell, 1974). Shrubs and trees that need protection from sand blast, such as *Leptospermum laevigatum* (tea tree), *Banksia integrifolia* (coast banksia), *Casuarina stricta* and *C. equisetifolia* (horsetail oak), were eventually transplanted on the dunes. Shrubs and trees are usually planted as container-grown plants raised in nurseries.

Control of invading alien species

Long-term management of restoration sites involves controlling and eradicating alien species. Serious attempts have been made to control the growth of *Hippophaë rhamnoides* by cutting the plants during the winter months, burning the cuttings, and by spraying new seedlings with herbicide (Ranwell & Boar, 1986). Alien species often lack the predators and pathogens that limit their growth in the native habitats. For instance, growth of *H. rhamnoides* is limited by soil nematodes (*Longidorus dunensis* and *Tylenchorhynchus microphasmis*) in its native habitat (Willis, 1989). Introducing these soil nematodes into the rhizosphere could therefore control populations of invasive *H. rhamnoides*. Natural control by fungal species and nematodes has been suggest for the control of *Ammophila arenaria* that was introduced from Europe to coastal sand dunes in South Africa (Lubke *et al.*, 1995).

CONCLUDING REMARKS

It is evident that coastal sand dune ecosystems will face serious threats in the near future because of sea level rise and increased impact from recreational activities. They are valuable as coastal defence against the sea and strong winds. Also, a remarkably high biodiversity is found on this relatively narrow strip of coastal land. Restoration should be an important part of the conservation strategy for this ecosystem. Restoration of deflated coastal sand dunes should consider the complex landscape of the ecosystem and should not only aim at establishing plant cover but rather at building a self-sustaining and diverse ecosystem. Studies are needed on the effect of biodiversity on ecosystem functioning and more knowledge is needed on the basic ecology of native sand dune plants. This information will allow greater use of a variety of native plants in the restoration work. The use of native populations of plants, harbouring authentic ranges of genetic variation, is important for restoration. The role of soil micro-organisms in restoration of sand dunes needs further study, especially in relation to how soil micro-organisms regulate ecosystem functioning. Restoration techniques aimed at sustaining succession are in great demand.

REFERENCES

Ahmad, J. & Wainwright, S. J. (1977). Tolerance to salt, partial anaerobiosis and osmotic stress in *Agrostis stolonifera*. *New Phytologist*, **79**, 605–612.

Andersen, U. V. (1995a). Invasive aliens: a threat to the Danish coastal vegetation? In *Directions in European Coastal Management*, eds. M. G. Healy & J. P. Doody, pp. 335–344. Cardigan, UK: Samara Publishing.

Andersen, U. V. (1995b). Succession and soil development in man-made coastal ecosystems at the Baltic Sea. *Nordic Journal of Botany*, **15**, 91–104.

Anderson, P. & Romeril, M. G. (1992). Mowing experiments to restore a species-rich sward on sand dunes in Jersey, Channel Islands, GB. In *Coastal Dunes*, eds. R. W. G. Carter, T. G. F. Curtis & M. J. Sheehy-Skeffington, pp. 219–234. Rotterdam: Balkema.

Aptekar, R. & Rejmanek, M. (2000). The effect of seawater submergence on rhizome bud viability of the introduced *Ammophila arenaria* and the native *Leymus mollis* in California. *Journal of Coastal Conservation*, **6**, 107–111.

Avis, A. M. (1995). An evaluation of the vegetation developed after artificially stabilising South African coastal dunes with indigenous species. *Journal of Coastal Conservation*, **1**, 41–50.

Ayyad, M. A. (1973). Vegetation and environment of the western Mediterranean coastal land of Egypt. 1: The habitat of sand dunes. *Journal of Ecology*, **61**, 509–523.

Barbour, M. G., de Jong, T. M. & Pavlik, B. M. (1985). Marine beach and dune plant communities. In *Physiological Ecology of North American Plant Communities*, eds. B. F. Chabot & H. A. Mooney, pp. 124–146. New York: Chapman & Hall.

Barbour, M. G., Rejmánek, D. M., Johnson, A. F. & Pavlik, B. M. (1987). Beach vegetation and plant distribution patterns along the northern Gulf of Mexico. *Phytocoenologia*, **15**, 201–233.

Bazzaz, F. A. (1990). The response of natural ecosystems to the rising global CO_2 levels. *Annual Review of Ecology and Systematics*, **21**, 167–196.

Becking, J. H. (1970). Plant–endophyte symbiosis in non-leguminous plants. *Plant and Soil*, **32**, 611–654.

Benecke, W. (1930). Zür biologie der strand- und dünenflora. l: Vergleichende Versuche über die Saltztoleranz von *Ammophila arenaria* Link, *Elymus arenarius* L. und *Agropyron junceum* L. *Bericht der deutschen botanischen Gesellschaft*, **48**, 127–139.

Bergen, M. & Koske, R. E. (1984). Vesicular-arbuscular mycorrhizal fungi from sand dunes of Cape Cod Massachusetts. *Transaction of the British Mycological Society*, **83**, 157–158.

Binggeli, P., Eakin, M., Macfadyen, A., Power, J. & McConnell, J. (1992). Impact of alien sea buckthorn (*Hippophaë rhamnoides* L) on sand dune ecosystems in Ireland. In *Coastal Dunes*, eds. R. W. G. Carter, T. G. F. Curtis & M. J. Sheehy-Skeffington, pp. 325–337. Rotterdam: Balkema.

Blaszkowski, J. (1994a). Polish Glomales. I: *Acaulospora dilatata* and *Scutellospora dipurpurascens*. *Mycorrhiza*, **4**, 173–182.

Blaszkowski, J. (1994b). Polish Glomales. II: *Glomus pustulatum*. *Mycorrhiza*, **4**, 201–207.

Bledsoe, C., Klein, P. & Bliss, L. C. (1990). A survey of mycorrhizal plants on Truelove lowland, Devon Island, NWT, Canada. *Canadian Journal of Botany*, **69**, 1848–1856.

Bliss, L. C. (1993). Arctic coastal ecosystems. In *Ecosystems of the World*, vol. 2A, *Dry Coastal Ecosystems*, ed. E. Van der Maarel, pp. 15–22. London: Elsevier.

Boorman, L. A. (1977). Sand-dunes. In *The Coastline*, ed. R. S. K. Barnes, pp. 161–197. New York: John Wiley.

Brown, A. C. & McLachlan, A. (1990). The *Ecology of Sandy Shores*. Amsterdam: Elsevier.

Brown, J. C. (1958). Soil fungi of some British sand dunes in relation to soil type and succession. *Journal of Ecology*, **46**, 641–664.

Brown, R. L. & Hafenrichter, A. L. (1948). Factors influencing the production and use of beachgrass and dunegrass clones for erosion control. 3: Influence of kinds and amounts of fertiliser on production. *American Journal of Agronomy*, **40**, 677–684.

Brundrett, M. (1991). Mycorrhizas in natural ecosystems. *Advances in Ecological Research*, **21**, 171–313.

Chapman, V. J. (1976). *Coastal Vegetation*. Oxford: Pergamon Press.

Christie, P. & Nicolson, T. H. (1983). Are mycorrhizas absent from the Antarctic? *Transactions of the British Mycological Society*, **80**, 557–560.

Connell, J. H. & Slatyer, R. O. (1977). Mechanisms of succession in natural communities and their role in community stability and organisation. *American Naturalist*, **111**, 1119–1143.

Corkidi, L. & Rincon, E. (1997a). Arbuscular mycorrhizae in a tropical sand dune ecosystem on the Gulf of Mexico. 1: Mycorrhizal status and inoculum potential along a successional gradient. *Mycorrhiza*, **7**, 9–15.

Corkidi, L. & Rincon, E. (1997b). Arbuscular mycorrhizae in a tropical sand dune ecosystem on the Gulf of Mexico. 2: Effects of arbuscular mycorrhizal fungi on the growth of species distributed in different early successional stages. *Mycorrhiza*, **7**, 17–23.

Cordazzo, C. V. & Davy, A. J. (1994a). Seed production and seed quality of the dune building grass *Panicum racemosum* Spreng. *Acta Botanica Brasilica*, **8**, 193–203.

Cordazzo, C. V. & Davy, A. J. (1994b). Seed production, seed size, and dispersal of *Spartina ciliata* Bronginart (Gramineae) in southern Brazilian coastal dunes. *Atlantica*, **16**, 143–154.

Cordazzo, C. V. & Seeliger, U. (1993). Zoned habitats of southern Brazilian coastal foredunes. *Journal of Coastal Research*, **9**, 317–323.

Costa, C. S. B., Cordazzo, V. & Seeliger, U. (1996). Shore disturbance and dune plant distribution. *Journal of Coastal Research*, **12**, 133–140.

Cowles, H. C. (1899). The ecological relations of the vegetation on the sand dunes of Lake Michigan. *Botanical Gazette*, **27**, 95–117.

Crawford, R. M. M. & Wishart, D. (1966). A multivariate analysis of the development of dune slack vegetation in relation to coastal accretion at Tentsmuir, Fife. *Journal of Ecology*, **54**, 729–743.

Crawley, M. J. (1987). What makes a community invasible? In *Colonisation, Succession and Stability*, eds. M. J. Crawley, P. J. Edwards & A. J. Gray, pp. 429–453. Oxford: Blackwell.

Curr, R. H. F., Koh, A., Edwards, E., Williams, A. T. & Davis, P. (2000). Assessing anthropogenic impact on Mediterranean sand dunes from aerial digital photography. *Journal of Coastal Conservation*, **6**, 15–22.

Daft, M. J. & Nicolson, T. H. (1974). Arbuscular mycorrhizas in plants colonising coal wastes in Scotland. *New Phytologist*, **73**, 1129–1138.

Davy, A. J. & Figueroa, M. E. (1993). The colonisation of strandlines. In *Primary Succession on Land*, eds. J. Miles & D. W. H. Walton, pp. 113–131. Oxford: Blackwell.

de Lacerda, L. D., de Araujo, D. S. D. & Maciel, N. C. (1993). Dry coastal ecosystems of the tropical Brazilian Coast. In *Ecosystems of the World*, vol. 2A, *Dry Coastal Ecosystems*, ed. E. Van der Maarel, pp. 322–338. London: Elsevier.

de Ruig, J. H. M. (1998). Seaward coastal defence: limitations and possibilities. *Journal of Coastal Conservation*, **4**, 71–78.

Dieckhoff, M. S. (1992). Propagating dune grasses by cultivation for dune conservation purposes. In *Coastal Dunes*, eds. R. W. G. Carter, T. G. F. Curtis & M. J. Sheehy-Skeffington, pp. 361–366. Rotterdam: Balkema.

DiTommaso, A. & Aarssen, L. W. (1989). Resource manipulation in natural vegetation: a review. *Vegetatio*, **84**, 9–29.

Doody, J. P. (1995). Information and coastal zone management: the final frontier. In *Directions in European Coastal Management*, eds. M. G. Healy & J. P. Doody, pp. 399–416. Cardigan, UK: Samara Publishing.

Eldred, R. A. & Maun, M. A. (1982). A multivariate approach to the problem of decline of *Ammophila*. *Canadian Journal of Botany*, **60**, 1371–1380.

Espejel, I. (1987). A phytogeographical analysis of coastal vegetation in the Yucatan Peninsula. *Journal of Biogeography*, **14**, 499–519.

Everitt, J. H., Escobar, D. E., Yang, C., Lonard, R.I., Judd, F. W., Alaniz, M. A., Cavazos, I., Davis, M. R. & Hockaday, D. L. (1999). Distingushing ecological parameters in a coastal area using a video system with visible/

near-infrared/mid-infrared sensitivity. *Journal of Coastal Research*, **15**, 1145–1150.

Fay, P. J. & Jeffrey, D. W. (1992). The foreshore as a nitrogen source for marram grass. In *Coastal Dunes, Geomorphology, Ecology and Management for Conservation*, eds. R. W. G. Carter, T. G. F. Curtis & M. J. Sheehy-Skeffington, pp. 177–188. Rotterdam: Balkema.

Filion, M., St-Arnaud, M. & Fortin, J. A. (1999). Direct interaction between the arbuscular mycorrhizal fungus *Glomus intraradices* and different rhizosphere microorganisms. *New Phytologist*, **141**, 525–533.

Firbank, L. G., Watkinson, A.R., Norton, L. R. & Ashenden, T. W. (1995). Plant populations and global environmental change: the effects of different temperature, carbon dioxide and nutrient regimes on density dependence in populations of *Vulpia ciliata*. *Functional Ecology*, **9**, 432–441.

Forster, S. M. & Nicolson, T. H. (1981). Microbial aggregation of sand in a maritime dune succession. *Soil Biology and Biochemistry*, **13**, 205–208.

Garcia-Novo, F. (1976). Ecophysiological aspects of the distribution of *Elymus arenarius* and *Cakile maritima* on the dunes at Tentsmuir Point (Scotland). *Oecologia Plantarum*, **11**, 13–24.

Garcia-Novo, F. (1979). The ecology of the dunes in Doñana National Park (southwest Spain). In *Ecological Processes in Coastal Environments*, eds. R. L. Jefferies & A. J. Davy, pp. 571–592. Oxford: Blackwell.

Gemma, J. N. & Koske, R. E. (1989). Field inoculation of American beachgrass (*Ammophila breviligulata*) with V–A mycorrhizal fungi. *Journal of Environmental Management*, **29**, 173–182.

Gemma, J. N. & Koske, R. E. (1997). Arbuscular mycorrhizae in sand dune plants of the North Atlantic coast of the U.S.: field and greenhouse inoculation and presence of mycorrhizae in plant stock. *Journal of Environmental Management*, **50**, 251–264.

Gemma, J. N., Koske, R. E. & Carreiro, M. (1989). Seasonal dynamics of selected species of V–A mycorrhizal fungi in a sand dune. *Mycological Research*, **92**, 317–321.

Gerlach, A., Alberts, E. & Broedlin, W. (1994). Development of the nitrogen cycle in the soils of a coastal succession. *Acta Botanica Neerlandica*, **43**, 189–203.

Giovannetti, M. (1985). Seasonal variations of vesicular-arbuscular mycorrhizas and endogonoaceous spores in a maritime sand dune. *Transaction of the British Mycological Society*, **84**, 679–684.

Giovannetti, M. & Nicolson, T. H. (1983). Vesicular–arbuscular mycorrhizas in Italian sand dunes. *Transaction of the British Mycological Society*, **80**, 552–557.

Goldsmith, V. (1989). Coastal sand dunes as geomorphological systems. *Proceedings of the Royal Society of Edinburgh*, **96B**, 3–15.

Greipsson, S. (1998). Ecological approaches to coastal sand stabilisation using dune-building grasses. *Journal of Coastal Research*, Special Issue 26, A11.

Greipsson, S. (2001). Seed coating with diatomaceous earth and GA3 improves establishment of *Leymus arenarius* a sand-stabilising grass. *Seed Science and Technology*, **29**, 1–10.

Greipsson, S. & Davy, A. J. (1994a). *Leymus arenarius*: characteristics and uses of a dune-building grass. *Icelandic Agricultural Sciences*, **8**, 41–50.

Greipsson, S. & Davy, A. J. (1994b). Germination of *Leymus arenarius* and its significance for land reclamation in Iceland. *Annals of Botany*, **73**, 393–401.

Greipsson, S. & Davy, A. J. (1995). Seed mass and germination behaviour in populations of the dune-building grass *Leymus arenarius*. *Annals of Botany*, **76**, 493–501.

Greipsson, S. & Davy, A. J. (1996a). Sand accretion and salinity as constraints on the establishment of *Leymus arenarius* for land reclamation in Iceland. *Annals of Botany*, **78**, 611–618.

Greipsson, S. & Davy, A. J. (1996b). Aspects of seed germination in the dune-building grass *Leymus arenarius*. *Icelandic Agricultural Sciences*, **10**, 209–217.

Greipsson, S. & Davy, A. J. (1997). Responses of *Leymus arenarius* to nutrients: improvement of seed production and seedling establishment for land reclamation. *Journal of Applied Ecology*, **34**, 1165–1176.

Greipsson, S. & El-Mayas, H. (1994). Coastal sands of Iceland. *Coastline*, **3**, 36–40.

Greipsson, S. & El-Mayas, H. (1996). The stabilisation of coastal sands in Iceland. In *Studies in European Coastal Management*, eds. P. S. Jones, M. G. Healy & A. T. Williams, pp. 93–100. Cardigan, UK: Samara Publishing.

Greipsson, S. & El-Mayas, H. (2000). Arbuscular mycorrhizae of *Leymus arenarius* on reclamation sites in Iceland, and response to inoculation. *Restoration Ecology*, **8**, 144–150.

Harris, D. & Davy, A. J. (1986). The regenerative potential of *Elymus farctus* from rhizome fragments and seed. *Journal of Ecology*, **74**, 1057–1067.

Harris, D. & Davy, A. J. (1987). Seedling growth in *Elymus farctus* after episodes of burial with sand. *Annals of Botany*, **60**, 587–593.

Harty, R. L. & McDonald, T. J. (1972). Germination behaviour in beach Spinifex (*Spinifex hirsutus* Labill). *Australian Journal of Botany*, **20**, 241–251.

Hesp, P. A. (1991). Ecological processes and plant adaptations on coastal dunes. *Journal of Arid Environments*, **21**, 165–191.

Hester, M. W. & Mendelssohn, I. A. (1987). Seed production and germination response of four Louisiana populations of *Uniola paniculata* (Gramineae). *American Journal of Botany*, **74**, 1093–1101.

Hillen, R. & Roelse, P. (1995). Dynamic preservation of the coastline in the Netherlands. *Journal of Coastal Conservation*, **1**, 17–28.

Hobbs, R. J., Gimingham, C. H. & Band, W. T. (1983). The effects of planting technique on the growth of *Ammophila arenaria* (L.) Link and *Leymus arenarius* (L.) Hochst. *Journal of Applied Ecology*, **20**, 659–672.

Houle, G. (1997a). No evidence for interspecific interactions between plants in the first stage of succession on coastal dunes in subarctic Quebec, Canada. *Canadian Journal of Botany*, **75**, 902–915.

Houle, G. (1997b). Interactions between resources and abiotic conditions control plant performance on subarctic coastal dunes. *American Journal of Botany*, **84**, 1729–1737.

Huiskes, A. H. L. (1977). The natural establishment of *Ammophila arenaria* from seed. *Oikos*, **29**, 133–136.

Huiskes, A. H. L. (1979). Biological flora of the British isles: *Ammophila arenaria* (L.) Link (*Psamma arenaria* (L.) Roem. et Shult.: *Calamagrostis arenaria* (L.) Roth). *Journal of Ecology*, **67**, 363–382.

Jehne, W. & Thompson, C. H. (1981). Endomycorrhizae in plant colonisation on coastal sand dunes at Cooloola, Queensland. *Australian Journal of Ecology*, **6**, 221–230.

Johnston, A. F. (1977). A survey of the strand and dune vegetation along the Pacific and southern Gulf coasts of Baja California, Mexico. *Journal of Biogeography*, **4**, 83–100.

Karl, T. R., Nichols, N. & Gregory, J. (1997). The coming climate. *Scientific American*, **276**, 55–59.

Kellman, M. & Roulet, N. (1990). Nutrient flux and retention in a tropical sand-dune succession. *Journal of Ecology*, **78**, 664–676.

Kirkpatrick, J. B. & Hassall, D. C. (1981). Vegetation of the Sigatoka sand dunes, Fiji. *New Zealand Journal of Botnay*, **19**, 285–297.

Kooijman, A. M. & de Haan, M. W. A. (1995). Grazing as a measure against grass encroachment in Dutch dry dune grassland: effects on vegetation and soil. *Journal of Coastal Conservation*, **1**, 127–134.

Kooijman, A. M., Dopheide, J. C. R., Sevink, J., Takken, I. & Verstraten, J. M. (1998). Nutrient limitations and their implications on the effects of atmospheric deposition in coastal dunes: lime-poor and lime-rich sites in the Netherlands. *Journal of Ecology*, **86**, 511–526.

Koske, R. E. (1975) Endogone spores in Australian sand dunes. *Canadian Journal of Botany*, **53**, 668–672.

Koske, R. E. & Halvorson, W. L. (1981). Ecological studies of vesicular–arbuscular mycorrhizae in a barrier sand dune. *Canadian Journal of Botany*, **59**, 1413–1422.

Koske, R. E. & Polson, W. R. (1984). Are V–A mycorrhizae required for sand dune stabilisation? *BioScience*, **34**, 420–424.

Koske, R. E., Sutton, J. C. & Sheppard, B. R. (1975). Ecology of *Endogone* in Lake Huron sand dunes. *Canadian Journal of Botany*, **53**, 87–93.

Lichter, J. (1998). Primary succession and forest development on coastal Lake Michigan sand dunes. *Ecological Monographs*, **68**, 487–510.

Lichter, J. (2000). Colonization constraints during primary succession on coastal Lake Michigan sand dunes. *Journal of Ecology*, **88**, 825–839.

Little, L. R. & Maun, M. A. (1996). The 'Ammophila problem' revisited: a role for mycorrhizal fungi. *Journal of Ecology*, **84**, 1–7.

Logan, V. S., Clarke, P. J. & Allaway, W. G. (1989). Mycorrhizas and root attributes of coastal sand dunes of New South Wales. *Australian Journal of Plant Physiology*, **16**, 141–146.

Lubke, R. A., Hertling, U. M. & Avis, A. M. (1995). Is *Ammophila arenaria* (marram grass) a threat to South African dune fields? *Journal of Coastal Conservation*, **1**, 103–108.

Marrs, R. H., Roberts, R. D., Skeffington, R. A. & Bradshaw, A. D. (1993). Nitrogen and the development of ecosystems. In *Nitrogen as an Ecological Factor*, eds. J. A. Lee, S. McNeill & I. H. Revison, pp. 113–136. Oxford: Blackwell Scientific Publications.

Martinez, M. L., Moreno-Casasola, P. & Vazquéz, G. (1997). Effects of disturbance by sand movement and inundation by water on tropical dune vegetation dynamics. *Canadian Journal of Botany*, **75**, 2005–2014.

Maun, M. A. (1985). Population biology of *Ammophila breviligulata* and *Calamovilfa longifolia* on Lake Huron sand dunes. 1: Habitat, growth form, reproduction and establishment. *Canadian Journal of Botany*, **63**, 113–124.

Maun, M. A. (1998). Adaptations of plants to burial in coastal sand dunes. *Canadian Journal of Botany*, **76**, 713–738.

Maun, M. A. & Baye, P. R. (1989). The ecology of *Ammophila breviligulata* on coastal sand dune systems. *C. R. C. Critical Review in Aquatic Sciences*, **1**, 661–681.

Maun, M. A. & Krajynk, I. (1989). Stabilisation of Great Lakes sand dunes: effect of planting time, mulches and fertilizer on seedling establishment. *Journal of Coastal Research*, **5**, 791–800.

Maun, M. A. & Lapierre, J. (1986). Effects of burial by sand on seed germination and seedling emergence of four dune species. *American Journal of Botany*, **73**, 450–455.

Mitchell, A. (1974). Plants and techniques used for sand dune reclamation in Australia. *International Journal of Biometeor*, **18**, 168–173.

Moore, P. D. (1971). Computer analysis of sand dune vegetation in Norfolk, England, and its implication for conservation. *Vegetatio*, **23**, 323–338.

Moreno-Casasola, P. (1986). Sand movement as a factor in the distribution of plant communities in a coastal dune system. *Vegetatio*, **65**, 67–76.

Newsham, K. K., Fitter, A. H. & Watkinson, A. R. (1994). Root pathogenic and arbuscular mycorrhizal fungi determine fecundity of asymptomatic plants in the field. *Journal of Ecology*, **82**, 805–814.

Nicolson, T. H. & Johnston, C. (1979). Mycorrhiza in the Graminae. 3: *Glomus fasciculatus* as the endophyte of pioneer grasses in a maritime sand dune. *Transactions of the British Mycological Society*, **72**, 261–268.

Olff, H., Huisman, J. & van Tooren, B. F. (1993). Species dynamics and nutrient accumulation during early primary succession in coastal sand dunes. *Journal of Ecology*, **81**, 693–706.

Olsauskas, A. (1995). Influence of recreation on flora stability on the Lithuanian coastal dunes. In *Directions in European Coastal Management*, eds. M. G. Healy & J. P. Doody, pp. 103–106. Cardigan, UK: Samara Publishing.

Olson, J. S. (1958). Rates of succession and soil changes on southern Lake Michigan sand dunes. *Botanical Gazette*, **119**, 125–170.

Olsson, H. (1993). Dry coastal ecosystems of southern Sweden. In *Ecosystems of the World*, vol. 2A, *Dry Coastal Ecosystems*, ed. E. van der Maarel, pp. 131–143. London: Elsevier.

Packham, J. R. & Willis, A. J. (1997). The *Ecology of Dunes, Salt Marsh and Shingle*. London: Chapman & Hall.

Pickett, S. T. A., Collins, S. L. & Armesto, J. J. (1987). Models, mechanisms and pathways of succession. *Botanical Review*, **53**, 335–371.

Puppi, G. & Riess, S. (1987). Role and ecology of V–A mycorrhizae in sand dunes. *Angewandte Botanik*, **61**, 115–126.

Randall, R. E. (1993). Dry coastal ecosystems of the eastern Mediterranean. In *Ecosystems of the World*, vol. 2A, *Dry Coastal Ecosystems*, ed. E. van der Maarel, pp. 463–473. London: Elsevier.

Ranwell, D. S. (1972). The *Ecology of Salt Marshes and Sand Dunes*. London: Chapman & Hall.

Ranwell, D. S. & Boar, R. (1986). *Coast Dune Management Guide*. Huntingdon, UK: Institute of Terrestrial Ecology.

Read, D. J. (1989). Mycorrhizas and nutrient cycling in sand dune ecosystems. *Proceedings of the Royal Society of Edinburgh*, **96**, 89–110.

Rhind, P. M. & Jones, P. S. (1999). The floristics and conservation status of sand dune communities in Wales. *Journal of Coastal Conservation*, **5**, 31–42.

Ritchie, W. & Gimingham, C. H. (1989). Restoration of coastal dunes breached by pipeline landfalls in north east Scotland. *Proceedings of the Royal Society of Edinburgh*, **96B**, 231–245.

Rose, S. L. (1988). Above- and belowground community development in a marine sand dune ecosystem. *Plant and Soil*, **109**, 215–226.

Rozema, J., Bijwaard, P., Prast, G. & Broekman, R. (1985). Ecophysiological adaptations of coastal halophytes from foredunes and salt marshes. *Vegetatio*, **62**, 499–521.

Sanjaume, E. & Pardo, J. (1992). The dunes of the Valencian coast (Spain): past and present. In *Coastal Dunes*, eds. R. W. G. Carter, T. G. F. Curtis & M. J. Sheehy-Skeffington, pp. 475–486. Rotterdam: Balkema.

Seliskar, D. (1995). Coastal dune restoration: strategy for alleviating die-out of *Ammophila arenaria*. *Restoration Ecology*, **3**, 54–60.

Seliskar, D. & Huttel, R. N. (1993). Nematode involvement in the die-out of *Ammophila breviligulata* (Poaceae) on the mid-Atlantic coastal dunes of the United States. *Journal of Coastal Research*, **9**, 97–101.

Seneca, E. D. & Cooper, A. W. (1971). Germination and seedling response to temperature, daylength, and salinity by *Ammophila breviligulata* from Michigan and North Carolina. *Botanical Gazette*, **132**, 203–215.

Shao, G., Young, D. R., Porter, J. H. & Hayden, B. P. (1998). An integration of remote sensing and GIS to examine the response of shrub thicket distribution to shoreline changes on Virginia Barrier Islands. *Journal of Coastal Research*, **14**, 299–307.

Sharma, A. K., Jori, B. N. & Gianinazzi, S. (1992). Vesicular–arbuscular mycorrhizae in relation to plant disease. *World Journal of Microbiology and Biotechnology*, **8**, 559–563.

Sidaway, R. (1995). Recreation and tourism on the coast: managing impacts and resolving conflicts. In *Directions in European Coastal Management*, eds. M. G. Healy & J. P. Doody, pp. 71–78. Cardigan, UK: Samara Publishing.

Siguenza, C., Espejel, I. & Allen, E. B. (1996). Seasonality of mycorrhizae in coastal sand dunes of Baja California. *Mycorrhiza*, **6**, 151–157.

Stahl, E. (1900). Der Sinn der mycorhizen Bildung: Eine vergleichend-biologische Studie. *Jahrbuch für wissenschaftliche Botanik*, **34**, 539–668.

Stewart, W. P. D. (1965). Nitrogen turnover in marine and brackish habitats. *Annals of Botany*, **29**, 229–239.

Sturgess, P. (1992). Clear felling dune plantations: studies in vegetation recovery. In *Coastal Dunes*, eds. R. W. G. Carter, T. G. F. Curtis & M. J. Sheehy-Skeffington, pp. 339–349. Rotterdam: Balkema.

Sutton, J. C. & Sheppard, B. R. (1976). Aggregation of sand dune soil by endomycorrhizal fungi. *Canadian Journal of Botany*, **54**, 326–333.

Sykes, M. T. & Wilson, J. B. (1990). Dark tolerance in plants of dunes. *Functional Ecology*, **4**, 799–805.

Sylvia, D. M. (1986). Spatial and temporal distribution of vesicular–arbuscular mycorrhizal fungi associated with *Uniola paniculata* in Florida fore dunes. *Mycologia*, **78**, 728–734.

Sylvia, D. M. & Burks, J. N. (1988). Selection of a vesicular–arbuscular mycorrhizal fungus for practical inoculation of *Uniola paniculata*. *Mycologia*, **80**, 565–568.

Sylvia, D. M. & Will, M. E. (1988). Establishment of V–A mycorrhizal fungi and other microorganisms on a beach replenishment site in Florida. *Applied Environmental Microbiology*, **54**, 348–352.

Tilman, D. (1985). The resource-ratio hypothesis of plant succession. *American Naturalist*, **125**, 827–852.

van der Maarel, E. (ed.) (1993a). *Ecosystems of the World*, vol. 2A, *Dry Coastal Ecosystems*. London: Elsevier.

van der Maarel, E. (ed.) (1993b). *Ecosystems of the World*, vol. 2B, *Dry Coastal Ecosystems*. London: Elsevier.

van der Putten, W. H. (1990). Establishment and management of *Ammophila arenaria* (marram grass) on artificial coastal foredunes. In *Proceedings of the Canadian Symposium on Coastal Sand Dunes*, ed. R. Davidson-Arnott, pp. 367–387. Ottawa: National Research Council of Canada.

van der Putten, W. H. & Peters, B. A. M. (1995). Possibilities for management of coastal foredunes with deteriorated stands of *Ammophila arenaria* (marram grass). *Journal of Coastal Conservation*, **1**, 29–39.

van der Putten, W. H., van Dijk, C. & Troelsra, S. R. (1988). Biotic soil factors affecting the growth and development of *Ammophila arenaria*. *Oecologia*, **76**, 313–320.

van der Putten, W. H., van der Werf-Klein Breteler, J. T. & Van Dijk, C. (1989). Colonisation of the root zone of *Ammophila arenaria* by harmful soil organisms. *Plant and Soil*, **120**, 213–223.

van der Putten, W. H., van Dijk, C. & Peters, B. A. M. (1993). Plant-specific soil-borne diseases contribute to succession in foredune vegetation. *Nature*, **362**, 53–55.

van der Veen, M. A. C, Grootjans, A. P., de Jong, J. & Rozema, J. (1997). Reconstruction of an interrupted primary beach plain succession using a Geographical Information System. *Journal of Coastal Conservation*, **3**, 71–78.

van Dijk, H. W. (1992). Grazing domestic livestock in Dutch coastal dunes: experiments, experience and perspectives. In *Coastal Dunes*, eds. R. W. G. Carter, T. G. F. Curtis & M. J. Sheehy-Skeffington, pp. 235–250. Rotterdam: Balkema.

van Zuidam, R. A., Farifteh, J., Eleveld, M. A. & Cheng, T. (1998). Developments in remote sensing, dynamic modeling and GIS applications for integrated coastal zone management. *Journal of Coastal Conservation*, **4**, 191–202.

Vardavakis, E. (1992). Mycorrhizal endogonaceae and their seasonal variations in a Greek sand dune. *Pedobiologia*, **36**, 373–382.

Vestergaard, P. (1997). Possible impact of sea-level rise on some habitat types at the Baltic coast of

Denmark. *Journal of Coastal Conservation*, **3**, 103–112.

Wagner, R. H. (1964). The ecology of *Uniola paniculata* L. in the dune strand habitat of North Carolina. *Ecological Monographs*, **34**, 79–96.

Walker, J., Thompson, C. H., Fergus, I. F. & Tunstall, B. R. (1981). Plant succession and soil development in coastal sand dunes of subtropical eastern Australia. In *Forest Succession: Concepts and Application*, eds. D. C. West, H. H. Shugart & D. B. Botkin, pp. 167–189. New York: Springer-Verlag.

Wallen, B. (1980). Changes in structure and function of *Ammophila* during a primary succession. *Oikos*, **34**, 592–614.

Warming, E. (1909). *Oecology of Plants: An Introduction to the Study of Plant Communities*. Oxford: Clarendon Press.

Watkinson, A. R., Huskies, A. H. L. & Noble, J. C. (1979). The demography of sand dune species with contrasting life cycles. In *Ecological Processes in Coastal Environments*, eds. R. L. Jefferies & A. J. Davy, pp. 95–112. Oxford: Blackwell.

Webb, C. E., Oliver, I. & Pik, A. J. (2000). Does coastal fore-dune stabilisation with *Ammophila arenaria* restore plant and arthropod communities in southeastern Australia? *Restoration Ecology*, **8**, 283–288.

Webley, D. M., Eastwood, D. J. & Gimmingham, C. H. (1952). Development of a soil microflora in relation to plant succession on sand-dunes, including a 'rhizosphere' flora association with colonising species. *Journal of Ecology*, **40**, 168–178.

Wiedermann, A. M., Dennis, L. R. J. & Smith, F. H. (1974). *Plants of the Oregon Coastal Dunes*. Corvallis, OR: O.S.U. Book Stores Inc.

Willis, A. J. (1985). Plant diversity and change in a species-rich dune system. *Transactions of the Botanical Society of Edinburgh*, **44**, 291–308.

Willis, A. J. (1989). Coastal sand dunes as biological systems. *Proceedings of the Royal Society of Edinburgh*, **96B**, 17–36.

Woodhouse, W. W. (1982). Coastal sand dunes of the USA. In *Creation and Restoration of Coastal Plant Communities*, ed. R. R. Lewis, pp. 1–44. Boca Raton, FL: CRC Press.

Wright, S. J. (1994). *Beach wild rye: Planting Guide for Alaska*. Fairbanks, AK: State of Alaska, Division of Agriculture, Plant Materials Center.

Yuan, T., Maun, M. A. & Hopkins, W. G. (1993). Effects of sand accretion on photosynthesis, leaf water potential and morphology of two dune grasses. *Functional Ecology*, **7**, 676–682.

11 • Saltmarshes

JOY B. ZEDLER AND PAUL ADAM

INTRODUCTION

Organisms that live on intertidal sheltered shores have to contend with the twin challenges of frequent inundation and variable salinity. Yet this seemingly inhospitable environment supports saltmarshes (and, in the tropics and subtropics, mangroves) and a great diversity of organisms. Unlike exposed shores, where waves, currents and coarse substrates preclude establishment of vascular plants, protected shorelines allow fine sediments to accrete and vascular plants to root. Rather than being constrained by high tides and saline soils, some saltmarshes achieve productivity levels that rank among the world's highest.

Saltmarshes are found in a variety of geomorphological settings (Allen, 2000) but are particularly extensive in estuaries or on open coasts sheltered by offshore islands or spits. These same settings are highly desirable for human settlement, and large areas of saltmarsh have been filled for industrial and urban development, as well as modified for agricultural use. In Europe, saltmarshes have been used for grazing of domestic livestock and for haymaking since at least 600 BC (Esselink, 2000), and embankments to create non-tidal, non-saline agricultural land date to Roman times.

In many parts of the world saltmarsh values are being recognised and more widely appreciated, and saltmarsh restoration is a growing science and practice. The public at large benefits from shoreline stabilisation (Brampton, 1992), biodiversity support and recreational uses, while those who earn their livelihoods harvesting products of saltmarshes benefit from ecosystem support of commercial and recreational fin- and shellfisheries. Increased appreciation has followed recognition of the enormous scale of past saltmarsh losses. Current conservation efforts aim to minimise future losses, to rehabilitate damaged sites and to create new saltmarshes to mitigate for past or intended future losses, although the focus of activity varies between regions. In northern Europe, major interests are in increasing the nature conservation value of marshes (Esselink, 2000), in managed retreat of shorelines to ameliorate the impacts of rising sea level and to utilise saltmarshes to reduce coastal erosion. In the United States, various mandates for mitigation (i.e. compensation for filling other saltmarshes, as required under the Clean Water Act, the National Environmental Policy Act and other state and local regulations) have resulted in a very large number of saltmarsh restoration activities. Several saltmarshes have also been planted in the United States to stabilise dredge spoils, and these efforts have been studied extensively (Streever, 2000). In many countries, *Spartina* marshes have been established for shoreline protection, channel stabilisation and pasture, and the practice continued until very recently in China.

Much of the literature on saltmarsh rehabilitation and restoration consists of development proposals and environmental assessments that are not peer-reviewed. Published accounts often deal with only a few attributes (e.g. vegetation cover) rather than the whole ecosystem (Zedler & Callaway, 2000). Given the very large number of projects carried out in the United States, we draw heavily on the few that have been studied scientifically. This chapter provides a broad overview of saltmarsh ecology before discussing various aspects of restoration. Our approach is cautious; we wholeheartedly support the objectives of saltmarsh restoration and restoration ecology, while calling for much more research in order to evaluate more thoroughly the outcomes of past efforts and to develop better protocols for restoring saltmarshes in the future.

ESSENTIALS OF SALTMARSH ECOLOGY

Saltmarshes are vegetated wetlands that are dominated by herbaceous and small shrubby halophytes (salt-tolerant plants) (Adam, 1990; Packham & Willis, 1997). Saltmarshes are geographically widespread, occurring from high latitudes to the tropics. Mangroves (salt-tolerant trees) overshadow and displace saltmarsh vegetation at lower latitudes, and the herbaceous species often occur inland of mangroves along tropical and subtropical shores, as in southern Australia and Florida. Extensive saltmarshes occur along temperate-zone sea coasts, especially within estuaries and embayments. Coastal saltmarshes can be fully tidal, with regular inundation by seawater, or irregularly flooded by seawater; or non-tidal, but saline, around the shoreline of closed lagoons. We focus here on tidal saltmarshes.

Tidal marshes

Saltmarshes occur within the intertidal zone where wave energy is low. Their vertical range is determined by the tidal regime (especially the maximum amplitude) and by tidal energy and currents, while horizontal extent is influenced by local topography and bathymetry and stage of maturity of marsh development. The seaward limit depends on the inundation tolerance of local halophytes. Although there are few data on absolute elevation limits, marshes with *Spartina alterniflora* may extend lower than other types (Lefeuvre & Dame, 1994). The seaward area is either vegetated with seagrasses or free of vascular plants (mudflat) and dominated by algae, either micro- or macro-algal species. The upper limit of salt influence often coincides with the extreme high water of a storm event that transports ocean salts inland. At the inland margin, saltmarshes grade into forest, shrubland, grassland, desert, or coastal scrub. The transition to these drier ecosystems supports upland halophytes that are not well adapted to inundation but do tolerate occasional wetting by seawater (James & Zedler, 2000).

The intertidal zone encompasses considerable variation in environmental conditions (Fig. 11.1). The lowest, most frequently flooded, elevations of the saltmarsh typically have little variation in topography and uniform vegetation. The tidal channels that meander across the marsh plain dissect the saltmarsh as they grade into creeks and then tiny rivulets. With increasing surface elevation, topographic variation increases and a mosaic of communities is found. Attributes of microtopography include creek banks of various slopes, levees, mounds and depressions (Fig. 11.2). The density and nature of the drainage system varies among sites (Allen, 2000), but with a correlation between complexity of creeks, sediment type and vegetation (Adam, 1990). Composition and structure often differ between creek levees and intercreek basins (Adam, 1990; Zedler et al., 1999).

Vegetation

The saltmarsh vascular flora is halophytic by definition. Halophytes have a range of morphological, physiological and biochemical adaptations for tolerating salt. However, most saltmarsh plants do not require salt; they are readily grown under non-saline conditions, although some show improved growth at moderate salinities. The restriction of many species to saline habitats reflects poor competitive ability under freshwater conditions. Although many saltmarsh species are restricted to saline habitats, the flora includes salt-tolerant ecotypes of more widespread species, such as the grasses *Festuca rubra* or *Agrostis stolonifera*, halophytic ecotypes of which dominate large areas of north European saltmarshes.

Broad patterns in the distribution of species and vegetation types have been recognised (Adam, 1990; Zedler et al., 1999). The vascular plant diversity of saltmarshes is very low in the tropics and highest in temperate regions. The composition and structure of saltmarsh vegetation varies with latitude and also with rainfall and river flows. The most extensive saltmarsh type is dominated by *Spartina alterniflora* along the Gulf and Atlantic coasts of the United States, and in many ways it can be viewed as a unique biome. On arid and semi-arid coasts,

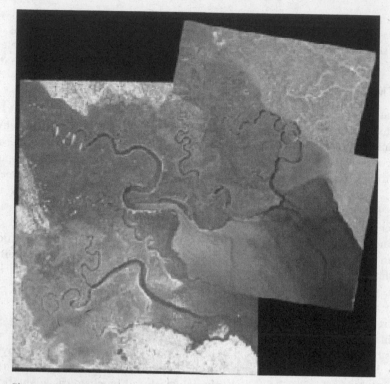

Fig. 11.1. Aerial image of a saltmarsh at San Quintin Bay, Baja California, Mexico. Mosaic by Bruce Nyden, Pacific Estuarine Research Laboratory, San Diego State University, San Diego.

subshrubs are more diverse and graminoids less dominant than in areas having more rainfall.

Within a region, the sequence of species' occurrences with elevation is generally predictable, but there is considerable site-to-site variation, and the sequence towards the head of estuaries is sometimes reversed. At any one site, an individual species oc-cupies a limited part of the intertidal zone (Zedler *et al.*, 1999) (Fig. 11.3) but species distributions overlap substantially and form a continuum of mean elevations. Still, there are large areas of recurrent assemblages that relate to microtopography, giving the marsh an overall patchy appearance (Fig. 11.1).

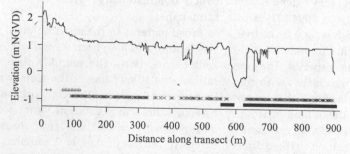

Fig. 11.2. Elevation profile at Volcano Marsh, San Quintin Bay, showing the occurrences of *Salicornia subterminalis* (+), *Salicornia virginica* (×) and *Spartina foliosa* (■). Elevations are relative to the lowest occurrence of *Spartina foliosa*. From Zedler *et al.* (1999).

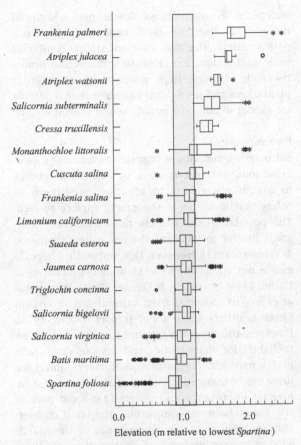

Fig. 11.3. Elevation ranges (boxes include upper and lower quartiles divided by the median; whiskers are quartiles; * and o are outliers) of saltmarsh plants at San Quintin Bay, Baja California Norte, Mexico. The 30-cm shaded elevation range denotes the marsh plain (modified from Zedler et al., 1999).

Temporal dynamics

Coastal saltmarshes are naturally dynamic systems. Saltmarsh species evolved over millions of years, and their position in coastal landscapes has varied over millennia with changing sea levels. Current coastline positions were largely set about 6000 years ago at the end of the last ice age, when the glaciers melted and global sea level reached approximately its present level. Isostatic changes in land level have continued to affect the land–sea boundary on some coastlines (Bird, 1993) while abrupt tectonic events,

although rare, have dramatically changed some salt-marshes, as was the case in the Alaska earthquake of 1964 (Ovenshine & Bartsch-Winkler, 1978). At present there is widespread concern about the potential for future sea level rise as a consequence of global warming, and its possible impact on the survival of many saltmarshes.

While the origin of some present saltmarshes may date back 6000 years, many have much shorter histories, in response to changing biological and geomorphological circumstances. The rate of saltmarsh development varies between sites but can be rapid (e.g. Packham & Liddle, 1971; Ward, 2000). Models of natural saltmarsh development start with the colonisation of sand or mudflats by pioneer vascular plants, a process that micro-algae facilitate by stabilising the sediment surface (Coles, 1979). Vertical accretion and horizontal expansion of the marsh is driven by the accumulation and trapping of sediment. Most of the sediment is inorganic material washed in by the tide or brought downstream by rivers, with smaller contributions of autochthonous organic matter (see Randerson, 1979; Adam, 1990). As the marsh surface rises, tidal influence declines, soils dry and salinity increases. The process of species replacement is often inferred from spatial distributions of species in relation to elevation, rather than direct observation.

Sedimentation may be both chronic and episodic. At Tijuana Estuary, in southern California, the last remaining mudflat has accreted sediment during several flood events over the past 10 years to the point (0.5 m National Geodetic Vertical Datum [NGVD is mean sea level as determined in 1929]) that it now supports large clones of the region's native *Spartina foliosa*. These clones are predicted to coalesce within 15 to 20 years, eliminating the mudflat's utility for shorebird feeding (Ward, 2000). Nearby, the older marsh plain has accreted about 30 cm of sediment (Weis, 1999), causing a shift from a diverse mixture of succulents at 60–80 cm NGVD to dominance by *Salicornia virginica* (J. B. Zedler, unpublished data). These short-term habitat conversions suggest how long-term patterns develop, but the concept of an orderly/predictable change in composition (i.e. succession) remains untested for these arid-region wetlands.

It would be simplistic to read current species occurrences × elevation as an exact reflection of the change in species composition during succession (Adam, 1990). Recruitment of species is often temporally and spatially variable, followed by vegetative expansion that creates a mosaic, rather than visible bands. Changes in environmental conditions and differences in the pool of available species may alter successional pathways for the current seaward edge relative to that of the present upper marsh. In some localities the pattern of marsh development is still poorly understood; for example, explanations of the zonation of mangrove and saltmarsh in temperate Australia remain controversial (Mitchell & Adam, 1989).

RATIONALE FOR RESTORATION

Functions and values

Saltmarshes are valued for a range of reasons: intrinsically for their support of biodiversity, aesthetically and economically as providing important resources, including waterfowl for hunting, forage for grazing, and plants that stabilise eroding shorelines. Habitat values are frequently recognised and the protection and restoration of intertidal marshes is often undertaken expressly for the purpose of conserving birds (Zedler, 1997) or sustaining coastal fisheries (Chamberlain & Barnhart, 1993; Minello et al., 1994; Simenstad & Thom, 1996; Streever, 1999a).

Productivity

Considerable importance has been attached to extremely high productivity levels found in saltmarshes, especially those dominated by Spartina alterniflora along the Atlantic coast of North America. Some studies have yielded productivity estimates that are among the highest known from natural systems. However, productivity figures are available from few marshes globally, and much discussion is based on extrapolation from Spartina alterniflora marshes, for which vascular plant productivity probably exceeds that of other marsh types.

High productivity derives in part from landscape position. Nutrients are collected from the entire watershed as water flows downslope. Additional nitrogen is fixed by both endophytic and soil cyanobacteria. The high nutrient status, combined with tidal action that alternately floods and drains the soil, supports high productivity of vascular plants. Live and dead plant matter is then available for export to adjacent estuarine and coastal waters.

Fish support

Saltmarshes contribute organic matter to the estuarine food web in two ways, by exporting detritus to coastal waters and by affording prey items to fishes that move on to the marsh surface to feed. The popular model depicts tidal transport of organic matter to adjacent shallow waters (Pomeroy & Wiegert, 1981). However, this 'outwelling hypothesis' is not applicable to all saltmarshes (Lefeuvre & Dame, 1994; Weinstein & Kreeger, 2001). Some estuaries export material from saltmarshes to coastal systems; others show a net import (Dame, 1994). However, the food-web-support function is well established for many fishes and invertebrates. Saltmarsh pans and creeks provide nursery habitat for juveniles of many species, including several of commercial importance and others that form part of the food chain of commercial species. Considerable research shows the importance of the marsh edge in supporting fish feeding (Minello et al., 1994; Zimmerman et al., 2001).

Bird use

Saltmarshes are important to migratory waders and waterfowl (ducks, swans and geese) and their predators, especially in the northern hemisphere. The intertidal mudflats provide food for waders, while the marsh offers cover and high-tide roosts for many species. A few species nest in saltmarshes, especially the high marsh, e.g. savannah sparrows (Passerculus sandwichensis) in the United States, shelduck (Tadorna tadorna) and redshank (Tringa totanus) in Europe. Because saltmarshes support waterbirds, and because people value recreational bird-watching, various governments support the conservation and restoration of coastal wetlands and endorse international agreements such as the Ramsar and Bonn conventions (Shine & de Klemm,

1999). Lost bird-support functions are difficult to replace, because preferred nesting and feeding requirements can be complex (Zedler, 1997).

Rare plants
Plant species with restricted ranges are also recognised as rare and of conservation concern. The majority of species in low and mid-marsh habitats are relatively widespread (Adam, 1990); towards the upper tidal limit, habitats and communities are more spatially heterogeneous, and more rare and local species tend to occur. The upper elevations are also the most susceptible to human encroachment and trampling.

Historical losses and kinds of damages

The history of human modification and destruction of saltmarshes dates to the earliest civilisations, which built major seaports on estuaries. The cities around them were developed by 'reclaiming' (land claiming) shallow water habitats. Throughout history, estuarine saltmarshes have been damaged by both large- and small-scale modifications. Many and diverse damages need to be repaired.

Structures restricting tidal flow
These include embankments (also known as levees or dykes) floodgates, culverts, weirs, causeways and bridges. In The Netherlands, whole estuarine systems have been dyked to allow desalinisation of lands for agricultural use. Tidal action is lost and salts are gradually diluted from the soil, permitting cropping and other agricultural use of former saltmarsh. Conditions in front of a new embankment are often conducive to new marsh formation outside the dyke. Over a long period of time the 'new' marsh develops to a stage where embankment is feasible and, eventually, a sequence of reclamation events can be recognised, with the most seaward being fringed by an actively growing marsh. Overall, the total area of intertidal saltmarsh declines. The recognition that such whole-scale habitat conversions have many negative impacts, including major economic losses to fisheries, has led to some efforts to renew tidal action and restore salt marshes (Smit *et al.*, 1997).

In arid regions, saltmarsh may be leveed to construct evaporation ponds for salt extraction, an industry with a very long history, now carried out on a very large scale. Salt production has been the cause of large-scale loss of saltmarsh in Australia and California (both San Francisco Bay and San Diego Bay). Saltmarshes at Laguna San Ignacio, in Baja California Sur, Mexico, have been targeted for replacement by salt-production ponds. This low-technology industry is argued as a way to make economic development feasible in remote areas. Aquaculture is also a threat to coastal saltmarshes, because shallow waters are readily converted to ponds for fish and shellfish (Streever, 1999b).

In New South Wales, a recent survey identified over 5300 tide-restricting structures, of which 1388 were considered modifiable to increase tidal flushing and rehabilitate tidal wetlands (Williams & Watford, 1997). In southern Australia, there is considerable interest in re-establishing tidal exchange to impounded estuarine wetlands (Streever, 1997), with a view to restoring habitat and function. Various schemes have been proposed and actively promoted by conservationists and the fishing industry, but few have been implemented. In Delaware Bay, the Public Service Electric and Gas Company is breaching dykes to restore 2500 ha of *Spartina alterniflora* marshes as part of their estuary enhancement and mitigation programme (Weinstein *et al.*, 1997).

Increasing the rate of deposition of fine sediments
Because estuaries collect materials from entire watersheds, they are subject to infilling over geological time. Humans accelerate the process by destabilising soils within estuarine watersheds. The Pacific coast of the United States has greatly accelerated rates of sedimentation due to inland soils that are prone to erode into estuaries when catastrophic flooding occurs. California's Mugu Lagoon, south of Santa Barbara, lost 40% of its low-tide volume during the floods of 1978 and 1980 (Onuf, 1987). Other sites have been deliberately filled with various materials, including trash and spoils dredged to maintain nearby navigation channels (Niering, 1997).

Pollution

Where urban saltmarshes remain, they are continually subject to degradation from runoff that is rich in nutrients or contaminated with heavy metals and/or toxic organic materials. Unfortunately, water-quality monitoring programmes are not a high priority for estuaries and saltmarshes, in part because of high cost and in part because estuarine waters do not provide drinking water. Among the more visible forms of pollution are oil slicks that occur near ports and marinas, following leakages of fuel and oil. Large oil spills can occur along both urban and rural coasts. Thick deposits of oil may persist for many years, inhibiting saltmarsh regeneration, as occurred following the *Metula* spill into the Strait of Magellan in 1974 (Baker *et al.*, 1994).

Salinity dilution

Freshwater can be a pollutant in saltmarshes when flows are augmented and soils are leached of salts. Especially in arid regions, the discharge of irrigation water or stormwater runoff introduces pulses of nutrients, litter, weed propagules, and other unwanted materials into saltmarshes. Glycophytes and brackish-marsh plants can invade the low-salinity soils and outcompete native plants. Reed (*Phragmites australis*) and reedmace or cattails (*Typha* spp.) are considered pests in many coastal areas, because these clonal species rapidly expand when excess freshwater flows into brackish or saline waters (Zedler *et al.*, 1990; Burdick *et al.*, 1997). Even annual grasses are a problem in high saltmarshes if they out compete native annuals, such as the endangered salt marsh bird's-beak (*Cordylanthus maritimus* ssp. *maritimus*) (Callaway *et al.*, 1997; Noe & Zedler, 2001).

Exotic species invasions

Given the adaptations required for species to thrive in the saltmarsh environment, the weed problems are much less extensive than in many terrestrial habitats. Nevertheless, invasive plant species are an increasing cause of concern in several parts of the world. *Spartina* species are, globally, the most serious problem and may fundamentally alter the composition and structure of invaded marshes (Spicher

& Josselyn, 1985; Callaway & Josselyn, 1992). In most cases, invasions have followed deliberate plantings for shoreline erosion control (in Europe) and stabilisation of dredge spoil islands (in Puget Sound, Washington). The hybridisation of the American *S. alterniflora* with the native *S. maritima* in southern England in the late nineteenth century produced the form *S. × townsendii*, which later underwent chromosome doubling to produce the fertile hybrid *S. anglica*. This species is now widespread in northern Europe, in China, southern Australia and New Zealand. In Washington and Oregon, *S. alterniflora* is now a major pest species, occurring over large areas of several estuaries, where it grows lower on the shore than native halophytes. Hence, it threatens the oyster industry and reduces mudflat habitat that is valued for foraging use by shorebirds. In Humboldt Bay, *S. densiflora* is the dominant plant. It was apparently brought in with ballast water by ships trading timber with Chile (Spicher & Josselyn, 1985). This caespitose species has a canopy structure that is very different from that of the native *S. foliosa*. In San Francisco Bay, both *S. alterniflora* and *S. densiflora* have invaded. The latter grows higher on the shore than the native species. *Spartina alterniflora* grows at both the lower and upper edges of the range of the native *S. foliosa*, and the two species have hybridised (Ayres *et al.*, 1999). In eastern Australia both the groundsel bush (*Baccharis halimifolia*) and the rush *Juncus acutus* are proving to be aggressive invaders. Ironically, in some areas where *J. acutus* is native, such as Britain and California, it is an uncommon component of saltmarshes, worthy of conservation attention (Ferren, 1985).

In contrast, the impact of invasive animal species has been substantial and widespread, owing to the global distribution of larvae and adult animals that are carried on the outsides of ships and in their ballast. Introducing invertebrates that can burrow into saltmarshes substantially alters geomorphology, and adding exotic predators that feed high in the food chain can completely change animal and plant populations through trophic cascades.

Grazing

Many of the world's coastal saltmarshes are grazed by domestic livestock, principally cattle, sheep and horses; grazing and its management may determine the structure and composition of the vegetation (Adam, 1990). Some species cannot survive grazing, either because the plants are particularly palatable or because they are damaged by trampling. Grasses become dominant in grazed saltmarshes (Zedler et al., 1995). On Kooragang Island, in New South Wales, Australia, the fencing of grazed saltmarshes allowed the rapid recovery of mangroves, where marsh vegetation had previously dominated (P. Nelson & J. Zedler, unpublished data). Thus, grazing might shift the balance between herbaceous and woody vegetation. Grazing management can influence the conservation value of a site. For example utilisation of a site by overwintering waterfowl may be enhanced by an appropriate level of summer grazing by stock. An issue in planning for restoration or recreation is whether the objectives will require the site to be grazed and, if so, at what stocking rate and what stage grazing could be introduced.

Physical disturbance

Saltmarsh surfaces are often disrupted by local engineering work. Pipelines are laid through marshes; borrow pits are dug to build embankments; recreational vehicles cut through the sod and soils are compacted. Human visitors are also of concern especially in small urban saltmarshes. In southern California, the plant species that occur adjacent to the upland are brittle succulents that are easily damaged by footsteps. Paths develop readily with only the decumbent and tough runners of a clonal grass, Monanthochloe littoralis, remaining. Spiny subshrubs, such as Lycium californicum, could be planted at the upper boundary of saltmarsh restoration projects to discourage public trespass (James & Zedler, 2000). Alternatively, boardwalks could be constructed to control access.

Upstream dams

Dams can benefit or threaten coastal saltmarshes that are fed by rivers. If excess sediments are filling in marshes, then the trapping of sediments by dams would benefit downstream wetlands. However, if sediments are needed to counteract sea level rise, a dam would negatively affect the process. River flows regulated by dams may change salinity regimes and reduce variability in estuaries. In Mozambique, the Kariba Dam, constructed in 1959, and the newer Cabora Bassa Dam have reduced flooding, decreased nutrient inflows, reduced sediment delivery, and increased coastal erosion, leading to lowered estuarine productivity, decreased fisheries and a loss of mangroves in the Zambezi River Delta (Beilfuss & Davies, 1999). In West Africa, loss of mangroves and their replacement by herbaceous halophytic vegetation has been reported from the Volta estuary as a consequence of upstream dams (Rubin et al., 1999).

When dams are regulated to reduce flood pulses, not only do the erosive forces change but salinity regimes are also modified. Saltmarshes that once experienced seasonal salinity dilution might shift toward a more uniform year-round salinity. Reduced flows can lead to desiccation of soils and invasion of woody plants, conditions that are less than ideal for wildlife and nesting waterbirds (e.g. wattled cranes [Bugeranus carunculatus]: Beilfuss & Davies, 1999). In this case, direct action is needed to restore floodplain vegetation and habitat value. Prescribed flooding is required to benefit downstream ecosystems (Beilfuss & Davies, 1999).

Fragmentation

Urban and agricultural encroachments have sequentially reduced the area and quality of tidal marshes, with habitat fragmentation and loss of connectivity between habitat units becoming a major issue. The degree of tidal influence is often reduced, allowing saltmarsh habitats to develop wider swings in environmental conditions. Sites can be come wetter or drier, saltier or fresher, depending on local circumstances. Given recognition of the historical and continuing loss of saltmarshes, and the widespread occurrence of degradation, there is considerable interest in rehabilitating, restoring and creating saltmarshes in many parts of the world (Thayer, 1992; Streever, 1999b).

Climate change and rising sea level

Climate change caused by the enhanced greenhouse effect is predicted to accelerate the rate of sea level rise. Over geological time, saltmarshes keep up with sea level rise by accreting sediments and migrating

inland. If the rate of sedimentation (and hence upward growth of the marsh surface) matches the rise
in sea level, the marsh will remain in its present
position. However, where sea level rise exceeds sedimentation, saltmarsh will survive only if it can
retreat landward. On many urbanised coasts, saltmarsh abuts embankments, sea walls or developed
land confining marshes to ever-shrinking strips between the city and the sea. Even where retreat is
possible, species that are poorly dispersed will probably be lost under increased rates of sea level rise if
habitat changes more rapidly than populations can
adjust.

Climate change may have other impacts on saltmarshes. Increased storminess may cause erosion or
sedimentation from stronger and/or more frequent
floods. Increased temperature may result in shifts
in species distributions, in subtropical areas possibly promoting mangrove spread at the expense of
saltmarsh.

CONSTRAINTS ON SALTMARSH RESTORATION

Urbanisation and fragmentation

Restoration often concerns urban saltmarshes,
which have borne the brunt of coastal development. Urban saltmarshes suffer reduced habitat
area, habitat fragmentation, impaired tidal flushing, altered freshwater inflows and reduced water
quality. Highly disturbed urban saltmarshes are
often targeted for restoration, especially to compensate (mitigate) for new developments. Many problems confront the manager who is charged with
restoring saltmarshes to achieve specific conservation targets. In urban areas, full restoration is not
really feasible, as neither the watershed nor the immediate surroundings of an urban estuary can be
restored. Rehabilitation and enhancement become
more feasible goals in such situations. The potential for restoring hydrology, reconnecting habitat
fragments, and restoring upland–wetland linkages
is often limited. Where buffers between the wetland and developments are inadequate, the connections with upland habitats are not restorable. Here,

we consider the environmental conditions that can
slow or prevent full restoration of saltmarshes.

Subsided sediments

All sites that have been dyked experience subsidence, and topography is not readily restored.
Opening dykes to tidal flow produces more open
water and deeper-water habitats than previously occurred on the site (Frenkel & Morlan, 1991). Dredge
spoils have been added to some dyked sites in San
Francisco Bay (e.g. the Sonoma Baylands project:
Marcus, 1994; US Geological Survey, 2001), raising
issues of appropriate particle size and potential
contaminants. Where dredge spoils are suitable for
restoring dyked, subsided wetlands, their use might
help solve two problems at once, as finding sites to
dispose of dredge spoils is difficult, and the cost
of transporting spoils to offshore dumping sites is
high. There may be problems with the high sand
content of dredge spoils.

Altered biogeochemistry

Drained saltmarshes undergo chemical changes
that are not readily reversible. Reflooding of dyked,
drained saltmarshes can lead to sulphide accumulation after the available iron is used up as an elec-
tron acceptor. Portnoy & Giblin (1997) collected sediment cores from two types of hydrologically disturbed saltmarshes on Cape Cod, Massachusetts. As
expected, dyked and drained marshes and seasonally flooded marshes responded differently to the
restoration of tidal flushing (Table 11.1). Their work
provides a caution that rewetting soils can cause
acidity (acid-sulphate soils). Acid may then drain
into adjacent estuaries, and the resultant chemical changes may kill fish and invertebrates. In such
cases, care would be needed to prevent acid release.

Impaired tidal flushing

Coastal saltmarshes are naturally inundated by
tides, but tidal flushing is commonly impaired by
roads, tide gates, culverts and other structures associated with development. Saltmarshes are not likely

Table 11.1. *Reintroducing seawater changes several components of drained marsh soil (sediment cores tested in the greenhouse)*

	Dyked and seasonally flooded	Dyked and drained
After dyking		
Organic content	65% dry mass	45%
Porewater pH	6.4	3.9
eH (mV)	300	500
21 months after seawater reintroduction		
Freshwater vegetation	Killed	Killed
Subsidence	6–8 cm	
pH	Initial decline, then ~6.5	Increase to ~6 at four months
Nutrients	Increased N and P mineralisation	Higher PO_4, NH_4, Fe(II)
Sulphides	Increased 10×	[precipitated by Fe(II)]
Sulphide toxicity	Likely	Unlikely
Alkalinity	Increased 10×	Increased
Revegetation potential	Limited due to sulphide toxicity	Less problem with sulphide toxicity

Source: Portnoy & Giblin (1997).

to be completely restorable without full tidal flushing. This may be difficult to achieve if the removal of structures is costly or would result in flooding of costly real estate. Often the agency in charge of such structures (highway departments, flood districts) is not the same as the one promoting saltmarsh conservation.

Even when tidal flushing is restorable, the natural links between tidal basins and marshes may not reform on their own. Tidal marshes are dissected by branched creek networks that form in the mud before vegetation develops. Plants then stabilise the sediment and allow steep banks to form. A non-tidal saltmarsh that is fully vegetated and then exposed to tidal action will already have stable sediments, and tides will not erode channels very quickly, if ever. Creeks are increasingly known to facilitate fish use of saltmarshes and to support a greater diversity of plant species (Zedler, 2001). Hence, the careful planning of restoring both tidal action and tidal creek networks is advised.

A large constraint on sediment excavation projects is the disposal of spoils. If overlying sediments are coarse enough, they can be fluidised and piped to a nearby shore, helping to sustain eroding beaches. In the United States, however, the Environmental Protection Agency will not permit the beach disposal of fine sediments. Hence, silt and clay spoils need to be trucked to a landfill. The high cost of trucking materials off site and the shortage of disposal sites add further constraints to excavation projects. Sediments that are contaminated with heavy metals and organic pollutants might not be welcome in landfills where materials could leach out and flow to adjacent lands. In between excavation and removal, additional space is needed for dewatering of wet spoils. Proposals to mitigate the loss of the Eve Street saltmarsh in Sydney, Australia through excavation of a nearby fill site were terminated when the degree of contamination of the fill was recognised.

Inadequate flood pulsing

Plans to reintroduce flooding to the Zambezi Delta offer possibilities for habitat restoration, as well as the challenge to identify the flooding regime that can best restore lost wetland values (Beilfuss

& Davies, 1999). The constraints on planning include: the difficulty of assessing historical flood
flows throughout the delta, given that records date
only to 1930; lack of knowledge of the minimum
flooding regime that will re-establish desired ecological conditions; the need to provide flood pulses
large enough to remove trees and other mature
vegetation in the delta; human settlements that
have moved into formerly flood-prone areas of the
delta; recent damaging floods that enhance the arguments for opposing planned floods; the public
perception that flooding generates mosquito habitat; and the design of the Kariba Dam, which has
sluice gates at the top of the dam wall, instead of
at the base, where greater flows could be released
(although cold water at the reservoir bottom might
be detrimental to river organisms).

Herbivores

Newly planted seedlings are attractive to many herbivores, including ground squirrels (*Spermophilus
beechyi*), rabbits (*Sylvilagus audoboni*) and coots (*Fulica
americana*) in southern California, nutria (*Myocaster
coypus*) in Louisiana, and geese along the United
States Pacific Northwest and Atlantic coasts. Herbivory 'eat-outs' are episodic and localised. While
the timing of migrating geese can be predicted,
the sites they select for feeding and the amount of
damage can be hard to predict. Fencing was employed at Tijuana Estuary to exclude rabbits and
coots after the first problems were noted (Zedler,
2001). However, not all projects are monitored often enough to identify problems in time to re-
act, and many lack response protocols to deal with
problems.

Exotic species

Restoring saltmarsh vegetation in hydrologically
modified sites requires knowledge of the salt tolerances of all life-history stages of both the native
and exotic species of the ecosystem in question. At
San Diego Bay, for example, exotic annual grasses
and an endangered annual plant compete for limited habitat in early spring, when both establish

seedlings after winter rainfall lowers soil salinity.
While salt additions could control the annual grass
(Kuhn & Zedler, 1997), an increase in salinity at the
wrong time would prevent the endangered plant
from germinating. Controlling soil salinity with salt
additions is especially challenging where rainfall is
unpredictable.

Insufficient topographic complexity

Saltmarshes are sites of considerable microtopographic diversity (Zedler *et al.*, 1999; Vivian-Smith,
2001) (Fig. 11.1). Few created marshes have the complexity of creek and pan systems of natural marshes.
Planting and rapid vegetative spread no doubt slow
erosion of creeks. Greater effort should be given to
designing and constructing creek systems, as the desired complexity might not develop once marshes
are fully vegetated.

THE IMPETUS FOR SALTMARSH RESTORATION

Restoration can be initiated by individual citizens,
undertaken by governmental agencies to enhance
wetland resources, or mandated by mitigation
agreements. In the United States, a permit is required to fill wetland; if filling cannot be avoided
or minimised, then the damages must be compensated before a permit to discharge fill material into
a wetland can be granted. The intent is to avoid a net
loss in wetland area and function (Streever, 1999*a*).

Private landowner efforts

In some cases, individual landowners have undertaken saltmarsh restoration projects. Funding is
usually a constraint in such operations, and the
scope of the project may be limited. However, there
are notable exceptions. In South Australia, the saltmarshes of the Murray River mouth disappeared following dyking that converted a 75 000-ha estuary
into a freshwater system (Denver, 1999). Wetlands
were lost to agriculture and grazing; wooded areas
were logged; and monotypic vegetation (*Typha domingensis*) invaded large tracts of former saline

wetland. Owners of a 1200-ha property known as Wyndgate are voluntarily excavating stream channels that connect the Murray River to the sea in order to encourage tidal flows and reduce *T. domingensis* infestations. Considerable dedication is needed when the costs include culverts under roads, maintenance dredging of channels, and problems of vandalism, equipment theft, and bogged machinery.

Governmental efforts

Many agencies within federal, state and local governments have found reasons to support saltmarsh restoration efforts. In the United States, the National Marine Fisheries Service recognises the value of saltmarshes for commercial fishes, and promotes habitat restoration (Thayer, 1992). The US Fish and Wildlife Service values wetlands for their support of endangered species; this agency and its state-level cooperators recently spent US$3.1 million to excavate accumulated sediments and restore just 8 ha of Tijuana Estuary (Entrix *et al.*, 1991).

Mitigation

In the United States, proposals to fill wetlands are subject to the provisions of Section 404 of the Clean Water Act (Box 11.1). The US Endangered Species Act also calls for mitigation of impacts to species on the federal list of endangered species (species whose populations are so depleted that they are in danger of extinction). A common concern with mitigation projects is that one type of valuable wetland is sometimes modified to create another. For example, port developers who damaged deep-water fish habitat were allowed to dredge a shallow-water lagoon and its fringing saltmarsh to replace lost fish habitat at Batiquitos Lagoon in San Diego County. In Florida, compensation for damages to saltmarsh often comes at the expense of uplands, which are highly valued for their support of rare plants. In the UK, saltmarshes that have been reclaimed and become freshwater systems have conservation value in their own right by providing habitat for plants, birds, frogs and other amphibians. Thus, enthusiasm for on-site, in-kind miti-

gation needs to be tempered by broader environmental perspectives. Proactive planning to protect a region's saltmarsh biodiversity and functions might include a portfolio of land purchase, preservation, restoration, and creation alternatives, which are prioritised in advance of funds being made available through the mitigation process (Zedler, 1996).

OBJECTIVES OF SALTMARSH RESTORATION

Restoration takes many forms and is conducted in many different contexts. Several examples illustrate the range of efforts that have taken place historically or are under way today (Table 11.2). In northern Europe the so-called 'farmers' method' of promoting sediment accretion and marsh expansion has been undertaken from at least the early eighteenth century (Dijkema *et al.*, 1990; Esselink *et al.*, 1998). Although more accurately termed creation, the management emphasis for many reclaimed marshes is now changing from productive agriculture to nature conservation and thus the ideals and practice of such projects may be similar to restoration efforts. Esselink *et al.* (1998) have demonstrated that terminating the maintenance of the drainage system allows levees to form, and some areas between ditches become poorly drained depressions. This increased topographic diversity is accompanied by increased biological diversity, which improves the nature conservation value of the marsh.

As in all wetland restoration efforts, the most important step is restoring natural hydrologic regimes. In addition to restoring water sources and hydroperiods, it is necessary to provide the appropriate soil, vegetation, and animals. Most saltmarsh restoration projects emphasise the manipulation of hydrology, often with vegetation planted, but soils and animals expected to develop unaided. It is likely that all four components will need attention. Although saltmarsh restoration can be very expensive, maintenance of sites is not always guaranteed. The Eve Street saltmarsh in Sydney was extensively rehabilitated (Stricker, 1995), but a major motorway was subsequently constructed over it.

Box 11.1   Mitigation for discharging fill to saltmarshes under the US Clean Water Act

Projects that involve filling of saltmarshes can be given permits by the US Army Corps of Engineers, but a sequence of alternatives must first be explored in a process termed 'mitigation'. Attempts are first made to (1) avoid or (2) minimise damages by redesigning the project. If damages cannot be avoided or reduced, then (3) unavoidable impacts must be compensated. The nature of the required compensation is indicated in the permit record. In most cases, degraded habitat must be restored or upland habitat converted to saltmarsh as compensation for destruction of natural saltmarsh habitat.

Initially, preference was given to on-site, in-kind habitat and to restoration rather than creation of wetland. Thus, when urban saltmarshes were to be filled, nearby degraded saltmarshes were the preferred site for restoration. In areas of rapid development, however, the demand for mitigation sites often exceeds the availability of former wetlands that need to be restored. When no degraded marshes are available, disturbed upland may be converted to wetland or the mitigation may be accomplished off site. Mitigation 'banks' were first developed in 1991 in Illinois by entrepreneurs, who began building wetlands and selling credits. Federal guidelines for mitigation banks were issued in 1995. More recently, a new practice has developed, namely charging a fee in lieu of identifying a specific project. The fee then goes to a responsible party, which uses discretion in preserving, restoring or creating wetland in exchange for permits to fill existing wetland. Federal guidelines for in-lieu fees have just been released (US Department of Defence, 2000). Unfortunately, many projects were never assessed post-construction so that neither the degree of ecological replacement nor the extent of compliance with permit conditions can be determined (Ambrose, 2000). Mitigation practices and outcomes of mitigation projects were recently reviewed by the National Research Council, under request from the US Environmental Protection Agency (National Research Council, 2001). Other restoration efforts (not mitigation) are aimed at expanding saltmarshes and/or enhancing their functioning. Usually, a specific set of objectives must be accomplished by the restored or created wetland, and these are set forth in the mitigation agreement, which also spells out the monitoring period and reporting requirements.

The Clean Water Act has an important loophole, namely that projects for the common good can proceed, even if they affect large areas of highly valued wetlands. In coastal urban areas, many projects have been deemed 'for the common good' including development of sewage-pumping stations, expansion of port facilities, construction of marinas, widening of highways, and construction of flood-control channels. Thus, compensatory mitigation projects are common in urban areas. The process can lead to much-needed restoration, rehabilitation and/or enhancement of degraded wetlands in or near the damage sites, but one project in southern California is 150 km from the damage site. In such cases, it is less clear that damages to specific wetlands are being compensated.

Some of the largest and most costly restoration programs are mitigation for damages to coastal fish habitat. Urban projects are often very costly. The restoration of Bolsa Chica Wetland near Long Beach, California (356 ha, purchased by the State of California in 1997 for US$25 million), will cost US$60 million and is intended to mitigate damages caused by the expansion of port facilities nearby. The high cost of the project derives from plans to excavate a new tidal inlet, add a highway bridge over the inlet, and install tide gates to control water levels in 'muted' tidal lagoons.

Restoring tidal flow to impounded saltmarshes

Dykes and levees that separate saltmarshes from tidal influence can be breached to return tidal flows and restore saltmarshes (Box 11.2). This may be the most straightforward type of saltmarsh restoration, because any increase in seawater influence will be likely to improve salinity regimes, provide clearer water, reduce the abundance of invasive plants, and restore marine animals, especially those with free-swimming larvae.

Many former saltmarshes are protected from the sea by embankments. The costs of providing

Table 11.2. *Examples of saltmarsh restoration*

Restoration challenge	Action(s) taken	Constraints	Reference
Reclaiming mudflats	Drainage, add groynes	Exotic plant invasions	Dijkema *et al.* (1990)
Impounded saltmarsh	Breach dyke	Subsided land	Frenkel & Morlan (1991)
			Weinstein *et al.* (1997)
		Loss of sediment	Flynn *et al.* (1999)
Subsidence behind dyke	Add dredge spoil	Coarse substrate	Marcus (1994)
		Contaminants?	
Drained saltmarsh	Backfill ditches	Acidified soil	White *et al.* (1997)
		Insufficient fill	
Sedimented wetland	Excavate sediments	Future sedimentation	Simenstad & Thom (1996)
and dredge spoil deposits	Excavate spoils	Coarse substrate	Langis *et al.* (1991)
		Erosion, accretion	Haltiner *et al.* (1997)
		Contaminants	
	Plant with *Spartina*	Erosion	LaSalle *et al.* (1991)
Shoreline erosion control	Plant with *Spartina*		Craft *et al.* (1999)
Freshwater flows reduced	Planned floods	Risks downstream	Beilfuss & Davies (1999)
by dams or increased	Enhance tidal action		Denver (1999)
by discharges			
Contamination, e.g. oil spills	Remove oiled soil	Elevation drop	Baker *et al.* (1994)
and acidified soils			
Exotic plant invasion	Salt addition	Risk to endangered plants	Kuhn & Zedler (1997)
			Callaway *et al.* (1997)
	Eradication efforts	Large populations	

drainage for the reclaimed land and of maintaining the embankments may be high (particularly when sea level is rising relative to the land) and may outweigh the value of the agricultural productivity of the reclaimed land. In these circumstances it may be more economic to accommodate the rising sea level by a process of managed retreat, in which embankments are breached to allow re-establishment of saltmarsh (Packham & Willis, 1997). The nature conservation benefits of such an approach are potentially large.

Historically, some sea defences have failed, and saltmarsh has re-established without assistance (Gray & Adam, 1974; Packham & Willis, 1997). However, managed retreat may not always be a simple matter of removing a sea wall and hoping for the regrowth of saltmarsh vegetation. Following reclamation, the soil of the former marsh may settle and shrink (about half a meter for Connecticut marshes

cut off from tidal flow for 25–50 years: Niering, 1997). Where the land surface has subsided, the breaching of dykes will inundate the former saltmarsh, and the result will be open water or mudflat. Sediment may need to be added to provide saltmarsh habitat. The breach needs to be carefully planned to avoid excessive scour and erosion. Weirs and sluices may be required to create an appropriate tidal regime and flow velocities (Packham & Willis, 1997). Former agricultural land may have a high nutrient status which can promote proliferation of green algae (e.g. *Enteromorpha*, *Ulva*) in developing saltmarsh.

In the United States, various structures have been used to protect or create coastal marshes (Sanzone & McElroy, 1998). In some cases, structures such as levees, tide gates and culverts have been installed to enhance waterfowl habitat and/or create nursery habitat for fish and invertebrates.

Box 11.2   Restoring tidal flushing

In Connecticut's Hammock River valley, a 100-ha marsh was directly connected to Long Island Sound until 1913, when a causeway dissected the marsh and reduced tidal flows. Later, tidal flows were further restricted by installation of four tide gates in 1947. The 32-ha area above the gates shifted to brackish marsh and dominance by *Phragmites australis*. In 1985, the opening of one tide gate reduced the growth of *P. australis* and allowed *Spartina alterniflora* to regain dominance and saltmarsh fauna to return (Niering, 1997).

In Oregon, the breaching of a dike on the Salmon River estuary rapidly returned tidal hydrology to the pasture, where salt-intolerant vegetation died and native saltmarsh plants recovered rapidly (Frenkel & Morlan, 1991). However, the locations of low- to high-elevation marsh habitats moved inland, due to subsidence of soils during the non-tidal period of agricultural land use. Wherever dykes have been constructed, soils are likely to be more aerated and decomposition rates elevated. Organic matter is likely to decompose and clay soils are likely to compress, acting together to lower surface elevations. Breaching dykes will thus restore less saltmarsh and more shallow water than was historically present.

In Connecticut's Wequetequock–Pawcatuck marshes, causeways were constructed to create waterfowl impoundments in the 1940s. Tidal flows were restricted by small-width culverts, and saltmarsh vegetation converted to *Typha angustifolia* and *P. australis*. The installation of large culverts in the 1970s substantially decreased *T. angustifolia* cover and allowed *Spartina alternifolia* to recover (Niering, 1997).

In all three cases, restoration efforts took place in areas adjacent to natural marshes. The increase in tidal influence and the removal of barriers to seed dispersal was sufficient to allow native plants and animals to colonise. In both cases, saltmarsh recovery was feasible where the topography had not substantially subsided (<0.5 m); it simply required patience.

In the exploitation of underlying oil deposits (e.g. in Louisiana), canals and levees have been constructed across saltmarshes. Various attempts have been made to restore hydrologic conditions to leveed saltmarshes, including the use of tide gates (Turner & Lewis, 1997). While such 'structural marsh management' may be of value in specific situations, it is not viewed as a universal solution to the problems faced by Louisiana's coastal wetlands (Flynn *et al.*, 1999). A national committee recently reviewed the benefits and problems of managing coastal marshes using structural marsh management. The consensus was that self-sustaining wetlands that provide a suite of functions should not be subject to structural marsh management. The use of structural marsh management should: (1) reflect scientific understanding of marsh degradation; (2) give preference to strategies that restore natural processes; and (3) be considered experimental and undertaken within appropriate experimental designs and assessment of outcomes (Sanzone & McElroy, 1998).

Removing accumulated sediments and spoils

Restoration of saltmarshes and tidal creeks with accumulated sediments involves recontouring by dredging and bulldozing (Box 11.3). Historical maps and photos are useful in locating former marsh and creek habitats, but soil cores are needed to determine how deeply the marsh soils are buried. If the weight of the overlying sediments has compressed the underlying marsh soil, a choice might need to be made between excavating to the natural marsh soil and excavating to the desired marsh elevation. Sediment texture is critical to both invertebrates and marsh vegetation, so the choice might well depend on how fine the texture of overlying sediments is. If primarily clay and silt, it may be suitable for the restored marsh. If sandy, it might not promote satisfactory plant growth (Box 11.4).

Shoreline erosion control

Spartina alterniflora has proven effective in stabilising shorelines in hundreds of sites along the United States Atlantic coast, where it is native (Niering,

Box 11.3   Excavation of flood-borne sediments to restore saltmarsh

Tijuana Estuary, in southwestern California, occurs downstream of a large, arid-region watershed that releases substantial sediments during heavy rainfall events. Over the past century, former saltmarsh areas in the path of sediment plumes have accreted >2 m of sand and silt. At this site, restoration of saltmarsh involves excavation and off-site disposal of sediments (Entrix *et al.*, 1991). The work is done with an experimental approach so that each restoration module provides information on how to restore the next module.

The first restoration site was designed to explore the need for planting marsh-plain species. The site supported 87 2 × 2-m plots, which were planted with 0, 1, 3 or 6 species (species randomly selected from 8 that are native to the marsh plain). Each plot was planted with 90 greenhouse-grown seedlings in a grid with 20-cm spacing. Thereafter, we counted seedlings of all species in all plots. Only 3 of the 8 planted species readily recruited seedlings; these were *Salicornia virginica* (long-lived perennial), *S. bigelovii* (annual) and *Suaeda esteroa* (short-lived perennial). The remaining 5 species recruited only rarely, but the species persisted where planted (Zedler, 2001). With no planting, a single species (*Salicornia virginica*) would probably have dominated the site. We also learned that the number of species in an assemblage affects canopy layering but not height or cover, while the species composition affects height but not layering or cover (Keer & Zedler, in press). More complex canopies are predicted to support more species of insects and to function better as habitat for marsh birds. Vegetation cover achieved 100% within two years, although height and layering still lagged behind the natural marsh after four years (Keer & Zedler, in press). The experimental results are still being analysed to determine effects on productivity and nitrogen accumulation (J. C. Callaway *et al.*, unpublished data).

The second, restoration module (8 ha) was excavated in February 2000 as part of a long-term adaptive management programme that would ultimately remove sediments from 200 ha. The original marsh surface was uncovered, as evidenced by the unearthing of Native American shell middens and patches of charcoal. (The project was interrupted for extensive archaeological study, which determined that the remains were not of major significance and the project could resume.)

Because earlier excavations of sediment in the region had not been planned to restore topographic heterogeneity, and because tidal creeks and channels were hypothesised to be important to marsh plants and animals, the 8-ha excavation was designed as an experiment to test the importance of excavating tidal creek networks instead of leaving a smooth marsh plain (Fig. B11.1). The site was divided into six equal-area 'cells', and three of the cells were left smooth, while a branched creek system was cut into

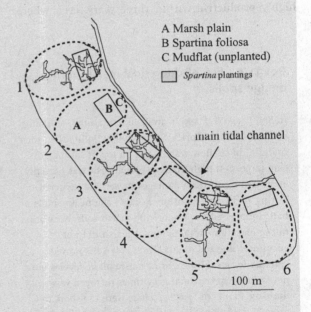

Figure B11.1. The 8-ha excavation at Tijuana Estuary, designed as an experiment to test the effects of tidal creek networks on ecosystem structure and functioning. Shown are the six experimental 'cells' with and without tidal creeks, the three habitats within each cell (mid-marsh plain, low marsh or *Spartina* zone, and mudflat) and plots with transplanted *Spartina foliosa* (cf. Zedler, 2001).

the remaining three. Most of the area was left to develop vegetation on its own; however, *Spartina foliosa* (with and without an organic [kelp-based] soil amendment) and six species of marsh-plain halophytes were planted in experimental arrays to test the importance of plant spacing (horizontal heterogeneity) along with the importance of creeks (vertical heterogeneity).

1997), as well as in Europe and China (Chung, 1994). Transplantation procedures were developed in North Carolina, where appropriate spacing for planting was experimentally established in the mid-1970s (Broome *et al.*, 1986). Hybridisation of introduced *S. alternifolia* and native *S. foliosa* has been recorded in the United States. Where hybridisation is actively occurring, the genetic integrity of any putative native stock should be ascertained before it is utilised in any restoration project (Ayres *et al.*, 1999).

North Carolina *Spartina* marshes can become highly productive within three years, even where sediments are coarse and nutrients are in short supply. Some are among the oldest saltmarsh restoration sites with long-term documentation; the most recent comparison was made at years 22–26 (Craft *et al.*, 1999). Where the constructed marshes receive nutrients and sediments from freshwater inflows, plant biomass is similar to that of reference sites. The benthic infauna becomes similar in composition and abundance to that at reference sites within 15–25 years; however, soil attributes are slow to match those of reference sites, and equivalency has not yet been attained (Craft *et at.*, 1999).

Box 11.4   Establishing tidal marsh from dredge spoils

In San Diego Bay, spoils dredged to maintain shipping channels were historically disposed by piling the sandy material in shallow waters and on top of saltmarshes. One large, 5-m high dredge spoil deposit was partially excavated to mitigate damages to wetland incurred during highway expansion. Because the heavy spoils had compressed the underlying marsh soils, excavation to the elevation suitable for saltmarsh did not encounter fine sediments. The sandy soils proved suitable for establishment of native plant species but not suitable for optimal growth, as nitrogen was limiting. Plants of *Spartina foliosa* were too short to provide nesting habitat for an endangered bird, the light-footed clapper rail (*Rallus longirostris levipes*); hence, conditions of the mitigation permit were not met (Zedler, 1993, 1997). Long-term research at the site allowed projections of soil conditions (organic matter and total Kjeldahl nitrogen) and plant heights into the future, and it was predicted that the site would never achieve the desired conditions (Zedler & Callaway, 1999). Short plants failed to support populations of the predatory beetle *Coleomegilla fuscilabris*, which is an important predator of scale insects (*Heliaspis spartina*). Scale-insect populations reached outbreak proportions in the planted marshes, such that short plants either senesced early or died (Boyer & Zedler, 1996).

Nitrogen addition was explored as a possible solution to the problem, but single soil amendments did not increase plant height enough to provide nesting habitat (Gibson *et al.*, 1994). Bi-weekly nitrogen amendments increased plant heights sufficiently but had to be repeated annually to sustain tall vegetation (Boyer & Zedler, 1998). Continual addition of nitrogen favoured a native annual halophyte (*Salicornia bigelovii*), which outcompeted *Spartina* for nitrogen (Boyer & Zedler, 1999); hence, the desired canopy type was not restorable under these field experimental conditions. Natural nitrogen supply regimes were not understood sufficiently in this restoration setting. Even if they were, the cost of management might be prohibitive. A better solution is to salvage and reuse fine sediments in marsh restoration.

Regulating freshwater inflows

The deliberate increase or decrease of freshwater flows to estuarine marshes can shift vegetation toward more or less salt-tolerant species. Where upstream dams are regulated for floodwater storage, managers will sometimes discharge large volumes of reservoir water over several months, decreasing soil salinities downstream and encouraging the growth of *Typha* species. Such was the case for the San Diego River in 1980, when discharges continued through March and April. The coastal salt marsh, dominated by *Salicornia virginica*, was drowned; *Typha domingensis* established and became dominant by June 1980, flourishing again with the flood of 1983 (Beare & Zedler, 1987). Restoration was passive; after several years without flooding or drawdown, the *Typha* plants gradually died back and saltmarsh vegetation re-established without assistance.

Treating contaminants and acidified soils

Restoration of polluted marsh soils is difficult, because both vegetation and soil may need to be removed, and damages may exceed benefits (see Hawkins *et al.*, this volume). Marshes in Brittany were heavily oiled following the *Amoco Cadiz* spill in 1978. Some of the marshes were cleaned by removing as much as 50 cm of sediment, as tidal creeks were straightened and widened at the same time. The lowering of the marsh to below Mean-High-Water Neap Tide prevented halophyte recolonisation. In contrast, the uncleaned sites recovered on their own (Baker *et al.*, 1994).

Multiple oil spills in New York Harbor have degraded stands of *Spartina alterniflora* and their associated fauna. Subsequent plantings of 200 000 seedlings of the same species (with 30–45 cm spacing and 30 g of slow-release fertiliser per plant) restored 2.4 ha of marsh (Niering, 1997), allowing invertebrates, fish and birds to return. It is plausible that the plantings accelerated the breakdown of oil, as microbial decomposers would benefit from both the fertilisers (a nitrogen source) and the soil aeration afforded by plants (oxygen leakage from aerenchyma in roots and rhizomes).

Some soils can become toxic when saltmarshes are drained, because the reduced forms of sulphur convert to sulphuric acid. In eastern Australia, White *et al.* (1997) suggested that reintroduction of tidal waters offers high potential for alleviating acidic discharges, as the twice-daily wetting enhances neutralisation, while algae in the tidal water provide organic matter to feed the microbes that reduce sulphate.

Controlling invasive plants

By far the most important invasive plants in saltmarshes have been species of *Spartina* (see above). Attempts to eliminate *Spartina* spp. have involved hand pulling, herbicides, and smothering by black plastic. None of these approaches is feasible for large patches, however. The best current approach is to halt invasions when the first clones appear. Potential biological controls are under active investigation.

In Perth, Western Australia, *Typha orientalis* from eastern Australia is invading saltmarshes that are naturally dominated by the native *Juncus krausii* (Zedler *et al.*, 1990). In this case, prevention of invasion is preferable to restoration of invaded marshes. The cutting and flooding of *Typha* stems is effective in freshwater marshes, but in tidal systems, there is no easy way to impound water to smother the rhizomes.

The deliberate planting of New Zealand stock of the mangrove *Avicennia marina* was undertaken at Mission Bay in San Diego, California, where it naturalised and expanded throughout the lower elevations of a remnant saltmarsh. There it altered the tall-grass nesting habitat of the light-footed clapper rail (see Box 11.4). Managers concerned about roosting by raptors, which prey on clapper rail chicks, organised an eradication programme, which involved cutting all stems and pulling seedlings for several years, until no further individuals were found. The escape of a mangrove at 32° N latitude was unexpected, as the nearest native mangroves occur at about 20° N in Baja California.

Offsetting sea level rise

The effects of a global sea level rise will not be felt equally by all saltmarshes. Tectonically rising coastlines and/or high rates of sedimentation could sustain saltmarshes, or even allow their expansion, while subsiding coastlines and saltmarshes with sparse sediment supplies will exacerbate the effects of sea level rise. In China, Chung (1989) long advocated the planting of *Spartina anglica*, and later *S. alterniflora*, to stabilise shorelines and keep up with sea level rise, as well as to trap enough sediment to increase land area. Where saltmarshes can migrate inland as sea level rises, wide buffers are needed to protect the full elevation range of anticipated sea level rise. Otherwise, future development could irrevocably eliminate the hinterland of saltmarshes. Managed retreat of coastal defences and saltmarsh re-establishment are feasible, as evidenced by historical examples where embankments have been breached by storms or deliberately damaged during wars (Allen, 2000). Managed retreat has been adopted in the United Kingdom as part of the government's strategic response to future sea level rise. A large-scale experimental study involving breaching of sea defences and re-establishment of saltmarsh is currently under way at Tollesbury, Essex (Packham & Willis, 1997).

RESTORATION IN PRACTICE

The status of the science of saltmarsh restoration

The role of restoration scientists is to understand how ecosystems develop under alternative management practices that seek to restore hydrology, soils, vegetation and fauna. From a thorough understanding, scientists hope to prescribe actions needed to achieve predictable outcomes for the many combinations of site type and degradation history. The ability to predict an outcome begins with an assessment of how much of the historical system might be restorable: Can the historical hydrology be returned? Is the historical soil type (especially particle size) still present or can similar soils be imported? Is the topography suitable for saltmarsh vegetation, or can it be suitably modified? Are animals likely to colonise the site and find the resources they need? If there are many constraints on restorability, then the goal should shift toward alternative reference systems, aiming to rehabilitate the site, rather than return it to its former condition. The overall goal can still be to restore some of the biodiversity and functioning of regional saltmarshes, but there is little reason to assume that any prior condition can be reactivated at any particular site. Watersheds that dictate inflowing water, sediments and nutrients might have been permanently altered; saltmarsh area might be permanently diminished and habitat distribution might be permanently fragmented. Science can help make these determinations, but the status of restoration science needs to be strengthened (Hobbs & Norton, 1996; Dobson *et al.*, 1997; Cole, 1999; Chapman & Underwood, 2000). Support for scientific studies of restoration efforts is growing. The National Oceanic and Atmospheric Administration (NOAA) has funded research in coastal restoration sites (Zedler, 1996, 2001) and a conference to review 'Concepts and controversies in tidal marsh ecology', including restoration (Weinstein & Kreeger, 2001). NOAA's National Marine Fisheries Coastal Service has long supported research in seagrass restoration (Thayer, 1992; see Fonseca *et al.*, this volume). There is also a long history of habitat restoration and creation of *S. alterniflora* marshes (Craft *et al.*, 1999), but techniques developed for this community do not necessarily translate to other habitats or regions. Work is needed on several aspects of saltmarsh restoration so that outcomes can be better predicted.

Useful conceptual models

A simple model (Bradshaw, 1987) suggests that restored sites will develop functional equivalency with reference ecosystems and that function (a process) develops linearly with structure (a condition) in restoration sites. A prediction from this model is that a saltmarsh with high biomass (a structural attribute) would have high primary productivity, support more animals, improve water quality, and reduce shoreline erosion (all functional attributes). Of these functions, productivity has received considerable attention in saltmarsh studies, and biomass has been evaluated in a few constructed saltmarshes (Zedler & Callaway, 2000). At San Diego Bay, end-of-season standing biomass was lower at constructed marshes than reference sites (Boyer & Zedler, 1998), but insect densities were much higher (Boyer & Zedler, 1996); hence, we could not predict productivity from biomass. The temptation to infer function from structure should be resisted in comparing restored and natural wetlands. There is room for improvement in conceptual models of restoration, i.e. what we expect of restoration efforts (Hobbs & Norton, 1996). We might, for example, consider a range of alternative stable states.

Utility of succession theory

Terrestrial ecosystems undergo successional development, with pioneer species establishing in open areas and longer-lived, shade-tolerant species invading later. Restoration efforts often seek to accelerate the process by planting shrubs or trees, thereby skipping the pioneer phase. Efforts to create saltmarshes, however, typically assume that target plant associations can be established on bare substrate. It is not clear if a low-elevation pioneer zone needs to be created before a higher-elevation assemblage can establish. An experimental planting of marsh-plain halophytes at Tijuana Estuary (Box 11.3) did not require prior establishment of the low-elevation *Spartina foliosa* vegetation. This experiment also showed that species-rich plantings develop attributes (often considered valuable) that are not present in monotypes.

Letting a marsh develop without intervention is a low-cost option, but specific elevations suitable for particular species of halophytes to establish are required and establishment is unlikely if erosive forces (from waves and/or currents) are strong. It might be necessary to build up the surface level or to overexcavate a marsh plain so that it can accrete fine sediments. Some form of protection from wave action or currents might be needed until vegetation can establish. Nurse plants might facilitate seedling establishment on a bare marsh plain. Without planting, a creek and salt pan system might develop before vegetation can stabilise the sediments. If vegetation is not planted, a salt crust

is more likely to form, especially in arid-region saltmarshes. Where surface soils are more saline than local species can tolerate, vegetation will be very difficult to establish (Zedler, 2001). Over time, assemblages of species might undergo succession as environmental conditions change. The sequences that would occur need to be understood so that plantings can be designed to accelerate the process. Exotic species might take hold where plantings do not cover the soil, but non-native plants (other than *Spartina* spp.) are generally restricted to the upper saltmarsh, which is more likely to have lower soil salinity. In the early years and decades, biodiversity might be limited without deliberate introductions. The utility of the succession model is not yet clear for saltmarsh restoration sites.

Assumptions about faunal colonisation

Although the emphasis of saltmarsh restoration and creation projects is usually on establishing plant communities, the major goal is often to support animals, such as birds, fish, and rare and endangered species (Pavlik, 1996). It is generally assumed that restoring the floristic framework will allow other components to colonise and 'natural' ecosystem functioning to develop. These assumptions are rarely tested, but, in the case of the light-footed clapper rail (Box 11.4), replanting the preferred vegetation type was not sufficient to support the endangered bird's nesting activities. Where specific animal populations are the restoration target, we recommend species-based research, including experimental introductions, to help achieve goals.

Research should be incorporated into restoration sites

More experimentation at a range of spatial and temporal scales is needed in order for scientists to predict outcomes of restoration projects. The ultimate merger between science and practice comes from designing entire restoration sites as ecosystem experiments. The recent 8-ha excavation of a marsh plain at Tijuana Estuary provides an excellent model (Entrix *et al.*, 1991) (Box 11.3). Managers did not know if they should spend extra money to have tidal creek networks added to this and future restoration modules; thus, this early module was designed to support the science necessary to answer the question. The site was excavated to have three replicate areas with tidal creek networks and three without. Researchers are exploring many attributes that are hypothesised to depend on the creek system.

Planting

Design details

Even though the driving force for saltmarsh restoration or creation is often habitat for wildlife (either in general or a particular rare species), the restoration activities concentrate on the plant community. We are not aware of any cases where saltmarsh animals have been deliberately translocated, although a range of invertebrates may be inadvertently introduced with planting material. Even where the plant community has been designed in detail, the assembly of the whole ecosystem has a large element of 'self-organisation' and few studies have investigated the progress of whole-ecosystem development in terms of structure, composition and function (see below).

W. J. Mitsch (personal communication) promotes the 'self-design' approach, wherein plant propagules may or may not be introduced but where no specific outcome is anticipated – the site is expected to sort out the species that can tolerate conditions. Some restorationists invest heavily in producing and implementing detailed plans for planting species in specific microhabitats, expecting the site to match the plan (Middleton, 1999). The question is not *whether* to plant but *which species* need to be planted and which will recruit on their own. Given sufficient time, the natural diversity of species may well colonise if the saltmarshes of the region are in good condition, if there are few barriers to dispersal, and if the sediment dynamics are suitable for saltmarsh development. In most instances, however, the initial vegetation and fauna will be low in diversity, with dominance by one or a few aggressive colonisers (Zedler, 2001). Additional questions concern how species that need to be introduced should be planted. Much of the restoration

work has focused on single-species plantings of *Spartina* spp. from vegetative material, rather than seeds. Obviously, there are many alternatives for restoring more species-rich wetlands; e.g. succulent-dominated wetlands in southern California have been planted with seeds, seedlings and/or rhizomes (Zedler, 2001). Region-specific information would assist in deciding which plants and animals should be introduced to achieve species-rich saltmarshes and fully functional ecosystems. Then, long-term observations of saltmarsh development would help to determine how long it takes for an unplanted site to achieve natural-marsh levels of species richness, plant canopy architecture (vertical structure), and processes, such as productivity and soil nitrogen and carbon accumulation rates.

We recommend experimentation with planted and unplanted plots (see Box 11.3 for a suitable model). Such experiments are needed for each habitat type and each biogeographic region. We recommend not planting colonisers that are readily available in a nearby saltmarsh, but we do recommend planting species that are unlikely to establish naturally. Planting with appropriate native species may be necessary to preempt invasion by aggressive exotics, such as *Spartina anglica* in northern Europe.

A major reason to favour planting over natural colonisation is to shorten the time-scale for marsh development. Planting holds the promise of a quick fix, whereas development of species-rich marshes may take decades if unaided. Where establishment of new marsh is a condition of a development approval, consent authorities need to see a new marsh early in the life of the project rather than to await the uncertain outcome of an uncontrolled process. Requiring a replacement marsh may also salve the bureaucratic conscience that accompanies the approval to destroy an existing saltmarsh.

Seeds versus plants

The range of options for establishing plants has been discussed by Guerrant (1996). Annual saltmarsh species flower and set seed every year, but seed set for perennial species may fluctuate annually. Collected seed can either be sown directly on to appropriate substrate or germinated under glasshouse or nursery conditions to produce stock for later planting.

Seeds of most saltmarsh species are not commercially available and collection of sufficient seed from existing saltmarshes to plant a large area would be labour-intensive. Trampling and depleting existing marshes of their seed stock may diminish disturbance–response functions and reduce granivory (passerine birds feed on the seed of saltmarsh plants). Seedling establishment is a hazardous phase in the life of any plant in any ecosystem. Saltmarsh seedlings face the possibilities of young seedlings being scoured by tidal currents, smothered in sediments or covered by algal blooms (Zedler, 2001). The windows of opportunity for successful seedling establishment between flooding tides are short and limited to particular seasons (Adam, 1990). While seeding to establish new marshes is often impractical, it is appropriate for the repair of degraded patches within established marshes.

Most saltmarsh species are perennial, and the dominants are often monocots (grasses, sedges, rushes) that can form large clones as a result of stolon or rhizome growth. It is possible (although disruptive) to cut sod from existing clones for transplantation to new marshes. In northwest England the cutting of 'sea-washed turf' from saltmarshes for use in lawns and sports fields was a considerable industry (Gray, 1972), although the practice has declined in recent years. Where an existing marsh is to be destroyed, the salvage and transplantation of vegetation can accelerate succession on the mitigation site. The removal of 'plugs' for vegetative propagation in a nursery can, with lead time, produce sufficient shoots to plant a large area. Planting of individual shoots is labour-intensive, but volunteers seem to enjoy such conservation projects.

Few saltmarsh plants have been subject to autecological studies, so knowledge of appropriate conditions for propagation and growth is limited. Much current practice is based on ad hoc empiricism; not only is there a need for studies of particular species, but all rehabilitation/creation projects should be fully documented and monitored so that the merits of alternative approaches can be assessed.

Biogeographic integrity

Saltmarshes have been deliberately created using species not indigenous to the site. During the twentieth century, extensive Spartina anglica-dominated marshes were planted in many parts of the world (Ranwell, 1967) (see above). Most creation and rehabilitation exercises now involve local species and, in many instances, attempts to eliminate alien species. The provenance of material needs to be addressed during planning.

Genetic integrity

Genetic considerations have long been important in the planning of conservation projects involving individual rare species (Fenster & Dudash, 1994; Gray, volume 1). With the increasing acceptance of the concept of biodiversity, there is wider recognition of the importance of maintaining genetic diversity in all species. One approach to maintaining genetic diversity is to require that plant material used in rehabilitation and creation projects be derived, as far as is possible, from parent material local to the planting area. This is likely to be better adapted to local environmental conditions, but the sites available for rehabilitation or creation may often be markedly atypical of the environment of surviving 'natural' examples of saltmarsh in the same general region. Even if these 'natural' marshes contain distinctive ecotypes of widespread species, these ecotypes may be ill-adapted to the conditions of degraded sites.

Saltmarsh plants have been classic material for genecological investigation (Adam, 1990). Examples are known of both variation over large geographical areas (e.g. Smith-White, 1988) and within single sites (Gray et al., 1979). Variation may be expressed in both morphological and physiological traits (Smith-White, 1981). Some variants may have very restricted distributions. For example, a viviparous pentaploid variant of the very widespread grass Sporobolus virginicus is known only from a single location in Western Australia (Smith-White & Adam, 1988). In sourcing plant material for restoration projects it would be unwise to treat saltmarsh species as if they were crops by planting commercially available material over wide areas without reference to local conditions. For rehabilitation of particular sites it is advisable to use plants from the same sites, planted in microhabitats similar to those at the source. Natural colonisation of new sites might include propagules from a large genetic catchment. Although the period of flotation and viability of propagules (fruit, seeds or vegetative fragments) after immersion in saltwater might limit the source area, species dispersed by birds can transcend national, or even continental, boundaries. For deliberate plantings, it is important to document and archive the sources.

The majority of saltmarsh restoration projects to date have been in the United States, and many have involved *Spartina alterniflora*. Phenotypic features of three *S. alterniflora* genotypes (from Massachusetts, Delaware and Georgia) persisted when planted to a common garden and to a large restoration site (Seliskar, 1995). Genotype mattered to many functions of the resulting restored marsh, including ramet density, detritus production, below-ground biomass, root and rhizome depth, recoverable reserves, soil respiration, and even fish use (J. Gallagher & D. Seliskar, personal communication). Parental height forms of *S. foliosa* did not persist in a restoration site where environmental conditions were important in determining plant growth (Trnka & Zedler, 2000). While little attention has been given to genotype in many plantings of *Spartina* species, future projects should give this topic greater consideration.

The progress of ecosystem development

Assessment strategies
The assessment of restoration progress requires agreement on the goals of the project, the reference sites that are suitable for comparison, and the methods of assessment. Too often, projects lack clear statements of their targets. When restoration is undertaken as mitigation, assessments can become highly contentious. Proponents rarely want their effort labelled a failure, even if the outcome falls wide of the proposed target. Insistence on 'success' often leads to pressure to revise the intent of the project to match the outcome or to evaluate only those areas that show the desired results or the few attributes that follow the desired trajectory. Four im-

provements are needed in assessment efforts (Zedler & Callaway, 2000), beginning with a change in terminology that better reflects the complex outcomes of restoration projects:

- The terms 'success' and 'failure' should be replaced with the term 'progress', which measures degrees of site development; the term 'compliance' should be used to indicate when mitigation criteria have been met (Quammen, 1986).
- More parameters should be assessed, including topography, hydrology, soil (soil total Kjeldahl nitrogen, soil organic matter, soil texture, soil salinity within the top 10 cm), vegetation (canopy architecture, cover from low-elevation remote sensing images) and animals (abundances of invertebrates and fish, size distributions of fish, gut contents of fish; nesting and production of fledglings by target bird species). Special attention should be given to the presence and abundance of rare species and of exotic species.
- The monitoring record should be evaluated and interpreted as the data are collected.
- Research should be integrated into the assessment programme to help identify and explain problems and predict if and when restoration goals might be reached.

Comparisons of restored and natural saltmarshes
The detailed assessments of restored saltmarshes, i.e. those that have appeared in the peer-reviewed literature, include comparisons of multiple sites of different age (space-for-time substitution) in North Carolina (Craft *et al.*, 1988; Sacco *et al.*, 1994), South Carolina (LaSalle *et al.*, 1991), Texas (Webb & Newling, 1985; Minello & Webb, 1997) and California (Haltiner *et al.*, 1997) with sites ranging in age from one to 17 years. These and other studies reveal differences between restored and natural saltmarshes, which suggest that:

- Restoration needs to be done in appropriate locations, e.g. where salinity and inundation regimes are suitable and where the region's biota can gain access once the site is available.
- Sites need to have sufficient topographic complexity, including tidal creek networks (Minello *et al.*, 1994; Zedler *et al.*, 1997).

- Soils need to have the appropriate sediment texture and nutrient supply rates, as well as sufficient soil organic matter for providing nutrients and retaining moisture (Craft *et al.*, 1988; Langis *et al.*, 1991).
- Sites need protection from excessive erosion and/or sedimentation (Simenstad & Thom, 1996; Haltiner *et al.*, 1997).
- Non-target vegetation will need to be monitored and controlled, especially invasive exotic species.
- Desired animal populations might need to be introduced if species are dispersal-limited (Sacco *et al.*, 1994; Minello & Webb, 1997).
- Contaminated soils should be ameliorated.

Two long-term data sets indicate that assessment periods should be extended beyond the typical five-year period, in order to document functional equivalency with natural saltmarshes (Simenstad & Thom, 1996; Minello & Webb, 1997; Zedler & Callaway, 2000). A third long-term study of soils (Craft *et al.*, 1999) indicates that soil carbon and nitrogen concentrations require up to 50 years to match levels in reference saltmarshes. Fish use seems to be restored quicker than other biological parameters (Burdick *et al.*, 1997), perhaps because of fish mobility. However, the tidal creek networks that are important to both fish (Desmond *et al.*, 2000) and plants (Zedler *et al.*, 1999) might need to be incised into saltmarsh restoration sites.

CONCLUDING REMARKS

Saltmarshes are complex and dynamic systems with halophytic vegetation that varies vertically and horizontally in relation to tidal regimes and topography. They also respond to sea level anomalies, floods, and sea storms. Productivity and sediment accretion rates are temporally variable. Urbanisation has substantially reduced saltmarsh quantity and quality, through dyking, pollution, altered salinity regimes, structures, dams and fragmentation. The high value of saltmarshes to waterbirds, fish and shellfish has led to protection and restoration efforts across the globe; restoration and creation of saltmarsh dates to the eighteenth century in Europe. Management of marshes using water-control structures is widespread, especially along the United States Gulf coast (Louisiana), but the

removal of dykes and tide gates does not necessarily allow marshes to re-establish. Subsidence and chemical changes occur during non-tidal periods, and such changes are not easily remedied. Because a saltmarsh is a product of its long-term history, its restoration involves more than the return of hydrologic conditions that preceded degradation. Ways in which a natural marsh develops are poorly documented; the same is true for restored and constructed sites. The settings in which restoration takes place are varied, and methods for restoration are far from standardised.

Efforts to date demonstrate that fine-textured, intertidal sediments can be vegetated through planting and subsequently used by a multitude of invertebrates, including shellfish, and by fish and waterbirds. Sites with greater topographic complexity, especially tidal creeks, are more likely to be accessible to aquatic organisms that use the saltmarsh as a foraging site. Although saltmarsh soils are slow to recover organic matter and nutrient levels, native plant species can establish and grow, even if height and biomass are lower than on natural saltmarshes.

Considerable progress has been made in increasing the area and quality of saltmarshes by:

- Encouraging sedimentation on mudflats to raise elevations enough to support halophytes
- Increasing tidal influence to impounded marshes by breaching dykes, opening tide gates and enlarging culverts
- Excavating fill to expose historical marsh plains or lower topography to saltmarsh levels
- Planting *Spartina* species to stabilise dredge spoil islands and shorelines
- Regulating salinity by controlling freshwater inflows, including releases from upstream reservoirs
- Decontaminating soils damaged by oil and acid accumulation
- Controlling exotic species.

In general, we believe that restoration of previous saltmarsh will progress more rapidly than the creation of saltmarshes by excavating uplands or dredge spoil deposits. Natural colonisation is more likely in places where local and regional saltmarshes are intact and rich in species. Animals are more likely to disperse into marshes if corridors are intact. Species will be more likely to persist where extremes in salinity are moderated by full tidal flushing. Projects that aim to restore specific plants and animals (e.g. rare or endangered species) experience greater difficulty. Species that have declined to rarity are often the most sensitive to human disturbance; because restoration sites have been degraded in the past and are disturbed in the process of restoration, very specific conditions might be needed before sensitive plants and animals will re-establish.

Restoration efforts can be improved by incorporating research into the projects, e.g. by adding experiments with different topography, soil amendments, plants, and species-reintroduction techniques. Where possible, we advocate designing the entire restoration site as an experiment ('adaptive restoration': Zedler, 2001). The recent excavation of an 8-ha restoration site at Tijuana Estuary to test the importance of adding tidal creek networks (Box 11.3) serves as an adaptive restoration model. Even if not treated as a full experimental project, restoration efforts will benefit from the use of science and scientists. Adaptive management, to be effective, requires ongoing involvement of scientists and a lead agency that fosters the use of research results.

ACKNOWLEDGMENTS

J. Zedler thanks the National Science Foundation for support of research to test the effects of planting diverse saltmarsh assemblages on ecosystem functioning (DEB 96-19875), the Earth Island Institute for support of long-term research to improve saltmarsh restoration, the California Department of Transportation for funding science-based assessments of their mitigation sites, the California Sea Grant Program for long-term support of research on coastal wetlands, and the National Estuarine Research Reserve Program for supporting long-term monitoring of Tijuana Estuary.

REFERENCES

Adam, P. (1990). *Saltmarsh Ecology*. Cambridge: Cambridge University Press.

Allen, J. R. L. (2000). Morphodynamics of Holocene salt marshes: a review sketch from the Atlantic and Southern North Sea coasts of Europe. *Quaternary Science Reviews*, 19, 1155–1231.

Ambrose, A. F. (2000). Wetland mitigation in the United States: assessing the success of mitigation policies. *Wetlands (Australia)*, 19, 1–27.

Ayres, D. R., Garcia-Rossi, D., Davis, H. G. & Strong, D. R. (1999). Extent and degree of hybridization between exotic (*Spartina alterniflora*) and native (*S. foliosa*) cordgrass (Poaceae) in California, USA determined by random amplified polymorphic DNA (RAPDs). *Molecular Ecology*, 8, 1179–1186.

Baker, J. M., Adam, P. & Gilfinan, E. (1994). *Biological Impacts of Oil Pollution: Saltmarshes*. London: International Petroleum Industry Environmental Conservation Association.

Beare, P. A. & Zedler, J. B. (1987). Cattail invasion and persistence in a coastal salt marsh: the role of salinity. *Estuaries*, 10, 165–170.

Beilfuss, R. D. & Davies, B. R. (1999). Prescribed flooding and wetland rehabilitation in the Zambezi Delta, Mozambique. In *An International Perspective on Wetland Rehabilitation*, ed. W. Streever, pp. 143–158. Dordrecht: Kluwer.

Bird, E. C. F. (1993). *Submerging Coasts: The Effects of a Rising Sea Level on Coastal Environments*. Chichester, UK: John Wiley.

Boyer, K. E. & Zedler, J. B. (1996). Damage to cordgrass by scale insects in a constructed salt marsh: effects of nitrogen additions. *Estuaries*, 19, 1–12.

Boyer, K. E. & Zedler, J. B. (1998). Effects of nitrogen additions on the vertical structure of a constructed cordgrass marsh. *Ecological Applications*, 8, 692–705.

Boyer, K. E. & Zedler, J. B. (1999). Nitrogen addition could shift species composition in a restored California salt marsh. *Restoration Ecology*, 7, 74–85.

Bradshaw, A. D. (1987). The reclamation of derelict land and the ecology of ecosystems. In *Restoration Ecology: A Synthetic Approach to Ecological Research*, eds. W. R. Jordan, M. R. Gilpin & J. D. Aber, pp. 53–74. Cambridge: Cambridge University Press.

Brampton, A. H. (1992). Engineering significance of British saltmarshes. In *Saltmarshes: Morphodynamics, Conservation and Engineering Significance*, eds. J. R. L. Allen & K. Pye, pp. 115–122. Cambridge: Cambridge University Press.

Broome, S. W., Seneca, E. D. & Woodhouse, W. W., Jr

(1986). Long-term growth and development of transplants of the saltmarsh grass *Spartina alterniflora*. *Estuaries*, **9**, 63–74.

Burdick, D. M., Dionne, M., Boumans, R. & Short, F. (1997). Ecological responses to tidal restorations of two northern New England saltmarshes. *Wetlands Ecology and Management*, **4**, 129–144.

Callaway, J. C. & Josselyn, M. N. (1992). The introduction and spread of smooth cordgrass (*Spartina alterniflora*) in south San Francisco Bay. *Estuaries*, **15**, 218–226.

Callaway, J. C., Zedler, J. B. & Ross, D. L. (1997). Using tidal saltmarsh mesocosms to aid wetland restoration. *Restoration Ecology*, **5**, 135–146.

Chamberlain, R. H. & Barnhart, R. A. (1993). Early use by fish of a mitigation salt marsh, Humboldt Bay, California. *Estuaries*, **16**, 769–783.

Chapman, M. G. & Underwood, A. J. (2000). The need for a practical scientific protocol to measure successful restoration. *Wetlands (Australia)*, **19**, 28–49.

Chung, C.-H. (1989). Ecological engineering of coastlines with saltmarsh plantations. In *Ecological Engineering: An Introduction to Ecotechnology*, eds. W. J. Mitsch & S. E. Jòrgensen, pp. 255–289. New York: John Wiley.

Chung, C.-H. (1994). Creation of *Spartina* plantations as an effective measure for reducing coastal erosion in China. In *Global Wetlands: Old World and New*, ed. W. J. Mitsch, pp. 443–452. Amsterdam: Elsevier.

Cole, C. A. (1999). Ecological theory and its role in the rehabilitation of wetlands. In *An International Perspective on Wetland Rehabilitation*, ed. W. Streever, pp. 265–276. Dordrecht: Kluwer.

Coles, S. M. (1979). Benthic microalgal populations on intertidal sediments and their role as precursors to salt marsh development. In *Ecological Processes in Coastal Environments*, eds. R. L. Jefferies & A. J. Davy, pp. 25–42. Oxford: Blackwell.

Craft, C. B., Broome, S. W. & Seneca, E. D. (1988). Nitrogen, phosphorus and organic carbon pools in natural and transplanted marsh soils. *Estuaries*, **11**, 272–280.

Craft, C., Reader, J., Sacco, J. & Broome, S. W. (1999). Twenty-five years of ecosystem development of constructed *Spartina alterniflora* (Loisel) marshes. *Ecological Applications*, **9**, 1405–1419.

Dame, R. F. (1994). The net flux of materials between marsh–estuarine systems and the sea: the Atlantic Coast of the United States. In *Global Wetlands: Old World and New*, ed. W. J. Mitsch, pp. 295–302. Amsterdam: Elsevier.

Denver, K. (1999). Rehabilitating Wyndgate: bringing back wetlands on a family property in South Australia. In *An International Perspective on Wetland Rehabilitation*, ed. W. Streever, pp. 107–112. Dordrecht: Kluwer.

Desmond, J., Zedler, J. B. & Williams, G. D. (2000). Fish use of tidal creek habitats in two southern California salt marshes. *Ecological Engineering*, **14**, 233–252.

Dijkema, K. S., Bouwsema, P. & van den Bergs, J. (1990). Possibilities for the Wadden Sea marshes to survive future sea-level rise. In *Saltmarsh in Wadden Sea Region*, ed. C. H. Ovesen, pp. 125–145. Copenhagen: National Forest and Nature Agency.

Dobson, A., Bradshaw, A. D. & Baker, A. J. M. (1997). Hopes for the future: restoration ecology and conservation biology. *Science*, **277**, 515–522.

Entrix, Inc., Pacific Estuarine Research Laboratory & Philip Williams and Associates, Ltd (1991). *Tijuana Estuary Tidal Restoration Program: Draft Environmental Impact Report and Environmental Impact Statement*, vols. 1–3. Oakland, CA: State Coastal Conservancy.

Esselink, J. W. P. (2000). *Nature Management of Coastal Saltmarshes: Interactions between Anthropogenic Influences and Natural Dynamics*. Haren: Koeman en Bijkerk.

Esselink, P., Dijkema, K. S., Reents, S. & Hageman, G. (1998). Vertical accretion and profile changes in abandoned man-made tidal marshes in the Dollard estuary, The Netherlands. *Journal of Coastal Research*, **14**, 570–582.

Fenster, C. B. & Dudash, M. R. (1994). Genetic considerations for plant population restoration and conservation. In *Restoration of Endangered Species: Conceptual Issues, Planning and Implementation*, eds. M. L. Bowles & C. J. Whelan, pp. 34–62. Cambridge: Cambridge University Press.

Ferren, W. R. (1985). *Carpinteria Salt Marsh: Environment, History, and Botanical Resources of a Southern California Estuary*. Publication no. 4, The Herbarium, Department of Biological Sciences. Santa Barbara, CA: University of California.

Flynn, K. M., Mendelssohn, I. & Wilsey, B. (1999). The effect of water level management on the soils and vegetation of two coastal Louisiana marshes. *Wetlands Ecology and Management*, **7**, 193–218.

Frenkel, R. E. & Morlan, J. C. (1991). Can we restore our salt marshes? Lessons from the Salmon River, Oregon. *Northwest Environmental Journal*, **7**, 119–135.

Gibson, K. D., Zedler, J. B. & Langis, R. (1994). Limited response of cordgrass (*Spartina foliosa*) to soil amendments in constructed salt marshes. *Ecological Applications*, **4**, 757–767.

Gray, A. J. (1972). The ecology of Morecambe Bay. 5: The saltmarshes of Morecambe Bay. *Journal of Applied Ecology*, **9**, 207–220.

Gray, A. J. & Adam, P. (1974). The reclamation history of Morecambe Bay. *Nature In Lancashire*, **4**, 13–20.

Gray, A. J., Parsell, R. J. & Scott, R. (1979). The genetic structure of plant populations in relation to the development of saltmarshes. In *Ecological Processes in Coastal Environments*, eds. R. L. Jefferies & A. J. Davy, pp. 43–64. Oxford: Blackwell.

Guerrant, E. O. (1996). Designing populations: demographic, genetic, and horticultural dimensions. In *Restoring Diversity: Strategies for Reintroduction of Endangered Plants*, eds. D. Falk, C. Millar & M. Olwell, pp. 171–207. Washington, DC: Island Press.

Haltiner, J., Zedler, J. B., Boyer, K. E., Williams, G. D. & Callaway, J. C. (1997). Influence of physical processes on the design, functioning and evolution of restored tidal wetlands in California (USA). *Wetlands Ecology and Management*, **4**, 73–91.

Hobbs, R. J. & Norton, D. A. (1996). Towards a conceptual framework for restoration ecology. *Restoration Ecology*, **4**, 93–110.

James, M. L. & Zedler, J. B. (2000). Dynamics of wetland and upland subshrubs at the saltmarsh–coastal sage scrub ecotone. *American Midland Naturalist*, **82**, 81–99.

Keer, G. & Zedler, J. B. (in press). Saltmarsh canopy architecture differs with the number and composition of species. *Ecological Applications*.

Kuhn, N. & Zedler, J. B. (1997). Differential effects of salinity and soil saturation on native and exotic plants of a coastal salt marsh. *Estuaries*, **20**, 391–403.

Langis, R., Zalejko, M. & Zedler, J. B. (1991). Nitrogen assessment in a constructed and a natural salt marsh of San Diego Bay. *Ecological Applications*, **1**, 40–51.

LaSalle, M. W., Landin, M. C. & Jerre, G. S. (1991). Evaluation of the flora and fauna of a *Spartina alterniflora* marsh established on dredged material in Winyah Bay, South Carolina. *Wetlands*, **11**, 191–208.

Lefeuvre, J. C. & Dame, R. F. (1994). Comparative studies of saltmarsh processes in the New and Old Worlds: an introduction. In *Global Wetlands: Old World and New*, ed. W. J. Mitsch, pp. 169–179. Amsterdam: Elsevier.

Marcus, L. A. (1994). Marriage made in mud. *Coast and Ocean*, **10**, 6–17.

Middleton, B. (1999). *Wetland Restoration, Flood Pulsing, and Disturbance Dynamics*. New York: John Wiley.

Minello, T. J. & Webb, J. W., Jr (1997). Use of natural and created *Spartina alterniflora* saltmarshes by fishery species and other aquatic fauna in Galveston Bay, Texas, USA. *Marine Ecology Progress Series*, **151**, 165–179.

Minello, T. J., Zimmerman, R. J. & Medina, R. (1994). The importance of edge for natant macrofauna in a created salt marsh. *Wetlands*, **14**, 184–198.

Mitchell, M. L. & Adam, P. (1989). The relationship between mangrove and saltmarsh communities in the Sydney region. *Wetlands (Australia)*, **8**, 37–46.

National Research Council Committee on Mitigating Wetland Losses (2001). *Compensating for Wetland Losses under the Clean Water Act*. Washington, DC: National Academy Press.

Niering, W. A. (1997). Tidal wetlands restoration and creation along the east coast of North America. In *Restoration Ecology and Sustainable Development*, eds. K. M. Urbanska, N. R. Webb & P. J. Edwards, pp. 259–285. Cambridge: Cambridge University Press.

Noe, G. B. & Zedler, J. B. (2001). Spatio–temporal variation of saltmarsh seedling establishment in relation to the abiotic and biotic environment. *Journal of Vegetation Science*, **12**, 61–74.

Onuf, C. P. (1987). *The Ecology of Mugu Lagoon: An Estuarine Profile*. U.S. Fish and Wildlife Service Biological Report no. 85(7.15). Washington, DC: U.S. Fish and Wildlife Service.

Ovenshine, A. T. & Bartsch-Winkler, S. (1978). Portage, Alaska: case history of an earthquake's impact on an estuarine system. In *Estuarine Interactions*, ed. M. L. Wiley, pp. 275–284. New York: Academic Press.

Packham, J. R. & Liddle, M. J. (1971). The Cefni saltmarsh and its recent development. *Field Studies*, **3**, 331–356.

Packham, J. R. & Willis, A. J. (1997). *Ecology of Dunes, Salt Marsh and Shingle*. London: Chapman & Hall.

Pavlik, B. (1996). Defining and measuring success. In *Restoring Diversity: Strategies for Reintroduction of Endangered Plants*, eds. D. A. Falk, C. I. Millar &

M. Olwell, pp. 127–155. Washington, DC: Island Press.

Pomeroy, L. R. & Wiegert, R. G.-(eds.) (1981). *The Ecology of a Salt Marsh*. New York: Springer-Verlag.

Portnoy, J. W. & Giblin, A. E. (1997). Biogeochemical effects of seawater restoration to diked salt marshes. *Ecological Applications*, **7**, 1054–1063.

Quammen, M. L. (1986). Measuring the success of wetlands mitigation. *National Wetlands Newsletter*, **8**, 6–8.

Randerson, P. F. (1979). A simulation model of saltmarsh development and plant ecology. In *Estuaries and Coastal Land Reclamation and Water Storage*, eds. B. Knights & A. J. Phillips, pp. 48–67. Farnborough, UK: Saxon House.

Ranwell, D. S. (1967). World resources of *Spartina townsendii (sensu lato)* and economic use of *Spartina* marshland. *Journal of Applied Ecology*, **4**, 239–256.

Rubin, J. A., Gordon, C. & Amatekpor, J. K. (1999). Causes and consequences of mangrove deforestation in the Volta Estuary, Ghana: some recommendations for ecosystem rehabilitation. *Marine Pollution Bulletin*, **37** (8–12), 441–449.

Sacco, J. N., Seneca, E. D.& Wentworth, T. R. (1994). Infaunal community development of artificially established salt marshes in North Carolina. *Estuaries*, 17, 489–500.

Sanzone, S. & McElroy, A. E. (1998). *Ecological Impacts and Evaluation Criteria for the Use of Structures in Marsh Management*. Washington, DC: Marsh Management Subcommittee, Ecological Processes and Effects Committee, Science Advisory Board, Environmental Protection Agency.

Seliskar, D. M. (1995). Exploiting plant genotypic diversity for coastal saltmarsh creation and restoration. In *Biology of Salt-Tolerant Plants*, eds. M. A. Khan & I. A. Ungar, pp. 407–416. Karachi, Pakistan: Department of Botany, University of Karachi.

Shine, C. & de Klemm, C. (1999). *Wetlands, Water and the Law: Using Law to Advance Wetland Conservation and Wise Use*. IUCN Environmental Policy and Law Paper no. 38. Gland, Switzerland: IUCN.

Simenstad, C. A. & Thom, R. M. (1996). Functional equivalency trajectories of the restored Gog-Le-Hi-Te estuarine wetland. *Ecological Applications*, **6**, 38–56.

Smit, H., van der Velde, G., Smits, R. & Coops, H. (1997). Ecosystem responses in the Rhine–Meuse Delta during two decades after enclosure and steps toward estuary restoration. *Estuaries*, **20**, 504–520.

Smith-White, A. R. (1981). Physiological differentiation in a saltmarsh grass. *Wetlands (Australia)*, **1**, 20.

Smith-White, A. R. (1988). *Sporobolus virginicus* (L.) Kunth in coastal Australia: the reproductive behaviour and the distribution of morphological types and chromosomes races. *Australian Journal of Botany*, **36**, 23–39.

Smith-White, A. R. & Adam, P. (1988). An unusual form of the saltmarsh grass *Sporobolus virginicus* (L.) Kunth. *West Australian Naturalist*, **17**, 118–120.

Spicher, D. & Josselyn, M. (1985). *Spartina* (Gramineae) in northern California: distribution and taxonomic notes. *Madroño*, **32**, 158–167.

Streever, W. (1997). Wetland rehabilitation in Australia. *Wetlands Ecology and Management*, **5**, 1–4.

Streever, W. (1999a). Performance standards for wetland creation and restoration under Section 404. *National Wetlands Newsletter*, **21**, 10–13.

Streever, W. (ed.) (1999b). *An International Perspective on Wetland Rehabilitation*. Dordrecht: Kluwer.

Streever, W. (2000). *Spartina alterniflora* marshes on dredged material: a critical review of the ongoing debate over success. *Wetlands Ecology and Management*, **8**, 295–316.

Stricker, J. (1995). Reviving wetlands. *Wetlands (Australia)*, **14**, 20–25.

Thayer, G. (ed.) (1992). *Restoring the Nation's Marine Environment*. College Park, MD: Maryland Sea Grant College.

Trnka, S. & Zedler, J. B. (2000). Site conditions, not parental phenotype, determine the height of *Spartina foliosa*. *Estuaries*, **23**, 572–582.

Turner, R. E. & Lewis, R. R. III (1997). Hydrologic restoration of coastal wetlands. *Wetlands Ecology and Management*, **4**, 65–72.

US Department of Defence (2000). Federal guidance on the use of in-lieu-fee arrangements for compensatory mitigation under Section 404 of the Clean Water Act and Section 10 of the Rivers and Harbors Act. *Federal Register*, **65**, 66914–66917.

US Geological Survey (2001). www.sfbay.wr.usgs.gov/ access/Dingler/home.html.

Vivian-Smith, G. (2001). Developing a framework for restoration. In *Handbook for Restoring Tidal Wetlands*, ed. J. B. Zedler, pp. 39–88. Boca Raton, FL: CRC Press.

Ward, K. (2000). Episodic colonisation of an intertidal mudflat by cordgrass (*Spartina foliosa*) at Tijuana Estuary. MSc dissertation, San Diego State University.

Webb, J. W. & Newling, C. J. (1985). Comparison of natural and man-made salt marshes in Galveston Bay complex, Texas. *Wetlands*, **4**, 75–86.

Weinstein, M. & Kreeger, D. (eds.) (2001). *Concepts and Controversies in Tidal Marsh Ecology*. Dordrecht: Kluwer.

Weinstein, M., Balletto, J., Teal, J. & Ludwig, D. (1997). Success criteria and adaptive management for a large-scale restoration project. *Wetlands Ecology and Management*, **4**, 111–127.

Weis, D. A. (1999). Vertical accretion rates and heavy metal chronologies in wetland sediments of Tijuana Estuary. MSc dissertation, San Diego State University.

White, I., Melville, M., Wilson, B. & Sammut, J. (1997). Reducing acidic discharges from coastal wetlands in eastern Australia. *Wetlands Ecology and Management*, **5**, 55–72.

Williams, R. J. & Watford, F. A. (1997). Identification of structures restricting tidal flow in New South Wales, Australia. *Wetlands Ecology and Management*, **5**, 87–97.

Zedler, J. B. (1993) Canopy architecture of natural and planted cordgrass marshes: selecting habitat evaluation criteria. *Ecological Applications*, **3**, 123–138.

Zedler, J. B. (1996). Coastal mitigation in southern California: the need for a regional restoration strategy. *Ecological Applications*, **6**, 84–93.

Zedler, J. B. (1997). Replacing endangered species habitat: the acid test of wetland ecology. In *Conservation Biology: For the Coming Decade*, eds. P. L. Fiedler & P. M. Kareiva, pp. 364–379. New York: Chapman & Hall.

Zedler, J. B. (ed.) (2001). *Handbook for Restoring Tidal Wetlands*. Boca Raton, FL: CRC Press.

Zedler, J. B. & Callaway, J. C. (1999). Tracking wetland restoration: do mitigation sites follow trajectories? *Restoration Ecology*, **7**, 69–73.

Zedler, J. B. & Callaway, J. C. (2000). Evaluating the progress of engineered tidal wetlands. *Ecological Engineering*, **15**, 211–225.

Zedler, J. B., Paling, E. & McComb, A. (1990). Differential salinity responses help explain the replacement of native *Juncus kraussii* by *Typha orientalis* in Western Australian saltmarshes. *Australian Journal of Ecology*, **15**, 57–72.

Zedler, J. B., Nelson, P. & Adam, P. (1995). Plant community organisation in New South Wales saltmarshes: species mosaics and potential causes. *Wetlands (Australia)*, **14**, 1–18.

Zedler, J. B., Williams, G. D. & Desmond, J. (1997). Wetland mitigation: can fishes distinguish between natural and constructed wetlands? *Fisheries*, **22**, 26–28.

Zedler, J. B., Callaway, J. C., Desmond, J., Vivian-Smith, G., Williams, G. D., Sullivan, G., Brewster, A. & Bradshaw, B. (1999). Californian saltmarsh vegetation: an improved model of spatial pattern. *Ecosystems*, **2**, 19–35.

Zimmerman, R., Minello, T. J. & Rozas, L. P. (2001). Salt marsh linkages to productivity of penaeid shrimp and blue crabs in the northern Gulf of Mexico. In *Concepts and Controversies in Tidal Marsh Ecology*, eds. M. Weinstein & D. Kreeger, pp. 293–314. Dordrecht: Kluwer.

12 • Rivers and streams

PETER W. DOWNS, KEVIN S. SKINNER AND G. MATT KONDOLF

INTRODUCTION

In their natural condition, the majority of rivers and streams, the 'catchment ecosystems' of Bormann & Likens (1979) or the 'fluvial hydrosystems' of Petts & Amoros (1996a), represent extremely complex, diverse and dynamic habitats for flora and fauna. Whilst research may focus on the characteristic ecology of a particular 'patch' within a river system, it is the high proportion of spatially transitional habitats or 'ecotones' (Naiman & Décamps, 1990) that are a particular distinguishing characteristic of fluvial ecosystems. The wide variety of ecotones results from the morphological diversity of the river–floodplain environment, but also the ecotones change in time according to the hydrogeomorphological dynamics of the river reach. Overall, river-channel-based ecosystems can be said to vary longitudinally, laterally, vertically and temporally (Petts & Maddock, 1994). Individual stretches of natural river–floodplain environment therefore present a complex backdrop of 'functional sectors' (Petts & Amoros, 1996b) for floral succession and competition, and a template within which fauna consistently adjust through their life cycle. However, only recently has causal understanding of the intimate links between the physical habitats and the resulting flora and fauna, linked by the key concepts of hydraulic stress and disturbance (Bravard & Petts, 1996), begun to be fully appreciated.

For a sustainable approach to ecological restoration, rivers must be viewed in context as dynamic ecosystems. River restoration projects based solely on recreating desired river morphology are unlikely to succeed. Full river restoration, classically defined by Cairns (1991) as 'the complete structural and functional return to a pre-disturbance state'

emphasises this fact. While most commentators would agree that the 'complete' return is neither achievable nor desirable as a result of changes in the river catchment since the pre-disturbance state, the 'functional return' demands that a multidisciplinary perspective is taken (e.g. Downs & Thorne, 2000). Although restoration targets are often ecological, for instance, the return of a floodplain forest or certain, valued 'charismatic megafauna' such as salmon (*Salmo* and *Oncorhynchus* spp.) or otters (*Lutra lutra*), the task centres on restoring the underpinning physical habitats. Further, the controls on healthy geomorphological functioning are related to factors far beyond the restoration reach. They extend to the catchment water quality, hydrology and sediment transport processes, and eventually to land use policies, plans for regional development, the organisation of the river management authority, available funding, favourable local community perception, and other political factors. Within the reach itself, effective river restoration may need a detailed understanding of flow hydraulics and sediment transport (see Newson *et al.*, volume 1), an inventive approach to engineering the solution, site managers with environmental science training, and an effective commitment to post-implementation maintenance. Therefore, river restoration is multidisciplinary in nature and all the more difficult to achieve because of it.

While no one chapter can possibly do justice to these varied facets (start instead with the compiled volumes of NRC, 1992; Brookes & Shields, 1996; FISRWG, 1998) we concentrate here on the functional environmental aspects of river restoration that increase the chance of a sustainable solution. These aspects include understanding the conceptual framework of river ecosystems and the ecological degradation of rivers that form the rationale for restoration, and developing objectives for river restoration in their multifunctional perspective. We outline and illustrate four basic approaches to river restoration, including a short-list of 'common fallacies in river restoration', and note several factors currently constraining ecological restoration of river systems. Throughout, we deal with restoration from a habitat-level perspective: specific issues relating to restoration policy and infrastructure, and to particular physical and biotic factors, are dealt with in greater detail elsewhere in these volumes.

ECOSYSTEM FUNCTIONING OF RIVERS

The arrangement and dynamics of river ecosystems stem from interactions among their hydrodynamic, hydrochemical and photosynthetic processes (Large & Petts, 1994); as conditions alter progressively through a catchment, some regularity may be expected in the downstream and lateral distribution and production of plants. In addition to this in-stream (autochthonous) biological production source, there are sources related to inputs from terrestrial vegetation (an allochthonous source) and the transport of organic matter from upstream. These latter two conditions will also change progressively downstream through the catchment resulting in distinct changes in available habitats related to physical conditions downstream.

While progressive changes in a downstream direction have been long recognised, the concept became popular as the 'river continuum concept' (RCC), a 'framework for integrating predictable and observable biological features of lotic systems according to the structure and function of communities' (Vannote et al., 1980) (Fig. 12.1). Understanding was promoted through the provision of generalisations on the magnitude and variation through time and space of organic matter supply, the structure of the invertebrate community, and the resource partitioning along the river length (Minshall et al., 1985). The RCC can be viewed as a first approximation model for subsequent revision according to local variability (Welcomme et al., 1989) and guided many studies of lotic function (Johnson et al., 1995).

However, the RCC was often deemed to be inadequate in describing events that occurred in the lower potamonic reaches of rivers (Welcomme et al., 1989), and applied only to permanent, lotic habitats (Junk et al., 1989). Extensive river–floodplain interactions can significantly modify the longitudinal patterns predicted by the RCC (Sedell et al., 1989). Moreover, in many rivers the 'continuum' is belied by significant 'steps' (Richards, 1973) in hydraulic geometry and habitats (Bruns et al., 1984) at tributary confluences. Overall, the RCC tends to be most appropriate for streams that are small and temperate (Statzner & Higler, 1985), lack river–floodplain interactions (Sedell et al., 1989) or are heavily regulated (Bayley & Li, 1996). Alternative longitudinal concepts of river ecosystems include the resource spiralling concept (Newbold et al., 1981, 1982a, b; Elwood et al., 1983) and serial discontinuity concept (Ward & Stanford, 1983, 1995).

For larger rivers with extensive interaction with their floodplains, the annual flood pulse, extending the river onto its floodplain, is the most important hydrologic feature (Johnson et al., 1995). The 'flood pulse concept' (Junk et al., 1989) (Fig. 12.2) recognises the hydraulic connectivity of a river and its floodplain as a key factor influencing productivity and species diversity (Scheimer & Zalewski, 1992; Welcomme, 1995). The moving littoral (the 'aquatic/terrestrial transition zone', ATTZ) between the water's edge to a few metres in depth (Junk et al., 1989) governs lateral transfers and recycling of biomass and nutrients. Aquatic organisms are adapted to feed and spawn on floodplain habitats during flood stages, whereas terrestrial organisms, which occupy non-flooded habitats along the borders, are adapted to exploit the floodplain at low flow (Petts & Maddock, 1994). The 'flood pulse advantage' is often conceptualised as the degree to which annual multi-species fish yield is higher in flooded rather than constant flow regimes.

Overall, the ecology of channel–floodplain as the transverse component of the river ecosystem can be seen to comprise a wide diversity of habitat patches at the meso-scale. These can be viewed geomorphologically as 'functional sets' (Petts & Amoros, 1996b) centred on the active channel, major floodplain, bars and abandoned channels. The truly 'special'

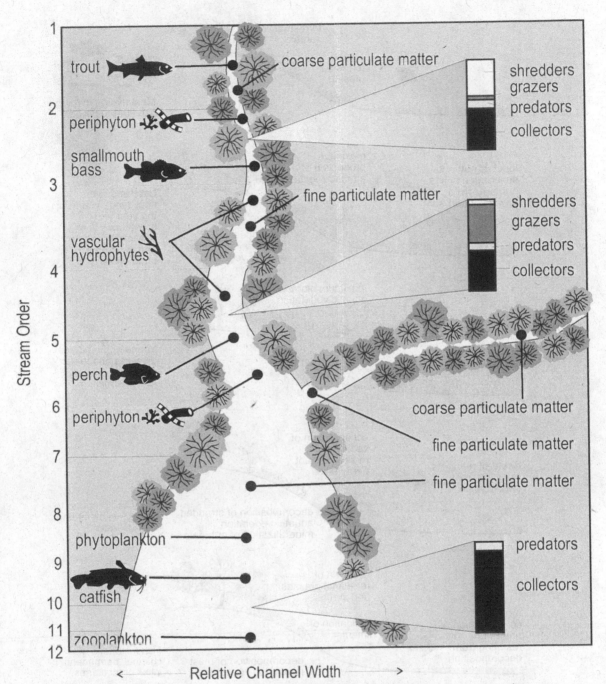

Fig. 12.1. Longitudinal ecological linkages in river systems illustrated by the river continuum concept. Adapted from Johnson *et al.* (1995) and FISRWG (1998).

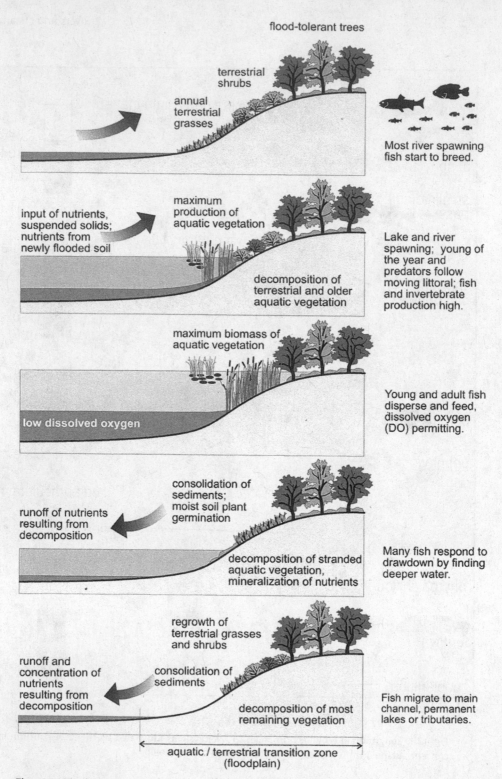

flood-tolerant trees

terrestrial shrubs

annual terrestrial grasses

Most river spawning fish start to breed.

input of nutrients, suspended solids; nutrients from newly flooded soil

maximum production of aquatic vegetation

decomposition of terrestrial and older aquatic vegetation

Lake and river spawning; young of the year and predators follow moving littoral; fish and invertebrate production high.

maximum biomass of aquatic vegetation

low dissolved oxygen

Young and adult fish disperse and feed, dissolved oxygen (DO) permitting.

consolidation of sediments; moist soil plant germination

runoff of nutrients resulting from decomposition

decomposition of stranded aquatic vegetation, mineralization of nutrients

Many fish respond to drawdown by finding deeper water.

regrowth of terrestrial grasses and shrubs

runoff and concentration of nutrients resulting from decomposition

consolidation of sediments

decomposition of most remaining vegetation

Fish migrate to main channel, permanent lakes or tributaries.

aquatic / terrestrial transition zone (floodplain)

Fig. 12.2. The importance of flood events in ensuring lateral ecological connectivity in river ecosystems illustrated by the flood pulse concept. Adapted from Bayley (1995) and FISRWG (1998).

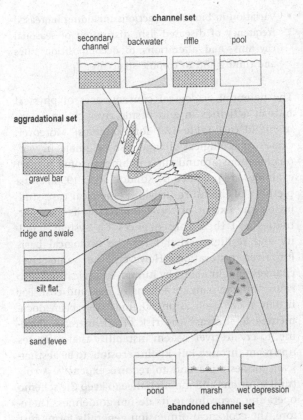

channel set

secondary channel backwater riffle pool

aggradational set

gravel bar

ridge and swale

silt flat

sand levee

pond marsh wet depression

abandoned channel set

Fig. 12.3. Detail from the fluvial hydrosystem concept wherein the river channel, riparian zone, floodplain and alluvial aquifer are viewed as four-dimensional system with longitudinal processes, lateral and vertical fluxes and strong temporal changes. 'Functional sectors' are distinguished within the catchment based upon channel form, flow, sediment transport and temperature, and their resultant habitat and relative stability. The illustration shows a 'functional set', a subset of a functional sector, in this case characterised by a laterally active channel. Adapted from Petts & Amoros (1996*b*).

characteristic of river ecosystems is the role played by periodic flooding, which provides the basis for the characteristic hydrodynamic and hydrochemical condition in each set and periodically resets the physical template on which the species develop. The 'fluvial hydrosystem concept' (Petts & Amoros, 1996*b*) (Fig. 12.3) was developed in an attempt to portray these features in terms of their structural and function connectivities. The fluvial hydrosystem is viewed as a four-dimensional system comprising the whole river corridor (river channel, riparian zone, floodplain and alluvial aquifer) that is influenced by longitudinal, lateral and vertical fluxes, as well as strong temporal changes (Petts & Maddock, 1994). The concept stresses a hierarchical view of river systems as a series of subsystems ranging from the drainage basin, down through functional sectors, sets and units to the smallest meso-habitat scale, with complex interactions between the hydrology, geomorphology and ecology at each of the scales.

By highlighting the ecosystem's driving forces, the fluvial hydrosystem concept points towards a functional notion of ecosystem health based on concepts of 'stability' and 'resilience' of the river system. Stability specifically refers to the ability of a system to return to an equilibrium level following a disturbance, whilst resilience is the ability of a system to maintain its structure and patterns of behaviour in the face of a disturbance (Common, 1995). In natural river channels, disturbances comprise a range of high flow events to which the channel adjusts over time.

RATIONALE FOR RESTORATION

While the ecological concepts outlined above have been developed to explain the natural functioning of rivers that are largely free from human disturbance, in reality, most of the world's river systems have been severely impacted by human modifications. Large-scale industrial, agricultural and mining practices as well as impoundments for flood defence and navigation have had significant, deleterious effects upon the health and integrity of river systems (Gore, 1996), to such an extent that by the beginning of the twentieth century few reaches of 'natural' river remained (Petts, 1989). In the UK this problem was recognised in the early part of the last century (Lamplugh, 1914).

Foremost among river modifications is channelisation for purposes of flood alleviation, agricultural drainage, navigation and the construction of roads and railways (Brookes *et al.*, 1983; see Brookes, 1988, for further details). The procedure often involved the complete straightening of the river, frequently accompanied by excavating an increased channel width in order to maintain flood conveyance, resulting in the imposition of increased geometric uniformity and implications for the resilience of the system. Consequently, following initial channelisation, the river would be of very low resilience since it would attempt to adjust to its new conditions by in-channel deposition of high-flow sediments that would formerly have been deposited overbank. This is indicative of recovery towards a natural system and suggests a move towards a more sustainable channel. However, many channelised rivers are also subject to regular 'maintenance', involving periodic dredging of the accumulated sediment to maintain flood conveyance. Consequently, natural system recovery is halted and the system returned to its condition of low resiliency, never approaching conditions of sustainability.

The nature of the ecological impacts caused by modification to river systems varies according to river and modification type. However, in a general sense, human modifications result in a series of physical 'river stresses' (NRC, 1992) that have repercussive biological stresses. Karr *et al.* (1986) suggests that the five most significant stresses on the biotic components of rivers arise from:

- Changes in food for organisms with respect to the quality, quantity and seasonal availability of particulate organic matter.
- Decline in water quality with respect to factors such as increases in temperature extremes, turbidity, nutrients and suspended solids and changes to the diurnal cycle of dissolved oxygen.
- Deterioration of habitat, including reductions in habitat area, habitat heterogeneity and in-channel shading and greater instability of channel bank and bed sediment.
- Deleterious changes in water quantity ('flow mistiming') such as altered flow extremes, greater extremes in flow velocities and reduced diversity of microhabitat velocities.

- Variation in biotic interactions including increased frequency of diseased fish, disruption of seasonal rhythms, and alterations to decomposition rates and timing and to trophic structure.

The potential ecological consequences of physical human activities in and beside river channels are manifold (see Table 12.1 for examples). Moreover, causes of ecological change also originate in various locations around the catchment (some further examples are provided by Kondolf *et al.*, 1996), and a plethora of literature provides empirical case studies on the extent of these impacts. It is these activities and their consequences for system stability and resilience that form the ecological basis for river restoration. However, the same river changes bought about by human modifications also cause 'human-scale' problems of risk and cost. For instance, channelisation schemes have been documented to increase the risk of downstream flooding, to create river system instability that migrates upstream through knickpoint erosion, to be aesthetically undesirable and to require expensive structural solutions and maintenance to keep the scheme operating according to its design guidelines. Therefore, the ecological dimension generally forms just one of a suite of objectives in undertaking river restoration.

OBJECTIVES FOR RIVER RESTORATION

Objectives for river restoration can be seen as community based and technically based as well as focusing on ecological improvement. Community objectives are driven by 'quality of life' concerns and may involve improving the aesthetics of degraded river environments or be part of a concern for preserving cultural heritage and nostalgic values, so that the 'non-use' value of the environment becomes an integral justification for the project. Intriguingly, perceived riverine aesthetic values and scientific conservation values of rivers do not necessarily coincide (Green & Tunstall, 1992; Gregory & Davis, 1993). This is often the case when restoration is undertaken for recreational benefit, another common objective of restoration schemes. Conversely, the overlap in

Table 12.1. *Examples of potential ecological consequences arising from selected human activity in and beside river channels*

Human activity and potential physical effect	Potential ecological consequences
Dams	
Reduced flood flows leading to reduction in channel migration	Reduced diversity of riparian habitats
Reduced flood flows eliminate frequent scour of active channel	Riparian vegetation encroaches into channel
Increased base flows and raised alluvial water table upstream of dam	Waterlogging of vegetation
Reservoirs drown existing vegetation	Longitudinal connectivity of riparian corridor interrupted
Base flows reduced or eliminated	Riparian and in-stream vegetation severely stressed or dies
Trapping of bed-load sediments behind dam, release of sediment-starved water downstream, channel incision	Reduction in water table, stressing or killing bankside and riparian vegetation, loss of connectivity in in-stream habitat, loss of upstream spawning areas for anadromous fish
Channelisation	
Channel excavation/dredging	Physical removal of fauna and flora, immediate reduction in number of fish present and species diversity, limited recovery in time, increase in non-native species suitability for fauna and flora
Silt released and redeposited through excavation/dredging	Increased turbidity kills macroinvertebrates and is unfavourable to game fish species, reduction in suitable nest sites for salmonid species
Smothering of channel bed gravels with sand/silt	Changes in standing crop numbers and species diversity, reduction in oxygen to buried fish eggs
Reduction in low-velocity habitat and slack-water channel edge habitat	Increased drift of macroinvertebrates, lack of shelter for fish fry, reduction in primary productivity
Elimination of pools, riffles, point bars	Reduction in macroinvertebrate diversity, reduction in fish number and biomass
Reduction in deep water habitat	Reduction in numbers of large fish, implications for rate of production
Increase in water temperature, and in temperature extremes	Direct impact on fish survival and production, reduction in macroinvertebrates
Reduction in volume of large woody debris	Direct loss of habitat, reduced cover for fish and reduction in nutrient status and trapping ability
Weed cutting	Removal of fauna in weeds, increased macroinvertebrate drift, reduction in overhanging food source, reduction in nursery habitat
Creation of unstable, silty banks	Reduction in number of fish through loss of habitat
Reduction in total bank length where straightened	Reduction in overall area of available habitat

Table 12.1. (*cont.*)

Human activity and potential physical effect	Potential ecological consequences
Forestry and timber harvest	
Removal of trees in riparian areas	Direct loss of trees, reduction in structural complexity for bird, mammal and insect habitat, partial elimination of large woody debris supply to the channel
Log transport on rivers leading to erosion of banks and simplification of channel geometry	Reduction in habitat complexity and consequent impact on fish, etc.
Removal of trees on hill slopes, resulting in increases in peak runoff and erosion	Bank erosion, conversion of vegetated bottomland and open gravel-bed channel, changes to species composition
Urbanisation	
Settlement along river banks and on bottomlands	Riparian habitat replaced by urban infrastructure
Increased impervious surface upstream increases peak runoff, induced channel widening and incision	Increased velocities causes additional stress for fish and macroinvertebrates, water table may lower stressing riparian vegetation
Land drainage to make land suitable for development	Flow quantity alteration from natural situation and impacts on in-stream fauna, desiccation of riparian vegetation

Source: Developed from Brookes (1988) and Kondolf *et al.* (1996).

interests between conservation needs and the needs of sport fisheries may be of benefit to ecological restoration initiatives. River restoration may also be in itself initiated by community action, and thus be an attempt to provide social cohesion or to raise local environmental awareness through 'pride in the river', as much as it is ecologically oriented. Nor should it be forgotten that many projects will continue to be the central 'deal' in which restoration is the mitigation device that allows development to proceed.

The technical objectives are mostly those initiated by the river management authority. Certainly these will, following the 1992 Earth Summit in Rio, include ecological and conservation perspectives under the guise of ensuring the environmental sustainability of rivers. But of increasing importance is the adoption of many of the more 'traditional' river management concerns under the 'umbrella' of restoration. A case in point is flood defence. Long viewed as fundamentally in conflict with river conservation, actions such as reflooding floodplains can both restore wetland habitats and provide flood protection to developed areas downstream by supplying 'off-line' flood storage. Other traditional concerns often now under the banner of restoration include measures to reduce systemic channel instability, approaches to reduce maintenance costs associated with sedimentation, and measures to improve water quality such as dissolved oxygen levels. These objectives may have related ecological benefits, such as new maintenance strategies that reduce fine sediment on the channel bed and also improve spawning habitat for salmonids, illustrating that ecological improvement may form only one of many objectives in river restoration.

Objectives for river restoration can be numerous and compromise may be required to achieve a balanced solution. From an ecological point of view, the overall aim is likely to revolve around improving the functional habitat of rivers so that there are wildlife improvements to in-stream or river corridor flora and fauna. Therefore, one key element in ecological objective setting is to specify the physical habitat requirements of the flora and fauna targeted (Downs & Kondolf, in press). This should involve identification of the target species, knowledge of their habitat requirements at different life stages, and knowledge of the habitat requirements of species with which they share a dependent or symbiotic relationship. Specifying these requirements will help geomorphologists and engineers translate the attributes of the current river ecosystem into designs for sustainable habitat improvements, which can be presented within the framework of the other river management objectives, such as flood defence, recreational improvements, etc.

APPROACHES TO RESTORATION

To meet the challenge of managing the river as an ecosystem, the overall approach of river restoration will be to 'undo stress' in terms of achieving structural and functional habitat improvements in the river channel, its riparian zone or corridor and its floodplain. Scheimer et al. (1999, p. 239) suggest that ecological restoration should be built on three basic principles:

- The approach should be based on common theoretical concepts in river ecology.
- It should be process- (ecosystem) orientated instead of species focused.
- It should primarily foster the hydrological and geomorphological functions of the river ('let the river do the work').

The first principle refers to concepts previously outlined such as the river continuum concept (Vannote et al., 1980), the flood pulse concept (Junk et al., 1989) and the fluvial hydrosystem concept (Petts & Amoros, 1996b). The key to these approaches is to understand the linkages between the geomorphology, hydrology and ecosystem development. The longer-term changes in both the morphology and hydrology of a river system progressively affect all the components of the community thus inducing major shifts in species dominance, relative and

total abundance and yield (Reiger et al., 1989). Over this period the stream fauna and flora will have evolved to survive any periodic disturbances without a significant major population change (Petts & Maddock, 1994). In disturbed catchments this situation will have changed considerably. The high biotic integrity (Reiger et al., 1989) and connectivity (Junk et al., 1989; Ward, 1998) associated with undisturbed catchments will have been significantly altered. Consequently, there is an immediate need to increase the river–floodplain interaction in order to improve the hydrological connectivity and ecological conditions along individual river segments (Scheimer et al., 1999).

The second principle outlined by Scheimer et al. (1999) refers to problems inherent in attempting to restore a particular habitat feature or the habitat of an individual fish species whilst ignoring its broader ecosystem setting. Historically, river restoration strategies, whether implicitly or explicitly, have been largely directed to sport and commercial fisheries on the assumption that this would be adequate for the habitat of other organisms (Petts & Maddock, 1994). However, this strategy is unlikely to prove successful (Reiger et al., 1989) as ecological restoration needs a more holistic perspective that focuses on improving the lateral and longitudinal connectivity of the system. If the functional integrity can be restored successfully then an increase in biodiversity should be expected to follow (Ward et al., 1999).

The final principle is directly linked to the need for an integrated approach since, in general, the most cost-effective procedure is to mimic natural processes (Bayley & Li, 1996). This requires an understanding of the geomorphic, hydrological and ecological functioning of a river system and is likely to be effective only through multidisciplinary approaches (Everest et al., 1991) that acknowledge that many habitat improvement schemes are, at present, experimental in nature.

Integrating these ecological principles for restoration into a framework for restoring the river requires the linked continuum from the target biota to the fundamental geomorphological processes controlling the river channel form to be understood. The continuum 'process–form–habitat–biota' depicts a hierarchical gradient for river restoration and is implicit in the foregoing reviews of the ecosystem functioning of rivers. It should also form the basis for specifying ecological objectives for restoration and is implicit in the NRC's (1992) setting of a hierarchy of objectives for sustainable approaches to river restoration, namely: (1) restore natural water and sediment regime, (2) restore natural channel geometry, (3) restore natural riparian plant community, and (4) restore native aquatic plants and animals.

The hierarchy acknowledges that restoration based on the lower-order objectives is unlikely to be sustained without attention to the upper order (NRC, 1992). For instance, restoring a native riparian plant community is unlikely to succeed in a highly channelised river channel because of the reduced extent of aquatic–terrestrial transitional bank habitat and the consequent disruptions to the near-surface water table in the bank zone. Likewise, restoration of physical habitats should provide the basis for landscape ecological improvements and thus the renaturalisation of plant and animal species indigenous to the area. However, structural habitat improvements based on 'naturalised' channel geometries are unlikely to be sustainable without functional improvements involving variable regimes of water discharge and sediment erosion, transport and deposition. Restoration of the upper-level objectives is expected to lead, over time, to improvements at the lower levels.

Approaches to the upper-level objectives are, thus, extremely important for guiding river restoration and restoration must have a core concern for the hydrogeomorphological interconnectivity within a river system. Related concerns include the need to understand zones of sediment supply, storage and transport within a catchment so that materials appropriate to a particular site are understood, to recognise whether a river channel is dynamically stable or unstable, to understand that channel adjustments are governed through spatial controls at a variety of scales and to gauge level of energy available for natural river recovery (Brookes, 1995; and see Newson *et al.*, volume 1). Approaches should also appreciate changes over longer time-frames and the symbiotic relationship of the geomorphological

features within the river system so that changes can be pre-empted (Sear, 1994). This should allow the context of the restoration to be understood in terms of the contemporary catchment sediment system and the direction of likely morphological change so that the most appropriate river morphology can be derived (Sear, 1994). In practice, several of the objectives may be tackled at the same time. The Clear Creek case study (Box 12.1) combined three of the objectives: restoration of process, and in some reaches of the stream, restoration of form and riparian vegetation. Despite the acknowledgement of the above concerns, restoration practices world-wide have often shown a marked departure from this understanding in practice. Four common fallacies are:

- That an inherently stable, static channel geometry exists for every stream. In reality, all river channels change in time, from imperceptibly slow change in some lowland clay streams to rapid avulsions during floods in high-energy channels. Many channels in drier climates or steeper, high-energy settings are inherently unstable and influenced by infrequent, high-magnitude events with channel widening in response to floods followed by gradual narrowing over subsequent low-flow years until the next big flood (Wolman & Gerson, 1978).

- That a static channel is ecologically preferable to an unstable one. However, the native flora and fauna of most stream systems are adapted to the periodic disturbances of floods (as discussed above) and even channel change, with an intermediate level of disturbance producing the greatest species richness (Connell, 1978; Pickett & White, 1985). Elimination of floods and substitution of steady, regulated flows for naturally variable flows facilitated establishment of exotic fish species that prey upon native salmon below dams in California (Baltz & Moyle, 1993); and, after construction of the Garrison Dam on the Missouri River in the United States, resulted in the virtual elimination of open gravel bars and early seral stages of riparian vegetation in the channel downstream (Johnson, 1992). Alternatively, actively migrating meandering rivers have very high ecological diversity related primarily to the length and diversity of river bank (Roux & Copp, 1996) and so

Box 12.1   Clear Creek channel–floodplain restoration project

RIVER DESCRIPTION

Clear Creek drains 590 km^2 of the northern Coast Ranges of California, and flows into the Sacramento River. The upper 520 km^2 are impounded by Whiskeytown Reservoir, below which Clear Creek flows 13 km through bedrock canyons, thence about 14 km through an alluvial valley with frequent bedrock control. Prior to dam construction and other alterations, the river provided habitat for anadromous salmonids: spring- and fall-run chinook salmon (*Oncorhynchus tshawytscha*) and steelhead trout (*O. mykiss*) (CDWR, 1986). Clear Creek has been profoundly modified by human actions since gold was discovered in the alluvial reach in 1850, and subsequently the entire floodplain was literally turned upside down as it was reworked into placer, hydraulic, and then dredger tailings, which are incapable of supporting vegetation. Since 1903, the 4.6-m high Saeltzer Dam diverted water into a irrigation ditch about 10 km upstream from the Sacramento River confluence and blocked fish passage.

Whiskeytown Reservoir (1963) impounds 370 million m^3, equivalent to about 81% of the average annual runoff; it has reduced the two-year return period flow (Q2) from 165 m^3 s^{-1} (pre-dam) to 88 m^3 s^{-1} (post-dam), a 47% reduction, and the Q5 from 277 m^3 s^{-1} to 188 m^3 s^{-1}, a 32% reduction. Elimination of flood scour led to encroachment of riparian vegetation (mostly *Alnus alba*) in the former active channel, which trapped overbank sediments and built up a berm along the channel margins. High flows still spilled over the dam every few years, incising the channel, washing out gravel bars and other features that would otherwise provide in-channel complexity. The end result was a simplified, 'bowling alley' channel, locked in place by dense root mats, and with little hydrological or ecological connection to the floodplain. In addition, Whiskeytown Dam traps bedload sediment supplied from the catchment, leading to 'sediment starvation' downstream.

Dam-induced sediment starvation was exacerbated by gravel mining (1950–78) that lowered parts of the floodplain down to a clay hardpan. During subsequent high flows, Clear Creek broke out of its channel and began to flow through a maze of ponds and across bedrock for much of its length thus eliminating spawning gravel, and stranding fish in the complex of shallow ponds (Fig. B12.1).

PROJECT DESCRIPTION

Despite the extent of its impacts, Clear Creek was viewed as a good candidate for restoration by virtue of the public ownership of most of the floodplain (USFWS, 1995; Calfed, 1999). The US Bureau of Land Management is now implementing a programme of land trades and acquisition to obtain remaining parcels from willing sellers in the Clear Creek corridor. The conceptual model underlying the Clear Creek restoration programme is to restore a dynamic flow regime and sediment load, and eliminate barriers to fish passage to restore ecosystem processes and increase habitat (particularly for anadromous salmonids).

The current controlled release capacity from Whiskeytown Dam is only 34 m^3 s^{-1} (far less than an annual flood on the pre-dam flood frequency curve), so the US Bureau of Reclamation is considering modification of the spillway, enlargement of low flow outlets, or changed operation of the existing reservoir to provide more frequent high flows downstream (USBR, 1999). In October 2000, the Bureau removed Saeltzer Dam, and out of concern for the potential effects of sand-sized sediment on restored spawning riffles downstream, mechanically removed most sediment stored behind the dam.

At the upper end of the alluvial reach, mounds of dredger tailings covering the surface of Reading Bar were removed, creating a lower floodplain surface hydrologically connected with the incised channel (i.e. flooded about every two years under the present flow regime). The riparian berm was removed and a gently sloping bank created. In the reach most affected by intensive gravel mining, a floodplain was reconstructed using dredger tailings from Reading Bar to fill floodplain gravel pits and create banks to confine the channel, and then revegetated. In effect, the gravel pits have been refilled with material similar to that removed by the miners, creating a topographic surface in connection with current hydrology.

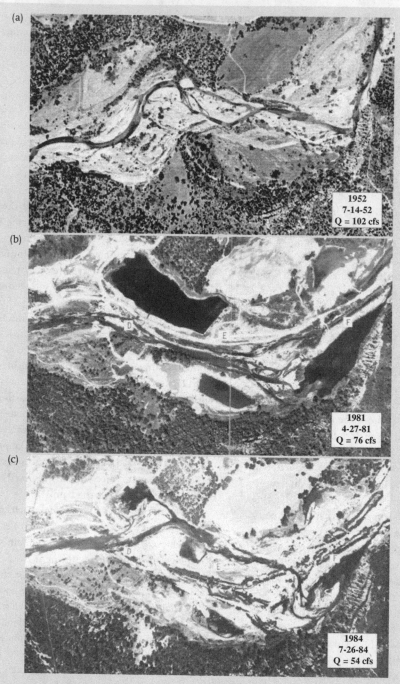

Fig. B12.1. Sequential aerial photographs of Clear Creek about 4–5 km upstream of the Sacramento River confluence preceding aggregate extraction (1952), during or immediately after extraction (1981), and after high flows (1984). North is up, flow is left to right. Adapted from McBain & Trush (2000).

DECISION ANALYSIS AND ADAPTIVE MANAGEMENT

To provide a systematic, rational basis to understand and quantify potential consequences of various choices on Clear Creek, the Calfed Bay–Delta Program (a multi-agency effort to restore ecosystem functions in the Sacramento–San Joaquin River system) has developed a decision-analysis and adaptive management model for Clear Creek. Increasingly, decision analysis is being used to quantify trade-offs in resource management decisions in general (Walters, 1986; Peterman & Peters, 1998), and for adaptive management experiments in river restoration in particular, including restoration of chinook salmon on the Snake River (Marmorek *et al.*, 1998), ecosystem restoration in the Colorado River (Collier *et al.*, 1997), and in-stream flow management in British Columbia (Higgins, 1999). The decision-analysis model for Clear Creek was designed to explicitly consider alternative flow regimes and their effects on water temperature and downstream flooding, gravel injection scenarios, and relations among physical habitats and biological populations (Alexander *et al.*, 2000). The performance measures are designed to serve as input to biological submodels.

OUTCOMES

Each component of the restoration programme has monitoring and evaluation built in, and there is a long data base of fish populations and redd counts, but the project's long-term effect on ecosystem health and target salmonid species will be unknown for some time. Clear Creek is notable because of the comprehensive approach to restoration, considering the entire catchment below Whiskeytown Dam, the effects of the dam on flow regime and sediment loads, as well as physical modifications to channel and floodplain from mining, and barriers to fish migration. The use of a decision-analysis tool represents an explicit attempt to rationally evaluate the trade-offs associated with range of possible management actions. Lack of human settlement along the channel and public ownership of most of the riparian zone meant that flood control was not an issue except for the timing of contributions of Clear Creek high flows to the Sacramento River when in flood.

channel resectioning and widening or the protection of meander bends against migration result in marked ecological degradation. Ecologically, there is nothing inherently desirable about a static channel, but in urban areas we are usually constrained by infrastructure and must often restrict channel movement for human reasons.

- That restoration of channel form without consideration of catchment processes will be sustainable. This is unlikely, because form follows function. The form and dimensions of alluvial river channels (i.e. channels whose bed and banks are composed of river sediments) reflect their flow and sediment transport regimes, the independent variables to which the dependent variables of channel geometry adjust. If flow or sediment load change, we can expect to see a corresponding change in channel form. For example, an increase in peak flows through an alluvial channel will typically cause an increase in channel dimensions through erosion of bed and banks. To restore requires that we first understand the processes: this requires looking upstream at a catchment scale (Kondolf & Downs, 1996), and looking back at historical changes that have led us to current conditions.

- That restoration projects can be designed solely from a channel classification system. Many restoration projects built this way have failed, but few have been documented, in part because most agencies prefer to spend their funds on building projects rather than post-project performance evaluation. One example is Deep Run, Maryland, where a 300-m long reach of symmetrical meanders corresponding to a Rosgen (1994) 'C-4' type were deemed to be inherently stable and were constructed in August 1995. The meanders were partially washed out in a flood in 1996, were repaired but failed again soon after (Smith, 1997). Building a 'stable' channel using a stream classification system remains popular among agency managers and other non-geomorphologists because it promises an easy shortcut to river restoration design.

There cannot be a simple and proven method for river restoration. Rivers differ too much in their geomorphological setting and ecological communities, their degradation history and their restoration objectives for this to be possible. Therefore, it is impossible to outline a single, best-practice approach. Instead, as a basis for comprehension, we offer four conceptual approaches to river restoration. An individual project may combine several of these approaches.

Non-structural techniques

This approach focuses on catchment and corridor measures for bringing about changes in catchment hydrology and sediment transport processes. Approaches may include: benign neglect in channels capable of natural recovery, catchment land and/or water management policies designed to reduce the extent of alteration to natural water and sediment regimes, the use of floodplain or corridor management policies (a 'semi-source' technique involving floodplain planning and the use of buffer strips), fencing to restrict livestock access to the channel, and aquatic marginal strategies such as strategic tree or reed planting. The inherent complexity of catchment-scale restoration is likely to restrict the use of this valuable approach, although non-structural elements such as land purchase for flood easements are being utilised in approaches to large rivers (e.g. SRAC, 2000). The Père Marquette catchment in Michigan, provides an example of the use of benign neglect (NRC, 1992). Following collapse of the timber industry, in the early 1900s, fewer stresses were applied to the river and it began to recover. By 1938, the Manistee National Forest was created over large parts of the catchment, aiding recovery further. The river is now designated as both a natural and scenic river and federal policies manage the sports fishing and canoe industries, while restricting commercial and residential development (NRC, 1992, pp. 211–212).

Improving network connectivity

This approach is not fully catchment-scale, but seeks to restore processes driven by the flood pulse through using measures such as flushing flow releases, the removal of small weirs and obstructions to flow and large dam removal. The approach is likely to involve process restoration to a 'less-modified' system state, rather than one that is fully restored. This approach is illustrated by the Clear Creek case study (Box 12.1), where sediment is injected at multiple points below the dam, and options to modify the dam to make higher flow releases are being evaluated. The approach is also central to the Danube Restoration Project in Austria (Box 12.2), the Kissimmee River restoration project in Florida (Box 12.3) and restoration of side arms of the River Rhône, France (Box 12.4). In these schemes, lateral connectivity is improved by increasing retention of the flood pulse onto the floodplain to achieve ecological benefits.

In-stream measures for prompted recovery

This principle usually involves the use of small structures designed to bring about a linked process–form improvement to the channel at the reach scale only, such as in the River Idle, England (Box 12.5). This scheme was designed to counter the prevailing trend of sedimentation and has been successful in developing a more heterogeneous channel by manipulating the geomorphological processes of erosion and deposition. The approach does not deal with fundamental links between geomorphology and the catchment hydrology but, if planned strategically, can instigate reach improvements that work within the context of an otherwise disturbed catchment (Downs & Thorne, 1998, 2000). Measures often include the use of flow deflectors, small weirs, sills and vanes, substrate reinstatement and pool–riffle re-creation (Hey, 1994), the provision of fish shelter through cover devices that are artificial or natural to the channel morphology (boulder clusters, large woody debris, etc.), the use of sediment traps and set-back embankments. The long-term ecological effects on fish biomass and diversity of such approaches will depend on the local impact of the scheme in combination with land use and channel management practices throughout the catchment as a whole.

Morphological reconstruction

This technique involves imposing a new morphology in the reach, usually with reference to the 'natural', pre-disturbance channel geometry, irrespective of the extent of process alteration since the pre-disturbance channel existed. However, it is a popular approach because restoration efforts are often focused in low-energy channels where rates of natural or prompted recovery will be slower than is politically acceptable for project funding. For these projects to succeed in anything other than very low-energy, static river environments, new approaches to channel morphology design are required that work within the context of the disturbed catchment to result in a channel designed without an inherent tendency for erosion or sedimentation. Recent advances in the sediment transport assessment needed to produce these 'zero sediment flux' designs gives hope that future morphological reconstruction designs will be more sustainable (see Hey & Heritage, 1993; Hey, 1994; Shields, 1996; FISRWG, 1998; Soar, 2000).

Box 12.2   The Danube floodplain restoration project

RIVER DESCRIPTION

The River Danube is one of the largest river systems in Europe draining 810 000 km^2 from its source in Germany to its mouth in the Black Sea. Most large European rivers have had a long history of regulation (Petts, 1989) and the River Danube is no exception. Within the last 50 years the effects of river regulation have been substantial. Over 90% of the headwater tributaries have been dammed for hydropower and much of the course of the river has also been modified (Scheimer et al., 1999). Recently, there has been recognition of the important role that flooding of the river's floodplain has in increasing ecological diversity (Junk et al., 1989; Scheimer & Zalewski, 1992; Petts & Amoros, 1996a). As a consequence of this, a 10-km reach of the Danube between Vienna and the Slovakian border is being used as a pilot scheme to investigate the ecological benefits of improving lateral connectivity. The Danube Restoration Project (DRP) is being undertaken on a reach that although partially regulated represents the largest remnant of an alluvial landscape in Europe (for further details see Heiler et al., 1995; Hein et al., 1999; Scheimer et al., 1999; Tockner et al., 1999; Ward et al., 1999) (Fig. B12.2). As a consequence of its importance it was designated a National Park in 1996.

PROJECT DESCRIPTION

The objectives of the project were to reduce channel incision by adding coarse-grained gravel, raise the water table by narrowing the shipping channel with a series of groynes, and improve the lateral migration and the exchange of processes between the river and its floodplain. Full-scale restoration was not possible because of alterations to sediment transport regimes, nutrient and pollution levels, as well as the constraints imposed by shipping and flood control (Scheimer et al., 1999).

These investigations were used to guide the management strategies that were adopted for the restoration programme. Engineering measures to improve the hydrological connectivity and dynamics of the river–floodplain system included lowering embankments where natural inflow channels currently exist from the main thalweg into the floodplain system, by constructing a series of 30-m wide overflow sills; adding extra openings at the inflow channels to ensure that a regulated inflow above the mean water level of –0.5m is maintained; lowering and widening weirs that currently control flows in the channels, within the floodplain system, to allow a shorter retention in the backwaters and to provide a more continuous flow of water.

OUTCOMES

A series of limnological investigations (begun two years before the planned scheme) concentrated on the Regelsbrunn floodplain segment to document changes in abiotic, biotic and functional properties (Hein et al., 1999). The scheme was implemented in the winter of 1997–8. Post-installation monitoring began soon after and is planned to continue for ten years. Through the monitoring programme, information on the role of hydrological connectivity in river–floodplain systems will help increase the chances of success in future attempts to restore such features.

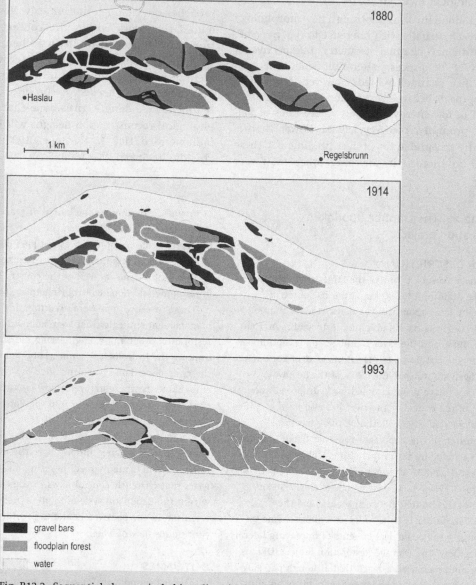

Fig. B12.2. Sequential changes in habitat diversity in the Regelsbrunn floodplain segment of the River Danube from 1880 to 1993, prior to restoration initiatives. Adapted from Scheimer *et al.* (1999).

Box 12.3   The Kissimmee River restoration project

RIVER DESCRIPTION

The Kissimmee River, central Florida historically had a 90-km long, 1.5–3-km wide river–floodplain ecosystem with over 20 lateral tributary sloughs (Toth, 1996) (Fig. B12.3a). The occurrence of flooding was an integral part of the Kissimmee River system. Precipitation in the wet seasons frequently caused large areas of the floodplain to be inundated with water for three to nine months per year (Warne et al., 2000). Floods typically receded at a slow rate, often less than 0.3 m a day, and as a result of the extent and duration of the flooding the Kissimmee was unique amongst rivers in North America (Toth, 1996).

Between 1962 and 1971 the river was channelised to provide flood protection in central Florida (Toth et al., 1993) (Fig. B12.3b). This project led to the diversion of 166 km of channel (US Army Corps of Engineers, 1985). As a result, the flood residence time on the Kissimmee decreased, on average, from 11.4 to 1.1 days (Toth, 1996). In the place of the meandering river, a 9 m deep, 64–105-m wide, 90-km long canal was excavated (US Army Corps of Engineers, 1985). The construction of the straight channel, and the destruction of the river–floodplain connectivity, meant that approximately 17 400 hectares of river–floodplain wetland habitat and 90% of the waterfowl were lost throughout the river valley (Shen et al., 1994). In addition, habitat for key-game species was also significantly reduced (Toth, 1996). Almost as soon as the project was completed in 1971 recommendations were made to restore the Kissimmee to its pre-channelisation status (Shen et al., 1994).

PROJECT DESCRIPTION

A principal goal of the proposed restoration was to restore the ecological integrity of the river (Warne et al., 2000). An important component of this was to restore the flood pulse so that the interaction between the river and its floodplain could be re-established. Following channelisation the Kissimmee was effectively converted from a well-connected river–floodplain ecosystem into a series of relatively

stagnant reservoirs (Toth, 1996). A series of post-channelisation hydrological manipulations had some success in improving the ecological condition of the Kissimmee (Toth et al., 1998). A more extensive demonstration project was undertaken between 1984 and 1989 to evaluate the possibility of restoring the Kissimmee to its historical condition (Toth et al., 1993). Three weirs were installed across the straightened channel to simulate the effects of diverting flow through the adjacent river channels and across the floodplain. The flow that was reintroduced was not as high as the historical discharges but nevertheless partially restored 9 km of remnant river channel (Toth, 1996) (Fig. B12.3c). After the reintroduction of flow, significant changes in channel morphology were observed and this was accompanied by an increase in the diversity and quality of the degraded habitat (Toth et al., 1993). Despite this improvement, restoration of key biological components was restricted by the existing discharge regimes in the upper basin and the regulation schedule.

As a response to the demonstration results a new restoration strategy was tested on another section of the Kissimmee to investigate whether more natural inflow characteristics could be achieved through modifying the discharge regime of the upper catchment (Toth et al., 1993, 1998). These studies concluded that to effectively restore the hydrological connectivity of the river–floodplain system would require some upper basin modifications to the flow regime and backfilling of 46 km of the straightened channel (Toth et al., 1993; Toth, 1996). This next stage, currently under construction, is expected to re-establish flow through 90 km of channel and provide interaction of the river and its floodplain over 11 000 hectares (Toth et al., 1993).

OUTCOMES

A key component throughout the various studies has been the flexible and adaptive approach that has been adopted throughout the planning, implementation and monitoring phases (Toth et al., 1998). This has increased the understanding of the system and has enabled further studies to be planned. By using monitoring, and subsequent evaluations, the

Kissimmee restoration project provides the opportunity to advance our understanding of river–floodplain interactions as well as providing valuable information on how to restore them.

Fig. B12.3. Temporal changes in the ecological complexity of the Kissimmee River: (a) river prior to channelisation 1961; (b) flood control channel created in the 1960s by dredging a straight channel and spreading the dredge spoil over the riparian wetland; (c) reflooding of a remnant backwater channel achieved following installation of demonstration project weirs. Photographs courtesy of the South Florida Water Management District (SFWMD, 2001).

Box 12.4   Side arms of the River Rhône

RIVER DESCRIPTION

The River Rhône originates in the Swiss and French Alps, flowing westward to Lyon, where it is joined by the Saône, and thence flows south to its delta on the Mediterranean coast just west of Marseilles. The Rhône has been extensively channelised, dammed and regulated throughout its length. As a result of sediment trapping by upstream dams, river training works, and extraction of gravel from the channel, the Rhône has incised along much of its length, resulting in the cutting off and/or drying out of side channels that were formerly connected with the main channel or at least experienced inundation during frequent floods. As a result, the ecological complexity of the riverine corridor has been reduced (Bornette & Heiler, 1994; Henry *et al.*, 1995).

On the Brégnier–Cordon Plain, about 80 km upstream (northeast) of Lyon, the Rhône had a braided channel until about 1800, when an embankment was built along the main channel, cutting off side channels from flow, except for seepage through the embankment and inundation during floods. At the heads of the cut-off channels, vegetation established and trapped fine sediment, creating plugs that terrestrialised and further isolated the downstream

reaches of the side channels. With construction of the Brégnier–Cordon hydroelectric project in 1982–4, the river was impounded and diverted through a canal, locally raising the alluvial groundwater table and inundating part of the floodplain, in response to which an emergency drainage canal was constructed, which lowered the water table. Isolation of the side channels by levees, armouring of main channel banks, and flow regulation eliminated much of the disturbance that formerly occurred on the floodplain, and the side channels had thus tended to become eutrophic and terrestrialised.

PROJECT DESCRIPTION

As described by Henry *et al.* (1995), one of the cut-off side arms, the Rosillon Channel, was treated, and it was compared with the Mortimer, a side channel unaffected by the hydroelectric project. On the Rosillon, a weir was installed to counteract the water-table lowering induced by the drainage canal, fine, organic-rich sediments were dredged, and coarse woody debris (that induced deposition) was removed. The conceptual model underlying the project was that by removing accumulated fine sediments, reducing hydraulic roughness in the channel to increase flood flow velocities, and increasing groundwater connectivity with the main channel, the now-eutrophic channel could be reset to a mesotrophic state, thereby increasing floodplain biodiversity (Fig. B12.4).

OUTCOMES

Much of the dredged fine sediment redeposited from suspension: pre-project fine sediment thickness ranged from 50 to 56 cm, post-project from 13 to 41 cm. Despite the persistence of fine sediment, the eutrophic aquatic plants that dominated the site pre-project (*Lemna minor*, *L. trisulca* and *Ceratophyllum demersum*) nearly disappeared, and were replaced with mesotrophic species (*Berula erecta*, *Callitriche platycarpa* and *Groenlandia densa*). Macroinvertebrate diversity increased, with the appearance of numerous lotic and oligotrophic species. Prior to the project, only one fish species was present, whereas afterwards numerous other species appeared.

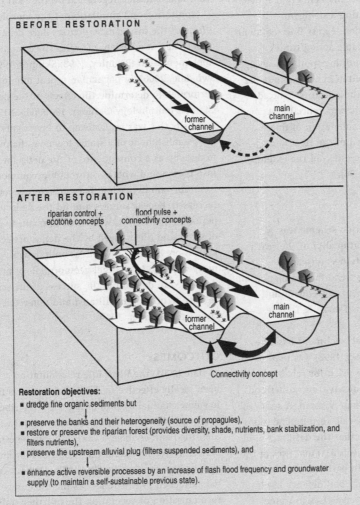

BEFORE RESTORATION

former channel

main channel

AFTER RESTORATION

riparian control + ecotone concepts

flood pulse + connectivity concepts

main channel

former channel

Connectivity concept

Restoration objectives:

■ dredge fine organic sediments but

■ preserve the banks and their heterogeneity (source of propagules),

■ restore or preserve the riparian forest (provides diversity, shade, nutrients, bank stabilization, and filters nutrients),

■ preserve the upstream alluvial plug (filters suspended sediments), and

■ enhance active reversible processes by an increase of flash flood frequency and groundwater supply (to maintain a self-sustainable previous state).

Fig. B12.4. Conceptual diagram showing increased surface and groundwater connectivity between side and main channels expected to result from restoration actions on the River Rhône. Adapted from Henry *et al.*, 1995.

Box 12.5   The River Idle rehabilitation project

RIVER DESCRIPTION

The River Idle is a low-gradient, mixed sand- and gravel-bed river draining an 842 km^2 catchment in the East Midlands, UK. The Idle was comprehensively channelised in the early 1980s for flood defence and land drainage (Downs & Thorne, 1998). This involved lowering the bed level, instating a trapezoidal channel cross-section, consolidating embankments over much of the Idle's length, and adopting a stringent maintenance regime. In addition, excess fine sediment supply derived locally from arable agricultural practices and upstream from mining spoil and sewage treatment works resulted in siltation throughout the course of the river. Together, these factors reduced habitat value; for instance, fishery potential had probably been reduced by a combination of fine sediment deposition on the bed of the river and steeply engineered banks that restricted the extent of shallow water habitat (Soar, 1996).

PROJECT DESCRIPTION

In the early 1990s a rehabilitation scheme was designed to improve the habitat quality of the reach through increasing habitat heterogeneity and accommodating the prevailing trend for sedimentation without a long-term commitment to frequent and costly operational maintenance (Downs & Thorne, 1998).

Initially, a geomorphological catchment baseline assessment (Environment Agency, 1998) was used to prioritise the most appropriate reach for rehabilitation based upon existing physical conservation values and on opportunities for riparian improvement. A series of catchment, corridor and reach-scale recommendations was made and tested to ensure that the designs would not deleteriously affect the multifunctional uses of the river (Downs & Thorne, 1998). In particular, a series of models was used to ensure that flood defence levels were not compromised by projected rises in water surface elevations consequent upon implementation of various elements of the scheme (Downs & Thorne, 2000). Some of the catchment-scale proposals were adopted as part of longer-term water resources

planning for the catchment (Environment Agency, 1999). Meanwhile, a series of river corridor and rehabilitation measures were implemented in two stages. Construction of the first stage of the scheme was undertaken in March 1996, followed by the second phase in late September 1996. Measures that were incorporated into the project included the installation of in-stream deflectors, riffles and fish shelters, reprofiling of channel banks as well as the planting of reeds and bankside vegetation over a 2-km reach.

Monitoring and simulation of the morphological effects of the instream structures have been undertaken at a variety of scales since project implementation (Bromley, 1998; Skinner, 1999; Swindale, 2000). In particular, repeat monitoring was performed to determine the effects of the deflectors on channel morphology (Skinner, 1999) (Fig. B12.5). Primarily this was undertaken to investigate whether the establishment of a stable low-flow channel started to develop as a consequence of the deflector installation and whether, through accommodating the trend for sedimentation, habitat heterogeneity had increased. Repeat surveys around the deflectors, over a period of 40 months, were used to quantify the degree of scour and fill induced by the deflectors relative to known pre-installation and 'as-built' reach morphologies. A 'channel geomorphology profiler' was developed to determine the success of the scheme relative to other rehabilitated and non-rehabilitated river sites (Skinner, 1999).

OUTCOMES

To date (2001) the River Idle rehabilitation scheme has been locally effective in developing a diversified low-flow channel by prompting the redistribution of channel sediments. Zones of scour and fill were identified and enabled a causal link to be established between river flows and the extent of morphological change between surveys. The formation of depositional bars attached to the deflectors has encouraged the recolonisation of vegetation thus further stabilising the new low-flow width. Planted bankside willows are beginning to provide valuable shade over the river. Using natural rivers as a basis for comparison, the combined riparian and in-stream effects should be beneficial for fish habitat in the Idle. Due to problems experienced with electro-fishing only visual evidence of fish behaviour was recorded. Numerous fry have been observed to shelter in the lee of the deflectors and consequently the presence of these low-flow zones (which had previously been absent) has mitigated one of the more apparent habitat deficiencies identified in the pre-rehabilitation assessments.

(a)

(b)

Fig. B12.5. Backfilled triangular deflector constructed from gabion baskets and installed on the River Idle in 1996: (a) shortly after construction, and (b) in 1997 during low flow illustrating the development of morphology diversity (sand bar and slack-water zone) in a previously plane bed channel. Vegetation has since colonised the deflector surface and channel margins upstream and downstream of the deflector. Flow is from left to right. (a) Photograph by PWD, (b) photograph by KSS.

CONSTRAINTS ON RESTORATION

There are numerous factors that exist to constrain ecological restoration, and many of them are not related to river ecology. In practical terms, this includes the primary requirement of obtaining floodplain land as either a permanent acquisition or under some co-operative agreement for 'set aside' land (such as under the Dutch programme of 'Space for rivers': Cals et al., 1998). There is also a host of potential institutional problems that have been previously noted as constraints on river basin management (Mitchell & Pigram, 1989; Mitchell, 1990), including administrative structures, overlapping jurisdictions of management authorities and boundary conflicts. There is a need for appropriate 'teams' to facilitate the restoration process (see FISRWG, 1998), for improved communication of benefits from practitioners to politicians, better ecosystem understanding by managers and, critically, supportive legislation. For instance, in Denmark, provision for stream restoration in the Danish Watercourse Act of 1982 (Iversen *et al.*, 1993) led to Denmark being a world leader in river restoration efforts. Some of these issues are dealt with elsewhere in these volumes.

Scientific constraints on river restoration are also numerous. They range from limited understanding of ecosystem functioning (NRC, 1992; Brookes, 1996), the lack of 'catchment historical' perspectives on river restoration (Kondolf & Downs, 1996), the need to improve communication of scientific information to managers and the public (Brookes, 1996) and the increasingly blurred distinction between scientists and managers (NRC, 1999). These issues are all worthy topics for discussion in themselves (see, for example, Adams & Perrow, 1999) and are beyond the scope of this chapter. Instead, we focus here on two specific constraints that are critical to the scientific improvement of ecological restoration of rivers. These relate to ecological target setting and monitoring and evaluation of restoration schemes.

Inability to set targets

At present, river restoration ecology suffers from the inability of ecologists to define their system requirements as closely as, for instance, engineers can specify and model (although not necessarily achieve) their flood defence requirements. Progress towards habitat requirement specification has generally involved the use of either species-specific habitat preference curves, reach-integrated template mapping of habitat or, in regulated rivers, setting of environmentally beneficial flow releases.

Habitat preference curves
One specific objective for river restoration is to restore 'natural' hydraulic habitats defined locally by the interaction of the flow with pool–riffle sequences, heterogeneity of the channel margins, planform sinuosity and temporary aquatic habitats on the floodplain. Accordingly, 'habitat preference

curves' for different species of fish and different life stages can act as a method of target-setting and are derived either from the literature, professional judgement or from detailed field observations (Petts & Maddock, 1994). They are usually expressed as a suitability index from 1 (suitable) to 0 (unsuitable) for flow depth, velocity and substrate. The most popular approach for simulating habitat preferences is via the 'instream flow incremental methodology' (IFIM) and its underlying model, the 'physical habitat simulation model' (PHABSIM) (Bovee, 1982), which computes the 'weighted usable area' (WUA) in a reach as a function of the river discharge, for different life stages and species of fish. Numerous procedural, biological and physical limiting assumptions prevent PHABSIM and IFIM from being reliable (e.g. Mathur *et al.*, 1985; Gore & Nestler, 1988; Gan & McMahon, 1990; Kondolf *et al.*, 2000). In response to these criticisms, a wide range of more complex approaches based on advanced understanding of the bioenergetics of fish biology and on more realistic representation and modelling of flow patterns is now at the research stage (Third International Symposium on Ecohydraulics, 1999).

Reach integrated template mapping
Rather than set a specific ecological target, an alternative is to set a target range of habitats based on the assumption that 'the river is the best model of itself' (Shields, 1996) and that natural river channel conditions provide the most favourable overall condition for the native flora and fauna. For example, the 'riverine community habitat assessment and restoration concept' (RCHARC) (Nestler *et al.*, 1993, 1996; Peters *et al.*, 1995) recognises the importance of variability in depth and flow velocity as the basis for fish habitat and is essentially a 'carbon copy' technique. A reference reach representing the target or ideal habitat conditions is used to develop a template of depth and flow velocity variability through field survey of at least 200 velocity–depth pairs, as the basis for hydraulic simulations at various discharges (Peters *et al.*, 1995). The approach depends on having a suitable reference reach, which is not possible in many degraded river environments (Hughes *et al.*, 1986), and involves subjective judgement of what is 'ideal'.

Another system of template mapping involves defining 'functional habitats' (Harper *et al.*, 1992) based on flow velocity–depth variability combinations associated with sampled mesohabitat types such as invertebrate assemblages found with aquatic macrophytes (e.g. 'fine-leaved emergent plants') or sediment conditions ('gravel', 'sand', etc.) (Harper & Everard, 1998). The approach is therefore a 'bottom–up' procedure for defining a series of 'building blocks' for river restoration, although there is no explicit connection between the building blocks and sustainable river channel configurations.

Another approach is a 'top–down' approach, mapping 'physical biotopes', identified primarily through a dominant flow type, as basic units of in-stream physical habitat (Padmore, 1997). Field survey in the UK identified nine distinct physical biotopes (waterfall, cascade, cascade rapid, riffle, run, boil, glide, pool, marginal deadwater, multiple biotopes), but the ecological significance of these various biotopes is only now being explored (Padmore, 1997).

A further alternative, the 'channel geomorphology profiler' (Skinner, 1999) assesses habitat quality by measuring habitat diversity from simple and quantifiable, geomorphologically significant, variables. Different rivers with different habitat qualities can be assessed on the same habitat diversity scale, and the results can be meaningful as a river configuration. The variables can be used to define three relationships outlining 'in-stream diversity', 'reach sustainability' and 'environmental suitability'.

Setting environmentally beneficial flow release

Downstream of dams, ecological targets for river restoration can also be specified in terms of pre-dam flow hydrology. These minimum 'environmental flows' are commonly set as statistical thresholds from the flow duration curve, such as the 95th percentile, a percentage of the average daily flow or the mean annual minimum seven-day flow frequency statistic (Petts & Maddock, 1994). However, this approach considers the hydrology without recourse to other ecological controls such as geomorphological processes, water quality issues, and the impact of other management factors such as channelisation, nor does it consider specific habitat requirements (Petts & Maddock, 1994). In addition to the minimum flows, deliberate high-flow discharges are released to mimic the effects of floods. These so-called 'flushing flows' can be categorised according to their general purpose as maintaining or restoring the habitat suitability of the channel bed, the channel perimeter or the floodplain (Petts & Maddock, 1994; Kondolf & Wilcock, 1996) but, without setting specific, geomorphologically based objectives, it is impossible to evaluate whether flushing flows succeed (Kondolf & Wilcock, 1996). Moreover, while flushing flows may result in a desirable deregulation of flow, the issue of restoring catchment sediment supply (trapped behind the dam) is not addressed. Therefore, flushing flows must interact with a highly modified sediment budget to produce a sustainable and ecologically acceptable river channel perimeter, but one that has few naturally formed counterparts. In part, this has led to increasing calls for dam removal as the ultimate re-creation of an environmental flow, despite the technical, environmental and social complexity of this procedure (Shuman, 1995; Born *et al.*, 1998; American Rivers *et al.*, 1999; Doyle *et al.*, 2000).

Lack of assessment

As a wide variety of natural river environments exists, underpinned by varying habitat dynamics driven from catchment-scale processes, and degraded to different extents, there can never a single approach to river restoration. Therefore, restoring the natural process dynamics is a difficult task and, along with numerous cases wherein the process factors are overlooked, a high rate of failure in river restoration schemes may be expected. There is some evidence that this is, indeed, the case. For example, Frissell & Nawa (1992) evaluated the performance of fish habitat enhancement projects in 15 streams in western Oregon and Washington and found the median success rate to be 40%. Miles (1998) evaluated enhancement projects after 8–14 years on the high-energy Coquihalla River in British Columbia and found 41% of structures were washed away or buried, and 87% of the structures were at least 50%

eroded away (as defined by loss of their material). On the nearby Coldwater River, Miles found 5% of structures were completely destroyed and 78% were at least 50% eroded. The success rates of these projects would become lower over time, as the projects are exposed to more high flows. Unfortunately, these figures are rarities because the vast majority of stream restoration projects are not evaluated objectively (Kondolf, 1995; Kondolf & Micheli, 1995).

CONCLUDING REMARKS

The ecological restoration of rivers has to be seen in the perspective of contemporary approaches to river channel management in general. Indeed, when a process-led approach to river restoration is adopted, incorporating non-structural measures for ecological improvement, then it becomes apparent that river restoration becomes the unifying practice that allows river basin management to become truly integrated. Seen in this light, the scientific context for the ecological restoration of river environments becomes one of a suite of practices for maintaining the ecological value of a river basin or catchment and the most appropriate approach to restoration depends upon the degree of river degradation (Boon, 1992). In this regard, we offer the following hierarchical perspectives on maximising the ecological value of river biomes (based on SRAC, 2000):

Where natural processes continue to function, preserve these processes

If the river continues to flood, if it still inundates its floodplain, moves sediment, and drives channel migration, preserve these processes. Protect the variable flow regime and use non-structural solutions to permit the river to continue to function dynamically.

Where natural processes can be restored, restore these processes

If, by non-structural means, it is possible to restore the characteristics of the pre-disturbance hydrograph and sediment transport processes, pursue this approach. In regulated rivers, a related approach is to use environmental flows such as flushing flows to simulate the effects of the flood pulse advantage (see section 'Ecosystem functioning of rivers' above and case studies). Sediment may need to be added to offset the tendency for 'hungry water' (Kondolf, 1997) (Box 12.1). A more comprehensive and sustainable approach is the removal of dams, especially when they are obsolete or structurally unsafe. Where large dams do not exist, and the project is only of local scale, in-stream 'prompted recovery' may be a means for creating a sustainable solution within the constraints of a disturbed catchment (Box 12.5). Alternatively, allowing the flood pulse to connect with previously disconnected floodplains may initiate process-based improvements in riparian habitat (Boxes 12.2–12.4).

Where natural habitats exist, preserve them

Some rivers still possess remnant stands of riparian forest, which were established by historical floods and channel processes. Even if the forest is not replicating itself naturally (because of reduced floods, sediment transport and channel migration), there may be habitat values worth preserving as relict stands. Existing intact habitat is usually better than can be re-created through restoration projects because time is one measure of ecological value that is inherently missing in 'faked nature' (Elliot, 1997).

Re-create natural habitats through restoration projects

This is essentially form-based restoration: for instance, regrading the channel and planting desired species to reproduce the natural river forms, or attempting to re-create habitat values on a small scale through installation of habitat enhancement structures. The problem is that these projects are not sustainable, unless dynamic river processes maintain the constructed forms. The NRC (1992) concluded that river restoration failures commonly resulted from a failure to take hydrology and natural processes into account.

In practice, the ecological restoration of river biomes cannot be seen free of its context as part of contemporary approaches to river channel (or basin) management. Therefore, not only does the science of ecological restoration continually test our capabilities in ecosystem understanding, but the process has to be compromised with the other community and technical objectives in undertaking the scheme. To an extent, therefore, all river restoration projects are experiments. These experiments can be approached in several ways. One way is to take a series of small, trial-and-error steps in pursuing ecological improvements with a high degree of security in the overall outcome. However, it has been argued that there are 'economies of scale' in river restoration (Brookes, 1996) such that small-scale, incremental, approaches are not effective in terms of cost or environmental improvement. An alternative is to acknowledge the experimental nature of river restoration and to adopt an 'adaptive management' approach (Holling, 1978; Walters, 1986) that allows uncertainty to be accommodated by using best-practice information-gathering and a 'direct feedback between science and management such that policy decisions can make use of the best available scientific information in all stages in its development' (Halbert & Lee, 1991, p. 138). Under this scenario, less conservative approaches are required in order to increase the chance of 'learning by surprise' in a rational way (McLain & Lee, 1996) as opposed to learning by trial and error. However, this approach puts an emphasis on post-project appraisal as the device for communicating both the 'success' of the project (requiring clearly stated ecological targets) and the increase in ecological understanding obtained in carrying out the experiment (Downs & Kondolf, in press). This is not a trivial task because, unlike short-term, construction-centred appraisals indicative of traditional river management approaches, river restoration demands an effect-centred and longer-term assessment that is suitable to the geomorphological context of the scheme (Downs et al., 1999; Skinner & Downs, 2000). Therefore, developing suitable approaches (and funding) for such appraisals and setting clearly stated target habitats are both critical in improving the future restoration of river ecology.

REFERENCES

Adams, W. R. & Perrow, M. R. (1999). Scientific and institutional constraints on the restoration of European floodplains. In *Floodplains: Interdisciplinary Approaches*, eds. S. B. Marriott & J. Alexander, pp. 89–97. London: Geological Society.

Alexander, C. A. D., Marmorek, D. R. & Peters, C. N. (2000). *Clear Creek Decision Analysis and Adaptive Management Model: Results of a Model Design Workshop held 24–26 January 2000*. Draft report prepared by ESSA Technologies Ltd. Vancouver, BC: Calfed Bay-Delta Program.

American Rivers, Friends of the Earth and Trout Unlimited (1999). *Dam Removal Success Stories: Restoring Rivers through Selective Removal of Dams that Don't Make Sense*. Washington, DC: American Rivers.

Baltz, D. M. & Moyle, P. B. (1993). Invasion resistance to introduced species by native assemblage of California stream fishes. *Ecological Applications*, **3**, 246–255.

Bayley, P. B. (1995). Understanding large river–floodplain ecosystems. *BioScience*, **45**, 153–158.

Bayley, P. B. & Li, H. W. (1996). Riverine fishes. In *River Biota, Diversity and Dynamics*, eds. G. E. Petts & P. Calow, pp. 92–122. Oxford: Blackwell.

Boon, P. J. (1992). Essential elements in the case for river conservation. In *River Conservation and Management*, eds. P. J. Boon, P. Calow & G. E. Petts, pp. 11–23. Chichester, UK: John Wiley.

Bormann, F. H. & Likens, G. E. (1979). *Patterns and Processes in a Forested Ecosystem*. New York: Springer-Verlag.

Born, S. M., Genskow, K. D., Filbert, T. L., Hernandez-Mora, N., Keefer, M. L. & White, K. A. (1998). Socio-economic and institutional dimensions of dam removals: the Wisconsin experience. *Environmental Management*, **22**, 359–370.

Bornette, G. & Heiler, G. (1994). Environmental and biological responses of former channels to river incision: a diachronic study on the Upper Rhône River. *Regulated Rivers: Research and Management*, **9**, 79–92.

Bovee, K. D. (1982). *A Guide to Stream Habitat Analysis using the Instream Flow Incremental Methodology.* Instream Flow Information Paper no. 12. Fort Collins, CO: US Department of the Interior Fish and Wildlife Service, Office of Biological Services

Bravard, J. P. & Petts, G. E. (1996). Human impacts on fluvial hydrosystems. In *Fluvial Hydrosystems,* eds. G. E. Petts & C. Amoros, pp. 242–262. London: Chapman & Hall.

Bromley, J. C. (1998). Hydraulic modelling and current deflectors: a study into the feasibility of using the CFX three-dimensional hydraulic model as a tool to help predict the effect of current deflectors on the River Idle. MSc dissertation, University of Nottingham, UK.

Brookes, A. (1988). *Channelised Rivers: Perspectives in Environmental Management.* Chichester, UK: John Wiley.

Brookes, A. (1995). River channel restoration: theory and practice. In *Changing River Channels,* eds. A. Gurnell & G. E. Petts, pp. 368–388. Chichester, UK: John Wiley.

Brookes, A. (1996). Floodplain restoration and rehabilitation. In *Floodplain Processes,* eds. M. G. Anderson, D. E. Walling & P. D. Bates, pp. 553–576. Chichester, UK: John Wiley.

Brookes, A. & Shields, F. D., Jr (eds.) (1996). *River Channel Restoration: Guiding Principles for Sustainable Projects.* Chichester, UK: John Wiley.

Brookes, A., Gregory, K. J. & Dawson, F. H. (1983). An assessment of river channelisation in England and Wales. *Science of the Total Environment,* **27,** 97–122.

Bruns, D. A., Minshall, G. W., Cushing, C. E., Cummins, K. W., Brock, J. T. & Vannote, R. L. (1984). Tributaries as modifiers of the RCC: analysis of polar ordinations and regression models. *Archiv für Hydrobiologie,* **99,** 208–220.

Cairns, J. (1991). The status of the theoretical and applied science of restoration ecology. *The Environmental Professional,* **13,** 186–194.

Calfed Bay–Delta Program (1999). *Strategic Plan for Ecosystem Restoration Program.* Sacramento, CA: Calfed Bay–Delta Program.

California Department of Water Resources (CDWR) (1986). *Clear Creek Fish Study.* Red Bluff, CA: CDWR Northern District.

Cals, M. J. R., Postma, R., Buijse, A. D. & Marteijn, E. C. L. (1998). Habitat restoration along the River Rhine in The Netherlands: putting ideas into practice. *Aquatic Conservation: Marine and Freshwater Ecosystems,* **8,** 61–70.

Collier, M. P., Webb, R. H. & Andrews, E. D. (1997). Experimental flooding in the Grand Canyon. *Scientific American,* **276,** 82–89.

Common, M. (1995). *Sustainability and Policy: Limits to Economics.* Cambridge: Cambridge University Press.

Connell, J. H. (1978). Diversity in tropical rainforests and coral reefs. *Science,* **199,** 1302–1310.

Downs, P. W. & Kondolf, G. M. (in press). Post-project appraisals in adaptive management of river channel restoration. *Environmental Management.*

Downs, P. W. & Thorne, C. R. (1998). Design principles and suitability testing for rehabilitation in a flood defence channel: the River Idle, Nottinghamshire, UK. *Aquatic Conservation: Marine and Freshwater Ecosystems,* **8,** 17–38.

Downs, P. W. & Thorne, C. R. (2000). Rehabilitation of a lowland river: reconciling flood defence with habitat diversity and geomorphological sustainability. *Journal of Environmental Management,* **58,** 249–268.

Downs, P. W., Skinner, K. S. & Soar, P. J. (1999). Muddy waters: issues in assessing the impact of in-stream structures for river restoration. In *The Challenge of Rehabilitating Australia's Streams: 2nd Australian Stream Management Conference,* eds. I. Rutherford & R. Bartley, pp. 211–217. Clayton, Australia: Co-operative Research Centre for Catchment Hydrology.

Doyle, M. W., Stanley, E. H., Luebke, M. A. & Harbor, J. M. (2000). Dam removal: physical, biological, and societal considerations. In *American Society of Civil Engineers Joint Conference on Water Resources Engineering and Water Resources Planning and Management,* 30 July–2 August 2000, Minneapolis, MN. New York: American Society of Engineers.

Elliot, R. (1997). *Faking Nature: The Ethics of Environmental Restoration.* London: Routledge.

Elwood, J. W., Newbold, J. D., O'Neill, R. V. & Van Winkle, W. (1983). Resource spiralling: an operational paradigm for analysing lotic ecosystems. In *Dynamics of Lotic Ecosystems,* eds. T. D. Fontaine & S. M. Bartell, pp. 3–27. Ann Arbor, MI: Ann Arbor Science.

Environment Agency (1998). *River Geomorphology: A Practical Guide.* National Centre for Risk Analysis and Options Appraisal Guidance Note 18, prepared by C. R. Thorne, P. W. Downs, M. D. Newson, M. J. Clark & D. A. Sear. London: Environment Agency.

Environment Agency (1999). *Idle and Torne Local Environment Agency Plan: Consultation Report.* Solihull, UK: Environment Agency Midlands Region.

Everest, F. H., Sedell, J. R., Reeves, G. H. & Bryant, M. D. (1991). Planning and evaluating habitat projects for anadromous salmonids. *American Fisheries Society Symposium*, **10**, 68–77.

Federal Interagency Stream Restoration Working Group (FISRWG) (1998). *Stream Corridor Restoration: Principles, Processes and Practices*. US National Engineering Handbook, Part 653. Washington, DC: US Department of Agriculture, Natural Resources Conservation Service.

Frissell, C. A. & Nawa, R. (1992). Incidences and causes of physical failure of artificial habitat structures in streams of Western Oregon and Washington. *North American Journal of Fisheries Management*, **12**, 187–197.

Gan, K. & McMahon, T. (1990). Variability of results from the use of PHABSIM in estimating habitat area. *Regulated Rivers: Research and Management*, **5**, 233–239.

Gore, J. A. (1996). Foreword. In *River Channel Restoration: Guiding Principles for Sustainable Projects*, eds. A. Brookes & F. D. Shields, Jr, pp. xiii–xv. Chichester, UK: John Wiley.

Gore, J. A. & Nestler, J. M. (1988). Instream flow studies in perspective. *Regulated Rivers: Research and Management*, **2**, 93–101.

Green, C. H. & Tunstall, S. M. (1992). The amenity and environmental value of river corridors in Britain. In *River Conservation and Management*, eds. P. J. Boon, P. Calow & G. E. Petts, pp. 425–441. Chichester, UK: John Wiley.

Gregory, K. J. & Davis, R. J. (1993). The perception of riverscape aesthetics: an example from two Hampshire rivers. *Journal of Environmental Management*, **39**, 171–185.

Halbert, C. L. & Lee, K. N. (1991). Implementing adaptive management. *Northwest Environmental Journal*, **7**, 136–150.

Harper, D. & Everard, M. (1998). Why should the habitat level approach underpin holistic river survey and management? *Aquatic Conservation: Marine and Freshwater Ecosystems*, **8**, 395–413.

Harper, D. M., Smith, C. D. & Barham, P. J. (1992). Habitats as the building blocks for river conservation and assessment. In *River Conservation and Management*, eds. P. J. Boon, P. Calow & G. E. Petts, pp. 311–320. Chichester, UK: John Wiley.

Heiler, G., Hein, T. & Scheimer, F. (1995). Hydrological connectivity and flood pulses as the central aspects for the integrity of a river–floodplain system. *Regulated Rivers: Research and Management*, **11**, 351–361.

Hein, T., Heiler, G., Pennetzdorfer, D., Riedler, P., Schagerl, M. & Schiemer, F. (1999). The Danube restoration project: functional aspects and planktonic productivity in the floodplain system. *Regulated Rivers: Research and Management*, **15**, 259–270.

Henry, C. P., Amoros, C. & Giuliani, Y. (1995). Restoration ecology of riverine wetlands. 2: An example in a former channel of the Rhône River. *Environmental Management*, **19**, 903–913.

Hey, R. D. (1994). Environmentally sensitive river engineering. In *The Rivers Handbook: Hydrological and Ecological Principles*, eds. P. Calow & G. E. Petts, pp. 337–362. Oxford: Blackwell.

Hey, R. D. & Heritage, G. L. (1993). *Draft Guidelines for the Design and Restoration of Flood Alleviation Schemes*. R&D Note no. 154. Bristol, UK: National Rivers Authority.

Higgins, P. (1999). Design of a large-scale experiment to estimate the functional relationship between steelhead smolt production and in-stream flow. Presentation at American Fisheries Society, North Pacific International Chapter meeting, February 1999.

Holling, C. S. (1978). *Adaptive Environmental Assessment and Management*. New York: John Wiley.

Hughes, R. M, Larsen, D. P. & Omernik, J. M. (1986). Regional reference sites: a method for assessing stream potentials. *Environmental Management*, **10**, 629–635.

Iversen, T. M., Kronvang, B., Madsen, B. L., Markmann, P. & Nielsen, M. B. (1993). Re-establishment of Danish streams: restoration and maintenance measures. *Aquatic Conservation: Marine and Freshwater Ecosystems*, **3**, 73–92.

Johnson, B. L., Richardson, W. B. & Naimo, T. J. (1995). Past, present, and future concepts in large river ecology. *BioScience*, **45**, 134–141.

Johnson, W. C. (1992). Dams and riparian forests, case study from the upper Missouri River. *Rivers*, **3**, 229–242.

Junk, W. J., Bayley, P. B. & Sparks, R. E. (1989). The flood pulse concept in river–floodplain systems. *Special Publication of the Canadian Journal of Fisheries and Aquatic Sciences*, **106**, 110–127.

Karr, J. R., Fausch, K. D., Angermeier, P. L., Yant, P. R. & Schlosser, I. J. (1986). *Assessing Biological Integrity in Running Waters: A Method and its Rationale*. Champaign, IL: Illinois Natural History Survey.

Kondolf, G. M. (1995). Five elements for effective evaluation of stream restoration. *Restoration Ecology*, **3**, 133–136.

Kondolf, G. M. (1997). Hungry water: effects of dams and gravel mining on river channels. *Environmental Management*, **21**, 533–551.

Kondolf, G. M. & Downs, P. W. (1996). Catchment approach to channel restoration. In *River Channel Restoration*, eds. A. Brookes & F. D. Shields, Jr, pp. 291–329. Chichester, UK: John Wiley.

Kondolf, G. M. & Micheli, E. M. (1995). Evaluating stream restoration projects. *Environmental Management*, **19**, 1–15.

Kondolf, G. M. & Wilcock, P. R. (1996). The flushing flow problem: defining and evaluating objectives. *Water Resources Research*, **32**, 2589–2599.

Kondolf, G. M., Kattelmann, R., Embury, M. & Erman, D. C. (1996). *Status of Riparian Habitat, in Sierra Nevada Ecosystem Project: Final Report to Congress*, vol. 2, *Assessment and Scientific Basis for Management Options*. Davis, CA: University of California, Centers for Water and Wildland Resources.

Kondolf, G. M., Larsen, E. W. & Williams, J. G. (2000). Measuring and modelling the hydraulic environment for assessing in-stream flows. *North American Journal of Fisheries Management*, **20**, 1016–1028.

Lamplugh, G. W. (1914). Taming of streams. *The Geographical Journal*, **43**, 651–656.

Large, A. R. G. & Petts, G. E. (1994). Rehabilitation of river margins. In *The Rivers Handbook*, vol. 2, eds. P. Calow & G. E. Petts, pp. 401–417. Oxford: Blackwell.

Marmorek, D. R., Peters, C. N. & Parnell, I. (eds.) (1998). *Plan for analysing and testing hypotheses (PATH)*. Final report for 1998. Prepared by ESSA Technologies. www.bpa.gov/Environment/PATH

Mathur, D., Bason, W. H., Purdy, E. J., Jr & Silver, C. A. (1985). A critique of the Instream Flow Incremental Methodology. *Canadian Journal of Fisheries and Aquatic Science*, **42**, 825–831.

McBain, S. & Trush, W. (2000). Lower Clear Creek floodway rehabilitation project: channel reconstruction, riparian vegetation, and wetland creation design document. Unpublished report, California, USA.

McLain, R. J. & Lee, R. G. (1996). Adaptive management, promises and pitfalls. *Environmental Management*, **20**, 437–448.

Miles, M. (1998). Restoration difficulties for fishery migration in high-energy gravel-bed rivers along highway corridors. In *Gravel-Bed Rivers in the Environment*, eds. P. C. Klingeman, R. L. Beschta, P. D. Komar & J. B. Bradley, pp. 393–414. Highlands Ranch, CO: Water Resources Publications.

Minshall, G. W., Cummins, K. W., Petersen, R. C., Cushing, C. E., Bruns, D. A., Sedell, J. R. & Vannote, R. L. (1985). Developments in stream ecosystem theory. *Canadian Journal of Fisheries and Aquatic Sciences*, **42**, 1045–1055.

Mitchell, B. (1990) Integrated water management. In *Integrated Water Management: International Experiences and Perspectives*, ed. B. Mitchell, pp. 1–21. London: Belhaven Press.

Mitchell, B. & Pigram, J. J. (1989) Integrated resource management and the Hunter Valley Conservation Trust, NSW, Australia. *Applied Geography*, **9**, 196–211.

Naiman, R. J. & Décamps, H. (eds.) (1990). *The Ecology and Management of Aquatic-Terrestrial Ecotones*. Paris: UNESCO.

National Research Council (NRC) (1992). *Restoration of Aquatic Ecosystems: Science, Technology and Public Policy*. Washington, DC: National Academic Press.

National Research Council (NRC) (1999). *New Strategies for America's Watersheds*. Washington, DC: National Academic Press.

Nestler, J. M., Schneider, L. T. & Latka, D. (1993). *Physical Habitat Analysis of Missouri River Main Stem Reservoir Tailwaters using the Riverine Community Habitat Assessment and Restoration Concept (RCHARC)*. US Army Corps of Engineers, Waterways Experimental Station, Technical Report EL-93-22. Vicksburg, MS: US Army Corps of Engineers.

Nestler, J. M., Schneider, L. T., Latka, D. & Johnson, P. (1996). Impact analysis and restoration planning using the Riverine Community Habitat Assessment and Restoration Concept. In *Proceedings of the 2nd IAHR Symposium on Habitat Hydraulics, Ecohydraulics 2000*, eds. M. Leclerc, A. Boudreault, H. Capra, S. Valentin & Y. Cote, pp. 871–876. Quebec: IAHR.

Newbold, J. D., Elwood, J. W., O'Neill, R. V. & Van Winkle, W. (1981). Measuring nutrient spiralling in streams. *Canadian Journal of Fisheries and Aquatic Sciences*, **38**, 860–863.

Newbold, J. D., Mulholland, P. J., Elwood, J. W. & O'Neill, R. V. (1982a). Organic carbon spiralling in stream ecosystems. *Oikos*, **38**, 266–272.

Newbold, J. D., O'Neill, R. V., Elwood, J. W. & Van Winkle, W. (1982b). Nutrient spiralling in streams: implications for nutrient limitation and invertebrate activity. *American Naturalist*, **120**, 628–652.

Padmore, C. L. (1997). Biotopes and their hydraulics: a method for defining the physical component of freshwater quality. In *Freshwater Quality: Defining the Indefinable*, eds. P. J. Boon & D. L. Howell, pp. 251–257. Edinburgh: Scottish Natural Heritage.

Peterman, R. M. & Peters, C. N. (1998). Decision analysis: taking uncertainties into account in forest resource management. In *Statistical Methods for Adaptive Management Studies*, Land Management Handbook no. 42, eds. V. Sit & B. Taylor, Victoria, BC: British Columbia Ministry of Forests. (www.for.gov.bc.ca/hfd)

Peters, M. R., Abt, S. R., Watson, C. C., Fischenich, J. C. & Nestler, J. M. (1995). Assessment of restored riverine habitat using RCHARC. *Water Resources Bulletin*, **31**, 745–752.

Petts, G. E. (1989). Historical analysis of fluvial hydrosystems. In *Historical Change of Large Alluvial Rivers*, ed. G. E. Petts, pp. 1–17. Chichester, UK: John Wiley.

Petts, G. E. & Amoros, C. (eds.) (1996a). *Fluvial Hydrosystems*. London: Chapman & Hall.

Petts, G. E. & Amoros, C. (1996b). The fluvial hydrosystem. In *Fluvial Hydrosystems*, eds. G. E. Petts & C. Amoros, pp. 1–12. London: Chapman & Hall.

Petts, G. E. & Maddock, I. (1994). Flow allocation for in-river needs. In *The Rivers Handbook: Hydrological and Ecological Principles*, eds. P. Calow & G. E. Petts, pp. 289–307. Oxford: Blackwell.

Pickett, S. T. A. & White, P. S. (1985). Patch dynamics: a synthesis. In *The Ecology of Natural Disturbance and Patch Dynamics*, eds. S. T. A. Pickett & P. S White, pp. 371–384. New York: Academic Press.

Reiger, H. A., Welcomme, R. L., Steedman, R. J. & Henderson, H. F. (1989). Rehabilitation of degraded river ecosystems. *Special Publication of the Canadian Journal of Fisheries and Aquatic Sciences*, **106**, 86–97.

Richards, K. S. (1973). Hydraulic geometry and channel roughness: a nonlinear system. *American Journal Science*, **273**, 877–896.

Rosgen, D. L. (1994). A classification of rivers. *Catena*, **22**, 169–199.

Roux, A. L. & Copp, G. H. (1996). Fish populations in rivers. In *Fluvial Hydrosystems*, eds. G. E. Petts & C. Amoros, pp. 167–183. London: Chapman & Hall.

Sacramento River Advisory Council (SRAC) (2000). *Sacramento River Conservation Area Handbook*. Sacramento, CA: California Department of Water Resources.

Scheimer, F. & Zalewski, M. (1992). The importance of riparian ecotones for diversity and productivity of riverine fish communities. *Netherlands Journal of Zoology*, **42**, 323–335.

Scheimer, F., Baumgartner, C. & Tockner, K. (1999). Restoration of floodplain rivers: the Danube restoration project. *Regulated Rivers: Research and Management*, **15**, 231–244.

Sear, D. A. (1994). River restoration and geomorphology. *Aquatic Conservation: Marine and Freshwater Ecosystems*, **4**, 169–177.

Sedell, J. R., Richey, J. E. & Swanson, F. J. (1989). The river continuum concept: a basis for the expected ecosystem behavior of very large rivers? *Special Publication of the Canadian Journal of Fisheries and Aquatic Sciences*, **106**, 49–55.

Shen, H. W., Tabios, G. & Harder, J. A. (1994). Kissimmee River restoration study. *Journal of Water Resources Planning and Management*, **120**, 330–349.

Shields, F.D., Jr (1996). Hydraulic and hydrological stability. In *River Channel Restoration: Guiding Principles for Sustainable Projects*, eds. A. Brookes & F.D. Shields, Jr, pp. 23–74. Chichester, UK: John Wiley.

Shuman, J. R. (1995). Environmental considerations for assessing dam removal alternatives for river restoration. *Regulated Rivers: Research and Management*, **11**, 249–261.

Skinner, K. S. (1999). Geomorphological post-project appraisal of river rehabilitation schemes in England. PhD dissertation, University of Nottingham, UK.

Skinner, K. S. & Downs, P. W. (2000). Monitoring AND evaluation: post-project appraisals – support for river restoration design. *American Geophysical Union Spring Meeting*, **81**, 19.

Smith, S. (1997). Changes in the hydraulic and morphological characteristics of a relocated stream channel. MS dissertation, University of Maryland, Annapolis.

Soar, P. J. (1996). In-stream physical habitat potential of the River Idle, Nottinghamshire, to evaluate the necessity for flow deflectors. BSc dissertation, University of Nottingham, UK.

Soar, P. J. (2000). Channel restoration design for meandering rivers. PhD dissertation, University of Nottingham, UK.

South Florida Water Management District (SFWMD) (2001). www.sfwmd.gov *and* www.sfwmd.ced.fau.edu/techpub/

Statzner, B. & Higler, B. (1985). Questions and comments on the river continuum concept. *Canadian Journal of Fisheries and Aquatic Sciences*, **42**, 1038–1044.

Swindale, N. R. (2000). Numerical modelling of river rehabilitation schemes. PhD dissertation, University of Nottingham, UK.

Third International Symposium on Ecohydraulics (1999). *Strategies for Sampling, Characterisation and Modeling of Aquatic Ecosystems in Applied Multidisciplinary Assessment Frameworks*. Salt Lake City, UT: Utah State University Extension. (CD-ROM)

Tockner, K., Schiemer, F., Bauggartner, C., Kum, G., Weigand, E., Zweimüller, I. & Ward, J. V. (1999). The Danube restoration project: species diversity patterns across connectivity gradients in the floodplain system. *Regulated Rivers: Research and Management*, **15**, 245–258.

Toth, L. A. (1996). Restoring the hydrogeomorphology of the channelised Kissimmee River. In *River Channel Restoration: Guiding Principles for Sustainable Projects*, eds. A. Brookes & F. D. Shields, Jr, pp. 371–380. Chichester, UK: John Wiley.

Toth, L. A., Obeysekera, J. T. B., Perkins, W. A. & Loftin, M. K. (1993). Flow regulation and restoration of Florida's Kissimmee River. *Regulated Rivers: Research and Management*, **8**, 155–167.

Toth, L. A., Melvin, S. L., Arrington, D. A. & Chamberlain, J. (1998). Managed hydrologic manipulations of the channelised Kissimmee River: implications for restoration. *BioScience*, **48**, 757–764.

US Army Corps of Engineers (1985). *Central and Southern Florida, Kissimmee River, Florida: Final Feasibility Report and Environmental Impact Statement*. Jacksonville, FL: US Army Corps of Engineers.

US Bureau of Reclamation (USBR) (1999). *Value Planning Report: Lower Clear Creek Hydraulic Analysis at Whiskeytown Dam*. Denver, CO: Technical Services Center.

US Fish and Wildlife Service (USFWS) (1995). *Anadromous Fish Restoration Plan: Working Paper*. Sacramento, CA: US Fish and Wildlife Service.

Vannote, R. L., Minshall, G. W., Cummins, K. W., Sedell, J. R. & Cushing, C. E. (1980). The river continuum concept. *Canadian Journal of Fisheries and Aquatic Sciences*, **37**, 130–137.

Walters, C. J. (1986). *Adaptive Management of Renewable Resources*. New York: MacMillan.

Ward, J. V. (1998). Riverine landscapes: biodiversity patterns, disturbance regimes, and aquatic conservation. *Biological Conservation*, **83**, 269–278.

Ward, J. V. & Stanford, J. A. (1983). The serial discontinuity concept of lotic ecosystems. In *Dynamics of Lotic Ecosystems*, eds. T. D. Fontaine & S. M. Bartell, pp. 29–41. Ann Arbor, MI: Ann Arbor Science.

Ward, J. V. & Stanford, J. A. (1995). The serial discontinuity concept: extending the model to floodplain rivers. *Regulated Rivers: Research and Management*, **10**, 159–168.

Ward, J. V., Tockner, K. & Scheimer, F. (1999). Biodiversity of floodplain river ecosystems: ecotones and connectivity. *Regulated Rivers: Research and Management*, **15**, 125–139.

Warne, A. G., Toth, L. A. & White, W. A. (2000). Drainage-basin-scale geomorphic analysis to determine reference conditions for ecologic restoration: Kissimmee River, Florida. *Geological Society of America Bulletin*, **112**, 884–899.

Welcomme, R. L. (1995). Relationships between fisheries and the integrity of river systems. *Regulated Rivers: Research and Management*, **11**, 121–136.

Welcomme, R. L., Ryder, R. A. & Sedell, J. A. (1989). Dynamics of fish assemblages in river systems: a synthesis. *Special Publication of the Canadian Journal of Fisheries and Aquatic Sciences*, **106**, 569–577.

Wolman, M. G. & Gerson, R. (1978). Relative scales of time and effectiveness of climate in watershed geomorphology. *Earth Surface Processes*, **3**, 189–208.

13 • Lakes

ERIK JEPPESEN AND ILKKA SAMMALKORPI

INTRODUCTION

Of the total water resource on earth, freshwater and saline lakes constitute less than 0.02% (Wetzel, 1983). In many parts of the world, however, freshwater lakes especially are a vitally important resource for humans. Lake water is used for potable supply and for irrigation, for harvesting fish and other food resources and for recreational activities such as boating, swimming and angling. Lakes contribute significantly to biodiversity on earth and act as important foraging areas for many terrestrial animals and waterfowl.

Although the total volume of lakes is dominated by a few large and deep ones, most lakes are small and shallow (Wetzel, 1983; Moss, 1998). There are significant differences in the trophic structure and dynamics of shallow and deep lakes as well as in their sensitivity to threats such as increasing nutrient loading or water abstraction. An essential difference is that, in summer, deep lakes often show thermal stratification that largely isolates the upper water layers (epilimnion) from the colder deep water (hypolimnion), preventing interaction with the sediment. Contrarily, in shallow non-stratified lakes there is direct contact between sediment and water and recycling of nutrients is then potentially rapid. Shallow water depth also enables much stronger interactions between the various elements of the food chain. Fish predation on zooplankton increases (Jeppesen et al., 1997a) and macrophytes and mosses have the potential to cover large areas. In contrast, macrophytes are limited to near-shore areas in deep lakes due to either the effect of light, wave action, the nature of the substrate or water pressure (Duarte & Kalff, 1986; Sand-Jensen & Borum, 1991). In shallow lakes and wetlands, submerged macrophytes are significant as both habitat and food for numerous consumers, including waterfowl, and have an essential structuring effect on the whole ecosystem (Carpenter & Lodge, 1986; Jeppesen et al., 1997a). The main emphasis of this chapter is on shallow lakes, which are particularly sensitive to the various anthropogenic impacts such as catchment drainage for agricultural purposes or nutrient loading and are often situated in areas with widespread human activities. Accordingly, shallow lakes have often been the objects of supplementary in-lake restoration measures. We will also concentrate on the eutrophication process and on how to restore eutrophied lakes. For further details, we refer to Sas (1989), Carpenter & Kitchell (1993), Cooke et al. (1993), Moss (1998) and Scheffer (1998). It is outside the scope of this chapter to discuss other restoration issues related to xenobiotics and acidification. For recent reviews of the latter, see Stoddard et al. (1999).

NATURE OF THE PROBLEM AND RATIONALE FOR RESTORATION

Eutrophication is the result of excessive nutrient loading to the receiving lake (Vollenweider & Kerekes, 1982), reflecting mainly human-related impacts such as poorly treated sewage water, industry and intensive agriculture. Since the 1970s, loading from the sources of sewage and industry has been reduced significantly in Europe, North America and other industrial countries. However, considering the continuing significant loading from agriculture and scattered settlements in these areas and the poor treatment of municipal and industrial sewage in other parts of the world, eutrophication remains a major global problem. The deterioration of water quality stands in contrast to the growing need for fresh water as 'the blood of the society' (Wetzel, 2000).

Increasing nutrient concentrations have led to a major increase in fish density and to a shift in dominance from piscivorous to plankti-benthivorous fish, such as cyprinids, with major implications for the lake ecosystem (Hrbácek, 1969; Persson et al., 1988; Jeppesen et al., 1991, 2000; Perrow et al., volume 1) (Fig. 13.1). In European lakes, an increase in roach (Rutilus rutilus) and bream (Abramis brama) abundance with total phosphorus (TP) and changes in the size structure of zooplanktivorous fish towards dominance by small specimens enhance the predation pressure on zooplankton. The biomass ratio of zooplankton to phytoplankton decreases from 0.5–0.8 in mesotrophic lakes to <0.2 when summer mean TP concentrations exceed 0.10–0.15 mg P l^{-1}. At the latter figure, zooplankton is not capable of controlling the phytoplankton, whose turnover time in eutrophic lakes may last 0.5–2 days. With decreasing grazing pressure by zooplankton and enhanced nutrient supply, phytoplankton biomass increases, resulting in reduced transparency (Secchi depth). An increase in fish predation may also reduce snail abundance and thus, grazing of epiphytes on plants, further impoverishing the growth conditions for submerged macrophytes (see Richardson & Jackson, volume 1). The plants disappear and the food source and feeding habitats of aquatic birds are diminished. The result is a lake with a large biomass of carp (Cyprinus carpio), roach and bream, high abundance of phytoplankton, few or no submerged macrophytes (Fig. 13.1) and a greatly reduced density of birds, dominated by fish-eating species (Jeppesen et al., 2000).

Resistance to increasing nutrient loading

Even though most multi-lake empirical equations and the data presented in Fig. 13.1 indicate a simple relationship between nutrient loading and the environmental state of lake ecosystems, this is not always true for an individual lake. Particularly in shallow lakes, the shift to the turbid state often occurs rather abruptly when a given lake-specific

nutrient threshold is reached (Scheffer et al., 1993). Thus, considerable resistance towards the shift is found in the early phase of nutrient loading which, as outlined below, may be attributed to a number of feedback mechanisms. One factor is retention of phosphorus by adsorption to unexploited binding sites in the surface sediment (as discussed in Søndergaard et al., volume 1). We will here focus on biological feedback mechanisms.

The role of macrophytes

Particularly in shallow lakes, submerged macrophytes may buffer in the following ways: (1) as nutrient loading increases, macrophyte biomass also increases causing enhanced fixation of nutrients in macrophytes and epiphytes, leaving less nutrients available to phytoplankton in summer; (2) increased abundance of submerged macrophytes may reduce sediment resuspension, which otherwise often entails increased nutrient release to the water (Carpenter & Lodge, 1986); (3) some investigations indicate that the roots and larger surface area of plants promote denitrification and, consequently, nitrogen loss from the lake (Eriksson & Weisner, 1999); (4) submerged macrophytes may locally diminish phytoplankton by shading (see also Weisner & Strand, volume 1).

The effect of plants on nutrients and light does not fully explain why submerged macrophytes promote clearwater conditions as, at the same P concentrations, lakes with high macrophyte coverage are more transparent than lakes without or only a low coverage of macrophytes (Canfield et al., 1984; Jeppesen et al., 1990; Faafeng & Mjelde, 1998; Scheffer, 1998). This is also true for lakes with macrophyte beds along the shore and open water offshore, implying that the effect must reach beyond the plant-covered area. A number of indirect effects have been offered as an explanation. First, by reducing wave forces, submerged macrophytes promote sedimentation and reduce resuspension with the effect that water is more transparent in shallow lakes where wind-induced resuspension may otherwise be quite significant (Hamilton & Mitchell, 1996; Barko & James, 1977). Second, resuspension may also be reduced via the effect of macrophytes upon fish community structure. For

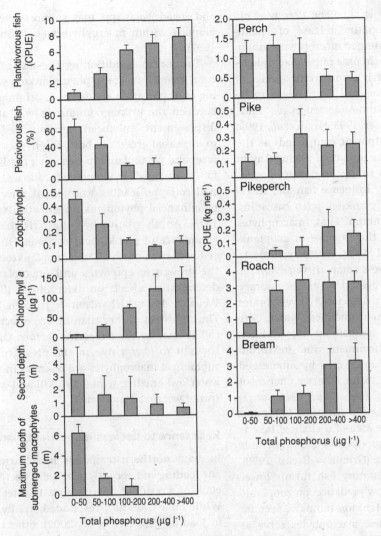

Fig. 13.1. (Left) August biomass of zooplanktivorous fish (measured as CPUE [catch per 18-hour night per net in kg wet weight] in late summer; catch in multiple mesh size gill nets, 14 different mesh sizes 6.25–75 mm) versus summer mean lake water total phosphorus (TP). Also shown are the percentage of carnivorous fish, summer mean (1 May–1 October) of zooplankton:phytoplankton biomass ratio, chlorophyll *a*, Secchi depth and the maximum depth of submerged macrophytes versus the lake water TP concentration. Mean ± S.E. of the five TP groups is shown. From Jeppesen *et al.* (1999). (Right) Biomass (CPUE) of various quantitatively important fish species in Danish lakes versus summer mean TP. The first three species are piscivorous, while the last two are plankti-benthivorous. Modified from Jeppesen *et al.* (2000).

example, benthivorous fish such as bream and carp stir up sediment when searching for food and this may substantially enhance the concentration of suspended sediment and nutrients (Meijer *et al.*, 1990;

Breukelaar *et al.*, 1994). Bream and carp are often abundant in macrophyte-free eutrophic lakes but rare or absent in macrophyte-rich lakes, being replaced by tench (*Tinca tinca*) or rudd (*Scardinius erythrophthalmus*) (Perrow *et al.*, 1999; Perrow *et al.*, volume 1). Third, allelopathy (release of chemical substances from submerged macrophytes inhibiting phytoplankton growth) may cause macrophyte-rich lakes to be particularly transparent. Species of *Ceratophyllum, Myriophyllum* and *Chara* seem to be potential candidates (Wium-Andersen *et al.*, 1982; Gross & Sütfeld, 1994; Jasser, 1995; Nakai *et al.*, 1999) with polyphenols and sulphuric compounds as the important substances. The impact of allelopathy in the field is, however, still subject to debate.

Macrophytes indirectly influence fish and invertebrates, particularly zooplankton, with cascading effects on the phytoplankton. First, macrophytes favour predatory fish at the expense of zooplanktivorous fish (Persson *et al.*, 1988; Perrow, volume 1). Perch (*Perca fluviatilis*) have a competitive advantage over roach within the vegetation, as they forage more efficiently on plant-associated invertebrates in a structured environment and clear water. Conversely, roach are better foragers on zooplankton, in an unstructured environment and in turbid water (Diehl, 1988). In lakes covered by submerged macrophytes, perch thus stand a better chance of reaching the piscivorous stage. Pike (*Esox lucius*) are associated with vegetation (Grimm, 1994) and if 30%–50% of the surface area of a shallow lake is covered by macrophytes, production of 0^+ cyprinids may be controlled by pike (Grimm & Backx, 1990). Higher abundance of predatory fish means lower abundance of cyprinids, low predation on zooplankton and thus low phytoplankton biomass. Second, at least in eutrophic lakes, macrophytes serve as a daytime shelter for pelagic zooplankton species (e.g. *Daphnia* spp.), enabling them to avoid fish predation or to persist longer in summer (Stansfield *et al.*, 1997). At night, when the predation risk is lower, the zooplankton migrate into open water (Box 13.1). Macrophyte refuges thus help augment the grazing pressure on phytoplankton, thereby enhancing water transparency and further improving plant growth conditions. Third, large mussels such as *Anodonta* and *Unio*, which depend on macrophytes in their early stages, may exert a high grazing pressure on phytoplankton in shallow lakes (Ogilvie & Mitchell, 1995). Fourth, macrophyte associated filter-feeding microcrustacea such as *Sida*

and *Simocephalus* spp. may suppress phytoplankton biomass within macrophyte beds (Stansfield *et al.*, 1997).

The factors conditioning the loss of plants and concordant increase in phytoplankton with increasing eutrophication are debated. Originally, the shift between the primary producers was attributed to displacement: enhanced nutrient loading leading to increased growth of both epiphytes on plant surfaces and phytoplankton, reducing the light climate for submerged macrophytes causing their eventual collapse, leading to nutrient release followed by enhanced phytoplankton growth (Phillips *et al.*, 1978). An alternative hypothesis is that increased abundance of plankti-benthivorous fish via enhanced predation on snails and zooplankton stimulates the growth of epiphytes and phytoplankton with deteriorating effects on macrophytes (Brönmark & Weisner, 1992; Richardson & Jackson, volume 1). Thus, indirect, rather than direct, effects of enhanced TP may be the triggering factors. Other factors thought to play a role in the presence/absence of submerged macrophytes are changes in water level, waterfowl grazing, winter fish kill and variations in spring weather conditions.

Resistance to decreasing nutrient loading

In many northern temperate lakes, external nutrient loading has recently markedly decreased, especially as a result of improved wastewater treatment. While some lakes have responded rapidly to changes in loading (Jeppesen *et al.*, 2002), others have been very resistant (Sas, 1989). For some lakes lack of improvement reflects insufficient reduction of nutrient input to trigger a shift to the clearwater state. For example, significant and sustaining changes in the biological community and water transparency of shallow temperate freshwater lakes cannot be expected unless the TP concentration has been reduced to below 0.05–0.1 mg P l^{-1} (Fig. 13.1).

Chemical resistance

Even when the P loading has been sufficiently reduced, resistance against improvements is often found. This resistance may be 'chemical': P concentrations remain high because of P release from

Box 13.1 The role of submerged macrophytes as a daytime refuge for zooplankton

Submerged macrophytes may act as a refuge for pelagic zooplankton during day when the risk of predation by visually hunting fish in the pelagial is high (Timms & Moss, 1984). A series of experiments in Lake Stigsholm, Denmark have shown that the accumulation per area unit is highest in small patches with high plant density (Fig. B13.1), as the zooplankton prefers to hide near the transitional zone between water and plants (Lauridsen & Buenk, 1996), as fish tend to avoid the beds unless being forced into them to avoid predators and as the foraging efficiency of several fish species declines in dense vegetation (Crowder & Cooper, 1979; Winfield, 1986; Jeppesen *et al.*, 1997a; Stansfield *et al.*, 1997).

At night, the zooplankton migrates into open water, probably in search of food since the concentrations of food are low within the macrophyte beds. Experiments have also evidenced that zooplankton accumulation is highest in macrophyte beds.

Calculations suggest that if small dense macrophyte-covered areas (2 m in diameter) were established in Lake Stigsholm, daytime accumulation of cladocerans in the vegetation would be so high that a coverage of only 3% of the lake area might cause a doubling of the whole-lake cladoceran density at night when they migrate into open water (Lauridsen *et al.*, 1996). Establishment of macrophyte refuges therefore considerably enhances zooplankton grazing on phytoplankton and thus indirectly encourages submerged macrophyte growth by improving the light conditions. Recent surveys, however, indicate that the role of zooplankton in

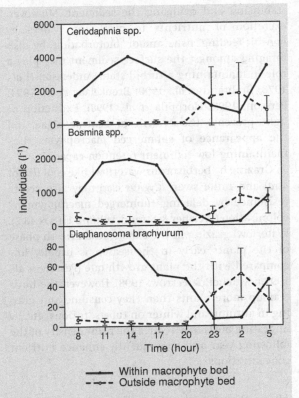

Fig. B13.1. Diel variations (mean ± SD) in the abundance of various cladocerans in a 2-m exclosure (open to small fish and zooplankton) with dense coverage of submerged macrophytes and at a reference station outside the macrophyte bed in Lake Stigsholm, Denmark, in August. Modified from Lauridsen *et al.* (1996).

controlling phytoplankton is less significant in oligomesotrophic lakes with abundant vegetation (Blindow *et al.*, 2000).

the sediment pool accumulated when loading was high (see Søndergaard *et al.*, volume 1). Several years may pass before the surplus pool is released or permanently buried. The duration of this transitional period depends on, for instance, the duration of the excessive nutrient loading, the residence time and the iron supply from the surroundings. As much as 20–40 years have been observed and the delay may be long even in lakes with a short residence time.

Biological resistance

Biological interactions also affect internal P loading and the physicochemical water quality. Planktivorous and benthivorous fish, particularly, contribute to biological resistance in shallow eutrophic lakes. Continuously high fish predation prevents both the appearance of large herbivorous zooplankton that would otherwise clear the water, and diminishes the number of benthic animals stabilising and oxidising the sediment. Moreover, excretion of nutrients to overlaying waters by benthic-feeding fish, and/or bioturbation by fish foraging amongst the surface sediment may play a role in maintaining a turbid state (Andersson *et al.*, 1978; Brabrand *et al.*, 1990; Breukelaar *et al.*, 1994; Persson, 1997; Horppila *et al.*, 1998). Extinction of light prevents the growth of benthic algae and the appearance of submerged macrophytes thus maintaining low sediment retention capacity.

Grazing by herbivorous waterfowl like coot (*Fulica atra*) and mute swan (*Cygnus olor*) may also create resistance by delaying submerged macrophyte re-colonisation (Weisner & Strand, volume 1) (Box 13.2). Waterfowl grazing in the exponential growth phase of the plants early in the season is usually low compared with the plant growth rate (Perrow *et al.*, 1997a; Mitchell & Perrow, 1998). However, the birds tear up more plants than they consume and grazing in autumn and winter on tubers, turions, etc. by migrating birds may reduce the plant density of the following year and consequently enhance nutrient concentrations.

Alternative states

Since resistance occurs both at increasing and decreasing nutrient loading, two alternative states of equilibria may occur within intermediate nutrient levels: the turbid and the clearwater states (Scheffer *et al.*, 1993). The decisive factors are the current nutrient level (the lower the nutrients the higher the probability of the clearwater state) (Scheffer *et al.*, 1993) and the state prevailing before the change in nutrient loading was initiated. The conceptual model described is founded on comprehensive data and theoretical deliberations. However, many debates concern the nutrient level at which alternative states can be expected to occur. Experience from Danish lakes indicates that alternate states typically appear between 0.04 and 0.15 mg P l^{-1}, although when N loading is low or in very small lakes with a large littoral zone or frequently low fish abundance, they may occur at higher nutrient levels. Also, seriously wastewater-impacted lakes may be artificially clear as high oxygen consumption or periodically high pH may lead to fish kill or poor fish recruitment (Beklioglu *et al.*, 1999). Finally, differences in water depth and temperature are likely to play a role.

PRINCIPLES OF RESTORATION AND THEIR APPLICATION

The acknowledgement of resistance to loading reductions and alternative states has, as outlined below, led to the development of a number of methods accelerating the recovery process after the reduction of external nutrient loading.

Reduction of internal loading by physicochemical methods

Various physicochemical methods have been used to reduce internal P loading, including sediment removal in shallow lakes and chemical treatment of the sediment with alum or iron salts in stratified lakes. In deep lakes, injections with oxygen or nitrate to the bottom layer or continuous destabilisation of the thermocline by effective circulation of the water column have been employed. A detailed description of these restoration measures is given in Søndergaard *et al.* (volume 1), and they will not be further elaborated here.

Water level alterations

Water level management has been used extensively as a tool to improve the habitat for waterfowl and promote game fishing and water quality (Cooke *et al.*, 1993). The method may, however, not always be feasible for recreational, nature conservation or agricultural reasons.

The net effect of natural or minor artificial alterations in water level is difficult to predict as multiple factors are involved, such as the degree of water level alterations, lake water turbidity, plant type and wave exposure (wind speed and exposure, sediment type and lake size). Both theoretical deliberations and empirical data suggest a unimodal relationship between water depth and submerged plant growth, with optimum conditions for plants at intermediate depths. If depth is high, plants may become light-limited and, conversely, if depth is

Box 13.2   Waterfowl exclosures as a restoration tool

Several reports have evidenced a major delay in the re-establishment of submerged macrophytes following clear water after fish manipulation. One of the resilience factors is grazing by herbivorous birds such as coots (*Fulica atra*) and mute swans (*Cygnus olor*). Results from an exclosure experiment in shallow Lake Stigsholm, Denmark (24 ha, mean depth 0.8 m) are shown in Fig. B13.2. A number of pots each containing one *Potamogeton crispus* shoot were placed inside and outside a chicken-wire fence at different locations in the littoral zone. Both total and mean shoot length, as well as number of shoots, became larger in the protected areas, while the percentage of stubble was higher in the unprotected areas showing signs of cutting.

Another experiment conducted in shallow Lake Væng, Denmark (16 ha, mean depth 1.2 m) using the same set-up evidenced significant differences in plant net growth between protected and unprotected plants in wind-sheltered areas near reed beds, while the difference was less pronounced at the exposed site. This difference can be ascribed to the higher abundance of coots in sheltered areas close to reeds that the birds used for nesting and hiding. Thus, the need for plant protection is greatest in sheltered areas, since the light conditions for plants, and consequently plant growth, are poorer at the sheltered sites in which the epiphyte biomass is typically greatest (Weisner *et al.*, 1997). Not all experiments have, however, demonstrated an effect of waterfowl grazing on plant growth and it may rightly be claimed that the experimental set-up overestimated grazer impact as the patches were small. Large-scale experiments in Lake Stigsholm did, however, also demonstrate

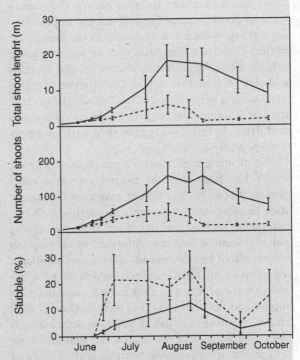

Fig. B13.2. Mean total shoot length, number of shoots and percentage stubble of *Potamogeton crispus* planted in pots in small-scale exclosure experiments. Unfenced exclosures are indicated by a dotted line, fenced exclosures by a solid line. Bars indicate standard error (*n* = 7). Modified from Søndergaard *et al.* (1996).

significantly higher plant growth in protected compared to unprotected areas (Søndergaard *et al.*, 1996).

The effect of waterfowl grazing is assumed to be greatest in the early colonisation phase and in years with sparse vegetation. Hence, M. R. Perrow (unpublished data) did not find significant grazer effects on plants in Lake Stigsholm in a year with high abundance of plants.

low sediment stability declines with frequent resuspension and inferior substrate conditions for the plants. Lake Tåkern, Sweden (Blindow *et al.*, 1993), and Hawksberry Lagoon, New Zealand (S. Mitchell, unpublished data) provide good examples of this phenomenon.

Changes in the water level may also influence lakes indirectly by affecting fish recruitment. Lack of flooding of marginal meadows in spring has been suggested as an important factor for poor recruitment of pike in the regulated Dutch lakes (Grimm, 1994). Conversely, shifting water levels, such as those found in reservoirs, may reduce the recruitment of planktivorous fish and thereby improve water quality (Kubecka & Duncan, 1994). Short-term partial drawdown has been used to improve game fishing (Cooke *et al.*, 1993), enhancing the biomass and size of predatory fish at the expense of planktivorous and benthivorous fish. This may be because the lower water table augments the predation risk or it dries the fertilised eggs in their early ontogeny (Seda & Kubecka, 1997).

The ultimate regulation is a complete drawdown, which has been used to control nuisance plant growth (Cooke *et al.*, 1993). It may also facilitate a shift to clearwater conditions in nutrient-rich turbid lakes, at least in the short term, as drying out may consolidate the sediment. Moreover, fish kill mediated by the drawdown enhances zooplankton grazing on phytoplankton, which in turn improves water clarity and thus growth conditions for submerged macrophytes. This phenomenon is well-known from re-established lakes on nutrient-rich soils, where the lakes during the first year following restoration are often clear with high abundance of submerged macrophytes despite high nutrient concentrations, but change to a turbid state after a few years, concurrently with immigration of planktivorous fish. Drawdown is also used to optimise conditions for waterfowl in wetlands as it often leads to a large-scale fish kill or dominance of piscivores (Cooke *et al.*, 1993).

Flushing to control phytoplankton growth

Lakes with low hydraulic retention time tend to be clearer than would be expected from their nutrient level, as the losses to flushing combined with other loss factors exceed the phytoplankton growth rate. Typically, however, less than 3–5 days are required for phytoplankton to regrow (Reynolds, 1984; E. Jeppesen, unpublished data) in summer, and the method is therefore often not applicable at this time. In contrast, flushing during winter when the phytoplankton growth rate is low may be a more realistic restoration tool, especially in lakes with relatively high concentrations of cyanobacteria that 'overwinter'. Winter flushing, which may wash out the cyanobacteria and precipitate a shift to the macrophyte state in early spring, has been considered beneficial in several Dutch lakes (Hosper & Meijer, 1986).

Fish manipulation

Various methods have been developed to overcome biological resistance, to enhance top–down control of phytoplankton and reduce resuspension and internal nutrient loading by plankti-benthivorous fish (Boxes 13.3–13.6). As reviewed by Shapiro (1990), the term 'biomanipulation' which he introduced in the 1970s covers all methods of reducing fish density to improve water quality inspired by early Czech experiments (Hrbácek *et al.*, 1961) and the size efficiency hypothesis of Brooks & Dodson (1965). As summarised by Drenner & Hambright (1999), the majority of successful biomanipulations thus far have been in lakes smaller than 25 ha and with a mean depth below 3 m. However, projects in deeper and stratified lakes (Lake Haugatjern: Reinertsen *et al.*, 1990) and in lakes larger than 1 km^2 (Boxes 13.3 and 13.4) have also been successful. The first successful biomanipulation in a large lake was carried out in shallow Lake Christina, Minnesota in 1987. It had shifted from a clearwater waterfowl lake to a turbid planktivore-dominated one. Following fish removal by rotenone treatment, stocking of walleye (*Stizostedium vitreum*) and largemouth bass (*Micropterus salmoides*) (see Table 13.2) and aeration to prevent winter fish kill of the piscivores, *Daphnia* and the clearwater state returned and macrophytes started to recolonise and waterfowl returned. *Potamogeton* spp. and *Myriophyllum* were followed by *Chara* dominance six years later (Hanson & Butler, 1990; Hansel-Welch *et al.*, in press).

The enhanced recruitment of planktivores, increased fecundity of the remaining adults and high survival and density of the 0^+ themselves are central problems of mass removal of planktivores. The problems are exacerbated at higher nutrient levels as 0^+ survival increases (Langeland & Reinertsen, 1982; Meijer *et al.*, 1999). Consequently, predation pressure on zooplankton may become even higher after an effective removal despite an overall reduction in fish biomass (Hansson *et al.*, 1998; Meijer *et al.*, 1999; Romare & Bergman, 1999). Enhancement of

Box 13.3   Lake Finjasjön: from sediment dredging to mass removal of fish

Lake Finjasjön (12 km²) in Sweden was clear with a transparency of about 2 m in the 1920s. External loading, particularly municipal sewage (up to 5.9 g P m⁻² y⁻¹, an order of magnitude higher than the critical level), shifted the lake to the turbid state. After tertiary wastewater treatment was implemented, the external loading decreased to 0.45 g P m⁻² y⁻¹, close to the acceptable level. Internal loading maintained high P concentrations and annual summer blooms of *Microcystis*. Restoration was first attempted by removing the nutrient-rich sediment by suction dredging. When 25% of the planned 6.6 km² was dredged, the management shifted to biomanipulation and the means of further decreasing external loading, since high P release also occurred from the dredged areas. A wetland was constructed between the sewage treatment plant and the lake and 5-m buffer zones were established along the most important inlet dykes. Removal of 430 tonnes (392 kg ha⁻¹) of plankti-benthivorous fish in 1992–4 (80% of the fish stock) caused a dramatic decrease in TP (from 0.2 to less than 0.05 mg l⁻¹) and chlorophyll *a* concentrations (from almost 100 to 20–30 µg l⁻¹) and a significant increase in transparency (from 0.4 to 1.5–2 m) in 1995–7 (Fig. B13.3). Submerged macrophyte coverage increased from 1% to 20% of the lake surface area and the percentage of piscivorous fish (which were returned to the lake) increased from 8% to 50%. The zooplankton:phytoplankton biomass ratio increased, suggesting a higher grazer control on phytoplankton. The substantial decrease of average P concentration reflected mainly reduction of the internal P loading maintained by the previously

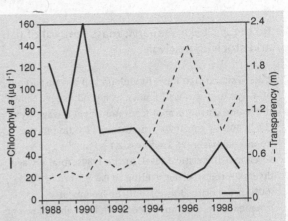

Fig. B13.3. Mean chlorophyll *a* (solid line) and transparency (dotted line) in Lake Finjasjön before and after fish removal carried out in 1992–1994 and continued since 1998. The years of removal are marked with horizontal bars.

dense cyprinid populations. Thus, biomanipulation had cascading effects throughout the food web down to the nutrient level. The inlet concentration was so low (0.04 mg P l⁻¹) that the lake could have maintained the clear state. However, destabilisation after a decline of *Elodea* from coverage of 30% to 18% as well as increased nutrient concentrations and phytoplankton biomass coinciding with higher numbers of cyprinids were observed three years after the fish removal. This could also have resulted from unrestricted catches of predatory fish. Additional control of the planktivorous fish (90 kg ha⁻¹) was undertaken in 1998/9 with a positive response in water quality in summers 1999 and 2000.

Sources: Persson (1997), Hansson *et al.* (1998), Annadotter *et al.* (1999), Nilsson (1999, and unpublished data).

piscivores is therefore an additional or alternative biomanipulation tool. Moreover, the growth of perch to predatory size is improved by removal of cyprinids (Søndergaard *et al.*, 2000).

Dramatic, cascading short-term effects are generally achieved in eutrophic lakes when the fish reduction has been effective (Boxes 13.3 and 13.4; Fig. 13.2). Reduced phytoplankton biomass and improved transparency typically follow. The chlorophyll *a*/TP ratio which is often high when small zooplankton dominates under intensive fish predation (Hrbácek *et al.*, 1978; Mazumder, 1994) may also be lowered by biomanipulation (Sarvala *et al.*, 2000) (Fig. 13.3). Fish manipulation may cascade to the nutrient level, leading to oligotrophication. A 30%–50% reduction in lake concentrations

Box 13.4 Lake Vesijärvi: mass removal of fish and stocking piscivores

Lake Vesijärvi (109 km²) in Finland had a transparency of 4–6 m and was well known as a good fishery before it became eutrophicated from municipal sewage from the 1950s. By the 1970s, the external loading to the Enonselkä basin (26 km²) was 2.1 g P m⁻² y⁻¹, seven-fold the critical level. In the 1980s, total sewage diversion reduced the loading to 0.2 g P m⁻² y⁻¹ (60% of the critical level), but the blooms of cyanobacteria continued. Moreover, high abundance of roach was recorded in regular gillnet monitoring, and a very dense population of smelt (*Osmerus eperlanus*) was found by acoustic recording. Values of recreation and fisheries collapsed in the 1960s and were still low 12 years after sewage diversion.

Restoration was carried out from 1987 to 1994 as a joint venture between municipal authorities, local NGOs, research institutes and university researchers. The measures included fish removal, large-scale stocking of zander (*Stizostedion lucioperca*, also known as pikeperch), diverting stormwaters to sewers in the city of Lahti, and management of shores for recreational purposes and enhancement of natural reproduction of pike. Intensive pelagic trawling was carried out in summer from 1989 to 1993. During those five years, 423 kg ha⁻¹ were removed. The roach biomass declined from 175 to 52 kg ha⁻¹ and that of smelt from 75 to 12 kg ha⁻¹. TP (from 0.05 to less than 0.03 mg l⁻¹) and chlorophyll *a* (from 23–28 to less than 10 μg l⁻¹) decreased markedly, while transparency increased (from 1.6 to 3 m) and the maximum depth distribution of submerged macrophytes increased (from 2.5 to 3.5 m) (Fig. B13.4). A reduction in cyanobacteria occurred from summer 1990 when the roach biomass had declined by almost 50%.

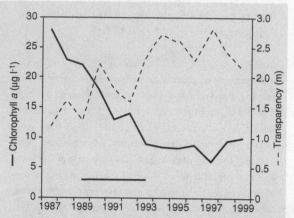

Fig. B13.4. Mean chlorophyll *a* (solid line) and transparency (dotted line) in the Enonselkä Basin of Lake Vesijärvi before and after fish removal carried out in 1989–1993 and during the maintenance fishing since 1994 (box).

According to cohort analysis of the trawl catches, *c*. 18 kg ha⁻¹ y⁻¹ roach ought to be removed to prevent the recovery of the roach stock. Follow-up fishing of this intensity has continued since 1994 with gear developed during the project. Massive zander introduction (700 000 fingerlings) in 1987–91 resulted in increased catches but not in top–down control of roach as zander preferred smelt and perch and large specimens were diminished by intensive gillnet fishing. Since 1996, a mesh limit of 100 mm (stretched) has been set to enhance the survival of, and predation by, zander. Roach has not recovered and the water has remained clear. The successful restoration activated local inhabitants, NGOs and municipalities. This co-operation has continued in 12 smaller lakes in the area.

Sources: Keto & Sammalkorpi (1988), Horppila & Peltonen (1994), Horppila *et al.* (1998), Malinen (1999), Peltonen *et al.* (1999).

of TP has been found in the most successful fish manipulation experiments in shallow and stratified eutrophic lakes (Fig. 13.2), even when macrophytes were absent (Søndergaard *et al.*, 2000). A significant contributory factor is increased growth of microbenthic algae owing to improved light conditions at the sediment surface stimulating fixation of inorganic N and P. More benthic algae, less sedimentation of phytoplankton due to higher grazing (particularly in deeper lakes) and more benthic animals due to lower predation may all result in higher redox potential in the surface sediment, which may reduce

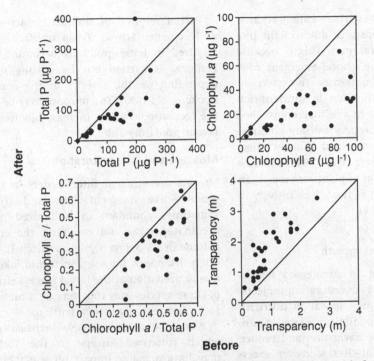

Fig. 13.2. Transparency, chlorophyll *a*, TP and chlorophyll/TP in lakes before and after an effective fish removal. Data (summer averages) compiled from Hansson *et al.* (1998), Søndergaard *et al.* (1998), Meijer *et al.* (1999) and Sammalkorpi (in press, *b*).

P release and stimulate the coupled nitrification-denitrification process, though the effect of benthic invertebrates is not unambiguous (Andersson *et al.*, 1988). Furthermore, benthic invertebrates and algae may stabilise the sediment, thereby reducing the risk of resuspension. Lower fish density reduces the enhancing effects of fish on nutrient concentrations (see above). Finally, nutrients are removed with the fish. The amount of P removed with fish has exceeded the average lake water P pool in some lakes. It may represent 20%–30% of the average external P loading, but it is usually low compared with the internal loading in eutrophic lakes.

The findings to date indicate that fish manipulation will have a long-term effect in shallow temperate lakes if P concentration is reduced to below 0.05–0.1 mg P l^{-1} during summer and in deeper lakes with a mean depth of 6–8 m (Hansson *et al.*, 1998). This threshold is in accordance with the major changes in lake biological structure which usually occur within this range (Fig. 13.1). It is unlikely

that the effect will prevail at higher nutrient concentrations in the long run unless periodical reduction in the abundance of planktivorous fish is conducted since cyprinid abundance will most likely rise again (Figs. 13.1 and 13.3). However, if N loading is low, fish manipulation may have a positive impact at higher TP.

Protection of submerged macrophytes and transplantation

The construction of exclosures to protect macrophytes against waterfowl grazing has been employed as an alternative or supplementary restoration tool to fish manipulation (Box 13.2). The exclosures enable the macrophytes to grow in a grazer-free environment from where they may spread seeds, turions or plant fragments augmenting colonisation. Moreover, they serve as a daytime refuge for zooplankton (Box 13.1). The usefulness of plant refuges as a restoration tool is probably particularly high in small lakes, and in lakes where colonisation is restricted to the near-shore area due to light limitation in a deeper central part. This is because waterfowl aggregate in the littoral emergent zone that provides shelter and nest sites. The effects of plants as a refuge seem greatest in nutrient-rich lakes where plant density in the macrophyte bed is often highest and where prey fish are confined to the pelagial to a higher extent than in oligo-mesotrophic lakes (Jeppesen et al., 1998; Blindow et al., 2000). Transplantation of plants or seeds is an alternative method, which is treated exclusively in Weisner & Strand (volume 1).

Combating nuisance plant growth

Although re-establishment of submerged macrophytes is the goal of many lake restoration projects, dense plant beds, which may appear in nutrient-enriched lakes, are occasionally considered a nuisance as they may impede navigation and reduce the recreational value for anglers. Moreover, excessive growth of invading species, like the Eurasian milfoil in many lakes in the United States as well as Africa, or the North American Elodea canadensis in Europe, may substantially alter lake ecosystems and constitute a serious threat to the native flora. Methods to combat such nuisance plant growth are manual harvesting, introduction of specialist phytophagous insects such as weevils (Room, 1990) or herbivorous grass carp (Ctenopharyngon idella), water level drawdown, coverage of the sediment with sheets or chemical treatment with herbicides (for a detailed review see Cooke et al., 1993). Often, harvesting and water level drawdown have only a short-term effect because of fast regrowth of the plant community and high external loading. Grass carp may have a strong effect on plant growth and are currently used in many parts of the world to reduce macrophyte abundance, but a shift to a turbid state may be caused by overstocking. In practice, it has proven difficult to find an optimal density promoting a diverse community with moderate biomass of submerged macrophytes. It is also difficult to compensate for an overdose since carp are difficult to catch (Leslie, 1988). The method should therefore be used with caution.

Partial removal of emergent macrophytes such as Phragmites, Scirpus, Typha or Equisetum may be required in lakes gradually becoming overgrown. Removal is carried out by cutting or dredging depending on the severity of the condition (see Cooke et al., 1993). The purpose may be to increase the recreational value, increase biodiversity and to favour predatory fish.

Mussels and lake restoration

Mussels are efficient filter feeders in lakes. Large unioids like Anodonta, Unio and Hyridella are sometimes abundant in well-mixed macrophyte-dominated lakes and can filter the entire water volume in 1.5–3 days (Ogilvie & Mitchell, 1995). They often disappear, however, in turbid lakes probably due to predation at the larval stage. Reintroduction of these species may therefore be a useful tool that so far has received little attention.

Also, the zebra mussel, Dreissena polymorpha, which colonized Europe in the 19th century, may have a major impact on water clarity when abundant. For example, field data from Polish lakes showed epilimnion filtering rates as high as 3–5 days (Stanczykowska, 1977). Significant effects on water clarity have also been shown in enclosure experiments (Reeders et al., 1992). The growing conditions for Dreissena may be enhanced through manipulation of substrate conditions. Whilst there is scope to use Dreissena as a management tool to improve water clarity and also enhance the food resource available to waterfowl in Europe, this may be problematic in other circumstances. For example, in North America, the lack of natural enemies in the rapid colonization phase allows Dreissena to reach enormous density. This has resulted in significantly reduced chlorophyll to TP ratios in nearshore areas of the Great Lakes (Nicholls et al., 1999) but also in fouling of water intakes in reservoirs and uncontrolled impacts on the entire lake ecosystem (Pace et al., 1999; Richardson & Jackson, volume 1).

PROPOSED STRATEGY FOR RESTORATION OF EUTROPHIC, TEMPERATE LAKES

Prior to initiating lake restoration, the strategy to be employed should be carefully considered. It is vital to know the present and past environmental state of the lake and its nutrient loading to ensure the optimum solution of the problem. The procedure set out below is recommended.

- Determination of annual P and N loading from direct measurements or area coefficient models. By means of the OECD-model (Vollenweider & Kerekes, 1982), the P level of the lake can be calculated and compared with actual measurements of average nutrient concentrations. Managers are recommended to employ the OECD-models of the lake type treated (shallow, deep or reservoirs) or to use equations developed on lakes from their own region.
- If the calculated annual mean concentration of TP based on the current external loading is higher than 0.05–0.1 mg l^{-1} for shallow lakes (mean depth <3 m) and, say, 0.01–0.02 mg l^{-1} for deep lakes (mean depth >10 m), the first step is to reduce the external P loading from point sources and/or the diffuse catchment loading. This may be achieved by establishing cultivation-free zones along inflow dykes and streams, changing manure practice, decreasing fertilising levels, improving wastewater treatment in scattered villages, and re-establishing or constructing wetlands as well as re-meandering of channelised streams to re-establish the lost retention capacity. If in-lake measures, described below, are implemented at too high an external P loading, there is a risk that their effects will be of only short duration unless the treatment is repeated continuously. This is treatment of the symptom rather than restoration. However, in shallow lakes with low TN loading but high TP concentrations, due to either past sewage loading or natural causes conditioned by the geology of the area, a clearwater state may prevail at higher TP concentrations. In deep lakes, N-fixation often seems to compensate for the N-deficit, leading to cyanobacteria dominance.
- If a sufficiently low external loading is achieved and the lake remains in the turbid state, additional measures could be applied, as a further reduction of external loading will improve the possibility of achieving permanent improvement.
- If the measured summer mean TP concentration is considerably higher than the critical values

calculated from OECD-models or local models and if there is a regular increase of TP in the growing season, internal loading is probably high. If the concentration is higher than, say, 0.25 mg P l^{-1} in shallow lakes and 0.05 mg P l^{-1} in deep lakes, achieving long-term effects by biomanipulation alone is uncertain. In such cases physicochemical methods should also be considered. For shallow lakes, these include removal of the sediment or treatment with iron or aluminium salts, and for deep lakes oxidation of the hypolimnion, perhaps in combination with chemical treatment (Søndergaard et al., volume 1).

- If the concentration is close to 0.1 mg P l^{-1} in shallow lakes and 0.02 mg P l^{-1} in deep lakes, if the fish density or CPUE is high and dominated by plankti-benthivorous fish, and if the chlorophyll a/TP ratio is high, biomanipulation may be used (see below). Particularly in shallow lakes, other biological measures may also be feasible. Where the recovery of submerged macrophytes is delayed, either introduction of seeds or plants of local provenance and/or protection against their grazing should be contemplated. If large mussels do not establish within the first few years despite earlier occurrence, it may be estimated whether mussels from nearby lakes or streams should be introduced.
- If external loading exceeds that specified above, its reduction is technically or economically difficult, but improvements in the environmental state of the lake are still desired, the restoration methods described above may be employed. However, frequent follow-up or continuous treatment of the lake may be necessary, and there is a risk that even extensive measures will not lead to an improvement.
- If an excessive biomass of submerged or emergent macrophytes is the matter of concern, partial manual harvesting is recommended. Biological control such as the introduction of grass carp or phytophagous insects such as weevils may offer an alternative solution.

PLANNING A BIOMANIPULATION

Prior to a biomanipulation it is essential to evaluate both the theoretical and practical options and to develop a suitable organisation of infrastructure and work out a detailed plan with clearly established objectives for the manipulation (see Holl & Cairns, volume 1). Consensus with the owners of fishing rights and public opinion must be thoroughly sought in the planning phase. Prevention of immigration from unmanipulated areas and an immediate release of piscivorous and other valuable species are crucial elements to be arranged. As often several tonnes of fish must be removed, the logistic preparations must be comprehensive and include handling, transport and final deposition or beneficial use of the catch. Active participation of local inhabitants or stakeholders may be important to projects in small lakes. In large lakes, it is recommended to include professional, experienced fishermen as they both possess the necessary skills for handling fish and also often are in possession of the necessary seines or trawls and equipment for transport.

It is important to undertake continuous monitoring, both of catches and of the effects on the fish community and the derived impacts on central biological and physicochemical variables. This will enable continuous adjustment of the intervention and the elaboration of a suitable strategy for a possible maintenance of the achieved target state. Also the presence of invertebrate predators. (e.g. *Chaoborus, Neomysis, Leptodora, Bythotrepes*) needs to be assessed, since a very strong reduction of planktivores may temporily increase their density and, thus, predation on *Daphnia* spp. A successful example of planning, implementation, adjustment and final results is described from Lake Wolderwijd (Backx & Grimm, 1994; Grimm & Backx, 1994; Meijer & Hosper, 1997).

Setting the targets for fish removal

The removal must be effective since inadequate stock reduction is considered as the main reason for a lack of short-term effects. In successful projects, the reduction has often been a minimum 70%–80% of the biomass of plankti-benthivores, often several hundreds of kg ha^{-1} in a short period (Perrow et al., 1997b; Hansson et al., 1998, Meijer et al., 1999). The target level to be left in the lake is c. 50 kg ha^{-1} or even less, particularly if the remaining fish are juveniles. However, improved water quality after a removal of c. 50% and remaining amounts

of 75–100 kg ha^{-1} were observed in, for instance, Lakes Væng and Vesijärvi (Søndergaard et al., 1990) (Box 13.4).

Catch in multimesh gillnets and point-abundance sampling by electro-fishing are efficient low-cost methods for analysis of proportional abundance of species, their temporal changes and spatial or between-lake differences (Perrow et al., 1996; Berg et al., 1997; Jeppesen et al., 1998; Peltonen et al., 1999). These methods are suitable for stratified random sampling and a high sample number which are advantages in the statistical testing of results. Fish density may be estimated by trawl surveys in the pelagial and seining in the littoral zones (Frankiewicz et al., 1986; Grimm & Backx, 1994; Perrow et al., 1998), by acoustic methods in deeper lakes (Kitchell, 1992; Horppila et al., 1996), or combining, for instance, vertical and horizontal echosounding in the pelagial and seining in the littoral (Kubecka et al., 1998). Density of some important pelagic planktivores, for instance smelt, can only be detected by trawl or echosounder. The traditional mark–recapture method is often used, particularly in smaller lakes with few species. Accuracy of the target estimates may be controlled by analysing the catch data of an intensive fish removal by virtual population analysis or depletion methods (Horppila & Peltonen, 1994; Peltonen et al., 1999). It is recommended to combine such monitoring methods to allow judgement of whether the change in CPUE is real or caused by an alteration in the behaviour and catchability when the aims of clear water, colonization of macrophytes and higher percentage of piscivores are achieved. For further details of fish stock assessment, see Perrow et al. (volume 1).

When the target catch for biomanipulation cannot be evaluated from quantitative determination of the fish stock, a tentative estimate of fish biomass (kg ha^{-1}) may be estimated from empirical relations to TP (µg l^{-1}). The approach was developed by Hanson & Legett (1982) (Fig. 13.3; solid line):

$$\text{fish biomass} = 2.17 \, TP^{0.78} \qquad \text{(Eq. 13.1)}$$

However, this equation underestimates the stock in shallow, nutrient-rich and cyprinid-dominated lakes. For Danish, primarily shallow, lakes crude

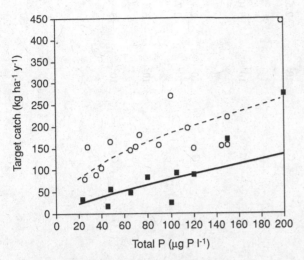

Fig. 13.3. The catch-need of fish removal vs. TP concentration in eutrophic European lakes dominated by planktivorous and benthivorous fish. White circles denote the annual catch in cases with effective fish removal and an improvement of water quality (increased transparency, decline in cyanobacteria) at least in the short term or increase in the numbers of piscivorous perch. Black squares denote cases in which the fish removal was too weak to have an effect on water quality or fish density. The solid and dotted lines are based on Eqs. 13.1 and 13.4, respectively.

estimates can be achieved from the following relationships (E. Jeppesen, unpublished data) based on gillnet catches in numerous lakes converted to biomass from mark–recapture experiments and capture figures from fish manipulation experiments (J. P. Müller & H. Jerl Jensen, unpublished data):

$$\text{fish biomass} = 9.42 \, TP^{0.62} \quad \text{(all fish)} \qquad \text{(Eq. 13.2)}$$

$$\text{fish biomass} = 1.46 \, TP^{0.93} \quad \text{(plankti-benthivores)} \qquad \text{(Eq. 13.3)}$$

Also, experience from successfully restored cyprinid dominated Finnish, Scandinavian and Dutch lakes suggests that the target catch (kg ha^{-1}) in one year is higher than the total biomass estimated from Eq. 13.1 (catch-need) (Fig. 13.3; dotted line):

$$\text{catch-need} = 16.9 \, TP^{0.52} \qquad \text{(Eq. 13.4)}$$

Techniques and strategies for fish manipulation

Detailed knowledge of fish species in the target lake is required to maximise the efficiency of fish manipulation. Special attention must be paid to the assessment and control of younger year classes as these are likely to have the greatest effect on water quality. Commercial fishing gear is usually inadequate as such, since smaller mesh sizes (10–20 mm, stretched mesh) are needed for removal of juveniles. Intensive stocking of piscivorous fish, which have decreased in eutrophic lakes (Fig. 13.1), is recommended as a follow-up measure. It may also be an alternative if the target fish are within the size range vulnerable to the gape-limited piscivores (see Perrow, volume 1). The dominant North American cyprinids do not typically attain the sufficient size to escape predation (Tonn et al., 1990), but several European cyprinids grow too large to be swallowed. Fish removal alone or combined with piscivore stocking has been the main strategy particularly in European lakes (Hansson et al., 1998; Perrow et al., 1998; Drenner & Hambright, 1999; Meijer et al., 1999). A flexible approach, using whatever gear to maximise results is needed (Perrow et al., 1998) (Table 13.1). A simple feasible strategy of fish removal is to catch passive fish with active gear and active fish with passive gear. Seasonally predictable behaviour such as spawning or foraging migration and shoaling of the target species increases the probability of catching a large part of the population aggregated in, or passing by, delimited areas (Fig. 13.4) (Sammalkorpi, 2000). For details of the methods and of average potential catches, see Table 13.1.

Active gear and methods

Autumn and winter fishing of aggregated fish is the most important removal method in temperate lakes (see Table 13.1). It is selective for the target species, and often for age classes too. Juvenile cyprinids, which are dispersed in the littoral in summer, either aggregate in the littoral margins, tributary streams, under boat bridges and other natural or man-made refuges in shallow lakes or shoal in the pelagial of deeper lakes in autumn and winter. Beach seine and electro-fishing are used in shallow lakes and pelagic

Table 13.1. *Applicability of methods used for fish removal, depending on lake type and age groups of target species, and an approximated potential daily catch per gear per day*

Method	Behavioural basis	Shallow lakes, dykes[a]	Deep/stratified lakes	Small lakes (<100 ha)	Large lakes (>100 ha)	Older fish (>2–3 years)	Juvenile cyprinids (0+ 1+)	Expenses	Release of piscivores	Local participation	Potential CPUE (kg gear^{-1}d^{-1})	References
Fykenets and traps at spawning time	SM	G	M	G	M	G	L	G	G	G	10^2	2,3,5,9
Large fykenets	DM, SM	G	M	G	M	M	M	G	G	M	10^2	9
Traps and small fykenets, in summer	WM	G	M	G	M	G	L	G	G	G	10^2	2,5,9
Pelagic seine, encircling shoals	SA	M	G	G	G	M	G	M	G	M	10^3	9,10
Beach seine	LA, SA	G	L	G	M	M	G	G	G	G	10^3	2,3,5,7
Large pelagic seines	SD	G	M	G	G	M	L	M	G	L	10^3	2,7
Sectorial total seining	SD	G	L	G	M	G	G	M	G	M	10^3	5,8
Scare line	SD	G	L	G	L	G	L	G	G	G	10^2	5

Method		Efficiency/selectivity				Fish density	References
Electro-fishing	LA, SA	G	G	G	M / L	10^2	1,2,3,5
Disturbance of spawning		M	M	G	G / G		1,5
Drawdown of water level	SD	G	L	G	G / M	10^3	1,12
Pelagic trawling	LA	L	G	M	M / L	10^3	2,4,11
Rotenone poisoning[b]	SD (LA)	G	G	G	M / L		1,4,6
Closing of immigration routes	SM	G	G	G	G / G	10^3	2,5,7,8
Gillnets	SD, LA	G	L	G	M / G	10^1	13
Piscivore stocking		G	G	G	G / G		1,6,12

Key: G: Potentially good efficiency or selectivity, low costs, easy release of piscivores and local participation. M: Moderate efficiency, selectivity or costs. L: Low efficiency or selectivity, high costs, difficult release of piscivores and local participation. DM: diel horizontal migrations between habitats (foraging migration of juveniles from littoral to pelagial). LA: local aggregation. SA: seasonal aggregation. SD: scattered distribution. SM: seasonal migrations between growing and overwintering areas. WM: within habitat movements.

[a] Large and pelagic gear are not used in dykes.

[b] The use of rotenone is limited by legislation or fisheries tradition in many countries.

References: 1. Cooke et al. (1993); 2. Backx & Grimm (1994); 3. Moss et al. (1996); 4. Hansson et al. (1998); 5. Perrow et al. (1996); 6. Drenner & Hambright (1999); 7. Meijer et al. (1999); 8. van Donk et al. (1990); 9. Sammalkorpi et al. (1999); 10. Turunen et al. (1997); 11. Horppila & Peltonen (1997); 12. Seda & Kubecka (1997); 13. Søndergaard et al. (1990).

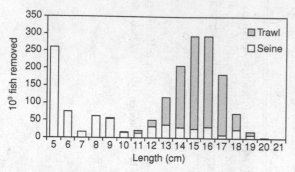

Fig. 13.4. Catch-weighed size distributions of roach caught with different gear in the biomanipulation of Lake Vesijärvi in 1993. Pelagic winter seining was carried out in the nursery area of Paimelanlahti Bay (see Turunen *et al.*, 1997) in 25.2–2.4. and pelagic trawling in the adjacent Enonselkä Basin in 21.6–26.8. (see Horppila *et al.*, 1996). Measurements: J. Horppila & I. Sammalkorpi, unpublished data.

seining or trawling in deeper lakes. Fish which seldom shoal as intensively as younger cyprinids, such as bream, may be collected with very large seines from large areas (up to 50 ha per day: Backx & Grimm, 1994) or with a scare line (a rope with coloured strings strung between boats) to herd the fish into a catching area (Perrow *et al.*, 1998). Particularly smaller lakes or channels may be divided by stopnets to sectors or lanes which are seined or electro-fished one by one (Perrow *et al.*, 1998). The mass removal in Lake Vesijärvi (Box 13.4) was based on summer foraging migration of 3+ and older roach and on the diel vertical migration of smelt. Trawling was effective since roach were in pelagic shoals and smelt aggregated in the deeper areas during day.

Passive gear and methods

Fish on their diurnal or seasonal migrations between habitats or lake basins or moving within littoral vegetation, are caught with passive gear. Adult fish with predictable spawning time and sites are caught on their way to and in the spawning grounds by fykenets or traps. Disturbance of spawning by artificial substrates to be removed after the spawning time, is used for cyprinids. Deliberate drawdown of water level, when possible, may be used to prevent spawning or development of fertilised eggs after spawning of the target species. Significant numbers of fish, including juveniles provided that the mesh size is small enough to retain

---

Box 13.5   Lake Wirbel: stocking 0+ pike to control young planktivores

Lake Wirbel (area 11 ha, maximum depth 4.4 m) is a eutrophic polymictic lake in Poland. Emergent plants covered 7% and floating-leaved plants 10% of the lake area, while submerged macrophyte abundance was low. The fish stock was numerically dominated by roach and *Leucaspius delineatus* with white bream (*Abramis bjoerkna*) and perch as subdominants. To enhance predation on 0+ planktivorous fish, juvenile pike, which are forced to feed on small prey because of gape limitation, were added in October 1987 (250 individuals m^{-2}). In each of the following years, pike were introduced mainly in May (when *c.* 3 cm long) at the time when newly hatched roach larvae appeared. The stocking in these years ranged between 730 and 3200 individuals ha^{-1}. Besides, in-lake transplantations of pike were performed to obtain higher densities in areas where cyprinids were abundant and pike density low. Finally, pike larger than 45 cm were reduced to minimise the predation upon the stocked 0+ pike. While the effect of low-density stocking in October was insignificant, substantial effects were recorded in the three succeeding years with spring stocking in high densities.

According to the authors, the implications for management are that in small shallow lakes yearly stocking of pike fry in high abundances can result in suppression of 0+ prey fish even when survival of the stocked pike at the end of the season is low. To optimise the effect, predator stocking should coincide with the appearance of the newly hatched larvae of the dominant prey species. Also, the authors recommend that predator access to densely vegetated habitats sheltering their potential prey be facilitated.
Source: Prejs *et al.* (1994).

Table 13.2. *Examples of piscivorous fish species and the annual stocking densities (D) used in biomanipulation of some eutrophic lakes in Europe (E) and North America (NAm). The numbers refer to fingerling densities unless otherwise stated*

Species	Lake	Area (ha)	Mean depth (m)	D stocked number ha^{-1} y^{-1}	Objective[a]	Fishing[b]	Effect[c]	Reference
Pike (E, NAm)	Wirbel	11	1.8	730–3200	R	–	S	1
	Lyng	10	2.4	500–3600	R	–	S	2
	Wolderwijd	2700	1.5	>200	F	p	N[h]	3
	Bleiswijkse Zoom	3.1	1.1	1100	F	–	N*	4
	Mendota	4000	12.7	6–14	F	+a, Re	L	5
Pike >2$^+$	Lake 227	5	4.4	40	R	–	S	6
Largemouth bass (NAm)	Wintergreen	15	2.4	47	R	–	S	7
	Christina	1619	1.3	50	F	Re	L*	8
Walleye (NAm)	Mendota	4000	12.7	137	F	+a, Re	(L)	5
	Christina	1619	1.3	25	F	Re	L*	8
Zander (E)	Bautzen	533	7.4	40–150	R	+a,g	(L)	9
	Enonselkä Basin	2600	5.8	40–60	F	Re	(S)	10

[a] R = restoration, F = follow-up.
[b] Fishing conditions: – = no fishing, + = intensive fishing (a = angling, g = gillnets, p = professional eel fishing), and Re = regulated fishing.
[c] Effects: S = short term, L = long term, N[h] = pike success was restricted by lack of suitable habitats, N* = unsuccessful pike stocking, but biomanipulation by fish removal succeeded, L* = signs of a slow return towards higher planktivore density and turbidity were observed during a ten-year period. (L) = fish densities declined but the cascading effects were less clear due to hypertrophy of the reservoir. (S) = the abundance and catches increased but foodweb effects were not observed.
References: 1. Prejs *et al.*, 1994, Box 13.5.; 2. Berg *et al.*, 1997; 3. Meijer & Hosper, 1997; 4. Meijer *et al.*, 1999; 5. Box 13.6; 6. Elser *et al.*, 2000; 7. Mittelbach *et al.*, 1995; 8. Hanson & Butler 1990; 9. Benndorf 1995; 10. Box 13.4.

them, may be caught both with small fykenets in the littoral area and with a few large fykenets located within the area of diurnal horizontal foraging migrations from the littoral to pelagic areas throughout the summer. Fykenets and traps are also used for removing fish aggregating in dykes or closing the immigration routes from other lakes. Gillnets may be used in size-selective removal of roach and other cyprinids in small lakes.

Enhancing populations of piscivorous fish
The basic tools in enhancement of populations of piscivorous fish are stocking with pond-raised

fingerlings, catch regulations and habitat management (e.g. aeration or shoreline management). Examples of stocking densities from different lakes are given in Table 13.2. Stocking with pelagic species (zander, walleye) should be accompanied by littoral species (pike, largemouth bass). Otherwise the nursery areas of the target species are not affected. In an optimal case, avoidance of the littoral piscivore by planktivores renders them vulnerable to the open-water predators (Berg *et al.*, 1997; Søndergaard *et al.*, 1997). Piscivore stocking is potentially more effective in North America where, compared with European cyprinids, few of the

Box 13.6   Lake Mendota: multiple intervention approach

Lake Mendota (catchment area 604 km^2, lake area 40 km^2, maximum depth 25 m) is located in a rich agricultural region near Madison, Wisconsin. The lake is eutrophic mainly due to agricultural runoff and historical sewage inputs from small communities in the catchment. The lake is used intensively for recreation, including fishing, boating, swimming and scenic enjoyment. Severe algal blooms first became problematic in the 1940s, but no remedial measures were implemented until 1971 when sewage was diverted. Algal blooms have continued to be a problem because of non-point pollution. Programmes to reduce these nutrient sources by encouraging voluntary 'best management practices' among the farmers followed. This led to the current comprehensive programme to reduce the loading from landowners and municipalities – the Lake Mendota Priority Watershed Project – which entered its implementation phase in 1998. It is supported by state funds (US$9 million) on a cost-sharing basis and requires clean-up of all 'critical sites' determined to be major sources of pollution. The action plan has included a detailed analysis of the P loading history and effects, and studies of food web interactions in the lake have been conducted. These studies suggested complex interactions between nutrient loading, food web structure and climate are influencing the production of cyanobacteria blooms and water quality.

Prior to the current non-point pollution abatement project, a large biomanipulation experiment was initiated in 1987 to reduce algal blooms and improve recreational fishing in the eutrophic lake. Piscivorous walleye (*Stizostedion vitreum*) and northern pike (*Esox lucius*) were heavily stocked to control planktivorous fish and hence increase algal grazing by *Daphnia*. A massive natural die-off of cisco (*Coregonus artedi*) in summer 1987, coupled with lower nutrient loadings from a prolonged drought and less internal mixing

from warm temperatures, produced exceptional water clarity in the summer of 1988. Planktivore biomasses have remained low throughout most of the 1990s and large-bodied *Daphnia* have remained abundant while nutrient levels have increased due to periods of high runoff. To protect the stocked piscivores, more restrictive fishing regulations had to be instituted. Another in-lake measure that has been active since the 1960s is mechanical harvesting to control the invading plant Eurasian water milfoil (*Myriophyllum spicatum*); harvesting continues but milfoil is less of a problem than in earlier years.

The Lake Mendota case study differs from the majority of lake restoration projects in that catchment action plans and in-lake measures are based on detailed knowledge obtained by: (1) intensive studies by scientists and managers for a prolonged period (many limnological data bases span much of the twentieth century), (2) palaeolimnology, (3) long-term monitoring, (4) development of simple and complex management models for the lake and its catchment, and (5) active co-operation between scientists, managers and the local population and interest groups. The catchment–lake models, including the effects of changing farming practices in the catchment on the storage and transport of nutrients to the lake, and the derived effects on lake trophic dynamics as well as lake water quality, have been valuable tools to guide managers in both the short and long term. This approach has resulted in a scientifically based solution to the eutrophication problem. The present average loading of ~0.85 g P m^{-2} y^{-1} is still too high. Modelling has shown that a P load reduction of 50% would decrease the probability of summer algal blooms >2 mg l^{-1} occurring from 6 out of 10 days to 2 out of 10 days. Further improvements in water clarity can be achieved through food web management. The effects of global climate change on Lake Mendota are more complex, but are being evaluated.

Sources: Kitchell (1992); Lathrop *et al.* (1998, in press); Carpenter & Lathrop (1999).

dominant planktivores attain the size refuge (Tonn *et al.*, 1990). Moreover, the largemouth bass has the relatively highest gape width of temperate piscivores (Mittelbach & Persson, 1998). There are

recent examples from European lakes showing that even a dominance of piscivores does not prevent cyprinids from becoming dominant (Annadotter *et al.*, 1999; Lammens, 1999) particularly in lakes without macrophyte dominance (Perrow et al., 1999). Stocking of zander or walleye needs to be accompanied also by catch and mesh size limits for fishing (see Boxes 13.4 and 13.6). Minimum mesh sizes of 100–120 mm stretched mesh (Bujse et al., 1992) are recommended for gillnets. This benefits both lake managers and fishermen as the average size of the fish caught increases and natural recruitment is boosted by a higher number or size (and fecundity) of spawning fish. For example in the hypertrophic Bautzen Reservoir in Germany, stocking with 40–150 zander fingerlings ha^{-1}, in combination with bag and size limits (>60 cm fish allowed) for zander and pike fishing, resulted in a higher piscivore biomass and a reduction of planktivorous fish (Benndorf, 1995). The opposite trend was found in two other lakes when fishing was not restricted (Table 13.2; see Boxes 13.4 and 13.6). In the small lakes Wirbel (Box 13.5) and Lyng, the introduction of pike fingerlings (3–4 cm) to control 0^{+} cyprinids has been successful in the short term, with excessive densities compensating for the high mortality caused by cannibalism (Table 13.2). In the large Lake Wolderwijd and in the small Lake Bleiswijkse Zoom, the number of refuges was considered too low for even a short-term effect. The introduction of older pike in Lake 227 was successful and after a natural reproduction of pike the planktivorous cyprinids were eliminated (Table 13.2). A dramatic increase of Daphnia density and cascading short-term effects in water quality were observed (Elser et al., 2000).

Aeration may sometimes be needed to prevent a selective winterkill of piscivores. Particularly the North American planktivores withstand oxygen depletion in winter better than their predators and may attain high densities in lakes with frequent oxygen depletion (Tonn & Magnuson, 1982; Hanson & Butler, 1990; Mittelbach et al., 1995). Since cannibalism of pike decreases with the refuges provided by emergent vegetation (Grimm & Backx, 1990), it may be valuable to establish artificial refuges, for example, sprigs of spruce (Skov & Berg, 1999). Also, mitigating water level fluctuations in spring (as discussed above) or increasing structural diversity by opening corridors and patches in large and dense reed beds (Grimm, 1989) may favour pike. The low percentage of piscivorous perch caused by a competitive bottleneck by dense roach populations (Persson et al., 1988) may be enhanced by mass removal of cyprinids (Søndergaard et al., 2000). However, a rapid increase in perch growth in the years following fish manipulation may also lead to higher mortality, implying that the desired enhanced predation on planktivorous fish is not always achieved in the long term.

In addition to the direct effect of piscivory, predator avoidance behaviour induced by piscivore introduction may reinforce the results due to reduced use of the pelagic habitat (Brabrand & Faafeng, 1993; He et al., 1993; Mittelbach et al., 1995; Eklöv & Persson, 1995; Horppila et al., 1998).

Costs of fish manipulation

The costs per weight of fish removed (Table 13.3) vary considerably due to factors affecting the catchability of fish (Table 13.1). The common experience from Lake Vesijärvi and Lake Wolderwijd, where several types of gear were used, was that the costs were lower for seine and fykenet fishing than for trawling. Fishing is usually more expensive in small lakes since the pre- and post-handling time of gear is about the same for large and small lakes. However, the role of less expensive gear such as traps, fykenets and locally developed applications can be significant, particularly in small lakes (Table 13.1). Local, previously inexperienced volunteers can be trained to use fykenets. By empowering local people, in some cases a remarkable percentage of the necessary removal can be done by locals, not least during follow-up maintenance fishing in the years after the mass removal.

CONCLUDING REMARKS

In recent years, a number of biological restoration methods have been developed. They have a large potential to substitute the traditional physico-chemical methods as primary in-lake methods when the external loading has been reduced. In shallow lakes especially, biological methods have yielded positive short-term results, although it must be acknowledged that long-term effects are poorly elucidated as yet. Several investigations have shown that efficient fish removal has positive effects, but a relapse often occurs a few

Table 13.3. *Examples of approximated costs of fish removal from some large lakes in the restoration phase*

Lakes	Fish removal ha^{-1}	Years of fishing	Total catch 10^3 kg	$ kg^{-1}	$ ha^{-1}	Reference
Lake Finjasjön[a,b,c]	392	2.5	425	2.1	*c.* 700	1
Lake Vesijärvi	423	5	1100	0.9[e]	*c.* 400	2
Paimela Bay	355	3	132	0.3[e]	*c.* 100	2
Lake Enäjärvi[d]	373	5	190	0.4[f]	*c.* 170	3
Lake Wolderwijd	154	1	425	0.9	*c.* 170	3
Lake Tuusulanjärvi	415	2.5	247	0.9	*c.* 250	5

[a] Fishing was continued until a decline in CPUE appeared.
[b] Decline was observed in connection with fish stock monitoring.
[c] Successful biomanipulation.
[d] Short-term effect.
[e] Maintenance fishing in Lake Vesijärvi has continued since 1994 with an average expense of *c.* 0.5$ kg^{-1} (J. Keto, unpublished data).
[f] The value of volunteer work, *c.* 30%, not included. *References:* 1. Annadotter *et al.*, 1999; 2. J. Keto & I. Sammalkorpi, unpublished data; 3. Lempinen, 1998; 4. Backx & Grimm, 1994; 5. M. Pekkarinen, unpublished data.

years after manipulation, indicating too high pre-manipulation nutrient levels and/or too low numbers of piscivores. Dominance of the macrophyte community by species showing marked fluctuations in abundance (like *Elodea* or *Ceratophyllum*) may also favour re-establishment of cyprinid dominance in years with low plant biomass (E. Jeppesen & M. Søndergaard, unpublished data). The experience gathered so far indicates that supplementary, but less intensive management to stabilise the lake system is needed in the years following the massive fish stock reduction.

An essential future challenge is optimisation of the different biomanipulation methods, including measures to stabilise the system, and to learn to combine biological and physicochemical methods in a cost-effective manner. For example, it is likely that hypolimnion aeration will strengthen the effect of fish removal in summer-stratified temperate eutrophic lakes. Besides reducing internal P loading, it may allow perch to forage over a larger sediment surface thereby making it easier for this species to pass through the benthic phase before becoming piscivores (Perrow *et al.*, volume 1). Management of eutrophic lakes also demands that the aims of both lake management and fisheries are considered.

Compared with the recently increased knowledge of the structure, function and restoration of temperate lakes, less information exists on subtropical and particularly tropical lakes, which differ in many ways. For instance, the faster nutrient cycling combined with higher predation pressure on zooplankton as a consequence of a lower percentage of pelagic piscivores, repeated fish recruitment over the season and an accordingly reduced grazer control of phytoplankton indicate a more direct link between nutrients and primary producers. There is also a risk of dominance by cyanobacteria owing to a higher probability of N limitation and the warm climate. To date, restoration by biomanipulation has been tested with mixed results in tropical lakes (Lazzaro, 1997) and as the improving economies of developing countries have led to increased nutrient loading of watercourses and thus elevated risk of lake deterioration in the future (Lewis, 2000), there is a great need for more information. Likewise, knowledge of the structure and function of brackish and saline lakes is poor. Lakes in different climatic regions and with different salinities represent a challenging area for applying and developing viable ecological restoration methods.

ACKNOWLEDGMENTS

Thanks to Anne Mette Poulsen for editorial assistance and to Kathe Møgelvang for layout and graphical assistance. Thanks also to Martin Perrow and Tony Davy for inviting us to write this chapter and to Martin Perrow for his very useful comments, and to Richard Lathrop for revision of the case study on Lake Mendota (Box 13.6). Mark Hanson, Jukka Horppila, Juha Keto and Per Åke Nilsson provided unpublished data from Lakes Christina, Vesijärvi and Finjasjön and useful comments.

REFERENCES

Andersson, A., Berggren, H. & Cronberg, G. (1978). Effects of planktivorous and benthivorous fish on organisms and water chemistry in eutrophic lakes. *Hydrobiologia*, **59**, 9–15.

Andersson, G., Granéli, W. & Stenson, J. (1988). The influence of animals on phosphorus cycling in lake ecosystems. *Hydrobiologia*, **170**, 267–284.

Annadotter, H., Cronberg, G., Aagren, R., Lundstedt, B., Nilsson, P.-Å. & Ströbeck, S. (1999). Multiple techniques for lake restoration. *Hydrobiologia*, **395/396**, 77–85.

Backx, J. J. G. & Grimm, M. P. (1994). Mass removal of fish from Lake Wolderwijd, The Netherlands. 2: Implementation phase. In *Rehabilitation of Freshwater Fisheries*, ed. I. G. Cowx, pp. 401–414. Oxford: Fishing News Books.

Barko, J. & James, W. F. (1998). Effects of submerged aquatic macrophytes on nutrient dynamics, sedimentation, and resuspension. In *The Structuring Role of Submerged Macrophytes in Lakes*, eds. E. Jeppesen, Ma. Søndergaard, Mo. Søndergaard & K. Christoffersen, pp. 197–214, New York: Springer-Verlag.

Beklioglu, M., Carvalho, L. & Moss, B. (1999). Rapid recovery of a shallow hypertrophic lake following sewage effluent diversion: lack of chemical resilience. *Hydrobiologia*, **412**, 5–15.

Benndorf, J. (1995). Possibilities and limits for controlling eutrophication by biomanipulation. *Internationale Revue gesamten Hydrobiologie*, **80**, 519–534.

Berg, S., Jeppesen, E. & Søndergaard, M. (1997). Pike (*Esox lucius* L.) stocking as a biomanipulation tool. 1: Effects on the fish population in Lake Lyng (Denmark). *Hydrobiologia*, **342/343**, 311–318.

Blindow, I., Andersson, G., Hargeby, A. & Johansson, S. (1993). Long-term pattern of alternative stable states in two shallow eutrophic lakes. *Freshwater Biology*, **30**, 159–167.

Blindow, I., Hargeby, A, Bálint, M. A., Wagner, A. & Andersson, G. (2000). How important is the crustacean plankton for the maintenance of water clarity in shallow lakes with abundant submerged vegetation? *Freshwater Biology*, **44**, 185–197.

Brabrand, Å. & Faafeng, B. (1993). Habitat shifts in roach (*Rutilus rutilus*) induced by pikeperch (*Stizostedium lucioperca*) introduction: predation risk versus pelagic behaviour. *Oecologia*, **95**, 38–46.

Brabrand, Å., Faafeng, B. & Nilssen, J. P. (1990). Relative importance of phosphorus supply to phytoplankton production: fish excretion versus external loading. *Canadian Journal of Fisheries and Aquatic Sciences*, **47**, 364–372.

Breukelaar, A. W., Lammens, E. H. R. R., Klein Breteler, J. P. G. & Tatrai, I. (1994). Effects of benthivorous bream (*Abramis brama* L.) and carp (*Cyprinus caprio* L.) on sediment resuspension and concentration of nutrients and chlorophyll *a*. *Freshwater Biology*, **32**, 113–121.

Brönmark, C. & Weisner, S. (1992). Indirect effects of fish community structure on submerged vegetation in shallow eutrophic lakes: an alternative mechanism. *Hydrobiologia*, **243/244**, 293–301.

Brooks, J. L. & Dodson, S. I. (1965). Predation, body size and composition of plankton. *Science*, **150**, 28–35.

Bujse, A. D., Pet, J. S., van Densen, W. L. T., Machiels, M. A. M. & Rabbinge, R. (1992). A size- and age-structured model for evaluating management strategies in a multispecies gillnet fishery. *Fisheries Research*, **13**, 95–117.

Canfield, D. E., Shireman, J., Colle, D. E., Haller, W. T., Watkins, C. E. & Maceina, M. J. (1984). Prediction of chlorophyll *a* concentrations in Florida lakes: importance of aquatic macrophytes. *Canadian Journal of Fisheries and Aquatic Sciences*, **44**, 497–501.

Carpenter, S. R. & Kitchell J. F. (eds.) (1993) *The Trophic Cascade in Lakes*. Cambridge: Cambridge University Press.

Carpenter, S. R. & Lathrop, R. C. (1999). Lake restoration: capabilities and needs. *Hydrobiologia*, **395/396**, 19–28.

Carpenter, S. R. & Lodge, D. M. (1986). Effects of submersed macrophytes on ecosystem processes. *Aquatic Botany*, **26**, 341–370.

Cooke, G. D., Welch, E. B., Peterson, S. A. & Newroth, P. R. (1993). *Restoration and Management of Lakes and Reservoirs*. Boca Raton, FL: Lewis Publishers.

Crowder, L. B. & Cooper, W. E. (1979). Structural complexity and fish–prey interactions in ponds: a point of view. In *Response of Fish to Habitat Structure in Standing Water*, eds. D. L. Johnson. & R. A. Stein. *American Fisheries Society Special Publications*, **6**, 1–10.

Diehl, S. (1988). Foraging effects of three freshwater fish: effects of structural complexity and light. *Oikos*, **53**, 207–214.

Drenner, R. & Hambright, D. (1999). Review: Biomanipulation of fish assemblages as a lake restoration technique. *Archiv für Hydrobiologie*, **146**, 129–165.

Duarte, C. M. & Kalff, J. (1986). Littoral slope as a predictor of the maximum biomass of submerged macrophyte communities. *Limnology and Oceanography*, **31**, 1072–1080.

Eklöv, P. & Persson, L. (1995). Species-specific antipredator capacities and prey refuges: interactions between piscivorous perch (*Perca fluviatilis*) and juvenile perch and roach (*Rutilus rutilus*). *Behavioural Ecology and Sociobiology*, **37**, 169–178.

Elser, J. J., Sterner, R. W., Galford, A., Chrzanowski, T., Findlay, D., Mills, K., Paterson, M., Stainton, M. & Schindler, D. W. (2000). Pelagic C:N:P stoichiometry in a eutrophied lake: responses to a whole-lake food-web manipulation. *Ecosystems*, **3**, 293–307.

Eriksson, P. G. & Weisner, S. E. B. (1999). An experimental study on effects of submerged macrophytes on nitrification and denitrification in ammonium-rich aquatic systems. *Limnology and Oceanography* **44**, 1993–1999.

Frankiewicz, P., O'Hara, K. & Peczak, T. (1986). Three small seine nets' method used for assessing the density of juvenile fishes in the Sulejów Reservoir. *Ekologia Polska*, **34**, 215–226.

Faafeng, B. & Mjelde, M. (1998). Clear and turbid water in shallow Norwegian lakes related to submerged vegetation. In *The Structuring Role of Submerged Macrophytes in Lakes*, eds. E. Jeppesen, Ma. Søndergaard, Mo. Søndergaard & K. Christoffersen, pp. 361–368. New York: Springer-Verlag.

Grimm, M. P. (1989). Northern pike (*Esox lucius*) and aquatic vegetation, tools in the management of fisheries and water quality in shallow waters. *Hydrobiological Bulletin*, **23**, 59–67.

Grimm, M. P. (1994). The characteristics of the optimum habitat of northern pike (*Esox lucius* L.). In *Rehabilitation of Freshwater Fisheries*, ed. I. G. Cowx, pp. 235–243. Oxford: Fishing News Books.

Grimm, M. P. & Backx, J. (1990). The restoration of shallow eutrophic lakes and the role of northern pike, aquatic vegetation and nutrient concentration. *Hydrobiologia*, **200/201**, 557–566.

Grimm, M. P. & Backx, J. (1994). Mass removal of fish from Lake Wolderwijd, The Netherlands. 1: Planning and strategy of a large scale biomanipulation project. In *Rehabilitation of Freshwater Fisheries*, ed. I. G. Cowx, pp. 390–400. Oxford: Fishing News Books.

Gross, E. M. & Sütfeld, R. (1994). Polyphenols with algicidal activity in the submerged macrophyte *Myrophyllum spicatum* L. *Acta Horticultura*, **381**, 710–716.

Hamilton, D. P. & Mitchell, S. F. (1996). An empirical model for sediment resuspension in shallow lakes. *Hydrobiologia*, **317**, 209–220.

Hansel-Welch, N., Butler, M. G., Carlson, T. J. & Hanson, M. A. (2000). Ten years of plant community change following biomanipulation of a large shallow lake. *Verhandlungen Internationale Vereinigung der Limnologie*, **27**.

Hanson, M. A. & Butler, M. G. (1990). Early responses to food web manipulation in a shallow prairie lake. *Hydrobiologia*, **200/201**, 317–328.

Hanson, J. M. & Legett, W. C. (1982). Empirical prediction of fish biomass and yield. *Canadian Journal of Fisheries and Aquatic Sciences*, **39**, 257–263.

Hansson, L.-A., Annadotter, H., Bergman, E., Hamrin, S. F., Jeppesen, E., Kairesalo, T., Luokkanen, E., Nilsson, P.-Å., Søndergaard, M. & Strand, J. (1998). Biomanipulation as an application of food chain theory: constraints, synthesis and recommendations for temperate lakes. *Ecosystems*, **1**, 558–574.

He, X., Wright, R. A. & Kitchell, J. F. (1993). Fish behavior and community responses to manipulation. In *The Trophic Cascade in Lakes*, eds. S. R. Carpenter & J. F. Kitchell, pp. 69–84. Cambridge: Cambridge University Press.

Horppila, J. & Peltonen, H. (1994). The fate of roach *Rutilus rutilus* (L.) stock under an extremely strong fishing pressure and its predicted development after the cessation of mass removal. *Journal of Fish Biology*, **45**, 777–786.

Horppila, J., Malinen, T. & Peltonen, H. (1996). Density and habitat shifts of a roach (*Rutilus rutilus*) stock assessed within one season by cohort analysis, depletion methods and echosounding. *Fisheries Research*, **28**, 151–161.

Horppila, J., Peltonen, H., Malinen, T., Luokkanen, E. & Kairesalo, T. (1998). Top–down or bottom–up effects by fish: issues of concern in biomanipulation of lakes. *Restoration Ecology*, **6**, 1–10.

Hosper, S. H. & Meijer, M.-L. (1986). Control of phosphorus loading and flushing as restoration methods for Lake Veleuwe, The Netherlands. *Hydrobiological Bulletin*, **20**, 183–194.

Hrbácek, J. (1969). On the possibility of estimating predation pressure and nutrition level of populations of *Daphnia* from their remains in sediment. *Verhandlungen internationale Vereinigung der Limnologie*, **17**, 262–274.

Hrbácek, J., Dvorakova, M., Korinek, V. & Prochazkova, L. (1961). Demonstration of the effect of the fish stock on the species composition of zooplankton and the intensity of metabolism of the whole plankton association. *Verhandlungen internationale Vereinigung der Limnologie*, **14**, 192–195.

Hrbácek, J., Desertová, B. & Popovský, J. (1978). Influence of the fish stock on the phosphorus–chlorophyll ratio. *Verhandlungen internationale Vereinigung der Limnologie*, **20**, 1624–1628.

Jasser, I. (1995). The influence of macrophytes on a phytoplankton community in experimental conditions. *Hydrobiologia*, **306**, 21–32.

Jeppesen, E., Jensen, J. P., Kristensen, P., Søndergaard, M., Mortensen, E., Sortkjær, O. & Olrik, K. (1990). Fish manipulation as a lake restoration tool in shallow, eutrophic, temperate lakes. 2: Threshold levels, long term stability and conclusions. *Hydrobiologia*, **200/201**, 219–227.

Jeppesen, E., Kristensen, P., Jensen, J. P., Søndergaard, M., Mortensen, E. & Lauridsen, T. (1991). Recovery resilience following a reduction in external phosphorus loading of shallow, eutrophic Danish lakes: duration, regulating factors and methods for overcoming resilience. *Memorie dell'Istituto italiano di Idrobiologia*, **48**, 127–148.

Jeppesen, E., Jensen, J. P., Søndergaard, M., Lauridsen, T. L., Pedersen, L. J. & Jensen, L. (1997). Top–down control in freshwater lakes: the role of nutrient state, submerged macrophytes and water depth. *Hydrobiologia*, **342/343**, 151–164.

Jeppesen E., Søndergaard, Ma., Søndergaard, Mo. &

Christoffersen, K. (eds.) (1998). *The Structuring Role of Submerged Macrophytes in Lakes*. New York: Springer-Verlag.

Jeppesen, E., Søndergaard, M., Kronvang, B., Jensen, J. P., Svendsen, L. M. & Lauridsen, T. (1999). Lake and catchment management in Denmark. In *Ecological Basis for Lake and Reservoir Management*, eds. D. Harper, A. Ferguson, B. Brierley & G. Phillips. *Hydrobiologia*, **395/396**, 419–432.

Jeppesen, E., Jensen, J. P., Søndergaard, M., Lauridsen, T. L. & Landekildehus, F. (2000). Trophic structure, species richness and biodiversity in Danish lakes: changes along a nutrient gradient. *Freshwater Biology*, **45**, 201–218.

Jeppesen, E., Jensen, J. P. & Søndergaard, M. (2002). Response of phytoplankton, zooplankton and fish to re-oligotrophication: an 11-year study of 23 Danish lakes. *Aquatic Ecosystem Health and Management*, **5**, 9–21.

Keto, J. & Sammalkorpi, I. (1988). A fading recovery: a conceptual model for Lake Vesijärvi management and research. *Aqua Fennica*, **18**, 193–204.

Kitchell, J. F. (ed.) (1992). *Food Web Management: A Case Study of Lake Mendota*. New York: Springer-Verlag.

Kubecka, J. & Duncan, A. (1994). Low fish predation pressure in the London reservoirs. 1: Species composition, density and biomass. *Internationale Revue der gesamten Hydrobiologie*, **79**, 143–155.

Kubecka, J., Seda, J., Duncan, A., Matena, J., Ketelaars, H. & Visser, P. (1998). Composition and biomass of the fish stocks in various European reservoirs. *International Review of Hydrobiology*, **83**, Special Issue, 559–568.

Lammens, E. H. H. R. (1999). The central role of fish in lake restoration and management. *Hydrobiologia*, **395/396**, 191–198.

Langeland, A. & Reinertsen, H. (1982). Interactions between phytoplankton and zooplankton in a fertilized lake. *Holarctic Ecology*, **5**, 253–272.

Lathrop, R. C., Carpenter, S. R., Stow, C. A., Soranno, P. A. & Panuska, J. C. (1998). Phosphorus loading reductions needed to control blue-green algal blooms in Lake Mendota. *Canadian Journal of Fisheries and Aquatic Sciences*, **55**, 1169–1178.

Lathrop, R. C., Johnson, B. M., Johnson, T. B., Vogelsang, M. T., Magnuson, J. H., Hrabik, T. R., Kitchell, J. F. & Carpenter, S. R. (in press). Stocking piscivores to improve fishing and water clarity: a synthesis of the

Lake Mendota biomanipulation project. *Freshwater Biology*, **47**.

Lauridsen, T. L. & Buenk, I. (1996). Diel changes in the horizontal distribution of zooplankton in the littoral zone of two shallow eutrophic lakes. *Archiv für Hydrobiologie*, **137**, 161–176.

Lauridsen, T., Pedersen, L. J., Jeppesen, E. & Søndergaard, M. (1996). The importance of macrophyte bed size for cladoceran composition and horizontal migration in a shallow lake. *Journal of Plankton Research*, **18**, 2283–2294.

Lazzaro, X. (1997). Do the trophic cascade hypothesis and classical biomanipulation approaches apply to tropical lakes and reservoirs? *Verhandlungen internationale Vereinigung der Limnologie*, **26**, 719–730.

Leslie, A. J., Jr (1988). Literature review of drawdown for aquatic plant control. *Aquatics*, **10**, 12–18.

Lewis, W. M. (2000). Basis for the protection and management of tropical lakes. *Lakes and Reservoirs: Research and Management*, **5**, 35–48.

Malinen, I. (1999). *The Condition of Lake Vesijärvi on the Basis of Water Quality Monitoring in 1998*. Mimeographed report. Lahti' Finland: Environmental Research Laboratory. (in Finnish)

Mazumder, A. (1994). Phosphorus–chlorophyll relationships under contrasting herbivory and thermal stratification: predictions and patterns. *Canadian Journal of Fisheries and Aquatic Sciences*, **51**, 390–400.

Meijer, M.-L. & Hosper, H. (1997). Effects of biomanipulation in the large and shallow Lake Wolderwijd, The Netherlands. *Hydrobiologia*, **342/343**, 335–349.

Meijer, M.-L., de Haan, W., Breukelaar, A. W. & Buiteveld, H. (1990). Is reduction of the benthivorous fish an important cause of high transparency following biomanipulation in shallow lakes? *Hydrobiologia*, **200/201**, 303–316.

Meijer, M.-L., de Boois, I., Scheffer, M., Portielje, R. & Hosper, H. (1999). Biomanipulation in shallow lakes in The Netherlands: an evaluation of 18 case studies. *Hydrobiologia*, **408/409**, 13–30.

Mitchell, S. F. & Perrow, M. R. (1998). Interactions between grazing birds and macrophytes. In *The Structuring Role of Submerged Macrophytes in Lakes*, eds. E. Jeppesen, Ma. Søndergaard, Mo. Søndergaard & K. Christoffersen, pp. 175–196. New York: Springer-Verlag.

Mittelbach, G. & Persson, L. (1998). The ontogeny of piscivory and its ecological consequences. *Canadian Journal of Aquatic and Fisheries Sciences*, **55**, 1454–1465.

Mittelbach, G., Turner, A. M., Hall, D. J., Rettig, J. E. & Osenberg, C. W. (1995). Perturbation and resilience: a long-term, whole-lake study of predator extinction and reintroduction. *Ecology*, **76**, 2347–2360.

Moss, B. (1998). Shallow lakes: biomanipulation and eutrophication. *Scope Newsletter* **29**.

Moss, B., Madgwick, J. & Phillips, G. (1996). *A Guide to the Restoration of Nutrient Enriched Shallow Lakes*. Norwich UK: Broads Authority, Environment Agency.

Nakai, S., Inoue, Y., Hosomi, M. & Murakami, A. (1999). Growth inhibition of blue-green algae by allelopathic effects of macrophytes. *Water Science and Technology*, **39**, 47–53.

Nicholls, K. H., Hopkins, G. J. & Standke, S. J. (1999). Reduced chlorophyll to phosphorus ratios in nearshore Great Lakes waters coincide with the establishment of dreissenid mussels. *Canadian Journal of Fisheries and Aquatic Sciences*, **56**, 153–161.

Nilsson, P. Å. (1999). *Finjasjön, Summer 1999: A Summary of Monitoring Data*. Hässleholm, Sweden. (in Swedish) Available at: http://www.hassleholmsvatten.se/

Ogilvie, S. H. & Mitchell, S. F. (1995). A model of mussel filtration in a shallow New Zealand lake, with reference to eutrophication control. *Archiv für Hydrobiologie*, **133**, 471–482.

Pace, M., Cole, J. J., Carpenter, S. R. & Kitchell, J. F. (1999). Trophic cascades revealed in diverse ecosystems. *Trends in Ecology and Evolution*, **14**, 483–488.

Peltonen, H., Ruuhijärvi, J., Malinen, T., Horppila, J., Olin, M. & Keto, J. (1999). The effects of food web management on fish assemblage dynamics in a north temperate lake. *Journal of Fish Biology*, **55**, 54–67.

Perrow, M. R., Jowitt, A. J. D. & Zambrano González, L. (1996). Sampling fish communities in shallow lowland lakes: point sample electric fishing versus electric fishing with stopnets. *Fisheries Management and Ecology*, **3**, 303–313.

Perrow, M. R., Schutten, J., Howes, J. R., Holzer, T., Madgwick, F. J. & Jowitt, A. J. D. (1997a). Interactions between coot (*Fulica atra*) and submerged macrophytes: the role of birds in the restoration process. *Hydrobiologia*, **342/343**, 241–255.

Perrow, M. R., Meijer, M.-L., Dawidowicz, P. & Coops, H. (1997b). Biomanipulation in shallow lakes: state of the art. *Hydrobiologia*, **342/343**, 355–365.

Perrow, M. R., Tomlinson, M. L., Phillips, G. L., Schutten, J. & Holzer, T. (1998). Fisheries related aspects of biomanipulation in the Norfolk Broads. In *Current Issues in Fisheries*, ed. R. Mann, A. Wheeler & I. Wellby, pp. 15–37. Cambridge: Institute of Fisheries Management.

Perrow, M. R., Jowitt, A. J. D., Leigh, S. A. C., Hindes, A. M. & Rhodes, J. D. (1999). The stability of fish communities in shallow lakes undergoing restoration: expectations and experiences from the Norfolk Broads (UK). *Hydrobiologia*, **408/409**, 85–100.

Persson, A. (1997). Phosphorus release by fish in relation to external and internal load in a eutrophic lake. *Limnology and Oceanography*, **43**, 577–583.

Persson, L., Anderson, G., Hamrin, S. F. & Johansson, L. (1988). Predation regulation and primary production along the productivity gradient of temperate lake ecosystems. In *Complex Interactions in Lake Communities*, ed. S. R. Carpenter, pp. 45–65. New York: Springer-Verlag.

Phillips, G., Eminson, D. F. & Moss, B. (1978). A mechanism to account for macrophyte decline in progressively eutrophicated fresh waters. *Aquatic Botany*, **4**, 103–126.

Prejs, A., Martyniak, A., Boron, S., Hliwa, P. & Koperski, P. (1994). Food web manipulation in a small, eutrophic Lake Wirbel, Poland: effect of stocking with juvenile pike on planktivorous fish. *Hydrobiologia*, **275/276**, 65–70.

Reeders, H. H., Bij de Vaate, A. & Noordhuis, R. (1992). Potential of the zebra mussel (*Dreissena polymorpha*) for water aquatic management. In *Zebra Mussels: Biology, Impacts, and Control*, eds. T. F. Nalepa & D. W. Schloesser, pp. 439–451. London: Lewis Publishers.

Reinertsen, H., Jensen, A., Koksvik, J. I., Langeland, A. & Olsen, Y. (1990). Effects of fish removal on the limnetic ecosystem of a eutrophic lake. *Canadian Journal of Fisheries and Aquatic Sciences*, **47**, 166–173.

Reynolds, C. (1984). *The Ecology of Freshwater Phytoplankton*. Cambridge: Cambridge University Press.

Romare, P. & Bergman, E. (1999). Juvenile fish expansion following biomanipulation and its effect on zooplankton. *Hydrobiologia*, **404**, 89–97.

Room, P. M. (1990). Ecology of simple plant–herbivore systems: biological control of *Salvinia*. *Trends in Ecology and Evolution*, **5**, 74–79.

Sammalkorpi, I. (2000). The role of fish behaviour in biomanipulation of a large lake. *Verhandlungen internationale Vereinigung der Limnologie*, **27**, 1464–1472.

Sammalkorpi, I., Ruuhijärvi, J. & Horppila, J. (1999). *Algal Bloom or Fisheries Boom? A Manual for Lake Management by Biomanipulation*. Helsinki: Finnish Environment Institute. (in Finnish)

Sand-Jensen, K. & Borum, J. (1991). Interactions among phytoplankton, periphyton and macrophytes in temperate freshwaters and estuarines. *Aquatic Botany*, **41**, 137–175.

Sarvala, J., Helminen, H. & Karjalainen, J. (2000). Chlorophyll-to-phosphorus relationship as a useful guide in lake restoration. *Verhandlungen internationale Vereinigung der Limnologie*, **27**, 1473–1479.

Sas, H. (ed.) (1989). *Lake Restoration by Reduction of Nutrient Loading: Expectation, Experiences, Extrapolatión*. Sankt Augustin, Germany: Richardz.

Scheffer, M. (1998). *Community Dynamics of Shallow Lakes*. New York: Chapman & Hall.

Scheffer, M., Hosper, S. H., Meijer, M.-L., Moss, B. & Jeppesen, E. (1993). Alternative equilibria in shallow lakes. *Trends in Ecology and Evolution*, **8**, 275–279.

Seda, J. & Kubecka, J. (1997). Long-term biomanipulation of Rimov reservoir (Czech Republic). *Hydrobiologia*, **291**, 95–108.

Shapiro, J. (1990). Biomanipulation: the next phase – making it stable? *Hydrobiologia* **200/201**, 13–27.

Skov, C. & Berg, S. (1999). Utilisation of natural and artificial habitats by YOY pike in a biomanipulated lake. *Hydrobiologia*, **408/409**, 115–122.

Stanczykowska, A. (1977). Ecology of *Dreissena polymorpha* (Pall.) (Bivalvia) in lakes. *Polska Archiv für Hydrobiologie*, **24**, 461–530.

Stansfield, J. H., Perrow, M. R., Tench, L. D., Jowitt, A. J. D. & Taylor, A. A. L. (1997). Submerged macrophytes as refuges for grazing Cladocera against fish predation: observations on seasonal changes in relation to macrophyte cover and predation pressure. *Hydrobiologia*, **342/343**, 229–240.

Stoddard, J. L., Jeffries, D. S., Luekeswille, A., Clair, T. A., Dillon, P. J., Driscoll, C. T., Forsius, M., Johannessen, M., Kahl, J. S., Kellogg, J. H., Kemp, A., Mannio, J., Monteith, D. T. & Wilander, A. (1999). Regional trends in aquatic recovery from acidification in North America and Europe. *Nature*, **401**, 575–578.

Søndergaard, M., Jeppesen, E., Mortensen, E., Dall, E., Kristensen, P. & Sortkjær, O. (1990). Phytoplankton biomass reduction after planktivorous fish reduction in a shallow, eutrophic lake: combined effects of reduced internal P-loading and increased zooplankton grazing. *Hydrobiologia*, **200/201**, 229–240.

Søndergaard, M., Olufsen, L., Lauridsen, T. L., Jeppesen, E. & Vindbæk Madsen, T. (1996). The impact of grazing waterfowl on submerged macrophytes: *in situ* experiments in a shallow eutrophic lake. *Aquatic Botany*, **53**, 73–84.

Søndergaard, M., Jeppesen, E. & Berg, S. (1997). Pike (*Esox lucius* L.) stocking as a biomanipulation tool. 2: Effects on lower trophic levels in Lake Lyng (Denmark). *Hydrobiologia*, **342/343**, 319–325.

Søndergaard, M., Jeppesen, E. & Jensen, J. P. (1998). *Sørestaurering i Danmark: Metoder, Erfaringer og Anbefalinger.* Copenhagen: Miljøstyrelsen. (in Danish)

Søndergaard, M., Jensen, J. P., Jeppesen, E. & Lauridsen, T. L. (2000). Lake restoration in Denmark. *Lakes and Reservoirs: Research and Management*, **5**, 151–159.

Timms, R. M. & Moss, B. (1984). Prevention of growth of potentially dense phytoplankton populations by zooplankton grazing in the presence of zooplanktivorous fish, in a shallow wetland ecosystem. *Limnology and Oceanography*, **29**, 472–486.

Tonn, W. & Magnuson, J. J. (1982). Patterns in the species composition and richness of fish assemblages in northern Wisconsin lakes. *Ecology*, **63**, 1149–1166.

Tonn, W., Magnuson, J. J., Rask, M. & Toivonen, J. (1990). Intercontinental comparison of small-lake fish assembleges: the balance between local and regional processes. *American Naturalist*, **136**, 345–375.

Turunen, T., Sammalkorpi, I. & Suuronen, P. (1997). Suitability of motorised under-ice seining in selective mass-removal of coarse fish. *Fisheries Research*, **31**, 73–82.

van Donk, E., Grimm, M. P., Gulati, R. D., Heuts, P. G. M., de Kloet, W. A. & van Liere, L. (1990). First attempt to apply whole lake food-web manipulation on a large scale in The Netherlands. *Hydrobiologia*, **200/201**, 291–301.

Vollenweider, R. & Kerekes, J. (1982). *Eutrophication of Waters: Monitoring, Assessment and Control.* Paris: OECD.

Weisner, S. E. B., Strand, J. A. & Sandsten, H. (1997). Mechanisms regulating abundance of submerged vegetation in shallow eutrophic lakes. *Oecologia*, **109**, 592–599.

Wetzel, R. G. (1983). *Limnology*, 2nd edn. Philadelphia, PA: W. B. Saunders.

Wetzel, R. G. (2000). Freshwater ecology: changes, requirements, and future demands. *Limnology*, **1**, 3–9.

Winfield, I. J. (1986). The influence of simulated aquatic macrophytes on the zooplankton consumption rate of juvenile roach, *Rutilus rutilus*, rudd, *Scardinius erythrophalmus*, and perch, *Perca fluviatilis. Journal of Fish Biology*, **29**, supplement A, 37–48.

Wium-Andersen, S., Anthoni, U., Christophersen, C. & Houen, G. (1982). Allelopathic effects on phytoplankton by substances isolated from aquatic macrophytes (Charales). *Oikos*, **39**, 187–190.

14 • Freshwater wetlands

BRYAN D. WHEELER, RUSS P. MONEY AND SUE C. SHAW

INTRODUCTION

In this chapter, the compass of 'wetlands' is restricted to telmatic wetlands, i.e. wet terrestrial freshwater habitats that are frequently referred to as mires, marshes, fens and bogs (Wheeler & Proctor, 2000) (Fig. 14.1).

Wetlands are often considered to be a single major habitat type comparable, say, with grasslands, heathlands or woodlands. However, wetlands encompass much of the environmental variation found across drier ecosystems, and form a series of habitats which broadly parallel these, differing primarily in their wetness. Differences in water sources and supply mechanisms further enhance their diversity. Wet conditions result from an interaction between landscape topography and sources of water supply and occur primarily because of impeded drainage or high rates of water supply (or both) (Fig. 14.2). The variability of wetlands has generated a rich terminology based on their broad environmental and hydrological subdivisions (Wheeler & Proctor, 2000) (Table 14.1). The diversity of 'wetlands' often makes generalisation difficult and many considerations of 'wetland restoration' have to be predicated with the question 'What type of wetland?' However, for all wetland types restoration consists of the identification of objectives, identification and reinstatement of appropriate habitat conditions, and facilitation of species recolonisation, and these form the main themes of this chapter.

The main variables that define the habitat of freshwater wetlands, and control the character of wetland vegetation, are water regime, nutrient richness, base richness, management status and successional status (Wheeler & Shaw, 1995a; Wheeler & Proctor, 2000) and these form the main components of habitat restoration in wetlands. However, these variables can show much variation amongst, and within, wetland sites, creating difficulties for both generalisation and the development of generic restoration strategies. Moreover, there are close and complex links between some of these variables, especially between water regimes and hydrochemistry, which can be difficult to disentangle for the practical purposes of restoration. Water supply is also the primary source of many of the chemical constituents of wetlands, and the specific behaviour of the water table can regulate the availability of some of them. For example, drying of wetland soils is often associated with increased availability of nitrogen through mineralisation processes. By contrast, waterlogging is usually associated with low oxidation–reduction (redox) potentials and, often, an increased availability of potential phytotoxins such as Fe^{2+}, Mn^{2+} and S^-; it may also lead to an increase in phosphorus availability (probably largely due to desorption processes) and high rates of denitrification. However, these relationships are complex and often far from exact and can be modified by conditions in individual sites (e.g. water movement and availability of reducible material can strongly affect the redox potentials that develop and the concentrations of reduced phytotoxins: Armstrong & Boatman, 1967; de Mars, 1996). These topics have been reviewed by various authors (e.g. Shotyk, 1988; Ross, 1995; Wheeler, 1999). Here it is possible only to draw attention to them.

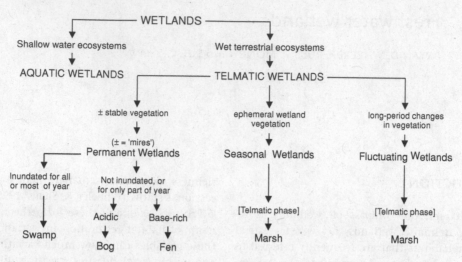

Fig. 14.1. Major wetland habitat categories and terms.

RESTORATION OBJECTIVES AND APPROACHES

Causes of modification ('damage') to wetlands

Wetlands have often been perceived as unremunerative, dangerous, disease-ridden or otherwise unpleasant tracts of land, and many attempts have been made to convert them into some other form of land use. A range of activities has modified the wetland habitat, with varying degrees of intensity (Table 14.2). Some other activities (e.g. urban development, industrialisation) have effectively destroyed the wetland habitat, although this sometimes reasserts itself temporarily through episodic flooding. Wildlife conservation is tabulated because it also often creates, or perpetuates, man-made modifications to the 'natural' wetland state. Conservationists usually consider 'damage' to be activities detrimental to conservational interest, whilst modifications they impose are called 'management' (or 'habitat creation').

Objectives of wetland restoration

Wetland restoration initiatives are mostly very recent, often driven primarily by wildlife conservation interests, but there is an increasing recognition, especially in larger-scale restoration projects, of the desirability of restoration of wider aspects of wetland 'function' (Wichtmann & Koppisch, 1998).

Restoration of wetland 'function'

Wetlands can perform various 'functions' for human society, particularly by helping to regulate ecological conditions at a catchment or even global level (Maltby, 1998). For example, they can help control water supply (Burt, 1995), and manage river flooding and coastal erosion (Williams, 1990). Peat-accumulating wetlands are a significant sink for atmospheric CO_2 and this may have important implications for global carbon cycling and climate change (Martikainen, 1996), not least on account of their enormous global area (Lappalainen, 1996). However, such functions may be strongly context-dependent and the actual environmental role of most wetland sites, and the implications of restoration, remain to be established.

Wetlands have served a number of economic functions. Efficient farming, forestry and peat extraction in wetlands usually demand their drainage, and create a condition that requires restoration, but some undrained or partly drained wetlands support low-intensity grazing, whilst others provide a renewable source of harvestable products (e.g. reed) and such 'traditional' activities sometimes form a focus of restoration initiatives. Some former damaging operations (e.g. peat extraction) have created

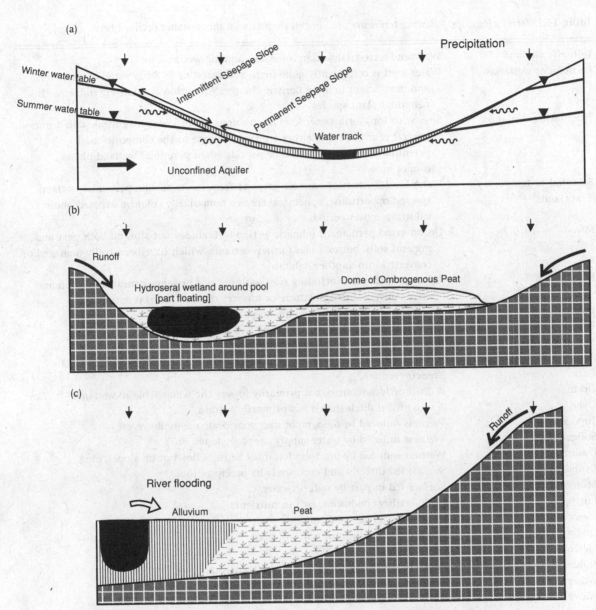

Fig. 14.2. Diagrammatic representations of three common, contrasting topographic situations in which wetlands develop. (a) Soligenous fen in a valley-head location fed by groundwater discharge on to the slopes. Water collects into an axial water track that drains the valley. (b) Hydroseral wetland in a runoff-fed basin upon a low permeability substratum. The deeper basin retains open water, with invasive peripheral rafts of vegetation. In the adjoining, shallower part of the basin, terrestrialisation is more advanced and a climax dome of ombrogenous peat has accumulated. (c) River floodplain wetland upon a low-permeability substratum. Water table is maintained by overbank flooding, surface runoff and precipitation.

Table 14.1. *Terminology: the following terms are used in this chapter with the meaning outlined here*

Telmatic wetland	Wet, semi-terrestrial wetlands (i.e. not aquatic wetlands).
Permanent wetlands	Water level is consistently quite high, and variation is fairly small or short term and insufficient to drive floristic change. Vegetation is relatively stable, with perennial plant species.
Fluctuating wetlands	Sites with long-term water level fluctuations of sufficient magnitude and duration (several years) to cause either a phased change in the composition of perennial wetland vegetation or periods when perennial wetland plants cannot grow.
Seasonal or temporary wetlands	Water level fluctuations are too great or frequent to sustain perennial wetland species; opportunist, ephemeral species temporarily colonise exposed, moist substrata in wet periods.
Mire	Unconverted permanent telmatic wetlands. Includes wet sites on both peat and mineral soils, but excludes former wetlands which have been badly damaged or converted into another habitat.
Peatland	All areas with peat, including sites with natural or semi-natural vegetation and areas converted to agriculture or forestry or used for peat extraction.
Bog[a]	Acidic (pH $< c$.5.5) mires (mainly on peat).
Fen[a]	Base-rich (pH $> c$.5.5) mires (peat and wet mineral soils).
Marsh	Seasonally dry wetlands on mineral soils.
Swamp	Wetlands with summer water table typically $> c$. 25 cm above ground level.
Carr	Tree-covered fen.
Drain	A ditch or watercourse that primarily lowers the water table in wetlands.
Dyke	A water-filled ditch that is not primarily a drain.
Topogenous	Wetness induced by topography and poor drainage (hollows, etc.).
Soligenous	Wetness induced by water supply (seepage slopes, etc.).
Ombrogenous	Wetness induced by precipitation (may be in hollows or on slopes, etc.).
Ombrotrophic	Surface fed directly and exclusively by precipitation.
Minerotrophic	Surface fed in part by telluric water.
Eutrophic	High fertility conditions, rich in nutrients.
Mesotrophic	Moderately fertile conditions.
Oligotrophic	Low fertility conditions, nutrient poor.
Meteoric water	Precipitation.
Telluric water	Water that has had some contact with the mineral ground.
Water table	Belowground free water surface.
Water surface	Surface of standing water.
Water level	Generic term for water table and water surface.

[a]These definitions follow Wheeler & Proctor (2000).

highly prized wildlife habitats, and the restoration of terrestrialising peat workings, to rejuvenate their hydrosere for wildlife, is sometimes an important restoration objective.

Restoration for wildlife conservation

Wildlife conservation and restoration objectives in wetlands can be grouped into three broad categories:

Table 14.2. *Causal factors of modification to the 'natural' wetland habitat*

	Management (cropping)	Drainage	Chemical enrichment	Peat wastage or removal	High water / inundation	Planting
Farming	√	√√√	√√	√		√√√
Forestry	√	√√	√	√		√√√
Turbary		√√	√	√√√	√	
Water supply	√				√	
Flood storage			√√		√√	
Sewage			√√√		√	
Industrial emission and effluent			√√		√	
Wildlife conservation	√√ (to maintain biodiversity and landscape)	√ (to control water levels)		√ (to create pools and scrapes)	√ (to create pools and scrapes)	√ (to reinstate former species; to provide habitats for new colonists)

Note: The number of √ represents an approximate intensity of the process that is typical for many regions. However, actual intensities show much variation.

Naturalness

'Naturalness' is often difficult to assess for many wetlands, not least because some appear to have been initiated through human disturbance (forest clearance) and because subsequent interference, such as periodic burning and light grazing, has often shaped their current character. It is not possible to restore 'naturalness', but man-modified wetlands can be restored to a condition similar to a former natural state. A complication is that many wetlands have had a complex developmental history so there may be a variety of former natural states to choose from.

The 'natural' state of a wetland, particularly in terms of its intrinsic environmental characteristics and water supply mechanisms, can help to determine restoration objectives and conditions for individual sites. For example, a floodplain wetland with a naturally fluctuating water table is not the most appropriate location for objectives that require permanently summer-wet conditions, even though, given suitable water engineering, this may technically be possible. Restoration initiatives that cut across the ecological 'grain' of wetlands are perhaps better seen as examples of water gardening for wildlife than of true wetland restoration.

Target species and community types

Many conservation, and some restoration, initiatives focus upon the protection, enhancement and reinstatement of specific target species and communities. For example, the European Union has identified a series of wetland 'habitats' in particular need of conservation and restoration (EEC, 1992). Some conflict of objectives can often occur in species-centred restoration initiatives, e.g. choosing between reedbeds for bird species such as bittern (*Botaurus stellaris*) versus a rare, species-rich type of fen vegetation.

Biodiversity

Wetland sites are often managed to maximise habitat and species diversity, to the extent that habitats are sometimes created (e.g. pools for wildfowl or Odonata) that are not necessarily part of the 'natural' state of particular sites. Unspecified 'biodiversity' is an appropriate objective where the former natural state or target species are either not known or not realistically restorable.

Restoration of traditional landscapes and land-use practices

Some wetlands owe much of their character to land-use practices that are nowadays largely archaic (e.g. summer-mowing for marsh hay); others have been even more modified and converted into low-intensity (wet) grasslands. Such landscapes are sometimes seen as a desirable focus for restoration as 'living museums', though in some examples only ditches represent their former wetland status.

Approaches to restoration in wetlands

Most restoration initiatives aim to reinstate a previous condition of a wetland site, by restoration of former habitats. However, sometimes 'restored' wetlands are very different from the original (e.g. reedbeds created over reflooded ombrogenous peatlands). This usually occurs when the former character of the wetland is not really known, when reinstatement of former conditions is not considered feasible or when creation of a specific wetland habitat is considered desirable by specific interest groups.

Many restoration initiatives essentially attempt to repair existing damage to re-create a former habitat. This approach is appropriate when salient components of the habitat already exist and where the damage is relatively minor – or is at least fairly easily repaired (e.g. ditch blocking to increase water tables, or cropping to remove nutrients). In other cases, direct repair may not be possible and 're-building' is necessary (e.g. where peat workings in ombrogenous bogs have become flooded with minerotrophic water so that ombrogenous conditions can only be re-established by seral development from recreated fen). Seral regeneration is sometimes seen as undesirable, because of its longer lead-in time for restoration, but in some cases it is unavoidable. Moreover, it can provide a more stable long-term foundation for restoration than attempted repair, because it provides scope for engineering an optimal restoration starting point. One

problem is that, as many of the seral phases thereby produced can have considerable wildlife value, seral regeneration can raise tricky questions about desirable restoration end-points.

An important constraint on restoration is the occurrence of damaging events external to wetlands. Because their water and nutrient supply may come from a large surrounding catchment, wetland sites can be much more influenced by external events than are many other habitats. Sometimes such effects (e.g. drainage of the surroundings, nutrient inflow) can be mitigated by on-site engineering or management actions, but in other cases effective restoration may require control over some or all of the catchment as well as the site itself (Grootjans & van Diggelen, 1995). In yet other cases, changes in their surroundings have helped produce the current wildlife value of wetlands and may be seen as desirable.

THEMES OF RESTORATION

Rewetting

Background

Although all wetlands are characterised by at least episodically high water levels, as a group they show enormous natural variation in their water regimes. In many cases water conditions have also been much influenced by human activities (e.g. drainage of the wetland and its surroundings, groundwater abstraction, and diversion and regulation of watercourses with a concomitant reduction of flooding). There is a widespread perception that many wetlands are drier than they 'should' be and rewetting is considered desirable. However, this presumption is not always correct: historical data on water levels are often sparse; drying sometimes reflects natural episodes of drought; and the high species richness of some wetlands is a consequence of partial drainage.

Whilst the reinstatement of high water tables to sites from which these have been lost is a prerequisite for their restoration, rewetting by itself does not necessarily provide complete habitat restoration because of other changes that may have occurred during the drained period. Nitrogen mineralisation and acidification can be consequences of drainage and rewetting is frequently associated with substantial release of N and P, especially in the early phases. Rewetting initiatives may therefore also need to deal with nutrient enrichment. Conversely, K deficiency has also been reported as a feature of drained peatland soils which may constrain successful restoration (van Duren et al., 1998).

Drying, sometimes reinforced by compaction caused by grazing animals or machinery, can also induce 'irreversible' changes to soil structure, leading to loss of macropores, lower water storage capacity, greater bulk density and smaller hydraulic conductivity (Schouwenaars & Vink, 1992; Schrautzer & Trepel, 1997). Such changes can be particularly critical in wetlands dependent upon intermittent water supply, such as precipitation-fed bogs, and mean that effective rewetting cannot always be achieved by such simple expedients as ditch blocking (Box 14.1).

For effective restoration, rewetting needs to be based on the water regime requirements of target species or communities, but these are generally complex and not well known. Extreme water level minima and maxima, average minima and maxima, 'average' water levels, the frequency and duration of fluctuations and the timing of these events can all influence the species composition and limits of wetland communities (Wheeler, 1999). Soil hydrophysical properties (von Müller, 1956; Gowing & Spoor, 1998), oxidation–reduction potentials, water flow, concentrations of reduced phytotoxins (especially Fe^{2+}, Mn^{2+} and S^-), availability of nutrients (N, P, K), competition with other plant species and facilitative oxygenation of the rooting zone by companion species can all modify the relationships of plants to water conditions (Wheeler, 1999). Moreover, some vegetation surfaces have hydroregulatory properties (acrotelms and rafts) which can modify both the water regimes experienced by plants and their response to them (Box 14.1). Thus although the effect of water tables on plant distribution is sometimes conceptualised as a single gradient, from water deficiency to excess (e.g. Gowing & Spoor, 1998), the reality is a great deal more complicated than this and appropriate conditions for restoration objectives can be difficult to specify.

Box 14.1   The acrotelm and restoration

The acrotelm is the thin surface skin (<1 m, often <0.5 m thickness) of peatlands, composed mainly of living plants and proto-peat (Ivanov, 1981). In natural mires this layer usually has hydrophysical characteristics (especially low bulk density, high hydraulic conductivity and high water storage capacity) that distinguish it from the underlying, more solid and permanently saturated, catotelm peats. The acrotelm often has expansible properties that help dampen the impact of water table fluctuations upon surface conditions by permitting rapid dissipation of water excess without a significant rise of water level and by reducing drying in dry periods (by reduction or cessation of horizontal seepage) (Ivanov, 1981). These properties can be seen particularly clearly in the spongy Sphagnum-dominated surfaces of some ombrogenous bogs, where they are enhanced by the capacity of Sphagnum to store water and to reduce evapotranspiration by 'mulching' and the albedo of bleached Sphagnum (Schouwenaars & Vink, 1992). The hydroregulatory properties of the acrotelm appear critical to the development of ombrogenous peatlands, particularly in regions subject to protracted periods without precipitation (Joosten, 1993).

The dependence of ombrogenous mires on conditions provided by the acrotelm is important in restoration, as this layer may be removed by operations such as peat extraction, or much modified and damaged by drainage. Where this is the case, operations such as simple ditch blocking often do not provide effective rewetting: this requires regeneration of the acrotelm and the trick of much bog restoration is to provide conditions that mimic the properties of an acrotelm and permit a new layer to develop.

Rewetting approaches and techniques

A prerequisite for all rewetting initiatives is to establish that a site is drier than it 'should' be and to identify the causes of this. Reasons for dryness are sometimes self-evident, but often hydrological measurements or modelling may be needed to identify them, especially where there are several possible contributory causes. Once identified, it may be possible to reverse the cause of dryness, but this is not always practicable. Sometimes, wet conditions can be engineered by another means, perhaps by utilising an alternative source of water or by lowering wetland surfaces, by excavation, to compensate for falling water tables. Such approaches can, of course, produce a rather different type of wetland to that which originally occurred, but this may be considered more acceptable than no effective rewetting.

Most wetlands are dependent on, or influenced by, hydrological processes within their catchments. Some are remnants of once-larger complexes, which may have had significant hydraulic interactions. Conservation managers sometimes identify a 'hydrological unit' around a site as a basis for water management within it, though often it is far from clear how, and with what validity, the boundaries of this are set. The extent to which control over a wider hydrological unit is required depends much upon the individual characteristics of wetland sites. Many wetlands that are remnants of larger complexes can, and do, survive as isolated units, though sometimes specific engineering or local water management are needed to maintain their characteristics.

In many situations, especially drained hollows or floodplains, rewetting can be achieved with surprising ease (e.g. by blocking drainage systems or disconnecting drainage pumps), but rewetting of sloping wetlands or wetlands with perched water tables can be more tricky. Even in topogenous circumstances, such as large, drained floodplains, rewetting is often deliberately limited (to prevent flooding of agricultural land, towns and villages) and this requires more elaborate engineering solutions than simply 'switching off the pumps'. Rewetting of upstanding peat remnants can be particularly difficult, especially sloping ombrogenous examples subject to the vagaries of precipitation supply. In some cases the most effective way of rewetting an upstanding remnant may be to lower its height and reconfigure the surface by removal of peat, but this solution is not always practicable or politically acceptable. However, without some such intervention approach remnants may just be left as rather dry, 'wasting assets'. In many cases rewetting initiatives are limited more by resource requirements, competing demands upon land and water resources or 'political' considerations than by technical constraints.

Ditch blocking

Blocking of drainage ditches, usually by dams of wood or metal or by plugs of soil (peat), is one of the most common, simple and important activities in wetland restoration, and considerable experience exists in cost-effective ditch-blocking technologies. However, in some situations, e.g. where near-surface soil physical properties have been much modified, ditch blocking *per se* may be insufficient to provide effective rewetting, and must be carried out in combination with other approaches.

The usual objective of ditch blocking is to reduce lateral drainage of water from the wetland, but in certain circumstances ditches also help drain wetlands by increasing vertical water flow (e.g. where they have been dug down into an underlying, more freely draining material). Whilst reduction of flow along a ditch can be achieved by one or more strategically placed dams, reduction of vertical loss may require that the entire ditch is infilled with a low-permeability material.

Dams can be particularly effective in topogenous wetlands, and can even maintain higher water levels than once occurred naturally. However, a potential problem with very high water levels is that, when ditch water is nutrient-enriched, they can enhance the penetration of nutrients into the adjoining wetland. They can also have directly adverse effects upon the vegetation. The value and deployment of dams has to be evaluated particularly carefully in soligenous wetlands, with a character maintained by lateral water flow, because dams can lead to stagnation, unnaturally low redox potentials and enhanced phytotoxicity. Dams can sometimes be used to mitigate dryness resulting primarily from reduced water supply. Again this may be suboptimal, because of changes to natural water supply mechanisms but, as is often the case in rewetting initiatives, sometimes the choice is between suboptimal rewetting and no rewetting at all.

Bunding

A 'bund' is effectively a long dam, designed to impound water over large areas by reduction of lateral flow. Bunds are often structures around all or part of the boundaries of a wetland or are used to subdivide wetland surfaces internally, often with a view to the creation of shallow lagoons (see below) (Fig. 14.3).

Bunds can be used to seal the margins of upstanding peat remnants, such as result from removal

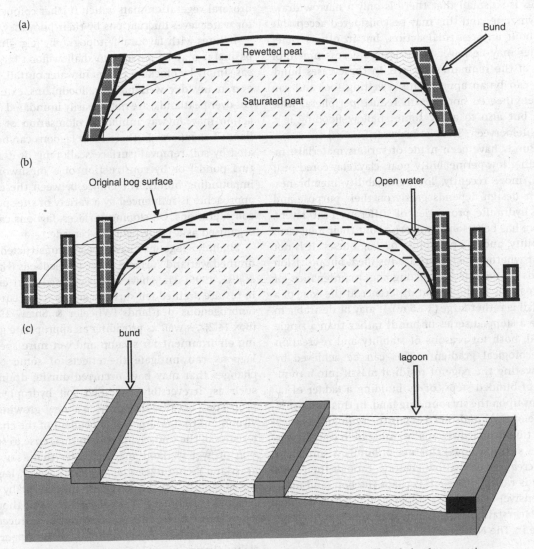

Fig. 14.3. Diagrammatic representation of some of the types and uses of bunds in the rewetting of drained remnants of ombrogenous peat and peat cutting surfaces. (a) Use of marginal parapet bunds to retain water within elevated peat remnant. (b) Use of stepped peripheral bunds located on the surroundings of the peat remnant, to re-create the original water mound and rewet the remnant; the lagoons created also form 'tanks' in which hydroseral regeneration can occur. (c) Use of a network of bunds to rewet a sloping peat-extraction by the creation of a cascade of shallow lagoons; a similar effect could be achieved by excavation of a series of scalloped depressions.

of peat, or drainage and subsidence, around them (Wheeler & Shaw, 1995*b*). The extent to which water drawdown occurs around the margins of such peat massifs depends primarily upon the hydraulic gradient and the hydraulic conductivity of the peat. Sometimes the value of one or both of these variables is so small that there is only a narrow drawdown zone, and this may be considered acceptable without any remedial action, but in others water tables may be lower than desired across much, or all, of the remnant's surface. In these cases bunding can be an appropriate rewetting tool and can be engineered not only to reduce peripheral leakage but also to allow some shallow flooding (e.g. Fochteloerveen, The Netherlands)

Bunds have been made of various materials, including low-permeability peat, clay, clay-cored peat and, more recently, low-permeability membranes. Their design depends partly on their purpose and the hydraulic properties of different peat layers. There has been some concern about their long-term stability, and the possibility of 'bog bursts'. It is not clear whether such fears are justified, but provision needs to be made for ongoing maintenance of bunds. Where the height difference at the edge of a massif is rather large (>1.5 m), it may be desirable to have a stepped series of bunds rather than a single bund, both for reasons of stability and re-creation of ecological gradients. This can be achieved by excavating the edge of residual massif into a number of bunded steps, or by building a ladder of lagoons upon the surrounding land. In this latter case none of the bunds needs abut the remnant, as they function primarily by maintaining high water levels in its surroundings. This use of bunds is similar to the creation of shallow lagoons, though their purpose is rather different. Perhaps the best (and most expensive) example of stepped bunding is around the Meerstalblok bog remnant in the Bargeveen reserve in The Netherlands.

Shallow lagoons

Large, shallow, flooded lagoons (depth normally *c*. 0.5–1 m) are sometimes constructed to rewet dry surfaces. They typically provide a swamp or wet mire environment, which may be either the main restoration objective (e.g. creation of reedswamp) or a starting-point for seral regeneration. Shallow lagoons effectively function as tanks that store water and help reduce water level fluctuations (relative to the unflooded state). They may also support the development of loose, floating or expansible, hydroseral vegetation mats which further compensate for water level fluctuations by their buoyancy.

In sites with favourable topography (e.g. shallow basins) it may be possible to shallow-flood large areas simply by blocking the main water outfall. However, many 'dry' wetlands (e.g. floodplains, extensive cut-over peatlands) are not so easily inundated (wave action may anyway inhibit recolonisation of large flooded expanses). In these cases lagoons can be created by soil removal (surface scalloping or digging turf ponds) or by construction of a meshwork of impounding bunds. The choice between these two approaches is influenced by a variety of site-specific considerations. On sloping surfaces, lagoons can be organised as a paddyfield-like cascade.

Shallow lagoons have been constructed on drained wetland surfaces to create reedbeds (Hawke & José, 1996). They have also been used in attempts to restore drained fen surfaces and to rewet cut-over ombrogenous peatlands (Wheeler & Shaw, 1995*b*) (Box 14.2). As well as providing an appropriate growing environment for swamp and wet mire species, lagoons can mitigate the effects of some other changes that may have occurred during drainage, such as 'irreversibly' modified soil hydrophysical properties, because they provide a 'new' growing environment that is largely independent of the characteristics of the underlying soil. In this sense, restoration of 'wet wetland' may be easier than, say, wet grassland (Schrautzer *et al.*, 1996). However, lagoon water quality can be adversely influenced by any residual enrichment of the underlying soil though, where practicable, this problem can be reduced by creating the lagoons by removal of the upper enriched layers of soil. Of course, if the principal water source for the lagoons is nutrient-rich surface water, consideration of nutrients residual in the soil assumes less importance. Also, some restoration objectives (e.g. creation of reedbeds) may be compatible with nutrient-enriched water.

Box 14.2   Restoration of peat cuttings in ombrogenous bogs

Peat extraction is an important commercial activity in many peat-rich landscapes and in some regions it is the main ongoing exploitation of peatlands. The physical removal of peat causes profound damage to peatlands, which may include: loss of species and seed banks; loss of acrotelm layer; drainage; mineralisation; humification and oxidation of residual peat; release of solutes and acids on rewetting. Such events may create considerable obstacles to the restoration of cut-over sites and perhaps make even more surprising the observation that some peat workings have recolonised and possess high conservation value. Older methods of peat harvesting were often rotational and permitted some spontaneous recolonisation with bog species (usually in temporarily abandoned trenches and pits), whereas modern commercial practices (usually sod cutting or milling) tend to remove smaller depths of peat over larger areas and with greater frequency. They also tend to produce more plane surfaces with less scope for spontaneous revegetation even if the return frequency of extraction permitted this.

Generalisation about restoration of cutting surfaces is difficult, because much depends upon the starting conditions – the dimensions and configuration of the peat deposit, the surface conditions and character of the peat and the climatic regime within which the mire is located. Wheeler & Shaw (1995b) provide a detailed review. In addition, somewhat different approaches have been followed in northwest Europe and North America.

APPROACHES IN NORTHWEST EUROPE

In northwest Europe, abandoned cutting surfaces, especially large, flattish expanses (as opposed to turf ponds or trenches) tend to experience strongly fluctuating water levels, with saturation or even shallow inundation in winter, but 'dry' conditions in summer. These conditions generally permit little regeneration towards 'good-quality' bog vegetation. Blocking of drainage ditches is a prerequisite for restoration but does not usually solve the problem of fluctuating water levels, which arise partly because the hydrophysical properties of the cut-over surface are usually very different from those of a natural bog acrotelm. However, reflooded depressions can act as tanks which help mimic the properties of a natural *Sphagnum*-based acrotelm, particularly by storing winter rainwater against subsequent summer deficit. They are also largely independent of the hydrophysical properties of the underlying peat, provide an environment in which a floating or 'swimming' raft of some aquatic *Sphagnum* species can develop, in effect acting as large bog pools, and apparently provide the best opportunity to rekindle *Sphagnum* growth on badly damaged bog surfaces (Schouwenaars & Vink, 1992). This has occurred spontaneously in many old, reflooded trenches and pits and can be engineered on flatter, modern cut-over surfaces, by scraping hollows in the residual peat, sealing existing baulks around abandoned cutting surfaces, building bunded lagoons on the peat surface, or by a combination of these approaches. Where the cut-over surface is sloping, a paddy-field-like cascade of lagoons may be necessary to ensure water retention (Wheeler & Shaw, 1995b). Such engineered surfaces have shown considerable restoration potential, and rapid development of extensive mats of *Sphagnum cuspidatum* or *S. recurvum* has occurred in shallow-flooded lagoons, followed by some vascular plants such as cotton grass (*Eriophorum angustifolium*). Of course, the lagoons are essentially just large bog pools, not 'new' raised bogs, but raised bog systems are known to have developed from comparable semi-aquatic starting-points. In some restoration sites (e.g. Lichtenmoor, northwest Germany), there has been limited colonisation by later-successional bog-building species, but invasion by these is generally not widespread. The reason for this is not known. Possibilities include the absence of propagules, the young age of the restoration sites and atmospheric N deposition (Money & Wheeler, 1999). In some parts of Europe (parts of Scandinavia, the Baltic States), bog-building species such as *S. magellanicum* rapidly colonise reflooded peat workings and it remains an open question why this does not occur so readily in much of northwest Germany, The Netherlands and England.

As a template for restoration, lagoons appear little influenced by the character of the underlying peat, and there is no evidence that lagoons on surfaces from which much peat has been removed necessarily recolonise less well than examples from where only a rather small amount has been taken. However, the depth of residual peat can be important in situations when water is able to seep downwards into unsaturated material beneath the bog. The rate of downward seepage from a deposit of saturated bog peat is controlled primarily by its hydraulic conductivity, not necessarily by its thickness, but when open water is impounded on top of the peat surface in lagoons, the thickness of residual peat has a more important control on vertical water seepage (Schouwenaars *et al.*, 1992). This is a problem in parts of The Netherlands and northern Germany where bogs are underlain by permeable deposits. Estimates of the peat thickness required to reduce vertical seepage to acceptable levels vary considerably (Blankenburg & Kuntze, 1987; Schouwenaars *et al.*, 1992), partly on account of variation in hydraulic conductivity. In some cases, 50 cm of ombrogenous peat is considered to be an acceptable minimum. Where bogs are located over crystalline rocks or thick clays such considerations have little relevance, though in all sites a minimum thickness of some 50 cm residual bog peat may be needed to prevent enrichment with bases and nutrients from underlying minerotrophic deposits.

APPROACHES IN NORTH AMERICA

Partly in recognition of the difficulties and expense associated with rewetting cut-over bogs, but also in response to the larger areas of 'undamaged' bogs, bog restoration in North America has proceeded along different lines to that in northwest Europe. Donor areas of intact bog are identified, from which the vegetation is stripped to a shallow depth (which can readily regenerate). The donor material is spread across cut-over surfaces and subsequently mulched with straw. The basis of this approach is the view that soil water tension at the cut-over surface is the critical factor influencing *Sphagnum* establishment. Mulches can generate water tensions above the level thought to be critical for sustaining *Sphagnum* inoculum without the need for expensive rewetting measures (Price, 1998). Once a *Sphagnum* layer is established it is postulated that it will become self-sustaining without the need for further mulching.

This approach has been tested widely in southeast Canada (Campeau & Rochefort, 1996; Rochefort & Campeau, 1997) where mulching has undoubtedly assisted establishment of vascular plants from the initial inoculum, in particular *Eriophorum spissum* and woody shrubs such as *Ledum palustre* and *Chamaedaphne calyculata*. *Polytrichum* also seems to establish well. However, once the mulch has deteriorated (after two to three years), survivorship of *Sphagnum* is generally poor, except in depressions and ditches with high water tables. Despite expectations, there is little empirical evidence to suggest that the cover of tussock-forming species and low shrubs improves microclimatic conditions sufficiently for the surfaces to remain dominated by *Sphagnum*; rather they appear to support wet heath and scrub comparable to spontaneously recolonised peat surfaces with subsurface water tables elsewhere (Money & Wheeler, 1999). Consequently the value of this approach over and above the approach already taken in northwest Europe is questionable, though the Canadian work has certainly demonstrated that mulches do improve conditions at the surface of cut-over bogs. On balance, this technique may be best suited to improving revegetation success to a 'drier' form of bog, in areas where it is not feasible to raise water tables.

The desirable depth of water in lagoons, and acceptable range of fluctuation, depends upon the restoration objectives. Reedswamp is generally tolerant of a quite wide range of water conditions, though reedbeds tend to be most vigorous and pure when summer water levels are well above surface, conditions that are also appropriate for some 'flagship' species such as bittern (Hawke & José, 1996).

In ombrogenous bogs, deep (>0.5 m) water is often less suitable for regrowth of aquatic *Sphagnum* species than shallower conditions, though the capacity of *Sphagnum* species to 'swim' in bog pools can vary amongst pools of similar depth, for reasons that are not well understood. Where storage of winter water surplus is an important function of lagoons, the depth of winter inundation needs to be sufficient to prevent the summer water tables falling substantially sub-surface. This may require a trade-off between depth of water optimal for vegetation re-establishment and that needed to sustain summer wetness. In some restoration initiatives in bogs, winter flooding depths of about 0.5 m appear to have been appropriate.

The depth of flooding for re-creation of floating fens is not well known, although rhizome mats over water and muds with depths between about 0.2 and 2 m have been reported. Nor is the mechanism of raft formation well understood, and may be variable (for example, by flotation of vegetation rooted on the bottom of the pit, possibly induced by a gradually rising water level, as well as by centripetal rafting from the margin: van Wirdum, 1995). Little guidance is available about constraints upon rafting and techniques that can be employed to facilitate this process. Rafting has occurred widely and spontaneously in some old, reflooded hollows, but can be difficult to induce in restoration initiatives: wave action, grazing by wildfowl and lack of patience are all possible reasons for this. There is scope for experimentation with provision of 'buoyancy aids', though the use of such structures over large areas of reflooded fen may be neither practically possible nor aesthetically pleasing.

Supplementary water supply

Supplementation of an existing water supply is sometimes used to restore wet conditions, particularly where dryness is considered to be due to reduction of supply. The success of supplementation depends primarily upon the extent to which the supply mimics the former hydrological regime of the wetland and its hydrochemical characteristics. Sometimes a supplementary source may be satisfactory in neither respect, presenting a choice between the desirability of the existing 'dry' state versus the establishment of wet conditions different to those

that once occurred. In some long-drained wetland sites, former water supply mechanisms and hydrochemical characteristics may be little known, and there is likely to be less concern about the use of supplementary water than in sites that have shown recent drying.

Perhaps the simplest supplementation circumstance is where the existing main source of water is distributed through a network of dykes. Although unnatural, dykes are often long established and can easily be topped up. The main issues are the rate at which supplementary water can be supplied and the suitability of its quality.

One important limitation of dyke-distributed supply is that lateral water seepage (sub-irrigation) from the channels is often insufficient to maintain high water tables except in their immediate vicinity, unless the peat is very transmissive. One solution is to excavate (or restore) a more intensive network of dykes, or a series of shallow 'foot drains' extending from them into the wetland. Another approach, where hydrologically feasible, is to keep the dykes brimming so that surface flow can occur into and across the wetland. Such 'border irrigation' (Dietrich *et al.*, 1998) can be used on sloping sites and may provide a suitable approach for rewetting soligenous wetlands. Richert *et al.* (2000) report that it can provide effective rewetting of sloping, deep peat of low hydraulic conductivity, though low summer water tables still occurred in dry years (Box 14.3).

Dyke distribution of supplementary water in groundwater-fed wetlands may be possible, but does not necessarily mimic their natural water regime and associated characteristics (e.g. high oxidation–reduction potential associated with water flow; calcite precipitation and associated co-precipitation of P).

In some sites, groundwater discharge has been simulated by installation of a buried system of perforated pipes or mole drains fed by water of appropriate quality, derived from boreholes (e.g. Newham Fen, Thriplow Meadows, UK). This approach is reported to provide effective rewetting, but is expensive; nor is it certain whether it results in the intended hydrochemical conditions. A problem here is that, whatever the quality of the

Box 14.3   Restoration of fens from farmland in the north German lowlands

Some 20% of the agricultural land of north Germany is estimated to be on fen peat (Wichtmann & Koppisch, 1998), many of the fens having been drained some 30 years ago mainly for conversion to intensive grassland. In recent years, various large fen restoration projects have been initiated. Many of the fens were formerly groundwater-fed, mesotrophic percolating wetlands (Michaelis, 1998). Hydrophysical changes in soil structure consequent on drainage (compaction, loss of macropores and storage capacity, increase in bulk density and decrease in hydraulic conductivity) mean that, even if restored, supplies of groundwater may not percolate through the soil as before (e.g. Schrautzer *et al.*, 1996). There is a general perception that the groundwater percolation components of these fens cannot readily be restored (Wichtmann & Koppisch, 1998), and rewetting has focused on surface water supply. Even with surface water sources rewetting can be difficult, on account of both low summer levels and limited recharge into the peat. Various workers have shown that normal dyke spacings are inadequate for effective rewetting of the peats by sub-irrigation. Hennings & Blankenburg (1994) concluded that dyke spacings of less than 10 m might be needed for rewetting, whilst Scholz *et al.* (1995) found that mole 'drains' connected to the dykes

could provide an effective, if expensive, rewetting solution. However, these workers concluded that flooding, or surface flow from ditches on gently sloping sites, provided the most cost-effective rewetting. Shallow flooding is often seen as the best solution for re-creating wet fen on these surfaces (Dietrich *et al.*, 1998). It provides both effective distribution of water across the peatland and a restoration environment that is substantially independent of the hydrophysical characteristics of the underlying 'drained' soils.

As the main sources of surface water available for these projects tend to be nutrient-rich, restoration objectives are usually floodplain reedswamp or fertile wet grassland (Koppitz *et al.*, 1998; Wichtmann & Koppisch, 1998) rather than mesotrophic fen. Reed production is seen as a desirable outcome, as the reedbeds have economic value as a crop, can be used for water purification and may provide a carbon sink (Schäfer & Wichtmann, 1998) as well as having some biological value. The long-term potential for regeneration of mesotrophic fen has received rather little consideration but, as examples are known of mesotrophic fen sourced primarily by enriched river water, with excess nutrients apparently stripped by passage through intervening 'filters' of vegetation and peat (Koerselman *et al.*, 1990), the possibility of re-creating this habitat on a river-fed floodplain should not be discounted.

source water, the wetland soils may be somewhat enriched in nutrients as a consequence both of past drying and the disturbance associated with installation of the sub-irrigation system. An alternative approach is to construct a series of shallow, open channels fed by groundwater across the slope, which partly mimic the runnels that naturally occur in some soligenous wetlands. This is cheaper and creates less disturbance than installation of a sub-irrigation system, but the practical outcomes of the two approaches do not seem to have been compared or evaluated.

In ombrogenous peatlands water supplementation is usually needed to mitigate water shortage

caused by drainage or insufficient water storage rather than reduction of supply, though it may be most necessary in relatively dry climates. Constructions designed to increase the water storage capacity of the surface of damaged bogs (such as shallow lagoons) usually accompany supplementation.

There are only limited options for supplementation using rainwater, especially in dry climates. Possibilities include the designation of parts of a peatland to become seasonally 'dry' water-donor areas, which feed other parts of the mire. Where the topography permits, this can be done by gravitational drainage, either by a terraced rewetting cascade or through a system of drains, otherwise pumping may be required. However, the use of pumped drain water to support water levels in lagoons on ombrogenous bogs may not always be satisfactory as there is some evidence that it can lead to an increase in solute loadings. Possible reasons for this are enhanced water throughflow rates and enrichment of the drain water. This latter may occur when ditches are cut into minerotrophic deposits, when they receive land drainage inputs or when there is solute release from adjoining disturbed peat. Ingress of enriched ditch water on to bog surfaces will preclude the immediate establishment of ombrotrophic conditions, but when it is the only supplementary source available this may be considered more desirable than the alternative of seasonally low water tables.

Re-establishment of former water supply mechanisms

Where dryness has resulted from modification to former water supply mechanisms, an obvious, and in some cases essential, restoration approach is to reinstate these. However, as wetlands may be dependent on, or influenced by, hydrological processes within large catchments, reinstatement of former mechanisms may require control over the catchment as well as the wetland. This may not always be feasible and reinstatement of former mechanisms is perhaps most appropriate where reduction of supply is attributable to a single 'problem', such as a groundwater abstraction point which can be relocated (Box 14.4). This approach is fundamentally simple but often expensive. Moreover, reinstatement of former water supply does not necessarily restore all of the former ecological conditions.

Nutrient reduction

Background

Wetlands vary very considerably in their fertility (capacity to support plant growth). Nutrient-rich wetlands occur naturally, particularly on river floodplains. However, in many regions input of nutrients into wetlands has increased in consequence of enrichment of their main water sources, particularly by agricultural runoff and leachate, and by sewage and other effluents. As their water supply may be sourced from a wide catchment, wetlands can be more subject to pervasive nutrient enrichment than many other habitats. However, some of the biggest inputs of nutrients are often associated with inwash of soil and deposition of alluvial solids. Water quality can also affect nutrient availability indirectly. For example, water rich in sulphate can help release some soil-bound nutrients and increase their availability to plants (Koerselman & Verhoeven, 1995). In addition, changes in water level can result in nutrient changes within the wetlands. Various workers have reported enhanced nitrogen mineralisation in drained wetland soils (e.g. Williams, 1974; Grootjans et al., 1986), though substantial N mineralisation can occur in waterlogged conditions (Williams & Wheatley, 1988), and the process is also regulated by the nutrient capital of the soils. By contrast, high water levels can result in phosphorus desorption and high rates of denitrification (Koerselman & Verhoeven, 1995).

Generalisation about the likely impact of nutrient input into wetlands is difficult, as it depends inter alia upon the starting conditions, the capacity of the substratum to sequester nutrients and the identity of the main growth-limiting nutrients. For example, Boyer & Wheeler (1989) measured concentrations of NO_3-N of some 30 mg l^{-1} in groundwater discharging into a spring-fed wetland, but the system was P limited and aboveground vegetation production rates were very small. Without detailed investigations, it is often difficult to predict the likely impact of, or thresholds for, particular nutrient inputs into specific wetland sites. Nor do high productivities necessarily result in increased crop mass and reduced species richness, though this is often the case (Wheeler & Shaw, 1994).

Techniques for reducing nutrients

Soil stripping

Where practicable, removal of the upper soil layers is an effective mechanism for reducing the nutrient status of enriched wetlands, especially where only the uppermost soil layers have become enriched. This is often the case in wetlands that have been converted to agricultural land and fertilised,

or where drying has led to the mineralisation of the
upper layers (Box 14.4). Soil stripping offers other
potential benefits, such as removing soil with
changed hydrophysical properties, and the seed
banks of potential weed species. In some cases it
may also help rewetting by reducing the elevation
of the wetland surface relative to the water table
though, conversely, sometimes this makes sites too
wet for direct re-establishment of the former vegetation, so that an indirect, hydroseral restoration
strategy becomes necessary.

The main problem with soil stripping is its expense, both of excavation and disposal. In some initiatives stripped spoil has been left as banks adjoining excavated areas. These can provide access tracks,
etc., but they may also lead to establishment of a
range of non-wetland species, and potential nutrient release back into the excavated areas. Soil strip-

ping may also remove any seed bank of wetland
plant species, though many drained and enriched
wetland soils have only a limited persistent seed
bank of wetland species.

Sorption and uptake

The quality of water feeding wetlands can often be
'improved' by natural processes of sorption and uptake. Wetlands are well known as sinks of nutrients
and other chemicals (Mitsch, 1995), and this has
been exploited in the development of artificial
wetlands for effluent treatment. In semi-natural
wetlands, nutrient inputs can be stored within the
peat, up to a point of saturation (Verhoeven *et al.*,
1983), and locations distant from eutrophic water
sources are often less enriched than proximate locations, because of uptake by plants and adsorption to peat and mud (Koerselman *et al.*, 1990). It is

therefore possible to design restoration strategies in which productive zones of vegetation bordering enriched surface-water sources help to reduce ingress of nutrients into the fen hinterland. This approach may be less suitable for enriched groundwater-fed wetlands on account of the more pervasive character of much groundwater input.

Cropping

There is considerable evidence that cropping of vegetation, particularly mowing with removal of the mown material, can lead to reduction of nutrients in fen meadows. However, as Bakker (1989) points out, it is possible to find evidence of mowing leading to nutrient decrease or having no effect; of reducing biomass with variable effects upon species richness; and of increasing species richness without affecting biomass. This probably reflects variation in starting conditions, the fact that cropping can have direct effects upon vegetation composition (by reducing dominance, etc.) independently of any effect upon soil nutrient status (Wheeler & Shaw, 1994), and the difficulty of relating measured soil nutrient concentrations to vegetation productivity and composition (Wheeler et al., 1992). These constraints need to be considered when interpreting different studies. Koerselmann & Verhoeven (1995) suggest that mowing may be appropriate for reducing P and K availability in some Dutch fens, but that it may have little impact upon N availability because estimated annual N input in precipitation may be similar to N export in harvested vegetation. However, workers are generally agreed that cropping is one of the most effective means of nutrient regulation in wetlands, even if reduction may be a slow, long-term and costly process.

Grazing is thought to be much less effective in nutrient removal (Bakker, 1989). In some situations herbivore activity may increase nutrient availability by recycling and returning nutrients sequestered in plant material and by promoting disturbance and mineralisation of the soil (Connors et al., 2000).

Denitrification

Elevation of water levels has been suggested as a possible procedure to remove nitrogen from eutrophic wetlands (Koerselman & Verhoeven, 1995), by stimulating denitrification and, perhaps, by reducing rates of mineralisation. Denitrification in wetlands is well documented and enormous rates of N loss have sometimes been reported (Guthrie & Duxbury, 1978), though they are by no means universal even in waterlogged sites. Denitrification occurs mostly in reducing conditions (Eh $<c.\,300$ mV), and is particularly associated with high water tables and base-rich conditions. It may be maximised by strongly fluctuating water levels but, except where they are natural, such regimes are likely to be detrimental to the established biota. However, sometimes there may be opportunities to use them as a pre-treatment to restoration on a badly damaged site. The main problem with the denitrification approach is that its general practical value is far from clear, not least because some studies suggest that the process can be limited and patchy even within waterlogged wetlands (Koerselman et al., 1989). Another problem, recognised by Koerselman & Verhoeven (1995), is that the low redox potentials appropriate for denitrification may increase phosphorus availability by desorption.

Amelioration of acidification

Background

'Acidification' refers to a reduction of base status (pH) at or near the surface of wetland soils. The acidity of wetlands depends on the balance of metallic cations and strong-acid anions, which in turn depends upon the composition of their water sources and the capacity of these to buffer acidity produced endogenously by plants, especially Sphagnum spp. (Clymo, 1984), imported in 'acid rain', or arising in other ways (Urban et al., 1986).

In some climatic regions acidification is a natural ontogenic process within wetlands, particularly associated with accumulation of peat above the influence of minerotrophic water. This process is expressed on a large scale in the development of ombrogenous bogs, but small, embryonic ombrogenous nuclei also occur within fens, sometimes associated with hummock-forming plant species. Floating mats of hydroseral vegetation in natural pools and reflooded peat workings can be particularly prone to surface acidification, even in calcareous wetlands, because their buoyancy helps reduce inundation by base-rich water (Giller & Wheeler, 1988).

Acidification has also occurred in some wetlands because of a reduction of base-rich water supply (e.g. lowering of groundwater tables or decreased flooding because of diversion or water management of watercourses) (Grootjans *et al.*, 1986). Drying of wetland soils also sometimes leads to a decrease of pH, in some cases through oxidation of sulphides giving rise to 'acid sulphate soils'. Haesebroeck *et al.* (1997) demonstrated drying-induced acidification of soil cores removed from the influence of groundwater supply, and found an associated increase in soluble reactive phosphorus concentrations. Acidification is often most evident when water is flushed from dried wetland soils during early phases of rewetting, when pulses of very low pH water can be released, often enriched with other solutes.

'Acid rain' is a potential source of surface acidification, but, in general, rather little is known about its effects within wetlands, partly because it is difficult to disentangle them from other potential causes. High pH, but weakly buffered, wetlands appear particularly susceptible to damaging acidification (Boeye & Verheyen, 1994), but even naturally acidic wetlands, such as ombrogenous bogs, can sometimes be too base-poor to support good growth of some of their most characteristic species. Richards *et al.* (1995) found that poor growth of *Eriophorum angustifolium* on ombrogenous hill peat, and the inability of this species to recolonise erosion surfaces, could be ameliorated (temporarily) by addition of lime.

The floristic effects of acidification depend upon its magnitude, the associated water regime and other starting conditions. In summer-'dry' wetlands, acidification may produce a highly impoverished vegetation, dominated by such species as purple moor grass (*Molinia caerulea*). In wetter conditions, it is often associated with the development of carpets of *Sphagnum*.

Techniques for restoring acidifying wetlands

Relatively few attempts have been made to restore acidifying wetlands. The approach used depends upon both the intensity and cause of acidification.

Where it has been induced by drying or a reduction in the proportionate contribution of base-rich water sources to a wetland, restoration of former water supply mechanisms can be considered, such as increasing groundwater levels or restoration of river flooding, though this latter may be inappropriate on account of high nutrient loadings.

Van Duren *et al.* (1998) report an attempt to restore fen meadow in The Netherlands where reduction of base-rich groundwater supply was thought to be the primary cause of degradation. The project included a variety of treatments, including flooding with surface water (supplied through a helophyte filter), top soil removal and liming. They found that removal of the top layer of acidified soil led to a small, but sustained, increase in pH whereas liming had no long-term beneficial effects. Some typical fen meadow species were introduced but did not survive well. These authors concluded that the restoration treatments were largely unsuccessful, probably because they had limited impact upon the acidification.

Dutch workers have also been much concerned with reversing the seral acidification of species-rich fen in terrestrialising turf ponds (see below). Two main restoration approaches have been used: excavation of irrigation channels, to assist the outflow of rainwater and inflow of base-rich water (van Wirdum, 1991), and removal of the top layers of acidifying peat. Beltman *et al.* (1995) report that a combination of both approaches could lead to an increase in base status and re-establishment of some characteristic species, though neither was effective on its own. However, it is not clear to what extent these manipulations have resulted in persistent mesotrophic fen communities comparable to those that formerly occurred. Moreover, in view of the patent suitability of some turf ponds for *Sphagnum* colonisation, turfing-out operations may have to be repeated periodically to maintain the desired state. Ironically, the *Sphagnum* surfaces that are removed can contain 'bog-building' species such as *S. magellanicum* and *S. papillosum*, which have proved to be elusive recolonists in attempts to restore ombrogenous bog in some ombrotrophic contexts. This suggests a need for holistic strategic planning of wetland restoration initiatives.

Controlling colonisation and stopping succession

Dealing with dereliction

Although some wetland habitats can have quite long-term stability, either because they are climax states (e.g. ombrogenous bog), subclimax states (e.g. some types of fen carr), or because successional change proceeds slowly (e.g. some types of swamp), many types of wetland vegetation are plagioclimax states and show substantial, and often quite rapid, change (dereliction) consequent upon abandonment of former management regimes. Many management practices (e.g. summer mowing, grazing) produce a fairly short, species-rich sward and abandonment often results in rank herbaceous vegetation and colonisation by woody plants. In both cases a loss of species richness often results.

Preventing, or reversing, the effects of dereliction is one of the main activities of conservation managers in many parts of Europe, but it is much less widespread elsewhere, partly because wetlands have traditionally not been managed or because the type of habitats that are maintained by management have naturally occurring, fairly stable, equivalents. Indeed, vegetation management is sometimes seen as the antithesis of conservation in some countries. Routine, regular management is not normally considered to constitute 'restoration', but one-off activities such as clearance of dense scrub, or burning of an overgrown reedbed, often are.

Mowing and grazing

Mowing is often a favoured management technique for maintaining species richness, partly because it helps remove nutrients from enriched fens, though its impact is determined by variables such as the timing of harvest (Rowell et al., 1985). Mowing can also reduce some 'desirable' species (e.g. hummock-forming bryophytes) as well as less desirable ones (e.g. Moen, 1995) and may sometimes promote acidification. Also, unless the harvest has commercial value, it is expensive. Mowing, with gathering and other maintenance, has been estimated to take about 30 person–days ha^{-1}, which in England in 1998 cost approximately US$2400 ha^{-1} (Wheeler, 1998).

The former role of natural grazers, such as elk and deer, has been removed or reduced in wetlands in many developed regions, sometimes to be replaced by cattle or sheep. Grazing is the primary usage of managed wetlands in Britain, but is not always seen as beneficial, particularly on account of disturbance (poaching), especially in upland systems. However, grazing has been introduced as a management tool in some wetlands, partly because it is cheaper than mowing (though real costs can be difficult to estimate). In The Netherlands, permanent populations of Heck cattle, Konik Polski (an old breed of horse) and deer have been established to manage the Oostvaardersplassen, a large (5600 ha) polder 'reclaimed' in the 1960s (Kampf, 2000).

Grazing may provide a cost-effective mechanism for managing or restoring derelict wetland vegetation, and can benefit certain avifauna, especially those associated with wet grassland (Fuller, 1982), but there has been little rigorous evaluation of its merits. These are likely to depend strongly upon context and intensity. High stocking rates can cause poaching and compaction of upper soil layers, reduce their hydraulic conductivity and water storage capacity, and increase water table fluctuations in systems where these are strongly regulated by soil physical properties (Schrautzer & Trepel, 1997). In other systems (e.g. spring-fed fens), grazing may have little influence on water table behaviour and moderate poaching may be beneficial in providing niches for low-growing species such as butterwort (*Pinguicula vulgaris*) (Magnusson & Magnusson, 1991). However, in less consistently wet contexts, poaching may mainly benefit weed species and enhance soil mineralisation, so that 'undesirable' species contribute to any observed increase in species richness. It is doubtful if anything but very light grazing provides appropriate management for deep-peat wetlands, unless these are to be kept partly drained, but it may be more valuable in some shallow-peat wetlands, or examples on mineral soils.

Burning

Controlled burning may provide a cheaper management option to mowing in wetlands. It has received only limited critical evaluation, but is not generally welcomed by conservationists. However, Curtis (1959) points out that fires are natural features which have probably helped prevent encroachment by woody plants in some Wisconsin wetlands. Also Ditlhogo *et al.* (1992) suggest that reedbed management by periodic burning may be no more damaging than regular harvesting, and may be less so for birds and invertebrates, if it takes place less frequently. The effects of burning are, however, context-dependent (Bowles *et al.*, 1996). Partly drained wetlands can be badly damaged by fire, especially if there is combustion of the peat. There is also a widespread view that burning is incompatible with the maintenance of species-rich fen vegetation, perhaps on account of pulses of nutrient enrichment, though Wheeler *et al.* (1998) documented the occurrence of fen orchid (*Liparis loeselii*) in species-rich fen vegetation which had been maintained by occasional winter burning. Burning is also used extensively in the UK for the management of moorland and blanket bog, especially for game birds, but this is generally not recommended by conservationists as it can be damaging to the habitat and the less mobile animals (Shaw *et al.*, 1998). Overall, burning is perhaps best seen as an occasional tool for restoring derelict wetlands, especially extensive examples, where it may be the only practicable management technique.

Turfing out

Hydroseral processes (Fig. 14.4) may also be the focus of restoration, usually because ongoing terrestrialisation leads to vegetation change and loss of species and communities confined to transient phases of the hydrosere. Peat accumulation and associated autogenic vegetation change can be rapid (Bakker *et al.*, 1997). The rate of hydroseral change can sometimes be reduced by regular vegetation management, but this does not usually prevent it: indeed, some hydroseral events, such as *Sphagnum* colonisation, may be accelerated by regular mowing. Periodic rejuvenation of the hydrosere, by 'turfing out' (peat and soil removal) is therefore sometimes considered to be a necessary approach to habitat restoration. Revegetation considerations are similar to those discussed for lagoons (above).

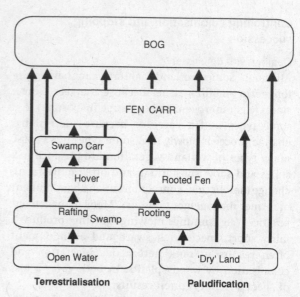

Fig. 14.4. Some main autogenic successional pathways in wetlands. The 'hydrosere' corresponds specifically to the terrestrialisation pathway and is not a generic term for all vegetation change in wetlands.

Turfing-out operations are resource-intensive (turf-pond excavation by machine has been estimated as requiring *c.*6 person–days ha^{-1}: Wheeler, 1998) and may be considered to be elaborately interventionist. However, in some regions certain species (such as fen orchid) are largely confined to artificially created or maintained hydroseres, possibly because natural disturbances (e.g. by large animals such as pigs: Grootjans & van Diggelen, 1995; Arrington, *et al.*, 1999) no longer occur.

Survival, recolonisation and reintroduction of species

Potential for natural recolonisation

Seed and spore banks

Undrained wetland soils may contain a rich seed bank, which can bear quite strong similarities to the extant vegetation (e.g. van der Valk & Verhoeven, 1988). However, there is considerable variation in the reported capacity of wetland species to form seed banks (Thompson *et al.*, 1997) (though this partly reflects differences in investigative methodology as well as natural variation amongst and within

Box 14.5   Restoration of 'prairie potholes' in North America

The Prairie Pothole region of North America contains a large number of shallow depressions (the 'prairie potholes') which contain wetland vegetation, typically showing a water depth zonation from open water surrounded by swamp, grading out into sedge fen and wet meadow. In the shallower examples, the central deeper water zones may be missing, and in examples surrounded by intensive farmland the outer zones may be truncated. Many examples (some 90% in certain regions) have been drained, using ditches or tile drains, but since the 1980s a number of these have been reflooded as part of a widescale restoration initiative (Galatowitsch & van der Valk, 1994).

Some wetland plant species, especially aquatic and swamp species, have rapidly re-established in reflooded potholes, leading to the impression that restoration was quick and effective (LaGrange & Dinsmore, 1989). However, analysis of the recolonist vegetation has shown that reflooded sites are less species-rich than 'natural' examples, particularly with respect to the sedge fen and wet meadow zones (Galatowitsch & van der Valk, 1995).

Certain types of plant species (submersed aquatics and mudflat annuals) may be better represented in reflooded potholes than natural examples, possibly due to the absence of competition. Galatowitsch & van der Valk (1995) attribute the rapid colonisation of some aquatics to their high vagility rather than persistent seed banks, but consider that seed banks may have been important for the annual species. The impoverished flora of the drier zones seems to reflect both absence of persistent seed banks and constraints upon dispersability. Various workers (e.g. Reinhartz & Warne, 1993) have observed that the number of recolonist species in reflooded potholes is inversely related to their distance from a colonist source in extant examples. However, Galatowitsch & van der Valk (1995) consider that the persistent patches of residual vegetation within the drained wetlands are also important. These are related to the method and efficiency of drainage (ditch-drained potholes tend to have richer and more rapid species recolonisation than the more efficiently tile-drained potholes, reflecting a greater number of wetland species surviving the drainage period within the ditches). Brown (1998) has also observed that spontaneous regeneration of vegetation in a wetland restored from former farmland showed more relationship to the composition of patches of residual vegetation than to the seed bank. Galatowitsch & van der Valk (1994, 1995) consider that recolonisation of the fen and wet meadow zones is likely to be very slow in isolated sites, and suggest that planting may be needed to re-establish them.

individual species). The importance of seed banks also varies in relation to wetland type.

Wetlands subject to cycles of drying and wetting, in which re-establishment of populations of wetland plants from seed is a recurrent event, typically have a well-developed persistent seed bank. This is often dominated by annual species in regularly dry wetlands (Haukos & Smith, 1993), whereas wetlands with longer-period, or less regular, fluctuations may contain a wider range of plant types with contrasting seed bank longevities and germination requirements (van der Valk & Davis, 1978) (Box 14.5).

Persistent seed banks are generally not well developed in wetlands with more stable water tables, presumably reflecting the lesser importance of regeneration by seed in such sites. In general, persistent seed banks tend to be produced by various graminoid monocotyledonous species (especially some sedges), and some dicotyledonous species, but rarer forbs tend to be poorly represented (Maas & Schopp-Guth, 1995). A few uncommon species (e.g. *Cyperus fuscus*) have been recorded from seed banks at sites where they are absent in the vegetation (Schneider & Poschlod, 1994), but these are exceptional. Many of the vascular plants of ombrogenous *Sphagnum* bogs do not form long-term persistent seed banks (though several form transient seed banks), but permanent spore banks have been found for many mosses and liverworts, including *Sphagnum* species (Poschlod, 1990).

Whilst some wetland species can undoubtedly form persistent seed banks, empirical investigations show that the soils of many drained wetlands contain few viable seeds of any wetland species

(Pfadenhauer & Maas, 1987; McDonald *et al.*, 1996), especially the less common taxa. Wienhold & van der Valk (1989) found that wetlands which had been drained for some 30 years had less than 10% of the seed density, and about half of the species, recorded from the seed banks of comparable extant wetlands. Thus in damaged wetlands, especially permanent wetlands, there is little reason to suppose that the majority of species will be able to regenerate from a persistent seed bank, in the event of favourable growing conditions being restored. Moreover, some damaging operations (most notably, but not exclusively, peat extraction) remove the upper layers of soil and any viable diaspores that might have persisted in them.

Seed dispersal

The incapacity of some wetland species to form persistent seed banks means that restoration projects may be critically dependent upon inward dispersal of seeds from external sources. The dispersal characteristics and agents of many wetland plants are not well known, though Middleton (1998) has tabulated much published information. Moore (1982) suggested that some bog plant species were well adapted for dispersal, and this is supported by some empirical observations. For example, Andreas & Host (1983) reported that the wet floor of an abandoned sandstone quarry was readily colonised by bog species, including *Sphagnum* spp., apparently from a source some 14 km distant. Likewise, various bog species have been recorded in the diaspore rain upon commercially cut peatlands (Salonen, 1987). However, whilst long-distance transport of seeds of wetland plants undoubtedly occurs, the limits of dispersability are not really known and various workers have reported limited seed vagility, even in wind-dispersed taxa (Poschlod, 1995; Vegelin *et al.*, 1997). Thus the recolonisation of isolated wetlands may be limited by constraints on dispersion, and the seed rain may be dominated by ruderal or dryland species of local provenance (Curran & MacNaeidhe, 1986).

Many wetland plants are well adapted to water dispersal (Poschlod, 1990; Skoglund, 1990), and have buoyant seeds. However, their floating time, which seems important to the effectiveness of water dispersal (Nilsson *et al.*, 1991), is very variable, typically between about a week and one or more years, though some scarcely float at all (Ridley, 1930; Coops & van der Velde, 1995). Long-distance seed dispersal by water can undoubtedly be important. For example, Skoglund (1990) concluded it was possible for seeds of most wetland species to reach all parts of his wetland study area by flooding. However, its significance for wetland restoration is far from clear, because many diaspores capable of floating are also dispersed by other mechanisms and because some restoration sites are largely isolated from seed-bearing water inputs. It is likely to be particularly important in river-flooded wetlands and for redistributing seeds within restoration sites. Middleton (1998, 2000) points to the need for a 'flood pulse regime', with high winter water levels to maximise dispersion followed by lower summer water conditions for germination and establishment.

Overall, the potential for inward invasion of wetland plants into restoration sites is not well known, but it is likely that propagule availability is often an important constraint upon recruitment of former species that have become extinct locally, especially in isolated sites. In such situations there may be little alternative to deliberate species reintroduction. In some mown restoration sites (fen meadows) mowing machinery could be an important 'accidental' vector that helps to achieve this (Bakker, 1989; Strykstra, 2000).

Dispersal and regeneration of vegetative fragments

Dispersal of vegetative fragments of wetland plants sometimes occurs. Poschlod (1995) reports that fragments of various bryophytes can be wind-dispersed to peat surfaces upon which they may regenerate. Many wetland plants can regenerate readily from small vegetative fragments (Grime *et al.*, 1988), which may be particularly important in assisting dissemination within restoration sites.

Assisting recolonisation

In view of the limited capacity for regeneration from seed banks, probable dispersal constraints, and the occurrence of site conditions that may be inconducive to the establishment of diaspores, consideration can be given to procedures that may assist recolonisation.

Seeding

The possibility of sowing large quantities of seeds of wetland species in restoration sites has been considered by various workers, especially in North America (e.g. Galatowitsch & van der Valk, 1994). In addition to cost and seed availability, potential limitations are poor germination (e.g. some sedge [*Carex*] species), unsuitable water conditions for germination and establishment, competition from seeds of ruderal species, and granivory. Many of these limitations can be controlled or mitigated by appropriate site preparation. It is also likely that spring sowing may optimise establishment, though there is little critical evidence for this.

Hydroseeding provides an efficient means for inoculating large areas of wetland, but attempts have been little documented. Other sorts of inoculum can also be applied this way. Money (1995) 'hydroseeded' experimental trenches in a degraded bog with *Sphagnum* fragments in water. The fragments could be applied more efficiently than whole plants, and produced equally good *Sphagnum* regeneration with much less starting material. In subsequent scaled-up inoculation trials, approximately 180 l of *Sphagnum cuspidatum* fragments were added to two large rectangular lagoons (total area *c*. 1.2 ha). Despite initial problems of water regulation, including both droughting and excessive inundation, after five years a cover of *Sphagnum* of 50%–70% has been observed in late summer (when water tables are at their lowest) (R. P. Money, unpublished data).

Transplants

Transplantation of individual species or turves can assist the recolonisation of restoration sites. Although it is sometimes deprecated on account of its 'unnatural' character, and the possibility of introducing new genotypes or species into the receptor sites, in some instances it may provide the only realistic option for species recruitment. Moreover, species introductions, both deliberate and accidental, have undoubtedly occurred widely in the past (Strykstra, 2000) and in some cases have created vegetation now considered to have high conservation value. For example, during the nineteenth and early twentieth centuries, when litter production had considerable economic value, in some regions litter fen meadows were deliberately created from such diverse starting-points as dry meadows and lakes using techniques which included planting or sowing of desired species in addition to water engineering. Stebler (1898) outlines the techniques used to create litter fen meadows in Switzerland.

Middleton (2000) observes that 'The planting of propagated nursery plants or vegetative fragments is a common practice at restoration sites', but there is considerably less published information about the transplantation of wetland plants and vegetation than attempts to do this. In Britain, extensive reedbeds have been created on rewetted drained wetlands by direct planting of *Phragmites* within prepared sites (Hawke & José, 1996). Translocations of turves, which may contain viable plants, a seed bank and some fauna, have also been made into restoration sites where appropriate donor material is available (sometimes rescued from a threatened wetland). Some of these attempts have not been very successful (Worthington & Helliwell, 1987; Yetka & Galatowitsch, 1999), others more so (Brown & Bedford, 1997; Bullock, 1998). Difficulties probably relate mainly to inappropriate environmental conditions (especially water regime and fertility) in the receptor sites as, in consequence of the regenerative capacity of wetland plants, wetland vegetation may be easier to transplant than some more terrestrial types, at least when suitably wet receptor conditions are provided. Translocation of invertebrate populations is always likely to be difficult (Bullock, 1998), though successful translocation of targeted rare species has been reported (e.g. *Vertigo moulinsiana*: Stebbings & Killeen, 1998).

Little attention has been given to the timing of translocation in relation to hydrological conditions. As a number of wetland plants appear to survive waterlogging-induced anoxia by aeration through 'snorkels', even during their dormant period, it may be expected that translocation into surface-flooded sites during winter may lead to the death of some individuals or species if aeration pathways become damaged by the operations. Transplanting in spring, when new growth is commencing, may well be optimal, though there is only limited empirical evidence for this. Yetka & Galatowitsch (1999) report that *Carex lacustris* rhizomes transplanted better in spring than in autumn, though they considered this was because new shoots were produced in autumn and were sensitive to damage then.

Careful assessment of the appropriateness of conditions at receptor sites, in terms of a suitable ecohydrological environment and the possible damaging impact of human and herbivore activity, is required for effective translocation. Would-be restorers sometimes have an optimistic view as to what constitutes 'suitable conditions'!

'Shoeing' and Bunkerde

The successful 'spontaneous' recolonisation of some old, abandoned peat workings may be partly attributed to the former practice of 'shoeing' the cuttings, i.e. the disposal of unwanted, newly dug living surface turves on the floor of the working. This practice is incompatible with current commercial peat extraction operations, but a related technique persists in the German practice of retaining the surface sods for use in restoration. This 'stripping spoil' or *Bunkerde* is normally stored for long periods and does not necessarily containing living remnants of wetland plants, but in some cases it may contain a seed or spore bank. It is usually composted material and its main value may be as a soil conditioner in restoration sites. However, rather little has been published about this material, which was originally retained to assist the conversion of cut-over peatlands to farmland, and its efficacy and basis of any beneficial effect in peatland restoration is not well documented. A problem of generalisation is that *Bunkerde* can be a very variable material, depending *inter alia* upon its provenance and storage conditions.

Nurse species and mulches

In general, rather little deliberate use has been made of nurse species in wetland restoration initiatives, but they may be useful in some circumstances. There is some evidence that they can assist recolonisation of damaged ombrogenous surfaces, especially under relatively 'dry' conditions. A cover of vascular plants (such as cotton grass, purple moor grass

[*Molinia caerulea*] or heather) can sometimes assist the survival of *Sphagnum* on damaged bog surfaces despite low water tables (Money, 1995; Sliva *et al.*, 1997). The effect appears to be achieved by modification of the surface microclimate, in particular by creating higher humidity and less extreme temperatures than on bare peat (Grosvernier *et al.*, 1995). A similar mulching effect was observed by Salonen (1992), using plastic plants to simulate a cover of *Vaccinium vitis-idaea*.

The capacity for *Sphagnum* to grow amongst nurse species has led some workers to propose this as a mechanism by which damaged bogs can be restored without the need for expensive rewetting operations of bunds and lagoons. For example, Grosvernier *et al.* (1995, 1997) describe the propensity for *Sphagnum fallax* dominated vegetation to colonise 'dried-out' bogs in Switzerland (despite 'unmeasurably' low water tables in some cases), facilitated by a cover of *Polytrichum alpestre* and *Eriophorum vaginatum*. In Canada, restoration trials have used a straw mulch to improve survival of *Sphagnum* transplants (Box 14.2). However, whilst nurse species may help *Sphagnum* to survive as a component of the vegetation at low water tables, it is less certain that, without a rise in the water table, they will result in an actively growing *Sphagnum* bog (Money & Wheeler, 1999).

Nurse species can also have an important role in the recolonisation of extensive shallow lagoons in bog restoration. A cover of emergent plants, including recolonist species such as *Eriophorum angustifolium*, and tussock-forming species established prior to reflooding, including (sometimes moribund) purple moor grass, can provide shelter and support for *Sphagnum* growth (Joosten, 1992; Sliva *et al.*, 1997). Such 'tussock buffering' both increases surface water storage and provides a more stable physical environment for *Sphagnum* expansion.

CONCLUDING REMARKS

Wetland restoration is still in its infancy, and many uncertainties remain. The variability of the habitat, starting conditions and problems make initiatives strongly context-dependent, and generalisation about likely success correspondingly difficult. Some habitats, with species-poor and fast-growing emergent vegetation, such as reedbeds, can be created or restored with considerable success, though not always with their key faunal components. The prospects for some other wetland types are less certain. In some cases, such as restoration of ombrogenous bogs on commercial peat workings, initial recolonisation appears favourable but subsequent development remains uncertain. Even more uncertainties attach to restoration of species-rich fen, partly on account of difficulties of identifying, or providing, appropriate starting conditions.

Restoration of appropriate habitat conditions is a prerequisite for effective restoration. In some, especially topogenous, situations rewetting *per se* can often be accomplished rather easily; in others, such as ombrogenous surfaces without a hydroregulatory acrotelm, it is more tricky and may require considerable engineering. In any case, restoration of appropriate hydrochemical conditions may be more difficult than rewetting. For example, reduction of on-site nutrients may be slow, difficult and sometimes impracticable, e.g. when available water sources (including precipitation) are enriched. Restoration of wetlands may often be more constrained by conditions outwith individual sites than those within them, and such problems are generally much more difficult to solve, though on-site 'workarounds' are sometimes available. When they are not, it may be possible to restore wet conditions but not a wetland of the type that once occurred. However, in many cases, unless there are proximate wetland habitats, the biggest constraint on effective restoration may be isolation of the site from recolonist species. Deliberate reintroduction of species may help circumvent this, but this depends on the availability of a supply of, sometimes rare, organisms and may only be really appropriate for vegetation, and perhaps a few, specifically targeted, animal species.

The limitations and uncertainties inherent in wetland restoration partly reflect deficiencies in knowledge and understanding of wetland ecology, and better characterisation of conditions appropriate for specific restoration targets would be desirable. There is a particular need to identify restoration strategies that are appropriate for contrasting starting-points and conditions. Another consideration is that some restoration practitioners prefer a more intuitive, empirical approach to one that is science-based. Of course, properly documented, all restoration ventures, empirical or not, can contribute considerably both to the disciplines of wetland restoration and wetland ecology, but often restoration initiatives are inadequately documented, monitored or evaluated, and their outcomes are not widely disseminated. Unsuccessful initiatives are often particularly poorly publicised and, whilst understandable, this leaves other restorers to repeat their mistakes. Effective monitoring and reporting of restoration attempts, successful or not, can contribute materially to the science and practice of wetland restoration and one of the biggest constraints on the present 'state of the art' is that very often both follow-up monitoring and reporting of restoration projects are inadequate, sometimes non-existent. Restoration projects are often costly, and it is unfortunate when, often at rather little extra expense, steps are not taken to maximise the amount of transferable information that can be gleaned from them.

REFERENCES

Andreas, B. K. & Host, C. E. (1983). Development of a *Sphagnum* bog on the floor of a sandstone quarry in NE Ohio. *Ohio Journal of Science*, **83**, 246–253.

Armstrong, W. & Boatman, D. J. (1967). Some field observations relating the growth of bog plants to conditions of soil aeration. *Journal of Ecology*, **55**, 101–110.

Arrington, D. A., Toth, L. A. & Koebel, J. W. (1999). Effects of rooting by feral hogs *Sus scrofa* L. on the structure of a floodplain vegetation assemblage. *Wetlands*, **19**, 535–544.

Bakker, J. P. (1989). *Nature Management by Grazing and Cutting*. Dordrecht: Kluwer.

Bakker, S. A., Jasperse, C. & Verhoeven, J. T. A. (1997). Accumulation rates of organic matter associated with different successional stages from open water to carr forest in former turbaries. *Plant Ecology*, **129**, 113–120.

Beltman, B., van den Broek, T. & Bloemen, S. (1995). Restoration of acidified rich-fen ecosystems in the

Vechtplassen area: successes and failures. In *Restoration of Temperate Wetlands,* eds. B. D. Wheeler, S. C. Shaw, W. J. Fojt & R. A. Robertson, pp. 273–286. Chichester, UK: John Wiley.

Blankenburg, J. & Kuntze, H. (1987) Moorkundlich-hydrologische Voraussetzungen der Wiedervernässung von Hochmooren. *Telma,* **17,** 51–58.

Boeye, D. & Verheyen, R. F. (1994). The relation between vegetation and soil chemistry gradients in a groundwater discharge fen. *Journal of Vegetation Science,* **5,** 553–560.

Bowles, M., McBride, J., Stoynoff, N. & Johnson, K. (1996). Temporal changes in vegetation composition in a fire-managed prairie fen. *Natural Areas Journal,* **16,** 275–288.

Boyer, M. H. L. & Wheeler, B. D. (1989). Vegetation patterns in spring-fed calcareous fens: calcite precipitation and constraints on fertility. *Journal of Ecology,* **77,** 597–609.

Brown, S. C. (1998). Remnant seed banks and vegetation as predictors of restored marsh vegetation. *Canadian Journal of Botany,* **76,** 620–629.

Brown, S. C. & Bedford, B. L. (1997). Restoration of wetland vegetation with transplanted wetland soil: an experimental study. *Wetlands,* **17,** 424–437.

Bullock, J. M. (1998). Community translocation in Britain: setting objectives and measuring consequences. *Biological Conservation,* **84,** 199–214.

Burt, T. P. (1995). The role of wetlands in runoff generation from headwater catchments. In *Hydrology and Hydrochemistry of British Wetlands,* eds. J. Hughes & A. L. Heathwaite, pp. 21–38. Chichester, UK: John Wiley.

Campeau, S. & Rochefort, L. (1996). *Sphagnum* regeneration on bare peat surfaces: field and greenhouse experiments. *Journal of Applied Ecology,* **33,** 599–608.

Clymo, R. S. (1984). *Sphagnum*-dominated peat bog: a naturally acid ecosystem. *Philosophical Transactions of the Royal Society of London,* **305B,** 487–499.

Coops, H. & van der Velde, G. (1995). Dispersal, germination and seedling growth of six halophyte species in relation to waterlevel zonation. *Freshwater Biology,* **34,** 13–20.

Connors, L. M., Kiviat, E., Groffman, P. M. & Ostfeld, R.S. (2000). Muskrat (*Ondatra zibethicus*) disturbance to vegetation and potential net nitrogen mineralisation and nitrification rates. *American Midland Naturalist,* **143,** 53–63.

Curran, P. L. & MacNaeidhe, F. S. (1986). Weed invasion of milled-over bog. *Weed Research,* **26,** 45–50.

Curtis, J. T. (1959). *The Vegetation of Wisconsin.* Madison, WI: University of Wisconsin Press.

De Mars, H. (1996). *Chemical and Physical Dynamics of Fen Hydro-Ecology.* Utrecht, The Netherlands: Koninklijk Nederlands Aardrijkskundig Genootschap and Faculteit Ruimtelijke Wetenschappen Universiteit Utrecht.

Dietrich, O., Dannowski, R. & Quast, J. (1998). Solutions of water supply for rewetting of degraded fen sites in northeastern Germany. In *Peatland Restoration and Reclamation,* eds. T. Malterer, K. Johnson & J. Stuart, pp. 220–228. Jyväskylä, Finland: International Peat Society.

Ditlhogo, M. K. M., James, R., Laurence, B. R. & Sutherland, W. J. (1992). The effects of conservation management of reed beds. 1: The invertebrates. *Journal of Applied Ecology,* **29,** 265–276.

EEC (1992). Council Directive 92/43/EEC of 21 May 1992 on the Conservation of Natural Habitats of Wild Fauna and Flora, O.J. L206, 22.07.92.

Fuller, R. J. (1982). *Bird Habitats in Britain.* London: Poyser.

Galatowitsch, S. M. & van der Valk, A. G. (1994). *Restoring Prairie Potholes: An Ecological Approach.* Ames, IA: Iowa State University Press.

Galatowitsch, S. M. & van der Valk, A. G. (1995). Natural revegetation during restoration of wetlands in the southern prairie pothole region of North America. In *Restoration of Temperate Wetlands,* eds. B. D. Wheeler, S. C. Shaw, W. J. Fojt & R. A. Robertson, pp. 129–142. Chichester, UK: John Wiley.

Giller, K. E. & Wheeler, B. D. (1988). Acidification and succession in a flood-plain mire in the Norfolk Broadland, UK. *Journal of Ecology,* **76,** 849–866.

Gowing, D. J. G. & Spoor, G. (1998). The effect of water table depth on the distribution of plant species on lowland wet grassland. In *United Kingdom Floodplains,* eds. R. G. Bailey, P. V. José & B. R. Sherwood, pp. 185–196. Otley, UK: Westbury Press.

Grime, J. P., Hodgson, J. G. & Hunt, R. (1988). *Comparative Plant Ecology.* London: Unwin Hyman.

Grootjans, A. P. & van Diggelen, R. (1995). Assessing the restoration prospects of degraded fens. In *Restoration of Temperate Wetlands,* eds. B. D. Wheeler, S. C. Shaw,

W. J. Fojt & R. A. Robertson, pp. 73–90. Chichester, UK: John Wiley.

Grootjans, A. P., Schipper, P. C. & van der Windt, H. J. (1986). Influence of drainage on N mineralisation and vegetation response in wet meadows. 2: *Cirsio-Molinietum* stands. *Oecologia Plantarum*, **7**, 3–14.

Grosvernier, P., Matthey, Y. & Buttler, A. (1995). Microclimate and physical properties of peat: new clues to the understanding of bog restoration. In *Restoration of Temperate Wetlands*, eds. B. D. Wheeler, S. C. Shaw, W. J. Fojt & R. A. Robertson, pp. 435–450. Chichester, UK: John Wiley.

Grosvernier, P., Matthey, Y. & Buttler, A. (1997). Growth potential of three *Sphagnum* species in relation to water table level and peat properties with implications for their restoration in cut-over bogs. *Journal of Applied Ecology*, **34**, 471–483.

Guthrie, T. F. & Duxbury, J. M. (1978). Nitrogen mineralisation and denitrification in organic soils. *Soil Science Society of America Journal*, **42**, 908–912.

Haesebroeck, V., Boeye, D., Verhagen, B. & Verheyen, R. F. (1997). Experimental investigation of drought induced acidification in a rich fen soil. *Biogeochemistry*, **37**, 15–32.

Harding, M. (1993). Redgrave and Lopham fens, East Anglia, England: a case study of change in flora and fauna due to groundwater abstraction. *Biological Conservation*, **66**, 35–45.

Haukos, D. A. & Smith, L. M. (1993). Seed-bank composition and predictive ability of field vegetation in playa lakes. *Wetlands*, **13**, 32–40.

Hawke, C. J. & José, P. V. (1996). *Reedbed Management for Commercial and Wildlife Interests*. Sandy, UK: RSPB.

Hennings, H. H. & Blankenburg, J. (1994). Investigations on the rewetting of fens in the Dümmer-Region, northwest Germany. In *Proceedings of International Symposium on Conservation and Management of Fens*, 6–10 July 1994, Warsaw, pp. 231–238. Falenty, Poland: International Peat Society and Institute for Land Reclamation and Grassland Farming.

Ivanov, K. E. (1981). *Water Movement in Mirelands*, transl. A. Thomson & H. A. P. Ingram. London: Academic Press.

Joosten, J. H. J. (1992). Bog regeneration in The Netherlands: a review. In *Peatland Ecosystems and Man: An Impact Assessment*, eds. O. M. Bragg, P. D. Hulme,

H. A. P. Ingram &, R. A. Robertson, pp. 367–373. Dundee, UK: Department of Biological Sciences, University of Dundee and International Peat Society.

Joosten, J. H. J. (1993). Denken wie ein Hochmoor: hydrologische Selbsregulation von Hochmooren und deren Bedeutung für Wiedernässung und Restauration. *Telma*, **23**, 95–115.

Kampf, H. (2000). The role of large grazing animals in nature conservation: a Dutch perspective. *British Wildlife*, October 2000, 37–46.

Koerselman, W., & Verhoeven, J. T. A. (1995). Eutrophication of fen ecosystems: external and internal nutrient sources and restoration strategies. In *Restoration of Temperate Wetlands*, eds. B. D. Wheeler, S. C. Shaw, W. J. Fojt & R. A. Robertson, pp. 91–112. Chichester, UK: John Wiley.

Koerselman, W., de Caluwe, H. & Kieskamp, W. (1989). Denitrification and dinitrogen fixation in two quaking fens in the Vechtplassen area, The Netherlands. *Biogeochemistry*, **8**, 153–165.

Koerselman, W., Claessens, D., ten Den, P. & van Winden, E. (1990). Dynamic hydrochemical gradients in fens in relation to the vegetation. *Wetlands Ecology and Management*, **1**, 73–84.

Koppitz, H., Kühl, H., Timmerman, T. & Wichtmann, W. (1998). Restoration and reed cultivation: first results of a multidisciplinary project in northeastern Germany – biotic aspects. In *Peatland Restoration and Reclamation*, eds. T. Malterer, K. Johnson & J. Stuart, pp. 235–243. Jyväskylä, Finland: International Peat Society.

LaGrange, T. G. & Dinsmore, J. J. (1989). Plant and animal community responses to restored Iowa wetlands. *Prairie Naturalist*, **21**, 39–48.

Lappalainen, E. (ed.) (1996). *Global Peat Resources*. Jyväskylä, Finland: International Peat Society and Geological Survey of Finland.

Maas, D. & Schopp-Guth, A. (1995). Seed banks in fen areas and their potential use in restoration ecology. In *Restoration of Temperate Wetlands*, eds. B. D. Wheeler, S. C. Shaw, W. J. Fojt & R. A. Robertson, pp. 189–206. Chichester, UK: John Wiley.

Magnusson, B. & Magnusson, S. H. (1991). Studies in the grazing of a lowland fen in Iceland. 1: The responses of the vegetation to livestock grazing. *Buvisindi*, **4**, 87–108.

Maltby, E. (1998). Peatlands: the science case for conservation and sound management. In *Conserving*

Peatlands, eds. L. Parkyn, R. E. Stoneman & H. A. P. Ingram, pp. 121–131. Wallingford, UK: CAB International.

Martikainen, P. J. (1996). The fluxes of greenhouse gases CO_2, CH_4 and N_2O in northern peatlands. In *Global Peat Resources*, ed. E. Lappalainen, pp. 29–36. Jyväskylä, Finland: International Peat Society.

McDonald, A. W., Bakker, J. P. & Vegelin, K. (1996). Seed bank classification and its importance for the restoration of species-rich flood-meadows. *Journal of Vegetation Science*, **7**, 157–164.

Michaelis, D. (1998). Palaeoecology and the challenge of restoration: the example of percolating mires. In *Peatland Restoration and Reclamation*, eds. T. Malterer, K. Johnson & J. Stuart, pp. 73–78. Jyväskylä, Finland: International Peat Society.

Middleton, B. (1998). *Wetland Restoration, Flood Pulsing, and Disturbance Dynamics*. New York: John Wiley.

Middleton, B. (2000). Hydrochory, seed banks and regeneration dynamics along the landscape boundaries of a forested wetland. *Plant Ecology*, **146**, 169–184.

Mitsch, W. J. (1995). Restoration of our lakes and rivers with wetlands: an important application of ecological engineering. *Water Science and Technology*, **31**, 167–177.

Moen, A. (1995). Vegetational changes in boreal rich fens induced by hay-making: management plan for the Sòlendet Nature Reserve. In *Restoration of Temperate Wetlands*, eds. B. D. Wheeler, S. C. Shaw, W. J. Fojt & R. A. Robertson, pp. 167–181. Chichester, UK: John Wiley.

Money, R. P. (1995). Establishment of a *Sphagnum*-dominated flora on cut-over lowland raised bogs. In *Restoration of Temperate Wetlands*, eds. B. D. Wheeler, S. C. Shaw, W. J. Fojt & R. A. Robertson, pp. 405–422. Chichester, UK: John Wiley.

Money, R. P. & Wheeler, B. D. (1999). Some critical questions concerning the restorability of damaged raised bogs. *Applied Vegetation Science*, **2**, 107–116.

Moore, P. D. (1982). How to reproduce in bogs and fens. *New Scientist*, 5 August 1982, 369–371.

Nilsson, C., Gardfjell, M. & Grelsson, G. (1991). Importance of hydrochory in structuring plant communities along rivers. *Canadian Journal of Botany*, **69**, 2631–2633.

Pfadenhauer, J. & Maas, D. (1987). Samenpotential in Niedermoorböden des Alpenvorlandes bei Grünlandnutzung unterschiedlicher Intensität. *Flora*, **179**, 85–97.

Poschlod, P. (1990). Vegetationsentwicklung in abgetorften

Hochmooren des bayerischen Alpenvorlandes unter besonderer Berücksichtingung standortskundlicher und populations-biologischer Faktoren. *Dissertationes Botanicae*, **152**, 1–331.

Poschlod, P. (1995). Diaspore rain and diaspore bank in raised bogs and implications for the restoration of peat-mined sites. In *Restoration of Temperate Wetlands*, eds. B. D. Wheeler, S. C. Shaw, W. J. Fojt & R. A. Robertson, pp. 471–494. Chichester, UK: John Wiley.

Price, J. S. (1998). Methods for restoration of a cutover peatland, Quebec, Canada. In *Peatland Restoration and Reclamation*, eds. T. Malterer, K. Johnson & J. Stuart, pp. 149–154. Jyväskylä, Finland: International Peat Society.

Reinhartz, J. A. & Warne, E. L. (1993). Development of vegetation in small created wetlands in southeastern Wisconsin. *Wetlands*, **13**, 153–164.

Richards, J. R. A., Wheeler, B. D. & Willis, A. J. (1995). The growth and value of *Eriophorum angustifolium* Honck. in relation to the revegetation of eroding blanket peat. In *Restoration of Temperate Wetlands*, eds. B. D. Wheeler, S. C. Shaw, W. J. Fojt & R. A. Robertson, pp. 509–521. Chichester, UK: John Wiley.

Richert, M., Dietrich, O., Koppisch, D. & Roth, S. (2000). The influence of rewetting on vegetation development and decomposition in a degraded fen. *Restoration Ecology*, **8**, 186–195.

Ridley, H. N. (1930). *The Dispersal of Plants Throughout the World*. Ashford, UK: Neeve & Co.

Rochefort, L. & Campeau, S. (1997). Rehabilitation work on post-harvested bogs in south eastern Canada. In *Conserving Peatlands*, eds. L. Parkyn, R. E. Stoneman & H. A. P. Ingram, pp. 287–294. Wallingford, UK: CAB International.

Ross, S. M. (1995). Overview of the hydrochemistry and solute processes in British wetlands. In *Hydrology and Hydrochemistry of British Wetlands*, eds. J. M. R. Hughes & A. L. Heathwaite, pp. 133–181. Chichester, UK: John Wiley.

Rowell, T. A., Guarino, L. & Harvey, H. J. (1985). The experimental management of vegetation at Wicken Fen, Cambridgeshire. *Journal of Applied Ecology*, **22**, 217–227.

Salonen, V. (1987). Relationship between the seed-rain and the establishment of vegetation in two areas abandoned after peat harvesting. *Holarctic Ecology*, **10**, 171–174.

Salonen, V. (1992). Effects of artificial plant cover on plant colonization of a bare peat surface. *Journal of Vegetation Science*, **3**, 109–117.

Schäfer, A. & Wichtmann, W. (1998). Restoration and reed cultivation: first results of a multidisciplinary project in northeastern Germany – economic aspects. In *Peatland Restoration and Reclamation*, eds. T. Malterer, K. Johnson & J. Stuart, pp. 244–249. Jyväskylä, Finland: International Peat Society.

Schneider, S. & Poschlod, P. (1994). Landshaftsökologisch-moorkundliche Untersuchungen im Osterried bei Laupheim. 3: Die generative Diasporenbank in untersuchiedlich genutzten Flächen. In *Hohenheimer Umwelttagung*, vol. 26, eds. R. Böcker & A. Kohler, Helmbach.

Scholz, A., Pöplau, R. & Warncke, D. (1995). Wiedervernässung von Niedermoor: Ergebnisse eines Versuches in der Friedländer Großen Wiese. *Telma*, **25**, 69–84.

Schouwenaars, J. M. & Vink, J. P. M. (1992). Hydrophysical properties of peat relicts in a former bog and perspectives for *Sphagnum* regrowth. *International Peat Journal*, **4**, 15–28.

Schouwenaars, J. M., Amerongen, F. & Booltink, M. (1992). Hydraulic resistance of peat layers and downward seepage in bog relicts. *International Peat Journal*, **4**, 65–76.

Schrautzer, J. & Trepel, M. (1997). Wechselwirkungen zwischen bodenphysicalischen Parametern, Grundwasserdynamik und der Vegetationzusammensetzung in unterschiedlich stark genutzen Niedermoor-Ökosystemen. *Feddes Repertorium*, **108**, 119–137.

Schrautzer, J., Asshoff, M. & Müller, F. (1996). Restoration strategies for wet grasslands in northern Germany. *Ecological Engineering*, **7**, 255–278.

Shaw, S. C., Wheeler, B. D. & Backshall, J. (1998). Review of the effects of burning and grazing of blanket bogs: conservation issues and conflicts. In *Blanket Mire Degradation: Causes, Consequences and Challenges*, eds. J. H. Tallis, R. Meade & P. D. Hulme, pp. 174–182. Aberdeen: Macauley Land Use Research Institute, on behalf of Mires Research Group (British Ecological Society).

Shotyk, W. (1988). Review of the inorganic geochemistry of peats and peat waters. *Earth Science Reviews*, **25**, 95–176.

Skoglund, J. (1990). *Seed Banks, Seed Dispersal and Regeneration Processes in Wetland Areas*. Uppsala: Acta Universitatis Uppsaliensis.

Sliva, J., Maas, D. & Pfadenhauer, J. (1997). Rehabilitation of milled fields. In *Conserving Peatlands*, eds. L. Parkyn, R. E. Stoneman & H. A. P. Ingram, pp. 295–314. Wallingford, UK: CAB International.

Smith, H. (2000). The status and conservation of the fen raft spider (*Dolomedes plantarius*) at Redgrave and Lopham Fen National Nature Reserve, England. *Biological Conservation*, **95**, 153–164.

Stebbings, R. E. & Killeen, I. J. (1998). Translocation of habitat for the snail *Vertigo moulinsiana* in England. *Journal of Conchology*, Supplement 2 June 1998, 191–204.

Stebler, F. G. (1898). *Die besten Streue-Pflanzen*, vol. 4, *Des schweizerischen Wiesenpflanzenwerks*. Berne, Switzerland: K. J. Wyss.

Strykstra, R. J. (2000). Reintroduction of plant species: s(h)ifting settings. PhD dissertation, University of Groningen, The Netherlands.

Thompson, K., Bakker, J. P. & Bekker, R. M. (1997). *Soil Seed Banks in North West Europe: Methodology, Terminology, Density and Longevity*. Cambridge: Cambridge University Press.

Urban, N. R., Eisenreich, S. J. & Gorham, E. (1986). Proton cycling in bogs: geographic variation in northeastern North America. In *Effects of Acidic Deposition on Forests, Wetland and Agricultural Systems*, eds. T. C. Hutchinson & K. M. Meena, pp. 577–598. New York: Springer-Verlag.

van der Valk, A. G. & Davis, C. B. (1978). The role of seed banks in the vegetation dynamics of prairie glacial marshes. *Ecology*, **59**, 322–335.

van der Valk, A. G. & Verhoeven, J. T. A. (1988). Potential role of seed banks and understorey species in restoring quaking fens from floating forests. *Vegetatio*, **76**, 3–13.

van Duren, I. C., Strijkstra, R. J., Grootjans, A. P., ter Heerdt, G. J. N. & Pegtel, D. M. (1998). A multidisciplinary evaluation of restoration measures in a degraded *Cirsio-Molinietum* fen meadow. *Journal of Applied Vegetation Science*, **1**, 115–130.

van Wirdum, G. (1991). Vegetation and hydrology of floating rich-fens. PhD dissertation, University of Amsterdam.

van Wirdum, G. (1995). The regeneration of fens in abandoned peat pits below sea level in the Netherlands. In *Restoration of Temperate Wetlands*, eds.

B. D. Wheeler, S. C. Shaw, W. J. Fojt & R. A. Robertson, pp. 251–272. Chichester, UK: John Wiley.

Vegelin, K., van Diggelen, R., Verweij, G. & Heinicke, T. (1997). Wind dispersal of a species-rich fen meadow (*Polygono-Cirsietum oleracei*) in relation to the restoration prospects of degraded valley fens. In *Species Dispersal and Land Use Processes*, eds. A. Cooper & J. Power, pp. 85–92.

Verhoeven, J. T. A., van Beck, S., Dekker, M. & Storm, W. (1983). Nutrient dynamics in small mesotrophic fens surrounded by agricultural land. 1: Productivity and nutrient uptake by the vegetation in relation to the flow of eutrophicated groundwater. *Oecologica*, **60**, 25–33.

von Müller, A. (1956). Uber die Bodenwasser-Bewegung unter einigen Grünland Gesellschaften des mittleren Wesentales und seiner Randgebiet. *Angewandte Pflanzensoziologie*, **12**, 1–85.

Wheeler, B. D. (1998). Conservation of Peatlands. In *Global Peat Resources*, ed. E. Lappalainen, pp. 285–301. Jyväskylä, Finland: International Peat Society.

Wheeler, B. D. (1999). Water and plants in freshwater wetlands. In *Hydroecology: Plants and Water in Terrestrial and Aquatic Ecosystems*, eds. A. Baird & R. L. Wilby, pp. 127–180. London: Routledge.

Wheeler, B. D. & Proctor, M. C. F. (2000). Ecological gradients, subdivisions and terminology of north-west European mires. *Journal of Ecology*, **88**, 1–21.

Wheeler, B. D. & Shaw, S. C. (1994). Conservation of fen vegetation in sub-optimal conditions. In *Conservation and Management of Fens*, eds. H. Jankowska-Huflejt & E. Golubiewska, pp. 255–265. Falenty, Poland: International Peat Society and Institute for Land Reclamation and Grassland Farming.

Wheeler, B. D. & Shaw, S. C. (1995a). A focus on fens: controls on the composition of fen vegetation in relation to restoration. In *Restoration of Temperate Wetlands*, eds. B. D. Wheeler, S. C. Shaw, W. J. Fojt & R. A. Robertson, pp. 49–72. Chichester, UK: John Wiley.

Wheeler, B. D. & Shaw, S. C. (1995b). *Restoration of Damaged Wetlands*. London: HMSO.

Wheeler, B. D., Shaw, S. C. & Cook, R. E. D. (1992). Phytometric assessment of the fertility of undrained rich-fen soils. *Journal of Applied Ecology*, **29**, 466–475.

Wheeler, B. D., Lambley, P. W. & Geeson, J. (1998). *Liparis loeselii* (L.) in eastern England: constraints on distribution and population development. *Botanical Journal of the Linnean Society*, **126**, 141–158.

Wichtmann, W. & Koppisch, D. (1998). Degraded fens in northeastern Germany: goals for cultivation and restoration. In *Peatland Restoration and Reclamation*, eds. T. Malterer, K. Johnson & J. Stewart, pp. 32–36. Jyväskylä, Finland: International Peat Society.

Wienhold, C. E. & van der Valk, A. G. (1989). The impact of drainage on the seed banks of northern prairie wetlands. *Canadian Journal of Botany*, **67**, 1878–1884.

Williams, B. L. (1974). Effects of water-table level on nitrogen mineralisation in peat. *Forestry*, **47**, 195–202.

Williams, B. L. & Wheatley, R. E. (1988). Nitrogen mineralisation and water table height in oligotrophic deep peat. *Biology and Fertility of Soils*, **6**, 141–147.

Williams, M. (ed.) (1990). *Wetlands: A Threatened Landscape*. Oxford: Blackwell.

Worthington, T. R. & Helliwell, D. R. (1987). Transference of semi-natural grassland and marshland onto newly created landfill. *Biological Conservation*, **41**, 301–311.

Yetka, L. A. & Galatowitsch, S. M. (1999). Factors affecting revegetation of *Carex lacustris* and *Carex stricta* from rhizomes. *Restoration Ecology*, **7**, 162–171.

15 • Polar tundra

BRUCE C. FORBES AND JAY D. McKENDRICK

INTRODUCTION

Terrestrial biologists define the Arctic as those lands poleward of the climatic limit of trees and agree that the flora is that of a single region (Walker *et al.*, 1994). Despite its occurrence over several disjunct landmasses, the Arctic is comprised of a single biome – the tundra. The vascular flora is depauperate and intercontinental similarity is extremely high, particularly in more climatically and edaphically severe high arctic zones (Young, 1971). Although similarity of the non-vascular flora is also high on a circumpolar basis, species richness of bryophytes and lichens tends to be higher than that of vascular plants within a given landscape (Longton, 1988). The animal biota also contains a high proportion of circumpolar species (Sage, 1986).

Summer warmth is considered the dominant macroenvironmental control delimiting gross distributions of vascular plants in the Arctic (Young, 1971; Edlund, 1990), although a mosaic rather than a zonal pattern is most helpful in understanding local biological diversity. Within a given bioclimatic zone (*sensu* Edlund, 1990), factors such as moisture, nutrient status, soil chemistry and wind become important controls, as evidenced by change along local catenas in a number of landscapes (Bliss & Matveyeva, 1992). Gross distributional patterns among arctic cryptogams, bryophytes and lichens are not so easily explained, although there can be prounounced differences between both cryptogamic and non-cryptogamic floras of adjacent areas where substrates vary even slightly in age since deglaciation or in chemical status. This is often the case among bryophytes, which may be sensitive to subtle spatial and temporal changes in factors such as soil texture, moisture and pH (Longton, 1988).

The division into Low and High Arctic has been mapped by Bliss & Matveyeva (1992), with phytogeographic subdivisions within each major zone (Walker *et al.*, 1994). The High Arctic, in general, differs from the Low Arctic in having very low-growing plant communities – polar desert and polar semi-desert – with only 1%–30% cover of vascular plants. Tundra heath/dwarf shrub and sedge-cryptogam communities with continuous vascular plant cover and a variety of growth forms, which are abundant in the Low Arctic, are very minor components and usually restricted to coastal lowlands (Bliss, 1997). Taken together, relatively productive ecosystems such as these, sometimes referred to as 'polar oases', comprise only a tiny fraction (2%–3%) of the surface area of the High Arctic, yet are critical to the region's terrestrial food web (Bliss, 1977). Cottongrass tussocks (*Eriophorum vaginatum*), which are important in terms of structure and function in many low arctic ecosystems, are absent in the High Arctic.

One of the central assumptions of tundra disturbance ecology is that natural regeneration from human impact in the Arctic is slow (Reynolds & Tenhunen, 1996), and in particular that disturbed high arctic tundra regenerates more slowly, but through similar processes, than similar ecosystems in the Low Arctic (Babb & Bliss, 1974). However, until recently there have been insufficient long-term data to test this assumption, to compare responses of different disturbance types within a given region, or to compare responses among different regions.

In a recent review (Forbes *et al.*, 2001), similarly created high arctic patches were indeed found to return more slowly toward the productivity and physiognomy of their respective control communities than they did in the Low Arctic. The exceptions to this trend provide important insights into factors controlling arctic vegetation responses to patchy disturbance: (1) in several cases (organic layer totally removed, multi-pass vehicle trails, and heavy trampling), vegetation on dry sites regenerated about equally slowly in both high and low arctic regions; (2) in sites that experienced nutrient enrichment due to deposition of nutrient-rich materials or accelerated decomposition, increased nutrient availability allowed high arctic sites to develop a vigorous plant cover similar to low arctic sites. However, species richness was severely reduced due to rapidly growing clonal species tolerant of severely compacted soils which were prominent on sites of modern housing and adjoining heavily trampled ground. In wet areas disturbed by vehicle tracks, the regeneration of vegetation cover was equally vigorous in both regions, although biomass and species richness remained significantly reduced in tracks in the High Arctic. This was attributed to the predominantly belowground allocation of biomass in tundra ecosystems. Particularly in communities characterised by rhizomatous graminoids, plants can easily resprout from intact tillers and then continue to spread vegetatively. Recruitment from the seed bank may also occur, but is not common in the High Arctic.

This chapter will provide an overview of the need for restoration and rehabilitation in tundra ecosystems today and provide concise examples of responses from a wide cross-section of the circumpolar North. The emphasis is necessarily on plants and soils, since vegetation establishment is the first step in stabilising severely disturbed terrain, but the role of animals will also be discussed. Terrestrial herbivores, in particular, exert important controls over vegetation composition and structure via a range of plant–animal interactions. The studies included here embrace everything from descriptions of natural plant regeneration (Box 15.13), to assisted revegetation with non-native species (Boxes 15.7, 15.9 and 15.11), to actual attempts at restoration with native species (Box 15.8), including both vascular and non-vascular plants (Box 15.6).

NATURE OF THE PROBLEM AND RATIONALE FOR RESTORATION

Ecological, conservation, sociological and commercial importance of tundra

Arctic tundra vegetation has low species diversity, simple structure, and low annual productivity. Nonetheless, tundra ecosystems support large populations of wild and semi-domestic animals highly valued by aboriginal and non-native peoples, and they supply critical nesting habitat for immense numbers of shorebirds, waterfowl and other birds (Chernov, 1995; Forbes & Kofinas, 2000). The vast petroleum resources of the Arctic have been known for many years (Crawford, 1997; Forbes, 1999a; Truett & Johnson, 2000). Until recently, development efforts were concentrated almost entirely on crude oil extraction with few new mega-projects since Prudhoe Bay, Alaska began in the 1970s. However, the price of oil tripled in the first 18 months of the twenty-first century and natural gas prices doubled within a year, making the high capital investments necessary for exploiting even remote, permafrost-bound arctic deposits seem reasonable (Varoli, 2000). The result is that there have been increasingly intensive efforts in the past few years to plan for and finance immense arctic petroleum developments in both North America and Russia, particularly as regards natural gas (James, 1998; Vuoria, 1999; Brooke, 2000; Varoli, 2000).

Causes and consequences of degradation

In addition to the more obvious and large-scale impacts associated with petroleum development, mining, and military activities (Crawford, 1997; Komárková & Wielgolaski, 1999), the explosive growth of so-called 'wilderness' or 'eco' tourism is affecting all sectors of the Arctic (Anonymous, 1990; Hamley, 1991; Christensen, 1992; Ilyina & Mieczkowski, 1992; Kaltenborn & Emmelin, 1993). The types of terrestrial degradation commonly associated with the petroleum industry have historically included: drilling pads and roads and the excavation of the gravel and sand quarries necessary for their construction; pipelines; rutting from tracked vehicles; and seismic survey trails. However, some of these impacts (e.g. vehicle tracks) have become relatively rare occurrences in the North American Arctic since the institution of strictly enforced mitigative regulations in the 1970s (Crawford, 1997; Truett & Johnson, 2000).

Recreation impacts tend to be more local (hiking/trampling, camping), but can have significant and long-lasting effects on vegetation and soils (Forbes, 1996; Wielgolaski, 1998), and at least short-term effects on wildlife (Aastrup, 2000). Motorised vehicles (ATVs, snowmobiles) can also have measureable effects on both vegetation/soils and wildlife (Greller et al., 1974; Ahlstrand & Racine, 1993; Wolfe et al., 2000).

Grazing and trampling by both wild and domestic herbivores can result in extensive local erosion. If sustained through many years and over large areas, these processes can lead, in the most extreme cases, to widespread desertification and trophic cascades. Examples include: domestic livestock in Iceland (Magnússon, 1997); the indirect influence of reindeer in association with the petroleum industry in northwest Siberia (Forbes, 1999a); and snow geese, which are increasing through declines in hunting combined with increased food provision in winter quarters (Jefferies, 1999).

Even moderate climatic warming is expected to cause a massive increase in thermokarst (subsidence of surface due to thawing), particularly in ice-rich permafrost regions like northern Alaska and northwest Siberia (Nelson & Anisimov, 1993; Billings, 1997). However, for the next few decades, direct human impacts on arctic ecosystems are expected to have a greater influence on tundra biodiversity and productivity than the projected warming temperatures (Chapin et al., 1997). Anthropogenic activities in the Alaskan Arctic up to this time have generally created small but locally intense disturbances, <1 km² and often <100 m², and in total they occupy small proportions, <10%, of oil and gas fields (Walker & Walker, 1991). These small-scale disturbances may affect wildlife out of proportion to their spatial extent by creating microscale heterogeneity with patches that can either attract (Truett & Johnson, 2000) or repel (Nellemann & Cameron, 1996; Volpert & Sapozhnikov, 1998) animals. In portions of arctic Russia, such as northwest Siberia,

the realised and potential impacts are greater by an order of magnitude, a result of little or no adherence to or enforcement of mitigative regulations (Forbes, 1999a).

Direct versus indirect or cumulative impacts

Disturbance often implies negative changes, such as a catastrophic reduction in species richness or biomass, but any sudden deviation from normal, or reference state, is considered disturbance (Hobbs & Huenneke, 1992). In our discussion, we will consider mostly direct disturbances, i.e. mechanical impacts such as anthropogenically bared patches. For a discussion of cumulative impacts see Walker et al. (1987a) and Forbes (1998).

PRINCIPLES AND CONSTRAINTS OF RESTORATION

To early observers, arctic ecosystems appeared to be so thoroughly affected by the natural disturbance regimes associated with frozen ground that 'stability', as represented by so-called 'climax communities', was simply absent (e.g. Raup, 1951). More recent thinking (sources in Salzburg et al., 1987; Rietkerk & van de Koppel, 1997) incorporates the various disturbance regimes into a theoretical framework in which the factors limiting directional succession and individualistic responses of species and communities are necessarily accounted for and alternative stable states are possible (see below). Periglacial erosional processes, such as thermokarst and shallow-layer detachment slides, are still considered important, particularly in regions with ice-rich permafrost (Leibman & Egorov, 1996; Osterkamp et al., 2000). However, herbivory is increasingly seen as a force to contend with: both as a form of disturbance in itself and as a potential limiting factor during succession (see below). This is particularly the case in the relatively lush lowland and coastal tundra ecosystems with high herbivore densities where petroleum development is now active (e.g. Prudhoe Bay, Alaska; Nenets Autonomous Okrug, Russia) and expected to spread to (e.g. Arctic National Wildlife Refuge, Alaska; Yamal-Nenets Autonomous Okrug, Russia).

While there are still important gaps in our understanding of natural tundra ecosystems, it is becoming increasingly clear that there are varied and critical functions to be served by restoration procedures in disturbed ecosystems. Among the most important applications are: (1) prevention of erosion; (2) rehabilitation of degraded wildlife habitat; and (3) protection of permafrost, and associated benefits from carbon sequestration. Most applications to date have taken place in subarctic and low arctic ecosystems. In the early years, research was necessarily concentrated on assisting recovery, i.e. establishing a simple plant cover as quickly as possible. More recently, however, ecological restoration has come to be seen as an aim in itself.

Limiting factors in the natural recolonisation of disturbed areas

In their synthesis on patch dynamics, Pickett & White (1985) draw attention to physiognomic system structure as a context for disturbance. Of the four contrasting sorts of community structure, tundra would be considered as 'root biased'. This distinction is important because the structure of a system will determine: (1) what sorts of disturbance may have an impact; (2) the threshold of intensity that is effective; and (3) the dependence of species coexistence on disturbance. They contend that disturbance of insufficient intensity to open the root mat in 'root biased' systems will have little impact on species coexistence, although there is evidence to the contrary in high arctic ecosystems (Forbes, 1992).

In addition to intensity, size and shape have proven to be useful predictors of the nature of regeneration within disturbed patches (Forman, 1981). This is due primarily to the alteration of the edge:interior ratio. In the case of indirect or cumulative impacts, patches may substantially change in size and shape over time, as in the case of a vehicle track (small strip patch *sensu* Forman, 1981), which may alter hydrology over an entire slope and/or lead to thermokarst erosion on level ground (large isodiametric patches). When we discuss response and the potential for recovery we are referring to the cumulatively affected area.

Walker *et al.* (1987b) emphasise that patch stability is a prerequisite for either assisted or unassisted vegetation regeneration. However, vegetation regeneration may, in fact, engender site stability. In other words, physical site stability and environmental stress comprise gradients along which suitable regeneration niches may or may not exist. These concepts are at the core of various theories of vegetation changes, i.e. safe site theory, facilitation, initial floristics (see below). The distinction between different states of recovery proposed by Walker *et al.* (1987b) is useful. However, it needs to be expanded to include not only vegetational characteristics but also hydrology, edaphic parameters (physical, chemical and thermal) and patch quality, such as above- and belowground nutrient pools which have implications for herbivory. More recent studies (Forbes, 1994; Strandberg, 1997) have focused on the importance of including bryophytes and lichens in the evaluation of recovery. The authors found that evaluation merely on the basis of vascular plants overestimated the degree of recovery in virtually every instance. On the other hand, if a manager simply wants a green, stable surface, then a measure of vascular cover, usually provided by graminoids, may be all that is feasible under the most severe conditions (Forbes & Jefferies, 1999).

The role of herbivores in the structure and function of regenerating plant communities

In general, arctic herbivores display sensitivity to variation in forage quality across a wide range of spatial scales (Batzli *et al.*, 1981; Post & Klein, 1996). Herbivores may, in turn, influence vegetation both directly, through consumption of plant tissues, and indirectly, via their effects on nutrient cycling and soil disturbance (Hulme, 1996). It is thus widely understood that grazing by vertebrate herbivores can have profound effects on dynamic processes in arctic ecosystems, particularly in successional communities (McKendrick *et al.*, 1980; Jefferies, 1999). In addition to favouring graminoids and ruderal bryophytes at the expense of lichens and certain selected dwarf shrubs (Thing, 1984; Lent & Klein, 1988), grazing is an ecologically important limiting factor in the regeneration of many vascular plant species.

Numerous arctic researchers have noted that herbivores ranging in size from lemming (*Lemmus* spp.) to caribou (*Rangifer tarandus*) are attracted to the plants growing on experimental fertilisation plots and that they can affect the structure, cover/abundance and successional trajectory of the affected communities (Schultz, 1969; McKendrick, 1996; Kidd *et al.*, 1997). Caribou, for example, may use sites with high forage nitrogen concentrations more intensely as a strategy of maximising nutrient intake, leading to a positive-feedback loop over the long-term (Post & Klein, 1996). The same pattern has been observed in the boreal zone, where selective grazing of vegetation plots has been reported for periods of up to five years after a one-time addition of NPK fertilizer (John & Turkington, 1997). This has serious implications for areas where assisted revegetation is attempted because, although plants are selected primarily to prevent erosion (Salzburg *et al.*, 1987; McKendrick & Masalkin, 1993; Martens, 1995), many of the chosen species also provide important food for wildlife (Prégent *et al.*, 1987; McKendrick, 1996, 2000b). Since it is standard practice to apply organic matter and/or chemical fertilisers to subsidise the initial stages of growth (Salzburg *et al.*, 1987; Martens & Younkin, 1997), managers need to be wary about the access of herbivores to sites either naturally recovering from disturbance or actively revegetated (Opperman & Merenlender, 2000).

Caribou are especially important to consider given their great numbers, circumpolar distribution and, in many cases, dependence on habitats partly or completely overlapping large-scale petroleum developments. For example, the number of reindeer on Yamal Peninsula in northwest Siberia (*c.* 200 000), while currently increasing, is already estimated to be 1.5 to 2 times greater than the optimum for the region (Martens *et al.*, 1996). Related points are that: ongoing petroleum exploitation is constantly reducing the area of tundra suitable for pasture; the number of animals and associated grazing pressures are increasing throughout much of the region and altering the vegetation and soils; and overall carrying capacity is constantly being reduced (Podkoritov, 1995).

In such situations, as land withdrawals for industry continue to increase and suitable habitat and/or habitat quality declines, reindeer may change their grazing habits from highly selected plant associations to an even more opportunistic utilisation (e.g. select populations of desirable plants) (Staaland *et al.*, 1993; Ouellet *et al.*, 1994). If grazed intensively and repeatedly, locally enriched patches or more extensive shrub–herb–moss habitats with sufficient organic material are likely to become graminoid-dominated lawns or herb–graminoid meadow–steppe, respectively, as is occurring already, albeit less intensively, at the landscape level in places like Yamal and Taimyr in northern Russia (i.e. 'grassification': Shchelkunova, 1993). Both vegetation types are characterised by low species diversity. However, less productive sites with poor or skeletal soils, such as naturally recolonising rubble/gravel/sand quarries or 'natural' (including zoogenic) sandy blowouts (deflation zones) (Forbes 1995, 1997), may simply revert to more or less primary surfaces indefinitely, thus exacerbating desertification.

Alternative stable states

The disturbance types discussed here tend to involve transitions between a number of different alternative, relatively stable vegetational states, rather than successional development *per se* (Svoboda & Henry, 1987). The transitions between states are caused by the different disturbance events. In a recent review of the topic, the results of many empirical studies of patchy 'natural' communities are interpreted in terms of the presence of alternative metastable states (Hobbs, 1994). At least one example comes from heavily disturbed (overgrazed) subarctic tundra ecosystems (Rietkerk & van de Koppel, 1997), with direct relevance to ecosystems ripe for restoration efforts (Handa & Jefferies, 2000).

A key difference between temperate zone ecosystems and arctic tundra is the ability of moderately disturbed communities of temperate systems to recover to their original state, or nearly so, relatively rapidly once pressure is removed, e.g. trampling (Hammitt & Cole, 1987). In the Arctic, it appears the 'early-successional' communities which occupy many patches may be self-perpetuating (Forbes *et al.*, 2001). The reason for the persistence of these communities is not clear. One factor may be the higher decomposition rates on some of these sites that favour fast-growing grasses which, in turn, prevent establishment by other colonisers. Once established, graminoids in northern ecosystems are able to form similarly dense swards and can benefit from repeated grazing by vertebrate herbivores. This can result in so-called 'grazing lawns' on disturbed patches, which may persist indefinitely (McKendrick et al., 1980; Thing, 1984).

Edaphic parameters can also contribute to community stability. Physical changes in the soil environment within disturbed patches can be persistent, e.g. soil compaction and increased thaw depths relative to undisturbed ground (Forbes, 1993; Kevan et al., 1995). Often, the soil pH of disturbed ground increases. This may limit the reinvasion of nonvascular species, which tend to be more sensitive to subtle changes in surface chemistry, texture, and moisture than vascular plants (Longton, 1988; Forbes, 1994). Higher levels of available nutrients are another factor which may help sward-forming graminoids to occupy a patch indefinitely (Forbes et al., 2001).

RESTORATION IN PRACTICE

At present, the evidence indicates that only the simplest (species-poor) wetland vegetation has been successfully restored, and then only at a small scale and generally lacking microtopography (see Boxes 15.1, 15.6 and 15.8). Habitats with any appreciable microtopography are much more complex and therefore present greater challenges for restoration ecology. The constraints that exist are not merely climatic, but also stem from our limited understanding of tundra ecosystems. That said, the few long-term studies which exist (McKendrick, 1997) have taught us many lessons, during the course of revegetation for its own sake, as well as to stabilise and control erosion. Here we demonstrate the ability to facilitate the regeneration of local species either directly or indirectly, and to rely on natural successional processes in the long run. After three decades, it may eventually be possible to attain native plant assemblages approximating those which were either originally displaced, or at least those which appear best

suited to the new substrate. The restoration of entire communities, including soil fauna, has yet to be addressed, although some baseline work is available to determine soil arthropod survival in various anthropogenic habitats (Meyer, 1993; Kevan et al., 1995; Hodkinson et al., 1996).

Even in the relatively 'mild' climate of the subarctic, it is necessary to take a long-term perspective for any restoration project. In northern Russia, the prevailing theoretical approach has been based on the principle of restoring an organic soil layer in which nutrients, soil fauna, and the bulk of the plant biomass are concentrated (Archegova, 1997). Restoration is conceived as a two-stage process. In the first phase (3–5 years), plant cover and the so-called 'fertile layer' are restored and accelerated soil erosion is prevented by means of intensive agricultural methods: planting perennial tundra grasses and herbs with fertiliser additions (see Boxes 15.3, 15.10, 15.11 and 15.12). It is understood that if the site is used for forage by reindeer then a meadow-like sward will develop (see *Role of Herbivores* and *Alternative Stable States* above). In the second phase (25–30 years), the planted community is ideally replaced by as many as possible of the species originally occurring on the site and a tundra soil develops. In practice, as discussed earlier, intensive grazing may delay or even prevent the realisation of the second stage.

Not only are the technologies for oil exploration and production changing, but also the level of understanding for how to use tundra plant species is improving. In Alaska, relying on secondary succession for tundra regeneration is gaining acceptance, because it results in the most natural-appearing type of plant cover. Leaving damaged sites with a soil that resembles the natural landscape in topography, hydrology and other physical and chemical properties is the most important practice for tundra restoration. Once the desirable soil conditions are established, plant restoration becomes a matter of allowing sufficient time for natural succession to proceed. If the damaged habitat differs markedly from the surroundings, reinvasion may be slow, because availability of propagules is limited. The process can be accelerated by introducing plant species adapted to the disturbed soil. For example, if gravel fill is left in place in the midst of a wet sedge meadow, there will be no nearby source of plant species adapted to dry gravel. There are indigenous tundra species adapted to gravel habitats (see Boxes 15.2, 15.4 and 15.13), and these can be introduced to accelerate revegetating such sites.

Selecting species to introduce requires an understanding of tundra plant recolonisation processes and knowing the habitat requirements (e.g. soil tolerances) for tundra plant species. Usually, natural recolonisation in arctic Alaska follows a process of original floristic composition, wherein the members of the climax community gradually invade and occupy the open ground, and the changes are simply shifts in species composition over time. When artificial revegetation has relied on seeding commercial grasses, it forces the natural recolonisation through the floristic relay process, wherein the species introduced must be eventually replaced by the natural tundra species. If the artificially applied species is a strong competitor, it will slow and possibly prevent the natural community of plants from developing within a reasonable time-frame. Acquiring seed of suitable species for revegetation, which do not outcompete local flora, is crucial for restoring natural tundra communities (see Boxes 15.2, 15.5, 15.7, 15.10 and 15.11). The alternative is laborious 'sprigging' with vegetative cuttings (see Boxes 15.1, 15.3, 15.4, 15.6, 15.8 and 15.12) or transplanting seedlings (Box 15.9). Currently, the demand for tundra revegetation seed is too low to support even one commercial seed-producer in the state of Alaska, and for some habitats, the most desirable species cannot be grown beyond the Arctic. As Alaska's oil exploration and production technologies continue reducing the extent of tundra damage, the prospects for future seed markets will likely decrease.

Alaska

Two commercial operations in Alaska that have damaged tundra are the exploration for, and production of, hydrocarbons. Prior to those hydrocarbon-related activities, the military, indigenous people and tundra researchers were the primary sources of human disturbances in the Alaska Arctic. During the 1940s through to the early 1990s

exploratory drilling affected tundra vegetation and soils at various locations in the National Petroleum Reserve in Alaska (NPR-A) and beyond (McKendrick, 2000a). Arctic scientists were supported out of the Naval Arctic Research Laboratory at Barrow, Alaska and often travelled across the tundra in track vehicles, leaving trails, some of which are still visible today. Until the 1970s, such disturbances were left to recover on their own. Natural tundra recolonisation vegetated most of those disturbances within 25–30 years. Generally, recovery time has been inversely related to site wetness.

Commercial oil production following the discovery of the Prudhoe Bay Oil Field in 1968 and the growing concerns for environmental considerations prompted industry-sponsored studies to address tundra rehabilitation. Based on early experiments, seeding of three grasses (*Arctagrostis latifolia* var. Alyeska, *Festuca rubra* var. Arctared and *Poa glauca* var. Tundra), became a standard practice. On medium texture, mesic soils, seeding and fertilising usually produced grass stands within three growing seasons. In the long term, seeding these commercial grasses has frequently delayed natural recolonisation, where seeded grasses established and persisted. Examples are overburden stockpiles at gravel mines. Where such stockpiles were seeded, the commercial grasses have resisted invasion by other tundra plant species for at least 15 years. It is believed that competition from the seeded grasses is responsible for the delay of indigenous plant invasion. The oil industry has created perhaps fewer than 20 gravel mine overburden stockpiles in northern Alaska. Stockpiles vary from about 0.4 ha up to 15 ha in area, and not all have been seeded and fertilised.

The most extensive tundra disturbance has been gravel fill for pads and roads. It has been calculated that 0.04% of Alaska's Arctic Coastal Plain (or 0.1% of the area between Colville and Canning Rivers) has been affected by gravel fill (Gilders & Cronin, 2000). Gravel fill creates a habitat that is drier, more compact, and lower in nutrient- and moisture-holding capacities than the common soils. The commercially available grass varieties are marginally suited to gravel fill habitat. Removal of gravel fill is a relatively recent practice in the Alaska oil

Box 15.1   Sprigging *Arctophila fulva* into flooded areas

Arctophila fulva is an important emergent aquatic grass associated with prime waterfowl habitat. It develops colonies along margins of ponds and lakes and slow-flowing streams. Gravel fill often creates impoundments and new open-water habitats. Excavations (reserve pits at exploration sites) and thermokarst may also produce new open-water habitats. In time, *Arctophila* often invades these habitats, increasing their usefulness to wildlife. To accelerate that process, the grass can be sprigged into those locations (McKendrick, 1993; Kidd *et al.*, 1997). Adding phosphorus fertiliser improves the chances for success (McKendrick, 1993), but can also attract vertebrate herbivores (Kidd *et al.*, 1997). Sprigs have been obtained from nearby natural stands and are planted at spaces ranging from 5 to 30+ cm along transects perpendicular to pond margins. The grass roots at stem nodes, and does not require rooted shoots for establishment in new locations.

To compensate for fluctuating water levels, transects begin above the waterline integrated into the water to a depth of approximately 10–15 cm. This technique has proven successful at other wetland restoration sites in the oil fields (Kidd *et al.*, 1997). The sprigs were not planted throughout pond basins because the water was too deep in the centre of the basins. Percentage survival of the planted sprigs was difficult to determine because, although some sprigs died after planting, new sprigs emerged in the following year, and it was not practical to tag each sprig that was initially planted to track its fate. However, the total number of sprigs within sampling contracts increased in one pond from 46 in 1992 to 177 in 1995 and in another pond, the total number sprigs increased from 20 in 1992 to 65 in 1995. Overall, mean sprig density increased from 4 sprigs m^{-2} in 1992 to 16 sprigs m^{-2} in 1995.

The area continues to be used by wildlife including shorebirds, geese and caribou, although grazing intensity has declined over time. The nutrients applied in 1995 appear to have been used up, possibly leaving the forage less nutritious and less attractive to herbivores. Vegetation adjacent to the created ponds, however, is still intensively grazed. Thus, the created ponds are serving as foraging habitat for waterbirds (Kidd *et al.*, 1997).

Overall, many ecological characteristics present at former Prudhoe Bay exploratory drill sites indicate that wetland functions are being established and that the initial response to the various treatments has been favourable. It is still uncertain as to whether plant growth at these sites can be maintained without continued fertilisation, and when the sites will become thermally stable. However, much has been learned about both the limitations associated with creating and restoring wetlands in disturbed environments in the Arctic and the state of rehabilitation strategies that best facilitate wetland recovery in a cost-effective and ecologically sensible manner.

fields, and vegetation responses to that are yet to be determined, but early results are promising.

Exploration and production technology is changing in Alaska. Currently, all exploration activities are conducted in winter, and drilling is done from ice pads. No reserve pits are created, and once the ice pad melts, the only visible sign of drilling is a small melt pool at the well itself. Thus, there is no need for tundra restoration at new exploration sites. New pipelines are also constructed from ice roads, and neither workpads nor gravel roads are built next to pipelines. Production facilities have been redesigned to reduce the footprint to half that which used to occur in the Prudhoe Bay Oil Field (Gilders & Cronin, 2000). Thus, future needs for tundra restoration will be much less than for production sites developed previously.

The following are examples of recent applications: (1) establishing commercial grasses on exploratory drilling pads; (2) sprigging *Arctophila fulva* into flooded areas (Box 15.1); (3) using indigenous species on overburden and gravel fill (Box 15.2); and (4) stabilising sand with *Elymus arenarius* (Box 15.3). They are not necessarily representative of future needs, owing to changing exploration and production practices.

The industry and federal government used two general types of drilling pads for exploration from

Box 15.2   Using indigenous species on overburden and gravel

The last two exploration wells on Alaska's North Slope (early 1990s), in which reserve pits were excavated and subsequently buried, were given applications of indigenous plant seed (hand harvested locally) and fertiliser (McKendrick, 2000b). The biennial *Descurainia sophioides* provided the greatest amount of cover immediately. *Puccinellia borealis* (mistakenly referred to as *P. arctica* in the article cited) was also an important revegetation component. *Puccinellia borealis* was used successfully in the 1970s to establish cover on an oil-damaged tundra at Prudhoe Bay (McKendrick & Mitchell, 1978). This grass species is adapted to a wide range of soil conditions and only persists in habitats unsuited to other tundra species, such as on drilling muds and saline soils. Both *D. sophioides* and *P. borealis* readily relinquish territory to new invaders. While it dominates a site, *D. sophioides* provides habitat useful for wildlife. *Puccinellia borealis* can be produced on farms outside the Arctic, making it a candidate for commercial seed production, should the demand warrant. The first commercially produced seed of this grass was used to establish vegetation at the Badami Oil Field gravel mine and exploration well in 1998. A grazing survey conducted on overburden stockpile sites in the Kuparuk Oilfield by Bishop *et al.* (1998) found that grazing was variable among years and appeared dependent upon time since fertilisation. Overall, *P. langeana* was the most selected species for grazers (primarily caribou), followed by *Arctagrostis latifolia*. *Deschampsia caespitosa* and *Festuca rubra* were heavily grazed in naturally colonised sites which had been

fertilised. These species are apparently selected by geese.

Forbs in the family Leguminosae are of particular interest for revegetation efforts because of their symbiotic relationship with *Rhizobium* bacteria, which form nodules on the legume roots. These bacteria convert atmospheric nitrogen into forms that plants can use, and the relationship allows legumes to grow on nitrogen-poor soils. Several species of legumes occur naturally on gravelly soils and are therefore good candidate species for gravel fill sites. Different techniques were tested on a gravel pad at the Kuparuk Oil Field for inoculation (three treatments), scarification (roughing of the seed coat; two treatments), and fertilisation (four treatments) (Bishop *et al.*, 1999). Species for all three experiments included *Oxytropis borealis*, *O. campestris*, *O. deflexa* and *O. viscida*. In addition, *Hedysarum mackenzii* (scarification) and *Astragalus alpinus* (fertilisation) were included. In the scarification experiment, mean survivorship in the unscarified treatment (19%) was more than five times as high as in the scarified treatment (3%), a significant difference. There was no significant effect of the inoculation treatments. Within the fertiliser experiment, mean survivorship ranged from 8% to 12%, with no significant differences among treatments. Available nitrate levels appeared higher in the high-N fertiliser treatments, but high within-treatment variability made interpretation of fertiliser treatment effects on nutrient levels difficult. Future monitoring will determine whether the presence of legumes improves the growth and persistence of grasses (or willows; see Box 15.4).

the 1940s to the 1960s. One was constructed from material excavated from reserve pits and mixed with snow and ice, and the other was constructed from gravel or coarse sands and gravel. During the late 1970s, the federal government explored NPR-A for the second time, using these types of drilling pads. Those drilling pads of the late 1970s were revegetated artificially using commercial grass seed and fertiliser (Smith, 1986).

Nearly 20 years has lapsed since these sites were rehabilitated. Results vary among locations due to

habitat conditions. Some sites with medium-texture soil and adequate moisture have been invaded by indigenous tundra species and now resemble the undisturbed tundra. At other locations, the seeded grasses still dominate plant communities. Where gravel fill was used and thermokarst created wet basins on drilling pads, the seeded grasses have been overtaken by accumulations of moss (as much as 5–7 cm) and indigenous vascular plant species (Box 15.4). If the gravel remained dry, *Poa glauca* has persisted on the horizontal surfaces of drilling pads with

Box 15.3   Using *Elymus arenarius* to stabilise sand at Prudhoe Bay

Wind erosion of the sandy soil cap on an abandoned landfill in the Sagavanirktok River Delta, Prudhoe Bay, Alaska required control. Several species of grasses were seeded: *Deschampsia beringensis* var. Norcoast, *Elymus arenarius* var. Reeve (from Norway), *Festuca rubra* var. Arctared, *Poa glauca* var. Tundra and *Puccinellia augustata*. Locally obtained sprigs of *E. arenarius* were sprigged at the site. All seeded grasses eventually established to varying degrees; however, the sprigs of *E. arenarius* proved most beneficial in stabilising the sand. Owing to the poor nutrient content of the sand, fertiliser applications greatly benefited grass establishment.

To date, stabilising sand has not been a significant problem related to oil production in the Alaska Arctic. However, there are extensive areas in NPR-A where sand predominates, and future development in those locations may require sand stabilisation. Also, production areas of western Siberia (Yamal-Nenets Autonomous Okrug) have sandy soils (see below), and erosion control is critical for those production sites, not only for protecting the habitat, but also to prevent sand from damaging equipment and facilities.

lesser amounts of *Festuca rubra*. Moss accumulation on dry habitats ranges from 0 to 1 cm. The side slopes of these exploratory gravel pads are usually dominated either by a mixture of *Arctagrostis latifolia* and *Festuca rubra* or *Arctagrostis* alone. Standing dead phytomass and litter, as opposed to mosses, have accumulated on side slopes. A number of indigenous plant species has been observed on these gravel fill pads (McKendrick, 1991), indicating secondary succession is occurring, but in many instances the invasion process is slow, due to competition from seeded grasses.

Canada

The earliest experiments with revegetation/restoration using tundra species in Canada date from the early 1970s. Both native and non-native species,

mainly grasses and sedges, have been grown in trials from the subarctic to the High Arctic (Boxes 15.5, 15.6, and 15.7). Areas of concern, due to the extent of terrain disturbance involved, include the valley and delta region of the Mackenzie River, where sizable petroleum reserves exist (Brooke, 2000), and along the western coast of Hudson Bay, where intensive grazing is resulting in the destruction of large areas of saltmarsh (Box 15.8).

Iceland

Degradation processes in Iceland have been intense with grazing, burning and clearing, all followed by erosion. The result is that totally barren land covers immense areas. The climate is cool year round with frequent strong winds and the principal soils are andosols of volcanic origin and with high glass content. It is estimated that the native mountain birch (*Betula pubescens*) and other vegetation covered c. 65 000 km^2 at the time of settlement, 1100 years ago (Magnússon, 1997). The current extent of birch is 1165 km^2, and 37 000 km^2 are now barren deserts with an additional 10 000 to 15 000 km^2 of disturbed areas with limited plant production. Magnússon (1997) has divided restoration efforts in Iceland into three main periods: (1) erosion control (1907–45) (Box 15.9); (2) cultivation period (1946–85); and (3) ecological revegetation (1986–present) (Box 15.10). Ecological methods are becoming increasingly important. They include the use of native and introduced legumes, and shrubs and trees, including birch. At the same time there is a growing concern and debate about the introduction and the use of exotic species for land reclamation.

Russia

Unlike Alaska, where there is relatively little need for tundra restoration at new exploration sites, the general lack of meaningful mitigation protocols in northern Russia means that there are now immense areas which require, at the very least, extensive assisted revegetation to control further erosion. Studies on sandy soils were initiated because this is a common soil type in large areas of northwestern Siberia and is a condition that makes revegetation more difficult. In 1992, AMOCO also tested various

Box 15.4   Use of native willows to control erosion on slopes of gravel fill

During early June 1998, a total of 570 cuttings of three common species of native willows (*Salix alaxensis, S. lanata* and *S. glauca*), were harvested from local populations in the Colville River Delta. These cuttings were 'inoculated' and transplanted into experimental plots where native grasses (*Deschampsia caespitosa, Poa glauca, Festuca rubra, Arctagrostis latifolia*) and a legume (*Astragalus alpinus*) were sown on the same day on the side slope of a gravel fill (Jorgenson *et al.*, 1999). Cuttings of *Salix alaxensis* and *S. lanata* were 0.4 to 0.5 m long and *S. glauca* cuttings were 0.2 to 0.4 m long. All three species were planted so that approximately 90% of their stems were below ground. Stem cuttings were inoculated with mycorrhizae by coating the cutting stems with a soil/water slurry using soil obtained from the shrub collection sites, because mycorrhizae have been found to be important to the growth of arctic shrubs. Of the 570 shrub cuttings that were transplanted, 389 were alive by late July (*c.* 50 days after planting) resulting in an overall mean

survivorship of 68%. Mean survivorship was similar in the upper (68%), middle (79%) and a lower (59%) slope height zones. Sprouting of many of the stems was robust, and survivorship appeared to be similar for *S. alaxensis* (75%), *S. lanata* (69%) and *S. glauca* (66%).

It is anticipated that the nitrogen-fixing capability of the legumes should reduce or replace fertilisation requirements over the long term. Many grass cultivars that were originally seeded in the mid-1970s along the Trans-Alaska Pipeline have died back to a sparse cover of widely spaced bunches of grasses and have been replaced with legumes. Jorgenson *et al.* (1999) hypothesise that the legumes are largely responsible for the long-term productivity of these sites. Since the initial survivorship of the shrub cuttings was relatively high, further growth and stem sprouting (coppicing) should continue to increase the shrub cover. A closed canopy of low shrubs may take five to ten years to develop however, and will have a relatively large nutrient requirement for sustained productivity. Planting of shrubs is also labour intensive, and the much higher costs of planting will need to be evaluated against the potential benefits.

techniques for establishing grasses on these soils (Martens, 1993) (Box 15.11). These included the use of a seed drill to plant seeds and apply fertiliser at controlled rates, and nurse crops to provide protection for slower-establishing but longer-lived grasses, and erosion control mats to hold the sand in place in highly erodable locations. Native willows have also been used to control erosion (Box 15.12). Other

Box 15.5   Pipeline trench revegetation using native subarctic species

The following revegetation treatments using locally collected native species and commercially available native cultivars were seeded with and without fertiliser on to exposed mineral soils in a simulated pipeline trench (see also Box 15.7). Local treatments were *Arctagrostis latifolia, Carex membranacea, Arctostaphylos rubra, Betula glandulosa, Empetrum nigrum, Ledum groenlandicum* and *Vaccinium uliginosum*. Commercial treatments were Alyeska polargrass (*Arctagrostis latifolia*), *Artemisia tilesii* and *Calamagrostis canadensis*, and Decora seed-mix (*Agropyron violaceum,*

Festuca ovina, Poa alpina and *P. glauca*) (Maslen & Kershaw, 1989). *Epilobium angustifolium* rhizomes were also cut to lengths of 25–60 cm and planted at depths of *c.* 5 cm. *Betula glandulosa* and *L. groenlandicum* seeds germinated well, but remained as seedlings during the first growing season. *Carex membranacea* and the other shrub species did not germinate. Based on aboveground production, fertilised Alyeska polargrass performed better than all other treatments and was the only treatment to respond to fertiliser, with a phytomass increase of 663% relative to the unfertilised control. The other cultivated and locally collected herbaceous species all established well and had similar, but low, aboveground production.

Box 15.6   High arctic wetland restoration

The study took place in the floristically impoverished wet meadows of Truelove Lowland, Devon Island (Forbes, 1999b). In gently sloping sedge-moss meadows, a single taxon, *Carex aquatilis* var. *stans*, was dominant with *Eriophorum angustifolium* as an important associate. These two plants appear morphologically similar but occupy different niches due to divergent rooting strategies and preferences for different peat depths. Since so few wet-meadow species produce viable seed in such extreme tundra environments (Bliss & Grulke, 1988), vegetative shoots and bryophyte sods were employed to restore wetlands which had suffered intense damage from vehicle activity. Clonal transplants of native *C. aquatilis* var. *stans* were planted with and without bryophytes into vehicle ruts and left unfertilised. After nearly two decades, *Carex* cover was somewhat less in ruts with flowing water and somewhat greater in restored ruts with standing water compared to controls. Reinvasion of *Eriophorum angustifolium* occurred in treated ruts, but cover was less in both treatments compared to controls. *Eriophorum scheuchzeri* was unexpectedly recruited from the seed bank in moss-sodded plots. Total plant cover in restored ruts was nearly comparable to controls, but biomass was somewhat less than control plots. Plots with bryophytes were environmentally distinct, due primarily to increases in organic mat depth relative to controls. Restoration efforts resulted in increased plant cover after 18 years compared to naturally recovering ruts.

Box 15.7   Agronomic and native plant production under various fertiliser and seed application rates

In spring 1987, a reclamation experiment was initiated on a simulated right-of-way in a permafrost-influenced subarctic *Picea mariana* forest near Fort Norman, Northwest Territories (Evans & Kershaw, 1989). The objective was to test the short-term effects of various rates of fertiliser and seed application on the productivity of agronomic and native plants. In the 25-m wide right-of-way, organic gleysolic turbic cryosol soils were disturbed by 1800 passes of an all-terrain-cycle (ATC) over a 70-m length. The site was seeded and fertilised with a commonly used commercial seed-mix (*Agropyron trachycaulum* 28%, *Festuca rubra* 20%, *F. ovina* 15%, *Alopecurus arundinaceus* 15%, *Phalaris arundinacea* 12%, *Phleum pratense* 5% and *Poa pratensis* 5%) at 30 kg ha^{-1} and fertiliser (NPK blend 17–25–15) at rates of 0 to 1000 kg ha^{-1}. Phytomass of native graminoids, native herbaceous and agronomic seed-mix species increased after the first and second growing seasons. The average increase in native species' phytomass from 1987 to 1988 was 573% in the unseeded treatments. Within seeded treatments the average increase in native plant production was 346%. The average increase in agronomic seed-mix phytomass was 454% over the two growing seasons of the study, with the highest productivity occurring in those treatments in which 500 and 1000 kg ha^{-1} of fertiliser were applied.

studies have concentrated on natural revegetation of primary surfaces such as sand, gravel and rubble quarries, roadside and railway verges and sandy deflation scars and aeolian sand surfaces.

CONCLUDING REMARKS

With the exception of Iceland, which has the longest history of attempting to rehabilitate severely degraded ecosystems, ecological restoration is still in its early stages in much of the circumpolar North. Restoration in the Arctic remains constrained not only by a severe climate, but also by a limited understanding of fundamental ecosystem processes. Despite this, assisted revegetation efforts have made great progress in identifying a range of techniques. An urgent need typically resulting from industrial and/or grazing impacts is that of rapidly establishing a plant cover for purposes of controlling erosion on nutrient-poor, well-drained soils (i.e. sands, loess, volcanic ash). Rehabilitated habitats, in particular gravel/overburden and flooded areas, are often

Box 15.8   Subarctic saltmarsh restoration

At present, 2500 ha of vegetation in tidal and freshwater marshes in the vicinity of La Pérouse Bay, Manitoba have been adversely affected by foraging activities of staging and breeding lesser snow geese, *Anser caerulescens caerulescens*. Intense grazing in summer has led to massive loss of saltmarsh vegetation via a positive-feedback mechanism that results in hypersaline soil. In the few intact areas of intertidal salt marsh that remain, the dominant graminoids are *Puccinellia phryganodes*, a stoloniferous grass, and *Carex subspathacea*, a rhizomatous sedge. Former *Puccinellia–Carex* swards have been replaced by relatively barren mudflats. In some inland saltmarshes and at some sites in the upper levels of the intertidal marshes, transient populations of the annual halophytes, *Salicornia borealis* and *Atriplex patula* var. *hastata* have naturally colonised mudflats (Handa & Jefferies, 2000).

In order to assess the potential for restoration of mudflats, soil plugs of the former dominant graminoids, *P. phryganodes* and *C. subspathacea*, were transplanted from an intact site into plots of hypersaline, bare soil in mid to late June 1996. Some plots were treated with fertiliser and peat mulch to determine if ameliorations enhance plant growth. Plants of *P. phryganodes* established in degraded sediments, but those that received an amelioration treatment showed significantly higher growth than those that were planted in bare soil. Plants of *C. subspathacea* did not establish readily in degraded sediments and no enhancement to plant growth was observed when amelioration treatments were applied to sediments. Growth rate and mortality of plants both showed high variation between sites and years (1996–7). The spatial and temporal variation in establishment and growth of plants reflected variation in edaphic and weather conditions that included high salinity and low water content of soils and hot, dry weather.

Box 15.9   Use of *Leymus* for controlling erosion

Leymus arenarius is the most important species used to stabilise active aeolian surfaces in Iceland (Aradóttir *et al.*, 2000). *Leymus* is typically seeded and fertilised with about 100 kg nitrogen per hectare on aeolian sands. All sites with the *Leymus* treatment have had active sand surfaces with sand sedimentation that can create sand dunes and ridges 2–3 m high. The stands are fertilized at both establishment and later stages, if needed, for maintaining or increasing plant cover and/or facilitating seed production. *Leymus* has been identified to have good potential for carbon sequestration, because roots and organic matter are continuously being buried under aeolian sediments (Arnalds *et al.*, 2000). This constitutes a significant additional benefit resulting from restoration practices in Iceland.

exploited by wildlife for forage and/or nesting, although this effect may not persist if available nutrients are depleted from the site. Nevertheless, managers need to be wary of access to recovering sites by large herbivores which can easily alter the intended successional trajectories via intensive grazing.

Use of agronomic species engineered for specific tasks is widespread given the difficulty of obtaining sufficient amounts of viable local seed. Long-term observations indicate that reinvasion by local indigenous species can and does occur, at least in the Low Arctic of Alaska. In some cases, such as in northwestern Siberia, both intentionally and unintentionally introduced species have demonstrated the ability to persist and spread in response to local disturbance regimes. As a result, extreme vigilance is advised when employing agronomic species, and road/railway corridors should be monitored for northward-migrating southern ruderals. It is important to note that during the same time period in which the above lessons have been learned, the nature and extent of impacts have changed, at least in North America. The main reason is that a less invasive oil industry leaves an increasingly smaller 'footprint' on the landscape. In northern Russia, however, a lack of meaningful mitigation protocols and lax enforcement means that the number of severely

Box 15.10   Restoration forestry in Iceland

In addition to preserving the remaining native birch woodlands, it is important to restore the original and productive birch ecosystems. The native birch can be an early coloniser in succession in Iceland, and its role for ecosystem restoration has often been underestimated. Increased emphasis has recently been placed on finding suitable means to exploit the colonising traits of birch to restore Icelandic woodland ecosystems (Aradóttir & Arnalds, 2001). Large-scale efforts to prevent erosion and reclaim severely degraded land began early in the twentieth century.

Since 1990, an average of 1 million tree seedlings has been planted annually as part of a national programme, at a total of 110 sites. Birch has been the most important species but larch (*Larix sukaczewii* and *L. siberica*) and pine (*Pinus contorta*) species are also important. The average survival in 1996 of seedlings planted in 1991 and 1992 was 70% for birch, 63% for pine and 44% for larch. The survival of birch at individual sites ranged from 32% to 94% for seedlings planted in 1991 and 55% to 89% for seedlings planted in 1992. This large variation reflects differences in environmental conditions, vegetation cover and surface characteristics of the planted areas, and different fertilisation treatments, among other factors. About 24% of the planted birch seedlings from 1991 and 1992 were reported as frost-heaved. Mortality of the frost heave to seedlings was relatively high, indicating that frost heaving was an important cause of mortality.

The use of nitrogen-fertilisers may increase mortality of birch seedlings in the first year after planting due to fertiliser-salt effects and increased competition by other vegetation. The fertiliser-induced mortality of young plantations can be avoided by using slow-release fertilisers. In poor and eroded soils the increased vigour of fertilised seedlings, compared to unfertilised ones, may compensate for the early mortality.

Box 15.11   Use of native and introduced grasses to control erosion on sandy soils

Site conditions on the Yamal Peninsula for revegetation are harsh due to severe climate exacerbated by poor soil conditions for plant growth. The sandy substrates have an inherently low capacity to retain plant nutrients and soil water. The growing season is short, precipitation is low, winter temperatures are extreme and the sandy nature of the soils makes them highly susceptibility to wind erosion. Strong winds in both summer and winter result in blowing sand and snow. In 1993, joint Russian/AMOCO revegetation trials were established at Bovanenkovo to test and evaluate revegetation species and methods at a common site with the objective of developing the most suitable revegetation prescriptions for conditions on the Yamal. In 1994, several new test sites were established to replace sites that were abandoned in 1993.

Based upon their performance in northern Canada and Alaska, varieties of 23 grasses and one legume were selected in 1991 for testing (Martens, 1995; see also Rozhdestvenskii & Sarapultsev, 1997). All but three of these species are circumpolar and are represented in the native vegetation of Russia as well as North America. A 5-m row of each species was planted at each location and growth of each species was examined in late August. Percentage of the row occupied by the seeded species was estimated visually, stage of plant development was assessed, plant heights were measured, and photographs were taken.

All of the species that maintained vigour through the fourth growing season are successful candidates for inclusion in seed mixtures for the Bovanenkovo area. Successful species were: *Arctagrostis latifolia* var. Alyeska, *Calamagrostis canadensis* var. Sourdough, *Deschampsia beringensis* var. Norcoast, *D. caespitosa* var. European, *Festuca ovina* var. Common, *F. ovina* var. Duriscula, *F. rubra* var. Arctared, *Poa alpina* var. Common, *P. glauca* var. Tundra and *P. pratensis* var. Nugget. All of these species also produced seed, further demonstrating their adaptation to this region and greatly improving their ability to maintain a long-term plant cover.

Although not tolerant of winter conditions at Bovanenkovo, grasses that are capable of rapid growth in the first growing season can be used very effectively in combination with slower-growing, longer-lived species, to stabilise a site where erosion is a concern or to provide physical protection in the first and second year while the slower-developing grasses are establishing. Species that produced abundant first-year cover but declined after the first or second winter included *Agropyron dasystachyum* var. Elbee and *A. trachycaulum* var. Revenue. *Bromus inermis* var. Manchar and *Deschampsia beringensis* var. Norcoast appear to have the ability to do both, providing a rapid first-year cover that is also winter-hardy. It should be noted that

A. trachycaulum (a Canadian cultivar) produced seed in the first or second year after its introduction, and thereafter invaded other test plots as a result of seed spread (Martens, 1995). Also, a heavy infestation of *Matricaria grandiflora* and *Stellaria media* occurred in the first year, but an application of herbicide eradicated the weeds. These results indicate that extreme caution should be exercised when using non-native species and native species should be employed whenever possible.

Box 15.12   Use of native willows to control erosion on sandy soils

In 1992, three approaches to establishing a shrub cover on an outer slope of a sand pad were tested on the Ob River floodplain near Nefteyugansk (Martens, 1995). Contour willow bundles (fascines) were installed on a 6-m portion of the pad slope. *Salix* spp. harvested nearby were pruned of all side branches and tied into bundles of approximately 3 m in length and 20 cm in diameter. Four successive horizontal rows of these bundles, spaced at approximately 1.2 m, were placed on the face of a slope in trenches approximately 20 cm in depth and 6 m in length, and securely anchored with willow stakes. Following staking, sand was filled in around the bundles and worked into the bundle.

Contour brush-layering consisted of embedding green branches of shrub or tree species on successive horizontal rows or contours in the face of the slope. Brush-layering could be incorporated for slope protection purposes during construction of a sand pad. Stem cuttings was the third technique tested. The sections of willows stem approximately 30 cm in length and 1–2 cm in diameter were pushed into the sand to a depth of approximately 20 cm in a 7-m section of slope. After one year, nearly 100% of the stem cuttings had sprouted, and the willow bundles and brush layers had developed continuous rows of shoots from the stock material. After three years, the willow bundles and brush layers continued to flourish. However, only 25% of the stems remained alive after the second year, and less than 10% after the third year.

Box 15.13   Natural regeneration by native and non-native plants on primary surfaces

Investigations of patterns of natural tundra regeneration are important both for identifying locally important candidate species for restoration efforts, in the short term, and for understanding regional prospects for recovery, over the long term (Kershaw & Kershaw, 1987). Surveys of successful plant colonisers on primary surfaces ranging in age from two to ten years (sand, gravel and rubble quarries; roadside and railway verges; sandy deflation scars and aeolian sand surfaces) were made in the vicinity of a major transportation corridor being constructed on southern Yamal Peninsula during the period 1993–6 (Forbes, 1995, 1997; Forbes & Sumina, 1999; B. C. Forbes & S.

Jonasson, unpublished data). On roadside verges, the grasses *Calamagrostis purpurea*, *C. neglecta* and *Puccinellia sibirica* dominated the lower, more stable portions of the slopes four years after road construction. *Rorippa palustris* and *Polygonum humifusum* dominated the upper, more active portions of the slopes (Forbes, 1997). The most abundant and frequent species on railway verges were *Festuca ovina*, *Poa alpigena* ssp. *colpodea*, *Oxytropis sordida*, *Rorippa palustris*, *Equisetum arvense* and *Dianthus repens* (B. C. Forbes & S. Jonasson, unpublished data).

The total list of taxa counted in quarries of various origins comprised four groups including 90 vascular plants, 31 mosses, 2 hepatics and 2 lichens (Forbes & Sumina, 1999). Group A includes very common, widespread species that were recorded in all study sites at three different locations and on many different substrates: *Festuca ovina, Poa alpigena, Bryum* spp., *Equisetum arvense, Tripleurospermum hookeri*. Somewhat less widespread species included: *Chamaenerium angustifolium, Funaria hygrometrica* and *Salix viminalis*. Group B species are normally associated with dry, rubbly slopes, screes, but sometimes moist and calcareous substrates: *Draba ochroleuca, Oxytropis sordida, Minuartia verna, Dianthus repens, Poa glauca, Luzula multiflora, Stellaria peduncularis, Androsace septentrionalis*. Others that were also characteristic in rubbly quarries included: *Thymus reverdattoanus, Saxifraga oppositifolia, Minuartia arctica, Papaver lapponicum* ssp. *jugoricum, Racomitrium lanuginosum, R. canescens*. Group C comprises species that were widespread mainly on fine-grained sandy substrates. In natural communities they are also associated with sandy, sometimes wet, surfaces: *Salix phylicifolia, Deschampsia obensis, Calamagrostis lapponica* and *Eriophorum scheuchzeri*.

Group D species occurred only on southernmost Yamal and are naturally associated with wet, often saline conditions: *Puccinellia hauptiana, Polygonum humifusum* and *Alopecurus geniculatus*.

Most of the 37 species counted as colonists in both quarries and along roadsides were vascular plants. Six were bryophytes, which generally occurred at high frequencies. No lichens were observed on any primary surfaces. In the southern Yamal, boreal and subarctic ruderals (*Rumex confertus, Polygonum aviculare, P. humifusum, Rorippa palustris, Stellaria media, Thlapsi arvense, Capsella bursa-pastoris* and *Chenopodium album*) have migrated north along the road/rail corridor built to facilitate petroleum development. At least two of these species (*P. humifusum* and *R. palustris*) have spread on to several types of disturbed surfaces, both primary and secondary, and are outcompeting native species (Forbes, 1997).

degraded ecosystems is likely to increase as industrial exploitation continues to expand.

An important avenue for future research would be to simply continue monitoring the results of the many short-term studies reported here. In addition, experiments with soil fauna could further the scope of restoration efforts and, in the process, help illuminate questions pertaining to their potential and realised niches as they colonise, or fail to colonise, rehabilitated substrates. Similarly, there is a need to know whether and how phytophagous arthropod communities are affected by restoration efforts.

ACKNOWLEDGMENTS

We would like to thank the following individuals and companies for their help, respectively, in providing source materials and allowing proprietary data to be published here: AMOCO-Eurasia Production Co., ARCO Alaska, Inc., Ása Aradóttir, Inna Archegova, CONOCO, Torre Jorgenson, Peter Kershaw, Margarita Magomedova and Harvey Martens.

REFERENCES

Aastrup, P. (2000). Responses of West Greenland caribou to the approach of humans on foot. *Polar Research*, **19**, 83–90.

Ahlstrand, G. M. & Racine, C. H. (1993). Response of an Alaska, USA, shrub-tussock community to selected all-terrain vehicle use. *Arctic and Alpine Research*, **25**, 142–149.

Anonymous (1990). Surging visitors stir fears for Arctic Refuge. *The New York Times*, 25 December, p. A8.

Aradóttir, Á. L. & Arnalds, Ó. (2001). Ecosystem degradation and restoration of birch woodlands in Iceland. In *Nordic Mountain Birch Ecosystems*, ed. F. E. Wielgolaski, pp. 293–306. Paris: UNESCO.

Aradóttir, Á. L., Svavarsdóttir, K., Magnússon, S. H., Gudmundsson, J., Sigurgeirsson, A. & Arnalds, A. (1999). *The Utilization of Native Willows in Reclamation of Degraded Areas*. Status Report 1997-8, SCS Report no. 1. Hella, Iceland: Gunnarsholt. (in Icelandic)

Aradóttir, Á. L., Svavarsdóttir, K., Jónsson, P. H. & Gudbergsson, G. (2000). Carbon accumulation in vegetation and soils by reclamation of degraded areas. *Icelandic Agricultural Research*, **13**, 99–113.

Archegova, I. B. (1997). Nature restoration strategy in the Far North. In *Proceedings of the International Symposium on Physics, Chemistry, and Ecology of Seasonally Frozen Soils*, eds. I. K. Iskander, E. A. Wright, J. K. Radke, B. S. Sharrat, P. H. Groenevelt & L. D. Hinzman, pp. 477–480. Hanover, NH: Cold Regions Research and Engineering Laboratory, US Army Corp of Engineers.

Arnalds, Ó., Gudbergsson, G. & Gudmundsson, J. (2000). Carbon sequestration and reclamation of severely degraded soils in Iceland. *Icelandic Agricultural Research*, **13**, 87–97.

Babb, T. A. & Bliss, L. C. (1974). Effects of physical disturbance on high arctic vegetation in the Queen Elizabeth Islands. *Journal of Applied Ecology*, **11**, 549–562.

Batzli, G. O., White, R. G. & Bunnell, F. L. (1981). Herbivory: a strategy of tundra consumers. In *Tundra Ecosystems: A Comparative Analysis*, eds. L. C. Bliss, O. W. Heal & J. J. Moore, pp. 359–375. Cambridge: Cambridge University Press.

Billings, W. D. (1997). Challenges for the future: arctic and alpine ecosystems in a changing world. In *Global Change and Arctic Terrestrial Ecosystems*, eds. W. C. Oechel, T. V. Callaghan, T. Gilmanov, J. I. Holten, B. Maxwell, U. Molau & B. Sveinbjörnsson, pp. 1–18. Berlin: Springer-Verlag.

Bishop, S. C., Kidd, J. G., Cater, T. C., Rossow, L. J. & Jorgenson, M. T. (1998). *Land Rehabilitation Studies in the Kuparuk Oilfield, Alaska, 1997*, 12th Annual Report prepared for ARCO Alaska Inc. Anchorage, AK: Alaska Biological Research Inc.

Bishop, S. C., Kidd, J. G., Cater, T. C., Rossow, L. J. & Jorgenson, M. T. (1999). *Land Rehabilitation Studies in the Kuparuk Oilfield, Alaska, 1998*, 13th Annual Report prepared for ARCO Alaska Inc. Anchorage, AK: Alaska Biological Research Inc.

Bliss, L. C. ed. (1977). *Truelove Lowland, Devon Island, Canada: A High Arctic Ecosystem*. Edmonton, Alberta: University of Alberta Press.

Bliss, L. C. (1997). Arctic ecosystems of North America. In *Ecosystems of the World*, vol. 3, *Polar and Alpine Tundra*, ed. F. E. Wielgolaski, pp. 551–683. Amsterdam: Elsevier Science.

Bliss, L. C. & Grulke, N. E. (1988). Revegetation in the High Arctic: its role in reclamation of surface disturbances. In *Northern Environmental Disturbances*, ed. G. P. Kershaw,

pp. 43–55. Edmonton, Alberta: Boreal Institute for Northern Studies.

Bliss, L. C. & Matveyeva, N. V. (1992). Circumpolar arctic vegetation. In *Arctic Ecosystems in a Changing Climate: An Ecophysiological Perspective*, eds. F. S. Chapin III, R. L. Jefferies, J. F. Reynolds, G. R. Shaver & J. Svoboda, pp. 59–89. New York: Academic Press.

Brooke, J. (2000). A big push is on for natural gas under the Arctic: environmental worry is taking back seat to rising prices for consumers in U.S. *The New York Times*, 28 September, pp. A1, C26.

Chapin, F. S. III, Hobbie, S. E. & Shaver, G. R. (1997). Impacts of global change on composition of arctic communities: implications for ecosystem functioning. In *Global Change and Arctic Terrestrial Ecosystems*, eds. W. C. Oechel, T. V. Callaghan, T. Gilmanov, J. I. Holten, B. Maxwell, U. Molau & B. Sveinbjörnsson, pp. 221–228. Berlin: Springer-Verlag.

Chernov, Yu. I. (1995). Diversity of the arctic terrestrial fauna. In *Arctic and Alpine Biodiversity: Patterns, Causes and Ecosystem Consequences*, eds. F. S. Chapin III & C. Körner, pp. 81–95. Berlin: Springer-Verlag.

Christensen, T. (1992). Greenland wants tourism. *Polar Record*, **28**, 62–63.

Crawford, R. M. M. (ed.) (1997). *Disturbance and Recovery in Arctic Lands: An Ecological Perspective*. Dordrecht: Kluwer.

Edlund, S. A. (1990). Bioclimatic zones in the Canadian Arctic Archipelago. In *Canada's Missing Dimension: Science and History in the Canadian Arctic Islands*, vol. 1, ed. C. R. Harington, pp. 421–441. Ottawa: Canadian Museum of Nature.

Evans, K. E. & Kershaw, G. P. (1989). Productivity of agronomic and native species under various fertiliser and seed application rates on a simulated transport corridor, Fort Norman, Northwest Territories. In *Proceedings of the Conference 'Reclamation: A Global Perspective'*, eds. D. G. Walker, C. B. Powter & M. W. Pole, pp. 279–287. Calgary, Alberta: Alberta Land Conservation and Reclamation Council.

Forbes, B. C. (1992). Tundra disturbance studies. 1: Long-term effects of vehicles on species richness and biomass. *Environmental Conservation*, **19**, 48–58.

Forbes, B. C. (1993). Aspects of natural recovery of soils, hydrology and vegetation at an abandoned high arctic settlement, Baffin Island, Canada. In *Proceedings of the 6th International Conference on Permafrost*, vol. 1, pp. 176–181. Wushan, Guangzhou, China: South China Institute of Technology Press.

Forbes, B. C. (1994). The importance of bryophytes in the classification of human-disturbed high arctic vegetation. *Journal of Vegetation Science*, **5**, 875–882.

Forbes, B. C. (1995). Tundra disturbance studies. 3: Short-term effects of aeolian sand and dust, Yamal Region, Northwest Siberia, Russia. *Environmental Conservation*, **22**, 335–344.

Forbes, B. C. (1996). Plant communities of archaeological sites, abandoned dwellings, and trampled tundra in the eastern Canadian Arctic: a multivariate analysis. *Arctic*, **49**, 141–154.

Forbes, B. C. (1997). Tundra disturbance studies. 4: Species establishment on anthropogenic primary surfaces, Yamal Peninsula, Northwest Siberia, Russia. *Polar Geography*, **21**, 79–100.

Forbes, B. C. (1998). Cumulative impacts of vehicle traffic on high arctic tundra: soil temperature, plant biomass, species richness and mineral nutrition. *Nordicana*, **57**, 269–274.

Forbes, B. C. (1999*a*). Reindeer herding and petroleum development on Poluostrov Yamal: sustainable or mutually incompatible uses? *Polar Record*, **35**, 317–322.

Forbes, B. C. (1999*b*). Restoration of high latitude wetlands: an example from the Canadian High Arctic. In *An International Perspective on Wetland Rehabilitation*, ed. W. J. Streever, pp. 205–214. Dordrecht: Kluwer.

Forbes, B. C. & Jefferies, R. L. (1999). Revegetation in arctic landscapes: constraints and applications. *Biological Conservation*, **88**, 15–24.

Forbes, B. C. & Kofinas, G. (eds). (2000). Proceedings of the Human Role in Reindeer/Caribou Systems Workshop. *Polar Research*, **19**, 1–142.

Forbes, B. C. & Sumina, O. I. (1999). Comparative ordination of low arctic vegetation recovering from disturbance: reconciling two contrasting approaches for field data collection. *Arctic, Antarctic and Alpine Research*, **31**, 389–399.

Forbes, B. C., Ebersole, J. J. & Strandberg, B. (2001). Anthropogenic disturbance and patch dynamics in circumpolar tundra ecosystems. *Conservation Biology*, **15**, 954–969.

Forman, R. T. T. (1981). Interaction among landscape elements: a core of landscape ecology. In *Proceedings of the International Congress of the Netherlands Society for Landscape Ecology*, eds. S. P. Tjallingii & A. A. de Veer, pp. 35–48. Veldhoven, The Netherlands: Netherlands Society for Landscape Ecology.

Gilders, M. A. & Cronin M. A. (2000). North Slope oil field development. In *The Natural History of an Arctic Oil Field: Development and the Biota*, eds. J. C. Truett & S. R. Johnson, pp. 15–33. San Diego, CA: Academic Press.

Greller, A. M., Goldstein, M. & Marcus, L. (1974). Snowmobile impact on three alpine tundra plant communities. *Environmental Conservation*, **1**, 101–110.

Hamley, W. (1991). Tourism in the Northwest Territories. *Geographical Review*, **81**, 89–99.

Hammitt, W. E. & Cole, D. N. (1987). *Wildland Recreation: Ecology and Management*. New York: John Wiley.

Handa, I. T. & Jefferies, R. L. (2000). Assisted revegetation trials in degraded salt marshes of the Hudson Bay lowlands. *Journal of Applied Ecology*, **37**, 944–958.

Hobbs, R. J. (1994). Dynamics of vegetation mosaics: can we predict responses to global change? *Ecoscience*, **1**, 346–356.

Hobbs, R. J. & Huenneke, L. F. (1992). Disturbance, diversity, and invasion: implications for conservation. *Conservation Biology*, **6**, 324–337.

Hodkinson, I. D., Coulson, S. J., Webb, N. R. & Block, W. (1996). Can high arctic soil microarthropods survive elevated soil temperatures? *Functional Ecology*, **10**, 314–321.

Hulme, P. E. (1996). Herbivory, plant regeneration, and species coexistence. *Journal of Ecology*, **84**, 609–615.

Ilyina, L. & Mieczkowski, Z. (1992). Developing scientific tourism in Russia. *Tourism Management*, **12**, 327–331.

James, B. (1998). Finland aims to give EU a 'Northern Dimension': its objective is coordination in strategy toward Russians. *International Herald Tribune*, 18 March, pp. 1, 7.

Jefferies, R. L. (1999). Herbivores, nutrients and trophic cascades in terrestrial environments. In *Herbivores: Between Plants and Predators*, eds. H. Olff, V. K. Brown & R. H. Drent, pp. 301–330. Oxford: Blackwell.

John, E. & Turkington, R. (1997). A 5-year study of the effects of nutrient availability and herbivory on two boreal forest herbs. *Journal of Ecology*, **85**, 419–430.

Jorgenson, M. T., Cater, T. C. & Roth, J. E. (1999). *Revegetation for Erosion Control on the Alpine Development Project, 1998*, 1st Annual Report prepared for ARCO Alaska Inc. Fairbanks, AK: Alaska Biological Research Inc.

Kaltenborn, B. P. & Emmelin, L. (1993). Tourism in the high north: management challenges and recreation opportunity spectrum planning in Svalbard, Norway. *Environmental Management*, **17**, 41–50.

Kevan, P. G., Forbes, B. C., Behan-Pelletier, V. & Kevan, S. (1995). Vehicle tracks on high arctic tundra: their effects on the soil, vegetation and soil arthropods. *Journal of Applied Ecology*, **32**, 656–669.

Kershaw, G. P. & Kershaw, L. J. (1987). Successful plant colonisers on disturbances in tundra areas of northwestern Canada. *Arctic and Alpine Research*, **19**, 451–460.

Kidd, J. G., Jacobs, L. L., Cater, T. C. & Jorgenson, M. T. (1997). *Ecological Restoration of the North Prudhoe Bay State No. 2 Exploratory Drill Site, Prudhoe Bay Oilfied, Alaska, 1995*, 4th Annual Report prepared for ARCO Alaska Inc. Fairbanks, AK: Alaska Biological Research Inc.

Komárková, V. & Wielgolaski, F. E. (1999). Stress and disturbance in cold region ecosystems. In *Ecosystems of the World*, vol. 16, *Ecosystems of Disturbed Ground*, ed. L. R. Walker, pp. 39–122. Amsterdam: Elsevier Science.

Leibman, M. O. & Egorov, I. P. (1996). Climatic and environmental controls of cryogenic landslides, Yamal, Russia. In *Landslides*, ed. K. Senneset, pp. 1941–1946. Rotterdam: Balkema.

Lent, P. C. & Klein, D. R. (1988). Tundra vegetation as a rangeland resource. In *Vegetation Science Applications for Rangeland Analysis and Management*, ed. P. T. Tueller, pp. 307–337. Dordrecht: Kluwer.

Longton, R. E. (1988). *Biology of Polar Bryophytes and Lichens*. Cambridge: Cambridge University Press.

Magnússon, S. H. (1997). Restoration of eroded areas in Iceland. In *Restoration Ecology and Sustainable Development*, eds. K. M. Urbanska, N. R. Webb & P. J. Edwards, pp. 188–211. Cambridge: Cambridge University Press.

Martens, H. (1993). *Revegetation Research Western Siberia: Year 3*. Report prepared for AMOCO Production Co. Calgary, Alberta: Harvey Martens & Associates Inc.

Martens, H. (1995). *Revegetation Research Western Siberia: Year 4*. Report prepared for AMOCO Production Co. Calgary, Alberta: Harvey Martens & Associates Inc.

Martens, H. & Younkin, W. (1997). Revegetation trials in Western Siberia: a five-year summary. In *Development of the North and Problems of Recultivation*, eds. E. G. Kuznetsova & I. B. Archegova, pp. 354–365. Syktyvkar, Russia: Komi Science Centre.

Martens, H., Magomedova, M. & Morozova, L. (1996). *Rangeland Studies in the Bovanenkovo Proposed Development Area: Year 3*. Report prepared for AMOCO Eurasia Production Co. Calgary, Alberta: Harvey Martens & Associates Inc.

Maslen, L. & Kershaw, G. P. (1989). First year results of revegetation trials using selected native plant species on a simulated pipeline trench, Fort Norman, N.W.T., Canada. In *Proceedings of the Conference 'Reclamation: A Global Perspective'*, eds. D. G. Walker, C. B. Powter & M. W. Pole, pp. 81–90. Calgary, Alberta: Alberta Land Conservation and Reclamation Council.

McKendrick, J. D. (1991). Colonising tundra plants to vegetate abandoned gravel pads in arctic Alaska. *Advances in Ecology*, **1**, 209–223.

McKendrick, J. D. (1993). *Transplanting Arctophila fulva to Create Emergent Vegetation Habitats in Arctic Alaska*. Anchorage, AK: BP Exploration (Alaska) Inc.

McKendrick, J. D. (1996). Vegetation recolonisation on salt-damaged soil in arctic Alaska. In *12th Biennial High Altitude Revegetation Workshop*, Information Series no. 83, pp. 1–13. Fort Collins, CO: Colorado Water Resources Research Institute.

McKendrick, J. D. (1997). Long-term recovery in northern Alaska. In *Disturbance and Recovery in Arctic Lands: An Ecological Perspective*, ed. R. M. M. Crawford, pp. 503–518. Dordrecht: Kluwer.

McKendrick, J. D. (2000a). Vegetative responses to disturbance. In *The Natural History of an Arctic Oil Field: Development and the Biota*, eds. J. C. Truett & S. R. Johnson, pp. 35–56. San Diego, CA: Academic Press.

McKendrick, J. D. (2000b). Northern tansy mustard fills a niche. *Agroborealis*, **32**, 15–20.

McKendrick, J. D. & Masalkin, S. D. (1993). Transferring U.S. rangeland technology assisting Russian gas and oil production in the arctic tundra. *Agroborealis*, **25**, 4–7.

McKendrick, J. D. & Mitchell, W. W. (1978). Fertilising and seeding oil-damaged tundra to effect vegetation recovery. *Arctic*, **31**, 296–304.

McKendrick, J. D., Batzli, G. O., Everett, K. R. & Swanson, J. C. (1980). Some effects of mammalian herbivores and fertilisation on tundra soils and vegetation. *Arctic and Alpine Research*, **12**, 565–578.

Meyer, E. (1993). The impact of summer and winter tourism on the fauna of alpine soils in western Austria (Oetzal Alps, Rätikon). *Revue suisse de Zoologie*, **100**, 519–527.

Nellemann, C. & Cameron, R. D. (1996). Effects of petroleum development on terrain preferences of calving caribou. *Arctic*, **49**, 23–28

Nelson, F. E. & Anisimov, O. A. (1993). Permafrost zonation in Russia under anthropogenic climatic change. *Permafrost and Periglacial Processes*, **4**, 137–148.

Opperman, J. J. & Merenlender, A. M. (2000). Deer herbivory as an ecological constraint to restoration of degraded riparian corridors. *Restoration Ecology*, **8**, 41–47.

Osterkamp, T. E., Viereck, L., Shur, Y., Jorgenson, M. T., Racine, C., Doyle, A. & Boone, R. D. (2000). Observations of thermokarst and its impact on boreal forests in Alaska, USA. *Arctic, Antarctic and Alpine Research*, **32**, 303–315.

Ouellet, J.-P., Boutin, S. & Heard, D. C. (1994). Responses to simulated grazing and browsing of vegetation available to caribou in the Arctic. *Canadian Journal of Zoology*, **72**, 1426–1435.

Pickett, S. T. A. & White, P. S. (eds.) (1985). *The Ecology of Natural Disturbance and Patch Dynamics*. San Diego, CA: Academic Press.

Podkoritov, F. M. (1995). *Reindeer Herding on Yamal*. Sosnovyi Bor, Russia: Leningrad Atomic Electrical Station. (in Russian)

Post, E. S. & Klein, D. R. (1996). Relationships between graminoid growth form and levels of grazing by caribou (*Rangifer tarandus*) in Alaska. *Oecologia*, **107**, 364–372.

Prégent, G., Camiré, C., Fortin, J. A., Arsenault, P. & Brouillette, J. G. (1987). Growth and nutritional status of green alder, jack pine and willow in relation to site parameters of borrow pits in James Bay Territory, Quebec. *Reclamation and Revegetation Research*, **6**, 33–48.

Raup, H. M. (1951). Vegetation and cryoplanation. *Ohio Journal of Science*, **51**, 105–116.

Reynolds, J. F. & Tenhunen, J. D. (eds.) (1996). *Landscape Function and Disturbance in Arctic Tundra*. Berlin: Springer-Verlag.

Rietkerk, M. & van de Koppel, J. (1997). Alternate stable states and threshold effects in semi-arid grazing systems. *Oikos*, **79**, 69–76.

Rozhdestvenskii, Yu. F. & Sarapultsev, I. E. (1997). Experimental results of testing plants for land reclamation in the Yamal Peninsula. *Russian Journal of Ecology*, **28**, 348–352.

Sage, B. (ed.) (1986). *The Arctic and its Wildlife*. New York: Facts on File.

Salzburg, K. A., Fridriksson, S. & Webber, P. J. (eds.) (1987). Restoration and vegetation succession in circumpolar lands. *Arctic and Alpine Research*, **19**, 339–586.

Schultz, A. M. (1969). A study of an ecosystem: the arctic tundra. In *The Ecosystem Concept in Natural Resource Management*, ed. G. M. Van Dyne, pp. 77–93. New York: Academic Press.

Shchelkunova, R. P. (1993). The effect of industry and transport on reindeer pastures: the example of Taymyr. *Polar Geography and Geology*, **17**, 252–258.

Smith, P. D. J. (1986). *Final Cleanup at Selected (1975–1981) Well Sites*. Anchorage, AK: Nuera Reclamation Co. and US Geological Survey.

Staaland, H., Scheie, J. O., Gròndahl, F. A., Persen, E., Leifseth, A. B. & Holand, Ò. (1993). The introduction of reindeer to Bròggerhalvòya, Svalbard: grazing preference and effect on vegetation. *Rangifer*, **13**, 15–19.

Strandberg, B. (1997). Vegetation recovery following anthropogenic disturbances in Greenland. In *Disturbance and Recovery in Arctic Lands: An Ecological Perspective*, ed. R. M. M. Crawford, pp. 381–391. Dordrecht: Kluwer.

Svoboda, J. & Henry, G. H. R. (1987). Succession in marginal arctic environments. *Arctic and Alpine Research*, **19**, 373–384.

Thing, H. (1984). Feeding ecology of the West Greenland caribou (*Rangifer tarandus groenlandicus*) in the Sisimiut-Kangerlussuaq region. *Danish Review of Game Biology*, **12**, 1–53.

Truett, J. C. & Johnson, S. R. (2000). *The Natural History of an Arctic Oil Field: Development and the Biota*. San Diego, CA: Academic Press.

Varoli, J. (2000). Energy on ice: rising oil and gas prices revive interest in Russia's arctic fields. *The New York Times*, 3 October, p. W1.

Volpert, Y. L. & Sapozhnikov, G. V. (1998). Responses of small mammalian fauna to different forms of technogenic influences on arctic landscapes. *Russian Journal of Ecology*, **29**, 133–138.

Vuoria, M. (1999). The Northern Dimension and energy. *New Northern Europe Business Magazine*, 1999, 38–39.

Walker, D. A. & Walker, M. D. (1991). History and pattern of disturbance in Alaskan arctic terrestrial ecosystems: a hierarchical approach to analysing landscape change. *Journal of Applied Ecology*, **28**, 244–276.

Walker, D. A., Cate, D., Brown, J. & Racine, C. (1987*a*). *Disturbance and Recovery of Arctic Alaska Tundra Terrain: A Review of Recent Investigations*. Hanover, NH: Cold Regions Research and Engineering Laboratory, U.S. Army Corp of Engineers.

Walker, D. A., Webber, P. J., Binnian, E. F., Everett, K. R., Lederer, N. D., Nordstrand, E. A. & Walker, M. D. (1987*b*). Cumulative impacts of oil fields on northern Alaskan landscapes. *Science*, **238**, 757–761.

Walker, M. D., Daniëls, F. J. A. & van der Maarel, E. (eds.) (1994). Circumpolar arctic vegetation. *Journal of Vegetation Science*, **5**, 758–920.

Wielgolaski, F. E. (1998). Twenty-two years of plant recovery after severe trampling by man through five years in three vegetation types at Hardangervidda. *NTNU Vitensk. Mus. Rapp. Bot. Ser.* 1998, **4**, 26–29.

Wolfe, S. A., Griffith, B. & Wolfe, C. A. G. (2000). Response of reindeer and caribou to human activities. *Polar Research*, **19**, 63–73.

Young, S. B. (1971). The vascular flora of St Lawrence Island with special reference to floristic zonation in the arctic regions. *Contributions from the Gray Herbarium*, **201**, 11–115.

16 • High-elevation ecosystems

KRYSTYNA M. URBANSKA AND JEANNE C. CHAMBERS

INTRODUCTION

High-elevation ecosystems are highly valued for their livestock forage, mineral and timber assets, and recreational opportunities. They also serve as vital watersheds for urban and agricultural areas, and include a unique array of animals and plants. High-elevation ecosystems are often among the most difficult to restore because of their slow recovery due to the severity of the environmental regime and the limited number of adapted species. For example, the effects of various mining-related disturbances in subalpine ecosystems may be still apparent after 100+ years (Veblin *et al.*, 1991), and in alpine ecosystems the effects of disturbances are evident for even longer. Other less visible impacts related to global change processes such as acid rain and nitrogen deposition are undoubtedly influencing recovery potentials and vegetation dynamics in many of these areas. In addition, current climate records indicate that future climate change will be more apparent at high than low elevations (Beniston *et al.*, 1997).

In this chapter, we discuss the unique environmental constraints of subalpine and alpine ecosystems, and outline methods for their restoration. We refer to alpine ecosystems as those areas that occur above the climatic limit of trees, although they may include krummholz and the associated subalpine herbs and shrubs. Subalpine ecosystems are those that occur in the zone between the alpine and montane or boreal forest zone. Our emphasis in on North American and European systems and the focus is on plants. The majority of principles and practices discussed should be widely applicable, but regional and local differences do exist and should always be taken into account.

CLIMATE AND PLANT ADAPTATIONS IN HIGH-ELEVATION ECOSYSTEMS

High-elevation ecosystems all over the world have in common: (1) short growing seasons and unpredictable weather patterns, (2) cold environments, (3) restricted water availability, both physical and physiological, (4) wind, and (5) potentially high radiant energy fluxes, especially in the damaging UV-B rays. These environmental constraints are accompanied by inherent constraints in the physiological traits and life histories of mountain plants. All these constraints must be considered when developing restoration schemes.

Growing periods in high-elevation areas are shorter than in lowlands. An average growing period in the subalpine belt of the Alps or the Rocky Mountains ranges from about five months at the lowest elevation to about three months at timberline. In contrast, plants occurring in the alpine belt may have only six weeks of growth. The stress related to short growing periods is further aggravated by unpredictable and often extreme weather.

High-elevation areas are cold, and the effect of individual thermal constraints is often compounded by other factors (Körner & Lärcher, 1988). The annual mean temperature of the air decreases with increasing elevation by about 0.55 °C per 100 m (Landolt, 1992). Consequently, plants at high elevations exhibit less annual growth than their lowland relatives (e.g. leaves of the alpine sedge *Carex curvula* extend on average only 1 mm per year). Nocturnal heat emission increases with altitude. Temperatures may drop by 20 °C in a single day and frost represents a constant danger in high-altitude sites. The freezing of saturated soils may directly kill seedlings or juveniles through the formation of needle ice. It may also cause soil movement (solifluction, frost heaving) which affects even established plants. Mortality of flowers or ripening fruits caused by frost may be less conspicuous, but also results in substantial damage.

Water availability in high-elevation areas is often limited. Absolute air moisture (g water vapour m^{-3}) decreases with increasing altitude so that the air may become exceedingly dry. Carbon dioxide content per volume unit also decreases with increasing elevation because the air pressure diminishes. Plant water stress can occur if (1) the area is generally arid, or (2) water is physically present but physiologically unavailable because of low soil temperatures. Even if solar radiation compensates locally for generally low temperatures, e.g. on southwest-facing slopes, sharp thermal contrasts often occur. Precipitation generally increases with elevation but there are considerable differences related to the geographical location of the mountains and the direction from which the humid air masses arrive. Since environmental changes at high elevations are often abrupt, some parts of a given mountain range may be considerably more arid than others.

Duration of snow cover generally increases with elevation but varies strongly with aspect and topography. Snow cover is a poor heat conductor and, consequently, provides some protection for plants. However, it also represents an environmental constraint because it limits light availability. For that reason, an intermittent snowfall in summer may result in decreased photosynthesis and growth in plants.

High-altitude areas are windy and this is a further constraint because wind substantially increases the chilling factor and reinforces air dryness. Wind may damage plants mechanically by breaking plant parts or tearing out whole plants. On a more local scale, it may cause snowdrifts and, thus, considerably shorten the growing period. It may also blow away the insulating snow cover and expose plants to frost.

Light intensity increases with altitude. The UV part of the spectrum (280–320 nm) is greater in high-altitude areas, especially when the skies are cloudless, but it may also be reinforced by diffuse radiation from thin clouds (see Körner, 1999 for more details). Seasonal variation in light intensity is often considerable (Reisigl & Keller, 1987). The light- and heat-energy available to plants depends not only on a direct solar radiation, but is also influenced by albedo and the heat emissions from the soil.

The severe environment places significant constraints on high-elevation plants. These constraints are evident in various adaptations. In the subalpine vegetation belt, dominant life forms include trees and shrubs, as well as both annual and perennial forbs and graminoids. In contrast, low-growing perennial cushion and rosette plants, deciduous and evergreen shrubs, forbs and graminoids dominate above the timberline. Annual species are extremely rare. Plants in alpine areas invest principally in their root system which may be about five times larger than that of related valley species. Their aboveground parts remain small but, in spite of lower partial carbon dioxide air pressure, average photosynthetic efficiency of alpine plants is often higher than that of plants in valleys (Körner & Diemer, 1994). The physiognomic convergence of alpine floras from various parts of the world is considerable (Körner, 1999).

Environmental constraints in high-elevation areas are not always compatible with plant growth and reproduction, especially above the timberline. The short growing periods can limit reproduction by seed and result in significant seedling mortality. Many mountain species, especially in the alpine belt, preform flower buds during the previous summer or autumn and this behaviour has considerable adaptive value. However, inclement weather conditions can prevent pollination and result in significant mortality of both flowers and developing seeds. Seed production in mountain species is accordingly characterised by strong spatial and temporal variation. Inherent constraints also influence other life history phases in plants (see section 'Population and community processes' below).

UNDERLYING ECOLOGICAL CONSIDERATIONS
Soils

Although physical and chemical soil properties are significantly influenced by geology, solar radiation, wind and precipitation interact at high altitudes

Table 16.1. *Generalised properties of high altitude disturbances with topsoils in place, topsoils removed or soils altered*

	Organic topsoil	Mineral topsoil	Topsoil removed	Soils altered
Topography	Slight–Moderate	Moderate–Extreme	Moderate–Extreme	Slight–Extreme
Soil nutrient retention and cycling capacity	High	Intermediate–Low	Low–Absent	Low
Propagule pool	High	High–Low	Absent	Absent
Productivity	High	Intermediate–Low	Low	Low
Recovery rate	Rapid	Intermediate-Slow	Very slow	Slow

Source: Modified after Chambers (1997).

with topographic gradients and determine both soil development and vegetation patterns (Billings, 1988; Grabherr *et al.*, 1995). Differences in soil moisture between dry, wind-swept knolls and topographic depressions on a single ridge are often larger than variation among level alpine areas along the gradient from the equator to the poles (Billings, 1974). Strong winter winds redistribute the snow cover and, thus, the soil development patterns. Snow cover depth determines the location of dry fellfield, dry meadow, moist meadow, and snowbed vegetation as well as that of krummholz and forest (May & Weber, 1982). Wind-exposed ridges and slopes that are snow-free in winter typically have soils with surface horizons formed by the physical mixing of mineral and organic materials due to soil movement (Eddleman & Ward, 1984). Soil texture and cation exchange capacity depend largely on lithology, and organic matter and soil nutrients levels are frequently low. In contrast, areas that receive moderate amounts of snow and occur on well-drained slopes, such as fellfields and dry meadows, accumulate large amounts of organic matter and as a result have high cation exchange capacities (Ratcliffe & Turkington, 1987) (Table 16.1). The low soil temperatures and generally slow decomposition often result in the accumulation of organic matter. This accumulation is more moderate in subalpine forests, where areas with maximum snow cover and the longest duration of snowbanks have saturated soils during the period of snowmelt. Soils exhibit increasing organic matter with decreasing snow duration (Stanton *et al.*, 1994), and are often mixed (Eddleman & Ward, 1984). Soils in areas with

elevated groundwater tables or that accumulate runoff (wet meadows) are often highly organic and peaty with mottled subsoils and dense silty layers deeper in the profile (Retzer, 1974).

Soil development at high elevations is influenced by low temperatures. In alpine areas, soil depths to fractured rock or bedrock range from only a few centimetres to about 1 m. Soil depth in subalpine areas can be somewhat greater. Levels of plant-available nitrogen (NH_4 and NO_3) and phosphorus (PO_4) are typically low. Decomposition rates and levels of nutrient availability vary over the growing season and are influenced by soil moisture, temperature, microbial activity and physiological responses of plants (O'Lear & Seastedt, 1994). Another important factor is mycorrhiza (see section 'Interactions' below). Incomplete oxidation of organic matter and accumulation of acidic end-products of microbial metabolism result in increased soil acidity. Nitrification is typically low with low pH values. All these differences influence plant productivity and also species composition.

Population and community processes

Knowledge of the population and community processes within a given ecosystem is essential for restoration. These processes are intertwined because the behaviour of populations influences their various interactions, and these in turn influence community dynamics.

Clonal growth is common in mountain plants (Hartmann, 1959; Stöcklin, 1992). Plants that exhibit

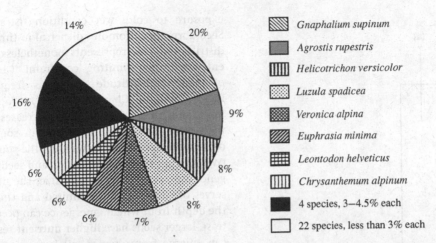

Legend:

- Gnaphalium supinum
- Agrostis rupestris
- Helicotrichon versicolor
- Luzula spadicea
- Veronica alpina
- Euphrasia minima
- Leontodon helveticus
- Chrysanthemum alpinum
- 4 species, 3–4.5% each
- 22 species, less than 3% each

Fig. 16.1. Species richness and contribution (%) of identified species to the seed rain on natural grassland at Jakobshorn Mountain (2500 m a.s.l., northeast Swiss Alps). Unpublished data from a three-year study period 1996–8.

clonal growth are valuable for restoration because (1) they are able to hold space, withstand damage well, and often respond with compensating regrowth, and (2) they have a high potential for vegetative reproduction by fragmentation, enforced (e.g. by soil movement) or self-induced, which results in population increase.

Seed production in high-elevation areas, especially above the timberline, was considered negligible by some earlier authors (Bliss, 1971; Billings, 1974), but more recent studies document its importance. Seed production depends on growing season conditions and can vary both among years and among species within years (Chambers, 1989). In order to collect the seeds required for restoration, it is necessary to locate plant populations on more favourable slopes or at slightly lower elevations to obtain enough seeds. Seed production is often episodic, so that it may be necessary to collect the seeds during one or more growing seasons and store them for later use. In mountainous ecosystems seeds often mature towards the end of growing period and are dispersed late in season, in winter, or shortly after spring snowmelt.

Most high-elevation plants use more than one dispersal mode. Dispersal distances in high-altitude areas are often very short (mostly less than 1 m), and long-distance dispersal represents isolated events (Spence, 1990; Chambers, 1993; Urbanska et al., 1998, 1999; Urbanska & Fattorini, 2000). Secondary dispersal of seeds on exposed soils can be significant. Relationships among soil characteristics and seed morphology greatly influence the spatial distribution of dispersed seeds and may be one of the primary determinants of colonisation patterns (Chambers et al., 1991; Chambers, 1995, 2000). Temporal and spatial variation in the seed rain is strong (Marchand & Roach, 1980; Stöcklin & Bäumler, 1980; Chambers, 1993; Urbanska et al., 1998, 1999). In most cases, a few species dominate the seed rain, and the others provide only minor contributions (Fig. 16.1). The species composition of seed rain is related to the resident vegetation or the communities occurring nearby. However, species abundance in the seed rain seldom reflects the actual species abundance in the aboveground vegetation (Chambers, 1993, 1995; Urbanska et al., 1999). Data on seed rain and in particular on distances separating degraded sites from intact vegetation are important for planning restoration schemes.

Soil seed banks in high-elevation ecosystems have been studied in only a few, different areas (Hatt, 1991; Chambers, 1993; Diemer & Prock, 1993; Semenova & Onipchenko, 1994; Urbanska & Fattorini, 1998a,b). Their common features are as follows: (1) spatial and temporal variation in seed

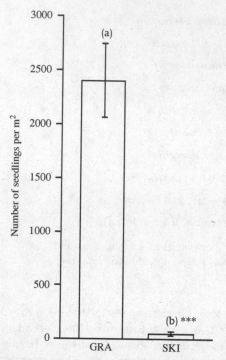

Fig. 16.2. Highly significant difference in the density of the germinable seed bank between a machine-graded ski run (SKI) and an adjoining natural grassland (GRA) at Jakobshorn Mountain (2500 m a.s.l., northeast Swiss Alps). *U*-test of Mann–Whitney. Reproduced by permission of the publisher from Urbanska & Fattorini (1998a).

density is very strong; (2) seed bank persistence is species-dependent, and (3) species composition and abundance of the seed bank seldom mirrors that of the aboveground vegetation. Differences between adjoining intact and disturbed sites may be pronounced (Fig. 16.2). Data on soil seed banks help to estimate the self-recovery potential of degraded sites.

Germination in mountain plants is strongly influenced by seed dormancy (Schütz, 1988). Dormancy is often enforced by temperatures falling below a certain threshold, so that germination is delayed until the proper environmental conditions occur (Chambers *et al.*, 1990). However, enforced dormancy is frequently accompanied by innate dormancy which may be broken only by e.g. mechanical damage to seed coat (scarification) or a prolonged

exposure to cold, wet conditions (stratification). Seed dormancy promotes dispersal in time and risk distribution, but represents nonetheless an inherent recruitment-limiting constraint. Germination of many high-altitude species is frequently enhanced by light (Chambers *et al.*, 1987a). Seed viability and, thus, germinability decreases generally with time (Chambers, 1989) but in some species older seeds germinate better than the younger ones (Weilemann, 1981). Germination and seedling emergence are also related to seed size. The nutrient reserves of small seeds are limited and this restricts the depth from which emergence can occur. In contrast, larger seeds have higher nutrient reserves and can emerge from deeper soil layers, but may desiccate if not covered by sufficient soil or organic matter.

Understanding of germination behaviour of high-altitude species is important to the timing and methods of restoration. Seeding should be done in the autumn to provide a natural stratification over winter and to avoid summer drought. Small seeds and seeds of species with a light requirement should be sown close to the soil surface.

Seedling establishment is strongly influenced by both soil characteristics and growing season conditions (Chambers *et al.*, 1990). It can be significantly higher on organic than mineral soils (Fig. 16.3). In areas where soil moisture is adequate, seedlings that emerge during longer and warmer growing seasons can have higher survival than those that emerge during short and cool growing seasons (Fig. 16.3). Seedling mortality on high-elevation sites is often considerable and can be attributed to slow seedling growth, needle-ice activity in saturated soils, late-season soil drought, or physical abrasion by wind. Other important hazards are grazing or destruction of seedlings by small mammals and other herbivores. However, given favourable conditions, seedling establishment can be higher than has generally been reported.

Safe sites are decisive in a successful establishment. Shelter from frost represents one of the most significant aspects of safe sites in high mountains (Urbanska & Schütz, 1986), and neighbouring plants often function as safe-site components. This 'nurse effect' may represent plant parental care (Urbanska,

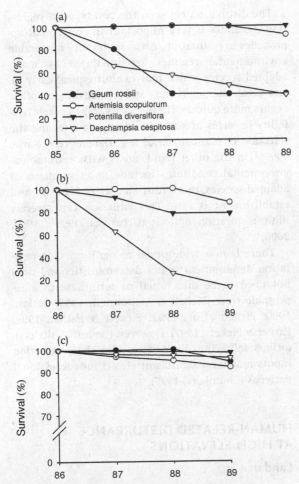

Fig. 16.3. Seedling survival of species with varying life histories, at Beartooth Plateau (2000 m a.s.l., southwest Montana, USA). Differences between a warm year 1985 (a) and a cool year 1986 (b) on the loamy sand soils of a gravel borrow area, as well as between the gravel borrow area (b) and the organic soils of a Geum turf (c) in 1986. After Chambers (1995).

1997a; Wied & Galen, 1998), be related to other conspecifics, or influenced by plants·representing different species. For example, cushion- or mat-forming species in the alpine vegetation belt protect seedlings from winter desiccation and summer transpirational losses, as well as abrasion from wind-blown snow, ice and soil. Because of increased boundary-layer resistances within cushions or mats,

convective heat transfer is reduced and temperatures are less extreme. Leaf temperatures may be 15–20 °C above ambient on sunny days, and even 5–8 °C on cloudy days (Bliss, 1985). Crowns and root systems of healthy cushions reduce wind erosion of soil and enhance accumulation of fine soil particles and organic matter resulting in favourable soil texture, increased moisture storage, and increased nutrient availability. Comparable conditions often occur within older clonal structures influenced by natural die-back of some ramets. For example, self-thinned tussocks of Carex curvula function as secondary safe sites for many dicotyledonous species (Urbanska, 1997a).

The conditions promoting seedling growth and survival should always be considered in restoration schemes. Nurse plants can greatly facilitate seedling establishment in extreme alpine and subalpine environments.

Interactions

Interactions occurring in high-elevation ecosystems are varied and involve associations ranging from conspecifics to species representing different ecosystem inhabitants (plants–micro-organisms, plants–fungi, plants–animals).

Mutualistic interactions are widely distributed in mountain biomes. Prime examples are pollinators and dispersers, vital to many plant species. Some of the most important mutualistic interactions involve mycorrhiza (ectomycorrhiza, ericoid mycorrhiza, vesicular-arbuscular mycorrhiza, orchid mycorrhiza), and dark-septate hyphae associations (Haselwandter & Read, 1982; Lesica & Antibus, 1986, Allen et al., 1987; Allen, 1989; Barnola & Montilla, 1997; Haselwandter, 1997). Mutualistic associations between plants and N-fixing organisms are also common in mountain ecosystems (Wojciechowski & Heimbrook, 1984; Holzmann & Haselwandter, 1988). In the alpine vegetation belt of the Alps, for example, Rhizobia were found in all legumes (Hasler, 1992). However, the relative contribution of legume di-nitrogen fixation seems to decrease with altitude (Körner, 1999).

Commensalism may be observed at any altitude. The nurse effect, which represents a particularly clear example of such interactions, may be recognised in many high-elevation sites, especially those exposed to frost and wind (Kikvidze & Nakhutsrishvili, 1990; Urbanska, 1997a).

Competition may be strong in densely vegetated mountainous areas situated above, below or at the timberline. Depending on the particular site or region, various forms may occur (both asymmetric and symmetric intraspecific competition, interspecific competition), and various resources may be involved. In high-alpine sites characterised by extreme environmental conditions, competition does not represent an overall community-shaping force, although it may be recognised in dense patches of grassland.

Plant–herbivore interactions are widely distributed in the mountains except for the subnival areas. Overgrazing by livestock or wild animals may have particularly damaging effects at local and larger scales (Wilhalm & Florineth, 1990).

Vegetation dynamics

Development of vegetation in high-elevation areas is influenced by: (1) the environmental attributes of the site, (2) the numbers and types or species adapted to establish on the site, and (3) the availability of adapted propagules. Limited numbers of species are adapted to survive and reproduce in subalpine, and especially alpine ecosystems, and the species that establish following disturbance are often members of the pre-disturbance or adjacent undisturbed community. On sites with severe environmental constraints (e.g. high altitude or exposed mineral soils), seral phases cannot be recognised and clear sequences of species replacement or of species dominance are absent. Vegetation development in such sites does not follow the classical successional pattern (Urbanska, 1994, 1997a, b). On the other hand, on sites with more moderate environmental regimes (e.g. lower elevations or organic soils), changes in species composition can occur over time and it may be possible to recognise distinct seral stages. However, both species establishment processes and species interactions are strongly influenced by each of the initial factors listed above and successional patterns are not always predictable.

The distinction between the two types of vegetation dynamics is very important in conceptual approaches to restoration. Areas with more favourable environmental regimes are inhabited by many adapted species which may exhibit typical early vs. late seral characteristics, and establishment often occurs more quickly. These factors allow greater flexibility in terms of both site manipulations and the selection of plant material (e.g. Densmore & Karle, 1999). On the other hand, areas with extreme environmental conditions include fewer numbers of adapted species, more restricted growth forms, and establishment is rare. This situation can severely limit restoration options (Urbanska, 1997b, 1999, 2000).

There is now widespread recognition that vegetation development is not deterministic, and does not need to be directional or terminate in a repeatable state (Pickett & McDonnell, 1989; Botkin, 1990; Pickett et al., 1992; Pickett & Parker, 1994; Parker & Pickett, 1997). However, the initial site conditions following disturbance and the restoration inputs may have significant effects on successional patterns (Chambers, 1997).

HUMAN-RELATED DISTURBANCE AT HIGH ELEVATIONS

Land use

The demands on mountain ecosystems are increasing as human populations grow and expand. Overgrazing of high-altitude areas by livestock has occurred for the last century in North America and significantly longer in Europe. In many areas of Europe, the treeline has been artificially lowered to facilitate livestock grazing and a high percentage of the lower elevation forests have been converted to pastureland. Timber harvest in subalpine conifer forests is a common and ongoing practice. In North America, recent advances in mineral exploration and mining technology have resulted in more numerous and larger mining-related disturbances. Increased interest in outdoor recreation has led to the development and expansion of ski areas (e.g. Meisterhans, 1988; Urbanska, 1997c), camp sites (e.g. Cole, 1995), and to numerous other human activities. These land uses have been accompanied by the construction of an ever-expanding system of roads and trails. As a result, disturbances occurring throughout the alpine and subalpine regions of the world represent a broad range of scales and intensities.

Types and characteristics of disturbance

Recognition of the differences that exist in the spatial and temporal scales as well as severity of disturbance is critical for determining the appropriate restoration approach.

Disturbance may be local (gap formation due to selective logging, establishment of a small camp site), relatively limited (construction of a single downhill ski run, a limited surface exploitation of mineral resources), or extensive (large-scale overgrazing by livestock, construction of whole road networks or skiing facilities, large open-pit mining operations). Disturbances can also differ in terms of temporal scales. The two obligate questions restorationists have to ask are: (1) is the disturbance a continuous long-term phenomenon or is it rather limited in time, and (2) does (did) the disturbance represent a single isolated event or is it recurrent? To answer these questions, the history of the disturbed area and the whole landscape context have to be considered (Parker & Pickett, 1997).

Disturbance severity may range from relatively minor damage to a single ecosystem component, e.g. primary producers, to complete alteration of the soils and hydrology in addition to destruction of the biotic components of an ecosystem. The relative severity of disturbance is influenced significantly by the soil characteristics and physical environment (Bradshaw, 1997a, b) and this influence is clearly recognisable in high-elevation areas. Although disturbance characteristics represent a continuum, they can generally be categorised according to increasing severity based on soil characteristics (Table 16.1). Organic topsoils have relatively high soil nutrient retention and cycling capacities, and contain plant propagules as well as much of the original microbial and mycorrhizal community. Recovery after disturbance is often relatively rapid even if the vegetation has been removed or otherwise damaged. Less-intensive restoration inputs are required and recovery processes can resemble secondary succession (Carghill & Chapin, 1987). In contrast, mineral topsoils or subsoils have lower nutrient retention and cycling capacities as well as more variable propagule pools. Recovery following disturbance can be a slow process and primary restoration concerns are amelioration of the extreme physical environment and plant establishment. On sites with irreversibly damaged soil (see Box 16.1), and in areas with topsoils removed (e.g. Urbanska 1997c), it may be necessary to improve the physical and chemical properties of the soils. Altered soils or mine spoils frequently represent the most severe disturbances. For example, in high-elevation areas of the western United States, mining activities can expose pyritic materials that initiate a cycle of sulphide oxidation and result in lower soil pH and increased bio-availability of heavy metals. Restoration often involves remediating the acidic conditions, preventing acid mine drainage, and stabilising the surface to decrease soil erosion (Brown & Johnston, 1979).

Many disturbances leave the biota in place and the effects may be recognised at various levels of biological organisation:

- Population-level changes resulting from disturbance are basically demographic. High-frequency, low-severity disturbances such as livestock grazing can prevent seedling establishment and result in populations dominated by older individuals, while low-frequency, high-severity disturbances such as avalanches can result in periodic stand renewal. In dense stands, population-level effects of disturbance reflected in individual species responses are influenced by competition for light or other resources. Species-specific potential for recovery after damage is very relevant, especially on extreme sites where abiotic factors are more important than competition.
- Community-level effects of disturbance are recognisable in the composition and architecture of stands. Successional pathways may be strongly affected or irreversibly altered; also, patterns of interactions within and among populations belonging to a given community may be changed.

Box 16.1 Restoration of subalpine camp sites in the Eagle Cap Wilderness, Oregon

David N. Cole

Camping often seriously degrades the environment. Wilderness managers sometimes close camping sites to facilitate a return to conditions approximating those prior to disturbance. Various site manipulations, often employed to accelerate recovery, are often costly, time-consuming and unsuccessful. This study assessed the effectiveness of restoration in closed subalpine camp sites in the Eagle Cap Wilderness, northeast Oregon. The six studied sites were located between 2170 m and 2320 m in a basin containing numerous lakes (Fig. B16.1). The subalpine forest in the study area consists of *Abies lasiocarpa, Picea engelmannii, Pinus contorta* and *P. albicaulis*. The most common groundcover plants in undisturbed areas are *Vaccinium scoparium, Phyllodoce empetriformis* and *Carex rossii*. Soils are derived from a granitic substrate.

DAMAGE ASSESSMENT

Each camp site was typically about 200 m² in size with about 100 m² completely devoid of vegetation. The soil compaction and minimal soil organic horizon in the sites indicated a severe, long-lasting (at least 50 years) disturbance. Unsuccessful attempts to close and restore camp sites began in the 1970s.

Fig. B16.1. View from a camp site area in the Eagle Cap Wilderness.

IMPLEMENTATION

Each camp site was divided into two plots, one without and one with a surface mulch application (a biodegradable erosion control blanket made of straw interwoven with cotton string and jute). Each plot was subdivided into subplots (Fig. B16.2) which received soil treatments (organic material + inoculum; organic material + inoculum + compost; no soil amendment), and planting treatments (transplanted + seeded; no plant material). Soil was scarified in all these plots to a depth of about 15 cm. The additional control plot received no treatment.

Subplots with the organics + inoculum treatment were covered with c. 2.5 cm thick layer of peat moss mixed with well-decomposed, locally collected organic matter. This material was then mixed with mineral soil to a depth of 7.5 cm. Inoculum came from the rooting zone of local transplants. About 1.2 l of soil were mixed with about 20 l of water, and 3 l of this slurry was sprinkled over each plot and raked into the soil. Organic matter and inoculum in the compost treatments were applied in an identical manner. Also, 2.5 cm of sewage sludge/log-yard waste compost was lightly watered and raked into the top 10 cm of organic and mineral soil.

Half of the plots were seeded and transplanted. Seeding involved local seed harvesting from several species, pinch-broadcasting over the plot, and

Fig. B16.2. Setting up the restoration plots.

raking seed into the upper 2.5 cm of soil. Direct transplanting involved digging up plugs in the vicinity, planting along with Vita-start (vitamin B1) to reduce transplant shock, and watering (0.6 l per transplant). Canopy cover and maximum height of each transplant were measured after transplanting in September 1995, then in September of 1996, 1997 and 1998. Seedling density was assessed at least twice a year, beginning in early July 1996. Also, height of ten, randomly selected individuals of each species was measured in each plot. In 1996 and 1997, another four individuals of the same species were carefully excavated, and dry biomass of their root and shoots determined.

POST-RESTORATION MONITORING

One year after restoration, an average of 26 seedlings m^{-2} had established on the plots (almost 20 000 total). More than 70% of these seedlings originated from the broadcasted seed. The remaining seedlings were either immigrants or possibly originated from the soil seed bank. Three years after restoration, 10 000 perennial

seedlings were alive. Seedling density differed significantly between treatments (Fig. B16.3). In 1998, scarified plots had a greater seedling density (31 seedlings m^{-2}) than control plots (7 seedlings m^{-2}). Seeded plots had over five times as many seedlings as unseeded plots. Plots that were amended with organic matter and compost had more seedlings than unamended plots, but the surface mulch treatment had no effect on seedling density. Seedling height on plots with the organic material and compost amendment was significantly greater than on plots receiving organic material only or no soil amendments. Seedling height was greater on mulched plots than on those without mulch. Seedlings on plots amended with organics and compost had significantly more biomass than seedlings on other plots. Mulching had no effect on biomass.

Transplant survival was high regardless of treatment. Three years after transplanting, 90% of the 206 plugs had survived and 40% had flowered. Total canopy cover had increased 33%. The increase in transplant canopy cover was significantly

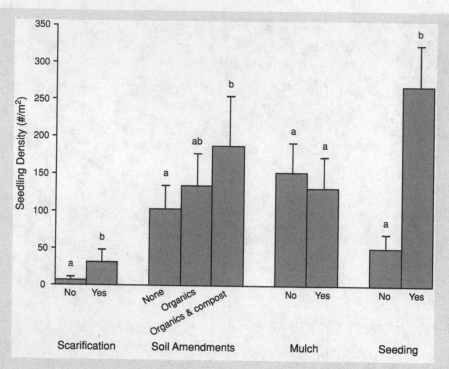

Fig. B16.3. Effect of scarification, soil amendments, mulch and seeding on seedling density. Status as of September 1998. The scarification treatment received no soil amendment, mulch or seeding. Means ± 1 S.E. Different letters indicate significantly different means ($p = 0.05$).

greater on plots with organic and compost soil amendments than on plots with no soil treatment.

Vegetation cover was negligible before restoration. Immediately after transplanting and seeding, mean total cover was 3.7%. It increased to 9.0% in 1996, 10.6% in 1997 and 13.2% in 1998. Initial cover was mostly provided by transplants, but in subsequent years more cover was provided by plants recruited from seed. The plots amended with organic matter and compost had twice the mean cover of plots that were not amended. Plots that were mulched and scarified also had more cover than non-treated plots, but the differences were not statistically significant.

CONCLUSIONS

Overall, the restoration techniques employed – planting, soil scarification and compost application – were highly effective. However, even with the most effective restoration techniques, hundreds of years will be required to eliminate the undesirable and unnecessary camp-site impacts in the area. The results demonstrate the importance of avoiding damage by implementing effective management programmes wherever regular recreation activities occur.

• Ecosystem-level effects of disturbance affect cycling and recycling of resources and energy influenced by abiotic factors such as altered geomorphology (erosion, solifluction), hydrology (water table depth, water-holding potential of soil, surfaceflow patterns) and/or biogeochemistry (soil nutrient availability, water chemistry modifications). These processes may also be affected by changes in biotic factors, e.g. plant species composition. Recognition of the abiotic and biotic effects of disturbance is the basis for selecting the most effective restoration material and restoration techniques as well as the best aftercare procedures.

RESTORATION IN PRACTICE

Assessment of site damage and recovery potential

Once the type of disturbance and its general characteristics have been identified (Table 16.1), specific assessments of both the abiotic and biotic factors are required. Abiotic assessments should include information on the hydrology and soils of the disturbed area, its large- and small-scale topographic features, and possible topography alterations. A basic assessment of the hydrology of high-altitude systems includes gathering data on the precipitation regime, the extent and duration of snow cover, the vegetation cover and the length of the growing season. It is also important to evaluate potential effects of the disturbance on soil infiltration and water storage characteristics as well as surface drainage patterns, stream flows, sediment discharge and water quality. An assessment of the soils should begin with a evaluation of the nature and distribution of the soil resource and, for severely disturbed sites, with identification of topsoil materials. The necessary analyses of the physical and chemical properties of the different types of topsoils and subsoils include soil texture, percentage coarse fragments, bulk density, pH, plant-available nitrogen, phosphorus and potassium, and cation exchange capacity. Soil microbial activity is another useful indication of disturbance (see e.g. Zabinski & Gannon, 1997). The establishment of self-regulating vegetation on degraded sites is influenced by the seed rain, seed bank, and ultimately by recruitment (Fig. 16.4). Combined with the abiotic data, these criteria are useful in deciding how the restoration may be achieved (Table 16.2).

An optimal restoration scheme should be a one-time intervention. Subsequent manipulations should be reduced to a bare minimum because they may interfere with the initiated natural processes. Restoration of high-elevation areas also requires understanding of specific problems related to altitude. Last, but not least, post-restoration monitoring

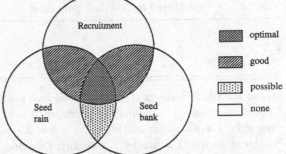

Fig. 16.4. Self-recovery potential of a degraded site based on seed rain, seed bank and recruitment.

should always be included in any restoration scheme and high-elevation sites are no exception.

Restoration techniques

Site manipulations and the use of plant material represent the two principal issues of restoration. They have to be carefully addressed if restoration is going to succeed.

Site manipulations

Site manipulations involved in restoration include landscaping, altering the soil structure, texture and chemistry, or both intervention types.

Shaping and stabilising severely disturbed areas is a critical component of many restoration projects (see Box 16.2). Severely disturbed areas should be shaped to conform as closely as possible to the original land contour (Brown & Johnston, 1979). Steep slope angles should be minimised and unsuitable topographic configurations should be avoided. If possible, the original drainage pattern should be re-established, and the necessary precautions (e.g. the use of erosion matting) taken to ensure that it can withstand spring runoff events.

High-elevation ecosystems require special considerations of bed preparation and seeding methods (Brown & Chambers, 1989). Topsoil salvage and replacement accelerate ecosystem recovery of severely disturbed high-elevation sites. This practice also represents a cost-saving element because topsoils are

Table 16.2. *Disturbance threshold rating and planning decisions*

Scale[a]	Topsoil/seed bank[b]	Safe sites[c]	Diaspore source[d]	Threshold[e]	Decision[f]
1	1	1	1	1	X
2	2	2	1–2	2	XX
2–3	3	3	1–3	3	XXX

[a] Scale: 1 = disturbance local; 2 = disturbed area medium-sized; 3 = disturbed area extensive.
[b] Topsoil/seed bank: 1 = present; 2 = damaged; 3 = none or virtually none.
[c] Safe sites: 1 = numerous and diverse; 2 = few; 3 = none.
[d] Diaspore sources: 1 = nearby; 2 = distant; 3 = none.
[e] Threshold: 1 = disturbance local and not intensive; 2 = deterioration considerable and/or extensive; 3 = site destroyed.
[f] Decisions; X = no intervention necessary, only diaspore entrapment may be improved; XX = some site manipulations needed (e.g., diversification of microrelief, site protection, diaspore entrapment); XXX = a full restoration programme required.
Source: Modified from Urbanska (1997a).

valuable sources of soil micro-organisms, mycorrhizal fungae and plant propagules. In most cases, even poorly developed topsoils or topsoils mixed with subsurface soils will provide a better plant growth medium than subsoils alone, and can be a first step towards re-creating microbial communities. The length of topsoil storage should be minimised because the biological activity of stockpiled soils decreases significantly as the length of storage increases (Kundu & Ghose, 1997). Specifically, the viability of most seeds and population densities of soil micro-organisms can decline dramatically over time. Should topsoils from areas other than surrounding natural areas be used, a possible involuntary introduction of invaders should be avoided at all cost. If present, soils and substrates containing acid-generating materials, heavy metals or other materials harmful to plant growth should be buried.

The soil amendments used are highly dependent on the characteristics of the soils and the severity of the disturbance (Chambers, 1997) (see also Box 16.1). Amendments on sites with well-developed topsoils are seldom necessary, but the final determination should be based on the assessment of physical and chemical properties. In contrast, on severely disturbed sites with the topsoils removed or on sites that are at high risk of surface erosion, it may be necessary to consider organic matter additions or fertiliser applications to increase the nutrient retention and cycling capacity of the soil (Chambers, 1997). Finally, on high-elevation sites with poorly developed topsoils or subsoils that exhibit little erosion or that have naturally low production potentials, soil amendments should be used only if deemed necessary for plant establishment.

Fertiliser application should be based not only on knowledge of the physical and chemical composition of the soil, but also on the effects of fertilisers on vegetation development. Nitrogen is highly mobile in soils and fertilisation often results in a short-term nitrogen pulse. In contrast, phosphorus is less mobile and is retained over time. The use of high fertiliser levels often requires substantial economic investment and has several negative aspects when restoring high-elevation ecosystems. For instance, grasses exhibit high growth rates in response to nitrogen but prevent establishment of other species because of competition (Chambers et al., 1987a). Also, leaching of water-soluble nutrients, especially nitrogen, during spring runoff can result in nitrification of ponds and streams. The primary benefit of fertiliser applications in high-elevation ecosystems may be in accelerating plant establishment on areas that lack topsoils or that are characterised by mineral topsoils generally low in nutrients. Low levels of slow-release,

Box 16.2   Restoring high-alpine social trails on the Colorado Fourteeners

James J. Ebersole, Robin F. Bay and
David K. Conlin

Hiking on the Colorado Fourteeners (peaks over 14 000 ft = 4268 m a.s.l.) has increased dramatically (c. 300%) in the last ten years. Hikers often ascend one to several unconstructed 'social' trails which become eroded. After construction of regular trails, the social trails are closed but they require restoration. Several restoration techniques were evaluated after one year on Humboldt Peak (Fig. B16.4) and after three years on Mount Belford, both located in south Colorado. Restoration sites are located at c. 3660–3960 m a.s.l. on the steep (30%–35%), south-facing slope of Humboldt Peak and on the broad, northwest-facing ridge crest of Mount Belford. Snowmelt at both mesic sites occurs relatively early. The turf vegetation is species-rich with the highest cover provided by *Geum rossii* and *Carex elynoides*.

DAMAGE ASSESSMENT

The social trail at Humboldt Peak was eroded in places to almost 1 m deep and 1–2 m wide with virtually no vegetation. The ridge of Mount Belford was less eroded, about 1 m wide and 20 to 30 cm deep, but it also lacked vegetation. The regular trail on Humboldt Peak was constructed in 1997, that on Mount Belford in 1996.

IMPLEMENTATION

Turf transplantation

Across the gully-like social trail on Humboldt Peak, rock walls were built and backfilled with soil. Turf blocks (c. 25–35 cm x 30–50 cm and 15 cm deep) left from the new trail construction were planted on the nearly level terraces (Fig. B16.5). On Mount Belford, similar-sized turf blocks were planted directly into the trail.

Single species transplantation

In early summer 1998, small (4–6 cm diameter) tussocks of *Carex scopulorum*, *C. elynoides* and *C. haydeniana* were harvested from adjacent undisturbed sites and divided into single tillers or small tiller groups following the technique of Urbanska *et al.* (1987). Transplants of *Geum rossii* included one ramet with rootstock each. The transplants were grown in a temporary greenhouse at 3660 m a.s.l. for 4–5 weeks, then planted onto the restoration site on Humboldt Peak and covered with Curlex erosion mat made of wood fibres. Additional tillers harvested in late August 1998 were planted directly on to the same site.

Seeding

Seeds collected in mid September 1998 in the surrounding undisturbed vegetation, were sown at 40g m^{-2} into 70 × 70 cm plots on Humboldt Peak, then covered with a thin layer of soil and the Curlex mat. Species seeded were *Polygonum bistortioides*, *Castilleja*

Fig. B16.4. The area of Humboldt Peak (4290 m a.s.l.) seen from Upper Colony Lake (3670 m a.s.l.).

Fig. B16.5. Closed social trail with turf pieces transplanted in 1998 on a horizontal terrace. Photograph taken in summer 1999.

occidentalis, Trifolium dasyphyllum, Silene acaulis, Poa alpina, Elymus scribneri, Trisetum spicatum ssp. *congdonii* and *Deschampsia cespitosa.*

POST-RESTORATION MONITORING

The transplanted turf blocks survived well (about 90%) but only when they were properly dug in. Their species richness was not different from control plots on either peak. Absolute covers of 35 species on Humboldt Peak, and of 38 species on Mount Belford, did not differ between transplants and controls. On Mount Belford, eight species had higher absolute covers in transplanted turf than in controls (p <0.05), whereas on Humboldt Peak this was not the case for any species. Eight species on Humboldt Peak and three species on Mount Belford had lower covers in the transplants. Most differences were small (1% to 5%), except for *Geum rossii* which decreased by 50% or more. On Mount Belford, three short-lived forbs and five graminoids had higher relative covers in transplant plots. Relative covers of the remaining species were similar in transplant and control plots. In both restoration sites, grasses increased moderately but

forbs still dominated. Survival of single transplanted species differed between species and apparently was influenced as well by the transplant preparation. Despite an extremely dry late summer and fall 1998, survival after one year was good for directly transplanted *Geum rossii* but only moderate for its pregrown transplants. An opposite pattern was found in *Carex scopulorum*. Survival of *Carex elynoides* and *C. haydeniana* was moderate in both transplant types (Table B16.1). Seedling emergence in the seeded plots was clearly influenced by the use of matting. On Mount Humboldt, total seedling density was 540 ± 33 m^{-2} by mid August 1999. *Polygonum bistortioides* (320 ± 21 seedlings m^{-2}) and unidentified grasses (228 ± 34 seedlings m^{-2}) dominated; in addition, moderate amounts (25 ± 3 seedlings m^{-2}) of species not used in restoration were present. In the plots non-covered by matting, seedling density was c. 80% lower.

In the non-seeded plots on Mount Belford, natural seedling emergence was apparently promoted by the transplanted turf. The highest seedling density (9.7 ± 1.4 m^{-2}) in 3 × 10 cm plots was found in places immediately adjacent to the turf blocks, but it

Table B16.1. *Humboldt Peak site: transplant survival one year after restoration, status as of August 1999*

Species	Direct transplants[a]	Percentage survival	Pregrown transplants	Percentage survival
Geum rossii	203(294)	69	90(294)	31
Carex elynoides	115(294)	39	132(294)	45
Carex scopulorum	35(98)	36	63(84)	75
Carex haydeniana	35(98)	36	17(98)	17

[a] Numbers of planted plugs are given in parentheses.

diminished by about a half already at 5 cm distance from them. A further decrease at 10 cm distance from the transplanted turf was, however, only slight.

CONCLUSIONS

Several approaches to restoring high-alpine vegetation on the Colorado Fourteeners appear likely to work. Whenever available, turf from nearby sources should be transplanted to partly re-establish vegetation in damaged areas, and to enhance natural seedling emergence in a close proximity of transplanted blocks. Survival of single species transplants was moderate to high through a dry late summer and the first winter; this method seems, thus, promising. Late summer seeding led to good seedling emergence in the following spring; however, the seeded plots must be protected. These results are encouraging, but long-term monitoring is indispensable.

organic fertilisers applied at the time of seeding or transplanting on these types of soils can increase the growth and survival of plant species adapted to both high and low nutrient levels and result in earlier and more successful reproduction (Chambers *et al.*, 1990; Chambers, 1995).

Human activities are altering the nutrient dynamics of many high-elevation ecosystems and these changes must be considered when developing restoration schemes. Deposition of soluble atmospheric nitrogen is increasing in both European and North American alpine ecosystems. For instance, annual–deposition rates in the Central Alps reach 0.5–$1.4 \, \mathrm{g \, m^{-2}}$ (Graber *et al.*, 1996). This increase could result in a shift from predominantly N- to more P-limited conditions (Körner, 1999) as the annual soluble P deposition by precipitation in the Central Alps is very low (7 mg $\mathrm{m^{-2}}$: Psenner & Nicklus, 1986).

Safe sites should always be provided for establishing seedlings and transplants (see Box 16.3). The use of biodegradable geotextiles or mulches is strongly recommended in high-altitude restoration because they: (1) decrease water loss and moderate soil temperatures and thus promote survival and recruitment, (2) stabilise the soil surface to some extent, and (3) trap wind-blown diaspores (see Boxes 16.1–16.3). In areas with extreme environmental conditions, transplants, which often function as nurses to the new recruits, can also form part of the safe site and should be locally established in clusters (see e.g. Urbanska, 1997*a* for the concept of safety islands).

Plant material

Plant material is a central issue in restoration, and offers a particular challenge in high-elevation areas. Native species should be given priority because they usually perform better than introduced, exotic taxa. Their use improves visual continuity with the surrounding local landscape, promotes biological interaction with natural communities, and often results in lower maintenance costs or no aftercare. The plant material used should originate

Box 16.3   Restoration trials on high-alpine ski runs in the Swiss Alps

Krystyna M. Urbanska

Downhill ski runs in the alpine surroundings of Davos, Switzerland were constructed in the early 1970s. Topsoil and large boulders were then removed and discarded, and the exposed mineral soil was machine-graded (Fig. B16.6). Even about 30 years after construction, the high-alpine ski runs are eroded and remain largely uncolonised by plants (Fig. B16.7).

DAMAGE ASSESSMENT

The coarse soil of the ski runs is unstructured, has an exceedingly low water-holding capacity, and is deficient in organic carbon, nitrogen and phosphate. Temperature fluctuations on the soil surface are large, with day/night amplitudes averaging 27 °C in the summer (Flüeler, 1992). Vegetation cover about ten years after the construction varied between <0.5% and 5% (Meisterhans, 1988). A decade later, it increased locally on the silicate (<10% to 15%), but remained

unchanged on dolomite. Seed rain on ski runs was significantly lower than on an adjacent natural grassland (96 vs. 930 seeds m^{-2}: Urbanska *et al.*, 1998). The soil seed reserve in the ski runs was virtually nil compared with that in the grassland (52 vs. 2401 seeds m^{-2}) (see Fig. 16.2). Thus, unassisted recovery of the graded high-alpine ski runs was unlikely.

RESTORATION CONCEPT

The principal aim of restoration was to initiate natural processes on the population and community levels after a single intervention. This would be achieved by transplanting native species onto the ski run, and creating the safe sites for plant colonisation and recruitment. The restoration scheme included light scarification of the soil without fertilisation followed by establishment of safety islands, i.e. plots with transplants covered by biodegradable erosion mats. The use of transplants in the restoration of harsh and windy high-alpine environments has several advantages: (1) transplants often survive better than seedlings and juveniles; (2) transplants reproduce by

Fig. B16.6. A machine-graded ski run in alpine area near Davos (2500 m a.s.l.) in winter. The quadrangle marks a restoration plot.

Fig. B16.7. The ski run in summer. A ten-year-old restoration plot seen as the vegetation stripe in the middle.

seed more quickly than seedlings; (3) transplant clusters act as mini-windbreaks improving entrapment of diaspores; and (4) transplants often function as nurses to their own offspring and/or seedlings of immigrant species. Biodegradable wood-fibre mats enhance diaspore entrapment and provide protection from frost and herbivores for about the first three years after restoration.

IMPLEMENTATION

Transplants obtained either from seed (genet transplants) or single-ramet cloning of plants collected in the wild (clonal transplants) were container-grown in garden soil for at least six weeks, and then acclimatised at timberline for about two weeks prior to the restoration work. At the time of planting, each transplant consisted of several ramets and had well-developed roots; all legume transplants had root nodules.

The restoration plots (4–6 m×1 m) were established in early autumn by pocket planting of 50 transplants m^{-2} (Fig. B16.8). After planting, plots were covered with a wood-fibre mat (Curlex) held in place with U-pins. In general, mixtures of graminoids, legumes and forbs were used. The plots formed part of the ski runs maintained and used throughout the winter (Fig. B16.6).

POST-RESTORATION MONITORING

Post-restoration monitoring throughout the first three years after restoration focused on survival and reproduction of the transplants, and on colonisation by other plant species. Eight to ten years after restoration the development of populations founded by transplants and immigrant species, the seed rain and soil seed bank were evaluated. Transplant survival was high, although some species not suited as restoration material suffered considerable mortality. Clonal transplants of a given species survived better than the genet transplants grown from seed. Many transplants produced flowers in the first year after restoration, but flowering was spatially and temporally variable. Transplant offspring often established at the border of or directly under the mother plant canopy. This was probably due to more favourable environmental conditions and perhaps higher colonisation by mycorrhizal fungi within the close proximity to the nurse plant.

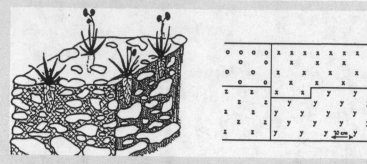

Fig. B16.8. Planting design for restoration of high-alpine ski runs. Left: three-dimensional plan with a planting depth of about 8 cm. Right: surface plan of a plot section with the distribution of transplants. Single-species sectors (marked with different symbols) were planted in random combinations.

Considerable colonisation was recorded in the first year after restoration (9–13 species per plot); it progressed more slowly in the following years. The immigrants represented various seral stages. Travel distances from the nearest possible diaspore source in the neighbouring natural vegetation were short. Many immigrants were found in the immediate vicinity or within the clonal structures of transplants, so that the nurse effect apparently promoted their establishment. Seed rain in two restoration plots about ten years old varied seasonally as well as between plots. The total seed rain was 1096–3557 seeds m^{-2} and totalled 18 species (Urbanska & Fattorini, 2000). The seed bank included eight species with seed densities of 611–916 seedlings m^{-2} in autumn, and of 698–3580 seedlings m^{-2} after the spring snowmelt. Plant cover in the plots was substantially higher than in the non-restored ski run nearby (30% to 60% vs. 5% to 15%)

(see Fig. B16.7). The number of species per plot totalled 22–24 in the aboveground vegetation. Of those, 12–16 species were immigrants, and the remainder were the transplant species initially used in restoration.

CONCLUSIONS
After ten years, the recovery in the restored plots was rapid for the severely degraded high-alpine ski runs, especially given that our restoration trials included no fertiliser amendments and repeated manipulations. Re-creation of functioning, self-supporting vegetation is possible within a reasonable time-span, even if the composition of the vegetation does not represent a carbon-copy of the adjacent natural plant cover. Positive interactions in plant communities, and specifically the nurse effect, should be given particular attention in restoring harsh environments.

from the area of restoration but generally does not need to represent the immediate surroundings of a restoration site. On the other hand, it is important to select plant species from a comparable elevation and type of soil (acidic vs. base-rich).

A correct use of biological resources in restoration implies biodiversity. This problem should be approached on three levels: (1) initial biodiversity refers to the plant material used in restoration, which should always include at least half a dozen species representing various seral stages and various

life forms; (2) spontaneous increase in biodiversity after restoration via natural immigration of seeds from outside may be enhanced by site manipulations that improve seed entrapment or attract pollinators and/or dispersers; and (3) age-state diversity of populations or whole stands is related to reproduction by seed and seedling recruitment. In the latter case, suitable plant material and site manipulations may be decisive (e.g. local use of transplants which usually flower faster than plants recruited from seed).

Depending on the site characteristics and damage type, seed material, transplants, or both will be required (see Boxes 16.1–16.3). Seed harvesting of local populations of high-elevation species is often restricted to good seed production years, and commercial seed availability is limited. Transplant material or containerised seedlings are often prepared and grown on an individual job basis. Seed production in the United States has been promoted by the US Department of Agriculture Plant Materials Centers which select broadly adapted ecotypes of various high-elevation species and release them to commercial growers. Also, various private companies collect and sell local seed, and prepare pregrown transplants for both general and specific restoration projects. Considerably less progress has been achieved in this area in Europe where the producers should be further encouraged to provide suitable materials.

Pieces of turf (e.g. from a trail construction: see Box 16.2) including several species, may be directly established on a restoration site. On the other hand, a direct transplanting of single species is not recommended, especially above the timberline, on account of considerable damage to natural populations and slow regrowth of donor plants. Transplant materials may be prepared *ex situ* (e.g. Urbanska *et al.*, 1987; Majerus, 1997) (see also Boxes 16.1–16.3). Plant symbionts need to be considered in restoration schemes and, in some cases, *Rhizobia* or mycorrhiza fungi are produced separately for inoculating the plants used in restoration. These procedures are rather expensive. More cost-effective approaches may involve adding to the containers topsoil from undisturbed areas (which often contains mycorrhizal spores) and small pieces of leguminose roots (*Rhizobia* nodules) left after preparing the transplants.

Specific problems related to high elevation

Timing, logistics and acclimatisation of the plant material are very relevant to restoration at high elevations. Experience shows that restoration work should be done at the end rather than at the beginning of the growing period. This strategy ensures higher survival of dormant transplants because of insulating snow cover, and provides a cold, wet stratification for seeds. Seeding or planting in spring results in substantial risk of drought and frost damage. Also, many high-altitude areas remain snow-covered until early summer.

Access to high-altitude sites is often difficult even in full summer, and often requires expensive transport measures. Establishing small 'field nurseries' e.g. at the timberline for the plants to be utilised in restoration of alpine disturbances reduces the transport distance and at the same time provides acclimatisation sites for transplants (see Box 16.2). Similar nurseries can also provide favourable conditions for seed production.

Post-restoration monitoring

It is critical to monitor the restoration results. Taking photographs is not sufficient, and a preferable monitoring approach is collecting data on plants, soils, and if at all possible, on various animal groups. Monitoring allows one to assess restoration success, and should therefore constitute an integral part of any restoration scheme.

The monitoring programme should be clearly defined as a function of time following the restoration work (Table 16.3). Monitoring during the first two years following restoration should provide information on the suitability of the plant material and the site manipulations involved in restoration, and possibly indicate early signs of restoration success such as increases of plant species richness. It should also include an assessment of soil stability (occurrence of surface erosion, rill or gully formation), and any potential soil chemistry or water quality concerns. Five years after restoration, the vegetation within the restored site should include both the species initially used in the restoration, as well as some immigrants. Soil erosion should be minimal or non-existent as should soil chemical or water quality problems. Population and community processes and the biological activity of the soil should be at least initiated at that time. More advanced development of plant cover and soil may take longer, but at least some of these processes should be recognisable within seven to ten years after restoration (Urbanska, 1997c; Urbanska & Fattorini, 2000) (see also Box 16.3). The details

Table 16.3. *Post-restoration monitoring in high-altitude sites*

Time after restoration	Monitoring object	Monitoring goal
1–2 years	Survival and reproduction of transplants, seedling emergence and establishment	Suitability of plant material used
	Plant species inventory	Suitability of abiotic site manipulations
	Pollinators/dispersers	
	Soil erosion	
	Soil and water chemistry	
5 years	Plant species inventory, cover %	Population and community processes
	Population size	
	Age–state structure (whole stands)	
	Seed rain and seed banks	
	Pollinators/dispersers	
	Soil microbial communities	Soil function and stability
	Soil erosion	
	Soil and water chemistry	
8–10 years	Plant species inventory, cover %	Restoration success
	Seed rain and seed bank	
	Pollinators/dispersers	
	Soil microbial comunities	
	Soil erosion	
	Soil and water chemistry	

of the monitoring programme proposed here will obviously differ between sites depending on disturbance extent, site location (above vs. below the timberline) and the distances separating the restored sites from the natural areas. The deadline for significant indications of restoration success should not exceed a decade. This deadline, suggested previously for more productive ecosystems (Jackson *et al.*, 1995) should be observed also in restoration of high-elevation biomes.

CONCLUDING REMARKS

The current state of knowledge demonstrates that the original restoration concept of recreating a carbon-copy of the past or reference conditions is not a viable proposition. It is generally accepted that the long-term objective of restoration should be establishment of self-sustaining ecosystems. Accordingly, restoration focuses on ecosystem function rather than its exact pre-disturbance structure

(Bradshaw, 1997a; Webb, 1997). Maintenance of biodiversity, which is an important ecosystem service, and site integration into the landscape represent further long-term objectives. These principles are particularly important in restoration of high-altitude sites.

Restoration goals have to be regarded in the context of environmental constraints, site damage and the surrounding natural landscape. In high-elevation ecosystems, the degree of stress has to be considered because there may be substantial differences in patterns of assisted recovery between low mountain areas which have more favourable environmental regimes and extreme high-alpine sites. Another obvious criterion is the extent of damage. For instance, a reasonably fast assisted recovery could be prognosed in a disturbed high-alpine site with topsoil remaining, but a much slower recovery will have to be expected for an adjoining site in which the topsoil was irretrievably lost (Table 16.1).

A realistic mid-term objective of restoration in high-elevation areas could be getting a site off the degradation trajectory and putting it into a condition from which it will develop towards a self-sustaining state. This means initiation of population and community processes. It is possible to reach this goal within a reasonable time-span even in severely degraded high-alpine sites if suitable plant material and site manipulations are used in a coherent design (Urbanska, 1997b; Urbanska & Fattorini, 1998b, 2000). Such an option may be preferable to setting a definite target community in restoration of high-elevation areas, expecially in the extreme high-alpine environments. On the other hand, species composition of the undisturbed neighbouring communities should be considered when the plant materials for restoration are being selected.

High-altitude ecosystems have a long history of land use and their restoration has to be considered in the historical context. However, the management of the mountain areas should focus on sustainability, and this includes an obligatory regulation of the land use to avoid, for example, overgrazing or over-exploitation of the alpine areas for recreational or other purposes.

Extensive 'developmental' projects at high elevations inevitably result in disturbance. We strongly recommend that an appropriate cost–benefit evaluation (financial and ecological) and a well-defined restoration scheme be integral parts of any such project.

REFERENCES

Allen, M. F. (1989). Mycorrhizae and rehabilitation of disturbed arid soils: processes and practices. *Arid Soil Research*, **3**, 229–241.

Allen, E. B., Chambers, J. C., Connor, K. F., Allen, M. F. & Brown, R. W. (1987). Natural re-establishment of mycorrhizae in disturbed alpine ecosystems. *Arctic and Alpine Research*, **19**, 11–12.

Barnola, L. G. & Montilla, M. G. (1997). Vertical distribution of mycorrrhizal colonisation, root hairs and belowground biomass in three contrasting sites from the tropical high mountains, Merida, Venezuela. *Arctic and Alpine Research*, **29**, 206–212.

Beniston, M., Diaz, H. F. & Bradley, R. S. (1997). Climate change at high elevation sites: an overview. *Climate Change*, **36**, 233–251.

Billings, W. D. (1974). Arctic and alpine vegetation: plant adaptations to cold summer climates. In *Arctic and Alpine Environments*, eds. J. D. Ives & R. G. Barry, pp. 403–444. London: Methuen.

Billings, W. D. (1988). Alpine vegetation. In *North American Terrestrial Vegetation*, eds. M. G. Barbour & W. D. Billings, pp. 391–420. New York: Cambridge University Press.

Bliss, L. C. (1971). Arctic and alpine plant life cycles. *Annual Reviews of Ecology and Systematics*, **2**, 405–438.

Bliss, L. C. (1985). Alpine. In *Physiological Ecology of North American Plant Communities*, eds. B. F. Chabot & H. A. Mooney, pp. 41–65. London: Chapman & Hall.

Botkin, D. B. (1990). *Discordant Harmonies: A New Ecology for the Twenty-First Century*. New York: Oxford University Press.

Bradshaw, A. D. (1997a). What do we mean by restoration? In *Restoration Ecology and Sustainable Development*, eds. K. M. Urbanska, N. R. Webb & P. J. Edwards, pp. 8–14. Cambridge: Cambridge University Press.

Bradshaw, A. D. (1997b). The importance of soil ecology in restoration science. In *Restoration Ecology and Sustainable Development*, eds. K. M. Urbanska, N. R. Webb & P. J. Edwards, pp. 33–64. Cambridge: Cambridge University Press.

Brown, R. W. & Chambers, J. C. (1989). Reclamation of severely disturbed alpine ecosystems: new perspectives. In *Reclamation: A Global Perspective*, eds. D. G. Walker, C. B. Powter & M. W. Pole, pp. 59–68. Calgary, Alberta: Canadian Land Reclamation Association and American Society for Surface Mining and Reclamation.

Brown, R. W. & Johnston, R. S. (1979). Revegetation of disturbed alpine rangelands. In *Special Management Needs of Alpine Ecosystems*, pp. 76–94. Denver, CO: Society for Range Management.

Carghill, S. M. & Chapin, F. S. III (1987). Application of succession theory to tundra restoration: a review. *Arctic and Alpine Research*, **19**, 366–372.

Chambers, J. C. (1989). Seed viability of alpine species: variability within and among years. *Journal of Range Management*, **42**, 304–308.

Chambers, J. C. (1993). Seed and vegetation dynamics in an alpine herbfield: effects of disturbance type. *Canadian Journal of Botany*, **71**, 471–485.

Chambers, J. C. (1995). Disturbance, life history strategies, and seed fates in alpine herbfield communities. *American Journal of Botany*, **82**, 421–433.

Chambers, J. C. (1997). Restoring alpine ecosystem in the western United States: environmental constraints, disturbance characteristics, and restoration success. In *Restoration Ecology and Sustainable Development*, eds. K. M. Urbanska, N. R. Webb & P. J. Edwards, pp. 161–187. Cambridge: Cambridge University Press.

Chambers, J. C. (2000). Seed movements and seedling fates in disturbed sagebrush steppe ecosystems: implications for restoration. *Ecological Applications*, **10**, 1400–1413.

Chambers, J. C., Brown, R. W. & Johnston, R. S. (1987a). A comparison of soil and vegetation properties of seeded and naturally revegetated pyritic alpine mine spoil and reference sites. *Landscape and Urban Planning*, **14**, 507–519.

Chambers, J. C., MacMahon, J. A. & Brown, R. W. (1987b). Germination characteristics of alpine grasses and forbs: a comparison of early and late seral dominants with reclamation potential. *Reclamation and Revegetation Research*, **6**, 235–249.

Chambers, J. C., MacMahon, J. A. & Brown, R. W. (1990). Alpine seedling establishment: the influence of disturbance type. *Ecology*, **71**, 1323–1341.

Chambers, J. C., MacMahon, J. A. & Haefner, J. A. (1991). Seed entrapment in disturbed alpine ecosystems: effects of soil particle size and diaspores morphology. *Ecology*, **72**, 1668–1677.

Cole, D. N. (1995). Disturbance of natural vegetation by camping: experimental applications of low-level stress. *Environmental Management*, **19**, 405–416.

Densmore, R. V. & Karle, K. F. (1999). Stream restoration at Denali National Park and preserve. *Colorado Water Resources Research Institute Information Series*, **89**, 174–187.

Diemer, M. & Prock, S. (1993). Estimates of alpine seed bank size in two Central European and one Scandinavian subarctic plant communities. *Arctic and Alpine Research*, **25**, 194–200.

Eddleman, L. E. & Ward, R. T. (1984). Phytoedaphic relationships in alpine tundra of north-central Colorado. *Arctic and Alpine Research*, **16**, 343–359.

Flüeler, R. P. (1992). Experimentelle Untersuchungen über Keimung und Etablierung von alpinen Leguminosen. *Veröffentlichungen des geobotanischen Institutes ETH Zurich*, **110**, 1–149.

Graber, W., Siegwolf, R., Nater, R. T. W. & Leonardi, S. (1996). Mapping the impact of anthropogenic depositions on high elevated alpine forest. *EnvironSoftware*, **11**, 29–64.

Grabherr, G., Gottfried, M., Gruber, A. & Pauli, H. (1995). Patterns and current changes in alpine plant diversity. In *Arctic and Alpine Biodiversity: Patterns, Causes and Ecosystem Consequences*, eds. F. S. Chapin III & C. Körner, pp. 167–182. New York: Springer-Verlag.

Hartmann, H. (1959). Vegetative Fortpflanzungs-möglichkeiten und deren Bedeutung bei hochalpinen Pflanzen. *Die Alpen*, **55**, 173–184.

Haselwandter, K. (1997). Soils microorganisms, mycorrhiza, and restoration ecology. In *Restoration Ecology and Sustainable Development*, eds. K. M. Urbanska, N. R. Webb & P. J. Edwards, pp. 65–80. Cambridge: Cambridge University Press.

Haselwandter, K. & Read, D. J. (1982). The significance of a root-fungus association in two *Carex* species of high-alpine plant communities. *Oecologia*, **53**, 352–354.

Hasler, A. R. (1992). Experimentelle Untersuchungen über klonal wachsende alpine Leguminosen. *Veröffentlichungen des geobotanischen Institutes ETH Zürich*, **111**, 1–104.

Hatt, M. (1991). Samenvorrat in zwei alpinen Böden. *Berichte des geobotanischen Institutes ETH Stiftung Rübel*, **57**, 41–71.

Holzmann, H.-P. & Haselwandter, K. (1988). Contribution of nitrogen fixation to nitrogen nutrition in an alpine sedge community (*Caricetum curvulae*). *Oecologia*, **76**, 298–302.

Jackson, L. L., Lopoukhine, N. & Hillyard, D. (1995). Ecological restoration: a definition and comments. *Restoration Ecology*, **3**, 71–75.

Kikvidze, Z. & Nakhutsrishvili, G. (1998). Facilitation in subnival vegetation patches. *Journal of Vegetation Science*, **9**, 261–264.

Körner, C. (1999). *Alpine Plant Life: Functional Ecology of High Mountain Ecosystems*. Berlin: Springer-Verlag.

Körner, C. & Diemer, M. (1994). Evidence that plants from high altitudes retain their greater photosynthetic efficiency under elevated CO_2. *Functional Ecology*, **8**, 58–68.

Körner, C. & Lärcher, W. (1988). Plant life in cold climates. In *Symposium of the Society for Experimental Biology*, vol. 42, *Plants and Temperature*, eds. S. F. Long & F. I. Woodward, pp. 25–57. Cambridge: Company of Biologists Ltd.

Kundu, N. K. & Ghose, M. K. (1997). Shelf life of stock-piled topsoil of an opencast mine. *Environmental Conservation*, **24**, 24–30.

Landolt, E. (1992). *Unsere Alpenflora*. Jena, Germany: Gustav Fischer Verlag.

Lesica, P. & Antibus, R. K. (1986). Mycorrhizae of alpine fellfield communities on spoils derived from crystalline and calcareous parent materials. *Canadian Journal of Botany*, **64**, 1691–1697.

Majerus, M. (1997). Restoration of disturbances in Yellowstone and Glacier National Parks. *Journal of Soil and Water Conservation*, **52**, 232–236.

Marchand, P. J. & Roach, D. A. (1980). Reproductive strategies of pioneering alpine species: seed production, dispersal, and germination. *Arctic and Alpine Research*, **12**, 137–146.

May, D. E. & Weber, P. J. (1982). Spatial and temporal variation of the vegetation and its productivity, Niwot Ridge, Colorado. In *Ecological Studies in the Colorado Alpine*, ed. J. C. Halfpenny, University of Colorado, Institute of Arctic and Alpine Research, Occasional Paper, **37**, 35–62.

Meisterhans, E. (1988). Vegetationsentwicklung auf Skipistenplanierungen der alpinen Stufe bei Davos. *Veröffentlichungen des geobotanischen Institutes ETH Zürich*, **97**, 1–169.

O'Lear, H. A. & Seastedt, T. R. (1994). Landscape patterns of litter decomposition in alpine tundra. *Oecologia*, **99**, 95–101.

Parker, V. T. & Pickett, S. T. U. (1997). Restoration as an ecosystem process. In *Restoration Ecology and Sustainable Development*, eds. K. M. Urbanska, N. R. Webb & P. J. Edwards, pp. 17–32. Cambridge: Cambridge University Press.

Pickett, S. T. U. & Mc Donnell, M. J. (1989). Changing perspectives in community dynamics: a theory of successional forces. *Trends in Ecology and Evolution*, **43**, 241–245.

Pickett, S. T. U. & Parker, V. T. (1994). Avoiding the old pitfalls: opportunities in a new discipline. *Restoration Ecology*, **2**, 75–79.

Pickett, S. T. U., Parker, V. T. & Fiedler, P. L. (1992). The new paradigm in ecology: implications for conservation biology above the species level. In *Conservation Biology: The Theory and Practice of Nature Conservation, Preservation and Management*, eds. P. Fiedler & S. Jain, pp. 65–88. New York: Chapman & Hall.

Psenner, R. & Nicklus, U. (1986). Snow chemistry of a glacier in the Central Eastern Alps (Hinterreisferner, Tyrol, Austria). *Zeitschrift für Gletscher Glazialgeologie*, **22**, 1–18.

Ratcliffe, M. J. & Turkington, R. (1987). Vegetation patterns and environment of some alpine plant communities on Lakeview Mountain, southern British Columbia. *Canadian Journal of Botany*, **65**, 2507–2516.

Reisigl, H. & Keller, R. (1987). *Alpenpflanzen im Lebensraum*. Stuttgart, Germany: Gustav Fischer Verlag.

Retzer, J. L. (1974). Alpine soils. In *Arctic and Alpine Environments*, eds. J. D. Ives & R. G. Barry, pp. 771–802. London: Methuen.

Schütz, M. (1988). Genetisch-ökologische Untersuchungen an alpinen Pflanzen auf verschiedenen Gestein-unterlagen: Keimungs- und Aussaatversuche. *Veröffentlichungen des geobotanischen Institutes ETH Zürich*, **99**, 1–148.

Semenova, G. V. & Onipchenko, V. G. (1994). Soil seed banks. *Veröffentlichungen des geobotanischen Institutes ETH Zürich*, **115**, 69–82.

Spence, J. (1990). Seed rain in grassland, herbfield, snowbank and fellfield in the Craigieburn Range, New Zealand. *New Zealand Journal of Botany*, **28**, 439–450.

Stanton, M. L., Rejmanek, M. & Galen, C. (1994). Changes in vegetation and soil fertility along a predictable snowmelt gradient in the Mosquito Range, Colorado, USA. *Arctic and Alpine Research*, **26**, 364–374.

Stöcklin, J. (1992). Umwelt, Morphologie und Wachstumsmuster klonalen Pflanzen: eine Übersicht. *Botanica Helvetica*, **102**, 3–21.

Stöcklin, J. & Bäumler, E. (1980). Seed rain, seedling establishment and clonal growth strategies on a glacier foreland. *Journal of Vegetation Science*, **77**, 45–56.

Urbanska, K. M. (1994). Ecological restoration above the timberline: demographic monitoring of whole trial plots in the Swiss Alps. *Botanica Helvetica*, **104**, 141–156.

Urbanska, K. M. (1997a). Safe sites: interface of plant population ecology and restoration ecology. In *Restoration Ecology and Sustainable Development*, eds. K. M. Urbanska, N. R. Webb & P. J. Edwards, pp. 81–110. Cambridge: Cambridge University Press.

Urbanska, K. M. (1997b). Restoration ecology of alpine and arctic areas: are the classical concepts of niche and succession directly applicable? *Opera Botanica*, **132**, 189–200.

Urbanska, K. M. (1997c). Restoration ecology research above the timberline: colonisation of safety islands on a machine-graded alpine ski runs. *Biodiversity and Conservation*, 6, 1655–1670.

Urbanska, K. M. (1999). Resilience, tolerance and threshold: implications from restoration ecology. In *Ecosystem Management: Questions for Science and Society*, eds. E. Maltby, M. Holdgate, M. Acreman & A. Weir, pp. 83–91. Virginia Water, UK: Royal Holloway Institute for Environmental Research.

Urbanska, K. M. (2000). Environmental conservation and restoration ecology: two facets of the same problem. *Web Ecology*, 1, 20–27.

Urbanska, K. M. & Fattorini, M. (1998a). Seed banks in the Swiss Alps. 1: Un-restored ski run and the adjacent intact grassland at high altitude. *Botanica Helvetica*, 108, 93–104.

Urbanska, K. M. & Fattorini, M. (1998b). Seed banks in the Swiss Alps. 2: Restoration plots on a high-alpine ski run. *Botanica Helvetica*, 108, 289–301.

Urbanska, K. M. & Fattorini, M. (2000). Seed rain in high-altitude restoration plots in Switzerland. *Restoration Ecology*, 8, 74–79.

Urbanska, K. M. & Schütz, M. (1986). Reproduction by seed in alpine plants and revegetation research above the timberline. *Botanica Helvetica*, 96, 43–60.

Urbanska, K. M., Hefti-Holenstein, B. & Elmer G. (1987). Performance of some alpine grasses in single-tiller cloning experiments and in the subsequent revegetation trials above the timberline. *Berichte des geobotanischen Institutes ETH Stiftung Rübel*, 53, 64–90.

Urbanska, K. M., Erdt, S. & Fattorini, M. (1998). Seed rain in natural grassland and adjacent ski run in the Swiss Alps: a preliminary report. *Restoration Ecology*, 6, 159–165.

Urbanska, K. M., Fattorini, M., Thiele, K. & Pflugshaupt, K. (1999). Seed rain on alpine ski runs in Switzerland. *Botanica Helvetica*, 109, 199–216.

Veblin, T. T., Hadley, K. S. & Reid, M. S. (1991). Disturbance and stand development of a Colorado subalpine forest. *Journal of Biogeography*, 18, 707–716.

Webb, N. R. (1997). The development of criteria for ecological restoration. In *Restoration Ecology and Sustainable Development*, eds. K. M. Urbanska, N. R. Webb & P. J., Edwards, pp. 133–158. Cambridge: Cambridge University Press.

Weilemann, K. (1981). Bedeutung von Keim- und Jungpflanzenphase für alpine Taxa verschiedener Standorte. *Berichte des geobotanischen Institutes ETH Zürich*, 48, 68–119.

Wied, A. & Galen, C. (1998). Plant parental care: conspecific nurse effects in *Frasera speciosa* and *Cirsium scopulorum*. *Ecology*, 79, 1657–1668.

Wilhalm, T. & Florineth, F. (1990). Revegetation of overgrazed alpine and subalpine areas in South Tyrol (Italy). *Colorado Water Resources Research Institute Information Series*, 63, 238–239.

Wojciechowski, M. F. & Heimbrook, M. E. (1984). Dinitrogen fixation in alpine tundra, Niwot Ridge, Front Range, Colorado, USA. *Arctic and Alpine Research*, 16, 1–10.

Zabinski, C. A. & Gannon, J. E. (1997). Effects of recreational impact on soil microbial communities. *Environmental Management*, 21, 233–238.

17 • Atlantic heathlands

NIGEL R. WEBB

INTRODUCTION

To the ecologist heathland is a type of vegetation in which evergreen, dwarf shrubs from the family Ericaceae are the dominant plants. This word can also be used to describe open treeless landscapes on poor soils and where agricultural productivity is low.

Heathlands form on chemically poor acidic soils in those parts of Atlantic zone of Europe with an altitude <300 m. This zone has a temperate climate with cool moist summers and warm moist winters. The mean temperature of the warmest month is <22 °C and of the coldest month >0 °C. Over most of this area precipitation is <1000 mm per year. Soils suitable for the formation of heathland may be derived from non-calcareous or siliceous substrates and are generally podsolic. Suitable soils also occur over the fluvio-glacial plains of northwestern Europe and on decalcified coastal sand dune systems. Heathland is confined to an area stretching from Portugal and Spain in the south to beyond the polar circle in western Norway and from Ireland in the west to Poland in the east (Fig. 17.1).

Formerly, heaths, which occurred widely over western Europe, were considered natural, but we now recognise that heathland development followed forest clearances which began 4–5000 years ago. The subsequent use of these lands for grazing of livestock, turf-cutting, and burning and cutting of vegetation for fuel and fodder arrested succession to forest and maintained a low nutrient capital. This ensured the persistence of extensive heathlands until the eighteenth century. Over most of the heathland area, a similar form of management evolved to enable crops to be grown on infertile soils. This activity has produced one of the principal cultural landscapes of the Atlantic region (Webb, 1998a).

Throughout the heathland region plant species composition varies with local topography and geographical location. Although Calluna vulgaris (hereafter called Calluna following convention) is ubiquitous, other ericaceous species are not. In the southwest up to seven species may be present, but this number diminishes northwards and eastwards and only Calluna persists in the interior of the region. However, other ericaceous species, such as Vaccinium, become more abundant. Of the associated species, species of Ulex are associated with heaths in the central and more oceanic areas. Southwards they are replaced by species of Genista and Cytissus, and by Juniperus communis to the north and east.

Locally within heathlands plants species composition depends on topography and soil moisture. There are distinct plant associations on dry heath, humid (mesophytic) heaths, wet heaths, and oligotrophic ponds and valley mires (Noirfalise & Vanesse, 1976). It is important to recognise this variation in plant species when planning restoration. To date, many heathland restorations have concentrated on dry heath and have neglected wet heaths and valley mires.

Heathland is a stage in the succession between pioneer vegetation and forest. It is, thus, unstable unless there is management, such as grazing, burning or cutting that depletes nutrients and arrests the succession. When unmanaged, open heathland may remain for some 30 to 40 years depending on its size and the nature of the surrounding vegetation (Webb & Vermaat, 1990). Generally, before that time succession will have proceeded sufficiently for management to be needed. The

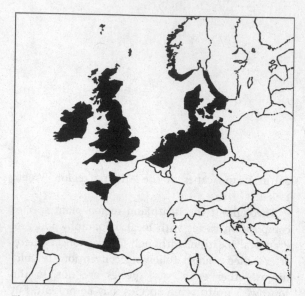

Fig. 17.1. The range of Atlantic heathlands.

management of heathland follows well-established practices (Gimingham, 1992; Michael, 1993).

Calluna has a life cycle extending over some 40 years on most sites, but this can be reduced to 20 on the more fertile sites or extended under stressed conditions (Gimingham, 1992). The life cycle of *Calluna* imparts a successional character to heathland itself, and four stages – pioneer, building, mature and degenerate – have been recognised (Watt, 1955). These descriptions may be applied to individual plants or to entire stands. Under a constant management, mixed-aged stands of heather may develop, but where the succession begins from fixed point disturbance, such as burning or cutting, then an entire stand will be even-aged and represent one of these successional stages. Under management the succession will be reset from time to time within the *Calluna* cycle.

Much of the foregoing is generally regarded as conservation management and not ecological restoration, although there is a close relationship between them. Urbanska (2000) has drawn a clear distinction between conservation, which aims to maintain ecological structures and services that still exist, and restoration, where the objective is to rebuild ecological structures and services that have been destroyed. This distinction is seen clearly on

heathands where there is a spectrum of management activities. Conservation management aims to control succession to scrub and woodland and maintain open dwarf shrub vegetation. On late successional heathland where scrub and trees have invaded, management can still be used to re-establish the dominance of dwarf shrubs; this is conservation. However, other types of vegetation may have replaced heathland, either through succession or the use of the land. When we convert this land to heathland again we are restoring. Finally, there is the creation of heathland on bare ground or on areas such as landfills and disused mineral workings, both of these last two activities being restoration. The first activity is closely related to normal conservation management and is concerned with the management of successional communities where many elements of the target community persist and it is a matter of manipulating the competitive interactions between species to restore the target community. In other instances, non-heathland communities may have replaced the target community. For instance, heathland may have developed into scrub or woodland. It may have been converted to plantation forestry, agricultural land or destroyed through activities such as mineral extraction. The likelihood that these latter areas will return to heathland within an acceptable time limit (see Jackson et al., 1995) is negligible if not non-existent.

There are thus four starting-points for heathland restoration: (1) restoration of heathland from late successional heathland vegetation; (2) restoration from other types of vegetation such as woodland, grassland and farmland; (3) restoration following disturbance or damage; and (4) restoration on bare ground.

RESTORATION PLANNING

Policy instruments

The cultural landscape began to break down from the mid eighteenth century as agricultural techniques improved. Formerly, the Atlantic heathlands extended over several million hectares, but today, it is estimated that only some 4000 km² remain. (Diemont et al., 1996; Webb, 1998b). Throughout the region, heathland has been converted to farmland

Table 17.1. *A summary of the environmental legislation affecting European heathlands*

Global

Ramsar Convention 1971 Conservation of Wetlands of International Importance

World Heritage Convention 1972 Protection of the World Cultural and Natural Heritage

Bonn Convention or CMS 1979 Conservation of Migratory Species of Wild Animals

Long Range Trans Boundary Air Pollution Convention 1979

Biodiversity Convention 1992

Pan-European

Berne Convention 1979 Conservation of European Wildlife and Natural Habitats

Espoo Convention 1991 Environment Impact Assessment in a Transboundary Context[a]

European Union

Birds Directive 1979

Habitats Directive 1992

Environmental Impact Assessment Directive 1985

Directive on Fire Control 1986

Directive on Freedom of Access to Information of the Environment 1990

Council Regulations of Structural Funds 1993

Council Regulations of Prevention of Forest Fires 1992

[a]Yet to come into force (as of 2002)
Source: Rebane *et al.* (1997).

or forestry, or used for urban and industrial development. Most remaining European heathland is now subject to legislative instruments to protect, conserve and manage biodiversity (Table 17.1). The most significant are the Ramsar Convention 1971 on wetlands of international importance, the Habitats Directive 1992 and the Birds Directive 1979 of the European Union (Rebane *et al.*, 1997). To promote biodiversity further, often to meet targets set by national and local biodiversity action plans, there is now much interest in restoring former heathland and creating new heathland. For instance, in the UK a target of 6000 ha of new heathland has been set (Anonymous, 1995).

Setting targets

We must decide between restoration to provide a heathland landscape and that to meet biodiversity targets. In the former case, restoration of the topography is important. The vegetation will be a mixture including scrub, acid grasslands, scattered trees and small woodlands together with heathland, most of which will be dominated by ericaceous shrubs especially *Calluna*. The emphasis is on appearance rather than the presence of certain species.

Restoration to meet biodiversity targets is more specific and has two approaches. First, the restoration of whole ecosystems or communities, and secondly, the provision of the habitats of species with a high conservation value (Tables 17.2 and 17.3). The emphasis is on ecological restoration in

Table 17.2. *Rare and notable heathland plants in Britain*

National Red Databook species	**Other species of note**
Eriophorum gracile	*Carex lassiocarpa*
Erica ciliaris	*Carex limosa*
Lobelia urens	*Chamaemelum nobile*
Pulicaria vulgaris	*Platanthera bifolia*
Nationally scarce species	
Cicendia filiformis	*Sparganium natans*
Crassula tillaea	*Whalenbergia hederacea*
Deschampsia setacea	*Anagallis minima*
Gentiana pneumonanthe	*Apium inundatum*
Hammarbya paludosa	*Baldellia ranunculoides*
Hypochaeris glabra	*Drosera longifolia*
Lotus subbiflorus	*Filago vulgaris*
Lycopodiella inundata	*Genista anglica*
Moenchia erecta	*Litorella uniflora*
Persicaria minor	*Pinguicula lusitanica*
Potentilla argentea	*Potentilla palustris*
Pilularia globulifera	*Radiola linoides*
Rhyncospora fusca	*Sagina subulata*
Trifolium glomeratum	*Utricularia intermedia*
Trifolium ornithopodioides	*Utricularia minor*
Trifolium suffocatum	*Veronica scutellata*
Viola lactea	

Source: Byfield & Pearman (1996).

Table 17.3. *Rare and notable heathland fauna*

Invertebrates	Vertebrates
Stethophyma grossum (large marsh grasshopper)	*Lacerta agilis* (sand lizard)
Metrioptera brachyptera (bog bush-cricket)	*Coronella austriaca* (smooth snake)
Ceriagrion tenellum (small red damselfly)	*Sylvia undata* (Dartford warbler)
Chorthippus vagans (heath grasshopper)	*Lullula arborea* (wood lark)
Coenagrion mercurale (southern damselfly)	*Caprimulgus europaeus* (nightjar)
Ischnura pumilio (scarce blue-tailed damselfly)	
Plebejus argus (silver-studded blue butterfly)	

which not only the species composition, but also the functional characteristics of the system are restored.

Whether to adopt the landscape or biodiversity approach will depend on the location. Many heathlands now exist as isolated fragments in a modified landscape. We may wish to increase the size of these fragments, in-fill irregularly shaped patches and link unconnected patches to maintain species population dynamics, and to increase the size and survival of key populations. Or we may wish to restore hydrological units and hence ensure ecosystem functions and services (Webb, 1997). Only in a few instances will large enough areas (>100 ha) be available to restore landscapes.

It is essential to set clear targets from the outset. First, suitable soil conditions, usually matching those of existing heathland (Table 17.4), must be established. Second, suitable vegetation targets must be set. In Britain the National Vegetation Classification (NVC) (Rodwell, 1991) might provide a basis. Targets based on the criteria used to determine the favourable conservation status under European Union Directives could also be used. Geographic location, the composition of surrounding heathland and local soil moisture status are factors in

selecting target communities. However, our ability to assemble these associations is limited by the lack of knowledge of the processes of community development. Most targets are plant-based and we lack the information to set targets for the fauna, particularly the invertebrates. Generally, it is assumed that if we can restore the vegetation then the fauna will follow. Dry heath is poor in higher plant species and here the emphasis is often on restoration for vertebrates, while on wet heaths and valley mires the emphasis is often on plants.

Site preparation

Site preparation is essential for successful restoration. The mosaic of heathland vegetation types depends on small-scale variations in topography that produce varying soil moisture conditions. It is a challenge to produce this topography by conventional civil engineering before restoration takes place, and this must be planned from the outset.

Monitoring

By setting targets there is an obligation to monitor. All too frequently restorations are not monitored, and therefore, the success of various techniques is not evaluated. Further more, fully replicated tests of different techniques are rare (e.g. Pywell *et al.*, 1995, 1996a; Dunsford *et al.*, 1998) causing information to accumulate slowly. Permanent quadrats or transects should be set up. On heathland it is sufficient to record annually. In experiments where a sound experimental design has been adopted then statistical procedures such as ANOVA and multivariate methods can be used to assess change. More usually in a restoration there will be no strict experimental design. Data will show the presence and absence of species and their abundances. Data of this kind can be compared using software such as Tablefit and Match to estimate the closeness of fit with target NVC classes. In dry heath vegetation, unless lower plants (bryophytes and lichens) are included in the recording programme, the target community can be reached quite quickly (five to eight years) as the community contains relatively few species of higher plants.

Table 17.4. *Typical soil chemical data for heathland in Dorset, comparing existing heathland farmland on former heathland and abandoned farmland on former heathland; ions are expressed as mg/100g of soil*

	pH	Percentage loss on ignition	NO$_3$	NH$_4$	Total N	P	Mg	Ca	K
Heathland	3.81	31.98	0.14	0.31	0.45	1.83	18.14	36.35	12.65
Abandoned farmland	5.21	6.75	0.12	0.17	O.29	0.27	7.31	98.36	2.99
Farmland	5.83	6.40	0.22	0.24	0.46	3.29	5.32	122.5	5.91

Source: Pywell *et al.* (1994).

Management

A commitment to aftercare and management is essential if the effort of restoring a heathland is not to be wasted. This is not usually emphasised sufficiently in restoration plans. Not only should there be aftercare in the years immediately following the restoration, but long-term conservation management must be planned to maintain the quality of the heathland and its biodiversity.

Invasive or undesirable species often establish in the early stages of a restoration and need controlling. These include scrub and tree species such as birch (*Betula* spp.), pine (*Pinus* spp.) *Rhododendron*, bracken (*Pteridium aquilinum*), gorse (*Ulex* spp.) seedlings and herbaceous species such as rosebay willowherb (*Chamerion angustifolium*), foxglove (*Digitalis purpurea*), sheep's sorrel (*Rumex acetosella*) and some mosses. Where heathland is restored on bare substrates, such as former mineral workings, invasive species are not usually a problem if soil fertility is managed appropriately. However, where heathland is restored on former areas of scrub these species may regrow. These species can be controlled by hand weeding, cutting and the application of herbicides. The procedures for chemical control of these species are well established (e.g. Marrs, 1987; Putwain & Rae, 1988; Gimingham, 1992).

In some instances, the establishment of heathland species may be inhibited by dense carpets of sheep's sorrel or mosses, particularly the introduced species *Campylopus introflexus*. These carpets are transitory, lasting for up to two years after which they break up enabling heathland species to establish. Grazing mammals especially rabbits and deer may damage or kill newly established plants. It may be necessary to control rabbits and fence against deer.

All heathland requires management to arrest succession to arrest to scrub and trees. Long-term management should be planned from the outset and should include provision for burning, cutting or grazing (see Gimingham 1992). Where a restored area is next to existing heathland its management may be incorporated into that of the adjoining heath.

BASIC TECHNIQUES

Soil conditions

The correct soil conditions are fundamental to the development of heathland. Heathland soils are acidic and poor in nutrients. Most heathland soils show some degree of podsolisation with a strongly leached upper horizon and well-developed pan layers. In these free-draining soils a well-developed series of horizons develops that may take many decades, even centuries, to form and once destroyed cannot be restored quickly. Where drainage is impeded or the soil is waterlogged or where the water table is rarely below 10 cm from the soil surface, gleyed podsols may form or peat may develop.

The main characteristic of heathland soils is their acidity which lies in the range 3.6 to 5.5 (Gimingham, 1972; Webb, 1986; Putwain & Rae, 1988). Associated with low pH values is the low availability of nitrogen, exchangeable calcium and extractable phosphorus (Chapman, 1967; Marrs, 1993; Pywell *et al.*, 1994; Owen & Marrs, 2000*a,b*) and the slow rates of mineralisation (Table 17.4). However many heathland species, especially *Calluna*, are tolerant of less acid conditions than are found on some heathlands and of higher concentrations of plant nutrients. Instead, these species are confined to the poorest soil by competition with other species (Ellenberg, 1974).

One of the most intractable problems in heathland restoration is the reduction of fertility and lowering of pH (Marrs & Gough, 1989). During conversion to agriculture there will have been additions of artificial fertilisers and lime or marl over varying periods of time. Where the fertiliser additions have been low in calcium, leaching will remove excess nitrogen and some phosphorus. This may take 12–15 years (Smith *et al.*, 1991; Pywell *et al.*, 1994) after which colonisation by heathland species may occur. Where the pH is higher and both calcium and phosphorus concentrations elevated, or when restoration is required more quickly, active steps must be taken to lower pH and fertility. If the elevated pH or nutrient content is concentrated in the uppermost soil horizons (this can be tested by analysis of soil samples taken down the horizon), topsoil stripping has been suggested (Marrs, 1985; Pywell *et al.*, 1994; Aerts *et al.*, 1995). Often, this has the advantage of removing the seed bank of competitive grasses and weeds.

The use of arable crops to impoverish soils is a possible method to restore heathland (Marrs, 1985, 1993; Marrs *et al.*, 1998). The crops are grown with controlled additions of fertilisers. Under these conditions more nutrients are removed from the system than are added. In a pioneering study at Roper's Heath in the East Anglian Breckland in the UK, cereal cropping and stubble burning were used to reduce fertility. Despite using additions of nitrogen to help deplete the soil phosphorus, this proved difficult (Marrs, 1985). It was essential to harvest all of the aboveground vegetation as harvesting only seed did not remove a significant quantity of nutrients.

Recent large-scale studies by Marrs *et al.* (1998) on abandoned farmland tested crops of linseed, spring and winter barley and rye. These were sown with additions of inorganic nitrogen and potassium. Yields were lower than with conventional farming and nutrient budgets showed that more nutrients were removed than added. Nevertheless, after seven years no appreciable reduction in pH and available nutrients was reported, and it was considered unlikely that heathland vegetation could be established rapidly. The use of arable crops, thus, appears to be essentially a long-term approach.

More rapid establishment of heathland requires acidification of the soil. Various materials have been mixed with soil to lower the pH. These include bracken litter, pine chippings, pyritic peat and elemental sulphur (Ford & Williams, 1994; Williams *et al.*, 1996; Davy *et al.*, 1998; Dunsford *et al.*, 1998; Owen & Marrs, 2000b). Experimental studies in East Anglia (Owen *et al.*, 1999) in agricultural soils with a pH of 6.0–7.0 showed that pine chippings reduced pH to 5.0–6.0, bracken litter to pH 4.0–4.5 and elemental sulphur to pH 2.0–4.0, depending on the application rate. It was concluded that pine did not reduce pH to the extent needed for heathland restoration. Bracken litter produced a satisfactory reduction in pH almost instantly, but there were problems with growth of ruderal species. Sulphur was the most effective material and when applied at a rate of 1–2 t ha^{-1} (higher rates of application are likely to lead to leaching into watercourses) reduced pH to about 4.0. However, this effect was only achieved after a few months. Because of a delay in the action of sulphur, Owen *et al.* (1999) suggest that a mixture of bracken litter and sulphur be added, the bracken providing a more rapid initial reduction and the sulphur enabling a lower pH to be achieved. Bracken litter may be difficult to obtain, but is sometimes available as a by-product of normal conservation management.

An effective method to establish *Calluna* is the addition of acidic peat (Davy *et al.*, 1998; Dunsford *et al.*, 1998). The supply of this peat is limited and other pyritic materials, such as colliery tailings, might be a substitute (Davy *et al.*, 1998). These materials must be incorporated in to the soil thoroughly to avoid creating areas of high acidity, and a mixture of about 50% peat gives best cover of desirable species. As with elemental sulphur, at high application rates, pulses of iron and sulphate can leach into surrounding watercourses (Dunsford *et al.*, 1998; Owen *et al.*, 1999).

In former farmland soils the seed bank of grasses and fast-growing ruderal species may cause problems (Pywell et al., 1997; Dunsford et al., 1998; Marrs et al., 1998; Owen et al., 1999; Owen & Marrs, 2000a). Even when soil conditions are suitable for the establishment of heathland species, fast-growing species will inhibit the slower-growing heathland species. One of the advantages of topsoil stripping to lower fertility is that the weed seed bank is also removed and any residual heathland seed bank exposed.

In rare instances it may be necessary to raise the pH. Gilbert & Anderson (1998) cite an example where lime needed to be added to promote the establishment of moorland vegetation. At high acidity, toxic ions, particularly aluminium, may be mobilised and reach concentrations that affect many heathland plants: Calluna, having the greatest tolerance to aluminium, is the least susceptible species to soil acidification. The decline in the diversity of heath vegetation in parts of The Netherlands has been attributed to these effects and successful restoration has been achieved by adding lime to raise the pH (Roclofs et al., 1996; de Graaf et al., 1998).

Sources of propagules

Seed
Heather seed is available from commercial sources. However, this is expensive for large restorations, and the seed may not be of an acceptable provenance. It is more usual to collect seed or seed-bearing materials for each restoration from nearby heathland.

Harvested shoots
Harvested heather shoots with seed capsules are the most abundant source of seed. They can be collected during normal heathland management using either a forage harvester (Gimingham, 1992; Pywell et al., 1996b) or brushing machines. Harvested shoots contain large quantities of germinable seed of heather and of some other heathland species (Table 17.5) and are best collected in the autumn from October to December. Using a double-chop forage harvester results in a mixture of capsules and heather shoots. These shoots act as a mulch that improves stability,

Table 17.5. *Germinable seeds per kg fresh material from Dorset heathland*

Species	Number of seeds
Calluna vulgaris	1538
Erica cinerea	82
Erica tetralix	117
Agrostis curtisii	2
Ulex minor	2
Molinia caerulea	12

Source: Pywell et al. (1996b).

conserves moisture, suppresses weeds and enhances germination.

Harvested material can be applied at rates of about 0.6–1.8 kg m^{-2} (Pywell et al., 1996b) enabling two to five times the harvested area to be restored. The shoots are rolled after application to improve retention and stability. These rates of application resulted in successful germination of seedlings on both abandoned farmland and mineral wastes (Pywell et al., 1996a). Generally, heather mowings are applied without storage. Where it is necessary to store the harvested material it must be dried first. Where erosion is likely application rates can be increased or companion species sown.

Bales
As an alternative to forage harvesting, heather plants can be cut and baled. This was a widespread practice when there was a demand for heather bales for road foundations and similar purposes. Bales can be stored and spread on areas to be restored: the woody material in the bales acting as a mulch. A deep layer of mulch should be avoided as this can suppress the germination.

Litter
Heather litter is a rich seed source not only of heather but also of associated species. It is difficult to collect and is only suitable for restoring small areas. Gimingham (1992) indicates that it is practicable to collect 30–40 kg dry weight by hand per person per day. For larger quantities a vacuum collector on wheels can be used, although this depends on the terrain. Only the litter containing seeds of

heathland species should be collected and areas of scrub, particularly birch, should be avoided.

Topsoil

Topsoil is generally only available where a site is to be destroyed. Topsoil is usually collected together with the litter. This material contains not only an abundant seed bank of heather and associated species but also root fragments and stem bases. In trials this material produced a more diverse heathland community than that produced from heather mowings (Pywell *et al.*, 1995). Topsoil should be collected only from heathland and areas of birch and scrub avoided.

The area of the donor sites should be rotovated to a depth of about 50 mm and the mixed litter and topsoil removed with light earth-moving machinery. Pywell *et al.* (1995) applied this mixture at a rate of 22–26 kg m^{-2} mixing it with the receptor site soil with a rotary cultivator and then rolling with a ring roller to make good contact with the seedbed. In some instances, a combined approach can be adopted, especially on sites to be destroyed. The heathland vegetation can be flailed, and this together with the litter and topsoil rotovated as before. If turf-cutting is instigated as conservation management for wet heath and valley mires, then the cut material could be used for restoration.

The seed content of collected materials varies from site to site and depends on season. Usually, seed capsules should be harvested between October and December as after this time most of the heather seed will have been shed. Samples of collected materials should be checked for seed content before they are used in restoration. By harvesting at other times of the year, seeds of associated species can be collected.

All of these seed sources are likely to produce seedlings in the first autumn or the following spring. However, there is evidence (Gilbert & Anderson, 1998) that freshly collected heather seed may not germinate until the spring following application. The survival of spring-germinated cohorts is often poor as the seedlings are not sufficiently established before the dry summer weather. Autumn-germinated cohorts survive better. Weather at the time of germination is an important factor in survival of seedlings and there can be much interannual variation. Where both cohorts fail in the first year there is usually sufficient seed for future years. It must be recognised that heathland restoration can be slow.

Turves

An effective, but expensive way of restoring heathland to produce an instant effect is to use turves. Normally this type of material is not available except where a site is to be destroyed. It is best to move turves that are as large as possible, as small turves tend to dry out. Specialised hydraulic equipment to lift turves 1.2 m × 2.3 m × 150 mm has been used in some instances (Pywell *et al.*, 1995). The receptor site should be prepared by removing topsoil. Care should be taken to ensure that the composition of the vegetation on the turves is compatible with that of nearby heathland. Usually, it is not necessary to irrigate the turves once laid except in very dry summers. Using this method, Pywell *et al.* (1995) created a mature heathland plant community over small areas with minor changes in the composition of the vegetation. Translocation of entire large turves has the advantage that many invertebrates are moved to the restoration site. Turves can be used in combination with the various seeding methods. In some cases, the turves can be broken up and the resulting material (clods) spread on the site; this is akin to spreading topsoil. The drawback of using turves is their scarcity and the high cost of lifting and moving them.

Nursery plants

In some instances, rooted seedlings or nursery-raised small plants can be used. This is generally expensive and time-consuming, but may be effective over small areas or in critical areas of a large restoration. Plants may be used in combination with various methods of seeding, the plants producing areas of established vegetation and a source of additional seed. Heathers can be grown from cuttings taken during the summer using standard horticultural techniques (Small, 1995). Where plants are needed for a restoration this should be planned well in advance and sufficient stock raised in time for planting out. Alternatively, seedlings can be raised by taking handfuls of heather litter and spreading on seed trays, although it takes several years before pot-sized plants are obtained. While there is a considerable background in the culture of heathers we know almost nothing about other heathland species. Almost certainly it will be easier to raise these from seed on the site of the restoration than to transplant. Species of *Ulex* are notoriously difficult to transplant and must be raised from seed *in situ*.

Companion species

In hostile situations, especially on slopes, fast-growing companion or nurse species may be required to stabilise the soil surface and to protect the slower-growing species. The need for companion species is greater in upland moorland areas than on lowland heathland. Companion species should establish quickly (within the first year) and not compete with the heathland species. Their requirement for fertiliser additions should be low and they should be short-lived, dying out of the vegetation after about five years. Several of the heathland grass species such as *Deschampsia flexuosa* and *Molinia caerulea*, while suitable in upland or wet areas respectively, tend to be too vigorous for most restoration purposes and are not suitable in the lowlands or under more fertile conditions. The same is true of some agricultural grasses such as *Lolium perenne*. Putwain & Rae (1988) provide a full review of companion grasses. The most useful cultivars are from species of *Agrostis* or *Festuca* with *Agrostis vienealis* and *A. castellana* cv. Highland being the most usual. In the southwest of England the native bristle bent (*Agrostis curtisii*) meets the requirements of a companion species.

Fertilisers

Fertiliser additions are not required when restoration is on former heathland. Where the land has been in arable production reducing and not increasing fertility is likely to be the problem. However, on mineral wastes, steep slopes and other hostile locations, rapid establishment of heathland species or support for companion species may require fertiliser applications (NPK and Mg). However, fertiliser applications should be added with care as inappropriate rates of application may lead to invasion by grasses and ruderal species.

There is a range of examples from upland areas in which rates of application have varied from 75 to 517 kg ha^{-1} (reviewed by Gilbert & Anderson, 1998). On mineral waste, rates of application vary from 100 to 300 kg ha^{-1} (Putwain & Rae, 1988; Pywell *et al.*, 1996a). In replicated experimental trials on mineral wastes in Dorset, UK, Pywell *et al.* (1996a) added NPKMg (6:19:10:9) and NPK (15:15:15) each at a rate of 100 kg ha^{-1}. Both these additions resulted in increased seedling establishment on plots on which heathland topsoil had been spread, but had no effect on plots treated with mown heather capsules.

Stabilisation

Both companion species and mowings collected with a double-chop forage harvester impart stability to restored areas. On steep slopes and in upland situations maintaining surface stability can be difficult. Other methods used include spreading of pine brashings or other woody materials, or chippings. However, these methods are difficult to apply over large areas and may in some circumstances inhibit establishment. Chemical stabilisers have been used in a few instances. Usually these are effective for fast-growing species but are less so for slow-growing heathland species (Putwain & Rae, 1988; Gilbert & Anderson, 1998).

RESTORATION PRACTICE

Restoration from different types of vegetation

Bracken

Bracken is a fern which has invaded considerable areas of heathland. In the past the traditional activities controlled its spread (Webb, 1998a). Bracken forms dense stands, which are of little conservation interest, with a deep litter layer in which nutrients accumulate, and where colonisation by heathland species is prevented.

Where bracken is beginning to invade heathland vegetation or exists as scattered fronds, it can be controlled by spraying with the herbicide asulam

(40%) between late June and the end of August; further treatment is usually required in subsequent years (Woodrow *et al.*, 1996).

Dense stands of bracken can be sprayed with herbicide, cut, rolled or ploughed (Snow & Marrs, 1997). Although these procedures kill or weaken the plants, they are unlikely to provide conditions for the re-establishment of heathland. To avoid a grass-dominated community and to favour *Calluna* establishment, the litter must be removed: it may be used in other restoration schemes. Bracken stands <50 years old may contain a residual heathland seed bank (Pakeman & Hay, 1996). If there is no residual seed bank of heathland species, heather seed or heather mowings must be added. Regeneration of heather in these circumstances may be variable (Snow & Marrs, 1997; Marrs *et al.*, 1998). Medium-term management may require the control of invasive species, and grazing by rabbits and deer. Where heathland is established successfully it will need to be brought in to a management programme.

Scrub

All heathlands have a tendency to be invaded by woody species particularly pine, birch, *Rhododendron*, willow and on occasions, gorse. Succession may go beyond that normally controlled by heathland management (Gimingham, 1992) and when this happens restoration is needed. Pine and *Rhododendron* present less of a problem than the other types of scrub as they alter the soil least. Birch, willow and gorse tend to increase the nutrient capital and raise the pH of the soil and make the re-establishment of heathland species more difficult.

Restoration from pine scrub follows procedures similar to those outlined for the restoration of heath from coniferous woodland (see below). Where there is relict heathland vegetation within the scrub it will be sufficient to cut the small trees and remove them. Where there is no ground vegetation then, as in coniferous woodland, the litter layer should be removed to expose the buried seed bank. Some pine trees may be retained to provide bird habitat (Auld *et al.*, 1992; Woodrow *et al.*, 1996).

Rhododendron can be cut with chainsaws. It is preferable to leave the stumps, as winching them out of the ground causes excessive disturbance. It is more satisfactory to kill them by painting the cut ends with herbicide. Any resprouting from the cut stumps is spot-treated with herbicide using a knapsack sprayer later in the first season. Further spot treatment in the following two seasons may be required during which any seedlings are also treated. Despite these measures, heathland species are often slow to regenerate on areas from which *Rhododendron* has been removed.

Birch and willow scrub present different problems. These species are of conservation importance particularly for insects. Scattered bushes or trees or small areas of these species are acceptable as part of the heathland community. Where there are more extensive areas or the area of dwarf shrubs needs to be increased, they must be removed by cutting and treating with herbicide. Follow-up spot treatment to kill regrowth and seedlings may also be required (Marrs, 1987). Where heathland vegetation remains beneath the scrub it can be allowed to regrow or regenerate. Where the scrub is dense then there may be a deep litter layer or the soil may have become less acid and more nutrient rich. Regeneration on this material is often of grasses and other non-heathland species. In this case, the litter should be removed and the lower layers of the soil exposed. This action may expose a residual seed bank of heathland species, but if seedlings do not establish after two seasons, seed or other propagules should be added. Once heathland vegetation is established it must be managed.

Conifers

During the late nineteenth and throughout the twentieth century large areas of heathland were planted with conifers. In the 1800s trees were planted without modification to the soils and when they were removed heathland re-established. In contrast to earlier practice, more recent plantations have been fertilised and the soil deep ploughed to break the iron pan (Box 17.1). A substantial seed bank of *Calluna* remains beneath conifers planted on heathland for as long as 70 years (Hill & Stevens, 1981; Pywell *et al.*, 1997); however, we know little about the longevity of the seed bank of other heathland species. Elsewhere, especially near plantations,

Box 17.1   Restoring coniferous plantations to heathland

Conifers are felled and the stumps cut close to the soil surface to allow the passage of machinery. The pine litter to a depth of 20–30 mm is then swept up using a tractor-mounted road sweeping brush, collected by a large vacuum and removed from the site. The timber is sold, and the litter and brash chipped and sold to offset cost of the restoration. This procedure exposes the mineral soil surface which contains an abundant seed bank of *Calluna*. Germination and establishment occurs within the first two years with >1000 seedlings per square metre. Where the litter is not removed seedlings number <10 per square metre and grasses may establish (Symes, 1999).

Where areas have been deep ploughed before afforestation it may be necessary to level the ground after felling to provide access for machinery. In these circumstances the disturbance, both during ploughing and restoration, is likely to damage the heathland seed bank. In these cases propagules, such as heather mowings, must be added and invasive weeds controlled.

During the first five years the vegetation may be dominated by bryophytes, which can achieve up to 20% cover. The introduced moss species *Campylopus introflexus* may produce a dense carpet which inhibits heather seedling regeneration. Only when this carpet breaks up does heather establish. Pioneer species, such as *Rumex acetosella*, can also achieve a high cover but are not persistent.

Results indicate that after five years *Calluna* attains 21.5% cover and *Erica cinerea* 18.3%. After ten years these species had covers of 72% and 11% respectively (Woodrow *et al.*, 1996). In this instance the aim of restoration was to enhance the populations of rare heathland birds. It resulted in a 23% increase in nightjar, a 36% increase in wood lark and a 108% increase in the number of Dartford warbler territories as these species responded to the changes in vegetation resulting from the restoration. We have less information on plants, but species such as *E. ciliaris* have established in some areas (Woodrow *et al.*, 1996).

conifers may invade the heathland, and if left, develop into woodland. Generally speaking, heathland regenerates following the removal of the trees and adjustments to fertility or pH are not required. In Dorset, the Royal Society for the Protection of Birds has restored extensive areas of heathland from beneath conifers (Auld *et al.*, 1992; Woodrow *et al.*, 1996). In some instances this requires a felling licence from the forest authority.

Broadleaves

In comparison with conifers, restoration on land with broadleaved trees is poorly known. Unlike conifers, which tend to leave the soil acid, broadleaves often result in soil changes which need to be reversed before conditions are suitable for heathland restoration. There are almost no examples of restoring heathland in these circumstances.

Farmland

Much of the former area of heathland in Britain has been converted to agriculture. For example in Dorset, Webb & Pywell (1992) estimated that between 1811 and 1990 some 50% (*c*. 125 km^2) of heathland was converted to farmland and of that 12% (33 km^2) had been converted since 1960. These conversions were encouraged by financial incentives from government. Agricultural policy has now changed and incentives focus on promoting biodiversity by reconverting farmland to dwarf shrub vegetation.

One of the main obstacles to restoring farmland to heathland is the elevated soil fertility, which depends on soil type, farming history and inputs of fertilisers and lime (Pywell *et al.*, 1994). The supply of propagules, especially where large areas are to be restored, and the high costs of the restoration, are further obstacles. Related to this type of restoration is the regeneration of dwarf shrubs on areas that have become dominated by grasses as a result of grazing (Bullock & Pakeman, 1997) or through increased atmospheric deposition of nitrogen (Bobbink & Heil, 1993).

The first step is to ascertain the past use of the land and where possible information on the rates of application of fertilisers and lime. Often, reliable information of this type is difficult to obtain. Surveys of soil nutrients and their distribution in the soil profile, the seed bank and the existing vegetation should then follow. The vegetation survey alone will provide many clues as to the soil conditions and the potential for restoring heathland.

In a few instances, the elevated nutrient content of the soil may decline through weathering and, where a seed bank remains, heathland species may begin to establish some 12–15 years after abandonment (Pywell et al., 1995). Intact heathland soils may have a seed bank of 6700–16 000 germinable seeds per m^2 concentrated in the top 40 mm. Seed of associated species such as Erica cinerea, E. tetralix, Ulex minor and grasses also occur in significant quantities in this horizon. Despite cultivation, the heathland seed bank can be remarkably persistent especially in the layers deeper than 40 mm (Pywell et al., 1997). In heathland soils converted to agriculture heathland seed has an estimated half-life of ten years (under conifers this is 13 years). The decline in heathland seeds is accompanied by a 200–800-fold increase in seeds of non-heathland species.

The principal methods for the restoration of farmland are addition of harvested heather shoots, the application of heathland topsoil and the translocation of heathland turves. A comparison of these three methods was carried out by Pywell et al. (1995). Successful establishment of dwarf shrubs occurred in all cases. The most effective was turfing, which resulted in a mature heathland community albeit with small vegetation changes between the donor and recipient sites. In topsoil applications heathland species were both more numerous and diverse than where heather mowings were applied (Pywell et al., 1995).

The recovery of heathland from grass-dominated areas may involve changes to the grazing regime, applications of herbicide and turf-stripping. Under grazing, heathland species persist at low densities or in the seed bank and relaxation in grazing can lead to their re-establishment. Grazing is generally regarded as a useful form of maintenance management rather than as a means reclamation management (Aerts & Heil, 1993; Bakker & Londo, 1998; WallisdeVries et al., 1998). In Britain, grazing management has been confined mainly to the upland heaths and moors (Bullock & Pakeman, 1997),

but elsewhere, especially in The Netherlands, Denmark and Belgium, there have been attempts to restore heathland from grassland using grazing animals. Success has been variable and depends on site differences, particularly soils and climate, land use history, grazing regime and the extent of nitrogen deposition from the atmosphere (Bokdam & Gleichman, 2000). In some instances, successful heather restoration has been achieved, for example in Denmark using cattle (Buttenschon & Buttenschon, 1982a, b), but more often the results using sheep, cattle and ponies have been less satisfactory (Bakker et al., 1983; van den Bosch & Bakker, 1990; Bokdam & Gleichman, 2000). Grazing can alter the composition of the vegetation and herding systems may be more effective in promoting dwarf shrub vegetation than free-range grazing. Free-range grazing produces dynamic mosaics between grass, heather and trees. To maintain at least a heather–grass mosaic trees must be cut (Bokdam & Gleichman, 2000). The use of grazing to restore heathland remains a subject for further research.

Turf-stripping, which removes nutrients and the seed banks of grasses and aggressive weeds, has been used successfully in The Netherlands to convert grass-dominated heaths to dwarf shrubs (Diemont & Linthorst Homan, 1989). In southern Britain, Smith et al. (1991) successfully increased the frequency and abundance of heathland species in old fields by stripping the top 50 mm. Other techniques, which included soil disturbance, cutting the grass and the addition of flailed heathland vegetation mixed with topsoil, were less successful. Using herbicides to encourage heathland species at the expense of grasses proved unsuccessful in trials in Dorset (Pywell et al., 1995).

The decline in fertility of old fields is often slow. There are three approaches to hasten the restoration of heathland vegetation: (1) removing the top soil, (2) diluting the topsoil with less fertile soil from deeper horizons or from other sites, and (3) inverting the soil profile by ploughing (Marrs, 1985, 1993; Pywell et al., 1994; Marrs et al., 1998). However, none of these techniques has been evaluated for conservation purposes (Marrs et al., 1998).

Damaged and disturbed areas

After fires

During prescribed burning the fire is controlled to avoid damage to the rootstock of the heather plants, which then resprout from the stem bases (Gimingham, 1972). However, during wildfires, the higher temperatures may kill rootstocks, and burn the litter layer including much of its seed bank. Usually, sufficient seed remains for heathland to regenerate, albeit slowly (Gimingham, 1972; Legg *et al.*, 1992; Pywell *et al.*, 1994). Mats of bryophytes, algae or lichens, which frequently colonise the soil surface following severe fires, may inhibit the germination of heather seeds. When these mats are broken up or removed germination rates are increased (Legg *et al.*, 1992). Following exceptionally hot fires restoration may be needed as almost all the seed bank will have been burnt and propagules in the form of mowings, cut heather litter or topsoil must be added. Often, heathland scrub regenerates as scrub and does not revert to heathland after a fire (Bullock & Webb, 1996), and in these circumstances restoration will be needed, and this should follow the procedures adopted for scrub.

Development: roads and buildings

Road embankments and disturbed areas such as building sites and parking lots frequently require restoration to blend with the landscape and to minimise their effects as barriers to species dispersal. In the past, topsoil from elsewhere was frequently spread over such areas with fertiliser application. Agricultural grass-seed mixes were often sown and inappropriate species of trees and bushes planted. This can be avoided if the restoration to heathland is planned from the outset.

Litter and topsoil should be removed during the development, and kept for use in the restoration. We have little information on the best conditions for storage to maintain the viability of the heathland seed bank, although Putwain & Rae (1988) state that when dry, heather litter can be stored for long periods. Access roads and hard standings should be made from acidic sands and gravels and the use of chalk or limestone hardcore must be avoided.

In some instances, heathland can be allowed to regenerate naturally through colonisation from the surrounding areas, although this is usually slow. More usually, stored or locally available soil of a suitable chemical type should be applied. In upland areas, peat and peat-rich soils are often unstable after disturbance and have a tendency to flow when saturated with water (Putwain & Rae, 1988). On slopes, stability can be achieved using nurse grasses. These are usually short-lived species of *Agrostis* such as *A. castellana* cv. Highland or in the southwest of Britain the native *Agrostis curtisii*. In a few cases in highly sensitive landscapes, pot-grown heather plants can be used to ensure rapid and effective restoration (Putwain & Rae, 1988).

Aftercare is particularly important. Restored road embankments are often managed inappropriately or even neglected. If a benefit from the restoration is to be achieved the right aftercare is essential. If soil fertility is too high then invasive species may establish, such as rosebay willowherb, creeping thistle (*Cirsium arvense*), purple moor grass (*Molinia caerulea*) or soft rush (*Juncus effusus*). These species may need control in the early stages.

Pipelines

Pipelines for oil, gas, water and chemicals now form an extensive network throughout the UK. They are usually laid through farmland, but on occasions the crossing of sensitive habitats such as heathland is unavoidable. The restoration of heathland in these cases has been well researched (Gillham & Putwain, 1977; Burden, 1979; Holliday *et al.*, 1979; Putwain *et al.*, 1982; Putwain & Rae, 1988; Rose & Webb, 1994). Formerly, few special procedures were adopted and the pipe trench was filled with the material dug out and agricultural seed mixes and fertilisers applied. The restored vegetation dominated by invasive species of grass, rushes or scrub and heathland regeneration was poor. The techniques have now been refined considerably (Box 17.2).

Mineral workings, landfills and quarries

Beneath many areas of heathland lie important deposits of minerals. In the past, small-scale workings left areas which, given time, regenerated as

Box 17.2   Laying pipelines across heathland

An easement of some 8 m width is fenced (Fig. B17.1). On either side of the proposed pipe trench sheets of geotextile membrane are laid over the vegetation. On one side ballast or wooden slats are laid to provide a roadway, while on the other side the soil removed from the pipe trench is stored. First, turves are cut from the pipe trench and stored behind the roadway. The soil is removed and laid on the other side in the order of its horizons. The pipe is assembled, welded and then lowered in to the trench. The soil is replaced in the reverse order to which it was removed;

finally the heather turves are replaced. The whole process often takes just a few days. The heather turves are stored for a minimum period of time and the covering of the heathland for the roadways and soil storage is short. The covering of heathland vegetation is best done early in the season for periods of no more than a month. If the vegetation is covered late in the season there will be insufficient time for the vegetation to recover before the winter (Rose & Webb, 1994). Results using these procedures are now very good and there a number of examples of high-quality restorations where pipelines cross heathland.

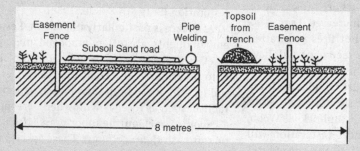

Fig. B17.1. The method used to lay pipelines across lowland heathland. After Burden (1979) and Putwain & Rae (1988).

heathland. Today, large-scale mechanised working leaves excavations which require restoration.

Restoring heathland on former mineral workings provides scope for the most imaginative restoration. It is important, however, to consider all aspects of the reinstatement and restoration at the planning stage because meeting specific restoration designs will influence the way in which the minerals are worked. There must be a clear plan of how the site will be restored and of the target vegetation, despite the fact that this may be some decades in the future. All too frequently restoring heathland has been an afterthought.

Central to this approach is the reconstruction of the topography with a range of hydrological conditions that matches the original or surrounding areas. This will enable dry heath, wet heath and mire communities to be restored. The restoration of wet heath may not always be possible once

minerals have been worked due to change in groundwater levels. Although this is primarily a civil engineering exercise, it requires ecological knowledge in its planning. Where the site is worked by backfilling it may be difficult to restore the final topography and vegetation, as the hydrology may not be established until the working is finished.

The topsoil and vegetation must be conserved for future use. Sometimes, vegetation, litter and topsoil can be stripped from an area to be worked and immediately spread on backfilled areas. Where turves, litter and topsoil have to be stored this must be for the shortest possible time. Finally, the aftercare and management of the restored heathland must be planned.

Putwain & Rae (1988) have reviewed the extensive body of earlier work on restoring vegetation on these areas. They recommend that heathland vegetation to be stripped be cut and lightly rotovated (50–100 mm depth). In some instances, the litter and topsoil can be stripped and spread on the area to be restored either in lumps or, where rotovated, spread as a thin layer (<50 mm thick). Results have been variable. Successful restoration was achieved on china clay sand wastes in Devon, UK by adding topsoil, the companion grasses *Agrostis castellana* cv. Highland, *Lolium perenne* cv. S24 and *Festuca tenuifolia*, together with moderate to high (<300 kg. ha^{-1} of NPK 17:17:17) fertiliser applications. After seven years this treatment gave a balanced heathland community. However, restoration failed if grazing animals were not excluded for the first five years (Putwain & Rae, 1988). Often unsuitable seed mixtures are applied to restorations on mineral workings, and many become poor grass swards with no heathland species and with a predominance of invasive species such as species of *Ulex*, *Rubus* and rushes (*Juncus* spp.) This is often the result of inappropriate soil conditions. Considerable care should be taken to ensure that if heathland topsoil and litter are not available from adjacent areas, then the soil applied conforms to that suitable for heathland, with a low pH and low calcium and phosphorus content.

Restoration of wet heath

In contrast to dry heathland communities, there has been almost no attempt to restore wet heath although there have been attempts to restore peat and blanket bog vegetation (reviewed by Putwain & Rae, 1988; Gilbert & Anderson, 1998). In the uplands, peat is easily eroded if damaged by fires, trampling or overgrazing. Because many of these areas are exposed or on steep slopes erosion can be difficult to control, and their remoteness and inaccessibility to machinery makes them difficult to restore. Experimental studies show that it is essential to exclude grazing animals if establishment of heather is to occur. Regeneration from seed banks is slow and seed should be added, usually in the form of heather mowings. Heather establishment is improved where the surface is stabilised using either conifer brashings or chemical stabilisers (Putwain & Rae, 1988; Gilbert & Anderson, 1998).

In other instances, especially in relation to developments, blanket peat has been translocated experimentally (Merrilees *et al.*, 1995; Standen & Owen, 1999). There may be problems with invasive species such as *Juncus effusus*, but in general, so long as the hydrological conditions can be met, translocation of large peat turves (300–1000 mm deep), where they are available, is reasonably successful.

Lowland wet-heath turves can also be translocated successfully provided that hydrological conditions are met at the receptor site. In an experimental translocation a small area of wet-heath vegetation (1000 m^2) in Dorset was taken from an area to be worked for ball clay to a specially constructed cell in which hydrological conditions could be controlled (Box *et al.*, 1998). Monitoring since 1993 has shown an increase in the abundance of species of *Sphagnum* and of mire species mainly because wetter conditions have been maintained at the receptor site (Hill & Box, 1999). This suggests that the composition of the vegetation can be manipulated when efficient control of the hydrology is possible.

CONCLUDING REMARKS

For the most part, heathland restoration means growing dwarf shrub vegetation, principally *Calluna*, at dry locations. This has been most successfully achieved when heather has been established following the spreading of mown seed capsules on suitably prepared soils and following the removal of conifers. In some instances, attempts have been made to provide a fuller flora, but there have been almost no attempts to restore wet heathland vegetation, and there have been almost no attempts to restore the fauna. There are a few instances where sand lizard (*Lacerta agilis*) and natterjack toad (*Bufo calamita*) have been translocated to heathlands on which they had become extinct There are also examples of rescue translocations of these and other herptiles (Hodder & Bullock, 1997). However, there are almost no other attempts to introduce other elements of the fauna, particularly invertebrates. Where turf, and in some cases topsoil, is used in restoration many components of the invertebrate fauna may be introduced. Where heathland is restored using seed material, it is normal to allow the fauna to colonise naturally. However, consideration should be give to introducing faunal elements at appropriate stages during a restoration.

The use of grazing stock to restore heathland, mainly from grasslands, is another area that can be developed further and for which research is needed. There have been almost no attempts to reconstruct a suitable topography and to restore ground water conditions. This presents the next challenge, as it will enable a whole range of heathland communities to be restored. Finally, much more thought needs to be given to planning restoration in the early stages. This is particularly the case in areas subject to development or mineral working. Plans for the restoration of heathland must include provision for, and identification of, the resources for conservation management in the long term.

REFERENCES

Aerts, R. & Heil, G. W. (eds.) (1993). *Heathlands: Patterns and Processes in a Changing Environment*. Dordrecht: Kluwer.

Aerts, R., Huiszoon, A., van Oostrum, J. H. A., van der Vijver, C. A. D. & Willems, J. H. (1995). The potential for heathland restoration on formerly arable land at a site in Drente, the Netherlands. *Journal of Applied Ecology*, **32**, 827–835.

Anonymous (1995). *Biodiversity: The UK Steering Group Report*, vol. 2, *Action Plans*. London: HMSO.

Auld, M., Davis, S. & Pickess, B. (1992). Restoration of lowland heaths in Dorset. *RSPB Conservation Review*, **6**, 68–73.

Bakker, J. P. & Londo, G. (1998). Grazing for conservation management in historical perspective. In *Grazing and Conservation Management*, eds. M. F. WallisdeVries, J. P. Bakker & S. E. Van Wieren, pp. 23–47. Dordrecht: Kluwer Academic Publishers.

Bakker, J. P., de Bie, S., Dalinga, J. H., Tjaden, P. & de Vrier, Y. (1983). Sheep grazing as a management tool for heathland conservation and regeneration in the Netherlands. *Journal of Applied Ecology*, **20**, 541–560.

Bobbink, R. & Heil, G. W. (1993). Atmospheric deposition of sulphur and nitrogen in heathland ecosystems. In *Heathlands: Patterns and Processes in a Changing Environment*, eds. R. Aerts & G. W. Heil, pp. 25–50. Dordrech: Kluwer.

Bokdam, J. & Gleichman, J. M. (2000). Effects of grazing by free-ranging cattle on vegetation dynamics in a continental north-west European heathland. *Journal of Applied Ecology*, **37**, 415–431.

Box, J., Coppin, N., Grigg, C. & Hill, A. (1998). Restoration of heathlands. *Mining and Environmental Management*, November 1998, 19–22.

Bullock, J. M. & Pakeman, R. J. (1997). Grazing of lowland heathland in England: management methods and their effects on heathland vegetation. *Biological Conservation*, **79**, 1–13.

Bullock, J. M. & Webb, N. R. (1996). Responses to severe fires in heathland mosaics in southern England. *Biological Conservation*, **73**, 201–214.

Burden, R. F. (1979). Landscape scientist in a county planning office. *Landscape Design*, **126**, 9.

Buttenschon, J. & Buttenschon, R. M. (1982a). Grazing experiments with cattle and sheep on nutrient poor acidic grassland and heath. 1: Vegetation development. *Natura Jutlandica*, **21**, 1–18.

Buttenschon, J. & Buttenschon, R. M. (1982b). Grazing experiments with cattle and sheep on nutrient poor acidic grassland and heath. 2: Grazing impact. *Natura Jutlandica*, **21**, 19–27.

Byfield, A. & Pearman, D. (1996). *Dorset's Disappearing Heathland Flora*. Sandy, UK: RSPB.

Chapman, S. B. (1967). Nutrient budgets for a dry heathland ecosystem in southern England. *Journal of Ecology*, **55**, 677–689.

Davy, A. J., Dunsford, S. J. & Free, A. J. (1998). Acidifying peat as an aid to the reconstruction of lowland heath on arable soil: lysimeter experiments. *Journal of Applied Ecology*, **35**, 649–659.

De Graaf, M. C. C., Verbeek, P. J. M., Bobbink, R. & Roelofs, J. G. M. (1998). Restoration of species-rich dry heath: the importance of the appropriate soil conditions. *Acta Botanica Neerlandica*, **47**, 89–111.

Diemont, W. H. & Linthorst Homan, H. D. M. (1989). Re-establishment of dominance by dwarf shrubs on grass heath. *Vegetatio*, **85**, 13–19.

Diemont, W. H., Webb, N. R. & Degn, H.-J. (1996). A pan-European view on heathland conservation. In *Proceedings of the National Heathland Conference 1996*, pp. 21–32. Peterborough, UK: English Nature.

Dunsford, S. J., Free, A. J. & Davy, A. J. (1998). Acidifying peat as an aid to the reconstruction of lowland heath on arable soil: a field experiment. *Journal of Applied Ecology*, **35**, 660–672.

Ellenberg, H. (1974). Indicator values of vascular plants in Central Europe. *Scripta Botanica*, **9**, 1–97.

Ford, M. A. & Williams, C. M. (1994). On establishment of heath vegetation on ex-arable land. *Aspects of Applied Biology*, **40**, 563–570.

Gilbert, O. L. & Anderson, P. (1998). *Habitat Creation and Repair*. Oxford: Oxford University Press.

Gillham, D. A. & Putwain, P. D. (1977). Restoring moorland disturbed by pipeline installation. *Landscape Design*, **119**, 34–36.

Gimingham, C. H. (1972). *Ecology of Heathlands*. London: Chapman & Hall.

Gimingham, C. H. (1992). *The Lowland Heathland Management Handbook*. Peterborough, UK: English Nature.

Hill, A. & Box, J. (1999). Experimental translocation of wet heath in Dorset, England: implications for restoration of mineral workings. In *Mineral Planning in Europe*, eds. E. E. Fuchs, M. R. Smith & M. J. Arthur, pp. 342–351. Nottingham, UK: Institute of Quarrying.

Hill, M. O. & Stevens, P. A. (1981). The densities of viable seed in soils of forest plantations in upland Britain. *Journal of Ecology*, **69**, 693–709.

Hodder, K. H. & Bullock, J. M. (1997). Translocation of native species in the UK: implications for biodiversity. *Journal of Applied Ecology*, **34**, 547–565.

Holliday, R. J., Gillham, D. A., Putwain, P. D. & Hogg, W. (1979). The restoration of heather moorland following severe disturbance: a case study of the installation of a gas pipeline in the Pentland Hills, near Edinburgh. *Landscape Design*, **126**, 33–36.

Jackson, L. L., Lopoukhine, N. & Hillyard, D. (1995). Ecological restoration: a definition and comments. *Restoration Ecology*, **3**, 396–397.

Legg, C. J., Maltby, E. & Proctor, M. C. F. (1992). The ecology of severe moorland fire on the North York Moors: seed distribution and seedling establishment of *Calluna vulgaris*. *Journal of Ecology*, **80**, 737–752.

Marrs, R. H. (1985). Techniques for reducing soil fertility for nature conservation purposes: a review in relation to research at Roper's Heath, Suffolk, England. *Biological Conservation*, **34**, 307–332.

Marrs, R. H. (1987). Studies on the conservation of lowland *Calluna* heaths. I. Control of birch and bracken and its effect on heath vegetation. *Journal of Applied Ecology*, **24**, 163–175.

Marrs, R. H. (1993). Soil fertility and nature conservation. *Advances in Ecological Research*, **24**, 241–300.

Marrs, R. H. & Gough, M. W. (1989). Soil fertility: a potential problem for habitat restoration. In *Biological Habitat Reconstruction*, ed. G. P. Buckley, pp. 29–44. London: Belhaven Press.

Marrs, R. H., Snow, C. S. R., Owen, K. H. & Evans, C. E. (1998). Heathland and acid grassland creation on arable soils at Minsmere: identification of potential problems and a test of cropping to impoverish soils. *Biological Conservation*, **85**, 69–82.

Merrilees, D. W., Tiley, G. E. D. & Gwyne, D. C. (1995). Restoration of wet heathland after open-cast mining. In *Restoration of Temperate Wetlands*, eds. B. D. Wheeler, S. C. Shaw, W. J. Fojt & R. A. Robinson, pp. 523–532. Chichester, UK: John Wiley.

Michael, N. (1993). *The Lowland Heathland Management Booklet*. Peterborough, UK: English Nature.

Noirfalise, A. & Vanesse, R. (1976). *Heathland of Western Europe*, Nature and Environment Series no. 12. Strasbourg: Council of Europe.

Owen, K. M. & Marrs, R. H. (2000a). Creation of heathland on former arable land at Minsmere, Sufffolk, UK: the effects of soil acidification on the establishment of *Calluna* and ruderal species. *Biological Conservation*, **93**, 9–18.

Owen, K. M. & Marrs, R. H. (2000b). Acidifying arable soils for the restoration of acid grasslands. *Applied Vegetation Science*, **3**, 105–116.

Owen, K. M., Marrs, R. H., Snow, C. S. R. & Evans, C. E. (1999). Soil acidification: the use of sulphur and acidic plant materials to acidify arable soils for the restoration of heathland and acidic grassland at Minsmere, UK. *Biological Conservation*, **87**, 105–121.

Pakeman, R. J. & Hay, E. (1996). Heathland seed banks under bracken *Pteridium aquilinum* (L.) Khun and their importance for revegetation after bracken control. *Journal of Environmental Management*, **47**, 329–339.

Putwain, P. D. & Rae, P. A. S. (1988) *Heathland Restoration: A Handbook of Techniques*. Southampton, UK: British Gas.

Putwain, P. D., Gillham, D. A. & Holliday, R. J. (1982). Restoration of heather moorland and lowland heathland with special reference to pipelines. *Environmental Conservation*, **9**, 225–235.

Pywell, R. F., Webb, N. R. & Putwain, P. D. (1994). Soil fertility and its implications for the restoration of heathland on farmland in southern Britain. *Biological Conservation*, **70**, 169–181.

Pywell, R. F., Webb, N. R. & Putwain, P. D. (1995). A comparison of techniques for restoring heathland on abandoned farmland. *Journal of Applied Ecology*, **32**, 400–411.

Pywell, R. F., Putwain, P. D. & Webb, N. R. (1996a). Establishment of heathland vegetation on mineral workings. *Aspects of Applied Biology*, **44**, 285–292.

Pywell, R. F., Webb, N. R. & Putwain, P. D. (1996b). Harvested heather shoots as a resource for heathland restoration. *Biological Conservation*, **75**, 247–254.

Pywell, R. F., Putwain, P. D. & Webb, N. R. (1997). The decline of heathland seed populations following conversion to agriculture. *Journal of Applied Ecology*, **34**, 949–960.

Rebane, M., Wynde, R., Diemont, W. H., Jensen, F. P., Pahlsson, L. & Webb, N. R. (1997). Lowland Atlantic heathland. In *Habitats for Birds in Europe: A Conservation Strategy for the Wider Environment*, eds. G. M. Tucker & M. I. Evans, pp. 187–202. Cambridge: Birdlife International.

Rodwell, J. S. (ed.) (1991). *British Plant Communities*, vol. 2, *Mires and Heaths*. Cambridge: Cambridge University Press.

Roelofs, J. G. M., Bobbink, R., Brouwer, E. & de Graff, M. C. C. (1996). Restoration ecology of aquatic and terrestrial vegetation on non-calcareous soils in the Netherlands. *Acta Botanica Neerlandica*, **45**, 517–541.

Rose, R. J. & Webb, N. R. (1994). The effects of temporary ballast roadways on heathland vegetation. *Journal of Applied Ecology*, **31**, 642–650.

Small, D. J. (1995). Propagation of hardy heathers in the garden. *Yearbook of the Heather Society*, 1995, 23–26.

Smith, R. E. N., Webb, N. R. & Clarke, R. T. (1991). The establishment of heathland on old fields in Dorset, England. *Biological Conservation*, **57**, 221–234.

Snow, C. S. R. & Marrs, R. H. (1997). Restoration of *Calluna* heathland on a bracken *Pteridium*-infested site in northwest England. *Biological Conservation*, **81**, 35–42.

Standen, V. & Owen, M. J. (1999). An evaluation of the use of blanket bog vegetation for heathland restoration. *Applied Vegetation Science*, **2**, 181–188.

Symes, N. (1999) Techniques for re-establishing heathland: examples from RSPB work in Dorset. In *Heathland Management of North-West Europe, 43rd Eurosite Workshop*, pp. 103–106. Sandy, UK: RSPB.

Urbanska, K. M. (2000). Environmental conservation and restoration ecology: two facets of the same problem. *Web Ecology*, **1**, 20–27.

Van den Bosch, J. & Bakker, J. P. (1990). The development of vegetation patterns by cattle grazing at low stocking density in the Netherlands. *Biological Conservation*, **51**, 263–272.

WallisdeVries, M. F., Bakker, J. P. & Van Wieren, S. E. (eds.) (1998). *Grazing and Conservation Management*. Dordrecht: Kluwer.

Watt, A. S. (1955). Bracken versus heather, a study in plant sociology. *Journal of Ecology*, **43**, 490–506.

Webb, N. R. (1986). *Heathlands*. London: Collins.

Webb, N. R. (1997). The development of criteria for ecological restoration. In *Restoration Ecology and Sustainable Development*, eds. K. M. Urbanska, N. R. Webb & P. J. Edwards, pp. 133–158. Cambridge: Cambridge University Press.

Webb, N. R. (1998a). The traditional management of European heathlands. *Journal of Applied Ecology*, **35**, 987–990.

Webb, N. R. (1998b). History and ecology of European heathlands. *Transactions of the Suffolk Naturalists' Society*, **34**, 1–8.

Webb, N. R. & Pywell, R. F. (1992). Heathland restoratoration: the potential of old fields. In *Heathland Habitat Creation*, eds. A. J. Free & M. T. Kitson, pp. 48–60. Gloucester, UK: Nuclear Electric.

Webb, N. R. & Vermaat, A. H. (1990). Changes in vegetational diversity on remnant heathland fragments. *Biological Conservation*, **53**, 253–264.

Williams, C. M., Ford, M. A. & Lawson, C. S. (1996). The transformation of surplus farmland into semi-natural habitat. 2: On the conversion of arable land to heathland. *Aspects of Applied Biology*, **44**, 185–192.

Woodrow, W., Symes, N., Auld, M. & Cadbury, J. (1996). Restoring Dorset's heathland: the RSPB Dorset heathland project. *RSPB Conservation Review*, **10**, 69–81.

18 • Calcareous grasslands

MICHAEL J. HUTCHINGS AND ALAN J. A. STEWART

INTRODUCTION

The conservation significance of calcareous grasslands lies in the remarkable richness and density of species which they support. Calcareous grasslands are among the most botanically diverse communities in Europe, often with between 20 and 50 vascular plant species per square metre (a figure which may rise to 80 species per square metre if lower plants are included). Some 700 species of plant have been recorded in calcareous grasslands, one-third of which are confined to this habitat type (Willems, 1990). Amongst this long list, a significant number of species are rare or threatened, and many carry Red Data Book status. In recognition of this, chalk grasslands are designated as priority habitats in the European Commission Habitats and Species Directive (92/43 EEC).

Calcareous grasslands also support very diverse assemblages of animals that include many rare species which are of conservation concern. Most of this diversity is to be found amongst the invertebrates, whereas the richness of the vertebrate fauna of calcareous grasslands tends to be low. Variations in the composition of invertebrate communities on calcareous grasslands reflect a range of underlying factors, especially the diversity of the vegetation (in terms both of species composition and physical structure), physical influences such as geographical location, soils and topography, and site management, both current and historical (Kirby, 1992; Morris, 2000).

Calcareous grasslands develop on soils derived from parent material composed primarily of calcium carbonate. The most familiar of such geological formations are the various types of limestone, but calcareous grasslands are also to be found on chalky boulder clays, loessic soils and calcium-enriched sand dunes. Amongst the lithologies known generically as limestones, a useful distinction can be made between the soft chalk formations from the Upper Cretaceous period (100 Ma) and the older and harder true limestones from the Jurassic and Carboniferous periods (190–390 Ma). Whilst limestones occur world-wide, the distribution of chalk is much more restricted, with a focus in northwest Europe. The most extensive chalk deposits are to be found in England and northern France, although smaller outcrops occur in Belgium, Denmark, The Netherlands, Germany and southern Sweden (Smith, 1980).

The soils, and therefore the vegetation types, that form on limestone geology share remarkable similarities, although of course subtle variants arise on different limestone types and under different climatic conditions. It is undoubtedly because of their considerable ecological interest that most research on the restoration of calcareous grasslands has been done on communities developing over chalk substrates. Consequently, this chapter draws heavily on the results of attempts to restore chalk grassland, since many of the principles have general applicability to all calcareous grasslands.

The physicochemical properties of soils that develop over calcareous bedrock are both distinctive and relatively constant between sites. With a chemical composition that is substantially calcium carbonate, soils are moderately to highly alkaline with a pH range from 6.5 to 8.5. The permeable nature of the bedrock means that most chalk soils are well drained (often excessively so) except where superficial clay deposits or other drifted material impede drainage locally. In consequence, chalk-based soils are generally well aerated and often reach high temperatures in summer, especially on south-facing aspects. The main soil type to develop under such conditions is the rendzina, which is characterised by very limited depth (often no more than 30 cm), and by a transition from the humic A horizon straight into the underlying shattered chalk without an intervening B horizon. Apart from the high calcium ion content, the most important chemical feature of chalk soils is that the major plant nutrients, particularly nitrogen and phosphorus, are only available in low concentrations.

An appreciation of the factors that have created and maintained the high diversity of calcareous grassland is essential if we are to attempt its restoration after the original habitat has been destroyed or degraded. The structure and composition of the vegetation of calcareous grasslands are determined by the combined influences of soil factors and both past and present management. The low nutrient status, shallowness and free-draining nature of rendzina soils present plants with a comparatively harsh edaphic environment. The latter factor favours drought-tolerant species, many of which are perennials with deep roots and xeromorphic foliage (i.e. leaves have structural adaptations that help to conserve water). The low nutrient status of the soil favours species with traits that generate slow dynamics, including low relative growth rates, slow turnover of individual plant parts and long life spans. These characteristics result in conservative use of mineral nutrient resources that are in short supply.

Another important set of plant traits is associated with adaptations to grazing. These include small stature, rosette or tillering habits and, in some species, chemical defences against herbivores. The combination of low soil nutrient status and repeated disturbance and defoliation by herbivores greatly restricts plant growth and often prevents establishment of more competitive plant species with fast growth rates and large biomass. When competitive species do establish, the same environmental characteristics prevent them from outcompeting slow-growing and less competitive species. The small stature of many of the latter species enables dense packing of species and hence greater species richness per unit area. If left ungrazed, most

calcareous grasslands would succeed to scrub, developing ultimately into climax woodland. Thus, grazing seeks to arrest this successional process by maintaining a dynamic plagioclimax grassland community.

Many of the best permanent calcareous grasslands are found on very steep slopes. In many cases these slopes have been cultivated at some time in the past, but were abandoned because of the difficulty of working on them, producing 'tumbled-down' land (Smith, 1980). At the end of the fifteenth century wool production became profitable enough to allow many of these steep slopes to be used once again, but as permanent pasture for sheep, while grain and winter fodder-crop cultivation was transferred to the valley bottoms and less steep slopes. It was common for sheep to be grazed on the steep slopes during the day, but at night they were herded into pens in the valleys, where they would urinate and defaecate. Thus, nutrients were exported via the grazing animals from the already nutritionally impoverished slopes to the areas cultivated for crops. The grazing activities of the sheep consequently reinforced the effects of the low soil nutrient status, promoting high species diversity and density, and also arrested succession.

Most studies of calcareous grassland restoration are both recent and of limited duration, so there is much ignorance about what may ultimately be achievable. Most information to date suggests that there are major problems to be overcome and that satisfactory results, if at all possible, cannot be achieved rapidly. Studies of the Porton Downs, UK, have demonstrated that ecologically valuable calcareous grassland communities develop only after 100–200 years without major habitat disturbances (Wells et al., 1976).

LOSSES AND DEGRADATION

Although calcareous grasslands have very high ecological value, several factors, both natural and anthropogenic, have caused a considerable decline in their area. Initially, the greatest threats came from wholesale habitat destruction, as many sites were lost to urban development, transport infrastructure or intensive cultivation. In the UK, much chalk grassland was ploughed during the two world wars and the land cultivated for food. From the 1940s the availability and use of cheap artificial fertilisers enabled crops to be grown repeatedly on many marginal soils on chalk, and fertilisers were even added to permanent chalk-based pastures to improve their productivity. Latterly, many remaining sites have been given formal protection and so the principal threat has become habitat degradation rather than destruction.

According to Blackwood & Tubbs (1970), sheep pasture remained on only 4% of the chalk of the South Downs in southern England. A quarter of this remnant was lost, largely to arable cultivation, by 1984 (Anonymous, 1984a). Comparable reductions in size and number of remaining fragments of calcareous grassland, and in their degree of isolation, have been reported from other parts of Europe (Wolkinger & Plank, 1981; Keymer & Leach, 1990).

Significant declines in species diversity have attended this habitat loss, to the extent that many local populations of a large number of species have gone extinct. It is particularly disturbing that such extinctions have taken place even in sites that have been afforded protection and where habitat quality and traditional forms of management have been maintained (Fischer & Stöcklin, 1997). For example, by the early 1980s, less than 20% of the chalk grassland that remained in England in 1939 retained its former floristic richness (Anonymous, 1984b: Ratcliffe, 1984). The remaining areas of unimproved chalk grassland have suffered the familiar consequences of progressive fragmentation, becoming smaller and more isolated from each other. Many relict sites are now confined to steeper slopes, these being the only areas to have escaped agricultural intensification.

As a consequence of their mid-successional position, remnant calcareous grasslands are now threatened, ironically, both by agricultural intensification and by neglect. The challenge for conservationists is to sustain the grassland in a dynamic balance by offsetting the natural forces of succession with appropriate management. In agriculturally intensive landscapes, the greatest threats have historically been either from attempts to increase the productivity of the grassland as forage for grazing animals or from wholesale conversion to arable production. In less intensively managed areas, and even on many nature reserves, the greatest threat has come from relaxation of grazing pressure, resulting in scrub invasion.

Scrub encroachment

If grazing or other management pressure is weak, it may be insufficient to prevent encroachment by invasive grasses (such as *Brachypodium pinnatum*) or woody perennials (principally hawthorn [*Crataegus monogyna*], but also gorse [*Ulex europaeus*], elder [*Sambucus nigra*] and blackthorn [*Prunus spinosa*]) into the species-rich grassland. This may happen because grazing has ceased or has been relaxed to a point where it can no longer arrest succession. Changes in grazing pressure may arise from the varying fortunes of native herbivores or from long-term changes in agricultural economics. The outbreak of myxomatosis in rabbits in the 1950s in Britain demonstrated how the removal of grazing pressure allows the invasion of scrub on many calcareous grassland sites. Additionally, many sites that were once sheep- or cattle-grazed have been abandoned by agriculture because livestock rearing on such poor pastures is no longer economically viable.

Scrub encroachment into grassland is generally viewed by botanists as detrimental, because the domination of the community by a few large woody species suppresses others in the field and ground layers and thereby reduces the overall plant diversity. However, scrub communities are not without their ecological interest, since they are important refuges for passerine birds and have a distinctive invertebrate fauna (Duffey *et al.*, 1974). For the more mobile insects, the ecotone between open grassland and dense scrub is especially valuable as it provides the juxtaposition of a warm flower-rich habitat for feeding with dense protective scrub for shelter (Kirby, 1992). For this reason, a more pluralistic management objective would be to create and maintain a spatially heterogeneous mixture of grassland and scrub (Jones-Walters, 1990).

Nutrient enrichment

The most common method used to improve the quality of pasture for domestic grazing animals involves elevation of soil nutrient levels through the addition of inorganic fertilisers. The dual intention is to raise productivity and to shift the balance of species within the community in favour of those that are most palatable to stock. It is a widely accepted principle of grassland management that there is a strong negative correlation between productivity and species richness. Raising soil nutrient levels provides most benefit to fast-growing competitive grass species at the expense of slower growing and less competitive species, with a resultant decline in overall species richness (Smith, 1994). This has been convincingly demonstrated on calcareous grasslands across a range of climatic conditions (Smith *et al.*, 1971; Willems & van Nieuwstadt, 1996). This method of grassland intensification has resulted in severe reductions in the conservation interest of many calcareous grassland sites across Europe.

Nutrient levels in the soils beneath many chalk grasslands have recently been increased by fertiliser runoff from adjacent arable land and by atmospheric deposition, particularly of nitrogen. This has changed both the species composition and productivity of the communities they support (Bobbink & Willems, 1987; Bobbink *et al.*, 1988; Willems, 1990). Species that survive under these conditions tend to be more typical of mesotrophic grasslands, such as dandelion (*Taraxacum officinale*), greater plantain (*Plantago major*), cocksfoot (*Dactylis glomerata*), meadow buttercup (*Ranunculus acris*) and daisy (*Bellis perennis*).

Reseeding

A more invasive method of pasture improvement involves the direct modification of plant species composition through the introduction into the sward of high-yielding grass cultivars. These are invariably varieties of species that are either not natural components of the calcareous grassland community, e.g. Italian ryegrass (*Lolium multiflorum*) or species that occur at low frequencies, such as perennial ryegrass (*L. perenne*), Yorkshire fog (*Holcus lanatus*) and bent grasses (*Agrostis* spp.). Seed is introduced into the community by oversowing or slot-seeding, with or without herbicide treatment to remove the dicotyledonous species.

CONSTRAINTS ON RESTORATION

Restoration of calcareous grassland on bare chalk by natural processes can take decades or longer. The chief limitations are lack of propagules and their slow rate of immigration, the physical hostility of the environment, and the deficiency of essential resources (Bradshaw, 1997). The following specific constraints need to be addressed when planning a restoration programme.

Soils and soil fertility

Without soil, ecosystem functions cannot be supported, so that in restoration on bare rock, it is usually necessary to kick-start the process by adding soil, soil organisms (particularly mycorrhizae), and some plants, particularly legumes for nitrogen fixation. Without such action primary succession will be a much slower process.

Atmospheric deposition of nutrients

Another source of nutrients that may cause significant eutrophication in environments where soils have naturally low nutrient contents, such as calcareous grasslands, is atmospheric deposition, particularly of nitrogen. Bobbink & Willems (1987) estimated that at least 50 kg ha^{-1} y^{-1} of nitrogen was deposited in South Limburg, The Netherlands, and that this was responsible for a significant increase in the dominance of the competitive tor grass (*Brachypodium pinnatum*) and a concomitant decline in numbers of associated species. Rates of deposition are lower in the UK (Hurst & John, 1999a), where increases in *B. pinnatum* are believed to be due more to a decline in grazing pressure (Wilson *et al.*, 1995). Management cannot prevent the addition of nutrients to chalk soils via atmospheric deposition, but the solution to changes in community composition will involve any of the techniques to lower productivity and reduce the ability of aggressive species to dominate the vegetation.

Availability of plant propagules

The availability of propagules of species characteristic of calcareous grassland communities is a major constraint in the restoration of these communities. There are two possible natural sources of suitable species, namely the seed bank and the seed rain.

Seed bank

It is well known that in most habitats there are considerable differences between the composition of growing vegetation and that of the seed bank in the underlying soil (e.g. Harper, 1977; Thompson & Grime, 1979). As succession proceeds, this discrepancy increases and the density and species richness of the seed bank declines. Beneath calcareous grassland there is little resemblance between the composition of the seed bank and that of the growing vegetation, and few calcareous grassland species accumulate persistent seed banks under encroaching scrub (Dutoit & Alard, 1995; Davies & Waite, 1998). Both number of seeds and number of species in the seed bank declines as the age of scrub increases (Box 18.1), a situation which appears to be similar in other grasslands. Bekker et al. (1997) demonstrate that restoration of species-rich grassland vegetation is easier following short rather than long periods of agricultural improvement or habitat degradation. In many sites, however, restoration may only be possible by deliberately sowing desirable species or introducing pot-grown transplants (Dutoit & Alard, 1995; Hutchings & Booth, 1996a; Davies & Waite, 1998). Several studies have shown that the seeds of many characteristic species of chalk grassland communities do not persist for long as viable, dormant seeds in soil, especially during cultivation (Graham & Hutchings, 1988a; Hutchings & Booth, 1996a). Thus, the seed bank appears to offer little potential for restoring species-rich calcareous grassland vegetation.

Compared with arable weed species and grasses with low habitat specificity, few seedlings of chalk grassland specialists germinate from the seed bank. Because of their slow growth rate and low stature,

many of these seedlings are likely to suffer competitive exclusion, especially as they will often be growing on sites where soil nutrient content has been elevated above its background level.

Seed rain

Few studies have explicitly investigated either the seed rain in chalk grassland habitats or the distances to which seeds of chalk grassland species can be dispersed. Verkaar et al. (1983) examined seed dispersal in some short-lived chalk grassland species, and estimated that maximum dispersal distances ranged from 0.3 to 3.5 m. Most seeds of all species were deposited very close to the parent plant. Consequently, as in many habitat types (e.g. Thompson, 1986), distribution of seeds of species in the soil seed bank showed strong spatial heterogeneity, and was closely associated with the sources of dispersing seeds. The dispersal range was also affected by the structure of vegetation surrounding source inflorescences. Inflorescence heights, and therefore the heights from which most seeds were released, were greater in plants in tall vegetation than in more open areas. For species with tall inflorescences, this gain in height more than offset the reduction in dispersal distances caused by lower wind velocity within the vegetation layer, leading to wider dispersal of their seeds. For species with short inflorescences, the gain in height of seed release in tall vegetation did not compensate for the reduction in wind velocity within the vegetation canopy, and dispersal ranges diminished. Consequently, the distribution in the soil of seeds of species with short inflorescences was more patchy, and that of species with taller inflorescences was less patchy, in taller vegetation.

Jefferson & Usher (1989) studied the composition of the seed rain in two disused chalk quarries. They showed that whereas seeds of nearly all species in the vegetation were present in the seed rain, few other species were recorded. In both quarries the seed rain was strongly dominated by a small number of species (85% of all seeds were from 6 of the 31 species recorded at one site and from 11 of the 34 species recorded at the other site), but the dominant species in the seed rain were different at the two sites. In both cases, there was a strong similarity

Box 18.1   Seed bank size and composition on soil beneath developing scrub vegetation on chalk

Davies & Waite (1998) investigated the size and composition of the germinable seed bank in soil on a transect running from ancient chalk grassland through developing scrub to mature woodland. Few species germinated from all sites along the transect and two species (perforate St John's wort [*Hypericum perforatum*] and common centaury [*Centaurium erythraea*]) together accounted for two-thirds of germinating seedlings. There was little correspondence between the composition of the seed bank and the overlying vegetation. Both number of species that germinated from the seed bank samples, and total number of seedlings that emerged, decreased from the grassland to the wooded end of the transect (Fig. B18.1). The

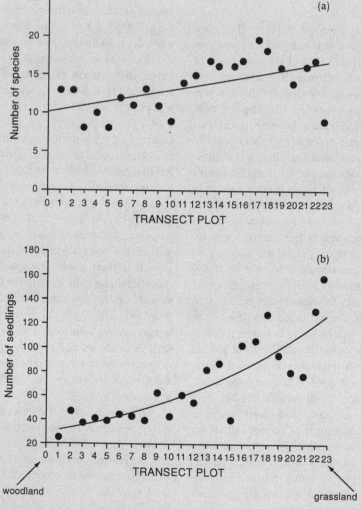

Fig. B18.1. The relationship between (a) species richness along a transect from rabbit-grazed calcareous and (b) size of seed bank, as measured by the number grassland (plot 23) to incipient woodland (plot 1). From of seedlings germinated from soil samples, at points Davies & Waite (1998).

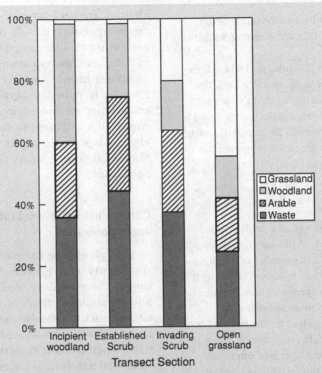

Fig. B18.2. The percentage contribution of species associated with four principal habitat types (grassland, woodland, waste, arable) to the germinable seed bank over four sections of the transect from open rabbit-grazed calcareous grassland to incipient woodland shown in Fig. B18.1. The four sections of the transect are of approximately equal length. From Davies & Waite (1998)

percentage contribution of grassland species in the germinable seed bank fell sharply under developing scrub, until these species represented approximately 1% of germinating seeds in soil beneath established scrub and incipient woodland (Fig. B18.2) Species characteristic of arable cultivation and of waste habitats dominated the seed bank over all parts of the transect.

between the species in the local vegetation and the dominant species in the seed rain. The mean density of the seed rain at the two sites was 540 and 420 m^{-2} per month. A study in an abandoned arable site in a valley surrounded by chalk grassland vegetation recorded both higher densities of seeds and wide variations in density from month to month (685–2321 seeds m^{-2} per week: Hutchings & Booth, 1996b). In this case, three grass species accounted for more than half of the seeds trapped, and grasses and sedges accounted for over 70%. The seed rain of most species displayed spatial patchiness, suggesting that the spatial distribution of adult plants was patchy. Analyses of the patterns for individual species suggested increasingly localised seed rain distribution patterns for grasses, weedy forbs and chalk grassland forbs, reinforcing the conclusions from other studies regarding the poor dispersal capabilities of many chalk grassland species (Box 18.2).

The distribution of dispersing seeds around their parent plants is also strongly influenced by the direction of the prevailing wind. Gibson et al. (1987) report colonisation by species of calcareous grassland against the direction of the prevailing wind. This also happened in the study by Verkaar et al.

Box 18.2   Seed rain in an ex-arable field bordering undisturbed calcareous grassland

Hutchings & Booth (1996b) recorded seed rain at 10 m intervals along transects in an ex-arable field. The tendency of different species to exhibit widespread or localized seed rain characteristics was analysed by applying Levin's (1968) formula for calculating niche breadths. For all commonly recorded species, dispersal breadth (B_{ij}) was calculated as:

$$B_{ij} = 1 / \left[\sum_{j=1}^{j=n} p_{ij}^2 \right]$$

where p_i is the proportion of all recorded seeds of species i recorded at each of the sampling points, j, on the transects, and n is the number of sampling points on the transects (7 in this case). Values of B_{ij} were calculated for each species on each transect using the numbers of seeds recorded at each sampling point, and the values for the two transects were averaged. Values could range from 1, when all the seeds of a species, no matter how numerous, were all recorded at the same sampling point, to 7, if the seeds of the species were evenly distributed between all sampling points. The mean dispersal breadths for grasses, 'weedy' forbs and calcareous grassland forbs were 4.13 ± 1.08, 2.87 ± 1.24 and 2.01 ± 0.95 respectively. These values were all significantly different.

These results indicate that, of these categories, grasses will disperse most widely, and calcareous grassland forbs least widely, into ex-arable sites following the cessation of cultivation. Taken together, (1) the relative absence of perennial forbs characteristic of calcareous grassland vegetation in the seed bank in soils of ex-arable sites, and (2) the poor dispersability of chalk grassland forbs, suggest that restoration of calcareous grassland vegetation on ex-arable habitats will be slow unless interventionist management is practised.

abilities are very limited. Thus, even a short distance separating islands of chalk grassland from each other in a landscape that has been converted to other uses, such as arable crop production, poses an almost insurmountable obstacle to colonisation when it is reliant on gravity or wind. Moreover, colonisation of unoccupied sites from adjacent occupied sites will often be short-lived, because it usually involves single plants or a few plants, rather than enough individuals to establish a sustainable population.

Constraints on invertebrate dispersal and colonisation

As with plants, the composition of the invertebrate community on a restoration scheme will be a function of the prevailing conditions on the site itself and the ability of individuals dispersing from other sites to reach it, colonise and build up a viable population. All invertebrates are likely to be very sensitive to the microclimatic conditions, especially temperature, which in turn are strongly influenced both by topography (especially aspect) and by management. For example, temperatures at the soil surface can be up to 10 °C hotter under short than tall vegetation. Successful colonisation by herbivorous insects, many of which are specialists on particular plant species, will of course be dependent upon the presence of their food plants. However, mere presence of the food plant does not guarantee its suitability for an invertebrate herbivore. Whilst many herbivorous insects use powerful olfactory senses to locate even single individuals of their host plant, isolated plants are less likely to be colonised than ones growing in greater abundance. Also, it is reasonable to assume that a certain minimum number or density of plants is needed to sustain a permanent herbivore population. Finally, many herbivorous insects have rather narrow niches, normally utilising only particular parts of the plant and often only when the plant is in an appropriate physiological condition.

Few insect species survive in the soil during cultivation (analogous to the seed bank in plants) and few are dispersed passively into a site (equivalent to seed rain) except members of the aerial plankton (e.g. aphids, some ballooning spiders). Communities

(1983), although dispersal was greater in the direction of the prevailing wind.

According to Verkaar et al. (1983) many short-lived species of chalk grassland are dispersed by gravity or wind. The low point of release of many of these species inevitably means that their dispersal of invertebrates on newly created sites develop almost entirely from whatever colonists actively immigrate from other populations. It is becoming increasingly apparent from studies on a wide range of invertebrate groups that many rare species have very limited powers of dispersal. This may be because they do not have the physical apparatus to disperse (e.g. wings) or because their behaviour is essentially sedentary. Butterflies have furnished the best examples of the latter limitation because they are easy to mark and to follow. Many rare species form small discrete colonies with limited dispersal between them (Thomas, 1984), maximal dispersal distances are modest (Thomas et al., 1992) and individuals are reluctant to cross more than certain distances of unsuitable habitat in order to get to new sites. In smaller and less mobile insect groups, dispersal patterns have to be deduced from the spectrum of successful colonists at new sites. These tend to share a number of life history traits; macroptery (fully winged), multivoltinism (several generations per year) and high fecundity, allowing rapid population increase, and host plant generalism (Morris, 1990). Probably because of these traits, most of these species, in turn, tend to be common and ubiquitous in a range of grassland types. Few restoration schemes have been monitored long enough to determine the time that needs to elapse before rarer invertebrate species can be expected to invade.

Invertebrates at higher trophic levels would be expected to colonise newly created habitats more slowly than their prey, in part due to the need for an adequate food supply to build up. Parasitic insects tend to be specialists on their particular hosts, thus compounding the specificity of their niche. Predatory arthropods, many of which hunt by sight (e.g. spiders, ground beetles), tend to have more generalist feeding habits and are more dependent upon the physical structure of the habitat. Finally, the assemblage of flower-visiting insects may provide essential pollinators for many native grassland plants. These need to include species in a range of size classes (and therefore proboscis lengths) in order to service all the insect-pollinated plant species. For this reason, Corbet (1995) points out that this essential ecological service cannot be replaced simply by bringing in honeybees, as they are not capable of pollinating the full spectrum of flower types. The ability of bees to colonise newly created grassland will depend not only on the presence of flowers with suitable corolla lengths but also on the bees' foraging range.

Conflicting objectives

Conflicting objectives are not uncommon in restoration programmes. However, the problems are particularly acute in the case of calcareous grasslands. First, these grasslands possess high floristic and faunistic diversity, so there are many species, each with particular, and sometimes conflicting, habitat and management requirements that need to be satisfied. Second, remnant calcareous grassland sites are often restricted in size, thus forcing the resolution of the conflicting objectives within a small area. Potential conflicts arise at a number of levels. Management for maximal species richness may be at the expense of species with particular habitat requirements that do not conform to the general pattern for the community. Use of grazing to maintain a species-rich plant community by keeping the vegetation short will reduce the diversity of invertebrates that depend on the three-dimensional structure of the habitat (Morris, 2000). Whilst some species require the warm microclimate provided by a closely cropped sward, maximal invertebrate species richness is attained in taller vegetation. There is even the potential for conflicting habitat requirements between different life history stages within a species. For example, the wart biter cricket (Decticus verrucivorus) requires the warm microclimate provided by short vegetation for maturation of eggs and young nymphs, but also needs coarse vegetation tussocks in which the later nymphal stages and adults can hide from predators (Cherrill & Brown, 1992). In this case, the management objective becomes the creation and maintenance of small-scale heterogeneity in vegetation height.

SPECIFIC RESTORATION TECHNIQUES

Reduction in soil nutrient status by removal of biomass

Since the introduction of subsidy schemes to remove land from cereal production in the early

1980s, there has been interest among ecologists and conservationists in re-establishing calcareous grassland communities on ex-arable land. The ease with which this can be done depends strongly on the length of time for which any site has been cultivated and on the quantity of nutrients that have been added to the soil. Seed banks of many typical calcareous grassland plant species are rapidly depleted during the first years of cultivation (e.g. Bekker *et al.*, 1997), and the nutrient status of the soil is increased by annual fertiliser additions. It is often asserted that soil nutrient status can be quickly reduced by cutting vegetation and removing the clippings. This view needs to be qualified, since its truth appears to depend on the fertility of the soil. In a long-term experiment to examine the effects on chalk grassland of different cutting regimes (cut once, twice or three times annually, with or without removal of clippings), Wells (1980) found no change in floristic composition after ten years. Careful calculations showed that this was because the biomass removed contained very small quantities of nutrients and that any reduction in nitrogen availability due to removal of vegetation and clippings would have been balanced by atmospheric deposition of nitrogen or by nitrogen fixation by legumes. In this experiment, the yields of the vegetation in the different treatments eventually diverged: yields on plots to which clippings were returned significantly exceeded those on removal plots in each of the last four years of the study. After eight years, soils in the treatments with cuttings removed had lower exchangeable Mg and, more importantly, lower soil extractable P levels. After 22 years, Wells's experiment showed that the concentration of P in the soil solution was lower where clippings were removed (Rizand *et al.*, 1989). It might be expected that these differences, particularly in availability of P, would eventually be reflected in changes in vegetation composition. However, nutrient depletion on already infertile soils is clearly a slow process, and fertility may already be so low that further reductions may lower rather than increase the diversity of species that can survive.

When attempting to restore chalk grassland vegetation on sites that have not been managed for a long period, it may be more important to remove cut material. Green (1980) describes an experiment at Lullington Heath (East Sussex, southern England) NNR, where 12-year-old gorse (*Ulex europaeus*) which had invaded chalk-heath was cut, and the cuttings either comminuted and left on site, or removed. The gorse litter contained 500 kg N ha^{-1}. Where cuttings were left to decay, the vegetation that developed included aggressive, fast-growing nitrophiles, such as rosebay willowherb (*Chamerion angustifolium*), and coarse, weedy grasses such as cocksfoot, Yorkshire fog and creeping bent (*Agrostis stolonifera*). Where cuttings were removed, much finer-scale vegetation developed that more closely resembled the original chalk-heath community. Thus, if calcareous grassland restoration is attempted on sites where productivity or standing crop is large, it may be essential to remove clippings in order to lower productivity and enable more diverse vegetation to develop.

Reducing phosphorus levels

Extractable P levels are greater in soil of arable and improved grassland habitats, including those based on calcareous substrates, than in soil of semi-natural grasslands (Gough & Marrs, 1990). Greenhouse studies showed that plant growth on these soils was strongly correlated with the level of extractable P, suggesting that plant growth in semi-natural grassland soils is limited by lower P levels. Thus, P content must be reduced before species-rich vegetation can be established on ex-arable and improved grasslands in set-aside and other extensification schemes. Soils from woodland sites also had higher extractable P levels than soils beneath semi-natural grassland, implying that the same may be true of former grassland sites that have developed to later successional stages. Measurements from an experiment in which crops were removed continuously for nearly 80 years, with no fertiliser inputs, showed that extractable P in the soil had been reduced by only 40% compared with controls (Gough & Marrs, 1990). This experiment was carried out on clay, where nutrient levels may fall more slowly than on chalk. However, the rate of soil nutrient depletion can clearly be very slow.

Nutrient addition

Where soil nutrient status is high, addition of nitrogen (a nutrient that can easily be controlled in soil), and sowing of a cereal crop can be beneficial in lowering soil fertility (Marrs, 1993; Marrs, volume 1). Such treatments can yield productive crops, which can take up problem nutrients, such as P, in considerable quantities. In a long-term experiment, more nutrients were removed from the soil in such N-treated plots than from control plots, and the available P in the soil eventually became virtually exhausted (Dyke *et al.*, 1983).

Burning and topsoil stripping

Marrs (1985) has reviewed other ways of lowering soil fertility as part of a management programme aimed at creating or restoring species-rich vegetation. Two of the most dramatic methods are burning of crops or stubble, and topsoil stripping (see Marrs, volume 1). Burning volatilises much of the N and between 20% and 25% of the P and K contained in the burnt material. The remainder is deposited as ash, and either leached, as in the case of much K, which quickly enters soluble states, or taken up by growing vegetation. The remaining P becomes fixed and unavailable to plants, at least temporarily.

Topsoil stripping has the dual advantage that both the highest fertility and the greatest density of buried, viable seeds, are found in the upper layers of the soil. On fertile ex-arable sites, many of these seeds will be weeds of arable cultivation (Graham & Hutchings, 1988*b*; Hutchings & Booth, 1996*b*). However, it has the disadvantage that the site may look very unsightly for a period. Relocating the topsoil elsewhere may be either a problem or an opportunity to sell a valuable resource, thus raising funds that can defray the costs of restoration work.

Grazing

Grazing by domesticated livestock is the traditional management technique that has helped to create and maintain calcareous grassland over the centuries. Within the overall technique, site managers have considerable flexibility for fine-tuning of grazing, to suit circumstances and to meet particular objectives. The variables that can be adjusted are: (1) the species of grazing animal used, (2) the stocking density, and (3) the seasonality, frequency and duration of the grazing events. There is a sizeable literature on the different effects produced by different grazer species on semi-natural grasslands (Gibson, 1995). Of the two most commonly used grazing animals, sheep crop the vegetation more evenly and to a shorter height than cattle. This tends to produce a sward that is richer in plant species. Cattle graze more heterogeneously and, being heavier animals, break up the sward more extensively. The small patches of bare ground produced by hoof-prints may provide opportunities for colonising plants and are known to be important for certain thermophilous invertebrates. The temporal aspects of the grazing events (especially the season in which grazing is imposed) have a considerable influence on the successional course which the plant community will take over time (Box 18.3) and on the immediate impact on invertebrate life cycles. Field experimental evidence suggests that the richness of both plants and invertebrates is maximised by the concentration of grazing in spring and autumn; relaxation of grazing in summer allows flowering and seed set and also coincides with the time of maximal invertebrate activity.

According to Marrs (1985) grazing is not a viable option for reducing soil fertility in chalk grasslands unless the efficiency of the process is increased by removing the animals at night, so that they urinate and defaecate elsewhere. This technique is employed in the 'plaggen' management system in The Netherlands. It has the double benefit that the animals' excreta can be used to fertilise areas where crops are to be grown and, in the process, nutrients are stripped from the grazed areas. The same principle was applied in traditional management of many English chalk grasslands, where sheep were grazed on the slopes during the day and then penned at night in the valley floors. However, this management is very labour-intensive and no longer economically viable even on conservation sites.

Box 18.3   The effects of grazing at different seasons on the course of natural succession on ex-arable land

The consequences of grazing by sheep at different times of year on colonisation and establishment of plant species, and on diversity of vegetation, was examined by Gibson *et al.* (1987) on a calcareous site that was abandoned from cultivation in 1981. In 1984, an experiment was established with four treatments involving grazing at different times, plus a control treatment without grazing. The effects on vegetation development were recorded at intervals until the end of 1996. Several fragments of old calcareous grassland remained within a short distance of the experimental site, from which species of permanent grassland communities might colonise the experimental plots. The recording of a single plant of a species within the experimental plots was used as a liberal criterion by which to judge colonisation. On this basis, colonisation by species was unaffected by grazing treatment, although greater establishment occurred in the grazed treatments. Colonisation occurred against the direction of the prevailing wind. By the end of 1986, 57% of the vascular species restricted to fragments of old calcareous grassland within 2 km of the ex-arable site had colonised the experimental plots. From the start of recording, measurements of diversity were higher in all grazed treatments than in the controls. Grazing allowed more early successional species, including annuals and short-lived perennials, to persist, whereas they were lost from control plots. Species characteristic of permanent grassland tended to increase in grazed plots, but not in controls. Short-term spring grazing produced vegetation of higher diversity, containing more species characteristic of permanent grassland, than short-term grazing in autumn.

In conclusion, appropriate grazing treatments by themselves enabled permanent grassland species to establish on the ex-arable site. These species had apparently arrived from small, local patches of old calcareous grassland. Short-term grazing in spring produced vegetation of higher diversity, containing more species characteristic of permanent calcareous grassland, than did short-term grazing in autumn.

Cutting

In some circumstances, cutting the vegetation followed by removal of the clippings may be an acceptable substitute for grazing. Cutting differs from grazing in two important respects: it is immediate (in that it happens suddenly rather than gradually) and is unselective with respect to plant species. The suddenness of the operation may be catastrophic for comparatively immobile invertebrates. It is also a technique that is difficult to apply on the steeply sloping terrain that typifies much of the remaining species-rich calcareous grassland.

Sowing seed mixtures

A faster and perhaps more reliable route to establishing the desired plant community is to sow a seed mixture, the composition of which reflects the target calcareous grassland community that has been chosen as appropriate for the local climatic and topographic conditions. A number of studies have tested the establishment success of a range of species and precise guidelines exist for the establishment of calcareous grassland communities under different background conditions (Wells *et al.*, 1981).

Wells (1987, 1989) has emphasised the importance of careful soil preparation by removing perennial weeds and producing a fine, stone-free tilth that will maximise seedling establishment. Appropriate site management after initial establishment is also vital. Recommended species mixtures and sowing rates for initiating the process of calcareous grassland re-creation are given in Wells *et al.* (1981) and Wells (1991) (Box 18.4). It is recommended that mixtures should consist (by weight) of 80% grass and 20% forb seeds. The cost per unit area will increase as more species are added to the mixture, placing constraints on what can be used in many re-creation projects.

It cannot be emphasised strongly enough that restoration schemes should only use seeds of local provenance, in order to maintain the integrity of local plant gene pools (see Gray, volume 1). Some wildflower seed companies harvest their material from widely separated populations, resulting in inappropriate genetic mixing. Amongst other problems, this could have serious consequences for insect herbivores that may have developed locally adapted gene complexes that cannot adapt quickly to feeding on host plants from a different area. However, being restricted to obtaining seed only from local populations raises two further dangers: local populations may provide insufficient supplies of seed for large restoration schemes and removal of large amounts of seed may pose risks to the plant populations on the donor sites. Stevenson *et al.* (1995) examined whether progressive reductions of the sowing rates of a chalk grassland mixture would affect either the establishment of the sown species or invasion by weeds. Species richness of chalk grassland species increased, and weeds decreased, over time and with the sowing rate, although it was possible to reduce the latter considerably (from 4.0 to 0.4 g seed m^{-2}) without affecting the outcome after two years in terms of similarity to a nearby undisturbed chalk grassland.

Slot seeding

There is considerable interest, although unfortunately rather little research information, on ways of enriching species-poor grasslands without resorting to destroying the existing vegetation and re-seeding with the desired species mixture. If successful, this would have a number of advantages over attempts to start a new community by sowing seed onto bare soil, not the least of which is the saving on the cost of destroying the previous community and preparing the ground for the new seed. Some species will exhibit low establishment or will fail to establish completely from seed. In such cases it may be more practical and economical to introduce them as pot-grown plants, perhaps a few

---

Box 18.4 Seed mixtures for establishing species-rich grassland on calcareous substrates

Wells (1983, 1989) has described several techniques for creating floristically rich grasslands on calcareous substrates, although none of these is presented as capable of re-creating the structure and proportional species composition of calcareous grassland communities, at least in the short to medium term. The techniques include sowing mixtures of seeds of suitable species, slot seeding into established grass swards, and transplanting pot-grown plants into established grass swards by using a bulb planter. In connection with the use of sown seeds, Wells stresses the importance of careful site and seed bed preparation, the use of nurse crops (see Boxes 18.5 and 18.6), and appropriate management after seedling establishment. Important qualities required in a nurse crop are fast germination and quick growth, the ability to suppress undesirable weed species, and rapid die-back within two years to be replaced by species of greater interest. Nurse crops also improve the aesthetic appearance of a site. Amongst the nurse crops Wells (1983) recommends on soils of lower fertility are the grasses Westerwolds ryegrass (*Lolium multiflorum*), rye (*Secale cereale*), sorghum (*Sorghum vulgare*) and common oat (*Avena sativa*), and annual or biennial legumes such as kidney vetch (*Anthyllis vulneraria*), black medick (*Medicago lupulina*) and hare's-foot clover (*Trifolium arvense*).

Wells *et al.* (1981) and Wells (1991) have published grass/forb seed mixtures that may be sown for the creation of chalk or limestone grassland (Table B18.1) or meadows on calcareous clay soils (Table B18.2). They emphasise that management after germination is essential to produce and maintain a suitable vegetation structure for the persistence of sown species. They also state that 'much more research is required on the effects of the range of techniques available for managing grasslands for nature conservation purposes'. Unfortunately, very little such research has been undertaken in the years since this statement was made.

Table B18.1. *Grass/forb mixture for creating a chalk or limestone grassland*

	Percentage by weight of grasses	kg ha^{-1}
Grasses (8 species)		
Bromus erectus	10	2.4
Cynosurus cristatus	15	3.6
Festuca ovina	25	6.0
F. rubra ssp. pruinosa	20	4.8
Helictotrichon pratense	5	1.2
Koeleria cristata	5	1.2
Poa pratensis ssp. angustifolia	10	2.4
Trisetum flavescens	10	2.4
Total	100	24

	Percentage by weight of forbs	g ha^{-1}
Forbs (25 species)		
Achillea millefolium	4	240
Anthyllis vulneraria	4	240
Centaurea nigra	5	300
C. scabiosa	2	120
Clinopodium vulgare	2	120
Filipendula vulgaris	2	120
Galium verum	3	180
Leontodon hispidus	2	120
Leucanthemum vulgare	15	900
Lotus corniculatus	2	120
Medicago lupulina	2	120
Onobrychis viciifolia	5	300
Ononis spinosa	2	120
Origanum vulgare	2	120
Pimpinella saxifraga	1	60
Plantago lanceolata	5	300
P. media	5	300
Primula veris	4	240
Prunella vulgaris	10	600
Ranunculus bulbosus	2	120
Reseda lutea	4	240
Rhinanthus minor	10	600
Sanguisorba minor	4	240
Silene vulgaris	1	60
Stachys officinalis	2	120
Total	100	6000

Notes: The mixture consists of 80% grasses to 20% forbs by weight. Grasses are sown at a rate of 3914 seeds m^{-2} and forbs at a rate of 894 seeds m^{-2}. Total sowing rate is 30 kg ha^{-1}. Westerwolds ryegrass is recommended as a nurse crop, sown at a rate of 10 kg ha^{-1}.
Source: Wells (1991).

Table B18.2. *Grass/forb mixture for creating a meadow community on calcareous clay soil*

	Percentage by weight of grasses	kg ha^{-1}
Grasses (9 species)		
Agrostis capillaris	10	2.4
Anthoxanthum odoratum	2	0.5
Cynosurus cristatus	15	3.6
Festuca ovina	18	4.3
F. rubra ssp. commutata	15	3.6
Hordeum secalinum	5	1.2
Phleum nodosum	10	2.4
Poa pratensis	10	2.4
Trisetum flavescens	15	3.6
Total	100	24

	Percentage by weight of forbs	g ha^{-1}
Forbs (24 species)		
Achillea millefolium	4	240
Centaurea nigra	5	300
C. scabiosa	2	120
Clinopodium vulgare	2	120
Filipendula vulgaris	2	120
Galium verum	3	180
Hypericum perforatum	2	120
Leontodon hispidus	2	120
Leucanthemum vulgare	15	900
Lotus corniculatus	5	300
Lychnis flos-cuculi	1	60
Malva moschata	4	240
Pimpinella saxifraga	1	60
Plantago lanceolata	5	300
P. media	5	300
Primula veris	4	240
Prunella vulgaris	10	600
Ranunculus acris	10	600
Rhinanthus minor	10	600
Rumex acetosa	3	180
Sanguisorba minor	1	60
Saxifraga granulata	1	60
Silene vulgaris	1	60
Stachys officinalis	2	120
Total	100	6000

Notes: The mixture consists of 80% grasses to 20% forbs by weight. Grasses are sown at a rate of 5795 seeds m^{-2} *and* forbs at a rate of 1133 seeds m^{-2}. Total sowing rate is 30 kg ha^{-1}. Westerwolds ryegrass is recommended as a nurse crop, sown at a rate of 10 kg ha^{-1}.
Source: Wells (1991).

years after an initial sowing of more tractable species. Among species that have been successfully established in this way are quaking grass (*Briza media*), clustered bellflower (*Campanula glomerata*), harebell (*C. rotundifolia*), horseshoe vetch (*Hippocrepis comosa*) and large thyme (*Thymus pulegioides*) (Wells 1991).

Wells *et al.* (1989) experimented with 'slot seeding' of various herbaceous species into mesotrophic grassland. Herbicide was used to clear the vegetation from a narrow strip of ground either side of the seed line, to encourage seedling growth in the absence of competition from the existing sward. A companion experiment examined the potential for introducing pot-grown species as plant 'plugs'. All 16 species of forb tested in these experiments established successfully, but there were significant differences in long-term survival and ability to spread into the adjacent sward. Neither technique is cheap and both would be prohibitively expensive except on comparatively small sites. Much more research is needed on this topic to explore ways of maximising the establishment rate of desirable species.

Control of weeds

Most of the species that constitute species-rich calcareous grassland communities are perennials, and their establishment rates from seed will be both slow and somewhat unpredictable. Undesirable species may invade and overtake the community in the crucial early stages of development before the more sensitive species have had a chance to become fully established. There is some experimental evidence to suggest that more diverse seed mixtures provide better protection against weed invasion in the early stages of succession than less diverse ones (van der Putten *et al.*, 2000), although this effect is highly dependent upon the identity of the species involved. Where weed species such as thistles (*Cirsium* spp.), ragwort (*Senecio jacobaea*) and fat hen (*Chenopodium album*) become established, it may be necessary to apply selective herbicides or to remove the aboveground parts of the problem species.

Use of nurse crops

One potential solution to the weed problem is to sow an annual grass species as a nurse crop, in order to provide an immediate 'green cover' that will suppress weeds while the slower-growing perennials become established. The assumption is that the species used for this will eventually be excluded from the community by the perennials, although the presence of the nurse crop will itself slow the rate of perennial establishment (Stevenson *et al.*, 1995). A further problem arises where nutrient-depleted soils are too infertile to allow good establishment of the nurse crop. Of course, this is unlikely on previously cultivated sites, but may be the case where calcareous soils are exposed after chalk extraction or engineering working. In such cases, an initial nutrient pulse may be required to assist in the establishment of the nurse crop (Mitchley *et al.*, 1996) (Box 18.5).

Several chalk grassland species have higher establishment when sown amongst short vegetation than on bare chalk soil (Hutchings & Booth, 1996a) (Box 18.6). When a nurse crop is used, however, it may be necessary to cut it once or more during the growing season, particularly on more fertile soils, to prevent it from growing so vigorously that it eliminates the less competitive sown species. It is also imperative to cut the nurse crop before it disperses its seeds to ensure its virtual disappearance within two years, enabling more desirable perennials to flourish.

Controlling dominant grasses

Formerly species-rich calcareous grasslands can become dominated by tall competitive grass species such as upright brome (*Bromus erectus*) and tor grass (*Brachypodium pinnatum*). Invasion and spread by the latter species has caused serious reductions in species-richness in many Dutch calcareous grasslands over the last 30 years (Bobbink & Willems, 1987). It remains unclear whether this has been due to the abandonment of grazing or to the long-term eutrophication effects of increased atmospheric nitrogen deposition. Similarly, whether or not a *B. pinnatum*-dominated stand is a semi-stable

Box 18.5   The use of a nurse crop and fertiliser application to establish maritime calcareous grassland on chalk spoil

Almost 4 million m^3 of chalk-marl spoil were excavated from the UK side of the English Channel tunnel workings. This material was used to create a reclamation platform 36 ha in area. Mitchley *et al.* (1996) undertook an experimental study of the establishment of vegetation on this spoil. Perennial ryegrass (*Lolium perenne*) was used as a nurse crop for a sown mixture of ten species characteristic of maritime chalk cliff communities (Table B18.3). Because the marl substrate had a very low fertility it was regarded as necessary to add fertiliser to ensure effective growth of the nutrient-demanding nurse species. However, it was also recognised that too vigorous a growth of the nurse crop might reduce the establishment of the other sown species. It was also a requirement that the nurse crop would quickly decline in vigour as the availability of added fertiliser fell, allowing the other species to expand in cover. A factorial experiment was conducted, using four levels of addition of each of nitrogen and phosphorus, to determine the growth of the nurse crop and establishment of the other species.

The experimental area was hydro-seeded in autumn 1992 with the seed mixture plus a cotton fibre mulch, specific rhizobia to inoculate the legumes, and a slow-release fertiliser added at a rate of 50 g m^{-2} (delivering 3 g m^{-2} N, 8.5 g m^{-2} P_2O_5 and 6 g m^{-2} K_2O). Four levels of nitrogen fertiliser, at rates of 0, 5, 10 and 20 g m^{-2}, and four levels of phosphate, at rates of 0, 3, 6 and 12 g m^{-2}, were added to experimental plots in all combinations in spring 1993.

Results obtained in summer 1993 showed that greater N and P applications produced higher nurse

Table B18.3. *Composition and sowing rate of the grassland seed mixture used by Mitchley* et al. *(1996), including the nurse species* Lolium perenne

Species	Sowing rate (seeds m^{-2})
Agrostis stolonifera	200
Festuca arundinacea	50
Festuca rubra	200
Anthyllis vulneraria	10
Lotus corniculatus	10
Ononis repens	15
Daucus carota	50
Echium vulgare	5
Plantago lanceolata	50
Silene nutans	25
Lolium perenne (nurse species)	2500

crop yields. However, at higher rates of N application the nurse crop lodged, and this, together with greater growth of the nurse crop, resulted in a negative effect of greater N addition on the density of establishment of other species. The legume species, which would themselves fix N, thereby improving soil nutrient status for other species, were particularly suppressed at high N.

In conclusion, for vegetating this reclaimed chalk spoil habitat, Mitchley *et al.* (1996) recommended low fertiliser applications to support low productivity of the nurse crop and maximum establishment of the wildflower mixture. It should be stressed that the vegetation produced by this approach has neither the species composition, nor the structure, of ancient calcareous grassland communities, at least in the medium term.

community that will persist indefinitely or one that is merely a transitional stage before scrub invasion occurs is still a matter of debate.

Brachypodium pinnatum is able to spread clonally through deep rhizomes, and it deposits a dense and persistent litter layer which suppresses seedling regeneration (Hurst & John, 1999a). Attempts

to restore species-rich calcareous grassland from *B. pinnatum*-dominated communities using traditional grazing management have had limited success, as grazing animals avoid its rough leaves and it is resistant even to severe trampling. Mowing as a substitute for grazing has some beneficial effect (Bobbink & Willems, 1991), but neither technique is

Box 18.6   Seed germination in response to vegetation management

Hutchings & Booth (1996b) examined the germination of seeds of four species of calcareous grassland (yarrow [*Achillea millefolium*], burnet-saxifrage [*Pimpinella saxifraga*], hoary plantain [*Plantago media*] and small scabious [*Scabiosa columbaria*]) in the laboratory, in chalk grassland, and in ex-arable sites where the vegetation was either: (1) fully cleared to leave a bare soil surface, (2) cut and maintained at a height of 3 cm, or (3) left intact. Survival of the seedlings in the field plots was also regularly monitored for a total of 22 weeks.

For all species, percentage germination was higher in the laboratory than under any field conditions, and field germination declined from the cleared to the cut to the uncut ex-arable in all species except for *Pimpinella saxifraga*, in which field germination was highest in the cut ex-arable plots. For all species, germination in the chalk grassland plots was lower than in cleared and cut ex-arable plots, but higher than in the uncut ex-arable plots (Table B18.4). Thus, to promote germination, clearance or cutting of vegetation is recommended, at least for these species.

Detailed analysis of the within-species survival of germinated seedlings showed several consistent patterns with significant differences between treatments. First, germination rates beneath uncut vegetation on ex-arable plots were so low that virtually no seedlings appeared throughout the study. Second, survival was greater on both cut and cleared ex-arable plots than on undisturbed chalk grassland. Lastly, and most importantly, survivorship on cut ex-arable plots was commonly greater than on cleared ex-arable plots, suggesting that, although it is important to relieve competition from other vegetation to promote the initial establishment of species, total removal of vegetation can be disadvantageous. It was suggested that a low cover of vegetation ameliorates environmental conditions by preventing the occurrence of extremes of temperature and drought. Thus, short vegetation may act as a nurse for sown seed.

It is important to emphasise that the results of this study indicate that on ex-arable sites the most suitable conditions for promoting germination are not also the best conditions for seedling survival.

Table B18.4. *Mean (± S.E.) percentage germination per plot of seeds of four different species under different conditions in the field, and in the laboratory*

	Achillea millefolium	Pimpinella saxifraga	Plantago media	Scabiosa columbaria
Ex-arable cleared	45.5 ± 8.4	13.6 ± 2.0	32.1 ± 2.5	7.0 ± 0.3
Ex-arable cut	7.6 ± 1.1	21.0 ± 3.0	17.6 ± 1.2	5.8 ± 0.5
Ex-arable uncut	0.8 ± 0.2	1.7 ± 1.0	1.8 ± 0.6	0.1 ± 0.1
Chalk grassland	4.0 ± 1.0	11.0 ± 1.3	8.5 ± 1.2	4.3 ± 0.4
Laboratory	96.8 ± 1.5	19.3 ± 2.5	81.8 ± 4.9	24.3 ± 1.6

able to restore the grassland to its previous composition and richness (Bobbink & Willems, 1993). The seasonality and frequency of management has an important influence on the outcome. Bobbink & Willems (1991, 1993) found that autumn mowing had limited benefits, whereas better results were obtained with bi-annual cuts in spring and summer. Chemical treatment of discrete *B. pinnatum* stands with a broad-spectrum herbicide (glyphosate) produced some increases in species typical of the surrounding chalk grassland (Hurst & John, 1999b). However, *B. pinnatum* subsequently reinvaded the treated patches and was expected eventually to return to its former dominance.

Passive and active dispersal of plant and animal material

Despite earlier comments about the low probability of species dispersing naturally between isolated fragments of calcareous grassland, long-distance dispersal is clearly often important in maintaining and extending species' ranges (Cain *et al.*, 2000). Indeed, Gibson *et al.* (1987) claim that most permanent grassland species that colonised their experimental plots on a calcareous site must have immigrated from elsewhere, as the soil seed bank only contained weeds of arable cultivation. Calcareous grasslands, which owe their existence and maintenance to human management of grazing animals over a very long period, may have acquired much of their species richness as a result of the movements of large flocks of sheep between summer and winter grazing sites. In mainland Europe, this transhumant shepherding often involved moving many thousands of animals over distances up to 100 km (Poschlod *et al.*, 1998). Seeds of chalk grassland species can be dispersed over wide distances on the fur and hooves of animals or, in the case of hard-coated seeds, within their guts. Small seeds in particular are transported by hoof, and considerable numbers of a very wide variety of the species of chalk grassland are transported on the animals' coats. Many seeds can remain attached to animals for long periods, giving the opportunity for very long-distance transport. Fischer *et al.* (1996) calculated that a flock of 400 sheep could transport at least 3 million seeds. Moreover, sheep are also potent agents for the dispersal of several types of invertebrate, particularly grasshoppers and bush crickets. Poschlod *et al.* (1998) argue that the movement of very large numbers of sheep over large distances between fragments of chalk grassland was one important mechanism enabling species to disperse between widely separated suitable patches of habitat. They suggest that the decline in such flock movements has had a significant impact upon the ability of chalk grassland fragments to maintain their species richness, and to gain species following the initiation of restoration programmes. They argue persuasively that programmes to restore and conserve communities should include restoration of the factors that created them, and that, in the case of calcareous grassland communities, this includes these important long-distance dispersal mechanisms. Wild animals, including birds, man and farmland machinery can also, to a more limited extent, act as agents of seed dispersal over smaller distances.

Seed dispersal and the process of recolonisation are clearly major constraints that limit our ability to restore species-rich chalk grassland vegetation (Dutoit & Alard, 1995). The decline in the economic viability of transhumant shepherding (Poschlod *et al.*, 1998) is unlikely to be reversed, but the potential importance of this route for long-distance dispersal and colonisation is clear. Small-scale reintroductions of movements of grazing animals between remnant calcareous grasslands and sites that are being restored may be possible in some cases, and may have considerable benefits.

Translocation of vegetation turves

There is increasing interest in the possibility of translocating grassland and other low vegetation communities by cutting intact turves and re-establishing them at new sites. This technique has invariably been employed as a last resort attempt to save communities that would otherwise be destroyed (e.g. by engineering work or housing development). Bullock (1998) has reviewed a range of cases, including three that refer to calcareous grasslands. At first sight, some of these projects appear to have been quite successful, with over 60%–70% similarity in plant species composition between the translocated turves and control plots. However, post-translocation changes in species representation invariably show rare plant species being lost and replaced by more common ones. Where monitored, invertebrate communities showed greater adverse changes than the plants. In most cases, post-translocation monitoring was too brief to test whether changes in translocated turf were distinguishable from changes over the same period of time in undisturbed turf. Nevertheless, several cases showed progressive divergence over time between the translocated communities and their controls, with the former invariably deteriorating in quality.

Thus, the value of community translocation as a technique for restoring calcareous grasslands is still unproven and should never be treated as a substitute for *in situ* preservation of intact sites.

ROUTES TO RESTORING CALCAREOUS GRASSLANDS

Figure 18.1 illustrates the different processes by which calcareous grasslands have been lost in the past, together with the possible routes towards restoration of the original species-rich grassland.

Reclamation from scrub

There are many instances where dense scrub has invaded species-rich calcareous grassland and has completely excluded all species associated with the original grassland. The question then becomes whether the original grassland community can be restored once the scrub itself has been physically removed. Scrub removal is normally done by cutting and removing the bushes, followed by chemical poisoning of the stumps and clipping of any regrowth.

The soil conditions immediately after scrub removal are likely to be very different from those under grassland. Nutrient enrichment, particularly where the scrub included nitrogen-fixing plants such as *Ulex europaeus*, will encourage colonisation by tall, fast-growing, competitive species such as blackberry (*Rubus fruticosus*), rosebay willowherb and stinging nettle (*Urtica dioica*), which may be difficult to suppress (Duffey et al., 1974). As stated before, the limited number of studies on the composition of seed banks under dense scrub suggest that they do not contain suitable species for restoration of species-rich calcareous grassland (Donelan & Thompson, 1980; Dutoit & Alard, 1995; Davies & Waite, 1998) (Figs. B18.1 and B18.2).

There has been rather little research in western Europe to determine how best to restore species-rich grassland on areas that have been cleared of scrub or woodland. However, some informative studies have been done on calcareous grasslands in northern and eastern Europe, including Estonian alvars (calcareous grasslands on thin soils), that have been overgrown by trees (Kiefer & Poschlod, 1996; Zobel et al., 1996; Partel et al., 1998). Partel et al. (1998)

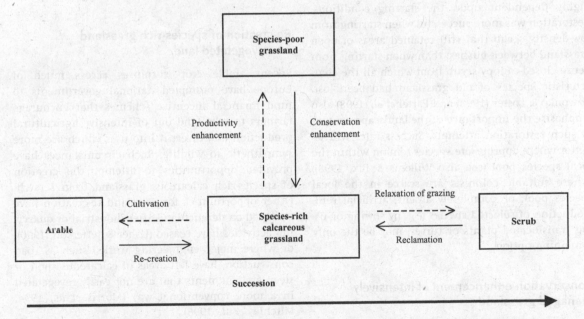

Fig. 18.1. Processes by which calcareous grasslands have been lost (dashed lines) and potential routes for restoration (solid lines). Adapted from Mortimer et al. (1998).

studied the effect of tree removal from a grassland area that had been overgrown by pine (*Pinus sylvestris*) but which still retained some of the original grassland species in the field layer. Seven years after the trees had been removed and the vegetation had been annually grazed or cut, they found that plant species richness resembled that in grassland areas that had remained open. Transplantation of turves from open grassland areas into the cleared plots did not further enhance the level of species richness finally achieved. Dzwonko & Loster (1998) followed the vegetation changes after clearance of a 35-year-old secondary pine forest that had developed on abandoned limestone grassland. After five years, the experimental plots were still significantly poorer in species than the old grassland, and contained a higher proportion of ruderal, annual, biennial and nitrophilous species. Thus, the communities achieved were compositionally very different from that of the old grassland which served as the target.

The most important general conclusion from these studies was that the degree of success in restoring species-rich calcareous grassland from communities dominated by woody plants was highly dependent upon the starting conditions. Restoration was more successful when starting from low-density scrub that still retained areas of open grassland between bushes, than when starting from dense closed-canopy scrub from which all the characteristic species of old grassland had been lost. Dzwonko & Loster (1998) and Partel *et al.* (1998) also emphasise the importance of the landscape context of such restoration attempts. Success rates will be better where appropriate species remain within the local species pool (see also Willems & Bik, 1998). Where suitable colonists are scarce in the local species pool, or completely absent, artificial reintroduction of selected species (e.g. by sowing or using translocated plants or turves) may be the only remaining option.

Conservation enhancement of intensively managed grassland

The first priority in any attempt to restore intensively managed grassland back to a species-rich community must be to lower soil nutrient levels and reduce the proportion of grasses in favour of smaller dicotyledonous species (Willems & van Nieuwstadt, 1996). The density of the vegetation will preclude natural regeneration of the more sensitive grassland species unless there is some disturbance of the sward, such as through intensive grazing or physical scarification. A number of studies have shown that seed will not germinate and seedlings can not establish in an intact sward or where a dense litter layer has accumulated (e.g. Silvertown, 1981).

Because most chalk grassland herbs have limited powers of dispersal and short life spans as viable seeds in soil, it will be necessary in many re-creation projects to introduce them directly into the site, either as sown seeds or as transplanted pot-grown seedlings or adult plants. Without such intervention, several important components of calcareous grassland plant communities can take well over 100 years to colonise sites (Wells, 1991). This applies to sensitive species such as squinancywort (*Asperula cynanchica*), spring sedge (*Carex caryophyllea*), dropwort (*Filipendula vulgaris*), common rockrose (*Helianthemum chamaecistus*), meadow oat-grass (*Helictotrichon pratense*), burnet saxifrage (*Pimpinella saxifraga*) and common milkwort (*Polygala vulgaris*).

Re-creation of species-rich grassland on unvegetated land

Recent arable crop surpluses across much of Europe have prompted national governments to fund financial incentive schemes that encourage farmers to take land out of intensive agricultural production and divert it into uses which are more sympathetic to wildlife. Such circumstances have provided opportunities to attempt the creation of species-rich calcareous grassland from scratch. Other opportunities for grassland re-creation have occurred on derelict land after industrial or quarrying activities have ceased (Usher & Jefferson, 1990), or where major engineering works, such as road construction, have left areas of calcareous spoil or steep embankments that are not easily revegetated in a more conventional way (Morris *et al.*, 1994; Mitchley *et al.*, 1996).

In all such cases, the challenge has been to establish the right initial mixture of species on bare soil and then guide natural succession towards the desired grassland community. This is a difficult exercise, as the initial plant communities are unstable and subsequent succession can easily become deflected in an undesirable direction (Gibson & Brown, 1992). This is the aspect of calcareous grassland restoration that has received the most attention (Gibson, 1995), and yet reliable principles remain elusive. Success or failure depends upon a number of interacting factors, including the available species pool, the nature of the underlying soils and the type of management imposed.

CONCLUDING REMARKS

Calcareous grassland is a diverse and highly complex habitat and it is therefore not straightforward to restore. All the evidence suggests that the species-rich grasslands that we witness today are the product of perhaps several hundred years' development without major interruption. It may therefore be unrealistic to assume that there are techniques that can be employed to short-circuit this lengthy process. Several restoration projects have demonstrated considerable success in producing aesthetically attractive grasslands that bear a reasonable superficial resemblance to undisturbed calcareous grassland and that contain a number of charismatic and uncommon plant species. However, detailed analysis of the vegetation invariably shows that the balance of species within the community differs significantly from that in ancient calcareous grassland and that the small-scale density of species is considerably less. Thus, species-rich vegetation achieved via a re-creation programme is a poor substitute, at least in the short term, for ancient calcareous grasslands, and therefore conservation of the latter should always be the highest priority.

At present, we have no single successful technique for restoration of calcareous grassland, whichever habitat is the starting point (see Fig. 18.1). To some extent, this lack of knowledge is a function of the short time for which most restoration schemes have been running. This means that we do not yet know the end points that can be achieved. In any case, such grasslands are highly dynamic and are therefore continually vulnerable to change. General prescriptions may be dangerous, since what works well in one situation may not be equally effective in another. This is partly because we can not rely on consistent responses to restoration in different geographical locations. However, it also reflects the importance of the starting conditions (soil nutrient status, seed banks, etc.). Seed banks in particular rapidly lose their value for restoration after the land use has changed, because species that are characteristic of calcareous grassland generally have short-lived seeds.

As with most habitats, restoration of calcareous grassland is an ongoing process which must continue even if a close approximation to the target habitat can be achieved. Follow-up management in the form of regular grazing or cutting will be needed to prevent competitive species from eliminating slower-growing species, and to prevent succession towards scrub and woodland.

Finally, species-rich grasslands present particular restoration problems because we are often attempting to establish many species in a very restricted area. This compounds already considerable problems, because we have to achieve multiple goals which may not be compatible. For example, the habitat requirements of rare plants may be in direct conflict with those of rare insects. To some extent, this problem may be circumvented by partitioning the site or by initiating rotational management. In general terms, the encouragement of spatial heterogeneity at a variety of scales will help to accommodate maximal ecological diversity within a confined area and should therefore be a major objective of the restoration process.

REFERENCES

Anonymous (1984a). *Wildlife of the South Downs.* Peterborough, UK: Nature Conservancy Council.

Anonymous (1984b). *Nature Conservation in Great Britain. Summary of Objectives and Strategy.* Peterborough, UK: Nature Conservancy Council.

Bekker, R. M., Verweij, G. L., Smith, R. E. N., Reine, R., Bakker, J. P. & Schneider, S. (1997). Soil seed banks in European grasslands: does land use affect regeneration perspectives? *Journal of Applied Ecology,* **34**, 1293–1310.

Blackwood, J. & Tubbs, C. R. (1970). A quantitative survey of chalk grassland in England. *Biological Conservation,* **3**, 1–5.

Bobbink, R. & Willems, J. H. (1987). Increasing dominance of *Brachypodium pinnatum* (L.) Beauv. in chalk grasslands: a threat to a species-rich ecosystem. *Biological Conservation*, **40**, 301–314.

Bobbink, R. & Willems, J. H. (1991). Impact of different cutting regimes on the performance of *Brachypodium pinnatum* in Dutch chalk grassland. *Biological Conservation*, **56**, 1–21.

Bobbink, R. & Willems, J. H. (1993). Restoration management of abandoned chalk grassland in the Netherlands. *Biodiversity and Conservation*, **2**, 616–626.

Bobbink, R., Bik, L. & Willems, J. H. (1988). Effects of nitrogen fertilisation on vegetation structure and dominance of *Brachypodium pinnatum* (L.) Beauv. in chalk grassland. *Acta Botanica Neerlandica*, **37**, 231–242.

Bradshaw, A. D. (1997). The importance of soil ecology in restoration science. In *Restoration Ecology and Sustainable Development*, eds. K. M. Urbanska, N. R. Webb & P. J. Edwards, pp. 33–64. Cambridge: Cambridge University Press.

Bullock, J. M. (1998). Community translocation in Britain: setting objectives and measuring consequences. *Biological Conservation*, **84**, 199–214.

Cain, M. L., Milligan, B. G. & Strand, A. E. (2000). Long-distance seed dispersal in plant populations. *American Journal of Botany*, **87**, 1217–1227.

Cherrill, A. J. & Brown, V. K. (1992). Ontogenetic changes in the microhabitat preferences of *Decticus verrucivorus* (Orthoptera: Tettigoniidae) at the edge of its range. *Ecography*, **15**, 37–44.

Corbet, S. A. (1995). Insects, plants and succession: advantages of long-term set-aside. *Agriculture, Ecosystems and Environment*, **53**, 201–217.

Davies, A. & Waite, S. (1998). The persistence of calcareous grassland species in the soil seed bank under developing and established scrub. *Plant Ecology*, **136**, 27–39.

Donelan, M. & Thompson, K. (1980). Distribution of buried viable seeds along a successional series. *Biological Conservation*, **17**, 297–311.

Duffey, E., Morris, M. G., Sheail, J., Ward, L. K., Wells, D. A. & Wells, T. C. E. (1974). *Grassland Ecology and Wildlife Management*. London: Chapman & Hall.

Dutoit, T. & Alard, D. (1995). Permanent seed banks in chalk grassland under various management regimes: their role in the restoration of species-rich plant communities. *Biodiversity and Conservation*, **4**, 939–950.

Dyke, G. V., George, B. J., Johnson, A. E., Poulton, P. R. & Todd, A. D. (1983). The Broadbalk Wheat Experiment 1968–78: yields and plant nutrients in crops grown continuously and in rotation. *Report of the Rothamsted Experimental Station, 1982*, Part 2, 5–44.

Dzwonko, Z. & Loster, S. (1998). Dynamics of species richness and composition in a limestone grassland restored after tree cutting. *Journal of Vegetation Science*, **9**, 387–394.

Fischer, M. & Stöcklin, J. (1997). Local extinctions of plants in remnants of extensively used calcareous grasslands 1950–1985. *Conservation Biology*, **11**, 727–737.

Fischer, S. F., Poschlod, P. & Beinlich, B. (1996). Experimental studies on the dispersal of plants and animals by sheep in calcareous grasslands. *Journal of Applied Ecology*, **33**, 1206–1222.

Gibson, C. W. D. (1995). *Creating Chalk Grasslands on Former Arable Land: A Review*, Oxford: Bioscan.

Gibson, C. W. D. & Brown, V. K. (1992). Grazing and vegetation change: deflected or modified succession? *Journal of Applied Ecology*, **29**, 120–131.

Gibson, C. W. D., Watt, T. A. & Brown, V. K. (1987). The use of sheep grazing to recreate species-rich grassland from abandoned arable land. *Biological Conservation*, **42**, 165–183.

Gough, M. W. & Marrs, R. H. (1990). A comparison of soil fertility between semi-natural and agricultural plant communities: implications for the creation of species-rich grassland on abandoned agricultural land. *Biological Conservation*, **51**, 83–96.

Graham, D. J. & Hutchings, M. J. (1988a). Estimation of the seed bank of a chalk grassland ley established on former arable land. *Journal of Applied Ecology*, **25**, 241–252.

Graham, D. J. & Hutchings, M. J. (1988b). A field investigation of germination from the seed bank of a chalk grassland ley on former arable land. *Journal of Applied Ecology*, **25**, 253–263.

Green, B. H. (1980). Management of extensive amenity grasslands by mowing. In *Amenity Grassland: An Ecological Perspective*, eds. I. H. Rorison & R. Hunt, pp. 155–161. Chichester, UK: John Wiley.

Harper, J. L. (1977). *Population Biology of Plants*. London: Academic Press.

Hurst, A. & John, E. A. (1999a). The biotic and abiotic changes associated with *Brachypodium pinnatum* dominance in chalk grassland in south-east England. *Biological Conservation*, **88**, 75–84.

Hurst, A. & John, E. A. (1999*b*). The effectiveness of glyphosate for controlling *Brachypodium pinnatum* in chalk grassland. *Biological Conservation*, **89**, 261–265.

Hutchings, M. J. & Booth, K. D. (1996*a*). Studies on the feasibility of re-creating chalk grassland vegetation on ex-arable land. 1: The potential roles of the seed bank and the seed rain. *Journal of Applied Ecology*, **33**, 1171–1181.

Hutchings, M. J. & Booth, K. D. (1996*b*). Studies on the feasibility of re-creating chalk grassland on ex-arable land. 2: Germination and survivorship of seedlings under different management regimes. *Journal of Applied Ecology*, **33**, 1182–1190.

Jefferson, R. G. & Usher, M. B. (1989). Seed rain dynamics in disused chalk quarries in the Yorkshire Wolds, England, with special reference to nature conservation. *Biological Conservation*, **47**, 123–136.

Jones-Walters, L. M. (1990). A new approach to the management of chalk grassland with particular reference to the integration of conservation measures for invertebrates. In *Calcareous Grasslands: Ecology and Management*, eds. S. H. Hillier, D. W. H. Walton & D. A. Wells, pp. 140–148. Huntingdon, UK: Bluntisham Books.

Keymer, R. J. & Leach, S. J. (1990). Calcareous grassland: a limited resource in Britain. In *Calcareous Grasslands: Ecology and Management*, eds. S. H. Hillier, D. W. H. Walton & D. A. Wells, pp. 11–17. Huntingdon, UK: Bluntisham Books.

Kiefer, S. & Poschlod, P. (1996). Restoration of fallow on afforested calcareous grasslands by clear-cutting. In *Species Survival in Fragmented Landscapes*, eds. J. Settele, C. R. Margules, P. Poschlod & K. Henle, pp. 209–218. Dordrecht: Kluwer.

Kirby, P. (1992). *Habitat Management for Invertebrates: A Practical Handbook*. Sandy, UK: Joint Nature Conservation Committee and RSPB.

Levin, R. (1968). *Evolution in Changing Environments: Some Theoretical Explorations*. Princeton, NJ: Princeton University Press.

Marrs, R. H. (1985). Techniques for reducing soil fertility for nature conservation purposes: a review in relation to research at Roper's Heath, Suffolk, England. *Biological Conservation*, **34**, 307–332.

Marrs, R. H. (1993). Soil fertility and nature conservation

in Europe: theoretical considerations and practical management solutions. *Advances in Ecological Research*, **24**, 241–300.

Mitchley, J., Buckley, G. P. & Helliwell, D. R. (1996). Vegetation establishment on chalk marl spoil: the role of nurse grass species and fertiliser application. *Journal of Vegetation Science*, **7**, 543–548.

Morris, M. G. (1990). The Hemiptera of two sown calcareous grasslands. 3: Comparisons with the Auchenorrhyncha faunas of other grasslands. *Journal of Applied Ecology*, **27**, 394–409.

Morris, M. G. (2000). The effects of structure and its dynamics on the ecology and conservation of arthropods in British grasslands. *Biological Conservation*, **95**, 129–142.

Morris, M. G., Thomas, J. A., Ward, L. K., Snazell, R. G., Pywell, R. F., Stevenson, M. J., & Webb, N. R. (1994). Re-creation of early-successional stages for threatened butterflies: an ecological engineering approach. *Journal of Environmental Management*, **42**, 119–135.

Mortimer, S. R., Hollier, J. A. & Brown, V. K. (1998). Interactions between plant and insect diversity in the restoration of lowland calcareous grasslands in southern Britain. *Applied Vegetation Science*, **1**, 101–114.

Partel, M., Kalamees, R., Zobel, M. & Rosen, E. (1998). Restoration of species-rich limestone grassland communities from overgrown land: the importance of propagule availability. *Ecological Engineering*, **10**, 275–286.

Poschlod, P., Kiefer, S., Tränkle, U., Fischer, S. & Bonn, S. (1998). Plant species richness in calcareous grasslands as affected by dispersability in space and time. *Applied Vegetation Science*, **1**, 75–90.

Ratcliffe, D. A. (1984). Post-medieval and recent changes in British vegetation: the culmination of human influence. *New Phytologist*, **98**, 73–100.

Rizand, A., Marrs, R. H., Gough, M. W. & Wells, T. C. E. (1989). Long-term effects of various conservation management treatments on selected soil properties of chalk grassland. *Biological Conservation*, **49**, 105–112.

Silvertown, J. W. (1981). Micro-spatial heterogeneity and seedling demography in species-rich grassland. *New Phytologist*, **88**, 117–128.

Smith, C. J. (1980). *The Ecology of the English Chalk*. London: Academic Press.

Smith, C. T., Elston, J. & Bunting, A. H. (1971). The effects of cutting and fertiliser treatment on the yield and botanical composition of chalk turfs. *Journal of the British Grassland Society*, **26**, 213–219.

Smith, R. S. (1994). Effects of fertilisers on plant species composition and conservation interest of UK grassland. In *Grassland Management and Nature Conservation*, eds. R. J. Haggar & S. Peel, pp. 65–73. Reading, UK: British Grassland Society.

Stevenson, M. J., Bullock, J. M. & Ward, L. K. (1995). Re-creating semi-natural communities: effect of sowing rate on establishment of calcareous grassland. *Restoration Ecology*, **3**, 279–289.

Thomas, C. D., Thomas, J. A. & Warren, M. S. (1992). Distributions of occupied and vacant butterfly habitats in fragmented landscape. *Oecologia*, **92**, 563–567.

Thomas, J. A. (1984). The conservation of butterflies in temperate countries: past efforts and lessons for the future. In *The Biology of Butterflies*, eds. R. I. Vane-Wright & P. R. Ackery, pp. 327–332. London: Academic Press.

Thompson, K. (1986). Small-scale heterogeneity in the seed bank of an acidic grassland. *Journal of Ecology*, **74**, 733–738.

Thompson, K. & Grime, J. P. (1979). Seasonal variation in the seed banks of herbaceous species in ten contrasting habitats. *Journal of Ecology*, **67**, 893–921.

Usher, M. B. & Jefferson, R. G. (1990). The concepts of colonisation and succession: their role in nature reserve management. In *Calcareous Grasslands: Ecology and Management*, eds. S. H. Hillier, D. W. H. Walton & D. A. Wells, pp. 149–153. Huntingdon, UK: Bluntisham Books.

van der Putten, W. H., Mortimer, S. R., Hedlund, K., Dijk, C. V., Brown, V. K., Leps, J., Rodriguez-Barrueco, C., Roy, J., Len, T. A. D., Gormsen, D., Korthals, G. W., Lavorel, S., Regina, I. S. & Smilauer, P. (2000). Plant species diversity as a driver of early succession in abandoned fields: a multi-site approach. *Oecologia*, **124**, 91–99.

Verkaar, H. J., Schenkeveld, A. J. & van de Klashorst, M. P. (1983). The ecology of short-lived forbs in chalk grasslands: dispersal of seeds. *New Phytologist*, **95**, 335–344.

Wells, T. C. E. (1980). Management options for lowland grassland. In *Amenity Grassland: An Ecological Perspective*, eds. I. H. Rorison & R. Hunt, pp. 175–195. Chichester, UK: John Wiley.

Wells, T. C. E. (1983). The creation of species-rich grasslands. In *Conservation in Perspective*, eds. A. Warren & F. B. Goldsmith, pp. 215–232. Chichester, UK: John Wiley.

Wells, T. C. E. (1987). The establishment of floral grasslands. *Acta Horticulturae*, **195**, 59–69.

Wells, T. C. E. (1989). The re-creation of grassland habitats. *The Entomologist*, **108**, 97–108.

Wells, T. C. E. (1991). Restoring and re-creating species-rich lowland dry grassland. In *The Conservation of Lowland Dry Grassland Birds in Europe*, eds. P. D. Goriup, L. A. Batten & J. A. Norton, pp. 125–132. Peterborough, UK: Joint Nature Conservation Committee.

Wells, T. C. E., Sheail, J., Ball, D. F. & Ward, L. K. (1976). Ecological studies of the Porton Ranges: relationships between vegetation, soils and land-use history. *Journal of Ecology*, **64**, 589–626.

Wells, T. C. E., Bell, S. & Frost, A. (1981). *Creating Attractive Grasslands using Native Plant Species*. Shrewsbury, UK: Nature Conservancy Council.

Wells, T. C. E., Cox, R. & Frost, A. (1989). Diversifying grasslands by introducing seed and transplants into existing vegetation. In *Biological Habitat Reconstruction*, ed. G. P. Buckley, pp. 283–298. London: Belhaven Press.

Willems, J. H. (1990). Calcareous grasslands in continental Europe. In *Calcareous Grasslands: Ecology and Management*, eds. S. H. Hillier, D. W. H. Walton & D. A. Wells, pp. 3–10. Huntingdon, UK: Bluntisham Books.

Willems, J. H. & Bik, L. P. M. (1998). Restoration of high species density in calcareous grasslands: the role of seed rain and the seed bank. *Applied Vegetation Science*, **1**, 91–100.

Willems, J. H. & van Nieuwstadt, M. G. L. (1996). Long-term after-effects of fertilisation on aboveground phytomass and species diversity in calcareous grassland. *Journal of Vegetation Science*, **7**, 177–184.

Wilson, E. J., Wells, T. C. E. & Sparks, T. H. (1995). Are calcareous grasslands in the UK under threat from nitrogen deposition? – an experimental determination of a critical load. *Journal of Ecology*, **83**, 823–832.

Wolkinger, F. & Plank, S. (1981). *Dry Grasslands of Europe*, Nature and Environment Series no. 21. Strasbourg: Council of Europe.

Zobel, M., Suurkask, M., Rosen, E. & Partel, M. (1996). The dynamics of species richness in an experimentally restored grassland. *Journal of Vegetation Science*, **7**, 203–210.

19 • Prairies

SCOTT D. WILSON

INTRODUCTION

North American prairie comprises a diverse array of grasslands and environments that cover a vast area, from Saskatchewan in the north to Texas in the south, from Colorado in the west to Illinois in the east (Coupland, 1992). Among the world's most diverse temperate habitats (Janssens *et al.*, 1998), prairie is dominated by grasses, sedges and forbs, with woody trees and shrubs mostly restricted to low areas such as swales or river valleys. Water availability decreases from east to west due to the rain shadow associated with the Rocky Mountains. In the east, tall C_4 grasses dominate and produce sufficient shoot mass to slow nutrient cycling and depress diversity. In the west, short C_4 grasses produce so little shoot mass that little litter accumulates (Fig. 19.1); bare ground and cryptogams are abundant. Aboveground disturbances which affect shoot mass, such as fire and vertebrate grazing, have the greatest effect on species composition and ecosystem function in the east, and less in the west. Belowground processes become more important as soil moisture decreases. Although seed production is copious in some years, establishment from seed is rare and small natural disturbances are colonised by vegetative growth.

RATIONALE FOR RESTORATION

Prairie restoration is needed in order to preserve native species in areas with relatively little prairie remaining. There is a gradient in prairie conservation, from very low (<1%) in the eastern parts of the Great Plains (Smith, 1998) to relatively high in the drier west (Epstein *et al.*, 1997). There is a parallel but opposite gradient in restoration efforts, being most common in the east (Illinois, Iowa) where little prairie is left, and least common in the west (Montana, Alberta) where native prairie is more abundant. In addition to providing habitat, restoration adds organic matter and nutrients to soils in abandoned agricultural fields, and native plants do this more effectively than introduced species (Christian & Wilson, 1999).

In spite of the fact that prairie species did not evolve with large-scale belowground disturbances like cultivation, there is abundant evidence that prairie can recover from this kind of disruption, given propagules and enough time (Burke *et al.*, 1995; Christian & Wilson, 1999). The most widespread disturbance of prairie results from agricultural cultivation, and areas recovering from cultivation are termed 'old fields'.

The restoration of North American prairie has a long history and includes some of the first ecological restorations (Kline & Howell, 1987). Much previous work has centred on prairie restoration with specific goals, such as erosion control, mine reclamation, and the creation of pasture or wildlife habitat (Table 19.1). Although these projects fulfil particular needs and provide valuable general insights, they can produce grasslands quite different from natural prairies in terms of species composition, diversity and structure. Ecological restoration (Jordan, 1988) aims to mimic natural processes and vegetation, and is most likely to succeed in producing sustainable restorations.

CONSTRAINTS

The main constraints to prairie restoration include a lack of seeds, among-year variability in establishment, and the persistence of introduced, non-native

Fig. 19.1. A 50-year-old stand of the introduced perennial grass *Agropyron cristatum* in Grasslands National Park in southwest Saskatchewan (see Box 19.1). The plot in the left foreground is dominated by native grasses sown as part of an experimental restoration. The plot had been treated with a general herbicide annually for three years, but care was taken to apply the herbicide only to the relatively tall *A. cristatum* using a wick. Seeds of native grasses were added as discarded cleanings three years earlier.

perennial species. The first constraint has been somewhat alleviated by commercial suppliers and efficient collection techniques, and the second is a largely uncontrollable characteristic of grassland environments. Much remains to be learned about the third constraint. Strategies for overcoming these constraints are discussed in this chapter.

Lack of a seed bank

Sites for grassland restoration typically contain few or no seeds of the desired species (Archibold, 1981; Bakker & Berendse, 1999). One reason that native prairie species have no seeds in disturbed soils is that they tend to occupy natural disturbances via vegetative spread rather than by seed establishment. For example, 99% of stems colonising experimental gopher (*Geomys bursarius*) mounds in mixedgrass prairie in South Dakota arose from vegetative growth (Umbanhowar, 1992). Similarly, soils from a grazed pasture in Saskatchewan produced 476 seedlings m^{-2}, but 2154 shoots m^{-2} from belowground parts (Archibold, 1981).

Strategies to store prairie seed banks are apparently unsuccessful. Undisturbed North Dakota prairie contained 2980 seeds m^{-2} in the top 7.5 cm, but fresh stockpiles of topsoil contained only 255 (Iverson & Wali, 1982). Similarly, moving the top 60 cm of prairie soil to cover mine spoils in Colorado and Wyoming produced only 1–4 plants m^{-2} from both seeds and rhizomes, clearly indicating the need for seed addition (Howard & Samuel, 1979). The reason that stockpiled soil contributes few germinable seeds to restorations is that most of the seeds in undisturbed prairie are within 7.5 cm of the surface (Iverson & Wali, 1982); machinery typically takes deeper soils, and dilutes the shallow seed-rich soil with deeper seed-free soil.

Restorations are occasionally undertaken on sites with some native species present. This occurs where native prairie has had introduced species added, or where native prairie has been invaded by introduced species, or where stands of introduced

Table 19.1. *Prairie restoration: syntheses, reviews and long-term studies of a practical or theoretical nature*

Reference	Area or system	Topic
Practical		
Anderson, 1995	European grasslands	general
Betz, 1986	Illinois	general
DePuit & Redente, 1988	Montana	mine wastes
Duebbert *et al.*, 1981	North Dakota	wildlife habitat
Harris & Dobrowolski, 1986	Washington	long term
Howe, 1994	tallgrass	grazing, fire
Jones & Hayes, 1999	UK	diversity and grazing
Jordan *et al.*, 1987	USA	various
Kline & Howell, 1987	USA	review
Lippitt *et al.*, 1994	USA	seed collection
Packard & Mutel, 1997	tallgrass	general
Redente & DePuit, 1988	Wyoming	mine wastes
Ringe & Graves, 1987	USA	mulches
Rock, 1975	Wisconsin	general
Roundy & Call, 1988	USA	rangelands
Schramm, 1978	tallgrass	general
Shirley, 1994	Iowa	general
Tyser *et al.*, 1998	Montana	roadsides
Wark *et al.*, 1995	Canada	duck habitat
Theoretical		
E. B. Allen, 1988a	USA arid	general
M. F. Allen, 1988	USA arid	soil
Archer & Pyke, 1991	USA	grazing
Bakker & Berendse, 1999	Netherlands	diversity
Bakker *et al.*, 1996	Netherlands	seed banks and dispersal
Call & Roundy, 1991	USA arid	general
Morghan & Seastedt, 1999	general	C amendments
Pyke & Archer, 1991	USA	competition
Workman & Tanaka, 1991	USA	economics (grazing)

perennials have been invaded by native species. In Saskatchewan, a 50-year-old field dominated by the introduced perennial grass crested wheatgrass (*Agropyron cristatum*) (Box 19.1) had an understorey of native plants (Bakker *et al.*, 1997), but the soil contained 930 seeds m^{-2} of *A. cristatum*, and only 1 seed m^{-2} of the most common native grass, blue grama (*Bouteloua gracilis*) (Ambrose, 1999). Thus, native seed is needed even where a few native adults are present.

The lack of seeds of species typical of undisturbed prairies led Lauenroth & Coffin (1992) to propose that classic old-field succession models developed for eastern North America do not describe prairie succession. Prairie succession may be better described using gap models which incorporate among-species differences in dispersal and establishment ability (e.g. Coffin & Lauenroth, 1990). These models point to the need for seed addition as part of restoration.

Box 19.1   Restoration strategies vary with initial conditions

The effects of neighbour control, seeding method and mulch were investigated in two contrasting fields in Grasslands National Park in southwest Saskatchewan. One was recently abandoned from cultivation and dominated by annual weeds (Christian, 1996). The second had been abandoned 50 years previously and planted with the introduced perennial grass *Agropyron cristatum* (Bakker, 1996; Bakker *et al.*, 1997) (Fig. 19.1). Both were sown with seeds of the native grasses blue grama (*Bouteloua gracilis*) and needle-and-thread grass (*Stipa comata*) at a very high rate (3.1 g m^{-2}) in May 1994, 1995 and 1996. All treatments and years were applied in a replicated, randomised design ($n = 5$) to 3×10 m plots.

In the recently abandoned field, neighbour control slightly increased seedling density in the first year, and had no significant effect in the following years (Bakker *et al.*, 1998). In the *A. cristatum* field, neighbour control increased native seedling density sixfold (Bakker *et al.*, 1997). Similar patterns were found for the cover of native species after four years of neighbour control: control had no effect on native cover in the recently abandoned field, but increased native cover fourfold in the *A. cristatum* field (Bakker *et al.*, 1998). Thus, neighbour control was far more important in the presence of introduced perennials, and was essentially unnecessary in the case of annual neighbours.

Neither tilling or herbicide provided good long-term control of *A. cristatum*. For example, tilling *A. cristatum* in May allowed seedlings of native grasses to establish, but the cover of *A. cristatum* in August had returned to 20%, compared with 45% in untilled plots (Bakker *et al.*, 1997). Similarly, applying herbicide to *A. cristatum* in spring produced only slight control in the following August (35% in herbicide plots vs. 45% in control plots).

In contrast, two years of simulated grazing decreased the cover of *A. cristatum* to 3% (vs. 35% in untreated controls) and increased the total cover of

native species to 38% (vs. 18% in controls: S. D. Wilson, unpublished data).

The two fields also differed in the number of seedlings established, and in their responses to among-year variation in weather. In the *A. cristatum* field, first-year native seedling density (in plots without neighbours and broadcast with native seed) in 1994, 1995 and 1996 was 150 m^{-2}, 450 m^{-2}, and 20 m^{-2} respectively. The same figures for the recently abandoned field were 100 m^{-2}, 70 m^{-2} and 5 m^{-2} (Bakker *et al.*, 1998). In spite of being separated by about 15 km, the fields showed a sixfold difference in establishment in 1995. Further, the rank order of 1994 and 1995 for establishment differed between the fields. There was agreement, however, in low establishment in 1996 for both fields, and in the enormous variation among years found in both fields.

The fields were also similar in their responses to seeding method and mulch. After four years, the density of native plants in the cultivated field (in plots without neighbours) was 68 m^{-2} in plots broadcast with 180 g m^{-2} of chaff discarded from cleaning native seed, 33 m^{-2} in plots broadcast with seed, 15 m^{-2} in plots drilled with seed, and 1 m^{-2} in plots broadcast with native hay at a rate of 200 g m^{-2}. The pattern was the same in the experiment dominated by *A. cristatum* (Fig. 19.1). Thus, the response of added seeds to seeding methods did not vary between annual and perennial neighbourhoods. Similarly, mulch (200 g straw m^{-2}) did not affect native seedling density in either field.

In summary, this pair of experiments showed that some aspects of restoration strategy depend on initial conditions: neighbour control was important in perennial vegetation but not in annual vegetation. Seeding techniques had similar rankings in the two fields, with broadcasting being the most effective of the traditional methods investigated. The enormous variation in first-year establishment between sites and among years signals the difficulty of predicting the success of restorations, and highlights the fact that successful strategies vary among initial conditions.

Lack of dispersal

Given the low reliance of prairie vegetation on seeds for recovery from disturbance, natural seed dispersal cannot be relied upon for prairie restoration, even with adjacent native vegetation. For example, the cover of native species in a three-year-old Kansas tallgrass restoration did not vary with distance from the edge of an undisturbed prairie (Kindscher & Tieszen, 1998). Even in fields abandoned for 50 years in northeast Colorado, the cover of the native dominant *Bouteloua gracilis* decreased significantly with distance from the edge of undisturbed prairie only in a few cases (Coffin *et al.*, 1996). Both studies suggest that native species are poor dispersers.

In contrast, low rates of dispersal may eventually bring in natives over very long periods: only a few native grasses were used in a Kansas restoration, but after 35 years, many other native species had dispersed into it from adjacent prairie (Schott & Hamburg, 1997). In the Curtis Prairie in Wisconsin, unplanted old fields near planted stands that were retired from cultivation in 1938 supported 42 native species in 1982 (Sperry, 1994). Similarly, the covers of common native species in Saskatchewan fields abandoned for 50 years without any restoration were similar to those in native prairie (Christian & Wilson, 1999).

The colonisation of restorations by native species may occur less often now than in the past for two reasons. First, newly abandoned fields are currently typically surrounded by introduced species, and there are few natives available for natural dispersal. Second, fewer dispersal propagules are probably available in the soil because fields abandoned now have been cultivated for much longer periods than fields abandoned in earlier times. Further, fields abandoned now are probably more deeply and frequently cultivated, using tractors, than those abandoned when horses were used.

The presence of introduced perennials

Introduced perennial species pose enormous challenges for restoration. Ironically, some of them, such as *Agropyron cristatum*, were introduced for restoration (Bock & Bock, 1995), and are still widely used today. They are used because their large, heavy seeds are well suited for agricultural seed drills, and the seed is commercially available. As a result, much of our knowledge about grass responses to revegetation techniques on prairie is based on data from introduced species.

Stands of introduced perennials are the biggest single factor causing failure of restoration (Duebbert *et al.*, 1981; Jordan, 1988; Roundy & Call,

1988). Late-successional perennials exert greater competitive effects than early-successional annual species (Wilson, 1999). Introduced perennials do far more than simply determine the species composition of a stand, however (M. F. Allen, 1988): their long-term effects determine ecosystem development (Walker & Smith, 1996). The invasiveness of many introduced perennials means that their effects extend from original points of introduction to other areas (D'Antonio *et al.*, 1999).

ESTABLISHING NATIVE PRAIRIE VEGETATION

Site preparation: soil effects

Soil structure (particle size, pore size, cracks) affects soil moisture and grass germination, so germination varies between old soils and tilled soils, and among soil types within these classes (Dobrowolski *et al.*, 1990). The presence of loamy soils was more important to establishment of *Agropyron cristatum* in Wyoming sagebrush stands than was site preparation (Cluff *et al.*, 1983). Salt in soil reduces the establishment of some species and prevents germination in others (Weiler & Gould, 1983).

Soil moisture generally decreases and particle size increases as elevation increases along hillsides, leading to calls for the use of different and appropriate seed mixes for hilltops, valley bottoms and intermediate locations (Duebbert *et al.*, 1981; Wark *et al.*, 1995). Further, topographical variation in soil types will lead to spatial variation in establishment success and in the need for reseeding.

Nutrient-driven diversity loss in grasslands is usually associated with increased standing crop and enhanced competition for light (Wilson & Tilman, 1991). Sewage sludge increased standing crop in a restoration in Oklahoma (Kessler & Kirkham, 1985), and fertiliser applied to mine spoils in southwest Montana increased the cover of *A. cristatum* and three other grasses (Holechek *et al.*, 1982). Nitrogen significantly increased standing crop in mixed stands of *A. cristatum* and smooth brome grass (*Bromus inermis*) in Saskatchewan (Peltzer *et al.*, 1998). In this experiment, however, nitrogen also decreased the density of native seedlings from 5 m^{-2} in unfertilised plots to 0 in plots receiving

15 g N m^{-2} y^{-1} (Wilson & Gerry, 1995). Neither fertiliser nor alfalfa (*Medicago sativa*) increased the yield of the native grass side-oat grama (*Bouteloua curtipendula*) sown into abandoned Texas cropland (Willard & Schuster, 1971). In summary, fertiliser might be appropriate for restorations where the goal is high standing crop, but this comes at the cost of lower diversity and loss of native species.

Controlling annual weeds

The control of annuals by tilling prior to sowing native seeds increases the establishment of native perennials only in some cases. Discing before seeding in Nebraska sandhills initially increased native grass establishment (4 plants m^{-2} in controls vs. 151 in disced plots), but the benefit was much reduced by the end of the second year (2 vs. 14: King *et al.*, 1989). A more complex procedure in the same study involved discing in March followed by sowing oats (*Avena sativa*) to stabilise soil and suppress weeds. The oats were killed with herbicide on 1 May, and natives were planted a few days later with no further weed control. This treatment, however, decreased weeds and increased native grasses in only one in four cases, suggesting that annuals have small negative effects.

Following seeding, hand weeding and the application of herbicides to annuals in new stands of introduced C$_4$ grasses in Texas significantly increased grass cover (Bovey *et al.*, 1986). Similarly, application of an herbicide that acts against forbs to a tilled field in Nebraska increased the density of native C$_4$ grass seedlings 2.5 times relative to controls (Cox & McCarty, 1958). The use of herbicide to reduce annuals in Nebraska promoted establishment of one native grass (big bluestem, *Andropogon gerardii*) but not another (sand reed grass, *Calamovilfa longifolia*: Masters *et al.*, 1990), possibly because native grasses vary in their sensitivity to herbicides used to suppress annuals (Bahler *et al.*, 1990). The effectiveness of herbicides on annuals may vary among years: spraying annual weeds in northeastern Colorado resulted in good stands of *Agropyron cristatum*, but only in an average rainfall year, not in a dry or wet year (McGinnies, 1968).

Annual weeds were also significantly reduced simply by sowing desired perennial species in Nevada (Young *et al.*, 1969). Sowing native grass seed in Arkansas at 0.43 g m^{-2} resulted in a weed cover of about 50% after two years, whereas sowing at 1.76 g m^{-2} produced a weed cover of 25% (Dale & Smith, 1986). Planting a short-lived perennial grass (*Secale montanum*) on coal-mine spoils in southeast Montana succeeded in significantly reducing the annual alien grass downy brome (*Bromus tectorum*), but it came back when *S. montanum* died off after two years (Andersen *et al.*, 1992).

Controlling introduced perennials

There are two classes of management for controlling existing or invading populations of introduced perennials as part of restoration. The first includes single treatments and treatments which are repeated occasionally, such as tilling, herbicide, mowing and fire. As discussed below these have little long-term effect on introduced perennials. The second includes continuous treatments, such as grazing and competition from natives: these may have the persistence required to negatively affect the undesirable species over the long term.

Tilling

Over the short term, tilling or ploughing typically decreases established perennials and improves the germination of added species (Cluff *et al.*, 1983; Dovel *et al.*, 1990). Site preparation probably never removes all of the individuals of undesirable species, however, and survivors grow back quickly (Roundy & Call, 1988) (Box 19.1). Thus, tilling provides only short-term control of introduced perennials, and may have undesirable side-effects such as promoting the germination of unwanted species, destroying aggregates and disrupting vertical soil structure. In contrast, other methods discussed below do not affect soil structure.

Herbicide

Like tilling, herbicide applied to stands of introduced perennials increases the short-term establishment of natives added as seed. For example, the density of native seedlings establishing from drilled seed in a central Saskatchewan *Bromus inermis* field was 0.1 m^{-2} in control plots, but 2.4 m^{-2} in plots treated with the herbicide glyphosate (Wilson & Gerry, 1995) (see also Box 19.1).

As in the case of tilling, however, herbicide provides only short-term control of introduced perennials. For example, herbicide was most effective at controlling *B. inermis* in central Saskatchewan after stands had been burned, allowing the herbicide good contact with green shoots. *Bromus inermis* grew back, however, resulting in non-significant differences between treated and control plots after two growing seasons (Grilz & Romo, 1995). Nearby, glyphosate decreased the total cover of *B. inermis* and *A. cristatum* from 68% to 13% (Wilson & Gerry, 1995), but the remaining mass of introduced grasses was sufficient to allow complete recovery after five years (S. D. Wilson, unpublished data). Glyphosate decreased *A. cristatum* cover in central Saskatchewan, but sprayed and control plots were not significantly different after the second growing season (Romo et al., 1994) (see also Box 19.1). Similarly, combinations of herbicide, burning and mowing applied to stands of the introduced grass Lehmann lovegrass (*Eragrostis lehmanniana*) in southeast Arizona allowed native species to establish, but *E. lehmanniana* was also promoted by all treatments, and was the most abundant species emerging from the seed bank (Biedenbender & Roundy, 1996). Spraying stands of leafy spurge (*Euphorbia esula*) in Nebraska allowed native grasses to establish, but the cover of *E. esula* quickly returned to levels similar to those in untreated plots (Masters et al., 1996). Applying selective herbicide to Montana fescue prairie annually for four years reduced the cover of the invasive introduced forb spotted knapweed (*Centaurea maculosa*) to 4% compared with 23% in controls, but did not eliminate it (Tyser et al., 1998).

Herbicides, therefore, do not provide long-term control of introduced perennials in cases where the objective is to remove them and restore native species.

Fire

Although fire affects the species composition of tallgrass prairie (Collins et al., 1995), and the number of exotic species in Kansas tallgrass prairie was lower in annually burned plots than in control plots (Gibson et al., 1993), there is no evidence that fire suppresses introduced grasses and favours native. In Saskatchewan, fire produced only short-term reductions in the introduced grasses *Bromus inermis* (Grilz & Romo, 1995) and *Agropyron cristatum* (Romo et al., 1994). *Bromus inermis* increased its cover in a northeastern Nebraska native prairie over 17 years, invading from roadsides even though the site had been burned seven times (Nagel et al., 1994). Fire provided no help for natives sown among Kentucky bluegrass (*Poa pratensis*) and quack grass (*Agropyron repens*) in Wisconsin (Robocker & Miller, 1955). A fire in a 45-year-old Wisconsin restoration reduced the cover of sweet clover (*Melilotus alba*) to 5%–26% compared with 93%–100% in control plots, but clearly left plenty of *Melilotus* to grow back (Kline, 1986). Fire may even promote introduced perennials. For example, the establishment of alien grass from the seed bank in Arizona was increased by fire (Biedenbender & Roundy, 1996).

Mowing, herbivory and grazing

Mowing the introduced *Agropyron cristatum* in Saskatchewan had no effect on its survival (Romo et al., 1994).

Herbivory by pocket gophers consumed 21% to 84% of the planted grass seedings in eastern Oregon (Garrison & Moore, 1956), and the effects varied among species (e.g. 38% for *A. cristatum*). All the predation was on planted or volunteer seedings. Thus gophers could contribute to the suppression of further recruitment to the population of introduced perennials, but would have little effect on established plants. Native perennial bunchgrasses recovered following the removal of the introduced perennial forb St. John's wort (*Hypericum perforatum*) by imported beetles (Huffaker, 1951).

Grazing decreases dominance and might increase diversity in stands of introduced perennials (Collins, 1987), but there is mixed evidence that grazing favours natives over introduced species. The cover of the introduced grass *Poa pratensis* in a Nebraska prairie heavily grazed by cattle until 1975 decreased from 28% to 20% during the following 17 years of light grazing (Nagel et al., 1994), suggesting that heavy grazing favoured introduced perennials. In contrast, two years of simulated heavy grazing was very effective in controlling 50-year-old stands of *A. cristatum* in southwest Saskatchewan (Box 19.1).

Competition from native species

Establishing dense stands of natives might provide sufficient competition to slow or prevent the return of introduced perennials (Betz, 1986; Roundy & Call, 1988). Tilling followed by prompt establishment of dense stands of natives nearly eliminated introduced perennial grasses in Kansas and Wisconsin (Box 19.2).

Mulches

Mulch can decrease evaporation, leaving more moisture available for seedling establishment. Mulch can add nutrients during decomposition, or decrease nutrients by immobilisation, or have no effect. A variety of mulches and their cost-effectiveness is reviewed by Ringe & Graves (1987).

Compost improved the establishment of prairie grasses on tailings of crushed rock with no organic N or C in northern Minnesota (Noyd *et al.*, 1996). Asphalt mulch in Colorado increased both soil temperature and moisture, increasing the establishment of *Bouteloua gracilis* tenfold, although it had no effect on two other grasses (Bement *et al.*, 1961). Similarly, straw mulch on Wyoming coal-spoil banks produced a tenfold increase in seedling numbers relative to controls (Jacoby, 1969).

Mulch can also have negative effects on restoration by promoting weeds. Adding sterile litter to cultivated soil greatly increased the establishment of annual weeds (Evans & Young, 1970). Because the litter was sterile, the results suggest that increased weed establishment resulted from improved environmental conditions alone. Similar results were obtained in Colorado (McGinnies, 1987). In Texas, cottonbur mulch prevented emergence of the native grass sand bluestem (*Andropogon hallii*), but raised soil moisture and also promoted plant pathogens (Stubbendieck & McCully, 1972). Thus the effects of mulch may vary among sites and years (Power, 1980).

Reducing soil nutrients

European approaches to nutrient reduction tend to be intensive, such as stripping the top and most

Box 19.2   Successful control of introduced perennial grasses

In northeast Kansas, Kindscher & Tieszen (1998) restored a 2-ha area (the 'recently untilled area' in the Fall Leaf site) dominated by the introduced perennial grasses *Poa pratensis* and *Bromus inermis*. The area was thoroughly tilled twice in April 1989 and broadcast with a mix of locally collected native seeds at 0.58 g m^{-2}. The area was mowed once in July.

By September 1992, introduced species accounted for only 0.3% of the total cover. Dominant species included the native grasses Indiangrass (*Sorghastrum nutans*, 49% cover), switchgrass (*Panicum virgatum*, 10%), little bluestem (*Schizachyrium scoparium*, 7%) and *Andropogon gerardii* (4%). There were still considerable differences between nearby native prairie and the restoration in terms of species composition and diversity: prairie supported an average of 14 species, whereas the restoration contained only 10.

A second example in southern Wisconsin used more intensive site preparation (Howe, 1999). Established stands of the Eurasian grasses *Bromus inermis*, *Poa pratensis* and *Agropyron repens* were sprayed with the herbicide glyphosate in August 1988 and burned two weeks later. The field was ploughed in November. The field was periodically disced and spot-sprayed the following summer (1989) and spring (1990). In July 1990 a native seed-mix was broadcast at 1000 seeds m^{-2}.

By July 1992, 95% of the cover was the native grass reed canary grass (*Phalaris arundinacea*), implying good control of the introduced perennials. All 19 of the sown species established.

In both cases, introduced perennials were rapidly replaced by native plants broadcast on tilled soil. In Wisconsin, this was helped by intensive site preparation, but natives also succeeded in Kansas with less intensive site preparation.

fertile soils, cutting and removing standing crop, cultivation to promote leaching, and carefully controlled grazing, with animals removed from the managed area each night (Anderson, 1995). Some North American workers have added carbon to soils in the hopes of immobilising nutrients and favouring native species over introduced species. Morghan & Seastedt (1999) provide a good review of recent studies.

Sugar applied at 45 g m^{-2} to a sagebrush steppe in southeast Oregon had no effect on soil available nitrogen, or on the growth of native grasses (Miller et al., 1991). Sawdust applied for three years at 400 g m^{-2} y^{-1} to a Saskatchewan grassland dominated by Agropyron cristatum and Bromus inermis significantly decreased soil available nitrogen (to 10% of controls) and increased bare ground, but did not affect the density of native seedlings (Wilson & Gerry, 1995). After five years, sawdust had no significant effect in this experiment, although species richness was about 50% higher in plots receiving sawdust than in control plots (Peltzer et al., 1998). A mixture of sugar (1 kg m^{-2}) and sawdust (650 g m^{-2}) applied for three years to a Colorado restoration dominated by introduced species produced significant short-term reductions in available nitrogen and cut standing crop nearly in half, but had no significant effect on the relative abundances of native and introduced species (Morghan & Seastedt, 1999). Mulch applied to the soil surface might also decrease available nitrogen by cooling the soil and decreasing decomposition rates (Power, 1981).

Carbon amendments have little effect for two reasons. First, grassland soil organic matter is a complex entity comprised of fractions that vary in turnover time. Fifty years of succession in Colorado may be sufficient for restoration of the most active fraction of carbon to its pre-disturbance values (Burke et al., 1995), but restoring the least active fractions may require over a millennium of root production and decomposition (Seastedt, 1995). Consequently the complete restoration of natural soil organic matter is clearly a long-term project and unlikely to respond to short-term carbon additions. Second, North American restorations are mostly in long-cultivated fields: cultivation reduces soil organic matter (Parton et al., 1989), so total and available nutrients may already be relatively low at the start of restoration. For example, available nitrogen (sum of ammonium and nitrate) in undisturbed prairie in southwestern Saskatchewan was about 2 mg N per kg soil, whereas in fields abandoned for 50 years and dominated by A. cristatum, available nitrogen was <1 mg N per kg soil (Christian & Wilson, 1999). While dominance of abandoned fields by introduced species can be symptomatic of high nutrient levels, introduced species can also dominate fields with low amounts of remaining organic matter simply because they have been sown there, and, as in Saskatchewan, occupy relatively poor soils. Thus, further decreasing nutrient availability in this case will not encourage native species.

Rain and irrigation

Available moisture is a complex variable because it varies with rainfall, evaporation, soil texture and storage from previous seasons (Aber & Melillo, 1991). Rainfall alone is complex, and should be characterised by amounts, timing, variability and predictability (Sala & Lauenroth, 1982).

As an example of the complexity caused by the timing of rain and drought, a glasshouse experiment (Frasier et al., 1987) with C$_4$ grasses showed that two consecutive days of rain caused no germination, allowing seeds to survive a subsequent drought. Three days of rain caused germination, but the seedlings died during a subsequent five-day drought. Five days of rain allowed germination and production of enough root mass for seedlings to survive the same drought.

More complexity arises from differences among species in their responses to water. Both the introduced Agropyron cristatum and the native Bouteloua gracilis rely on adventitious roots for growth soon after germination. When both species were planted 18 mm deep, adventitious roots of A. cristatum emerged near the seed, whereas those of B. gracilis emerged only 2 mm beneath the soil surface. Consequently, B. gracilis was much more susceptible to desiccation as the top of the soil dried (Hyder et al., 1971), and smaller seedlings were more susceptible than larger ones (Briske & Wilson, 1980). This may explain why seedlings of B. gracilis established at rates only 10%–50% of those for A. cristatum in Colorado (Bement et al., 1965). In summary, experiments suggest that A. cristatum should be less sensitive to drought than B. gracilis. In contrast, water applied to a field experiment in Saskatchewan increased the germination of both *A. cristatum* and *B. gracilis* to about the same extent (Ambrose, 1999).

The complex nature of germination responses is well suited to exploration using simulation models. Simulations of natural establishment of *Bouteloua gracilis* stands in Colorado predicted that 20 years would pass before *Bouteloua* dominated, if seed is always present (Lauenroth & Coffin, 1992). If no seed is added, and if seed availability varies with the previous year's rainfall, 65 years will be required. Similar models suggest that a suitable rainfall for recruitment occurs on average every 30–50 years on silty clay and loam soils, but less than once each 5000 years on sandy soils (Lauenroth *et al.*, 1994). The models imply that a long commitment to reseeding may be necessary (Duebbert *et al.*, 1981).

First-year establishment of *B. gracilis* varied 20-fold over three years of repeated plantings in Saskatchewan (Box 19.1). Experiments in dry Texas grassland showed that the rank order of patch favourableness for establishment varied between native grass species, years, sample dates and germination cohorts (Fowler, 1988). In total, the complex interaction of factors determining available moisture results in unpredictable germination if rainfall is the only source of water.

Unpredictable temporal variation in water availability can be overcome with irrigation in some situations. Irrigation has been used for revegetating mine spoils in southeast Montana with a mix of native and introduced species. Irrigation had no effect during a wet year, but increased standing crop threefold in a dry year (DePuit *et al.*, 1982). Positive effects were greatest for cool-season grasses and the introduced legume *Melilotus officinalis*. In contrast to shoot results, irrigation significantly reduced community root mass, and roots in irrigated plots were strongly skewed towards the surface. Higher shoot mass and fewer and more shallow roots might make irrigated plots more susceptible to damage from future droughts.

An agricultural irrigation system was used in a Nebraska old-field restoration. Irrigation had no effect during a year with above-average rainfall. Irrigation did increase establishment during a dry year, but had no effect on cover the following year (Oldfather *et al.*, 1989). Irrigation experiments across Nebraska showed that more water does not always have positive effects, since it can increase weed competition and fungal pathogens (Masters *et al.*, 1990).

Saline ground water can be used for irrigation. A glasshouse experiment showed that *B. gracilis* can germinate when supplied with moderately saline water, but not at the highest level of salinity (Weiler & Gould, 1983).

Capturing winter snows might enhance soil moisture. An annual nurse crop was planted in rows running in four directions in southeast Wyoming, but neither the presence of stubble or its direction improved the establishment of native and introduced grasses (Hart & Dean, 1986).

High soil moisture may enhance recovery rates. Faster succession on abandoned old fields in Nebraska relative to Montana was attributed to higher rainfall (Judd, 1940). Similarly, succession in northeast Colorado was considered to be faster in six wet years following 1938 than in the 15 preceeding dry years (Costello, 1944). The recovery of prairie vegetation on experimental gopher mounds in South Dakota was faster at low, wetter sites than at high, drier sites (Umbanhowar, 1995). Using irrigation to enhance recovery rates, however, might produce species mixes unable to cope with subsequent drought. For example, fertilising Minnesota old fields caused native grasses to be replaced by introduced perennials such as *Agropyron repens*. The introduced species, however, were much more sensitive to a natural drought year than were the native species (Tilman & El Haddi, 1992).

Mycorrhizae

Mycorrhizae are prerequisite for restoration (E. B. Allen, 1988*b*). For example, mycorrhizae were associated with 99% of plants in undisturbed Colorado sagebrush grassland, but only 1% of plants on an unrevegetated roadside nearby (Reeves *et al.*, 1979). Agricultural fields have lower rates of mycorrhizal infection than restorations, but infection rates did not vary with restoration age between two and 11 years in Illinois (Cook *et al.*, 1988). Soils under Ohio agricultural fields have mycorrhizae-free patches, and mycorrhizal infections were more homogeneous in fields undergoing natural succession or planted with native prairie species (Boerner *et al.*, 1996). Thus, the prevalence of mycorrhizae increases over time.

Mycorrhizae were manipulated in a Minnesota roadside restoration for 15 months: mycorrhizae had no effect on total cover, ruderal species or a cover crop, but significantly increased the cover of native species from 70% to 100% (Smith *et al.*, 1998). Mixes of native grasses and forbs were grown for six weeks in pots filled with Kansas tallgrass prairie soil: those without mycorrhizae had 31% less mass, less C_4 mass (e.g. *Andropogon gerardii*) and more C_3 mass (e.g. *Poa pratensis*: Wilson & Hartnett, 1997).

A glasshouse experiment with three native grasses showed complex responses to mycorrhizae, including positive, none, and negative, with the effects depending on phosphorus availability (Noyd *et al.*, 1995). Although species differ in their responses to mycorrhizae, the vast majority perform better in the presence of mycorrhizae, and adding mycorrhizae may help restorations on heavily disturbed soils.

Seeds

Provenance
Seed provenance has long been a concern (Rock, 1975), partly because ecotypes from far away might not grow well in new environments. A comparison of 175 progeny lines of the widely introduced pasture grass *Agropyron cristatum* found significant differences in emergence, seedling height and seedling mass. Genetic variation accounted for about 50% of total phenotypic variance, and most variables were positively correlated with seed mass (Asay & Johnson, 1983). Grasses indeed exhibit considerable genetic variation.

The suitability of local and distant genotypes was compared over two years for five species of forbs added to a UK grassland dominated by agricultural grasses. Overall, there were no consistent differences between non-local, commercially obtained genotypes and local genotypes collected within 8 km (Jones & Hayes, 1999). There are few examples of geographic variation in fitness for grassland species; large-scale, long-term trials would help, as have been done for commercially valuable trees.

A second general concern about the use of few, commercial ecotypes is that their use would result in the loss of local genetic diversity (Jacobson *et al.*, 1994). Ecological restorations should use locally collected seed (Lippitt *et al.*, 1994).

Seed collection and storage
Lippitt *et al.* (1994) provides an extensive, practical review of prairie seed collection, handling and storage. Seeds are typically cleaned before use to remove awns, but a restoration of tilled land in Illinois obtained good establishment with seed collected by a combine and broadcast without cleaning (Betz, 1986). Cleanings contain many viable propagules that can be broadcast; doing this in Saskatchewan provided far higher establishment than either drilling or broadcasting cleaned seeds at high rates (Box 19.1, Fig. 19.1).

An even coarser method than using uncleaned seed is the scattering of hay, as recommended for Kansas (Wenger, 1941) and Europe (Bakker *et al.*, 1996). In contrast, adding native hay to Saskatchewan restorations at rates comparable to native prairie standing crop (*c.* 200 g m^{-2}) produced almost no seedlings (Box 19.1).

Seed density
The density of seeds under natural prairie falls within the range of 1000–3000 seeds m^{-2} (Iverson & Wali, 1982; Grilz *et al.*, 1994), but these figures are not directly relevant to restoration because of the differences between restorations and undisturbed prairies in terms of the origin and function of the seed bank.

Duebbert *et al.* (1981) recommend densities of 40 seeds m^{-2} for tallgrass prairie restorations and 30 seeds m^{-2} for mixed grasses in North Dakota, and provide tables listing seed weights of common restoration species.

Stands can be established with higher rates, e.g. 750 seeds m^{-2} in Montana fescue grassland (Tyser *et al.*, 1998), 1000 m^{-2} in Wisconsin (Box 19.2), and 1450 m^{-2} in Illinois (Betz, 1986). To some extent, higher densities yield more standing crop. Population shoot mass of *Bouteloua gracilis* seedlings grown for six years in 62-cm-deep pipes increased with planting density over the range of 6–123 plants m^{-2}

(McGinnies, 1971). In contrast, yield did not increase with density for the larger *Agropyron cristatum*.

Although the relationship between planting density and stand yield is not well known for perennials (Pyke & Archer, 1991), it is clear that competition among seedlings means that high planting densities are wasteful. For example, planting *Agropyron cristatum* in Idaho at 12 times greater density than controls produced only 25% higher yield after three years, and 20% lower yield after eight years (Mueggler & Blaisdell, 1955). An Arkansas old field sown with native species at half, standard and double rates (0.4, 0.9, 1.7 g m^{-2}) produced covers of native grass after eight years of 83%, 82% and 89%; the lowest rate gave satisfactory results (Dale & Smith, 1986). Similarly, a reduced seeding rate (66%) on Nebraska sandhill cropland had no effect on establishment density (Oldfather et al., 1989).

Seed-mix diversity

Increased diversity of mixes enhances stand success by increasing the chance that some species are present that are capable of establishment. The least diverse seed-mix (two species) in a Nebraska sandhill restoration on cropland failed, whereas some of the species in more diverse mixes established (Oldfather et al., 1989). Similarly, more diverse native seed-mixes produced more diverse vegetation after two years in sandy Minnesota old fields (Tilman et al., 1996). The diverse seed-mixes also produced higher standing crop, because they tended to contain larger and faster-growing species otherwise lacking from simpler mixes (Huston, 1997).

Seeding method

Both drilling and broadcasting can work (Schramm, 1978). Drilling gives better results in some cases (Roundy & Call, 1988). Drilling native grasses into heavily grazed Iowa pasture produced a cover of natives of 56%, compared with 37% in broadcast plots (Jackson, 1999). Germination was far higher for seeds buried 1 cm deep (80%) than for surface-sown seeds (20%) for both *Agropyron cristatum* and *Bouteloua gracilis* in Saskatchewan (Ambrose, 1999), suggesting that drilling is more effective than broadcasting.

A cost of drilling, however, is the production of lines of grass which persist for decades (Bleak et al., 1965). This can be partially mitigated by repeated drilling in perpendicular directions (Rock, 1975). In contrast, natural prairie succession can attain natural patterns of heterogeneity (Burke et al., 1995).

Although it has been said that 'broadcasting of seed is generally a waste of time on semiarid rangeland' (Hyder et al., 1975), a review found that broadcasting can be as good as drilling (Duebbert et al., 1981). Native seeds sown into a southwest Saskatchewan old field had significantly higher establishment when broadcast as opposed to drilled (Christian, 1996) (Box 19.1). Broadcast produced threefold higher establishment rates than drilling in a nearby field dominated by *A. cristatum*, although the opposite result occurred in a second field about 200 km away (Bakker et al., 1997). Broadcasting native grasses in Kansas produced stands of native grass dense enough to eliminate introduced perennials (Kindscher & Tieszen, 1998). All 18 species broadcast on tilled soil in a Wisconsin restoration germinated (Howe, 1999) (Box 19.2).

Broadcasting is commonly performed on tilled soils. For restorations on untilled sites, even light tilling as part of site preparation can allow seeds to contact the soil (Bakker et al., 1997). Broadcast seeds can also be encouraged to germinate by applying gravel, soil cracks, litter, or heavy cattle trampling (e.g. five cattle in a 6×6 m plot for 20 minutes) (Winkel et al., 1991). Packing with a heavy roller pulled by a tractor also increases germination by ensuring good contact between seed and soil (McGinnies, 1962). Packing, however, may depress natural amounts of topographical variation (e.g. Kleb & Wilson, 1997) required for normal ecosystem function.

Most importantly, broadcasting allows the development of a natural distribution of plants and, eventually, patterns of soil heterogeneity (E. B. Allen, 1988a), and should always be selected over drilling in ecological restorations.

Time to plant

Spring seeding was found to be superior to fall seeding in southeast Wyoming (Hart & Dean, 1986) and

Wisconsin (Rock, 1975), but fall seeding was better in Idaho (Douglas *et al.*, 1960). Fall-planted seeds of *A. cristatum* in British Columbia that failed to establish in fall emerged the following spring (McLean & Wikeem, 1983); similar results were found for native cool-season grasses in Saskatchewan (Bakker *et al.*, 1998). Seeding season (fall or spring) had little effect on the mean emergence date of a mix of native and introduced grasses planted in southeast Wyoming (Hart & Dean, 1986).

The interpretation of all such experiments is difficult because seasonal effects such as rainfall, soil moisture, relative humidity, and degree–days also vary among years, so that a season that provides good establishment in one year may provide poor establishment in another. This problem was partly overcome in an experiment in Saskatchewan in which eight introduced and native species were sown repeatedly at fortnightly intervals over five years (Kilcher, 1961). Two grasses (*Agropyron cristatum* and green needle grass [*Stipa viridula*]) had highest germination in fall, two in spring (intermediate wheat grass [*A. intermedium*] and Russian wild rye [*Elymus juncaeus*]), and the rest showed no preference. Thus, there appears to be no ideal season for the planting of diverse mixtures.

Another factor to consider is seed predation. Seeds broadcast in fall may be exposed to granivores for a longer period than those sown in the spring.

Seed and seedling predation
The influence of seed predators and grazers on restorations has received little attention, but experimental exclosures examined in an Illinois restoration showed that seed-eating birds reduced plant density by 20%, and grass mass by 24% (Howe & Brown, 1999). Further, voles reduced forb mass by 35%. Voles browsed selectively, particularly reducing some dicots. Similarly, seed predation by rodents in Colorado shortgrass steppe was higher on the larger seeds of buffalograss (*Buchloe dactyloides*, 70%) than on the smaller seeds of *Bouteloua gracilis* (30%) (Hoffmann *et al.*, 1995). The well-known among-year variability of small mammal population size may cause similar variations in the magnitude of the impact of herbivory.

Post-establishment losses
Successful restorations can be characterised by long-term increases in native species. A mix of native C_4 grasses sown in Arkansas increased from 23% after one year to 82% after eight years (Dale & Smith, 1986). Such successes unfortunately seem to be the exception.

Density typically decreases following establishment. Plantings of *B. gracilis* in Colorado lost 80% of the population in the first year (Bement *et al.*, 1961). Native grasses decreased from 163 to 6 m^{-2} over one year in a Nebraska sandhills restoration (King *et al.*, 1989). Mixes of introduced and native grasses on tilled ground in eastern Washington declined from 28 seedlings m^{-1} of row to 5 m^{-1} after two years; on untilled ground, density decreased from 21 m^{-1} to 0 (Gates & Robocker, 1960). Stands of *A. cristatum* in Utah were initially successful but died out after 17 years (Bleak *et al.*, 1965), suggesting that native grass restorations could also disappear over the long term. Diversity also decreases over time. Only 28 out of 47 species initially established in the Curtis Prairie in Wisconsin remained after 50 years (Sperry, 1994).

Vegetative propagules
The reliance of prairie plants on vegetative growth to recolonise small disturbances (Umbanhowar, 1992) suggests that vegetative propagules should be useful for restoration. Sods of *B. gracilis* were successfully transplanted in Colorado (McGinnies & Wilson, 1982), and soil collected by earth-moving machinery contained rhizomes which established on mine spoils (Howard & Samuel, 1979).

Collecting individual ramets causes less harm to source areas than sod-stripping (Pywell *et al.*, 1995). Seedlings can also be started in the greenhouse (Wilson & Tilman, 1991). Transplants of Mead's milkweed (*Asclepias meadii*) established better than seedlings in hay meadows in Kansas (Bowles *et al.*, 1998). Transplant growth is much reduced by established vegetation, but transplant survival is often facilitated by neighbours (Wilson & Tilman, 1991), suggesting that missing species can be transplanted into restorations. Greenhouse-grown transplants are widely used for experimental research in prairie and for reforestation, and might be more widely used to overcome the establishment filter in prairie restorations (Anderson, 1995). Vegetative propagules are usually not used for restoration because of the labour involved in starting or collecting transplants, and in planting them over large areas.

CONTINUING MANAGEMENT

There are many papers concerning the establishment of restorations, and many others on the management of natural grasslands, but relatively few address the continuing management of established restorations. Management can be as important as the species composition of the seed-mix in determining the character of a restoration (Jones & Hayes, 1999). Given that prairie ecosystems arrived at their current condition after thousands of years, long-term management is important. Because of the time involved, restorations must be regarded as ongoing endeavours, not one- to two-year engineering projects (Tyser et al., 1998).

Fire

In eastern tallgrass prairie with high standing crop, fire removes litter and increases nutrient mineralisation, light penetration, productivity and diversity (Collins et al., 1995). To the west and north, however, where standing crop is lower, fire has little effect on species composition (Wilson & Shay, 1990). The usefulness of fire in restoration management probably follows the same pattern.

The lack of impact of fire on introduced perennials was discussed above. Burning had no positive effect on the relative abundance of native species in a Wisconsin restoration (Robocker & Miller, 1955).

In eastern prairies, where fire influences species composition, easily controlled fires in early spring or late fall may have little negative effect on large, dominant C_4 grasses, and may allow them to increase in abundance at the cost of a loss of diversity. Instead, a mixture of fire (and grazing) seasons could be used to reduce grass dominance and promote diversity (Howe, 1994). Experiments in Arkansas found significant differences in fire effects between seasons: dormant-season fire increased

grasses, and decreased forbs and diversity, whereas growing-season fire had the opposite effect (Sparks et al., 1998). Likewise, August burning or mowing in a Wisconsin restoration doubled the density of the perennial native forb golden alexander (Zizia aurea) relative to unburned controls and to May burning (Howe, 1999). Spring-flowering forbs in an earlier restoration had a total cover of 6% in spring-burned plots, and a cover of 46% in plots burned in summer; fall-flowering forbs had a cover of 92% in spring-burned plots, but 47% in plots burned in summer (Howe, 1995). Fire season clearly affects species composition in tallgrass prairie.

Increased plant diversity should affect other trophic levels: a mathematical population model of a butterfly (Fender's blue butterfly, Icaricia icarioides fenderi) that specialises on prairie plants predicted maximum population growth rate in cases where one-third of the prairie was burned annually (Schultz & Crone, 1998).

Grazing

Grazers could be used to disperse desired species from native vegetation to restorations (Archer & Pyke, 1991; Bakker & Berendse, 1999). Evidence for this comes from Montana, where abandoned fields supported only scattered individuals of Bouteloua gracilis if they were ungrazed or lightly grazed, but were dominated by B. gracilis if heavily grazed (Judd, 1940).

As in the case of fire, grazing can decrease dominance and increase diversity (Grime, 1973). Mixes of native prairie species can tend towards low diversity over time as standing crop accumulates (Pyke & Archer, 1991). Grazing may be more effective than fire on dry prairies because fire causes only an occasional removal of litter and standing crop. Grazing, in contrast, can constantly remove leaves, resulting in decreased root growth (Roundy et al., 1985). Because grassland competition is primarily belowground (Wilson, 1998), reduced root mass decreases competition. Native prairie in Nebraska that previously had been heavily grazed showed an increase in the tallgrass Andropogon gerardii from 4% to 26% over 17 years when grazing was restricted to light grazing every four years; shorter species declined (Nagel *et al.*, 1994). Many of the benefits of grazing may be simulated by mowing: monthly hay removal from a Wisconsin restoration increased short grasses such as *Bouteloua curtipendula* and decreased taller grasses such as *Andropogon gerardii* (Robocker & Miller, 1955).

Adding more species

Restorations initially need seeds, and this need may persist. Tallgrasses were seeded into a 90-year-old cultivated field in Kansas. Thirty-five years later, the seed bank of the restoration was only 33% as dense as that of adjacent undisturbed prairie, and the seed rain was only 14% (Schott & Hamburg, 1997). Adding a high rate of seed (4.5 g m^{-2}) to an undisturbed tallgrass understory of a Minnesota oak savanna increased richness by as much as 83%, and increased total vegetation cover by 31% (Tilman, 1997).

Rejuvenation

The standing crop and shoot density of restorations sometimes decrease over time, resulting in calls to 'rejuvenate' such stands in order to create cover for duck nests (Duebbert *et al.*, 1981; Wark *et al.*, 1995). This is accomplished by tilling, which kills plants, reduces nutrient uptake, allows nutrients to pool temporarily in the soil, and increases shoot productivity over the short term. Old prairie successions, however, are characterised by low shoot mass and high root:shoot ratios (Burke *et al.*, 1995; Christian & Wilson, 1999). Rejuvenation by cultivating the stand introduces annual and perennial weeds, increases soil nutrient availability and leaching, and decreases diversity, and thus has no role in the management of a restoration for ecological purposes.

CONCLUDING REMARKS

Grasslands can occur in alternative stable states (Westoby *et al.*, 1989), such as native prairie, communities of introduced forage grasses (Christian & Wilson, 1999) or stands of unwanted introduced perennials such as *Euphorbia esula* (Belcher & Wilson, 1989) or *Centaurea maculosa* (Callaway & Aschehoug, 2000). Instead of attempting to move restorations toward some idealised dynamic equilibrium, it might be more realistic and practical to recognise that some alternative stable states are more desirable than others, and: (1) to exploit opportunities to move towards them when possible, and (2) to avoid moving towards less desirable states if possible (Westoby *et al.*, 1989). For example, our ability to restore stands of introduced perennials to native vegetation remains doubtful. Until this ability is better developed, management efforts might be more usefully addressed to preventing native prairie from being invaded and dominated by introduced perennials, through minimising establishment opportunities afforded by disturbance, and the intensive control of small patches of aliens.

Recommendations for restoration

Restoration procedures vary with initial conditions (Fig. 19.2), but all procedures are predicated on the following principles:

- A gradient of increasing soil moisture occurs from west to east, with implications for establishment and productivity.
- Burning, mowing, and grazing have interchangeable effects to a certain extent, so that less controversial methods, such as mowing, can be employed where more controversial methods, such as fire, must be avoided.
- Seeds of local provenance should be used, with species mixes appropriate for landscape position.
- Seeds should be added by broadcasting.
- Seed addition may need to be repeated due to weather-related establishment failure.
- Fertilisation decreases diversity and increases exotics.
- Soil disturbance after establishment increases exotics and nutrient availability.

Restoration procedures for different initial conditions can be assigned to the following four categories.

Stands of introduced perennials without native species

The main aim of site preparation is to greatly reduce the population of introduced perennials and

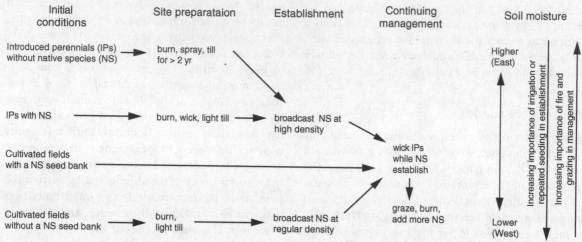

Fig. 19.2. Recommendations for restoring North American prairie starting from four initial conditions. Stands of introduced perennials are more common in the west; cultivated fields are more common in the east. In addition, the following recommendations apply to all situations: burning, mowing, and grazing have largely interchangeable effects; seeds of local provenance are added by broadcasting, with species mixes appropriate for landscape position; seed addition may need to be repeated due to weather-related establishment failure; fertilisation is inappropriate; soil disturbance after establishment is inappropriate. IPs: introduced perennials; NS: native prairie species.

prevent them from returning (Box 19.2). Burning (or mowing or grazing) will reduce standing crop and litter and provide better contact of herbicide with target vegetation. Herbicide spraying is more effective than wicking. The soil should be tilled to kill rhizomes and to encourage germination of buried seeds of introduced perennials. These should be killed in a second round of burning, spraying and tilling.

Seeds of native species should be broadcast at high density to provide competitive suppression of regrowing introduced perennials.

Selective control of introduced species during the initial phases of establishment should avoid the desired species by using techniques such as hand wicking. Established restorations should be managed with appropriate combinations of grazing, fire, and the addition of more native species.

Stands of introduced perennials mixed with native species

This initial condition occurs where native prairie has had introduced perennials added, or where native prairie has been invaded by introduced perennials, or where stands of introduced perennials have been invaded by native species (Box 19.1). In all cases, the extant native populations represent genetic diversity that deserve preservation, so the aim is to control the introduced perennials and increase and supplement the natives.

Burning or mowing will remove litter and standing crop, providing good contact for herbicide. In shortgrass prairie, a general herbicide such as Round-up, applied with a wick, will selectively affect tall, introduced species. A second round of burning followed by very light tilling will allow added seeds to reach the soil. Seed addition and further management should follow the recommendations given above.

Cultivated fields with a seed bank of native species

This situation may occur in a prairie that has been only temporarily disturbed. There is abundant evidence that prairie can re-establish, so, apart from selective control of invading introduced perennials, no further management may be necessary.

Cultivated fields without a seed bank of native species

Burning followed by light tilling will allow broad-cast seeds to reach the soil. There is no need for weed control or especially high densities of native seeds because alien weeds will disappear through succession (Box 19.1). The established stand can then be managed as outlined for stands of introduced species.

ACKNOWLEDGEMENTS

I thank Grasslands National Park and the Natural Sciences and Engineering Research Council of Canada for support, and Mike Christensen for li-brary help.

REFERENCES

Aber, J. D. & Melillo, J. M. (1991). *Terrestrial Ecosystems.* Philadelphia, PA: W. B. Saunders.

Allen, E. B. (1988a). Some trajectories of succession in Wyoming sagebrush grassland: implications for restoration. In *The Reconstruction of Disturbed Arid Lands: An Ecological Approach*, ed. E. B. Allen, pp. 89–112. Boulder, CO: Westview Press.

Allen, E. B. (ed.) (1988b). *The Reconstruction of Disturbed Arid Lands: An Ecological Approach*. Boulder, CO: Westview Press.

Allen, M. F. (1988). Below-ground structure: a key to reconstructing a productive arid ecosystem. In *The Reconstruction of Disturbed Arid Lands: An Ecological Approach*, ed. E. B Allen, pp. 113–135. Boulder, CO: Westview Press.

Ambrose, L. G. (1999). Seed persistence of an introduced and a native grass species in a prairie old field. MSc dissertation, University of Regina, Saskatchewan.

Andersen, M. R., Depuit, E. J., Abernathy, R. H. & Kleinman, L. H. (1992). Value of mountain rye for suppression of annual bromegrasses on semiarid mined lands. *Journal of Range Management*, **45**, 345–351.

Anderson, P. (1995). Ecological restoration and creation: a review. *Biological Journal of the Linnean Society*, **56**, 187–211.

Archer, S. & Pyke, D. A. (1991). Plant–animal interactions affecting plant establishment and persistence on revegetated rangeland. *Journal of Range Management*, **44**, 558–565.

Archibold, O. W. (1981). Buried viable propagules in native prairie and adjacent agricultural sites in central Saskatchewan. *Canadian Journal of Botany*, **59**, 701–706.

Asay, K. H. & Johnson, D. A. (1983). Genetic variability for characters affecting stand establishment in crested wheatgrass. *Journal of Range Management*, **36**, 703–706.

Bahler, C. C., Moser, L. E., Griffin, T. S. & Vogel, K. P. (1990). Warm-season grass establishment as affected by post-planting atrazine application. *Journal of Range Management*, **43**, 421–424.

Bakker, J. D. (1996). Competition and the establishment of native grasses in crested wheatgrass fields. MSc dissertation, University of Regina, Saskatchewan.

Bakker, J. P. & Berendse, F. (1999). Constraints in the restoration of ecological diversity in grassland and heathland communities. *Trends in Ecology and Evolution*, **14**, 63–68.

Bakker, J. P., Poschland, P., Strykstra, R. J., Bekker, R. M. & Thompson, K. (1996). Seed banks and seed dispersal: important topics in restoration ecology. *Acta Botanica Neerlandica*, **45**, 461–490.

Bakker, J. D., Christian, J., Wilson, S. D. & Waddington, J. (1997). Seeding blue grama in old crested wheatgrass fields in southwestern Saskatchewan. *Journal of Range Management*, **50**, 156–159.

Bakker, J. D., Christian, J., Wilson, S. D., Li, X., Ambrose, L. & Waddington, J. (1998). *The Establishment of Native Grasses in Crested Wheatgrass and a Cultivated Field in Grasslands National Park: Final Report*. Regina, Saskatchewan: Grasslands National Park and Canada–Saskatchewan Agricultural Green Plan Agreement.

Belcher, J. W. & Wilson, S. D. (1989). Leafy spurge (*Euphorbia esula* L.) and the species composition of mixed-grass prairie. *Journal of Range Management*, **42**, 171–175.

Bement, R. E., Hervey, D. F., Everson, A. C. & Hylton, L. O., Jr (1961). Use of asphalt-emulsion mulches to hasten grass-seedling establishment. *Journal of Range Management*, **14**, 102–109.

Bement, R. E., Barmington, R. D., Everson, A. C., Hylton, L. O., Jr & Remmenga, E. E. (1965). Seeding of abandoned croplands in the central Great Plains. *Journal of Range Management*, **18**, 53–59.

Betz, R. F. (1986). One decade of research in prairie restoration at the Fermi National Accelerator Laboratory (Fermilab), Batavia, Illinois. In *Proceedings of the 9th North American Prairie Conference*, eds. G. K. Clambey & R. H. Pemble, pp. 179–185. Fargo, ND: Tri-College University Center for Environmental Studies.

Biedenbender, S. H. & Roundy, B. A. (1996). Establishment of native semidesert grasses into existing stands of *Eragrostis lehmanniana* in southeastern Arizona. *Restoration Ecology*, **4**, 155–162.

Bleak, A. T., Frischknecht, N. C., Plummer, A. P. & Eckert, R. E., Jr (1965). Problems in artificial and natural revegetation of the arid shadscale vegetation zone of Utah and Nevada. *Journal of Range Management*, **18**, 59–65.

Bock, J. H. & Bock, C. E. (1995). The challenges of grassland conservation. In *The Changing Prairie: North American Grasslands*, eds. A. Joern & K. H. Keeler, pp. 199–222. New York: Oxford University Press.

Boerner, R. E. J., Demars, B. G. & Leicht, P. N. (1996). Spatial patterns of mycorrhizal infectiveness of soils along a successional chronosequence. *Mycorrhiza*, **6**, 79–90.

Bovey, B. W., Meyer, R. E., Merkle, M. G. & Bashaw, E. C. (1986). Effect of herbicides and handweeding on establishment of kleingrass and buffelgrass. *Journal of Range Management*, **39**, 547–551.

Bowles, M. L., McBride, J. L. & Betz, R. F. (1998). Management and restoration ecology of the federal threatened Mead's milkweed, *Asclepias meadii* (Asclepiadaceae). *Annals of the Missouri Botanical Garden*, **85**, 110–125.

Briske, D. D. & Wilson, A. M. (1980). Drought effects on adventitious root developing in blue grama seedlings. *Journal of Range Management*, **33**, 323–327.

Burke, I. C., Lauenroth, W. K. & Coffin, D. P. (1995). Soil organic matter recovery in semiarid grasslands: implications for the Conservation Reserve Program. *Ecological Applications*, **5**, 793–801.

Call, C. A. & Roundy, B. A. (1991). Perspectives and processes in revegetation of arid and semiarid rangelands. *Journal of Range Management*, **44**, 543–549.

Callaway, R. M. & Aschehoug, E. T. (2000). Invasive plants versus their new and old neighbors: a mechanism for exotic invasion. *Science*, **290**, 521–523.

Christian, J. M. (1996). Revegetation of abandoned cropland in southwestern Saskatchewan using native species, alien species and natural succession. MSc dissertation, University of Regina, Saskatchewan.

Christian, J. M. & Wilson, S. D. (1999). Long-term ecosystem impacts of an introduced grass in the northern Great Plains. *Ecology*, **80**, 2397–2407.

Cluff, G. J., Young, J. A. & Evans, R. A. (1983). Edaphic factors influencing the control of Wyoming big sagebrush and seedling establishment of crested wheatgrass. *Journal of Range Management*, **36**, 786–792.

Coffin, D. P. & Lauenroth, W. K. (1990). A gap dynamics simulation model of succession in a semiarid grassland. *Ecological Modelling*, **49**, 229–266.

Coffin, D. P., Lauenroth, W. K. & Burke, I. C. (1996). Recovery of vegetation in a semiarid grassland 53 years after disturbance. *Ecological Applications*, **6**, 538–555.

Collins, S. L. (1987). Interaction of disturbances in tallgrass prairie: a field experiment. *Ecology*, **68**, 1243–1250.

Collins, S. L., Glenn, S. M. & Gibson, D. J. (1995). Experimental analysis of intermediate disturbance and initial floristic composition: decoupling cause and effect. *Ecology*, **76**, 486–492.

Cook, B. D., Jastrow, J. D. & Miller, R. M. (1988). Root and mycorrhizal endophyte development in a chronosequence of restored tallgrass prairie. *New Phytologist*, **110**, 355–362.

Costello, D. F. (1944). Natural revegetation of abandoned plowed land in the mixed prairie association of northeastern Colorado. *Ecology*, **25**, 312–326.

Coupland, R. T. (1992). *Natural Grasslands*. Amsterdam: Elsevier.

Cox, M. L. & McCarty, M. K. (1958). Some factors affecting establishment of desirable forage plants in weedy bluegrass pastures of eastern Nebraska. *Journal of Range Management*, **11**, 159–164.

D'Antonio, C. M., Dudley, T. L. & Mack, M. (1999). Disturbance and biological invasions: direct effects and feedbacks. In *Ecosystems of Disturbed Ground*, ed. L. R. Walker, pp. 413–452. Amsterdam: Elsevier.

Dale, J. E. E. & Smith, T. C. (1986). The effects of differing seeding densities on establishment of grasses in a restored prairie at Pea Ridge National Military Park, Arkansas. In *The Prairie: Past, Present and Future*, eds. G. K. Clambey & R. H. Pemble, pp. 195–196. Fargo, ND: Tri-college University Center for Environmental Studies.

DePuit, E. J. & Redente, E. F. (1988). Manipulation of ecosystem dynamics on reconstructed semiarid lands.

In *The Reconstruction of Disturbed Arid Lands: An Ecological Approach*, ed. E. B Allen, pp. 162–204. Boulder, CO: Westview Press.

DePuit, E. J., Skilbred, C. L. & Coenenberg, J. G. (1982). Effects of two years of irrigation on revegetation of coal surface-mined land in southeastern Montana. *Journal of Range Management*, **35**, 67–74.

Dobrowolski, J. P., Caldwell, M. M. & Richards, J. H. (1990). Basin hydrology and plant root systems. In *Ecological Studies: Analysis and Synthesis*, vol. 80, *Plant Biology of the Basin and Range*, eds. B. Osmond, G. Hidy & C. Pitelka, pp. 243–292. New York: Springer-Verlag.

Douglas, D. S., Hafenrichter, A. L. & Klages, K. H. (1960). Cultural methods and their relation to establishment of native and exotic grasses in range seeding. *Journal of Range Management*, **13**, 53–56.

Dovel, R. L., Hussey, M. A. & Holt, E. C. (1990). Establishment and survival of Illinois bundleflower interseeded into an established kleingrass pasture. *Journal of Range Management*, **43**, 153–156.

Duebbert, H. F., Jacobson, E. T., Higgins, K. F. & Podoll, E. B. (1981). *Establishment of Seeded Grasslands for Wildlife Habitat in the Prairie Pothole Region*, Scientific Report – Wildlife no. 234. Washington, DC: US Department of the Interior, Fish and Wildlife Service.

Epstein, H. E., Lauenroth, W. K., Burke, I. C. & Coffin, D. P. (1997). Productivity patterns of C_3 and C_4 functional types in the U.S. Great Plains. *Ecology*, **78**, 722–731.

Evans, R. A. & Young, J. A. (1970). Plant litter and establishment of alien weed species in rangeland communities. *Weed Science*, **18**, 697–703.

Fowler, N. L. (1988). What is a safe site? Neighbors, litter, germination date, and patch effects. *Ecology*, **69**, 947–961.

Frasier, G. W., Cox, J. R. & Woolhiser, D. A. (1987). Wet-dry cycle effects on warm-season grass seedling establishment. *Journal of Range Management*, **40**, 2–6.

Garrison, G. A. & Moore, A. W. (1956). Relation of the Dalles pocket gopher to establishment and maintenance of range grass plantings. *Journal of Range Management*, **9**, 181–184.

Gates, D. H. & Robocker, W. C. (1960). Revegetation with adapted grasses in competition with dalmatian toadflax and St John's wort. *Journal of Range Management*, **13**, 322–326.

Gibson, D. J., Seastedt, T. R. & Briggs, J. M. (1993). Management practices in tallgrass prairie: landscape and small-scale experimental effects on species composition. *Journal of Applied Ecology*, **30**, 247–256.

Grilz, P. L. & Romo, J. T. (1995). Management considerations for controlling smooth brome in fescue prairie. *Natural Areas Journal*, **15**, 148–156.

Grilz, P. L., Romo, J. T. & Young, J. A. (1994). Comparative germination of smooth brome and plains rough fescue. *Prairie Naturalist*, **26**, 157–170.

Grime, J. P. (1973). Competitive exclusion in herbaceous vegetation. *Nature*, **242**, 344–347.

Harris, G. A. & Dobrowolski, J. P. (1986). Population dynamics of seeded species on northeast Washington semiarid sites, 1948–1983. *Journal of Range Management*, **34**, 46–51.

Hart, R. H. & Dean, J. G. (1986). Forage establishment: weather effects on stubble vs. fallow and fall vs. spring seeding. *Journal of Range Management*, **39**, 228–230.

Hoffmann, L. A., Redente, E. F. & McEwen, L. C. (1995). Effects of selective seed predation by rodents on shortgrass establishment. *Ecological Applications*, **5**, 200–208.

Holechek, J. L., Depuit, E. J., Coenenberg, J. & Valdez, R. (1982). Long-term plant establishment on mined lands in southeastern Montana. *Journal of Range Management*, **35**, 522–525.

Howard, G. S. & Samuel, M. J. (1979). The value of fresh-stripped topsoil as a source of useful plants for surface mine revegetation. *Journal of Range Management*, **32**, 76–77.

Howe, H. F. (1994). Managing species diversity in tallgrass prairie: assumptions and implications. *Conservation Biology*, **8**, 691–704.

Howe, H. F. (1995). Succession and fire season in experimental prairie plantings. *Ecology*, **76**, 1917–1925.

Howe, H. F. (1999). Response of *Zizia aurea* to seasonal mowing and fire in a restored prairie. *American Midland Naturalist*, **141**, 373–380.

Howe, H. F. & Brown, J. S. (1999). Effects of birds and rodents on synthetic tallgrass communities. *Ecology*, **80**, 1776–1781.

Huffaker, C. B. (1951). The return of native perennial bunchgrass following the removal of Klamath weed (*Hypericum perforatum*) by imported beetles. *Ecology*, **32**, 443–458.

Huston, M. A. (1997). Hidden treatments in ecological experiments: re-evaluating the ecosystem function of biodiversity. *Oecologia*, **110**, 449–460.

Hyder, D. N., Everson, A. C. & Bement, R. E. (1971). Seedling morphology and seedling failures with blue grama. *Journal of Range Management*, **24**, 287–292.

Hyder, D. N., Bement, R. E., Remmenga, E. E. & Hervey, D. F. (1975). *Ecological Responses of Native Plants and Guidelines for Management of Shortgrass Range*, Technical Bulletin no. 1503. Washington, DC: US Department of Agriculture.

Iverson, L. R. & Wali, M. K. (1982). Buried, viable seeds and their relation to revegetation after surface mining. *Journal of Range Management*, **35**, 648–652.

Jackson, L. L. (1999). Establishing tallgrass prairie on grazed permanent pasture in the Upper Midwest. *Restoration Ecology*, **7**, 127–138.

Jacobson, E. T., Wark, D. B., Arnott, R. G., Haas, R. J. & Tober, D. A. (1994). Sculptured seeding: an ecological approach to revegetation. *Restoration and Management Notes*, **12**, 46–50.

Jacoby, P. W., Jr (1969). Revegetation treatments for stand establishment on coal spoil banks. *Journal of Range Management*, **22**, 94–7.

Janssens, F., Peeters, A., Tallowin, J. R. B., Bakker, J. P., Bekker, R. M., Fillat, F. & Oomes, M. J. M. (1998). Relationship between soil chemical factors and grassland diversity. *Plant and Soil*, **202**, 69–78.

Jones, A. T. & Hayes, M. J. (1999). Increasing floristic diversity in grassland: the effects of management regime and provenance on species introduction. *Biological Conservation*, **87**, 381–90.

Jordan, W. R. III (1988). Ecological restoration. In *Biodiversity*, eds. E. O Wilson & F. M. Peter, pp. 311–316. Washington, DC: National Academy Press.

Jordan, W. R. III, Gilpin, M. E. & Aber, J. D. (1987). *Restoration Ecology: A Synthetic Approach to Ecological Restoration*. Cambridge: Cambridge University Press.

Judd, B. I. (1940). Natural succession of vegetation on abandoned farmlands in Teton County, Montana. *Journal of the American Society of Agronomy*, **32**, 330–336.

Kessler, E. & Kirkham, M. B. (1985). Restoration of eroded prairie with digested sludge. *Proceedings of the Oklahoma Academy of Science*, **65**, 25–34.

Kilcher, M. R. (1961). Fall seeding versus spring seeding in the establishment of five grasses and one alfalfa in southern Saskatchewan. *Journal of Range Management*, **14**, 320–322.

Kindscher, K. & Tieszen, L. L. (1998). Floristic and soil organic matter changes after five and thirty-five years of native tallgrass prairie restoration. *Restoration Ecology*, **6**, 181–196.

King, M. A., Waller, S. S., Moser, L. E. & Stubbendieck, J. L. (1989). Seedbed effects on grass establishment on abandoned Nebraska Sandhills cropland. *Journal of Range Management*, **42**, 183–187.

Kleb, H. R. & Wilson, S. D. (1997). Vegetation effects on soil resource heterogeneity in prairie and forest. *American Naturalist*, **150**, 283–298.

Kline, V. M. (1986). Response of *Melilotus alba* and associated prairie vegetation to seven different burning and mowing treatments. In *Proceedings of the 9th North American Prairie Conference*, eds. G. K. Clambey & R. H. Pemble, pp. 149–152. Fargo, ND: Tri-College University Center for Environmental Studies.

Kline, V. M. & Howell, E. A. (1987). Prairies. In *Restoration Ecology: A Synthetic Approach to Ecological Research*, eds. W. R. Jordan III, M. E. Gilpin & J. D. Aber, pp. 75–83. Cambridge: Cambridge University Press.

Lauenroth, W. K. & Coffin, D. P. (1992). Belowground processes and the recovery of semiarid grasslands from disturbance. In *Ecosystem Rehabilitation*, vol. 2, *Ecosystem Analysis and Synthesis*, ed. M. K Wali, pp. 131–150. The Hague: SBP Academic Publishing.

Lauenroth, W. K., Sala, O. E., Coffin, D. P. & Kirchner, T. B. (1994). The importance of soil water in the recruitment of *Bouteloua gracilis* in the shortgrass steppe. *Ecological Applications*, **4**, 741–749.

Lippitt, L., Fidelibus, M. W. & Bainbridge, D. A. (1994). Native seed collection, processing, and storage for revegetation graphics in the western United States. *Restoration Ecology*, **2**, 120–131.

Masters, R. A., Vogel, K. P., Reece, P. E. & Bauer, D. (1990). Sand bluestem and prairie sandreed establishment. *Journal of Range Management*, **43**, 540–544.

Masters, R. A., Nissen, S. J., Gaussoin, R. E., Beran, D. D. & Stougaard, R. N. (1996). Imidazolinone herbicides improve restoration of Great Plains grasslands. *Weed Technology*, **10**, 392–403.

McGinnies, W. J. (1962). Effect of seedbed firming on the establishment of crested wheatgrass seedlings. *Journal of Range Management*, **15**, 230–234.

McGinnies, W. (1968). Effect of post-emergence weed control on grass establishment in north-central Colorado. *Journal of Range Management*, **21**, 126–128.

McGinnies, W. J. (1971). Effects of controlled plant spacing on growth and mortality of three range grasses. *Agronomy Journal*, **63**, 868–870.

McGinnies, W. J. (1987). Effects of hay and straw mulches on the establishment of seeded grasses and legumes on rangeland and a coalstrip mine. *Journal of Range Management*, **40**, 119–121.

McGinnies, W. J. & Wilson, A. M. (1982). Using blue grama sod for range revegetation. *Journal of Range Management*, **35**, 259–261.

McLean, A. & Wikeem, S. J. (1983). Effect of time of seeding on emergence and long-term survival of crested wheatgrass in British Columbia. *Journal of Range Management*, **36**, 694–700.

Miller, R. F., Doescher, P. S. & Wang, J. (1991). Response of *Artemisia tridentata* spp. *wyomingensis* and *Stipa thurberiana* to nitrogen amendments. *American Midland Naturalist*, **125**, 104–113.

Morghan, K. J. R. & Seastedt, T. R. (1999). Effects of soil nitrogen reduction on non-native plants in restored grasslands. *Restoration Ecology*, **7**, 51–55.

Mueggler, W. F. & Blaisdell, J. P. (1955). Effect of seeding rate upon establishment and yield of crested wheatgrass. *Journal of Range Management*, **8**, 74–76.

Nagel, H. G., Nicholson, R. A. & Steuter, A. A. (1994). Management effects on Willa Cather Prairie after 17 years. *Prairie Naturalist*, **26**, 241–250.

Noyd, R. K., Pfleger, F. L. & Russelle, M. P. (1995). Interactions between native prairie grasses and indigenous arbuscular mycorrhizal fungi: implications for reclamation of taconite iron ore tailing. *New Phytologist*, **129**, 651–660.

Noyd, R. K., Pfleger, F. L. & Norland, M. R. (1996). Field responses to added organic matter, arbuscular mycorrhizal fungi, and fertiliser in reclamation of taconite iron ore tailing. *Plant and Soil*, **179**, 89–97.

Oldfather, S., Stubbendieck, J. & Waller, S. S. (1989). Evaluating revegetation practices for sandy cropland in the Nebraska Sandhills. *Journal of Range Management*, **42**, 257–259.

Packard, S. & Mutel, C. F. (1997). *The Tallgrass Restoration Handbook: For Prairies, Savannas, and Woodlands*. Washington, DC: Island Press.

Parton, W. J., Cole, C. V., Stewart, J. W. B., Ojima, D. S. & Schimel, D. S. (1989). Simulating regional patterns of soil C, N, and P dynamics in the U.S. central grasslands region. In *Ecology of Arable Land: Perspectives and Challenges*, eds. M. Clarholm & L. Bergstrom, pp. 99–108. Dordrecht: Kluwer.

Peltzer, D. A., Wilson, S. D. & Gerry, A. K. (1998). Competition intensity along a productivity gradient in a low-diversity grassland. *American Naturalist*, **151**, 465–476.

Power, J. F. (1980). Response of semiarid grassland sites to nitrogen fertilisation. 1: Plant growth and water use. *Soil Science Society of America Journal*, **44**, 545–550.

Power, J. F. (1981). Nitrogen in the cultivated ecosystem. In *Terrestrial Nitrogen Cycles*, eds. F. E. Clark & T. Rosswall, pp. 529–546. Stockholm: Swedish Natural Science Research Council.

Pyke, D. A. & Archer, S. (1991). Plant–plant interactions affecting plant establishment and persistence on revegetated rangeland. *Journal of Range Management*, **44**, 550–557.

Pywell, R. F., Webb, N. R. & Putwain, P. D. (1995). A comparison of techniques for restoring heathland on abandoned farmland. *Journal of Applied Ecology*, **32**, 400–411.

Redente, E. F. & DePuit, E. J. (1988). Reclamation of drastically disturbed lands. In *Vegetation Science Applications for Rangeland Analysis and Management*, ed. P. T. Tueller, pp. 559–589. Dordrecht: Kluwer.

Reeves, F. B., Wagner, D., Moorman, T. & Kiel, J. (1979). The role of endomycorrhizae in revegetation practices in the semi-arid west. 1: A comparison of incidence of mycorrhizae in severely disturbed vs. natural environments. *American Journal of Botany*, **66**, 6–13.

Ringe, J. M. & Graves, D. H. (1987). Economic factors affecting mulch choices for revegetating disturbed land. *Reclamation and Revegetation Research*, **6**, 121–128.

Robocker, W. C. & Miller, B. J. (1955). Effects of clipping, burning and competition on establishment and survival of some native grasses in Wisconsin. *Journal of Range Management*, **8**, 117–120.

Rock, H. W. (1975). *Prairie Propagation Handbook*, 4th edn. Milwaukee, WI: Boerner Botanical Gardens, Whitnall Park, Milwaukee County Park System.

Romo, J. T., Grilz, P. L. & Delanoy, L. (1994). Selective control of crested wheatgrass (*Agropyron cristatum* [L.] Gaertn. and *A. desertorum* Fisch.) in the northern Great Plains. *Natural Areas Journal*, **14**, 308–309.

Roundy, B. A. & Call, C. A. (1988). Revegetation of arid and semiarid rangelands. In *Vegetation Science Applications for*

Rangeland Analysis and Management, ed. P. T. Tueller, pp. 607–635. Dordrecht: Kluwer.

Roundy, B. A., Cluff, G. J., McAdoo, J. K. & Evans, R. A. (1985). Effects of jackrabbit grazing, clipping and drought on crested wheatgrass seedlings. *Journal of Range Management*, **38**, 551–555.

Sala, O. E. & Lauenroth, W. K. (1982). Small rainfall events: an ecological role in semiarid regions. *Oecologia*, **53**, 301–304.

Schott, G. W. & Hamburg, S. P. (1997). The seed rain and seed bank of an adjacent native tallgrass prairie and old field. *Canadian Journal of Botany*, **75**, 1–7.

Schramm, P. (1978). The 'do's and don'ts' of prairie restoration. In *Proceedings of the 5th Midwest Prairie Conference*, eds. D. C. Glenn-Lewin & R. Q. Landers Jr, pp. 139–150. Ames, IA: Iowa State University.

Schultz, C. B. & Crone, E. E. (1998). Burning prairie to restore butterfly habitat: a modeling approach to management tradeoffs for the Fender's blue. *Restoration Ecology*, **6**, 244–252.

Seastedt, T. R. (1995). Soil systems and nutrient cycles of the North American prairie. In *The Changing Prairie: North American Grasslands*, eds. A. Joern & K. H. Keeler, pp. 157–174. New York: Oxford University Press.

Shirley, S. (1994). *Restoring the Tallgrass Prairie*. Iowa City, IA: University of Iowa Press.

Smith, D. D. (1998). Iowa prairie: original extent and loss, preservation and recovery attempts. *Journal of the Iowa Academy of Science*, **105**, 94–108.

Smith, M. R., Charvat, I. & Jacobson, R. L. (1998). Arbuscular mycorrhizae promote establishment of prairie species in a tallgrass prairie restoration. *Canadian Journal of Botany*, **76**, 1947–1954.

Sparks, J. C., Masters, R. E., Engle, D. E., Palmer, M. W. & Bukenhofer, G. A. (1998). Effects of late growing-season and late dormant-season prescribed fire on herbaceous vegetation in restored pine–grassland communities. *Journal of Vegetation Science*, **9**, 133–142.

Sperry, T. M. (1994). The Curtis Prairie restoration, using the single-species planting method. *Natural Areas Journal*, **14**, 124–127.

Stubbendieck, J. & McCully, W. G. (1972). Factors affecting germination, emergence and establishment of sand bluestem. *Journal of Range Management*, **25**, 383–385.

Tilman, D. (1997). Community invasibility, recruitment limitation, and grassland biodiversity. *Ecology*, **78**, 81–92.

Tilman, D. & El Haddi, A. (1992). Drought and diversity in grasslands. *Oecologia*, **89**, 257–264.

Tilman, D., Wedin, D. & Knops, J. (1996). Productivity and sustainability influenced by biodiversity in grassland ecosystems. *Nature*, **379**, 718–720.

Tyser, R. W., Asebrook, J. M., Potter, R. W. & Kurth, L. L. (1998). Roadside revegetation in Glacier National Park, USA: effects of herbicide and seeding treatments. *Restoration Ecology*, **6**, 197–206.

Umbanhowar, C. E., Jr (1992). Early patterns of revegetation of artificial earthen mounds in a northern mixed prairie. *Canadian Journal of Botany*, **70**, 145–150.

Umbanhowar, C. E., Jr (1995). Revegetation of earthen mounds along a topographic–productivity gradient in a northern mixed prairie. *Journal of Vegetation Science*, **6**, 637–646.

Walker, L. R. & Smith, S. D. (1996). Impacts of invasive plants on community and ecosystem properties. In *Assessment and Management of Plant Invasions*, eds. J. O. Luken & J. W. Thieret, pp. 69–86. New York: Springer-Verlag.

Wark, D. B., Poole, W. R., Arnott, R. G., Moats, L. R. & Wetter, L. (1995). *Revegetating with Native Grasses*. Stonewall, Manitoba: Ducks Unlimited.

Weiler, G. & Gould, W. L. (1983). Establishment of blue grama and fourwing saltbush on coal mine spoils using saline ground water. *Journal of Range Management*, **36**, 712–717.

Wenger, L. E. (1941). *Re-Establishing Native Grasses by the Hay Method*, Circular no. 208. Washington, DC: US Department of Agriculture.

Westoby, M., Walker, B. & Noy-Meir, I. (1989). Opportunistic management for rangelands not at equilibrium. *Journal of Range Management*, **42**, 266–273.

Willard, E. E. & Schuster, J. L. (1971). An evaluation of an interseeded sideoats grama stand four years after establishment. *Journal of Range Management*, **24**, 223–226.

Wilson, G. W. T. & Hartnett, D. C. (1997). Effects of mycorrhizae on plant growth and dynamics in experimental tallgrass prairie microcosms. *American Journal of Botany*, **84**, 478–482.

Wilson, S. D. (1998) Competition between grasses and woody plants. In *Population Biology of Grasses*, ed. G. P Cheplick, pp. 231–254. Cambridge: Cambridge University Press.

Wilson, S. D. (1999). Plant interactions during secondary succession. In *Ecosystems of Disturbed Ground*, ed. L. R Walker, pp. 629–650. Amsterdam: Elsevier.

Wilson, S. D. & Gerry, A. K. (1995). Strategies for mixed-grass prairie restoration: herbicide, tilling and nitrogen manipulation. *Restoration Ecology*, **3/4**, 290–298.

Wilson, S. D. & Shay, J. M. (1990). Competition, fire and nutrients in a mixed-grass prairie. *Ecology*, **71**, 1959–1967.

Wilson, S. D. & Tilman, D. (1991). Components of plant competition along an experimental gradient of nitrogen availability. *Ecology*, **72**, 1050–1065.

Winkel, V. K., Roundy, B. A. & Blough, D. K. (1991). Effects of seedbed preparation and cattle trampling on burial of grass seeds. *Journal of Range Management*, **44**, 171–175.

Workman, J. P. & Tanaka, J. A. (1991). Economic feasibility and management considerations in range revegetation. *Journal of Range Management*, **44**, 566–573.

Young, J. A., Evans, R. A. & Eckert, R. E., Jr (1969). Wheatgrass establishment with tillage and herbicides in a mesic medusahead community. *Journal of Range Management*, **22**, 151–155.

20 • Semi-arid woodlands and desert fringes

JAMES ARONSON, EDOUARD LE FLOC'H AND CARLOS OVALLE

INTRODUCTION

In the semi-arid and arid drylands, which cover one-third of the emerged surface of the earth, ecosystems and ecological resource bases for human use have been dramatically altered and diminished by mismanagement, short-sightedness and overexploitation. Recent human population increase is the underlying problem, along with ill distribution of resources, and insufficient consideration of the limited carrying capacity of the drylands.

Under demographic, agro-industrial, military and other pressures, at least 12 million km² of drylands have become 'damaged beyond the repair capacity of individual farmers' (Whisenant, 1999). Governments, and grass-roots associations, must therefore intervene, and, as we will argue in this chapter, rural development in drylands should henceforth be combined with biodiversity conservation, ecosystem management, and ecological restoration or rehabilitation of damaged ecosystems.

The Sahel and other dry parts of Africa, along with the Mediterranean Basin, are perhaps the best-known victims of desertification or, more specifically, 'desertisation' (i.e. 'a combination of processes which result in more or less irreversible reduction of the vegetation cover, leading to the extension of new desert landscapes to areas which were formerly not desert': Le Houérou, 1969). It goes without saying that humans have been responsible for most of those processes, at least in the last few thousand years (Stebbing, 1953; Reifenberg, 1955; Thirgood, 1981).

Central Asia is another region characterised by massive deforestation and desertisation of great antiquity. Guo *et al.* (1989) cite 120 000 km² on the Ordos Plateau where deforestation and desertisation began more than 1500 years ago (Zhu *et al.*, 1986). Similar processes of ancient, anthropic transformation of natural vegetation in drylands is in fact quite widespread, for example, in South America (Ellenberg, 1979) and Australia (Jones, 1969), as well as in Eurasia and Africa (Kuhnholz-Lordat, 1938; Mikesell, 1960). More recently, in the various 'Neo-Europes' of the Americas, South Africa and Oceania, profound transformations have been inflicted on ecosystems in the space of just a few centuries (Bahre, 1979, 1991; Crosby, 1985; Hobbs & Hopkins, 1990).

In countries of the North, concern for biodiversity conservation, habitat for wildlife, and 'heritage landscapes' in drylands are increasingly active concerns, along with the 'problems of plenty' which lead, in some cases, to the desire for ecological and biological restoration. National monuments, national parks and 'Man in the Biosphere' reserves (MAB–UNESCO) are being created and reinforced, in rich and poor countries alike, as national and international patrimony. But the conceptual foundations for what should be done therein, and how, is not always well developed.

In the developing countries of the South, rehabilitation of ecosystems to improve productivity and security for use by local people is socially and politically more pressing than any notion of 'restoration', in the prevailing 'northern' sense of the term. In the South, notions related to rehabilitation, such as soil and water conservation, erosion control, watershed protection, etc. are far more relevant than restoration *sensu stricto* (see Boxes 20.1 and 20.2). Costly and difficult repair operations are now necessary in a huge number of countries, just to maintain existing agriculture and forestry, and to prevent catastrophic floods and mudslides, and the silting up of dam reservoirs.

But how are we to restore biodiversity, ecodiversity, ecosystem dynamism, and productivity in areas where rainfall is erratic, and devastation so widespread and so grave? As we have argued elsewhere (Aronson *et al.*, 1993*a*), restoration and rehabilitation of ecosystems share much ground, and should take inspiration from similar sources in ecology, history and human geography, even if their philosophical underpinnings contrast in many cases. Clearly formulated conceptual models are also a must.

Conceptual models

In addition to numerous studies carried out in ecosystem ecology and landscape ecology, conceptual models and syntheses specific to ecosystem management (e.g. Luken, 1990) and restoration ecology can help guide both restoration and rehabilitation efforts in drylands. In particular, Hobbs & Norton (1996) presented an attractive, three-part state-and-transition model in which an hypothetical ecosystem can migrate (or be 'pushed') to four possible states, with five possible transitions and a single threshold (Fig. 20.1). This model builds on earlier ecological state-and-transition models, e.g. those of Westoby *et al.* (1989) and Allen (1989), both of which focused on arid and semi-arid lands. Like Westoby *et al.* (1989) and Westman (1990), Hobbs & Norton (1996) seek to caution restorationists and ecosystem managers in semi-arid regions against unrealistic goals such as striving to establish a permanent equilibrium at some state of presumed succession. As Victor Hugo said, 'It is the unforseeable we must always anticipate.'

In our view, the major shortcoming of the Hobbs & Norton model was its underemphasis on historical reconstruction of specific dynamical processes and the trajectory of an ecosystem in the past (Aronson & Le Floc'h, 1996*a*). Additionally, we would note that, *in situ*, a rather more complicated situation usually exists in terms of ecosystem development than the idealised case presented by Hobbs & Norton's model.

In Fig. 20.2, a different set of predictions from Hobbs & Norton's is presented concerning (A) the most common trajectory of ecosystem transformations,

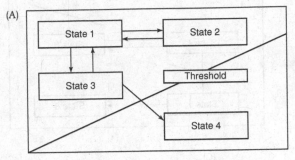

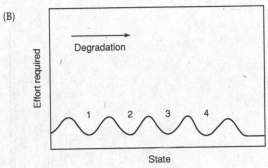

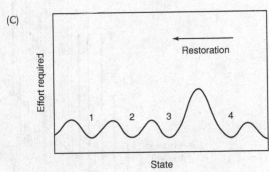

Fig. 20.1. A generalised state-and-transition model showing (A) four states, and five possible transitions, plus one threshold for an hypothetical ecosystem; (B) the relatively low and steady degree of effort predicted to be required to force transitions in a trough-and-wave model during the process of ecosystem degradation; and (C) the amount of hypothetical effort required to achieve various successive phases of restoration. The exceptional peak between States 4 and 3 in (C) corresponds to the threshold indicated in (A). From Hobbs & Norton (1996); reproduced with permission from Blackwell Publications, Inc.

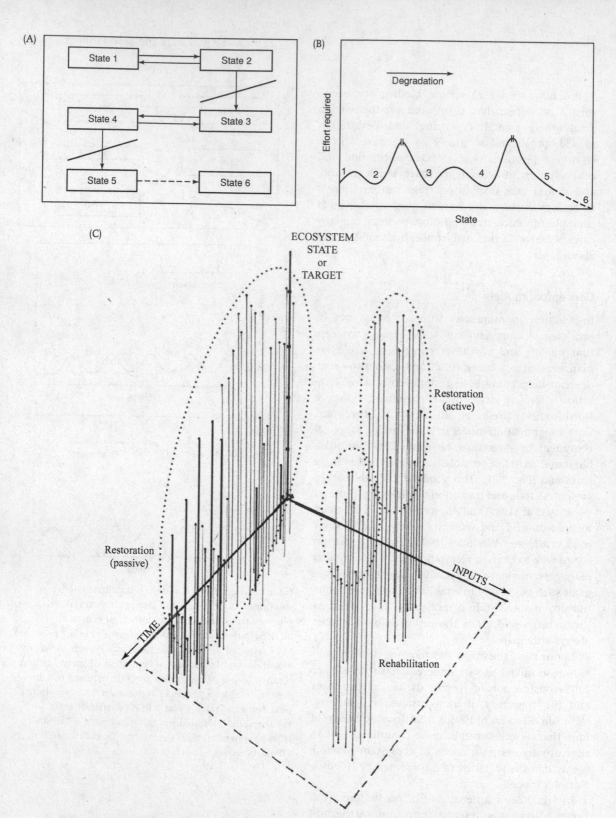

Fig. 20.2. A detailed response to the scheme shown in Fig. 20.1, with slight or major modifications proposed for all three parts. For discussion, see text. For explanation of dotted-line segments, see text.

(B) the amount of effort, or energy, that is likely to be required in various stages of ecosystem reparation and restoration, and (C) a more holistic approach to the modelling of restoration and rehabilitation trajectories. In parts A and B, the transition or trajectory between states 5 and 6 is indicated by a dotted line to suggest that beyond a certain point of desertification or denudation, further degradation is theoretically possible but will not normally proceed, without human intervention. Instead, at this point it is mostly abiotic vectors or determinants of degradation that continue to proceed, slowly but surely, even after people have abandoned the site and what little is left of its resources.

The ruptures in the highest peaks, in the effort required curve in part B, correspond to the two thresholds indicated in part A. This curve in Fig. 20.2B is inspired from Godron & Forman (1983), who compared the stability and 'potential energy' of successive ecosystem states in the course of a theoretical degradation process. In light of H. T. Odum's work, cited later in this section, quantifiable energetic terms (e.g. energy *per se*, or the more general 'inputs') should ultimately replace the more vague term 'effort' in models of this kind, as well as those concerning the response to degradation, i.e. restoration or rehabilitation, as in Fig. 20.2C.

We would quibble with the shape and, especially, the two-dimensionality of the third part of Hobbs & Norton's figure (Fig. 20.1C). Thus, in Fig. 20.2C, we propose an alternative, three-dimensional model for predicting, or indeed, planning trajectories and work agendas in both restoration and rehabilitation projects in drylands. Depending on the project target, and the means available, very different scenarios need to be considered, *vis-à-vis* not only the effort, or *energy*, required inputs, but also the *time* required to achieve the desired ecosystem state or 'target'.

Where 'passive' restoration is indicated, i.e. where functional damage is relatively limited and ecosystem resilience is thought to be high, only small inputs may be required, even if relatively long periods of time are needed. As indicated in Fig. 20.2C, however, there are cases where passive restoration can be quite rapid. Conversely, in situations where 'active' restoration is undertaken, larger inputs are required initially, but often for a short period only. In the case of rehabilitation projects, still greater initial effort and energy may be required, in the hope that large changes (or 'jumpstarts') can be achieved quickly, and that heavy investments will also be limited in time. However, given the disastrous state of many rehabilitation target ecosystems, a long, sustained effort may be required.

Rationale for restoring tree canopies in drylands

One of the critical challenges facing those attempting to restore or rehabilitate ecosystems in drylands is the loss of former tree canopies. The massive removal of trees, and the replacement of multi-tiered ecosystems by double-tiered shrublands or single-tiered herbaceous formations has led not only to diminished biotas, but also to profound changes in ecosystem and landscape dynamics. Fortunately, surviving trees in natural dryland ecosystems can often serve as bio-indicators to elucidate the distribution, structure, and the arboreal composition of former woodlands and desert fringe ecosystems.

As a network and multi-tiered canopy, together with a panoply of microsymbionts, trees provide an ecological, economic and cultural framework for smaller plants, animals and people alike. In drylands, trees also form 'islands of fertility' (Garner & Steinberger, 1989) which other organisms can benefit from. For example, in semi-arid woodlands of northern Mexico, Búrquez & Quintana (1994) found twice as many perennial plant species in the shade of ironwood (*Olneya tesota*) than outside the trees' shade and canopy.

Yet dryland trees not only confer a framework to natural and agro-forestry systems, and buffer the effects of climatic variability for understory soils and herbaceous plants, they also produce many other amenities for people and livestock (Ovalle & Avendaño, 1987; Joffre & Rambal, 1993; Busatto, 1998). In the drier parts of the 'South', planting trees with economic uses for local people is usually considered as essential.

FIVE SITUATIONS TO BE CONFRONTED IN DRYLANDS

This chapter covers a wide range of regions and biomes: semi-arid areas with 250–600 mm annual

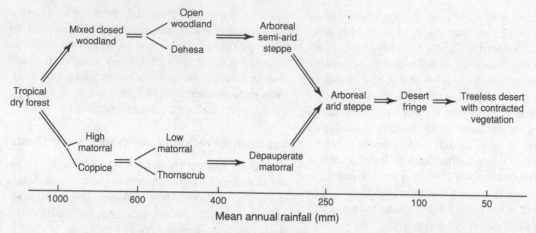

Fig. 20.3. A generalized mesic–xeric gradient, from tropical dry forest to treeless desert, also showing the two ways in which people have historically intervened to divert ecosystems' development. Upper branch: Selective removal and thinning of canopy trees leads to managed, open, more or less mixed woodlands (or 'dehesas') of a density and composition suitable for agro-forestry or other 'sustainable' land use systems. Lower branch: Excessive exploitation and transformation via 'artificial negative selection' and/or 'frenzies' of resource mining lead to shrublands or rangelands where trees and tall shrubs are considered undesireable, or else lost through artificial negative selection.

rainfall, where open woodlands once occurred, as well as more arid regions (50–250 mm), where trees are restricted to specific habitats, and finally disappear for lack of rainfall, soil types and appropriate topography. We also consider the transition zones at the wetter end of the gradient (600–1000 mm isohyets), where semi-arid woodlands merge into tropical dry forest (TDF) (Fig. 20.3).

Various secondary ecological situations (often called 'degradation phases') are encountered in these warm, mostly tropical semi-arid regions. They can be conveniently divided into five situations or categories, all of which arise from TDF or Mediterranean woodland ecosystems that have undergone 'human-mediated vegetation switches' (Barstow-Wilson & McG.-King, 1995) and thus have lost some or all of their resilience and initial structure and composition.

Thornscrub and coppices

In dry, warm temperate and dry tropical regions on all continents, various forms of secondary formations occur in areas formerly occupied by warm temperate, semi-arid woodlands or else seasonally dry tropical forests (Bullock *et al.*, 1995). In many cases they have been so transformed that their origins are nearly unrecognisable. They are typically 2–7 m tall in stature: shorter in Mediterranean climate areas, and taller in tropical areas. Their biodiversity is reduced, compared to the vegetation types from which they are derived, just as is their economic value to people. They tend to be rather homogeneous, spiny, and more or less inpenetrable. They are also fire-prone, which is a disadvantage in an increasingly crowded world.

The accompanying herbaceous layer is highly variable, from one geographical region to another. In the Sahel and northeast Brazil (the Caatingas), for example, the herbaceous layers of thornscrub are almost entirely indigenous, even if depleted and rarified. In southwestern Madagascar, by contrast, the herb 'community' is nearly entirely exogenous to the local, pre-existing ecosystems. Similar situations of synanthropic herbaceous floras prevail in the areas of former woodlands in the Neo-European Mediterranean-climate regions of Chile, California and Australia.

Dehesas

In southern Spain and Portugal, as well as North Africa, people have, over five centuries or more, fashioned or 'designed' a two- or multi-tiered ecosystem from within the former, highly complex framework or matrix provided by the natural woodlands (Joffre et al., 1988, 1999). These silvo-pastoral or agro-silvo-pastoral systems (dehesa in Spanish, and montado in Portugese) have evolved, with varied uses of trees (shade, fruits, bark, firewood, etc.) and the herbaceous stratum. Grazing is invariably an important element in dehesa management, and shrubs and small trees are regularly eliminated. Of particular interest are the 2.2 million hectares of cork oak (Quercus suber)-dominated dehesa woodlands in northwestern Africa and southwestern Europe (Montoya Oliver, 1988). The endemic Argania spinosa-based systems in arid southwest Morocco are also noteworthy (Benchekroun & Buttoud, 1989).

Dehesas are thus artificially opened and managed woodlands that are simplified as compared to TDF, natural woodlands, or even thornscrub. However, they have the virtue of 'mimicking natural ecosystems' (Lefroy et al., 2000) and thus are a valuable compromise as compared to other dry farming systems where trees are eliminated or treated solely as a periodic source of fuelwood. Unfortunately, these formations are widely threatenened with extinction (e.g. Mellado, 1989), either through intensification or extensification of their usage. They require new inputs and new ideas to promote natural regeneration and to reintroduce lost species. At the same time, they should be studied as models for the next category considered below.

Dehesas manqués

In parts of South America and northern Mexico, there are failed, never completed or degraded dehesas, that we would call dehesas manqués. They are notable for the large quantity and poor quality of woody biomass of intermediate height, and for the paucity and poor quality of the herbaceous and the tall canopy layers. These dehesas manqués occur largely in Latin America, where they are named

espinales, because of their spiny physiognomy. In these formations, as in thornscrub, the top canopy layer has been eradicated, and a lower-stature, secondary formation is managed as a long-term coppice, punctuated by livestock grazing and periodic plough agriculture. This constitutes a positive feedback cycle since only those tree and shrub species which regenerate readily after cutting, intensive grazing or fire survive.

In semi-arid regions where open woodlands once dominated, both excessive intensification and extensification of traditional land use practices can have negative effects, at least short-term, on biodiversity and ecosystem function. Restoration, in a strict sense, is therefore a very problematical proposition. In any case, the pre-colonial vegetation and ecosystems are often entirely or nearly eradicated. The best alternative in such cases would be to attempt to 'restore' the semi-cultural ecosystems and landscapes characterised by the best of the Iberian dehesas (see Box 20.2).

Tiger bush

Tiger bush or 'banded' vegetation is found in a number of semi-arid or subhumid regions, aligned perpendicular to slope (Tongway et al., 2000). It represents only a very small area of drylands worldwide, and therefore is not mentioned on Fig. 20.3. However, it is a fascinating and instructive avatar of natural dryland vegetation.

Capture and storage of limited water resources appears to be optimised in tiger bush through a near total spatial partitioning between runoff and runon/infiltration. The bare interband acts as a source while the vegetation band acts as a sink (Valentin & d'Herbès, 1999). In effect, tiger bush is a natural water-harvesting system whose productivity of woody biomass and rain-use efficiency are high, as compared to mean rainfall, as predicted by Noy-Meir (1973).

Desert fringes and other transition zones

As a result of increasing aridity, there is a point on woodland–desert gradients (e.g. Fig. 20.3), where woodland, and indeed all woody vegetation, becomes naturally constricted rather than being diffuse across the landscape (Khillian, 1925; Monod, 1954). In these transition zones (a less ambiguous term than ecotone or ecocline: van der Maarel, 1990) woody plants are progressively restricted by limited rainfall to those runon areas where surface runoff following rain accumulates in a concentrated fashion. Here only deep-rooted trees and shrubs have a chance to get established and survive. Given the paucity of woody vegetation and the relative fragility of these formations, desert fringe ecosystems are highly vulnerable to transformation, or outright destruction. Their restoration will demand much fine-scale work as well as a highly developed overview of the region and its landscapes. In addition, appropriate models and references are rare. In this context there is special need for studies seeking to elucidate surviving transition zone structure and dynamics, e.g. in the zone between open woodland and steppe in Inner Mongolia (Liu et al., 2000).

FIVE CONSTRAINTS

Severely disturbed ecosystems in arid and semi-arid lands are the rule rather than exception (Westoby et al., 1989; Whisenant, 1999). Due to canopy removal, reduced and altered species interactions, and physical degradation such as reduced infiltration or nutrient depletion (Milton et al., 1994) these systems are not going to be easy to repair. Moreover, drylands present a series of special constraints to restoration or rehabilitation, four of which are abiotic, and the fifth is anthropogenic. In addition to clear conceptual models and historical geographical understanding of the causes of desertification and desertisation, it should help to bear the following constraints in mind.

Scant, unpredictable rain and prolonged dry seasons

Variability of rainfall must be taken into account when planning a restoration project in drylands. Rainfall variability generally increases with aridity (Le Houérou & Popov, 1981), and variability

in annual primary production is even greater than variability in the amount of annual rains (Le Houérou, 1984). Long-scale events, cycles or disturbances can also have significant impact in these regions (Hobbs & Mooney, 1995), and should be taken into account (Dale et al., 1998), to avoid risks and take advantage of unusual opportunities, such as unusually wet years suitable for tree-planting. Duration of the dry season is also a critical factor in determining periods of physiological drought for plants, and ecosystem management strategies. Unexpected drought periods for people in semi-arid regions have frequently led to large changes in human use of resources and consequent impact on vegetation (Bahre, 1991). This in turn can lead to a new context, or trajectory, where disaster both socio-economic and environmental, can strike (e.g. Wainwright, 1994).

The oldest strategies whereby humans coped with life in deserts and desert edges were nomadism or transhumance (semi-nomadism), the two major means for human populations to adapt, ecologically, to the spatial and temporal variability in rainfall, and rangeland production in drylands. This necessitated integrated co-operation between people thoughout the regions visited by the nomadic populations, and provided a form of interweaving and interdependency that has now nearly vanished.

Heterogeneity of water and nutrient resources

Primary productivity and vegetation structure in arid and semi-arid ecosystems are determined, in large part, by water availability (Noy-Meir, 1973); both the maps and calendars of life in these regions reflect the presence or absence of water at a given spot. With redistribution of surface-flowing rainwater over uneven landscapes, plants experience even greater variability in soil moisture conditions, in space and time, than in annual rainfall. Precisely because soil water is heterogeneously distributed in arid regions, higher production per unit area can be achieved than if incoming water was evenly spread out (Noy-Meir, 1973, 1981, 1985). The same process holds true for nutrients (Tongway, 1990; Tongway & Ludwig, 1990; Ludwig & Tongway, 1995).

Irregular and unpredictable pulses of water and nutrients are followed by periods of consumption, recycling and conversion. Practical consequences of these observations can be directly applied to rehabilitation (Whisenant et al., 1995; Ludwig & Tongway, 1996, 1997), and to restoration. For example the simulation of gap dynamics in space and time can be used to favour regeneration and the establishment of tree seedlings in degraded woodlands (Yates et al., 2000).

Heterogeneity of energy resources

Like water inputs, energy inputs and use in ecosystems are also pulsed, perhaps especially in drylands. Sometimes the pulses are generated by the consumer animals and in other cases by long-range oscillations from even larger systems such as floods, landslides, economic oscillations, and/or harvest cycle (Odum, 1988). Pulsed consumption of phytomass by ruminant livestock has often been an effective management tool in rangelands. On the other hand, periods of frenzied, i.e. uncontrolled, consumption, such as the historic mining 'frenzies' of fuelwood, minerals, and other resources, have had huge destructive impact on vegetation, ecosystems and landscapes in deserts and desert fringes around the world. These frenzied attacks on ecosystem health and integrity accentuate another, long-term, and largely unconscious human process, described below.

Artificial negative selection

As Arturo Burkart (1976) pointed out, people in many semi-arid regions have practiced an 'artificial negative selection', with adverse effects on energy, water, and nutrient use efficiency, reduced genetic and specific diversity and loss of ecosystem resilience. This term describes the short-sighted practice whereby people selectively remove the most useful phenotypes and genotypes of woody plants, both within and among species and genera, in a progressive fashion generated by some kind of positive feedback cycle. Only inferior genotypes, phenotypes and, ultimately, species are left to reproduce and

contribute to the seed bank in the areas subject to this mismanagement. Consequences for restoration begin at the level of seed and germplasm collection for propagation and reintroduction. Often, the most abundant and readily available species, genotypes and phenotypes are precisely not the ones most suitable for restoration work. Ecologically and economically 'elite' individuals most suitable as sources of seeds, or cuttings, have to be looked for. The extra work this implies in the early stages of a project should be well rewarded by the long-term impact on genetic make-up of the restored plant and animal communities.

Growing demographic pressure

As mentioned at the outset of this chapter, demographic pressures, both rural and urban, are increasing rapidly in drylands around the world. Although outside the scope of this chapter, we wish to emphasise yet again the overriding importance of this constraint on the management and restoration of semi-arid woodlands and other drylands.

A NEW APPROACH

We have seen that ecosystems in dryland biomes are very often disturbed at the most fundamental ecological and genetic levels, such that the prospects for restoration, or even simply reintroducing key species are much reduced. What then can be done? In this section, we review five aspects of a new approach.

Reference information and areas

We argue in favour of establishing references – areas, ecosystems, landscapes – or even a collation of reference information from various sources (Aronson et al., 1993a, 1995; cf. Aronson & Le Floc'h, 1996b; White & Walker, 1997; Allen et al., 2001; Egan & Howell, 2001). This process or approach is useful, if not essential to ecological restoration, even if it is well understood that we are dealing with 'moving targets' and that blindly seeking to recreate a reference of some sort can be an obvious pitfall to avoid.

In drylands, moreover, as we have shown, there exists an array of forms, phases or avatars in which vegetation can be seen, most of which are rather removed from a pristine, 'undisturbed' state. Accordingly, several types of references and sources of reference information (White & Walker, 1997) can be suitable for restoration projects, e.g. an historically functional and 'healthy' dehesa can serve as a reference or yardstick with which to evaluate repair to a dehesa manqué. Furthermore, in most drylands, especially in the South, the line between restoration *sensu stricto* and rehabilitation is often a rather fuzzy one. This will be illustrated below by examples and case studies.

Means of evaluation

A second aspect of, or approach to, ecological restoration that is much debated among ecologists and practitioners is how to evaluate what has been done; which ecological indicators are the most pertinent? For example, Aronson *et al.* (1993*a, b*) introduced a series of 21 quantifiable 'vital ecosystem attributes' thought to be interrelated and helpful in formulating predictions and designing experiments in both restoration and rehabilitation. These included attributes or traits related to ecosystem composition, structure and functioning useful in charting ecological degradation and evaluating the results of attempted restoration or rehabilitation. Their key features are that they are all quantifiable and transferable, i.e. amenable to synchronic and diachronic comparisons, and of more or less universal relevance to ecosystem ecology, at least in arid and semi-arid lands. Since large numbers of attributes can rarely be monitored in a single project, it is important to identify an affordable suite of the most pertinent, sensitive and reliable attributes for each case study (Box 20.1).

Passive vs. active restoration

Relatively inexpensive, 'passive' restoration techniques are preferable where ecosystem structural/functional damage is relatively limited and resilience is high. They are also obviously preferable in situations where extensive land areas require treatment. For example, the exclusion or severe restriction of livestock grazing for a few years is sometimes sufficient to promote the self-recovery of dryland ecosystems where pasture plants have co-evolved with domestic grazing animals, e.g. in southern Tunisia (Floret, 1981; Ayyad & El-Khadi, 1982; Wesstrom & Steen, 1993) (Box 20.1). However, in such areas, outside of the rare national parks or private hunting reserves, revised management techniques, including controlled grazing, must be transmitted and inculcated in local populations if the effects of a temporary exclusion are not to be lost.

In most cases, however, in most drylands, simple exclusion or even tighter control of grazing will evidently not be sufficient to achieve restoration or rehabilitation of ecosystems, especially if one of the targets is to restore, or at least enrich, the tree canopy of a system. In such situations, 'active' restoration is required, with large 'inputs', and typically a number of synergistic techniques should be applied concurrently. We shall now briefly review some of the most prominent and promising of these physical and biological tools being tested and employed in drylands worldwide. The intelligent mixing, adapting and applying of these techniques will of course vary from one context to another.

Nuts and bolts

In all five situations considered above, the soils where diverse woodlands or TDF once occurred have become radically altered structurally, and depleted functionally. The reintroduction of trees and shrubs, or other presumed keystone or priority plants, often proves difficult without physical and/ or biotic amendments and reparations. The situation for most animal reintroductions is probably much the same.

The establishment or spontaneous re-establishment of trees, shrubs and other perennial plants can be aided by the use of carefully selected nurse plants and various techniques which mimic large- and especially medium-scale disturbances which reduce soil compaction and restore soil water infiltration to suitable rates (Ludwig & Tongway, 1996; Carillo-Garcia *et al.*, 1999; Yates *et al.*, 2000). A number of mechanical techniques, including 'pitting',

Box 20.1   Southern Tunisia

Over past millennia, vegetation in the Saharan desert fringe of northern Africa has become adapted to grazing (i.e. 'perturbation dependent' *sensu* Vogl, 1980). In the past century, however, the intensity and timing of grazing have become maladaptive, and the vegetation is now in a state of progressive degradation (Le Floc'h *et al.*, 1995, 1999). Trees and tall shrubs have been entirely removed, through artificial negative selection, and soil surface conditions have been modified through a general loss of fine-scale rugosity The least disturbed steppes in southern Tunisia are those dominated by a canopy *c.* 20–50 cm high, consisting largely of the Compositae subshrub *Rhanterium suaveolens,* and such formations were taken as a short-term reference for the experiments described below. The only significant remnant of the arboreal steppe known to have existed in the region (Le Houérou, 1959; Floret *et al.*, 1978) is in the 60-year-old Bou Hedma National Park, dominated by the native *Acacia tortilis* ssp. *raddiana.*

In a representative site in the desert fringe area (mean annual rainfall 175 mm), where vegetation and soils have been degraded, we sought to test the hypothesis that re-creation of a simplified version of the previously existing ecosystem could facilitate further reconstruction of the target, pre-existing system. The establishment of a simplified two-tiered plant community, with concurrent restoration of soil surface roughness or rugosity, should aid in trapping wind-borne diaspores and sand, leading them to reactivate soil water functions and, ultimately, prepare the ground for spontaneous reintroduction of additional native shrubs and trees.

We adopted also as a second hypothesis, based on the known life history traits of plants (Chaieb *et al.*, 1992), that a pertinent combination of native perennial herbs, grasses and shrubs could utilise the available water resources in an efficient and complementary fashion, while also showing resistance and sustained productivity in the presence of moderate livestock grazing. In this way soil erosion could be halted, and phytomass and livestock productivity reanimated. It has already been observed in long-term grazing exclosures near the study site that *Acacia tortilis* ssp. *raddiana* does reappear spontaneously in a healthy steppe dominated by *Rhanterium suaveolens.*

Four years after sowing of various native and presumed-to-be complementary species, the densities of *R. suaveolens* and the perennial bunch grass (*Stipa lagascae*) were similar to those found in our 'ecosystem of reference'. By contrast, in the unsown control plots, no individuals of either species occurred. Vegetation cover values of *R. suaveolens* and *Plantago albicans* in the sown plots were also similar to those of the ecosystem of reference four years after sowing (Le Floc'h *et al.*, 1995, 1999; Aronson *et al.*, 1993*b*).

By comparing a series of four biotic ecosystem attributes (Table B20.1) for three 'stages' of degradation and the experimental rehabilitated plot, the feasability of reintroducing a certain number of complementary pastoral species was demonstrated. Furthermore, the basis for a long-term comparison was established. Unable to measure directly water flux dynamics, we used soil surface state (Casenave & Valentin, 1989) as an abiotic indicator, relevant for the monitoring of hydrological functioning (infiltration, leaching and evaporation) (Table B20.2). In this context, sown plots showed an increase of litter and sand cover which are known to favour the hydrological functioning of a soil profile. This was a direct consequence of the increase in vegetation cover.

Although these results are preliminary, the working hypothesis appears to have been supported, i.e. that installing a simplified ecosystem with an enriched shrub layer could improve conditions for further reintroductions even as primary productivity was increased and stablised (provided co-operation of farmers was forthcoming to better manage grazing livestock).

Table B20.1. *Evolution of four biotic ecosystem indicators or attributes of the typical steppe, in late spring, in a southern Tunisian study site. Attributes were compared for four stages of degradation, and three years after the beginning of an experimental rehabilitation based on artificial sowing and reintroduction of native perennial species*

Ecosystem indicators	Stages of ecosystem development				
	RS1[a]	RS2[b]	RS3[c]	SP1[d]	Rehabilitated[e]
Number of annual species	41	13	6	2	21
Number of perennial species	23	18	3	1	12
Total plant cover (%)	60	20	2	0.5	37
Aboveground biomass (kg dry matter ha^{-1} y^{-1})	1800	600	200	100	1200

[a]RS1: *Rhanterium suaveolens*-dominated steppe on deep, sandy soils in relatively undisturbed conditions.

[b]RS2: same steppe, moderately degraded through overgrazing and woodcutting.

[c]RS3: same steppe, badly degraded and with truncated soils.

[d]SP1: *Stipagrostis pungens*-dominated steppe, with all shrubs removed.

[e]Rehabilitated steppe after reseeding of selected native species, including legume, grasses, and the dominant shrub, *Rhanterium suaveolens*.

Source: Le Floc'h *et al.* (1995).

Table B20.2. *Relative contribution of seven different components to soil surface cover (%) in control plots, sown plots and the 'ecosystem of reference' (April 1994)*

Surface characteristic	Control		Mixed sown plots	Ecosystem of reference
	External	Internal		
Battance	77.3	63.5	56.5	≅4
Litter	8.0	19.6	28.9	≅5
Gravel	6.6	6.2	1.5	≅1
Bare soil	6.3	2.6	2.1	≅4
Snail shells[a]	1.4	1.0	1.0	0
Animal excrement	0.4	0.2	0.0	0
Sand	0.0	6.9	10.0	≅90

[a]After soil deflation, snail shells present in the profile remain at the soil ground surface.

Source: Le Floc'h *et al.* (1999); reproduced with permission of Taylor & Francis Publishers.

can also substantially help increase rugosity of soil surfaces over large areas, and thereby help the establishment of new seedings or plantings by increasing the infiltration of rainfall and slowing the speed of surface runoff (Gintzburger & Skinner, 1985; Gintzburger, 1987). In Australia, and California especially, commercial firms and large equipment are available for this type of intervention. Elsewhere, the means are often lacking.

Microcatchments made by hand or machine can also make a huge difference for sowing and, especially, tree plantings (Table 20.1). The extra work and expense is usually justified by a dramatic increase in tree survival, especially if time and care has been spent in selecting the initial propagation material. In addition, in hilly or mountainous regions, microcatchments yield considerable soil conservation benefits at the field, slope and watershed levels.

Table 20.1. *Biomass and establishment success of two native woody species transplanted into microcatchment basins or unmodified surface in a semi-arid (400 mm) region of Texas*

Species	Planting location	1990		1992	
		Aboveground biomass (g plant^{-1})	Survival (%)	Aboveground biomass (g plant^{-1})	Survival (%)
Leucaena retusa	Basin	22[a]	100	49[a]	98[a]
	Unmodified	4[b]	50[b]	10[b]	45[b]
Atriplex canescens	Basin	89[a]	100[a]	502[a]	100[a]
	Unmodified	7[b]	100[a]	182[b]	100[a]

Note: Data pertain to all microcatchments (ratio of catchment basin to collection area varied from 1:1 to 30:1). Means within a column and species followed by similar letters do not differ according to LSD ($p = 0.05$).
Source: Whisenant *et al.* (1995); reproduced with permission of Blackwell Publications Ltd.

At levels of landscapes and watersheds, tree-planters and restorationists in drylands are increasingly attempting to harness and harvest rainfall, on medium to large scales, in imitation of the Nabateans, Persians and other ancient desert-dwellers of the Near and Middle East (Evenari *et al.*, 1982; Tongway & Ludwig, 1996). Applying old techniques to new problems is often efficacious in drylands, given that the so-called new problems are not so new after all.

In the plant nursery, appropriate containers for trees should be used, sufficiently large and structured so as to avoid the development of spiralling taproots which transplant poorly. Mycorrhization, rhizobial inoculation and other treatments can also dramatically increase chances for positive results, e.g. in planting trees (Querejeta *et al.*, 1998) and promoting the regeneration of dryland trees and shrubs (Montoya Oliver, 1988; Vallejo, 1996; Carrillo-Garcia *et al.*, 1999). There is an extensive literature on the use of tree–microsymbiont 'couples' and we will not review it here.

Fire management is also a biological tool of widespread importance and potential in restoration and especially rehabilitation of semi-arid multi-tiered formations. Numerous and extensive experiments and retrospective studies in this domain are under way in Australia (see Tongway & Ludwig, this volume). In Fig. 20.4, we illustrate a case study from the Sahel (Mali), in an agro-forestry system,

approximately 60 to 100 years old, which is derived from tropical dry forest with six to seven months absolute annual drought. Many of the woody elements of the pre-existing ecosystem are still present, especially along watercourses. More appropriate management of native woody elements in field–fallows rotation systems and in the surrounding area is crucial. To begin with, the results of this experiment along with many others for which results are not shown, have been used to persuade local farmers to adapt an early-burn schedule in replacement of the traditional late-burn practice. Favourable effects on biodiversity also favour soil fertility, stability and profitability of the farming system. As in so many other cases, however, throughout drylands of the South, accelerating human population densities in rural areas are *the* fundamental problem to be overcome.

Like fire, the ancient human practice of livestock raising in drylands must also be revised and adapted to current conditions. In the South, new strategies must be devised that will allow local populations to maintain ruminant animals on ecologically rational grounds, and participate in restoration-oriented activities. At the least, they must be helped to stop contributing to desertisation. In rich countries, ruminants also need to be managed in new ways, i.e. as a tool in preventing or limiting the spread of wildfires (Perevelotsky & Haimov, 1992; Etienne *et al.*, 1998; Vallejo *et al.*, 1999). In southern Europe

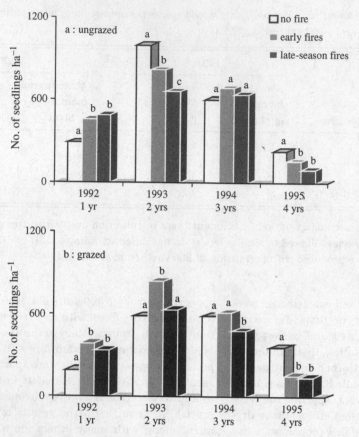

Fig. 20.4. Density of woody seedlings in fallows, with and without grazing, under three different controlled fire regimes: no fire (= once every three years), early fires, and late-season annual fires. Site: heavy silty soils at Missira, in semi-arid northern Mali. Histogram bar values for a given year and treatment with a different letter above them are significantly different ($p < 0.05$) by Duncan's multiple range test. After Dembele (1996).

and other semi-arid regions, horses, sheep, donkeys and even llamas are now being used in wildfire-control programmes, and for the maintenance of multi-purpose and landscape mosaics containing open areas as well as woodlands (Etienne *et al.*, 1998), as corresponds to most historic references for most Mediterranean climate and dryland regions generally.

Finally, we must mention the use of nitrogen-fixing legumes in dryland restoration and rehabilitation, especially in association with nursery inoculation of roots by selected vesicular–arbuscular mycorrhizal fungi and rhizobia (Herrera *et al.*, 1993).

Despite the fact that some woody legumes are sometimes invasive and potentially noxious (Hughes & Styles, 1987), the hypothesis is worth testing that multi-purpose legumes capable of biological nitrogen fixation can be useful in helping to restore, rehabilitate and otherwise repair damaged and desertified ecosystems, even as they provide shade, fuelwood, fodder and other renewable natural products (Box 20.2). Caution is necessary, however, as unfortunate experiences with introduced legumes in Australia, India, East Africa and elsewhere have shown. Native species present far less risk, as a rule.

Conservation, management and restoration combined

In developing countries, where most of the drylands occur, biological conservation, ecosystem management and ecological restoration (CM&R) should be approached simultaneously (Aronson *et al.*, 2000). All of these activities need to be oriented and guided by an understanding of: (1) the historical causes of ecosystem degradation, (2) ecological theory and data concerning ecosystem and landscape dynamics, and (3) biotic and abiotic constraints on dryland

---

Box 20.2   Central Chile

In the four and a half centuries since the arrival of Europeans in Chile, natural semi-arid woodlands and desert fringe ecosystems have declined and regressed steadily (Bahre, 1979). Productivity, biodiversity and ecosystem integrity are visably declining or else blocked in a highly limited, low-energy formation called *espinales* (Ovalle *et al.*, 1990). Ecoclines and zonation related to geophysical and climatic gradients have been wiped out, and indigenous sclerophyll woodlands have been destroyed, except for scattered patches, all under 50 ha in size (Aronson *et al.*, 1998; C. Lusk, unpublished data).

Throughout the unirrigated portions of central Chile, landowners and tenants practise low-energy agro-forestry systems based on the espinales, which is a poor imitation of the dehesas of southern Spain, where the first European colonists of Chile came from. Instead of the dehesas' productive and protective canopy of oaks, a generally scrubby espino (*Acacia caven*) is almost the only tree present in Chilean espinales, apart from a handful of ranches and farms where it is managed as part of a more 'intelligent', two-tiered silvo-pastoral system (Ovalle *et al.*, 1996a). Elsewhere it is cut and/or burned every 10–15 years, and no tree canopy is preserved at all. A near total 'artificialisation' has taken place, and integrated redevelopment programmes are essential, along with ecological restoration and rehabilitation.

In this context, we are testing the hypothesis that nitrogen-fixing legumes of woody and annual growth forms can be useful in the dual task of economic and ecological rehabilitation of biologically enriched and sustainably exploitable woodland systems. In accordance with ecological realities and socio-economic imperatives in the region, the model or reference system we employ is not the indigenous sclerophyll woodlands, but rather the mixed, biologically rich dehesas formed by early colonists in the seventeenth and eighteenth centuries (Aronson *et al.*, 1993b, 1998; Ovalle *et al.*, 1999). In that sense, we are seeking to restore dehesas.

For various multi-purpose native and introduced perennial legumes, we have begun to quantify biological nitrogen fixation over space and time, using various planting schemes and management systems (Aronson *et al.*, 1992). This is not easy in dryland conditions, but still more difficult is the task of assessing actual impact of N_2-fixing trees on soil biota and fertility, when they were planted in a typical degraded field. For this purpose we compared two slow-growing, long-lived native N_2-fixing legume trees (*Acacia caven* and *Prosopis chilensis*) to the very fast-growing but short-lived N_2-fixing legume tagasaste (*Chamaecytisus proliferus* ssp. *palmensis*) from the Canary Islands. For purposes of calculating nitrogen fixation, a non-fixing reference tree was required, and we used the naturalised European ash (*Fraxinus excelsior*) for this purpose (Ovalle *et al.*, 1996b) (Table B20.3). This kind of study is important both in a restoration or a rehabilitation context where native or carefully selected introduced tree species can provide fodder and forage for landowners' livestock, and thereby elicit co-operation and investment by those landowners, provided the cost–benefit analysis is positive. Such measurements are needed to test the hypothesis that legume–microorganism 'couples' can have positive effects on ecosystem dynamics and 'autogenic restoration' (Ovalle *et al.*, 1990; Aronson *et al.*, 1993a; Herrera *et al.*, 1993). Long-term studies are under way, in central Chile and elsewhere. Meanwhile, just as in the Tunisian study area, co-operation of local people will be essential to the long-term success of restoration and rehabilitation.

Table B20.3. *Soil rehabilitator effects produced after five years' growth by two native and one non-native nitrogen-fixing legume trees planted in a single old field as compared to a non-fixing reference species and untreated control sites (without trees) in the same field*

Species	pH	Organic matter (%)	Available N (mg kg^{-1})	Available P (mg kg^{-1})	Microbial N (mg kg^{-1} soil)
N$_2$-fixing trees					
Prosopis chilensis	5.8^a	1.0^b	8.3^b	3.8^a	3.8^b
Acacia caven	5.8^a	1.0^b	8.5^b	3.5^a	1.0^b
Tagasaste (Chamaecytisus proliferus ssp. palmensis)	5.4^a	3.2^a	13.3^a	2.3^a	25.3^a
Non-fixing trees					
Fraxinus excelsior	5.6^a	1.3^b	5.3^b	4.0^a	3.8^b
Control	5.7^a	1.1^b	7.7^b	4.0^a	1.3^b

Note: Soils were sampled at 0–40 cm depth. Figures in a column followed by a different letter are significantly different ($p < 0.05$) by Duncan's multiple range test.

Source: Ovalle *et al.* (1999); reproduced with permission of Taylor & Francis Publishers.

restoration and rehabilitation. Obviously, such a synergistic, holistic approach cannot be achieved easily. Both a 'reintegration of fragmented landscapes' (Hobbs & Saunders, 1992), and a 'reweaving' of local and regional communities appear to be essential to achieve successful ecological restoration and rehabilitation goals.

Ecological and economic rehabilitation of productive and profitable agro-silvo-pastoral systems which mimic natural woodland ecosystems should be seen as an integral part of the combined CM&R approach (Box 20.2). Specifically, rehabilitation on appropriate landscape units in semi-arid lands should promote productivity for local populations and, backed by environmental re-education, reduce pressure on sites or areas that local people agree to set aside for nature conservation and long-term restoration or rehabilitation. Ideally, restoration and rehabilitation should be pursued together, but in different portions of landscapes and watersheds.

CONCLUDING REMARKS

In most developing countries, like Tunisia, Chile and Mali, and in many rich, industrialised nations

as well, the notion of ecological restoration as a societal, individual and governmental priority has just recently emerged. This is perhaps especially true in areas where governmental investments in ecosystem management and development and in conservation have heretofore been disproportionately low and short-sighted. Until recently, human population levels in these regions was also much lower than they are today.

Drylands are also characterised by extreme variability and heterogeneity of resources both in space and in time. Extant ecosystems are, generally, badly disturbed, and lacking in resilience. Ecologically founded ecosystem management and, by extension, ecological restoration, should be planned in accordance with actual levels of ecosystem resilience, in order to set realistic goals in time and budget, keep options open, view succeeding events, and patterns of events, in a regional rather than a merely local context, and never to underestimate heterogeneity and unpredictability (Holling, 1973). Nowhere is this list of advice more pertinent than in drylands.

Furthermore, most large or even small-scale ecological restorations in drylands of the South cannot be seriously envisaged, however, unless they are embedded in a larger CM&R programme. Such programmes should aim at revising land, livestock and resource management strategies of local populations as well as repairing the already damaged wildlands and semi-wildlands. In the North, where 'problems of plenty' occupy some people and administrations as much or more than repairing damaged wildlands, CM&R still seems an attractive conceptual combination. Four of the five constraints listed above existed before human perturbations of ecosystems became widespread. The fifth, human population trends, serves to exacerbate the effects of those four, both in time and in space, and makes it imperative quickly to achieve more rational, sustainable resource and land use strategies.

In the broad CM&R context, in the South or in the North, the role of trees, which provide the framework of dryland ecosystems, cannot be overemphasised. Both for repairing damage done in the past, and for building robust, resilient systems for the future, trees are perhaps *the* essential building block for drylands, despite the long periods required initially for their establishment. Rational, ecological management and conservation of water are the essential complement to this approach. Multi-dimension conceptual models, multiple-input and transdisciplinary information systems, data bases and evaluation criteria are crucial too.

Finally, from researchers and students, there is a crying need, in drylands and elsewhere, for large-scale, long-term experiments to test the hypothesis that designed, restored, or 'natural' multi-tiered ecosystems are more likely to be functionally healthy and integrated than artificial, single-tiered systems. This, among many other live research questions from ecosystem ecology and field-oriented restoration ecology, should help advance the art and science of ecological restoration in semi-arid woodlands and other drylands including desert fringes.

ACKNOWLEDGMENTS

Martin Perrow's and David Tongway's penetrating criticisms of an earlier draft helped immeasurably. Edie Allen, Andre Clewell and Richard Hobbs kindly provided useful information and constructive commentary. We heartily thank Michel Grandjanny for technical support, and ECOS Sud-CONICYT (No. C95B05) and CNRS-CONICYT (No. 8734) for financial support.

REFERENCES

Allen, M. F. (1989). Mycorrhiza and rehabilitation of disturbed arid soils: processes and practices. *Arid Soil Research and Rehabilitation*, 3, 229–241.

Allen, E. B., Brown, J. S. & Allen, M. F. (2001). Restoration and Biodiversity. In *Encyclopedia of Biodiversity*, ed. S. Levin, pp. 185–202. San Diego, CA: Academic Press.

Aronson, J. & Le Floc'h, E. (1996a). Hierarchies and landscape history: dialoging with Hobbs and Norton. *Restoration Ecology*, 4, 327–333.

Aronson, J. & Le Floc'h, E. (1996b). Vital landscape attributes: missing tools for restoration ecology. *Restoration Ecology*, 4, 377–387.

Aronson, J., Ovalle, C. & Avendaño, J. (1992). Early growth rate and nitrogen fixation potential in 44 multipurpose legumes grown in an acid and a neutral soil in central Chile. *Forest Ecology and Management*, 47, 225–244.

Aronson, J., Le Floc'h, E., Floret, C., Ovalle, C. & Pontanier, R. (1993a). Restoration and rehabilitation of degraded ecosystems in arid and semi-arid regions. 1: A view from the South. *Restoration Ecology*, 1, 8–17.

Aronson, J., Le Floc'h, E., Floret, C., Ovalle, C. & Pontanier, R. (1993b). Restoration and rehabilitation of degraded ecosystems in arid and semi-arid regions. 2: Case studies in Chile, Tunisia and Cameroon. *Restoration Ecology*, 1, 168–187.

Aronson, J., Dhillion, S. & Le Floc'h, E. (1995). On the need to select an ecosystem of reference, however imperfect: a reply to Pickett & Parker. *Restoration Ecology*, 3, 1–3.

Aronson, J., Del Pozo, A., Ovalle, C., Avendaño, J., Lavin, A. & Etienne, M. (1998). Land use changes and conflicts in central Chile. In *Landscape Degradation and Biodiversity in Mediterranean-Type Ecosystems*, eds. P. W. Rundel, G. Montenegro & F. Jaksic, pp. 155–168. Berlin: Springer-Verlag.

Aronson, J., Li, J., Le Floc'h, E., Romane, F. & Vallauri, D. (2000). Combining biodiversity conservation, management and ecological restoration in China (with special reference to arid and semi-arid regions). In *The Conservation and Utilization of Biodiversity*, ed. J. Jianming, pp. 279–301. Beijing: CBHT.

Ayyad, M. A. & El-Khadi, H. F. (1982). Effects of protection and controlled grazing on the vegetation of a

Mediterranean desert ecosystem in northern Egypt. *Vegetatio*, **49**, 129–139.

Bahre, C. J. (1979). *Destruction of the Natural Vegetation of North-Central Chile*. Berkeley, CA: University of California Press.

Bahre, C. J. (1991). *A Legacy of Change: Historic Human Impact on Vegetation of the Arizona Borderlands*. Tucson, AZ: University of Arizona Press.

Barstow-Wilson, J. & McG.-King, W. (1995). Human-mediated vegetation switches as processes in landscape ecology. *Landscape Ecology*, **10**, 191–196.

Benchekroun, F. & Buttoud, G. (1989). L'arganeraie dans l'économie rurale du sud-ouest marocain. *Forêt méditerranéenne*, **11**, 127–136.

Bullock, S. H., Mooney, H. A. & Medina, E. (1995). *Seasonally Dry Tropical Forests*. Cambridge: Cambridge University Press.

Burkart, A. (1976). Monograph of the genus *Prosopis. Journal of the Arnold Arboretum*, **57**, 219–249, 450–525.

Búrquez, A. & Quintana, M. A. (1994). Islands of diversity: ironwood ecology and the richness of perennials in a Sonoran Desert biological reserve. In *Ironwood: An Ecological and Cultural Keystone of the Sonoran Desert*, eds. G. P. Nabhan & J. L. Carr, pp. 9–28. Washington, DC: Conservation International.

Busatto, B. (1998). The use of cork oak in the ecological and landscape restoration of the Latium hills. *Acta Horticulturae*, **457**, 47–54.

Carrillo-Garcia, A., León de la Luz, J.-L., Bashan, Y. & Bethlenfalvay, G. J. (1999). Nurse plants, mycorrhizae, and plant establishment in a disturbed area of the Sonoran Desert. *Restoration Ecology*, **7**, 321–335.

Casenave, A. & Valentin, C. (1989). *Les états de surface dans la zone sahélienne: Influence sur l'infiltration*. Paris: Office de Recherche Scientifique et Technique Outre-Mer; Didactiques.

Chaieb, M., Floret, C., Le Floc'h, E. & Pontanier, R. (1992). Life history strategies and water allocation in five pasture species of Tunisian arid zones. *Arid Soil Research and Rehabilitation*, **6**, 1–10.

Crosby, A. (1985). *Ecological Imperialism: The Biological Expansion of Europe, 900–1900*. Cambridge: Cambridge University Press.

Dale, V. H., Lugo, A. E., MacMahon, J. A. & Pickett, S. T. A. (1998). Ecosystem management in the context of large, infrequent disturbances. *Ecosystems*, **1**, 546–557.

Dembele, F. (1996). Influence du feu et du pâturage sur la végétation et la biodiversité dans les jachères en zone soudanienne-nord. PhD dissertation, Université de Droit, d'Economie et des Sciences d'Aix-Marseille III.

Egan, D. & Howell, E. A. (2001). *The Historical Ecology Handbook: A Restorationist's Guide to Reference Ecosystems*. Washington, DC: Island Press.

Ellenberg, H. (1979). Man's influence on tropical mountain ecosystems in South America. *Journal of Ecology*, **67**, 401–446.

Etienne, M., Aronson, J. & Le Floc'h, E. (1998). Abandoned lands and land use conflicts in southern France: piloting ecosystem trajectories and redesigning outmoded landscapes in the 21st century. In *Landscape Degradation and Biodiversity in Mediterranean-Type Ecosystems*, eds. P. W. Rundel, G. Montenegro & F. Jaksic, pp. 127–140. Berlin: Springer-Verlag.

Evenari, M., Shanan, L. & Tadmor, N. (1982). *The Negev: The Challenge of a Desert*. Cambridge, MA: Harvard University Press.

Floret, C. (1981). The effects of protection on steppic vegetation in the Mediterranean arid zone of southern Tunisia. *Vegetatio*, **46**, 117–129.

Floret, C., Le Floc'h, E., Pontanier, R. & Romane, F. (1978). *Modèle écologique régional en vue de la planification et de l'aménagement agro-pastoral des régions arides: Application à la région de Zougrata*, Document Téchnique no. 2. Tunis, Tunisia: Ministère de l'Agriculture.

Garner, W. & Steinberger, Y. (1989). A proposed mechanism for the formation of 'fertile islands' in the desert ecosystem. *Journal of Arid Environments*, **16**, 257–262.

Gintzburger, G. (1987). The effect of soil pitting on establishment and growth of annual *Medicago* spp. on degraded rangeland in Western Australia. *Australian Rangeland Journal*, **9**, 49–52.

Gintzburger, G. & Skinner, P. (1985). A simple single disc pitting and seeding machine for rangeland revegetation. *Australian Rangeland Journal*, **7**, 29–31.

Godron, M. & Forman, R. T. T. (1983). Landscape modification and changing ecological characteristics. In *Disturbance and Ecosystems*, eds. H. Mooney & M. Godron, pp. 12–28. Berlin: Springer-Verlag.

Guo, H., Wu, D. & Zhu, H. (1989). Land restoration in China. *Journal of Applied Ecology*, **26**, 787–792.

Herrera, M. A., Salamanca, C. P. & Barea, J. M. (1993). Inoculation of woody legumes with selected arbuscular mycorrhizal fungi and rhizobia to recover desertified Mediterranean ecosystems. *Applied Environmental Microbiology*, **59**, 129–133.

Hobbs, R. J. & Hopkins, A. J. M. (1990). From frontier to fragments: European impact on Australia's vegetation. In *Australian Ecosystems: 200 Years of Utilisation, Degradation and Reconstruction*, eds. D. A. Saunders, A. J. M. Hopkins & R. A. How, pp. 93–114. Chipping Norton, NSW: Surrey Beatty.

Hobbs, R. J. & Mooney, H. A. (1995). Effects of episodic rainfall events on Mediterranean-climate ecosystems. In *Time Scales in the Response to Water Stress: The Case of the Mediterranean Biota*, eds. J. Roy, J. Aronson & F. di Castri, pp. 71–85. Amsterdam: SPB Academic Publishers.

Hobbs, R. J. & Norton, D. A. (1996). Towards a conceptual framework for restoration ecology. *Restoration Ecology*, **4**, 93–110.

Hobbs, R. J. & Saunders, D. (eds.) (1992). *Reintegrating Fragmented Landscapes: Towards Sustainable Production and Nature Conservation*. New York: Springer-Verlag.

Holling, C. S. (1973). Resilience and stabilty of ecological systems. *Annual Review of Ecology and Systematics*, **4**, 1–23.

Hughes, C. E. & Styles, B. T. (1987). The benefits and potential risks of woody legume introductions. *International Tree Crops Journal*, **4**, 209–248.

Joffre, R. & Rambal, S. (1993). How tree cover influences the water balance of Mediterranean rangelands. *Ecology*, **74**, 570–582.

Joffre, R., Vacher, J., de los Llanos, C. & Long, G. (1988). The dehesa: an agrosilvopastoral system of the Mediterranean region with special reference to the Sierra Morena area of Spain. *Agroforestry Systems*, **6**, 71–96.

Joffre, R., Rambal, S. & Ratte, J. P. (1999). The dehesa system of southern Spain and Portugal as a natural ecosystem mimic. *Agroforestry Systems*, **45**, 57–79.

Jones, R. (1969). Fire-stick farming. *Australian Natural History*, **16**, 224–228.

Khillian, C. (1925). *Au Hoggar: Mission 1922*. Paris: Société d'Edition Géographique Maritime et Coloniale.

Kuhnholz-Lordat, G. (1938). *La Terre incendiée: essai d'agronomie comparée*. Nîmes, France: Maison Carrée.

Le Floc'h, E., Neffati, M., Chaieb, M. & Pontanier, R. (1995). Un essai de réhabilitation en zone aride: le cas de Menzel Habib (Tunisie). In *L'Homme peut-il refaire ce qu'il a défait*, eds. R. Pontanier, A. M. Hiri, N. Akrimi, J. Aronson & E. Le Floc'h, pp. 139–160. Paris: John Libbey Eurotext.

Le Floc'h, E., Neffati, M., Chaieb, M., Floret, C. & Pontanier, R. (1999). Rehabilitation experiment at Menzel Habib, southern Tunisia. *Arid Soil Research and Rehabilitation*, **13**, 369–381.

Lefroy, E. C., Hobbs, R. J., O'Connor, M. H. & Pate, J. S. (eds.) (2000). *Agriculture as a Mimic of Natural Ecosystems*. Amsterdam: Kluwer.

Le Houérou, H. N. (1959). *Recherches écologiques et floristiques sur la végétation de la Tunisie méridionale*. Algiers, Algeria: Institut de Recherches Sahariennes.

Le Houérou, H. N. (1969). *Végétation de la Tunisie Steppique (arec références au Maroc, à l'Algérie et à la Libye)*. Tunis, Tunisia: Annales de l'Institut National de Recherche Agronomique.

Le Houérou, H. N. (1984). Rain-use efficiency: a unifying concept. *Journal of Arid Environments*, **7**, 213–247.

Le Houérou, H. N. & Popov, G. F. (1981). *An Ecoclimatic Classification of Intertropical Africa*. Rome: FAO.

Liu, H., Cui, H., Pott, R. & Speier, M. (2000). Vegetation of the woodland–steppe transition at the southeastern edge of the Inner Mongolian Plateau. *Journal of Vegetation Science*, **11**, 525–532.

Ludwig, J. A. & Tongway, D. J. (1995). Spatial organisation of landscapes and its function in semi-arid woodlands, Australia. *Landscape Ecology*, **10**, 51–63.

Ludwig, J. A. & Tongway, D. J. (1996). Rehabilitation of semi-arid landscapes in Australia. 2: Restoring vegetation patches. *Restoration Ecology*, **4**, 398–406.

Ludwig, J. A. & Tongway, D. J. (1997) Landscape function. In *Landscape Ecology Function and Management: Principles from Australia's Rangelands*, eds. J. A. Ludwig, D. J. Tongway, D. Freudenberger, J. Noble & K. Hodgkinson, pp. 1–12. Melbourne: CSIRO.

Luken, J. L. (1990). *Directing Ecological Succession*. New York: Chapman & Hall.

Mellado, J. (1989). S O S Souss: Argan forest destruction in Morocco. *Oryx*, **23**, 87–93.

Mikesell, M. W. (1960). Deforestation in northern Morocco. *Science*, **132**, 441–448.

Milton, S. J., Dean, W. R. J., du Plessis, M. A. & Siegfried, W. R. (1994). A conceptual model of arid land degradation. *BioScience*, **44**, 70–76.

Monod, T. (1954). Modes 'contractés' et 'diffus' de la végétation saharienne. In *Biology of Deserts*, ed. J. L.

Cloudsley-Thompson, pp. 35–44. London: Institute of Biology.

Montoya Oliver, J. M. (1988). *Los alcornocales*. Madrid: Ministerio de Agricultura, Manuales Técnicos.

Noy-Meir, I. (1973). Desert ecosystems: environment and producers. *Annual Review of Ecology and Systematics*, **4**, 25–52.

Noy-Meir, I. (1981). Spatial effects in modelling of arid ecosystems. In *Arid Land Ecosystems*, eds. D. Goodall & R. Perry, pp. 411–432. Cambridge: Cambridge University Press.

Noy-Meir, I. (1985). Desert ecosystem structure and function. In *Ecosystems of the World*, vol. 12A, *Hot Deserts and Arid Shrublands*, eds. M. Evenari, I. Noy-Meir & D. Goodall, pp. 93–103. Amsterdam: Elsevier.

Odum, H. T. (1988). Self-organisation, transformity, and information. *Science*, **242**, 1132–1139.

Ovalle, C. & Avendaño, J. (1987). Interactions de la strate ligneuse et de la strate herbacée dans les formations d'*Acacia caven* (Mol.) Hook. et Arn. au Chili. 1. *Oecologica Plantarum*, **8**, 385–404.

Ovalle, C., Aronson, J., Del Pozo, A. & Avendaño, J. (1990). The espinal: agroforestry systems in the Mediterranean-type climate region of Chile. *Agroforestry Systems*, **10**, 213–239.

Ovalle, C., Avendaño, J., Del Pozo, A. & Aronson, J. (1996a). A study of the anthropogenic savannas in the subhumid region of central Chile: land occupation patterns and vegetation structure. *Forest Ecology and Management*, **86**, 129–139.

Ovalle, C., Longeri, L., Aronson, J., Herrera, A. & Avendaño, J. (1996b). N$_2$-fixation, nodule efficiency and biomass accumulation after two years in three Chilean legume trees and tagasaste (*Chamaecytisus proliferus ssp. palmensis*). Plant and Soil, **179**, 131–140.

Ovalle, C., Aronson, J., Del Pozo, A. & Avendaño, J. (1999). Restoration and rehabilitation of mixed espinales in central Chile: 10-year report and appraisal. *Arid Soil Research and Rehabilitation*, **13**, 369–381.

Perevelotsky, A. & Haimov, Y. (1992). The effect of thinning and goat browsing on the structure and development of Mediterranean woodland in Israel. *Forest Ecology and Management*, **49**, 61–74.

Querejeta, J. I., Roldán, A., Albaladejo, J. & Castillo, V. (1998). The role of mycorrhizae, site preparation, and organic amendment in the afforestation of a semi-arid Mediterranean site with *Pinus halepensis*. *Forest Science*, **44**, 203–211.

Reifenberg, A. (1955). *The Struggle between the Desert and the Sown*. Jerusalem: Jewish Agency.

Stebbing, E. P. (1953). *The Creeping Desert in the Sudan and Elsewhere in Africa*. Khartoum, Sudan: McCorquodale.

Thirgood, J. V. (1981). *Man and the Mediterranean Forest*. New York: Academic Press.

Tongway, D. J. (1990). Soil and landscape processes in the restoration of rangelands. *Australian Rangeland Journal*, **12**, 54–57.

Tongway, D. J. & Ludwig, J. A. (1990) Vegetation and soil patterning in semi-arid mulga lands of eastern Australia. *Australian Journal of Ecology*, **15**, 23–34.

Tongway, D. J. & Ludwig, J. A. (1996). Rehabilitation of semi-arid landscapes in Australia. 1: Restoring productive soil patches. *Restoration Ecology*, **4**, 388–397.

Tongway, D. J., Valentin, C. & Seghieri, J. (2000). *Banded Vegetation Patterning in Arid and Semi-Arid Environments*. New York: Springer-Verlag.

Valentin, C. & d'Herbès, J. M. (1999). Niger tiger bush as a natural water harvesting system. *Catena*, **37**, 231–256.

Vallejo, R. V. (ed.) (1996). *La restauración de la cubierta vegetal en la comunidad Valenciana*. Valencia, Spain: Centro de Estudios Ambientales del Mediterranean.

Vallejo, R., Bautista, S. & Cortina, J. (1999). Restoration for soil protection after disturbances. In *Life and Environment in the Mediterranean*, ed. L. Trabaud, pp. 301–343. Southampton, UK: WIT Press.

van der Maarel, E. (1990). Ecotones and ecoclines are different. *Journal of Vegetation Science*, **1**, 135–138.

Vogl, R. J. (1980). The ecological factors that produce perturbation-dependent ecosystems. In *The Recovery Process in Damaged Ecosystems*, ed. J. Cairns Jr, pp. 63–94. Ann Arbor, MI: Ann Arbor Science.

Wainwright, J. (1994). Anthropogenic factors in the degradation of semi-arid regions: a prehistoric case study in southern France. In *Environmental Change in Drylands: Biogeographical and Geomorphological Perspectives*, eds. A. C. Millington & K. Pye, pp. 285–304. New York: John Wiley.

Wesstrom, I. & Steen, E. (1993). Reconstitution de la végétation a la suite de mesures de protection des sols en terrains montagneux de Tunisie centrale. *Ecologia méditerranea*, **19**, 99–109.

Westman, W. E. (1990). Managing for biodiversity: unresolved science and policy questions. *BioScience*, **40**, 26–33.

Westoby, M., Walker, B. & Noy-Meir, I. (1989). Opportunistic management for rangelands not at equilibrium. *Journal of Range Management*, **42**, 266–274.

Whisenant, S. (1999). *Repairing Damaged Wildlands: A Process-Oriented, Landscape-Scale Approach*. Cambridge: Cambridge University Press.

Whisenant, S., Thurow, T. L. & Maranz, S. J. (1995). Initiating autogenic restoration on shallow semi-arid sites. *Restoration Ecology*, **3**, 61–67.

White, P. S. & Walker, J. L. (1997). Approximating nature's variation: selecting and using reference information in restoration ecology. *Restoration Ecology*, **5**, 338–349.

Yates, C. J., Hobbs, R. J. & Atkins, L. (2000). Establishment of perennial shrub and tree species in degraded *Eucalyptus salmonophloia* (Salmon Gum) remnant woodlands: effects of restoration treatments. *Restoration Ecology*, **8**, 135–143.

Zhu, Z., Liu, S., Wu, Z. & Di, X. M. (1986). *Deserts in China*. Langzhou, China: Institute of Desert Research, Academia Sinica.

21 • Australian semi-arid lands and savannas

DAVID J. TONGWAY AND JOHN A. LUDWIG

INTRODUCTION

The Australian semi-arid lands, occupying nearly two-thirds of the continent, about 5.5 million square kilometres, have a relatively short history of use by European pastoral systems (Noble & Tongway, 1986). Settlement began about 150 years ago, but some areas were only developed in the 1960s. Prior to that, the continent had been occupied by Aboriginal hunter–gatherer societies, whose major land management tool was fire. Analysis of the role of Aboriginal fire use in shaping the nature and distribution of the Australian flora and fauna is an active research topic (Nicholson, 1981; Bowman, 1998; Noble & Grice, 2001). It is certain that fire shaped the vegetation in a major way, but this chapter is concerned with the stress and disturbance due to pastoral use of the semi-arid lands and savannas, of which fire is just one part (Dyer *et al.*, 1997). Here, we briefly describe the impacts of pastoral land use and some efforts to rehabilitate these semi-arid lands and savannas using case studies in eastern and northwestern Australia (Fig. 21.1). We also present an alternative approach to such rehabilitation based on recently articulated principles of landscape ecology and evaluate the case studies against these principles.

NATURE OF THE PROBLEM

Australian semi-arid lands have amongst the world's least reliable rainfall regimes both in terms of spatial distribution and quantity (Fleming, 1978). Coupled with this is low soil fertility, due to the long history of soil weathering unrelieved by glacial stripping and rejuvenation, common in the northern hemisphere. Landscape relief is very low. The availability of surface water, prior to settlement,

was both sparse and temporary (Landsberg *et al.*, 1999; Freudenberger & Landsberg, 2000). These general conditions restricted the range, biomass and numbers of native herbivores. Their physiology suggests that they were adapted to survival, rather than production (Frith, 1970).

Occupation of the semi-arid lands expanded in good seasons when sheep and cattle could be droved long distances. Artesian water was exploited from 1879 and there was a rapid proliferation of permanent surface water storage in tanks and bores to sustain herbivores in drought times. The settlers did not understand the climate variability and stock numbers reflected expectations derived in good seasons. Set numbers of stock per paddock were typical and sheep and cattle were often retained long into a drought, due to low prices, transport difficulties and uncertainty as to when drought would end.

It is typical of the Australian semi-arid lands that degradation of the plant and soil resources was clear within about 20 years of settlement (Royal Commission, 1901). Eruptions of feral rabbits exacerbated the effect of domestic herbivores, and largely uncontrolled numbers of feral horses, camels and donkeys occurred in different biomes, competing with native grazing animals for herbage (Rolls, 1969, 1981; Low, 1999). Soil erosion surveys began as early as 1903 (Maiden, 1903) and the need for rehabilitation was recognised from this time forward. The cycle of drought, degradation, review and rehabilitation has been repeated several times in the ensuing century (Commonwealth & State Governments, 1978; Noble & Tongway, 1986). The outlook for pastoralism in the 1930s and 1940s was bleak (Pick, 1942).

The effect of overgrazing was the loss of perennial palatable grasses, their replacement by

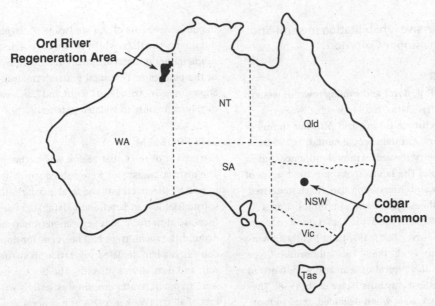

Fig. 21.1. Location of the case studies on the Australian continent.

short-lived herbage, soil erosion and in some land-scapes the increase in density of native unpalatable shrubs (Noble, 1997). All of these factors impinged on the capacity of land managers to maintain viable businesses. Dust storms were regular occurrences, which threatened both infrastructure and living conditions (Noble, 1904).

APPROACHES TO REHABILITATION

The past

The vast land areas affected by landscape deterioration and the relatively low financial returns per unit area militate against expensive and complex rehabilitation efforts. The procedures used can be divided into two classes: (1) passive, the removal of the disturbing influence and waiting for nature to effect a cure (Box 21.1) and (2) active, where efforts to re-establish pastures, often with exotic species, arresting erosion and ameliorating the soil are purposefully pursued (Box 21.2). Generally speaking, the passive procedures have had very slow and incomplete responses; often, critical events or stages in rehabilitation were missed due to lack of committed monitoring. Many active procedures, whilst technically successful at least in part, have been

very expensive and not economic (Friedel *et al.*, 1996*a, b*). Some have contributed to a subsequent weed problem (Lonsdale, 1994). Most of this work did not provide adequate explanations for observed system responses, whether favourable or not, so that generic principles have not emerged. Generally, this work focused too exclusively on the desired outcome ('perennial grasses have gone, let's plant more') rather than working out the most effective means to achieve the outcome. The mind-set behind these procedures has been called 'agronomic engineering' (Malcolm, 1990; Kearns, 1996), reflecting the capacity and desire to take charge of the needs of a piece of land, supplying any needs by management intervention (for example, artificial fertilisers, improved seeds, irrigation water and weed control). The two case studies review examples of the passive and active rehabilitation approaches.

Revised paradigm

In the process of studying semi-arid woodlands in western New South Wales, Australia, Ludwig & Tongway (1996, 1997) developed a conceptual framework by which to better understand the functioning of landscapes (Box 21.3). In relatively

Box 21.1   Passive rehabilitation in semi-arid woodland: the Cobar Common

THE SETTING

Cobar is a small pastoral and mining town in western New South Wales, about 700 km west of Sydney (Fig. 21.1). The climate is semi-arid with hot summers and cool winters. Annual average rainfall is 250–300 mm with tendency towards winter dominance, but is highly unreliable. The landscapes comprise a series of subdued ridges and intervening flats that are derived from Silurian slates, shales and sandstones. Slopes rarely exceed 3%. Soils are typic haplargids and durargids (Soil Survey Staff, 1975) and locally known as 'red earths' (Stace *et al.*, 1968). This landform/soil type combination is also typical of semi-arid woodlands in central and Western Australia (Stace *et al.*, 1968). The soils of the Cobar area, when degraded, tend to form hard surface seals with very low infiltration rates (Mucher *et al.*, 1988). The surface is also smooth, restricting seed lodgement and militating against organic matter accumulation (Greene & Tongway, 1989). Vegetation has been impacted by grazing for such a long period that shrub and tree species are the major components of the landscape. *Eucalyptus populnea* dominates the flats, whilst *Eucalyptus intertexta* and *Acacia aneura* (mulga) are co-dominants in other parts of the landscape. Perennial grasses remaining include *Stipa variabilis*, *Aristida jericoensis* and *Danthonia caespitosa*, mainly in isolated patches.

THE PROBLEM

Settlement of the Cobar region was in the mid-1870s. The district landscapes were subsequently heavily grazed by sheep, rabbits and feral goats, resulting in the substantial loss of perennial palatable grasses and the increase of native woody vegetation (Anonymous, 1969). Around the town, trees had been cut for domestic use, dairy herds had depleted vegetation to virtually bare soil, and dust storms affecting the town population were frequent. Water erosion was serious with soil loss of 10 cm (Walker, 1976). The general appearance of the landscape was as depicted in Fig. B21.1.

REHABILITATION ACTIVITIES

In 1959, a total area of 4811 ha around the town was fenced. Domestic stock were removed and not permitted to return (Walker, 1976). This was unpopular

Fig. B21.1. Cobar Common site, prior to rehabilitation.

among local graziers who unsuccessfully tried to reverse this decision by a political process. Although a small area was ripped to a depth of 45 cm, destocking represents the typical passive approach to rehabilitation. Natural recovery was monitored annually at a total of 14 photo-points. The objectives in undertaking this work were: (1) to abate the dust nuisance in the town, (2) to arrest soil erosion to protect the town water supply from siltation, and (3) to promote vegetation regeneration.

MONITORING REHABILITATION PROGRESS

The 16 years subsequent to 1959 included a severe and prolonged drought (1964–6) as well as two consecutive years (1973 and 1974) of more than double the long-term average rainfall. Photographic records show that mulga increased in number and size, as did turpentine (*Eremophia sturtii*), a 'woody weed' (Noble, 1997). However, the ground remained bare of herbage until the wet period of 1973–5. At an experimental area close by, mulga tree numbers rose from 33 to 71 and their mean height increased from 0.3 to 2.7 m in

the same time interval. Turpentine increased from 8 to 110 plants. There were several perennial grass germination events in intervening years where above-average rain fell, but no plants survived the ensuing dry seasonal conditions, even though grazing stress was absent. Although there were no official records of dust storms, local impressions are that the previously frequent dust nuisance had not occurred since 1965 (Walker, 1976). Casual observations suggested that the developing tree-belt dispersed approaching dust storms originating beyond the rehabilitated area. The monitoring record appeared to cease after 1976.

The years 1983 and 1984 had well above-average rain. Tongway & Smith (1989), working on the Cobar Common in this period, found patchy perennial grass production varied from 13.5 to 231 g dm m^{-2} (Fig. B21.2), the higher value representing excellent grass productivity. Plots with higher grass production had elevated soil organic carbon, nitrogen and infiltration values, compared to plots with lower production (Greene & Tongway, 1989), implying that natural rehabilitation processes had eventually created

Fig. B21.2. Cobar Common site in 1984, after 24 years of passive rehabilitation. Photograph by D. Tongway.

a soil with appropriate properties to support perennial grasses. This elevated soil condition and plant productivity also survived the prolonged drought from 1991 to 1994.

EVALUATION

It is now over 40 years since the initial passive stock exclusion treatment was commenced and although there are no true benchmarks available, full rehabilitation has probably not been achieved, even though the original objectives are largely satisfied. The length of time for a persistent ground layer to establish is of particular interest. The degraded soil surface was a smooth crust, with few obstructions to the transport of materials in runoff. These conditions provided poor habitat for grass establishment, with low seed lodgement capacity and infertile xeric soils. Rill and sheet erosion remains weakly active in some areas of the Common (D. Tongway, personal observation), implying that landscape instability persists. The rehabilitated land has not been subject to managed grazing over this time period, so the degree to which the rehabilitation has become resilient to the stress and disturbance effects of domestic grazing animals has not been assessed. The monitoring procedures concentrated on vegetation assessment, but not on the environmental requirements of 'missing' vegetation. This body of work suggests that degradation had caused this semi-arid ecosystem to fall below a critical threshold for the re-establishment of perennial grasses, though removal of the local dust hazard was effective in a reasonably short time, due to shrub growth. In summary, passive treatment in the simple form of removal of stress and disturbance due to grazing as a rehabilitation measure takes too long to achieve an outcome for pastoral production in Australian semi-arid lands.

EVALUATION AGAINST ECOLOGICAL PRINCIPLES

The Cobar Common was one of the sites used to develop the principles articulated in the section 'Revised paradigm' above (Tongway & Smith, 1989). It was clear that existing knowledge was insufficient in the 1970s and 1980s to determine the extent to which

any desertified landscape had lost productive potential and hence suggest the most efficacious rehabilitation procedure. Removal of the stress and disturbance was insufficient in itself to achieve full rehabilitation of the ground layer of vegetation, and seasonal conditions rather better than average were needed to re-establish resource regulating systems in the landscape.

Active rehabilitation activities that trapped water and organic matter or roughened the surface or eradicated inedible shrubs have also been used in similar landscapes to the Cobar Common. Water-ponding (Rhodes & Ringrose-Voase, 1987) effectively dealt with those soils with clay B horizons that have swell/shrink properties. Ponding eventually caused cracking of the B horizon, permitting water and root entry and leaching of soluble salts into the deeper horizon. Most semi-arid woodland soils in Australia do not have swelling B horizons. Blade ploughing aimed at killing shrubs by severing their roots to remove competition for water and facilitate grass growth failed (Robson, 1995). Shrubs re-established rapidly but grass establishment was so weak that grazing rapidly caused the total demise of palatable perennial grasses. In terms of our revised landscape function paradigm, the scale at which vital resources were regulated in the environment was too coarse for grasses (Tongway & Ludwig, 1994).

Soil surface treatments with mechanical implements to increase infiltration and trap seeds at another site near Cobar were partially successful (Cunningham et al., 1976). In central Australia in landscapes similar to Cobar, furrowing and banking were initially successful in establishing pasture grasses (Keetch, 1979). This was eventually shown to be due to the coincidence of the rehabilitation work and extremely favourable seasonal conditions. Later attempts to repeat the treatment were only partially successful in more normal or poor seasonal conditions (R. L. Keetch, personal communication). Friedel et al. (1996a, b) analysed the outcome from an enterprise point of view and showed that economic returns on outlays failed to break even. None of these trials made any overt attempt to capture organic matter in runon zones: water infiltration was the only factor considered essential.

Lack of herbivory control on establishing herbage, whether domestic, feral or wild, is a common thread in failed rehabilitation. Plant introductions, using sown exotic grass species in western New South Wales, have all failed (W. E. Mulham, personal communication). All accessions failed to survive two seasons. On the other hand, sporadic aerial seeding of the exotic buffel grass (*Cenchrus ciliaris*) over many decades has become highly successful to the point that it is invading native

pastures, and locally eliminating local species (Silcock, 1991). Buffel grass has moderate herbivore palatability, enabling pastoralism to persist. Weediness remains an issue, as buffel tends towards a monoculture that causes concern for biodiversity (Lonsdale, 1994). The loss of ecosystem services provided by native species has not yet been fully documented, but changes to granivorous avian fauna are a concern (Garnett, 1992).

Box 21.2   Active rehabilitation in semi-arid savanna: Ord River Regeneration Area

THE SETTING

The Ord River is a 55 000 km^2 catchment located in the East Kimberley region of northwestern Australia (Fig. 21.1). The climate is hot and semi-arid, where summer rainfall dominates with an annual average across the region of about 500–750 mm (de Salis, 1993). In higher rainfall zones, natural vegetation is semi-arid savanna or low, open, tallgrass woodland (Hacker, 1982, 1989), with an overstorey of eucalypt trees such as inland bloodwood (*Eucalyptus terminalis*) and an understorey of grasses such as whitegrass (*Sehima nervosum*) and black speargrass (*Heteropogon contortus*). These savannas become more open and grass-dominated at lower rainfalls, with arid shortgrasses such as limestone grass (*Enneapogon polyphyllus*) and kerosene grass (*Aristida contorta*) replacing tallgrasses. On rocky and sandy areas, limestone spinifex (*Triodia wiseana*) is common, and on areas of clay soil, nutwood trees (*Terminalia arostrata*) and grasses such as ribbon grass (*Chrysopogon fallax*) and barley Mitchell grass (*Astrebla pectinata*) dominate. The vegetation and soils in the region have been described and mapped into 'land systems' (Stewart *et al.*, 1970; Ryan, 1981). Demographic attributes of the region have also been described (Novelly & Gooding, 1997).

THE PROBLEM

Pastoralism in the East Kimberley region was initiated in 1884 when cattle (from Queensland) were

established on Ord River Station (Hacker, 1989). As now evident in hindsight, these first pastoralists of European origins did not understand the limits of these semi-arid tropical lands. By 1944, massive degradation was recorded due to overgrazing by cattle (Teakle, 1944; Medcalf, 1944). Loss of perennial plant ground cover and subsequent soil erosion was particularly severe on certain land systems. For example, much of the Nelson Land System was in poor range condition in 1981 (de Salis, 1993). This land system is an erosional plain formed from interbedded limestones, shales and mudstones (Stewart *et al.*, 1970). The arid short-grass vegetation on these lands was rapidly removed by grazing animals, leaving powdery calcareous loam soil surfaces exposed to massive sheet and gully erosion and to wind erosion (Payne *et al.*, 1979) (Fig. B21.3).

In 1960, the Ord River Irrigation Scheme, Stage 1, was initiated, and a Stage 2 has been proposed as the Kimberley Region continues to grow in population (Novelly & Gooding, 1997; Kinhill Pty Ltd, 2000). This irrigation development involved the construction of the Ord River Dam to form Lake Argyle and the Kununurra Dam below this lake is used to provide regulated levels of irrigation water to fields below this dam. Gauging station flow and sediment data at the dam site from 1962 to 1968 indicated that 29.8 million tonnes of sediment per year would be discharged into Lake Argyle (Kata, 1978). Most of these sediments were coming from erodable land systems in the catchment such as the Nelson Land System (Wasson *et al.*, 1994). In the 1960s, these land systems were in poor rangeland condition (Fitzgerald, 1967; de Salis, 1993),

Fig. B21.3. Ord River Regeneration Area. Gully erosion remains after 34 years of active rehabilitation. Photograph by N. Hindley.

with an estimated 3600 km² of the upper catchment being massively eroded where up to 30 cm of surface soil had been stripped (Medcalf, 1944; Teakle, 1944). This rate of soil sediment deposition into Lake Argyle would clearly threaten the long-term capacity of the lake to deliver irrigation water.

REHABILITATION ACTIVITIES

To reduce sediment losses from the upper Ord catchment, the Ord River Regeneration Area (ORRA) (Fig. 21.1) was established in 1960 (de Salis, 1993). The ORRA included about 10 000 km² out of about 46 000 km² located above the Ord River Dam. The objectives of rehabilitation on the ORRA (Hacker, 1989; Wasson et al., 1994) were to:

- Reduce sediment flows into Lake Argyle by treating surfaces to increase infiltration and decrease runoff and erosion
- Control the level of grazing
- Establish pastures with high seed producing plants.

The first objective involved using a number of mechanical implements to break the soil crust open,

thereby allowing water infiltration higher up in the catchment to reduce runoff and soil erosion (de Salis, 1993). Opposed disc and chisel ploughs, and disc pitters, were used to create seed beds and control runoff water. Bands of cultivation were created, mostly located on the interfluve zones between gullies. The complex network of rills and gullies made this mechanical rehabilitation difficult.

The second rehabilitation objective was supposed to reduce grazing by removal of domestic stock and feral animals. However, this proved to be difficult because it required massive co-operative efforts between many pastoral and governmental land management groups (de Salis, 1993). Costs were also great because about 700 km of fencing was required to help muster cattle and control feral animals (Hacker, 1989). Thus, effective cattle control was not obtained until about 1975, and feral donkey numbers remained high until 1990.

The third objective was to establish vegetation on these treated areas using pasture plants known to be prolific seed producers (Hacker, 1989). Three exotic species were selected and sown: buffel and birdwood grass (Cenchrus ciliaris and C. setigerus, respectively) and kapok bush (*Aerva javanica*). By 1968, 80 000 km of furrows had been ploughed, and altogether about 10 000 ha of land had been furrowed and seeded with these three plant species.

MONITORING REHABILITATION PROGRESS

Assessing this extensive rehabilitation required an active monitoring program (Hacker, 1989). Rangeland condition and soil erosion surveys in the ORRA began in the 1960s with a number of field-based studies and monitoring activities being conducted (e.g. Fitzgerald, 1968*a*, *b*; Ryan, 1981). A comprehensive assessment of range condition in the ORRA was conducted in 1981 (de Salis, 1993), which found that vegetation re-establishment was poor on areas where soils had been massively stripped, mainly on lower slopes. However, upper interfluve slopes responded well to rehabilitation treatments (Fig. B21.4). Unsown native grasses were the first to establish, as was the exotic kapok bush. Birdwood grass gradually increased over time, eventually dominating and forming a monoculture.

A grazing trial was conducted on one of these regenerated interfluve areas from 1983 to 1988 (Hacker & Tunbridge, 1991). This trial found that cattle selectively grazed sites within the regeneration area where perennial grasses had poorly established. Cattle liveweight gains in trial paddocks were most strongly related to the biomass of native arid short-grasses (e.g. *Enneapogon polyphyllus*), not exotic perennials (e.g. *Cenchrus setigerus*).

The cultivated strips formed an intricate mosaic, regulating the rate of runoff, which had formerly been excessive (Fitzgerald, 1968*a*). Overall, with these surface treatments and grass establishment, water was more competently retained in interfluve landscapes (Fitzgerald, 1968*b*). This improved water retention also favoured trees, both in the establishment of new trees and in the recovery of trees previously thought to be dead.

EVALUATION

In relation to the three objectives for the ORRA, rehabilitation was partially successful (Hacker, 1989; Wasson *et al.*, 1994). Mechanical treatment of crusted

soil surfaces on red soil plains increased infiltration rates (Fitzgerald, 1968*a*). This treatment promoted the establishment of seeded exotic grasses that themselves produced large amounts of seed (de Salis, 1993). Areas of black clay plains regenerated without seeding or mechanical treatment. Cattle and feral grazing animals have been controlled, although it took 30 years to effectively control donkeys (de Salis, 1993), and helicopter mustering and shooting must continue. However, this improved stock control and increased cover of vegetation in the upper catchment only reduced the estimated flow of sediment into Lake Argyle from 29.8 million tonnes per year (1962–8) to 23.5 million tonnes per year (1972–85) (Wasson *et al.*, 1994). About 90% of this sediment came from gully erosion occurring in areas that occupy less than 10% of the catchment. Natural and radioactive isotope tracer studies indicate that most of these gullies are relatively young, being formed since the arrival of pastoralists (Wasson *et al.*, 1994).

Thus, the use of the ORRA lands by domestic stock is still problematic (Hacker & Tunbridge, 1991). Patch grazing habits tend to disturb vulnerable areas such as lower slopes, rendering them subject to rain-splash erosion. Heavy rainstorms during the wet season are typical for this part of Australia. Prescribed fire is an option to balance the species composition of the woody and grassy components of these savannas (Craig, 1997; Dyer *et al.*, 1997), but the outcome of such fires on these recently rehabilitated lands is difficult to predict given rainfall variability in the region.

In relation to the paradigm of landscape function (Ludwig & Tongway, 1997), rehabilitation of the ORRA involved a project that was extensive enough for a whole-of-landscape approach from its beginning. Fragile landscape types (e.g. Nelson Land System) had a partial response to mechanical treatments, implying that they may be close to exceeding a critical threshold (see Fig. 21.3), this being due to improved control of surface hydrological flows. The black clay soils may never need rehabilitation as their initially low cover was purely a grass–herbivore interaction, and their soil productive potential was never affected by grazing. These soils are classic representatives of the 'robust' type of landscape (Tongway & Hindley, 2000).

(a)

(b)

Fig. B21.4. Ord River Regeneration Area. (a) Furrows created in 1961 to initiate regeneration on scalded interfluves (photograph by A. Payne). (b) The same area in 1966: response to ponding bank rehabilitation treatment.

The footslopes show the complexities of applying rehabilitation to semi-arid savanna landscapes. The first priority was to reduce sediment flows by improving hydrological processes by increasing infiltration rates and controlling erosion. Although infiltration was improved, gully cutting continues (Fig. B21.3) as revegetation efforts have been only partially successful (Wasson *et al.*, 1994). The case as we have presented it implies that the footslope landscapes are not yet ready to be subjected to the stress and disturbance of extensive cattle grazing (Hacker, 1989).

Purvis (1986) described his successful techniques for dealing with gully erosion, but these are somewhat more involved than what was attempted in the ORRA, and were developed over a 20-year period. Purvis aimed to control runoff water by initially installing banks at the top of the watershed and progressively moving downslope as control was gained over the amount, rate and pattern of upslope runoff. Flow into gullies was reduced to very low levels. This is not a trivial task,

but the expectation that gullies will spontaneously improve without major works to control upslope surface runoff leaves too much to nature when landscapes are already in a highly dysfunctional state.

Dealing with hydrological processes is not the only priority when improving landscape function (Tongway & Ludwig, 1996). Besides re-establishing vegetation, organic matter cycling processes also need to improve. For example, soil organic matter decomposition is needed to improve soil aggregate stability, which will reduce soil erosion. Successful rehabilitation of many upslope interfluves in the ORRA may be limited by low levels of these processes.

Box 21.3 Experimental rehabilitation based on landscape function

The authors tested the landscape function paradigm based on the trigger–transfer–reserve–pulse resource regulation framework (Fig. 21.2) by setting out to create perennial grass patches on bare stony runoff slopes of a heavily grazed paddock on soils and landscapes similar to those in the Cobar Common (Fig. B21.5). The method involved setting up patches

Fig. B21.5. Landscape function rehabilitation experiment: prior to treatment, 1988. Photograph by D. Tongway.

Fig. B21.6. Landscape function rehabilitation experiment: ten years after treatment application. Typical herbivore stocking levels were maintained throughout, and a three-year drought (1991–4) overlaid the herbivory. Photograph by J. Ludwig.

comprised of tree branches, with their long axes aligned with the contour so as to trap resources carried in runoff water, topsoil, litter, dung and seeds. The branch mounds were about 45 cm high, enabling them to also capture aeolian materials such as saltating sand grains and organic matter. The soil was analysed at commencement and after three years, measuring physical, chemical and biological soil properties (Tongway & Ludwig, 1996). Vascular plant germination and establishment were recorded at three-monthly intervals (Ludwig & Tongway, 1996). Perennial palatable grasses established within about 18 months of the experiment commencing, and survived the ensuing 11 years (Fig. B21.6). The soil properties after three years showed major developments in the accumulation of soil carbon and nitrogen, improvements in soil respiration and infiltration rate. Sheep and kangaroo were maintained on the site at levels similar to heavy commercial stocking throughout the experiment. After 11 years, all those soil variables continue to improve (Ludwig & Tongway, 1998) (Table B21.1); the experiment

also survived a major drought between 1991 and 1995 (Fig. B21.6). In summary, this experiment provided adequate means of capturing and retaining scarce, vital resources and encouraged a full range of soil processes to develop in concert. The resulting robust, productive patches supporting palatable perennial grasses can be maintained in the face of typical stress and disturbance regimes.

Table B21.1. *Changes in mean % soil organic carbon over a period of 11 years in rehabilitated and un-rehabilitated landscapes, semi-arid woodlands* (n = 20)

Depth (cm)	Rehabilitated			Un-rehabilitated		
	1988	1991	1999	1988	1991	1999
0–1	0.73	1.00	1.07	0.60	0.71	0.72
1–3	0.61	0.60	0.96	0.55	0.53	0.57
3–5	0.59	0.53	0.83	0.49	0.44	0.51
5–10	0.55	0.50	0.60	0.48	0.44	0.47

undisturbed semi-arid woodlands distinctive runoff and runon patterns were shown to provide enhanced water resources to woodland patches or groves, so as to support dense, long-lived vegetation (Tongway & Ludwig, 1990; Ludwig & Tongway, 1995). The runoff zone was substantially bare of vegetation, except for sparse ephemerals after rains. The runoff zone had a physical crust on the soil surface with few signs of accelerated erosion. On the upslope edge of the runon patches were perennial, palatable grasses, making use of runon water from more frequent small rain showers. A grove of mulga shrub (*Acacia aneura*) occupied the core of the runon zone. The whole landscape was a system comprised of conjugate units of runoff and runon. The scale and effectiveness of the various processes resulted in a system where a range of vascular plants had favourable long-term edaphic environments distributed as patches in the landscape.

Ludwig & Tongway (1997) formalised this ecosystem function in a framework, called trigger–transfer–reserve–pulse. This framework accounted for the spatial patterning in terms of resource regulation processes (Fig. 21.2). Feedback loops involving recycling and physical flow modification stabilised the overall functioning. Biota are both contributors to, and beneficiaries of, landscape processes. In particular, the approach is focused on the processes which retain, utilise and cycle scarce resources and less on simply listing the organisms or properties present or missing from a given landscape. These factors are important in infertile semi-arid lands because of the scarcity of vital resources (Noy-Meir, 1973; Ludwig & Tongway, 1997). This approach takes a whole-of-landscape view, recognising the need for both spatial heterogeneity in semi-arid lands and the processes by which this is maintained in space and time (Ludwig & Tongway, 2000). In degraded or desertified landscapes, the balance between resource retention and loss can be depicted as potentially unbalanced (Fig. 21.2). Functional landscapes are styled as 'resource conserving' whereas dysfunctional landscapes 'leak' resources to some degree.

This framework suggests that rehabilitation should more directly address the improvement of the edaphic habitat for species of concern. Four ecological principles need to be addressed in rehabilitation:

- The end land use must be specified in advance in order to design appropriate rehabilitation procedures.
- The current functional status of the degraded land needs to be understood.
- It is vital to set up processes to benefit the edaphic habitat for the specified biota.

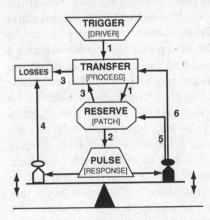

Ref	Processes Involved	
1	• Runon • Storage/Capture	• Deposition • Saltation capture
2	• Plant germination growth • Nutrient mineralisation	• Uptake processes
3	• Runoff into steams • Rill flow and erosion	• Sheet erosion out of system • Wind erosion out of system
4	• Herbivory • Fire	• Harvesting • Deep drainage
5	• Seed pool replenishment • Organic matter cycling/Decomposition processes • Harvest/Concentraion by soil micro-fauna	
6	• Physical obstruction/Absorption processes	

Fig. 21.2. The trigger–transfer–reserve–pulse framework. Examples of processes inferred by the framework are tabulated. Landscape degradation (processes 3 and 4) would tip the balance to the left (excessive resource loss), whilst adequate feedback mechanisms (processes 5 and 6) tend to balance inputs and outputs from the system.

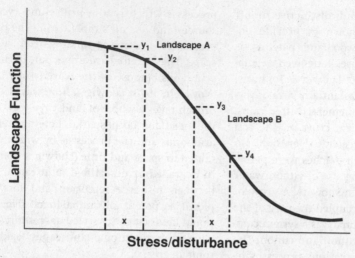

Fig. 21.3. The effect of similar quanta of applied stress/disturbance and landscape function on landscapes with differing initial function.

- A full range of ecosystem processes (physical, chemical and biological) needs to be improved, in order to minimise the time needed for rehabilitation.

CONCLUDING REMARKS

At present, there is a lack of connectivity between the three phases of rehabilitation: (1) assessment of landscape degradation, (2) the technical means of achieving rehabilitation, and (3) the monitoring of progress towards a satisfactory outcome within a reasonable time-frame. The informing science in the Australian semi-arid lands has been largely vegetation science, together with soil erosion information from the conservation technology perspective (Noble *et al.*, 1984). Recently, Ludwig & Tongway (1997, 2000) proposed that landscapes be treated as systems (sequences of processes stabilised by feedback loops), to facilitate a holistic assessment of ecosystem function. They proposed a framework summarising how landscapes work, with the potential to have detailed simulation models embedded within it (e.g. Ludwig *et al.*, 1994), but also with the capacity to identify which parts of a landscape were dysfunctional. Tongway & Hindley (1995, 2000) proposed field methodologies to provide appropriate information to the framework at low cost. These methods use a series of simple indicators of

processes to produce information at both patch and whole landscape scale. These summarise the productive potential of the soil and identify processes related to the regulation of scarce, vital resources in space and time. The data are broad-based, covering biological, biophysical and physical processes, facilitating an interpretation based on a function–dysfunction continuum, rather than on lists of missing species.

Tongway & Hindley (2000) also suggested a means of interpreting these monitoring data to identify critical thresholds that would signify several states related to self-sustainability, risk and potential to rehabilitate. This interpretational framework uses a sigmoidal shaped response curve, linking attributes of landscape function (y axis) to attributes of stress or disturbance (Fig. 21.3). Degradation/rehabilitation scenarios are depicted by curves. Stress and/or disturbance may be long term effects such as grazing pressure and droughts or short term stochastic events like fire. If a landscape with good functional status at location A undergoes a quantum of stress and disturbance x, the system has a very small functional response ($y1$ to $y2$). However, if a landscape with some dysfunction, B, undergoes the same quantum of stress/disturbance, x, it loses a large amount of function (from $y3$ to $y4$). This function loss may take the landscape below a critical threshold where spontaneous recovery is very slow. The states identified by this analytical approach are compatible with state-and-transition models (Westoby *et al.*, 1989) in principle, but use different information sources.

This landscape function approach enables the identification of processes under threat and also the means by which those processes can be restored or ameliorated, thus potentially supplying the needed connectivity between assessment and rehabilitation. The approach has been implemented on rehabilitating mine sites in the Australian semi-arid and savanna landscapes (Tongway *et al.*, 1997).

Related to these issues is the nature of the stress/disturbance regime that the rehabilitated landscape will be subjected to on completion. In our case studies, stress and disturbance regimes remain deliberately lower than the pre-rehabilitation situation. In that sense, rehabilitation success can only be gauged by subjecting the landscape to 'real world' stresses and by demonstrating little functional change.

REFERENCES

Anonymous (1969). *Report of the Inter-Departmental Committee on Scrub and Timber Regrowth in the Cobar–Byrock District and other Areas of the Western Division of NSW*. Sydney: Government Printer.

Bowman, D. M. S. (1998). Tansley Review 101: The impact of Aboriginal landscape burning on the Australian biota. *New Phytologist*, **140**, 385–410.

Commonwealth & State Governments (1978). *A Basis for Soil Conservation Policy in Australia: Commonwealth and State Governments Collaborative Soil Conservation Study 1975–77*, Report 1. Canberra: Australian Government Printing Service.

Craig, A. B. (1997). A review of information on the effects of fire in relation to the management of rangelands in the Kimberley high-rainfall zone. *Tropical Grasslands*, **31**, 161–187.

Cunningham, G. M., Walker, P. J. & Green, D. R. (1976). *Rehabilitation of Arid Zones: 10 years Research at Cobar, New South Wales*. Sydney: Soil Conservation Service of New South Wales.

de Salis, J. (1993). *Resource inventory and condition survey of the Ord River Regeneration Reserve*, Miscellaneous Publication no. 14/93. Perth, WA: Department of Agriculture.

Dyer, R., Craig, A. & Grice, A. C. (1997). Fire in northern pastoral lands. In *Fire in the Management of Northern Australian Pastoral Lands*, Occasional Publication no. 8, eds. A. C. Grice & S. M. Slatter, pp. 24–40. Brisbane: Tropical Grassland Society of Australia.

Fitzgerald, K. (1967). The Ord River catchment regeneration project. 1: The nature, extent and causes of erosion in the Ord River catchment area. *Journal of Agriculture of Western Australia*, 4th Series, **8**, 446–452.

Fitzgerald, K. (1968a). The Ord River catchment regeneration project. 2: Dealing with the problem. *Journal of Agriculture of Western Australia*, 4th Series, **9**, 90–95.

Fitzgerald, K. (1968b). The Ord River catchment regeneration project. 3: Eight years of progress. *Journal of Agriculture of Western Australia*, 4th Series, **9**, 398–405.

Fleming, P. M. (1978). Types of rainfall and local rainfall variability. In *Studies of the Australian Arid Zone*, vol. 3, *Water in Rangelands*, ed. K. M. W. Howes, pp. 18–28. Perth, WA: CSIRO.

Freudenberger, D. & Landsberg, J. (2000). Management of stock watering points and grazing to maintain landscape function and biological diversity in rangelands. In *Management for Sustainable Ecosystems*, eds. P. Hale, A. Petrie, D. Molonely & P. I. Sattler, pp. 71–77. Brisbane: University of Queensland.

Friedel, M. H., Kinloch, J. E. & Muller, W. J. (1996a). The potential of some mechanical treatments for rehabilitating arid rangelands. 1: Within-site effects and economic returns. *Rangeland Journal*, **18**, 150–164.

Friedel, M. H., Muller, W. J. & Kinloch, J. E. (1996b). The potential of some mechanical treatments for rehabilitating arid rangelands 2. Identifying indicators from between-site comparisons. *Rangeland Journal*, **18**, 165–178.

Frith, H. J. (1970). The herbivorous wild animals. In *Australian Grasslands*, ed. R. M. Moore, pp. 74–83. Canberra: Australian National University Press.

Garnett, S. (1992). *The Action Plan for Australian Birds*. Canberra: Australian National Parks and Wildlife Service.

Greene, R. S. B & Tongway, D. J. (1989). The significance of (surface) physical and chemical properties in determining soil surface condition of red earths in rangelands. *Australian Journal of Soil Research*, **27**, 213–225.

Hacker, R. B. (ed.) (1982). *The Problems and Prospects of the Kimberley Pastoral Industry*, Technical Report no. 6. Perth, WA: Department of Agriculture.

Hacker, R. B. (1989). An evaluation of range regeneration programs in Western Australia. *Australian Rangeland Journal*, **11**, 89–100.

Hacker, R. B. & Tunbridge, S. B. (1991). Grazing management strategies for reseeded rangelands in the East Kimberley Region of Western Australia. *Rangeland Journal*, **13**, 14–35.

Kata, P. (1978). *Ord River Sediment Study*. Perth, WA: Water Resources, Department of Public Works.

Kearns, A. (ed.) (1996). *Yesterday, Today and Tomorrow: The Changing Role of Engineers in Planning and Implementing Sustainable Development*. Darwin, NT: Institution of Engineers, Australia.

Keetch, R. L. (1979). *Rangeland Rehabilitation in Central Australia*, Publication no. LC/79/7. Alice Springs, NT: Land Conservation Unit.

Kinhill Pty Ltd (2000). *Ord River Irrigation Area Stage 2: Proposed Development of the M2 Area*, Draft Environmental Impact Statement, prepared for Wesfarmers Sugar Company Pty Ltd. Perth, WA: Marubeni Corporation and Water Corporation of Western Australia.

Landsberg, J., Craig, J. D., Morton, S. R., Hobbs, T. J., Stol, J., Drew, A. & Tongway, H. (1999). *The Effects of Artificial Sources of Water on Rangeland Biodiversity*, Biodiversity, Technical Paper no. 3. Canberra: Environment Australia.

Lonsdale, M. W. (1994). Inviting trouble: introduced pasture species in northern Australia. *Australian Journal of Ecology*, **19**, 345–354.

Low, T. (1999). *Feral Future: The Untold Story of Australia's Exotic Invaders*. Melbourne: Viking-Penguin Books.

Ludwig, J. A. & Tongway, D. J. (1995). Spatial organisation and its function in semi-arid woodlands, Australia. *Landscape Ecology*, **10**, 51–63.

Ludwig, J. A. & Tongway, D. J. (1996). Rehabilitation of semi-arid landscapes in Australia. 2: Restoring vegetation patches. *Restoration Ecology*, **4**, 398–406.

Ludwig, J. A. & Tongway, D. J. (1997). A landscape approach to rangeland ecology. In *Landscape Ecology Function and Management: Principles from Australia's Rangelands*, eds. J. Ludwig, D. Tongway, D. Freudenberger, J. Noble & K. Hodgkinson, pp. 1–12. Melbourne: CSIRO.

Ludwig, J. A. & Tongway, D. J. (1998). Ten years on, created landscape patches are still functioning. *Range Management Newsletter*, **98**, 1–7.

Ludwig, J. A. & Tongway, D. J. (2000). Viewing rangelands as landscape systems. In *Rangeland Desertification*, eds. O. Arnalds & S. Archer, pp. 39–52. Dordrecht: Kluwer.

Ludwig, J. A., Tongway, D. J. & Marsden, S. G. (1994). A flow-filter model for simulating the conservation of limited resources in spatially heterogenous, semi-arid landscapes. *Pacific Conservation Biology*, **1**, 209–213.

Maiden, J. H. (1903). The sand-drift problem in New South Wales. *Journal of the Proceedings of the Royal Society New South Wales*, **37**, 82–106.

Malcolm, C. V. (1990). Rehabilitation agronomy: guidelines for revegetating degraded land. *Proceedings of the Ecological Society of Australia*, **16**, 551–556.

Medcalf, F. G. (1944). *Soil Erosion Reconnaissance of the Ord River Valley and Watershed*. Perth, WA: Lands and Survey Reports.

Mucher, H. J., Chartres, C. J., Tongway, D. J. & Greene, R. S. B. (1988). Micromorphology and significance of the surface crusts in Rangelands near Cobar, Australia. *Geoderma*, **42**, 227–244.

Nicholson, P. H. (1981). Fire and the Australian Aborigine: an Enigma. In *Fire and the Australian Biota*, eds. A. M. Gill, R. H. Groves & I. R. Noble, pp. 55–76. Canberra: Australian Academy of Science.

Noble, A. (1904). Dust in the atmosphere during 1902–3. *Monitoring Weather Review*, **32**, 364–365.

Noble, J. C. (1997). *The Delicate and Noxious Scrub*. Canberra: CSIRO Wildlife and Ecology.

Noble, J. C. & Grice, A. C. (2001). Fire regimes in semi-arid and tropical pastoral lands: managing biological diversity and ecosystem function. In *Flammable Australia: The Fire Regimes and Biodiversity of a Continent*, eds. R. A. Bradstock, J. Williams & A. M. Gill, pp. 373–400. Cambridge: Cambridge University Press.

Noble, J. C. & Tongway, D. J. (1986). Pastoral settlement in arid and semi-arid rangelands. In *Australian Soils: The Human Impact*, eds. J. S. Russell & R. F. Isbell, pp. 8–14. Brisbane: University of Queensland Press.

Noble, J. C., Cunningham, G. M. & Mulham, W. E. (1984). Rehabilitation of degraded land. In *Management of Australia's Rangelands*, eds. G. N. Harrington, A. D. Wilson & M. D. Young, pp. 171–186. Melbourne: CSIRO.

Novelly, P. & Gooding, J. (1997). Kimberley Region, WA: case study – the present. In *Sustainable Habitation in the Rangelands*, eds. N. Able & S. Ryan, pp. 36–51. Canberra: Centre for Resource and Environmental Studies, Australian National University.

Noy-Meir, I. (1973). Desert ecosystems: environment and producers. *Annual Review of Ecology and Systematics*, **4**, 25–51.

Payne, A. L., Kubicki, A., Wilcox, D. G. & Short, L. C. (1979). *A Report on Erosion and Range Condition in the West Kimberley Area of Western Australia*, Technical Bulletin no. 42. Perth, WA: Department of Agriculture.

Pick, J. H. (1942). *Australia's Dying Heart: Soil Erosion and Station Management in the Inland*. Melbourne: Melbourne University Press.

Purvis, J. R. (1986) Nurture the land: my philosophy of pastoral management in central Australia. *Australian Rangeland Journal*, **8**, 110–117.

Rhodes, D. W. & Ringrose-Voase, A. J. (1987). Changes in soil properties during scald reclamation by waterponding. *Journal of Soil Conservation, New South Wales*, **42**, 84–90.

Robson, A. D. (1995). The effect of grazing exclusion and blade-ploughing on semi-arid woodland vegetation in north western New South Wales over 30 months. *Rangeland Journal*, **17**, 111–127.

Rolls, E. C. (1969). *They All Ran Wild: The Story of Pests on the Land in Australia*. Sydney: Angus & Robertson.

Rolls, E. C. (1981). *A Million Wild Acres*. Melbourne: Nelson.

Royal Commission (1901). *Report of Royal Commission to Enquire into the Condition of the Crown Tenants, Western Division of New South Wales*. Sydney: Government Printer.

Ryan, W. J. (1981). *An Assessment of Recovery and Land Capability of Part of the Ord River Catchment Regeneration Project*, Technical Bulletin no 53. Perth, WA: Department of Agriculture.

Silcock, R. G. (1991). *The Role of Pastures, Trees and Shrubs in Rehabilitating Mined Land in Queensland: Impediments to their Use on Opencut Coal and Alluvial Mines*, QO91018. Brisbane, Qld: Queensland Department of Primary Industries.

Soil Survey Staff (1975). *Soil Taxonomy: A Basic System for Making and Interpreting Soil Surveys*, Handbook no. 436. Washington, DC: U.S. Department of Agriculture.

Stace, H. C. T., Hubble, G. D., Brewer, R., Sleeman, J. R., Mulchcahy, M. J. & Hallsworth, E. G. (1968). *A Handbook of Australian Soils*. Glenside, SA: Rellim.

Stewart, G. S., Perry, R. A., Paterson, S. J., Traves, D. M., Slatyer, R. O., Dunn, P. R., Jones, P. J. & Sleeman, J. R. (1970). *Lands of the Ord–Victoria Area, Western Australia and Northern Territory*, Land Research Series no. 28. Canberra: CSIRO.

Teakle, L. J. H. (1944). Soil erosion and its relationship to soil type and geological formations of the Upper Ord Basin. In *File 23/44*, vol. 1, pp. 174–182. Perth, WA: Department of Lands and Surveys.

Tongway, D. J. & Hindley, N. (1995). *Manual for the Assessment of Soil Condition of Tropical Grasslands*. Canberra: CSIRO Wildlife and Ecology.

Tongway, D. J. & Hindley, N. (2000). Assessing and monitoring desertification with soil indicators. In *Rangeland Desertification*, eds. O. Arnalds & S. Archer, pp. 89–98. Dordrecht: Kluwer.

Tongway, D. J. & Ludwig, J. A. (1990). Vegetation and soil patterning in semi-arid mulga lands of Eastern Australia. *Australian Journal of Ecology*, **15**, 23–34.

Tongway, D. J. & Ludwig, J. A. (1994). Small-scale resource heterogeneity in semi-arid landscapes. *Pacific Conservation Biology*, **1**, 201–208.

Tongway, D. J. & Ludwig, J. A. (1996). Rehabilitation of semi-arid landscapes in Australia. 1: Restoring productive soil patches. *Restoration Ecology*, **4**, 388–397.

Tongway, D. J. & Smith, E. L. (1989). Soil surface features as indicators of rangeland site productivity. *Australian Rangeland Journal*, **11**, 15–20.

Tongway, D. J., Hindley, N., Ludwig, J. A., Kearns, A. & Barnett, G. (1997). Early indicators of ecosystem function on selected mine sites. In *Proceedings of the 22nd Annual Minerals Council of Australia Environmental Workshop*, pp. 495–505. Canberra: Minerals Council.

Walker, P. J. (1976). Cobar Regeneration Area: the first sixteen years. *Soil Conservation Journal of New South Wales*, July 1976, 119–130.

Wasson, R. J., Caitcheon, G., Murray, A. S., Wallbrink, P., McCulloch, M. & Quade, J. (1994). *Sources of Sediment in Lake Argyle*, Report CLW 48750 to Western Australia

Department of Agriculture. Canberra: CSIRO Land and Water.

Westoby, M., Walker, B. & Noy-Meir, I. (1989). Opportunistic management for rangelands not at equilibrium. *Journal of Rangeland Management*, 42, 266–274.

PETER BUCKLEY, SATOSHI ITO AND STÉPHANE McLACHLAN

INTRODUCTION

Broadleaved and mixed broadleaf–conifer forests occupy the temperate region, comprising much of eastern North America, northern and western Europe, eastern China and Japan, with parallel distributions in South America, southeast Australia and New Zealand in the southern hemisphere. In warmer and drier climates these forests merge into xeric, evergreen scrub of the Mediterranean zone, and into boreal, conifer-dominated forest at higher latitudes or greater altitude. Histories of the attrition of virgin, old-growth forest vary considerably in different countries. Over the past 8000 years, most intact, closed forests in Europe and Asia have disappeared, together with 55% of those in North America (Bryant *et al.*, 1997). However, even in Europe, much semi-natural and some old-growth forest, although modified by cutting, burning and hunting practices, survived well into the post-medieval period, and much later in pre-settlement North America (Peterken, 1996).

Rates of loss of old-growth and semi-natural forest have accelerated sharply in the past two centuries, owing to the increased demands of population growth, industrialisation, modern agriculture, logging and the rise of silvicultural management. Following the European tradition of production forestry dating from the early nineteenth century, plantations, often of non-native species, were substituted for the original forest cover, causing significant further fragmentation. In Britain, where semi-natural forest comprises only 13% of the total forest area, some 38% of semi-natural, ancient woodland was converted to plantation forestry during the period since 1930 (Spencer & Kirby, 1992), a proportion similar to

that in both Japan and South Korea (40% and 32%, respectively). The comprehensive removal, modification and substitution of all but a few remaining pockets of original forest cover provides a strong justification for developing new restoration strategies and targets.

It is a premise that at the outset of any restoration project, the structure, composition, dynamics and features of 'pristine' or near-natural forest *within that region* should be understood. This is easy to say, but difficult to unravel where forests have long since been fragmented and exploited, altered in composition, or entirely destroyed through human impact. However, no matter how unpromising the starting-point may appear, some attempt to envisage and reconstruct the natural forest state will provide important clues when deciding whether restoration is a feasible aim in the first place, the limitations of such an exercise, and the degree of compromise required.

In this chapter we consider three different contexts in which reconstruction of the original semi-natural forest has been attempted: modified semi-natural, hardwood forest (Canada), plantations set within semi-natural, broadleaved evergreen forest (Japan), and the expansion of remnant populations of pine (Scotland, UK). In each study both active and passive solutions to restoration are examined.

Features of 'natural' forests

Mature, temperate forests share a number of common features, including irregular canopy structures, the presence of native tree species, old-growth stands consisting of veteran trees, large volumes of dead wood, relatively undisturbed soils and sustained natural regeneration in gaps. In eastern

North America and continental Europe it is not unusual for canopy trees to live for 200–500 years, or even longer, in undisturbed conditions, while in the Pacific Northwest, stands of 800–1200 years have been reported in Douglas fir (*Pseudotsuga menziesii*) (Fowells, 1965). On good sites the trees may grow up to 35 m tall, with trunk diameters of 100 cm at breast height, although in old-growth stands of coastal Douglas fir and its associates heights of 50–90 m and diameters of 100–200 cm are typical (Franklin *et al.*, 1981). In Bialowieza Forest, Poland, old-growth stands frequently have tree diameters of 200 cm (lime, *Tilia cordata*), 230 cm (oak, *Quercus* species) and 130 cm (ash, *Fraxinus excelsior*), with stand basal areas of 25–40 m^2 ha^{-1}, tree densities of 120–500 trees ha^{-1} and living tree volumes of 200–500 m^3 ha^{-1} (Falinski, 1986; Martin, 1992). Crucially, plantations lack the multi-layered canopy structure, spatial variation and varied sizes, spacings and species typical of trees in natural stands (Franklin *et al.*, 1981; Spies & Franklin, 1988).

Old growth

The term 'old growth' is commonly applied to the mature phase of natural forest stands, describing not only old, large trees, but also the quantity of fallen and dead wood and species characteristic of these late-successional habitats. Where the forest has been modified but considerable biodiversity still remains, it may be expedient to employ relative definitions of old growth. Thus, where the evidence points to a long, continuous period under woodland cover of at least 400 years (but probably managed for much of this period and sometimes lacking old trees), the term 'ancient woodland' has been coined (Spencer & Kirby, 1992). However, only forests that have been protected from human disturbance are likely to have retained a full range of specialist species which depend on old stands of trees. For these species, restoration is an unlikely prospect unless relict populations remain *in situ* or are present in stands nearby.

Dead wood

Dead and decaying wood is a major component of natural forests, and an important host medium to decay fungi as well as to a range of specialist invertebrates, cavity nesting birds, bats, etc. Fallen trees can physically occupy large areas of the forest floor (commonly 5%–10%: Falinski, 1986), where they influence nutritional gradients, the distribution of the ground flora, tree regeneration and patterns of grazing. Interactions between fallen logs and woodland watercourses are of critical ecological importance (Lienkaemper & Swanson, 1987), the logjams reducing channel erosion and breaking the watercourse into a series of pools, riffles and falls, with slack water creating suitable cover and conditions for fish breeding.

In mature forests the accumulation of coarse woody debris (i.e. material exceeding 15 cm diameter) as snags, fallen wood, stumps and branch wood can amount to 50–150 m^3 ha^{-1}, with additional columns of decay in living tree trunks and main branches (Peterken, 2000). Coarse woody debris takes time to accumulate and rots down slowly, thus providing a stable and reliable habitat for long periods. Whereas forests under conventional management retain less than 5 m^3 ha^{-1}, the proportion of standing dead wood increases as the forest ages or is less intensively managed (e.g. Hodge & Peterken, 1998) (Table 22.1). Although there is usually no alternative but to allow sufficient time for dead wood to accumulate, restoration management can retain 'islands' of unharvested trees, mature individuals, snags and fallen timber within a production forestry context. Some traditional management practices actually produce appreciable quantities of decayed timber, as for example in wood-pasture stocked with ancient, pollarded trees, or ancient coppice stools (Ferris-Kaan *et al.*, 1993).

Key species

Old forest sites frequently have numbers of species not found in younger forest or scrub habitats. Through fragmentation, the areas remaining may simply be too small to support species with large home ranges, or have become increasingly unsuitable due to edge-effects. Other species with limited powers of dispersal tend to be restricted to areas under virtually continuous forest cover, or even within islands of old growth. Such species are often taken

Table 22.1. *Dead wood volumes recorded in managed and neglected British woods*

	Dead wood m³ ha⁻¹	Percentage fallen
Managed systems		
Ancient, semi-natural coppice	1–12	100
Conifer plantation[a]	4–20	100
Old, broadleaved plantation[b]	17–31	c. 60
Unmanaged systems		
Neglected coppice	7–46	c. 90
Young, semi-natural high forest[a]	20–31	c. 30
Native pinewoods	40–50	c. 33
Unmanaged broadleaved woodland	50–100	c. 80
Blowdown of high forest	300–400	100

[a]Less than 100 years old.
[b]More than 100 years old.
Source: After Hodge & Peterken (1998).

as indicators or symbols of the mature habitat. They may literally be keystone species – crucial to the integrity of the forest ecosystem – or significant simply because of their relative rarity or intrinsic appeal (Simberloff, 1997). Some categories may be useful as indicators of ecosystem recovery, at a range of complementary scales: for example, large-bodied species may indicate changes in features at the landscape level (e.g. elk [*Cervus elaphus canadensis*] and wolves [*Canis lupus*]), while small-bodied species with narrower niches (e.g. bats or woodpeckers) are fine-grain indicators of forest condition (Thompson & Anglestam, 1999).

Species requiring large forest areas
Several species require large areas of forest as territory or home range. This is particularly true of mammal predators, such as the wolf, and several large herbivores: extreme examples are the mountain lion (*Felis concolor*), requiring a minimum of 12 900 ha (Patton, 1992), and wintering caribou

(*Rangifer tarandus*), for which the minimum patch size may be several thousand square kilometres of forest (Chubbs *et al.*, 1993). Fragmentation increases the vulnerability of such species and their likely extinction from smaller patches. The absence of predator regulation on populations of herbivores has a profound influence on forest vegetation and processes of regeneration, while the extinction of large herbivores may reduce the amount of open space within the forest (Kirby *et al.*, 1994).

Saproxylic species
Saproxylic species include fungi, invertebrates, hole-nesting birds, epiphytic lichens, and mammals using hollow trees as shelters or roosts. Undisturbed forests with standing dead trees provide nesting sites for raptors such as the goshawk (*Accipter gentilis*) (Mannan & Meslow, 1984), while the holes and cavities are breeding sites for more than one-third of all British woodland birds (Fuller, 1995). The type of dead wood present as snags, fallen trunks, rot-holes in standing trees, etc., is critical to some specialist species. Elton (1966) estimated that 20% of the invertebrate fauna depended on the niches provided by dead and dying wood. Of 771 scarce woodland invertebrates in Britain, about one-third require dead wood in one form or another and of these, roughly 25% are listed as Red Data Book species (Kirby & Drake, 1993).

Veteran trees also support lichens, many of which are specialists of old-growth stands, with apparently only limited ability to colonise new areas (Gustafsson *et al.*, 1992). Disjunct distributions of both lichens and saproxylic invertebrates have been interpreted as relics of former natural forest, now surviving in fragments of old-growth forest or wood-pasture and parkland. Restoration strategies must take account of this by expanding woods at, or in close proximity to, concentrations of these specialist populations.

'Interior' forest species
Closed forest conditions are used by some species to avoid predation. A *cause célèbre* is the northern spotted owl (*Strix occidentalis caurina*), a 'flagship' species of old-growth conifer forest in the northwestern

United States, which became the subject of detailed demographic studies and conservation planning aimed at limiting the extent of commercial logging (Wilcove, 1993). Similarly, many plant species are also regarded as 'indicators' of ancient woods, owing to their relatively greater tolerance of shade and stress-tolerant strategies (Peterken, 1974; Whitney, 1986; Whitney & Foster, 1988), along with other key groups such as lichens, bryophytes, fungi, beetles and molluscs. All have potential as indicators of ecosystem recovery and are natural targets for restoration management.

Other features

Decades of relative stability will tend to result in closed-canopy forests dominated by shade-tolerant species (for example beech, lime and elm in Europe). However, the evidence of pollen analysis from the pre-Neolithic period suggests that provision should also be made for open-ground components within temperate mesic forests. These open habitats would have occurred on river floodplains, mires and wetlands, where they would have been maintained by herbivores, and on sites regularly disturbed by fire, flooding and landslips.

Forest soils can also be expected to show features of antiquity after generations of forest cover, progressively developing in structure while they accumulate organic matter, nutrients and a specialised soil biota. At the same time, periodic natural disturbance will create considerable spatial complexity and microtopographic gradients in soil profiles (Lutz, 1940; Beke & McKeague, 1984; Johnston *et al.*, 1986). Studies of forest succession on former agricultural soils generally show increasing levels of organic carbon and declining pH values (Schoenau *et al.*, 1989). Ancient woods in England have also been shown to have significantly more carbon and organic phosphorus than recent woods on equivalent soil (Wilson *et al.*, 1997). Proposals to 'restore' lost forest cover on sites cleared for agriculture would therefore do well to consider the likely timescale of soil maturation against the benefits of remnant sites.

Finally, field evidence of former woodland management, indicated by archaeological and historical features, may be considered suitable not only for protection, but even for restoration and renewal. This could include wood banks, tracks, saw-pits, charcoal hearths, ponds, etc. as well as old logging platforms, coppice stools, stubs and pollards.

NATURAL FOREST PROCESSES

Natural disturbance

Natural forests are shaped by the interaction of local climate and soil conditions with various forms of natural disturbance, resulting in a mosaic of stand types and seral species through space and time. Intermittent damage by wind, fire, drought or flooding creates heterogeneity within the forest structure and profoundly influences the processes of regeneration and succession. These effects may be unseen between one generation of observers and the next, fuelling controversies about the course of past events and frustrating attempts to prescribe accurate, authenticated restoration targets. Restoration managers therefore need to grasp the fundamental nature of natural disturbance events, and understand how their scale, frequency and intensity interact to produce a variety of characteristic forest types (Fig. 22.1).

The physical intensity of disturbance influences the rate of ecosystem recovery. Catastrophic events not only remove trees, but also soil layers during landslides, volcanic eruptions, flooding, debris flows in river channels and severe crown fires. Less severe disturbances include hurricanes and storms, lower-intensity fires, drought and disease. These have their impact mainly on the forest canopy, leaving intact areas with existing trees and shrubs, resprouting individuals, advanced regeneration and buried seed, all of which accelerate recovery.

Gap size and gap regeneration rate

Characterising or classifying natural disturbance is fraught with difficulty, but generally it is reasonable to expect a continuum from endogenous, small-scale events to the large-scale, wholesale destruction of forest tracts (Attiwill, 1994). This modern approach reconciles earlier views of the natural forest as an irregular, shifting mosaic within a 'steady

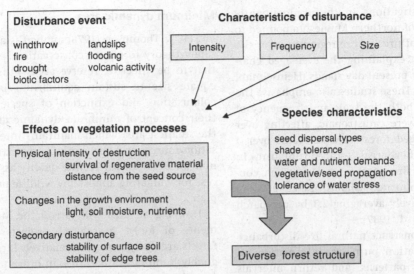

Fig. 22.1. The influence of the disturbance regime in diversifying forest structure.

state' (e.g. Bormann & Likens, 1979) with consistent observations of even-aged elements present within it (e.g. Jones, 1945). A dynamic view of forest processes and succession is the key to an understanding of ecological functioning in forests, providing measures of the effectiveness of restoration and silvicultural practices, respectively (Parker & Pickett, 1997).

The scale, distribution and timing of natural disturbance in forests are at least as important as the frequency and intensity of disturbance. Studies of natural disturbance in mesic, old-growth forests of eastern North America indicate an annual turnover, through gaps, of the order of 0.5%–2% of the forest area, equivalent to return times of 50–200 years (Runkle, 1982). In practice, some dominant trees may live for 300–500 years or even longer in places protected from disturbance, whereas areas subject to more frequent disturbance will typically result in shorter life spans.

Windthrow

Ecologists have consistently distinguished between: (1) small-scale tree falls and (2) catastrophic blowdowns of extensive forest areas. Generally in mesic, temperate forests windthrow gaps are small: for example, Runkle's studies of deciduous stands in eastern North America indicate predominantly small gap sizes of <0.1 ha, with recognisable gaps occupying less than 10% of the overall area (Runkle, 1982). In contrast, the incidence of tropical cyclones (hurricanes) and intense windstorms along the eastern coast of the United States and Canada is also well documented (Foster, 1988a, b). Meteorological records, together with historical and vegetation studies in the region, suggest a pattern of hurricane strike every 20–40 years, with catastrophic storms such as those of 1635, 1788, 1815 and 1938 occurring every 100–150 years (Foster, 1988a). Such records of gap size, gap creation rate and the pattern of disturbance, combined with windthrow risk analysis, provide the baseline for restoration managers seeking to mimic natural forest structures.

Burning

Fire is a major and widespread agent of forest disturbance, whether by natural means or human-assisted. A classic event was the fire of 1988 in which almost half of the 8000 km^2 area of Yellowstone National Park in Wyoming was burned (Romme & Despain, 1989). Much of it was damaging-intensity crown fires, drawing attention to the whole question of periodicity and the 'naturalness' of such events.

Historical investigations in the predominantly coniferous forests of northern Minnesota pointed to a high frequency of fires in the region prior to the twentieth century, explaining the even-aged characteristics of many present-day stands (Heinselman, 1973; Clark, 1990). These studies also emphasise the wide variation in both the scale and frequency of the burns. Some were catastrophic, affecting over 50% of the unlogged forest area in events over a century, while other areas escaped unburnt. In northern Alberta, fire patches of 41–200 ha contained scattered unburned 'islands' of trees of more than 1 ha, collectively averaging 2.3 ha per patch (Eberhart & Woodard, 1987).

In order to reconstruct natural fire disturbance regimes for restoration purposes, information on patch dimensions, patterns, and return intervals are required. Return times of 6–300 years have been estimated for small stands of conifer forests in the northern Rocky Mountains (Arno, 1980). In contrast, deciduous, mesic woodland has relatively low flammability (Rackham, 1980) and intervals between fires of more than 1000 years have been calculated for the hemlock–pine–hardwoods of Michigan (Whitney, 1986). For both safety and production reasons, forest management in the twentieth century has sought to regulate both the frequency and intensity of previous fire disturbance regimes. To avoid the high risk of catastrophic wildfire, in 1999 over 931 000 ha of prescribed burning was carried out by the US Federal agencies (Kloor, 2000). Such management implies shorter disturbance intervals and a greater uniformity of stand structure than might be expected for the same forests under natural conditions.

Other disturbance events

Other major abiotic disturbances are episodes of flooding, causing erosion of sediment and even debris flows, landslips and long drought periods. Together with biotic disturbances due to browsing mammals, defoliating insects and disease epidemics, the frequency and extent of each event must be anticipated by restoration managers and the assumptions, within stated limits, justified with proposals for action.

Minimum dynamic area

Pickett & Thompson (1978) applied island biogeography theory to nature conservation, pointing out that to be effective, reserve areas must be of adequate size to sustain equilibrium between the colonisation and extinction of species. Applying their concept of minimum dynamic area (MDA) – the smallest area which maintains internal colonisation sources and hence minimises extinctions – is now an essential part of the quality assurance process for validating almost any wildlife management plan.

How big an area will satisfy the MDA requirements of forest restoration projects? Fire-prone forests are likely to require relatively large areas, arguably hundreds of hectares, in order to accommodate the large grain size of individual disturbances and short return intervals. In comparison, mesic, temperate forests that turn over slowly in small gaps should require smaller areas to achieve a similar equilibrium. For much northern European woodland, Peterken (1996) suggested that MDAs of less than 100 ha would be satisfied by wind-disturbance regimes, with somewhat smaller areas sufficient to maintain a variety of canopy species.

In reality, however, applying minimal areas based on 'normal' disturbance effects to forest conservation is only part of the equation. Spatial models in conservation biology indicate that, for a given species, the outcome of interactions of population dynamics, dispersal abilities and habitat patch availability and suitability are by no means certain (Harrison, 1994). To provide a sufficient safety margin would ideally require the MDA to be increased so as to accommodate periodic, catastrophic disturbance. This principle is enshrined in the protection of large, non-intervention areas, as in national parks such as Yellowstone in the United States and Bayeriswald and Bialowieza in Europe, but in practice even these examples cannot be regarded as truly self-regulating, closed ecosystems. Ultimately the view of MDA is heavily dependent on the species to be conserved. Pickett & Thompson (1978) suggest that these 'core' species should determine area requirements: the corollary is that some species must be excluded from the restoration project if the proposed site is too small and fragmented.

SETTING A RESTORATION FRAMEWORK

During the planning of a restoration project the feasibility of the desired end point – how good a match it will be to the 'natural' forest state, and the technical difficulty of achieving it – will require careful consideration. Former management practices such as the planting of economic species, site cultivation and drainage, thinning, felling, the removal of predators and the encouragement of game species, can be addressed by direct action. Other, indirect effects are more intractable, including forest fragmentation (leading to isolation and edge-effects), species extinctions, invasions by non-native species, air pollution and climate change.

Fixed points in restoration

'Restoration' implies knowledge of the previous habitat condition or long-lost *status quo*. However, the lack of baseline data – what the 'original' forest was like and how it changed over centuries and millennia – militates against setting overprecise targets. Even when comprehensive records exist, it is optimistic to expect that modern conditions can be altered to resemble the past. If, at a stroke, all human influence could suddenly be removed, the legacy of external influences of both a biotic and abiotic nature would still continue to influence and alter the restoration site. How, then, can a consensus be arrived at for restoration?

Peterken (1996, 2000) listed five states of naturalness that might be used to formulate objectives in forest conservation (Table 22.2). A purist objective to re-create a prehistoric forest, uninfluenced by humans, (the original–natural forest) would almost certainly require the reintroduction of lost native species, implying an investment in sizeable preserves. However, this vision of 'true' wilderness ignores irreversible changes in forest condition which have taken place meantime, such as fragmentation, climate change and introductions of non-native species.

Less ambitious would be to protect the forest components inherited from ancient, near-natural forests (past-natural or inherited–natural forest), accepting previous species losses and other overriding changes. Unfortunately, advocates of this approach may interpret the past–natural forest as being in some 'equilibrium' state prior to the modern period. An example is the manipulation of western yellow pine (*Pinus ponderosa*) forest cover at Coconino Natural Forest, Flagstaff, Arizona to a level predating European settlement in 1876. Here the premise was that post-settlement controlled burning and selective logging had artificially increased the density of pine forest, thus reducing the diversity of open-ground species (Kloor, 2000). In fact, close examination of tree-ring data revealed a much more dynamic picture, indicating earlier pine regeneration flushes and suggesting that a wider range of forest cover was more authentic than the 'snapshot' of 1876.

A still more liberal approach, ignoring considerations of the past, is to adopt the *laissez-faire* equivalent of shutting the forest gate and allowing nature to take its course in whatever way it will, embracing both native and non-native elements as well as all previous changes, creating a future–natural forest.

'Traditional' versus 'natural' forest management

When few fragments of genuine old-growth forest remain in a region, attempts to re-create a past–natural wilderness may appear heroic rather than practical. On the other hand, arguments for restoring former forest management systems, based on archaeological, documentary and contemporary evidence, have their own justification. Forests managed traditionally, such as coppice, wood-pasture and wood-meadow systems in Europe and Scandinavia, are modified descendants of former forest and contain 'natural' features in their own right. In the case of coppice, cutting poles on short cycles produces a young growth structure that encourages specialist groups of plants, insects and birds (Buckley, 1992; Fuller & Warren, 1993), while similar practices in wood-pastures preserve old-growth features in veteran trees.

Table 22.2. *States of naturalness in forest restoration*

State of naturalness	Forest origin and influences	Action required
Original–natural	Primeval forests not significantly affected by people, reconstructed from palaeoecological evidence such as pollen diagrams and existing, relict original stands	Reconstruct as far as possible from original stands, reintroduce lost species where possible and control herbivores to facilitate natural regeneration
Present–natural	Forests that would be present today had they not been affected by humans, but acknowledging intervening (non-human-influenced) climatic changes, natural species migration, and other natural events since prehistoric times	(hypothetical only)
Past– or inherited–natural	Forests based on the remaining components inherited from virgin forest, but accepting intervening modification and change, including past species losses	Allow free development of locally native species but exclude non-native species and control herbivores to a point allowing natural regeneration
Potential–natural	Forest types that would develop now if human influence were removed completely, assuming instantaneous succession from available colonising sources of all species	(hypothetical only)
Future–natural	Forest that would eventually develop over centuries from existing forests if human influence were removed; will include both native and non-native elements	Allow free development of both native and non-native species, except possibly to prevent overgrazing and lack of regeneration

Source: After Peterken (1996, 2000).

Short rotations in traditional management demand heavy and continuous inputs – i.e. active, rather than passive restoration. The effort required may be too great to justify in the absence of ready markets. In Britain, more than 65% of the previously recorded coppice area is now neglected or 'stored' as high forest (Evans, 1992) and much regular management now takes place on behalf of nature conservation organisations. Wood-pasture is similarly threatened through the cessation of pollarding, resulting in the collapse of old tree specimens and encroachment by secondary woods and plantations.

Attempts to restore traditional management systems have been heavily criticised on the grounds that they do not deliver the potential biodiversity of natural forests. In their polemic on traditional practices in conservation, Hambler & Speight (1995) queried whether 'the niches suitable for 1,500 plant species and 58 butterfly species indicate the requirements for 28,500 invertebrates and 15,000 fungi', implying that intensive management regimes encourage species groups with more obvious appeal at the expense of less glamorous, old-growth specialists. However, the specialised conservation, historical and educational assets of traditional management systems must be balanced against the lack of natural forest alternatives in heavily human-modified landscapes.

Agencies of change

Pollution and climate change

Evidence of past climate change exposes the frailty of taking reference points in restoration. Packrat (*Neotoma* spp.) midden chronologies completed at study sites from New Mexico to Montana illustrate how climate fluctuation has caused the balance between mesic and xeric species to alter over past millennia (van Devender, 1986). Investigations into the distribution of the Utah juniper (*Juniperus osteosperma*) reveal that its abundance across Utah, Wyoming and Montana has fluctuated significantly over the past 4500 years. Climate cooling around 2500 years ago destroyed many juniper stands, but subsequent warming allowed the species to recapture lost ground. Recently observed accelerated migration of juniper has probably been facilitated by a combination of further warming, grazing and fire suppression. Evidence of even more rapid, contemporary climate change is also evident in phenological studies of tree leafing times, bird migratory arrivals and butterfly emergence in spring (Pollard & Yates, 1993; Fitter *et al.*, 1995; Sparks & Carey, 1995). In common with several other tree species, archive records of oak leafing times in southern England over 250 years indicate a trend towards earlier leafing of about a week since records began in 1736 (Sparks, 2000).

Pollution studies also indicate rapidly accelerating environmental changes that even large, remote forests cannot escape. Increases in industrial emissions of carbon dioxide, ozone, and oxides of sulphur and nitrogen have implications for forest canopies and understorey species, as indicated by studies in forest health monitoring in Europe (Innes, 1993). Taken together, these recent changes in the wider forest environment are still small relative to fluctuations in past millennia, suggesting that both 'past' and 'future–natural' restoration models fit the present time-frame.

Fragmentation and commercial exploitation

As the forest becomes more fragmented, island biogeography theory predicts that smaller patches will increase the isolation of remaining pieces. This in turn increases the vulnerability of 'interior' forest species to reduced dispersal and immigration, pollination and nutrient cycling (Wilcox & Murphy, 1985). Fragmentation also increases the relative proportion of edge to interior (Saunders *et al.*, 1991), resulting in higher radiation levels, ambient temperatures and temperature fluctuations, soil fertility and greater wind turbulence, while decreasing soil moisture and relative humidity (Matlack, 1993; Esseen, 1994; McCollin, 1998). Modified edge conditions affect species directly or indirectly, for example by reducing bird nesting success due to the increased risk of predation or infestations of brood parasites in the exposed forest perimeter (Murcia, 1995).

Although several empirical studies agree that direct edge-effects penetrate perhaps 50–150 m into the forest interior, impacts on species of the order of 1–5 km away from forest edges have been reported in tropical forests (Laurance, 2000). Species with extensive home ranges, particularly large mammalian carnivores and herbivores, are especially vulnerable to the fragmentation process, carrying future implications for ecosystem integrity. This suggests that not only is the protection of the forest core required, but also the provision of buffer areas, along with strategies to reverse fragmentation by planting or natural regeneration. Core areas, assuming edge-effects of varying intensity, can be simply calculated using a shape index based on the length of the woodland perimeter and its area (Laurance & Yensen, 1991). Assuming a circular wood of 100 ha and penetration of edge-effects to 50 m, it can be shown that the core area is effectively only 83 ha: conversely, to maintain 100 ha of core forest would require a total area of 118 ha.

PRINCIPLES OF AND APPROACHES TO RESTORATION

Pragmatically, restoration principles can be extended to already heavily modified forest ecosystems, such as the 'traditional' forest management practices already discussed, or even further to modern forestry practices, where the aim is to introduce elements of naturalness which 'add value' to the environment. Whatever the objective – natural restoration, traditional restoration or ecological enhancement – the restoration approach will require *active* or *passive* inputs, or combinations of both (Fig. 22.2).

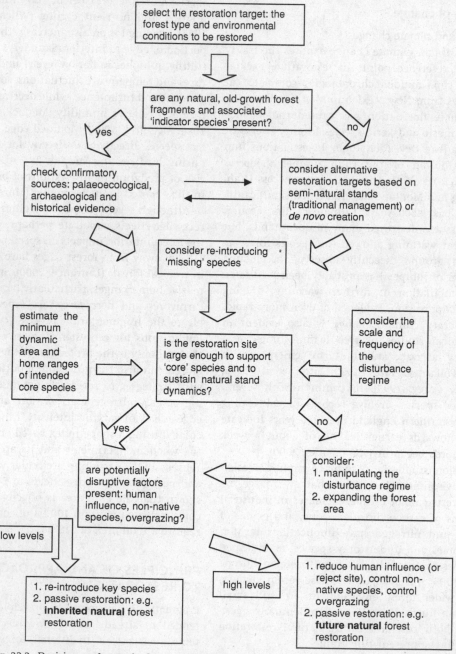

Fig. 22.2. Decision pathways in forest restoration.

Table 22.3. *Sizes of forest reserves (ha) in selected European countries*

Country	Mean size of strict forest reserves	Minimum size	Maximum size	Forest reserves as a proportion of semi-natural forest area (%)
Czech Republic	243	2	2 500	5.9
Finland	4 180	63	71 171	7.3
Netherlands	46	4	700	4.4
Slovakia	203	4	1 800	1.2
Russia	76 750	100	721 322	<0.01

Source: After Parviainen *et al.* (1999)

Natural restoration

Site selection and suitability

The appropriateness of selecting a particular forest type is clearly demonstrated when the evidence is in the form of remnant forest areas nearby, or failing that, an accurate historical record. Better still, some premier sites earmarked for restoration will already have *de facto* recognition through nature conservation protection policies, given that selection of reserves is guided by criteria such as size, naturalness, representativeness and recorded history (e.g. Ratcliffe, 1977). However, for non-designated sites, selection criteria are needed to target the best of a number of potential restoration sites. These include the following:

- The presence of old growth features, such as large and mature trees, dead wood accumulation, etc.
- A varied canopy composition, where possible confirming previous pollen records
- Presence of 'indicator' species of old growth which have poor colonising ability
- Large, compact sites which can satisfy MDA requirements of core species and sustain natural disturbance dynamics
- Sites with a good recorded history, confirming the extent of human modification
- Sites remote from human settlement
- Few or no exotic species
- Good integration and contiguity with adjacent forest areas.

Minimum intervention

In Europe, many forest reserves operate within a strict policy of non-intervention. Parviainen *et al.* (1999) defined these as 'forests left free for development without human influence', but interpretations in different countries vary considerably. Individual reserves generally are small compared with their equivalents in the United States, in total representing less than 2% of all natural and semi-natural forest. Average reserve sizes are generally less than 100 ha, but range up to 700 000 ha in countries with remote, boreal forests (Table 22.3).

One problem is that even forest reserves contain a legacy of non-natural features, such as the presence of introduced, non-native species and/or an overabundance of herbivores. Where active management is required for conservation purposes, minimum or limited intervention may be a more realistic option. Minimum intervention is more purist in motivation, the intervention being an attempt to restore the original–natural dynamic state by eliminating non-native species, reintroducing formerly extinct species and reorganising the age structure. However, there are usually pragmatic elements, such as carrying out 'safety felling' along public access routes (Table 22.4). For example, passive restoration at Point Pelée National Park, Canada involved not only the phased removal of human settlement, followed by natural forest regeneration, but also active restoration policies such as forest replanting and culling deer (Box 22.1).

Table 22.4. *Minimum and limited intervention strategies for forest restoration*

Nature of intervention	Non-intervention	Minimal intervention
Felling and harvesting	No felling	Safety felling along access routes only
Public access	Exclude public access	Limit access to selected routes and areas
Non-native species	No action	Eliminate or restrict where feasible
Reintroductions of native species	No active reintroductions, but encourage colonization	Introduce semi-wild grazing animals and carnivores
Control of grazing and browsing	No action	Culling of herbivore populations to 'natural' levels

Size of site

It has been stressed that the ideal restoration site should be large enough to accommodate the disturbance regime of the region. Fortunately the extensive body of literature on natural disturbance dynamics in forests provides general clues which can be used in a predictive sense. For example, the yardstick of a 1% per annum turnover in mesic, deciduous forest canopies suggests that average annual disturbances of 1 ha (probably as several small gaps) can be accommodated by 100 ha of this type of forest. However, even if edge-effects are ruled out, the turnover may be too dispersed and infrequent to maintain species with very specific growth-stage requirements indefinitely. The situation is made worse if past management has created a relatively stable, uniform forest structure.

For forests in which the scale of disturbance is extensive, for example in fire- and flood-affected forests, relatively larger areas will be needed to accommodate the increased patch size or severity of disturbance. In these cases, minimum areas based on the 'average' or normal events are unlikely to be adequate unless catastrophic events are also fed into the equation. Indeed, restoration management may itself increase the requirement for larger areas, as in the Yellowstone case where the 'no burn' policy allowed fuel loads to build, resulting in the severe fires of 1988 which devastated enormous tracts of forest. However, forest managers already have available to them risk-minimising protocols that are capable of predicting both windthrow and wildfire risk (e. g. Rothermel, 1983; van Wagner, 1987a; Quine *et al.*, 1995). Restoration managers can use a similar approach to risk analysis either to control or encourage a fuller range of natural disturbance probability at the selected site. Modern, GIS-linked fire models, for example, provide decision support for managers dealing with landscape-level forest dynamics at spatial resolutions as low as 1 ha (Li, 2000).

If, having considered the evidence, the restoration site is considered to be below optimum size, or in a too uniform and stable condition to permit 'natural' turnover, more active measures will be required. These include expanding the forest area to spread the opportunities and risks of natural disturbance, or managing *in situ* to regulate turnover to a desirable rate.

Mimicking natural processes, especially in plantations

Manipulating the disturbance regime

If the site cannot be expanded, or the forest age structure is too uniform, active restoration management can achieve a more balanced distribution, for example by creating artificial gaps by felling, poisoning or winching down trees. For broadleaved, temperate forests in the United States, Runkle (1991) recommended cutting 1% of the canopy over the first 20 years, creating gaps of 0.25–1 ha to simulate disturbance. The size of the gaps and their distribution in the landscape can be adjusted to other types of disturbance. For fire-disturbed forests, prescribed burning of patches can partly achieve the necessary structure, but not necessarily the intensity of disturbance of wildfire events.

Box 22.1 Restoration of the Carolinian Forest at Point Pelée National Park, southwestern Ontario, Canada

BACKGROUND

The Carolinian Forest Zone of southwestern Ontario represents the northern fringe of deciduous forest, extending into Canada from the northeast United States (Fig. B22.1). Prior to European settlement, this zone was dominated by closed-canopy forest, together with extensive wetlands, marshes, tallgrass prairies and alvars. It has the highest species richness of any life zone in Canada, and exhibits more than 2200 herbaceous plant and 400 bird species (Allen *et al.*, 1990). However, 250 years ago forests were cleared and converted into intensive agricultural, and, most recently, urban use. In some regions forest cover is less than 4%, and continues to decline (Riley & Mohr, 1994). This zone currently contains over half of Canada's endangered species and the integrity of the remaining forest is further threatened by introductions of exotic species such as garlic mustard (*Alliaria petiolata*) and overgrazing by white-tailed deer (*Odocoileus virginianus*).

Point Pelée National Park is one of Canada's highest-quality Carolinian preserves and is an internationally renowned bird-watching site. A sandspit projecting south in Lake Erie, it represents the southernmost tip of mainland Canada. The area was cleared of white pine (*Pinus strobus*) and planted with hackberry (*Celtis occidentalis*) before receiving park designation in 1918. About one-third of the park area (1100 ha) now consists of upland forest, dominated by hackberry, white pine, white ash (*Fraxinus americana*) and black walnut (*Juglans nigra*). Following designation, land use shifted towards fruit and vegetable production, recreation and settlement. By 1960, 600 cottages and numerous roads were located in the park and upwards of 700 000 people visited annually. Despite its containing at least 70 nationally rare vascular plant species, non-native species account for 40% of the total flora (Dunster, 1990).

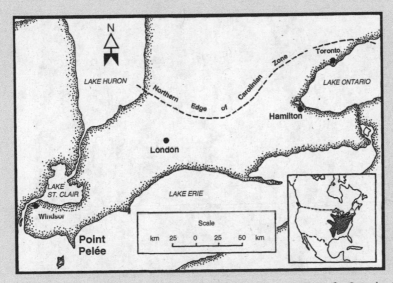

Fig. B22.1. Map of the Carolinian Zone or Deciduous Forest Zone for Ontario, Canada, showing major cities, and for North America (inset).

RESTORATION STRATEGY

In 1963, park staff initiated a radical new policy of restoration management in an attempt to mitigate the adverse effects of human use within the park. Over the last 35 years, the great majority of cottages have been acquired and, along with many roads, dismantled. Initially, former cottage and road sites were passively restored, simply allowing for natural colonisation

(a)

(b)

Fig. B22.2. An example of successful passive restoration. This site exhibited the greatest similarity to reference sites at Fish Point Nature Preserve of all restored sites. (a) Pre-restoration, *c.* 1961; (b) post-restoration, in 1995.

(Fig. B22.2). Over the last decade, sites have been actively restored, by planting shrubs, modifying the topography and hydrology, and controlling non-native plants (Fig. B22.3). Deer overgrazing also adversely affected the understorey, and in the early 1990s, herds were culled, reducing densities from 40 individuals km^{-2} to 11 km^{-2}, which has facilitated the regeneration of herbaceous and shrub communities (Koh, 1995). In 1994 and 1995, the effectiveness of this restoration was examined by comparing 28 former cottage and road sites with high quality 'reference' sites of upland forest located both within the Point Pelée National Park and the less-disturbed Fishpoint Nature Preserve on nearby Pelée Island (McLachlan, 1997).

Fig. B22.3. A road that was actively restored six months before. Note return to previous dune topography, removal of non-native vegetation and planting of early successional shrubs.

RESULTS

Results showed that the understorey plant communities at restored sites gradually became more similar in composition to those of the reference sites over time (Fig. B22.4). The degree of restoration was also positively associated with soil moisture, canopy cover and proximity to continuous forest cover. As regeneration occurred, non-native species declined (Fig. B22.5). Non-native ruderals showed the greatest decline, in part due to the cessation of disturbance and increased canopy cover. Species perceived as displacing native species also declined, with the exception of garlic mustard which occurred in all sites. Former lawn species such as Canada blue grass (*Poa compressa*), Kentucky blue grass (*P. pratensis*) and red fescue (*Festuca rubra*), that tend to inhibit successional change (Hiebert, 1990), only declined in the oldest cottage sites. Road sites tended to recover more rapidly than cottage sites, in part because of the absence of these former lawn grasses, and the relative proximity of forest cover. Actively restored sites were dominated by non-native ruderals including shepherd's purse (*Capsella bursa-pastoralis*) and horseweed (*Erigeron canadensis*), in large part because of the intensive and relatively recent disturbance associated with active restoration (McLachlan, 1997).

Overall, the trajectories suggested that highly disturbed deciduous forest sites, bearing in mind their already reduced species complement compared with reference sites, could show considerable improvement within 30–50 years. These relatively rapid recovery rates might be expected for small-scale disturbances compared with the larger disturbances of forest clearance (Peterken & Game, 1984; Duffy & Meier, 1992; Meier *et al.*, 1995) and flooding (Bratton *et al.*, 1994), where dispersal distances required for immigration and colonisation would be substantially greater. Even so, after 40 years of regeneration, restored sites were still missing a whole complement of the forest understorey found at the reference sites. These vulnerable herbaceous species tend to be spring-flowering ephemerals that are ant- or gravity-dispersed, and include wild leek (*Allium tricoccum*), white trillium (*Trillium grandiflorum*) and dutchman's breeches (*Dicentra cucullaria*) (McLachlan &

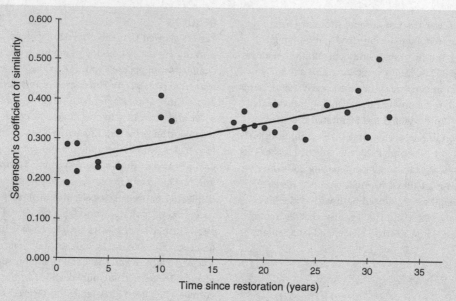

Fig. B22.4. Relationship between and time since restoration and Sørensen's coefficient of similarity for all understorey herbaceous species in restored sites at Point Pelée. Regression is $y = 0.0053x + 0.238$, $F_{1,27} = 29.23$, $p < 0.0001$.

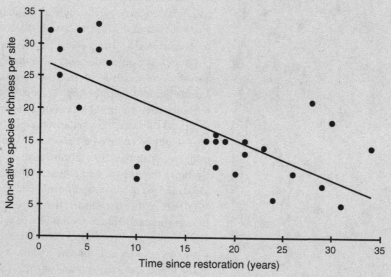

Fig. B22.5. Relationship between time since restoration and non-native species richness. Regression is $y = 27.666 - 0.628x$, $F_{1,26} = 35.89$, $p < 0.0001$.

Bazely (2001). It is likely that such species will have to be manually reintroduced, and experiments are now being conducted in restored sites to assess the effect of canopy shading and fertilisation on the re-establishment of selected vulnerable species (D. Bazely, personal communication)

In general, forests in dry sites seemed to recover more slowly than those in mesic and wet sites (McLachlan, 1997). Many of these dry areas were dominated by red cedar (*Juniperus virginiana*) savanna at the time of European colonisation (Reive *et al.*, 1992). The decline of this early-successional system in the park is associated with deer overgrazing and the suppression of natural disturbance regimes, including wave activity and fire. In 1995 and 1996, the effects of prescribed burns, red cedar density, brush clearance and soil discing on plant communities was determined at two in-park sites. Native species diversity increased in unburned plots that were cleared of brush. Burning was ineffective, in part because of the paucity of the native seed bank, suggesting that early-successional native species also will have to be reintroduced for recovery to occur (Falkenberg, 2000).

WIDER LANDSCAPE RESTORATION

Most of the restoration conducted by park management has focused on Point Pelée itself. However, the park has become highly isolated by intensive agriculture, and the nearest sizable forest remnant is more than 10 km away. Over the last decade, native trees have been planted extensively in the surrounding landscape matrix. While much of this planting aimed at mitigating the soil erosion associated with intensive agricultural use, participating landowners were also concerned to increase the amount and connectivity of wildlife habitat, aesthetics and revenue values of their properties (McLachlan, 1997). One notable project, the Natural Habitat Restoration Program (NHRP), a multi-stakeholder team consisting of Parks Canada, Ontario Ministry of Natural Resources, researchers, and environmental groups, has also established many plantations of native species grown from locally collected seed stock.

Despite the undoubted positive benefits of these and similar plantations across the region, current restoration: (1) focuses on forest creation rather than rehabilitating existing but degraded forests; (2) is restricted to the planting of trees, and (3) has yet to be co-ordinated at the landscape level. A recent vegetation classification for Point Pelée and the surrounding region included categories such as high- and low-quality forest, grassland (both native and exotic) and shrub cover (Pearce, 1996). These cover data could be used to identify potential sites for forest restoration which would not only increase regional forest cover, but could also improve connectivity between highly isolated, remnant patches, thereby increasing overall forest integrity.

STAKEHOLDER CONSERVATION

Most of the restoration activity in this human-dominated landscape is habitat-focused, expert- and agency-driven, and infrequently incorporates other stakeholders. Additionally, the majority of the remaining forest cover in this region is privately owned. Conservation efforts tend to be viewed as most stable when groups including Nature Conservancy purchase these forest remnants, and when human use becomes severely curtailed or eliminated. However, the effectiveness of more flexible, community-oriented, and multiple-stakeholder conservation strategies including conservation easements and land trusts also needs to be further explored (e.g. Hilts, 1985). The importance of partnering with other stakeholders also has been recognised recently by Parks Canada (Parks Canada, 2000). As with individual forest fragments, most Canadian national parks are too small and too isolated to be managed as 'islands' and many of the greatest threats to park integrity are transboundary in nature. Thus, a process of 'regional integration' is recommended, whereby parks are managed at the landscape level in partnerships with business, other governmental agencies, non-governmental organisations and local communities, especially Aboriginal First Nations (Parks Canada, 2000).

Ideally, these stakeholder-oriented conservation efforts would recognise and incorporate into future management efforts the rich 'ecological' knowledge that many of these rural communities and long-time residents have about the forests (Holling *et al.*, 1998). Ironically, many of these communities are as endangered as the forests they incorporate. As century-old family farms go bankrupt, they are replaced by large-scale, input-and machinery-intensive agricultural operations and, in communities close to urban centres, by ex-urbanites without the close, long-standing knowledge about these habitats that is so fundamental to effective restoration. Unfortunately, the decline of these rural communities and the local knowledge they reflect is occurring across Canada, and is probably permanent.

FUTURE PLANNING

The study illustrates many of the complexities surrounding forest restoration. Passive restoration, and the concomitant decrease in park visitors and deer numbers, has facilitated the regeneration of these diverse upland forests. Active restoration, though still positively associated with ruderal species, should eventually accelerate regeneration. However, plant reintroductions, which now emphasise the plantings of mid-successional trees and shrubs such as black raspberry (*Rubus occidentalis*) and white ash (*Fraxinus americana*), might be better served by focusing on dispersal-restricted ephemerals that are still almost entirely absent from restored sites, despite 35 years of regeneration. These vulnerable species as well as non-native ruderals would be effective indicators of forest recovery and should be monitored in the future. While successful at the site level, restoration has failed to address the large-scale degradation of the remaining forests that, due to deer overgrazing, agriculture-associated impacts of herbicide drift and soil erosion, and urban development, will almost certainly continue to decline in the future. Restoration and protection efforts need to be co-ordinated at the landscape level, incorporate privately owned forest remnants, and both involve and support the communities and long-term residents who have been effective stewards of these forests for centuries.

Similar principles can be applied to floodplain forests by recreating the original hydrology and natural watercourse configuration, followed by the reinstatement of appropriate forest vegetation. As a minimalist measure, Petts (1990) advocated buffer strips at least 20 m wide along the river corridor in order to separate it from adjacent intensive agricultural land production and nitrate runoff, allowing forest vegetation to develop. Alternatively, tree seeding or planting has been used in Denmark to reduce sediment erosion on floodplains (Petersen *et al.*, 1992). In The Netherlands and Germany, large areas of floodplain forest have been created in conjunction with low-intensity grazing by wild and semi-domestic breeds of pony, sheep and cattle.

Manipulation at the stand level

No single silvicultural system can exactly mimic the random scale and nature of natural disturbances. However, some systems are closer to the notional 'natural' state in terms of canopy structure and age-class representation (Matthews, 1989). Thus clear-felling, in which the felling unit varies conventionally from a few to hundreds of hectares, superficially resembles the large disturbance gaps and short return intervals associated with fire-prone natural forests, while group-felling and shelterwood systems incorporate features of small-gap disturbance moderated by windthrow.

Briefly, selection forestry uses small felling coupes (0.01–0.05 ha) to remove single individuals or a few trees, resulting in a fine patchwork that effectively maintains a closed and continuous canopy. Group selection systems fell larger patches at a time (e.g. 0.25 ha or more), thus favouring more light-demanding trees, while the coupe sizes employed in shelterwoods tend to be larger still. The relative benefits to wildlife of different silvicultural systems have been comprehensively reviewed by Mitchell & Kirby (1989), Peterken (1996) and Seymour & Hunter (1999) (Table 22.5). High-frequency disturbances, generated by fires or insect defoliations, tend to produce single or two-cohort structures, analogous to clear-felling, shelterwood (seed-tree) and two-storied high forest systems, often with survivors or 'legacy' trees from the original stand. On the other hand

Table 22.5. *Contrasts in structure, dynamics and composition between managed silvicultural systems and natural, broadleaved mesic woodland*

	Natural woodland	Group selection forest	Selection forest	Clear-felling forest	Traditional wood-pasture	Traditional coppice
Maximum tree age (years)	300–500	–	–	––	o	–––[a]
Average final tree age (years)	c. 200	–	–	––	o	–––[a]
Tree species diversity	mixed	––	–	–––	––	O
Gap size	mainly small	o	o	+++	n/a	++
Gap creation rate per year	1%	o	o	o	n/a	+++[a]
Open space	limited	+	+	++	+++	++
Structural diversity (stand)	high	o	–	–––	–	––
Structural diversity (whole forest)	patchy	o	–	++	++	++
Dead wood	abundant	–	–	–	–	–––[a]
Soil turnover rate	high	––	––	––	–––	–––[a]

Notes: +, ++, +++: attributes increased by management. o: no change. –, ––, –––: attributes reduced by management. n/a: not applicable.
[a]Old coppice stools and standard trees, the latter grown on long rotations, may counterbalance management trends.
Source: Modified from Peterken (1996).

infrequent disturbance, typical of old-growth, mesic forests where small-gap disturbance operates primarily through windthrow, leads to the development of multi-cohort structures emulated by group selection forestry systems. Compared with single-tree selection, group selection offers potentially greater structural diversity at the stand scale.

Many forests consist of plantations of non-native species, where restoration management can achieve only limited ends. However, when a mosaic of managed and semi-natural stands is present, a balance may be achievable where the commercial operations can be integrated to maintain moderate species diversity. The well-forested landscapes of southern Japan are a case in point (Box 22.2).

Since timber is the saleable commodity, all commercial forestry lacks the volumes of dead wood present in natural forests: trees are harvested before biological maturity and self-thinning is often pre-empted by interim harvesting. Short rotation stands and especially coppice have little opportunity to accumulate large, old trees. In a forestry context, blanket recommendations to retain 10 m³ or 10 large 'legacy' trees in clear-cut areas for conservation purposes are typical; for example Ferris-Kaan *et al.* (1993) suggest a formula of 6–8 mature standing trees, 6–8 snags, and 4–5 downed trees per hectare. Hodge & Peterken (1998) argued that high levels of dead wood retention would be difficult to justify in the absence of old growth, as in the case of commercial and secondary forests, whereas reserving 80% or more of the harvestable timber may be appropriate for ancient, semi-natural woodland.

Manipulation at the landscape scale

In practice the situation on the ground will determine how and where to employ 'silvicultural'

Box 22.2   Maintaining species richness within commercially managed sugi (*Cryptomeria japonica*) plantations, Kyushu, Japan

BACKGROUND

Seventy per cent of the land area of Japan is forested, of which 40% comprises plantations of sugi (*Cryptomeria japonica*) and hinoki (*Chamaecyparis obtusa*). The establishment of these extensive, even-aged monocultures has displaced much of the semi-natural, broadleaved evergreen forest in the warm–temperate zone of southern Japan. Fortunately, a number of factors ensure that the prospects for maintaining biodiversity in relatively unmodified forest habitat fragments remain relatively good. These include the small-scale nature of forestry management units or compartments in Japan and the high proportion of interconnected forest cover, which increases the prospects of colonisation and dispersal in the wider landscape. The presence of nearby or adjacent semi-natural stands and intermittent disturbances through natural events such as typhoons, volcanic activity, landslides and flooding also tend to reinforce an irregular and semi-natural forest structure.

RESTORATION STRATEGIES

Given the circumstances, the balance between commercial forestry and conserving or restoring the semi-natural forest cover of the region could follow one of three basic strategies:

(1) 'Business as usual', i.e. continued commercial management relying on 'fringe' areas and unconverted forest to maintain a more natural forest habitat
(2) Cease commercial production altogether (passive restoration)
(3) Diversify the plantation stand structure so as to protect and restore remnant native species populations within the commercial core (active restoration).

Nakagawa & Ito (1997) compared the occurrence of 144 major trees and shrubs within sugi plantations and native evergreen broadleaved forest respectively in southern Kyushu, southwestern Japan, a region in which there has been progressive conversion of native forest cover to plantations during the past 50 years. They reported that plantations had half the species richness of native evergreen forest, depending on stand age and area. Other factors thought likely to affect species richness were the management history (i.e. the status of the plantation as a first- or second-generation crop), the proximity of native forest patches, and variations in topography and soil type. Each factor was tested further in a smaller-scale study carried out within the same region in the 500-ha Miyazaki University Forest.

THE STUDY SITE

Miyazaki University Forest consists of 278 compartments, of which half are plantations of sugi and hinoki, replacing former evergreen broadleaved forest dominated by Fagaceae and Lauraceae species. Forty-one sugi forest compartments, differing in stand age and origin, were selected for the study, in which all woody species (trees, shrubs and climbers) were recorded along regularly spaced transects. These compartments covered 6.8% of the forest area and were chosen both to represent the full range of stand ages and to include a proportion of known first- and second-generation stands. The topographical position of each stand was noted as well as the adjacent stand types (whether plantation or native forest). From the full species list compiled, a group of 145 'less frequent' species (defined as those present in less than 50% of the 41 stands) was examined in more detail. This 'less frequent' group was in turn subdivided into those with significantly greater occurrences in 'older' (>45 year) or 'younger' (<35 year) stands.

Species richness (*d*) was evaluated by Gleason's formula (Whittaker, 1975) as:

$$d = S/\log_{10} A$$

where S and A were the number of species and stand area (m^2), respectively.

RESULTS

Larger numbers of 'less frequent' species (49) were found in older stands than younger ones (28). The former also contained a greater proportion of evergreen species and trees, while the latter had relatively more deciduous species and climbing plants. Species richness of trees and shrubs was positively correlated with stand age up to 40–50 years, but no trend was detected in climbing plants. There was also an increase in animal-dispersed and gravity-dispersed species richness over time, although not in wind-dispersed species.

Species richness in the species group characteristic of 'older' and 'younger' stands, respectively, showed strong and opposing correlations with age, as would be expected from the method of subgrouping (Fig. B22.6). The range of richness in 'older' stand species was far larger than for 'younger' stand species and this increased up to 60 years old, notably in herbivore- and gravity-dispersed trees and shrubs (Fig. B22.7). In both 'older' and 'younger' stands, compartments that were completely surrounded by conifer plantations also tended to have less species richness than those directly adjacent to evergreen,

broadleaved forest, but only in climbing plants was this significant. First- and second-generation sugi stands were generally similar in their overall species richness; however wind-dispersed species in the 'older' vegetation group were more abundant than in first-generation plantations. In evergreen, shrub and herbivore-dispersed species overall species richness of the 'younger' species groups was greater in second-generation stands. Overall, however, total species richness was not significantly affected between successive crop generations.

In contrast, the effect of topographical position was evident in 'older' stand vegetation, where species richness was greater in the valley stands than in those on ridges. 'Younger' stand types showed no significant differences in either situation.

RESTORATION MANAGEMENT IMPLICATIONS

Infrequent species progressively increased in older stands, indicating their vulnerability to clear-felling

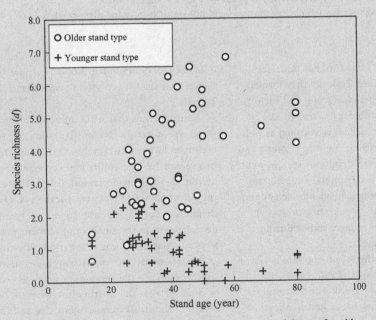

Fig. B22.6. Relationship between stand age and species richness for older and younger stand vegetation types respectively, based on infrequent species. Regression equations are $y = 0.06x + 1.52$, $r = 0.58$, $p < 0.001$ (older stands) $y = -0.02x + 1.90$, $r = 0.57$, $p < 0.001$ (younger stands)

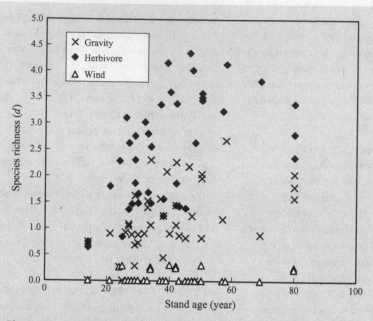

Fig. B22.7. Relationship between stand age and species richness in older stand vegetation types according to seed dispersal type. Regression equations are: $y = 0.02x + 0.40$, $r = 0.47$, $p < 0.01$ (gravity-dispersed) $y = 0.03x + 1.13$, $r = 0.52$, $p < 0.001$ (herbivore-dispersed) $y = 0.002x - 0.01$, $r = 0.30$, not significant (wind-dispersed)

and subsequent plantation establishment practices such as weeding and cleaning. Typically they formed part of the 'older' stand vegetation group and included many evergreen trees, e.g. live oak (*Quercus gilva*), *Q. hondae* and witch hazel (*Distylium racemosum*) as well as the shrubs *Lasianthus japonicus*, *Neolitsea aciculata* and *Maesa japonica*. In contrast, the deciduous species were typically pioneers of disturbed habitats, e.g. *Mallotus japonicus*, devil's pepper tree (*Fagara ailanthoides*), Konara oak (*Quercus serrata*) and yedo hornbeam (*Carpinus tschonoskii*) which thrived in younger stands during the early stages of development.

As rotations for sugi in the region commonly last 35–40 years, the infrequent species associated with older stands are clearly more at risk. For maximum benefits in terms of species richness, and to allow recovery of the native woody vegetation of older-growth stands, longer rotations of at least 60 years are indicated. This of course would represent

a loss in economic timber production that must be weighed against ecosystem restoration benefits.

A surprise was that little influence of the adjacent stand type, whether conifer plantation or broadleaved evergreen forest, was found in this study. This points to a reasonable availability and dispersal of propagules within the forest environment, originating either from the surroundings or from within the plantations themselves. Such apparently good dispersal over short distances within the small compartment matrix of Miyazaki may also account for similarities in species richness between first- and second-generation stands. On this evidence the restoration manager would have to conclude that the existing forest layout, both with regard to the small compartment size and the existing proportion of native forest, is probably adequate to maintain the present vegetation species richness over the area. However, this 'business as usual' stance also implies tacit acceptance of earlier species losses due to plantation conversion.

Stand topography had a strong and overriding influence on the abundance of many late-successional tree species. This has been corroborated by several other reports of topography-dependent distributions of native species in evergreen, broadleaved forests (Tanouchi & Yamamoto, 1995; Matsuda et al., 1998). Native forests in valley situations have traditionally been preferred for conversion to sugi plantations, owing to their high productive potential and the ease of forest road construction. Such intensive utilisation of valleys in the study region is likely to have been

responsible for losses in the original species richness, including endangered and infrequent species such as *Quercus hondae*, for which the riparian zone is their natural habitat. Restoration management should therefore prioritise the protection of unconverted valley forests, or cease commercial production altogether where remnant species are identified. Elsewhere, where conversion to plantation has already taken place, diversification measures such as lengthening rotations, retaining patches of evergreen, broadleaved forest and perpetuating the small stand structure should maintain current levels of plant species diversity.

restoration. One policy might be to segregate the site into areas of high and low potential disturbance, operating larger and more frequent felling coupes in the former (for example on floodplains subject to inundation, or windthrow-limited sites). On protected sites management could take the form of small, irregular coupes on long rotations, including a proportion of the area given over to minimum intervention zones. This raises a real conflict, as sheltered and undisturbed locations are precisely those with the greatest productive potential. Compromises may be possible where some limited old-growth stands are maintained on stable sites, in exchange for a more rapid, commercial turnover elsewhere (Peterken et al., 1992), perhaps operating variable-length rotations in different management compartments.

To increase landscape heterogeneity, harvesting can be done at a range of scales. One convenient method, which assumes a typical logarithmic gap-size distribution, apportions equal felling areas to each in a range of different gap-size classes, such that at the lower end of the scale there will be many small gaps, but only a few (covering an equivalent area) at the upper end (Hunter, 1990). This approach can be further refined by allocating different proportions of the forest area to increasing rotation intervals, based on the age structure found in natural forests (van Wagner, 1987b; Seymour & Hunter, 1999).

Reversing fragmentation: expanding existing woodlands

Expanding out from existing woodland areas is one alternative, but very long-term strategy for tackling the problem of small restoration sites. Natural regeneration of the canopy into a surrounding buffer area (i.e. through passive restoration) is likely to be very slow beyond the immediate forest edge, but has the advantage of having been achieved through natural succession. The alternative, planting, is less natural but accelerates the process of canopy formation. In East Sussex, England, Whitbread & Jenman (1995) advocated the creation of large, continuous forest areas to reverse fragmentation and to avoid the tendency of conservation managers to become preoccupied with maintaining small-scale diversity in small, isolated reserves. This neat solution included extensive planting on adjacent agricultural land, with the resulting large woodland districts being allowed to turn over naturally, without intervention, servicing the minimum dynamic area requirements of most species. However, several crucial assumptions are made:

- Planting land is available
- The new forest area will rapidly develop its own dynamic structure while maintaining continuity of habitat for key species of specific growth stages
- Newly created forest areas will be rapidly colonised by the target species

- The composition of the new woodland will be roughly similar to that of the original.

Woodland expansion on former crop or pasture land can also be criticised on the grounds that both the site and soil conditions are liable to be very different from those found in the original forest, especially after a long history of cultivation, drainage and fertiliser use. Clearly, very long time-scales indeed are needed before these fundamental differences reduce significantly.

Expansion strategies

Woodland expansion can be approached in a number of different ways, the two key variables being the relative increase in forest cover achieved and the distribution or pattern of that cover. Small woodlands can be individually expanded, as discussed previously, or linked together within a wider landscape perspective. The main, generic strategies for locating new forest areas can be summarised as (1) buffering, (2) linking, and (3) 'envelope' or proximity planting (Kirby et al., 1999) (Fig. 22.3). In the latter, planting is targeted towards concentrations of existing woodland in 'envelopes' sized according to ease of implementation.

Each different layout might be expected to favour some species groups more or less, the eventual choice hinging on the local conservation status of the most important individual species and taxa. This point was examined by Buckley & Fraser (1998), who compared the effect of different planting scenarios in four contrasting areas of lowland England on the habitat requirements of key species listed in local Biodiversity Action Plans. In these predominantly agricultural landscapes, modest expansion targets of 5% woodland cover were considered to benefit most key species groups without compromising individuals preferring open, non-forest conditions. Planting strategy made little difference to the outcome in the areas studied, with buffering, envelope and random planting strategies all producing a similar effect.

At high levels of landscape fragmentation, siting expansion areas close to old forest fragments might be expected to create sinks for dispersal and new available habitat for key species to expand their populations. This approach has been almost universally adopted in the restoration of the highly dispersed native pinewoods of Scotland (Box 22.3).

Predicting the potential distribution of new native woodland

At the extreme of forest fragmentation, as illustrated by the Scottish pinewoods, an altogether different approach is needed to restore forest cover. Here, palaeobotanical studies indicating the original–natural forest distribution have only limited relevance to efforts to restore new native woodland. Outside the context of existing woodland remnants, an alternative is to employ GIS-based expert systems, using spatial data on soils, climate zone and topography. Ecological site classifications (e.g. Pyatt & Suarez, 1997) can then be used to link the site data at a given location to an indicative forest type, enabling maps of potential forest vegetation to be produced at a regional level. The approach has been used successfully to model the potential distribution of natural forest types in the Cairngorm region of Scotland (Macmillan et al., 1997). The zoning produced can then be used to prioritise new area initiatives in forest restoration.

CONCLUDING REMARKS

In the foregoing sections we have stressed the importance of setting realistic restoration targets, avoiding where possible narrow 'fixed point' interpretations of some previous forest scene or purist effort to exactly remodel 'past–natural' vegetation composition. Basing the restoration project on and around sites containing surviving fragments of old forest seems most likely to give the best wildlife conservation return for effort, but long-term reconstructions of former forest cover, where this has disappeared, can also be worthwhile when good historical and ecological evidence is available to guide the process.

It may not be sufficient merely to close the forest gate and let nature get on with it, especially if

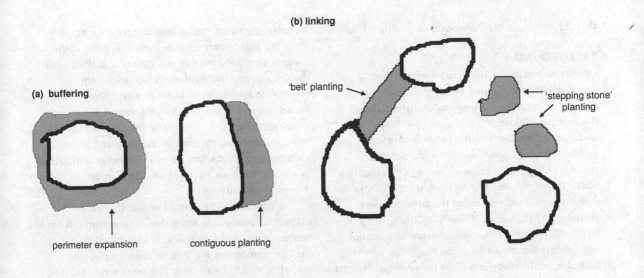

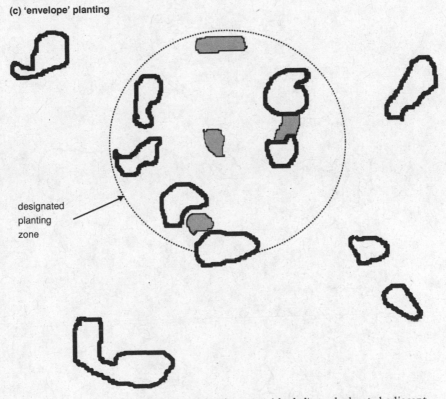

Fig. 22.3. Strategies for forest expansion. New planting areas (shaded) can be located adjacent to existing woods, either by (a) buffering or (b) linking woods together as continuous or discontinuous units. Either parent strategy can be employed within designated planting zones or 'envelopes' (c), coinciding with dense concentrations of semi-natural woodland.

Box 22.3   Native pinewoods of Scotland, UK

BACKGROUND

From their expansion after glacial retreat to 4000 years ago, pine–birch-dominated forests formed an incomplete cover over approximately three-quarters of the Scottish Highlands (more than 1.5 million hectares). Palynological studies show a consistent decline of pine pollen from this point, reflecting climatic change and increasing levels of precipitation which led to peat and bog formation, especially in the western Highlands, restricting the forest to lower altitudes or better-drained sites (O'Sullivan, 1977). Modification through human influence also became increasingly significant from the Mesolithic period, with fire, cattle, sheep and goats the principal reasons for forest clearance. By the Middle Ages probably only 4% of the forest cover remained, contradicting earlier theories that iron-smelting and timber exploitation in the seventeenth and eighteenth centuries were responsible for wholesale losses of forest (Smout, 1997). Rather, sheep-farming and, latterly, deer-stalking and grouse-shooting continued inexorably, preventing forest regeneration through overgrazing and muir-burning. That which remains, however, still ranks as some of the least modified woodland in Britain (Forestry Authority, 1994).

The impoverished situation of the Scottish or Caledonian pinewoods was reviewed in inventories by Steven & Carlisle (1959) and McVean & Ratcliffe (1962), the latter producing the first detailed distribution maps (Fig. B22.8). From these and more recent surveys,

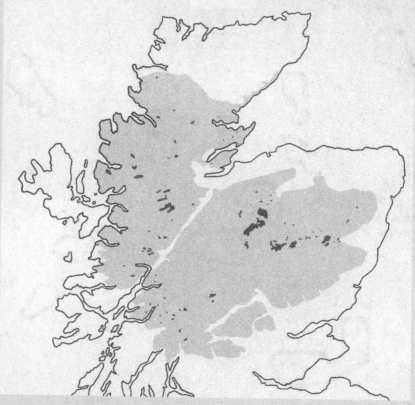

Fig. B22.8. Current areas of remnant Scots-pine-dominated forest in the Scottish Highlands compared with its reconstructed former extent (shaded), the latter based on pollen core analysis. After Royal Society for the Protection of Birds (1993).

it is estimated that the pinewoods today occupy only 1% of their original range, i.e. some 16 000 ha on 77 separate sites. Apart from indigenous Scots pine (*Pinus sylvestris* var. *scotica*), associated species consist of birch (*Betula* spp.), rowan (*Sorbus aucuparia*) and juniper (*Juniperus communis*), typically forming open woods with less than 70% canopy cover (Rodwell, 1991). The field layer comprises the dominant wavy hair grass (*Deschampsia flexuosa*), with a range of ericaceous subshrubs, e.g. blaeberry (*Vaccinium vitis-idaea*), cowberry (*V. myrtillus*) and heather (*Calluna vulgaris*), and large bryophytes, especially *Hylocomium splendens* and *Dicranum scoparium*. This habitat is associated with specialist and declining bird species, including the Scottish crossbill (*Loxia scotica*) and capercaillie (*Tetrao urogallus*, reintroduced since its extinction around 1770), and similarly reduced populations of mammals such as the wild cat (*Felis sylvestris*), pine marten (*Martes martes*) and red squirrel (*Sciurus vulgaris*). Plant rarities include the twinflower (*Linnaea borealis*), several wintergreen species (*Moneses uniflora*, *Orthilia secunda* and *Pyrola* species) and orchids such as *Goodyera repens*. Several insect species are restricted to relict pinewoods, including 44 species of beetle, many of which depend on old, dead or dying trees (Hunter, 1977) and the specialist hoverfly *Callicera rubra*, which breeds in wet rot holes.

LOCAL PINE PROVENANCES

The native pinewoods of the Scottish Highlands are distinctive, western outliers of the European range, separated by at least 500 km from the main distributions of continental Europe and Scandinavia. Biochemical analyses, based on the monoterpene composition of shoot resins, demonstrate that pine populations within Scotland show distinctive differences from each other as well as from other regions (Forrest, 1980, 1982). This work, supported by studies using isozyme and mitochondrial DNA markers, suggests that some of the western populations originated in southern Europe, recolonising Scotland after the last glaciation via refugia in western France or southern Ireland instead of via continental Europe (Kinloch *et al.*, 1986; Sinclair *et al.*, 1998). Genetic differences between regions are

further underpinned by ecological classifications of sites types based on vegetation communities (Bunce, 1977). The Forestry Commission, the government body dealing with forestry in Britain, has compiled a Caledonian Pinewoods Inventory of sites thought to be genuinely native (Forestry Commission, 1994). To qualify, these sites must contain more than 30 trees at minimum densities of 4 individuals ha^{-1} and support vegetation characteristic of native pinewood on undisturbed, semi-natural soil profiles.

The Forestry Commission also maintains a register of native Scots pine seed collection areas within seven seed zones, based on established genetic differences between populations (Fig. B22.9), which are used for planting in the appropriate areas. These must be sufficiently isolated from non-local origins of Scots pine (normally by at least 400 m) and must contain at least 200 trees capable of producing seed (Forestry Authority, 1994). 'Exclusion zones' are recognised for both the NW and SW regions where, to qualify for grant-aid, new planting must be derived from seed sources exclusive to those regions. Similar restrictions apply in the remaining five regions, but some transfer may be accepted near the boundaries.

RESTORATION STRATEGIES

Restoring the native Scottish pinewoods currently follows three strategic initiatives:

- Within existing pinewood remnants, facilitating natural regeneration in order to expand the tree population, redress the imbalance in age structure and increase its area by excluding grazing stock
- Where pine forest is known or thought to have existed in the past, reintroducing native trees to areas in which the grazing pressure is controlled
- Removing introduced conifer plantations such as sitka spruce (*Picea sitchensis*), Japanese larch (*Larix kaempferi*), lodgepole pine (*Pinus contorta*) and non-native Scots pine from remnant native pinewood stands.

Expansion through natural regeneration

Regaining control of deer and sheep populations is often the key to successful restoration of the surviving

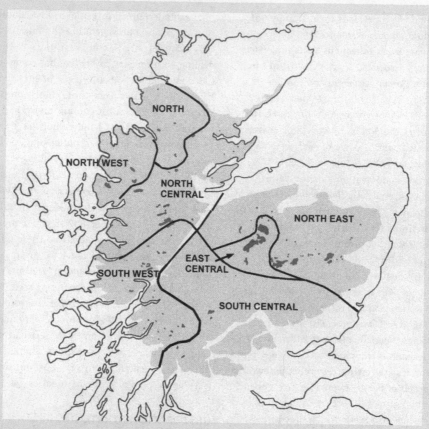

Fig. B22.9. Broad divisional boundaries between native Scots pine populations, based on their biochemical similarity. To maintain genetic integrity, these seven regions are currently adopted as seed collection and planting zones.

pinewood remnants. Red deer (*Cervus elaphus*) numbers have risen consistently over the last century, encouraged on sporting estates where land values often reflect the density of the deer populations they support. Aided by agricultural support policies, sheep numbers are also maintained at artificially high levels. The problem is vividly illustrated by the presence at several sites of repeatedly browsed individuals less than 1 metre tall but frequently more than 20 years old, beneath veteran populations of parent trees 300–400 years old (Edwards, 1998). Culling to reduce the deer population to a threshold allowing spontaneous natural regeneration is one approach, but requires strict co-operation between adjacent landowners. An alternative, such as reintroducing the wolf (*Canis lupus*), extinct since the seventeenth century, has yet to find political and public support. Fencing currently attracts some grant-aid and is therefore the most practicable means of excluding animals, but care is needed to avoid restricting the movement of non-target animals. Marking of fencelines is also necessary to prevent black grouse (*Tetrao tetrix*) and capercaillie from flying into them (Royal Society for the Protection of Birds, 1993).

Since most seed falls within 30 m of mature trees, regeneration zones can be designated around each pinewood by extending out from the existing forest boundaries to *c.* 100 m, depending on wind direction and topography (Forestry Commission, 1994). As regeneration occurs, further expansion of the area can be achieved incrementally, creating a varied age structure over time.

Planting

In areas where native pinewoods no longer exist but the vegetation, soils and previous site history indicate suitable conditions, restoring forest cover by planting native trees is a second option. To be consistent, all tree species should originate from nearby native populations, based on a framework of seed collection zones, as suggested for pine. Grant-aid is currently available towards the costs of such new native woodland planting from the Forestry Commission, including provision for shrubs and open space. Guidelines for planting mixed-species pine forest use the UK National Vegetation Classification as a framework to indicate suitable site types, species composition and planting patterns (Rodwell & Patterson, 1994). A 'recipe' approach can be avoided by varying the density and spacing of planting, leaving open areas and grouping species to mimic secondary succession, preferably without strict adherence to minor changes in soil and vegetation pattern (Peterken, 1998). Planting can also take place in small 'islands' or nuclei from which to expand a natural population, forming eventual stepping stones to remnant forest populations.

Removal of non-native plantations

Many native pinewoods are surrounded by plantations of non-native species, including non-native pine, creating problems of regeneration encroachment and gene flow between sites. Opportunities to reverse the situation may arise on restocking of the plantations, or through earlier efforts to remove them. In some areas the Forestry Commission operates a programme of removing underplanted non-native species, together with reduction of browsing pressures. For example, in 1994 the Commission declared 9000 ha of Glen Affric a Caledonian Forest Reserve, within which plantation species are to be gradually phased out and commercial management abandoned. Elsewhere, the restructuring of commercial plantations of local origins of Scots pine by retaining some stands to old-growth maturity, leaving areas unplanted and encouraging natural regeneration is being undertaken to create a more diverse forest environment.

PROTECTION AND RESTORATION

More than half of the native pinewood area is currently in private ownership (Royal Society for the Protection of Birds, 1993). Exceptionally for semi-natural woodlands in Britain, up to 80% of the native pinewoods are notified as Sites of Special Scientific Interest by Scottish Natural Heritage, the government body responsible for conservation. Caledonian forest is further recognised as a priority habitat under the 1992 Directive on the Conservation of Natural and Semi-natural Habitats and of Wild Flora and Fauna. In addition, the Forestry Commission owns large areas (3000 ha) as do several non-government organisations, such as the Royal Society for the Protection of Birds (RSPB). However, while such levels of conservation recognition are impressive, protection is not necessarily guaranteed, particularly from overgrazing.

Since 1988 a number of schemes have been put into action: over 3000 ha of regeneration of existing woods has been achieved and more than 11 000 ha of new native woodland planted. A programme co-ordinated by the Millennium Forest for Scotland Trust comprises over 70 individual projects, working on nearly 400 woodland sites (not all pinewoods) throughout Scotland, representing a total investment of £27 million up to 2001. Co-ordination of such initiatives is important for the future of the pinewoods. The RSPB, for example, envisages the expansion of pine forest in three core areas – the Beauly catchment, Strathspey and Deeside (Fig. B22.10), regions already containing extensive areas of native pinewood and including populations of some of the key specialist species. In the first of these areas a charitable trust, Trees for Life, has adopted a 155 000-ha area within which to carry out restoration projects. Co-ordination of policies between government departments, voluntary organisations and private owners on this scale is important to maximise conservation benefits in future.

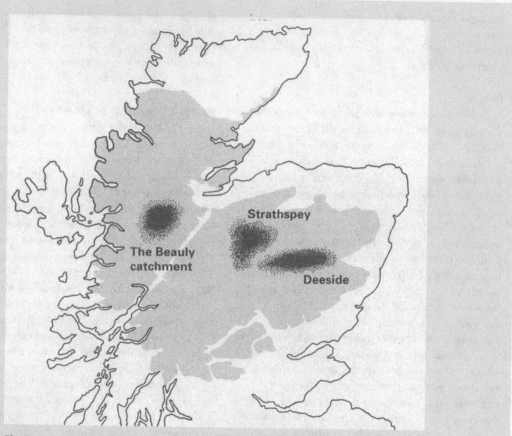

Fig. B22.10. Three potential expansion areas for Scots pine forest suggested by the Royal Society for the Protection of Birds (1993). These areas already contain some extensive areas of native pinewood, together with important key specialist species.

key elements of biodiversity are constrained either by small forest areas or missing growth stages. Active restoration may be necessary to increase the diversity of the forest structure by manipulating silvicultural regimes, extending rotations and allowing some areas to develop into old growth. The guiding principle, to mimic the amplitude of natural disturbance regimes of the region, may be aided by expert system modelling of events such as windthrow and fire. At higher levels of sophistication, different growth stages could be spatially targeted to match the predicted disturbance regimes to local site conditions and topography, or perhaps to develop

habitat configurations designed to accommodate key species.

Conserving *in situ*, expanding and creating new habitat are all valid strategies in forest restoration. The importance of maintaining areas of semi-natural forest is well illustrated both by Miyazaki and Point Pelée. Neither were pristine habitats, but conserving fragments of semi-natural woodland within the working forest appeared to maintain good plant colonising sources at Miyazaki, some of which also survived well in plantations. However, the need strictly to conserve the best fragments associated with particular topographical locations and rare trees and shrubs was well illustrated. A further case for reverting some plantation area to semi-natural forest and planning habitat corridors and networks could also be made.

To safeguard and consolidate surviving forest fragments, similar targeting will be necessary in native forest expansion initiatives of the future. These will involve planning at the landscape level, using expert systems to identify potential forest vegetation at both regional and local levels, while at the same time devising planting scenarios that will deliver the optimum biodiversity. Strong regional policies, declaring exclusion zones for local provenances and species, backed by planting grants and local planning initiatives, are needed to avoid well-intentioned but inappropriate planting of forest cover. Adding to this the facilitation of stakeholder activity and participation, the quality of the restoration effort should be vastly improved.

Relative success, measured as increasing species diversity over time, has been illustrated by the examples given of both active and passive restoration. Non-native species populations rapidly declined at Point Pelée, although, paradoxically, disturbances created by active restoration measures temporarily added to the problem. However, the highest plant species similarity coefficient achieved here, compared with the reference site, was still only 0.5 after more than 30 years, with 144 years projected to achieve the full species complement. Although in ecological terms these time-scales may seem reasonable, we should not be blind to the strong possibility that some 'lost' species will never make it back, even with assistance. Whether to reintroduce the vulnerable understorey plants and native woody pioneers at Point Pelée raises an ethical question familiar in wildlife conservation. If some disparity between aspiration and performance in restoration objectives is only to be expected, this not should deflect us from the wider benefits.

REFERENCES

Allen, G. M. P., Eagles, J. F. & Price, S. D. (eds.) (1990). *Conserving Carolinian Canada*. Waterloo, Ontario: University of Waterloo Press.

Arno, S. F. (1980). Forest fire history in the northern Rockies. *Journal of Forestry*, **78**, 460–465.

Attiwill, P. M. (1994). The disturbance of forest ecosystems: the ecological basis for conservative management. *Forest Ecology and Management*, **63**, 247–300.

Beke, G. J. & McKeague, J. A. (1984). Influence of tree windthrow on the properties and classification of selected forested soils from Novia Scotia. *Canadian Journal of Soil Science*, **64**, 195–207.

Bormann, F. H. & Likens, G. E. (1979). Catastrophic disturbance and the steady state in northern hardwood forests. *American Scientist*, **67**, 660–669.

Bratton, S. P., Hapeman, J. R. & Mast, A. R. (1994). The lower Susquehanna river gorge and floodplain (USA) as a riparian refugium for vernal, forest-floor herbs. *Conservation Biology*, **8**, 1069–1077.

Bryant, D., Nielsen, D. & Tangley, L. (1997). *The Last Frontier Forests*. Oxford: World Resources Institute and Oxford University Press.

Buckley, G. P. (ed.) (1992). *Ecology and Management of Coppice Woodlands*. London: Chapman & Hall.

Buckley, G. P. & Fraser, S. (1998). *Locating New Lowland Woods*, English Nature Research Report no. 28. Peterborough, UK: English Nature.

Bunce, R. G. H. (1977). The range of variation in pinewoods. In *Native Pinewoods of Scotland*, eds. R. G. H. Bunce & J. N. Jeffers, pp. 10–25. Cambridge: Institute of Terrestrial Ecology.

Chubbs, T. E., Keith, L. B., Mahoney, S. P. & McGrath, M. J. (1993). Responses of woodland caribou to clearcutting in east-central Newfoundland. *Canadian Journal of Zoology*, **71**, 487–493.

Clark, J. S. (1990). Fire and climate change during the last 750 yr in northwestern Minnesota. *Ecological Monographs*, **60**, 135–159.

Duffy, D. C. & Meier, A. J. (1992). Do Appalachian herbaceous understories ever recover from clearcutting? *Conservation Biology*, **6**, 196–201.

Dunster, K. (1990). *Exotic Plant Species Management Plan, Point Pelée National Park*. Ontario Region, Ontario: Parks Canada.

Eberhart, K. E. & Woodard, P. M. (1987). Distribution of residual fires associated with large fires in Alberta. *Canadian Journal of Forest Research*, **117**, 1207–1212.

Edwards, C. (1998). The secret of the pinewoods. *Tree News*, Spring 1998.

Elton, C. S. (1966). *Dying and Dead Wood: The Pattern of Animal Communities.* New York: John Wiley.

Esseen, P. (1994). Tree mortality patterns after experimental fragmentation of an old-growth conifer forest. *Biological Conservation,* **68**, 19–28.

Evans, J. (1992). Coppice forestry: an overview. In *Ecology and Management of Coppice Woodlands,* ed. G. P. Buckley, pp. 18–27. London: Chapman & Hall.

Falinski, J. B. (1986). *Vegetation Dynamics in Temperate Lowland Primeval Forests.* Dordrecht: Dr W. Junk.

Falkenberg, N. (2000). Restoration of red cedar savanna communities in Point Pelée National Park. MSc dissertation, York University, Ontario, Canada.

Ferris-Kaan, R., Lonsdale, D. & Winter, T. (1993). *The Conservation Management of Deadwood in Forests,* Research Information Note no. 241. Farnham, UK: Forestry Commission.

Fitter, A. H., Fitter, R. S. R, Harris, I. T. B. & Williamson, M. H. (1995). Relationships between the first flowering date and termperature in the flora of a locality in central England. *Functional Ecology,* **9**, 55–60.

Forestry Authority (1994). *The Management of Semi-natural Woodlands: Native Pinewoods,* Forestry Practice Guide no. 7. Edinburgh, UK: Forestry Authority.

Forestry Commission (1994). *Caledonian Pinewood Inventory.* Edinburgh, UK: Forestry Commission.

Forrest, G. I. (1980). Genotypic variation among native Scots pine populations in Scotland based on monoterpene analysis. *Forestry,* **53**, 101–128.

Forrest, G. I. (1982). Relationship of some European Scots pine populations to native Scottish woodlands based on monoterpene analysis. *Forestry,* **55**, 19–37.

Foster, D. R. (1988*a*). Disturbance history, community organisation and vegetation dynamics of the old-growth Pisgah Forest, southwestern New Hampshire, USA. *Journal of Ecology,* **76**, 105–134.

Foster, D. R. (1988*b*). Species and stand response to catastrophic wind in central New England, USA. *Journal of Ecology,* **76**, 135–151.

Fowells, H. A. (1965). *Silvics of Forest Trees in the United States,* Division of Timber Management Research, Agriculture Handbook no. 271. Washington, DC: US Department of Agriculture Forest Service.

Franklin, J. F., Cromack, K., Denison, W., McKee, A., Maser, C., Sedell, J., Swanson, F. & Juday, G. (1981). *Ecological Characteristics of Old-Growth Douglas-Fir Forests,* General Technical Report no. PNW-118. Portland, OR: Northwest Forest and Range Experiment Station.

Fuller, R. J. (1995). *Bird Life of Woodland and Forest.* Cambridge: Cambridge University Press.

Fuller, R. J. & Warren, M. S. (1993). *Coppiced Woodlands: Their Management for Wildlife.* Peterborough, UK: Joint Nature Conservation Committee.

Fuller, R. J. & Warren, M. S. (1995). Management for biodiversity in British woodlands: striking a balance. *British Wildlife,* **7**, 26–37.

Gustafsson, L., Friskesjo, A., Ingelog, T., Petterson, B. & Thor, G. (1992). Factors of importance to some lichen species of deciduous broadleaved species in southern Sweden. *Lichenologist,* **24**, 255–266.

Hambler, C. & Speight, M. R. (1995). Biodiversity conservation in Britain: science replacing tradition. *British Wildlife,* **6**, 137–147.

Harrison, S. (1994). Metapopulations and conservation. In *Large-Scale Ecology and Conservation Biology,* eds. P. J. Edwards, R. M. May & N. R. Webb, pp. 111–128. Oxford: British Ecological Society and Blackwell.

Heinselman, M. L. (1973). Fire in the virgin forests of the Boundary Waters Canoe Area, Minnesota. *Quaternary Research,* **3**, 329–382.

Hiebert, R. D. (1990). An ecological restoration model: application to razed residential sites. *Natural Areas Journal,* **10**, 181–186.

Hilts, S. (1985). Private stewardship. *Seasons,* Summer 1985, 47–49.

Hodge, S. J. & Peterken, G. F. (1998). Deadwood in British forests: priorities and a strategy. *Forestry,* **71**, 99–112.

Holling, C. S., Berkes, F. & Folke, C. (1998). Science, sustainability and resources management. In *Linking Social and Ecological Systems: Management Practices and Social Mechanisms for Building Resilience,* eds. F. Berkes & C. Fole, pp. 342–362. Cambridge: Cambridge University Press.

Hunter, F. A. (1977). Ecology of pinewood beetles. In *Native Pinewoods of Scotland,* eds. R. G. H. Bunce & J. N. Jeffers, pp. 42–55. Cambridge: Institute of Terrestrial Ecology.

Hunter, M. L. (1990). *Wildlife, Forests and Forestry: Principles of Managing Forests for Biological Diversity.* Englewood Cliffs, NJ: Prentice-Hall.

Innes, J. L. (1993). *Forest Health: Its Assessment and Status.* Wallingford, UK: CAB International.

Johnston, A. E., Goulding, K. W. T. & Poulton, P. R. (1986). Soil acidifcation during more than 100 years under permanent grassland and woodland at Rothamsted. *Soil Use and Management*, **2**, 3–10.

Jones, E. W. (1945). The structure and reproduction of the virgin forest of the northern temperate zone. *New Phytologist*, **44**, 130–148.

Kinloch, B. B., Westfall, R. D. & Forrest, G. I. (1986). Caledonian Scots pine: origins and genetic structure. *New Phytologist*, **104**, 703–729.

Kirby, K. J. & Drake, C. M. (eds.) (1993). *Dead Wood Matters: The Ecology and Conservation of Saproxylic Invertebrates in Britain.* Peterborough, UK: English Nature.

Kirby, K. J., Michell, F. J. & Hester, A. J. (1994). A role for large herbivores (deer and domestic stock) in nature conservation management in British semi-natural woods. *Arboricultural Journal*, **18**, 381–399.

Kirby, K. J., Buckley, G. P. & Good, J. E. G. (1999). Maximising the value of new farm woodland at a landscape scale. In *Farm Woodlands for the Future*, eds. P. J. Burgess, E. D. R. Brierly, J. Morris & J. Evans, pp. 45–55. Oxford: Bios Scientific.

Kloor, K. (2000). Returning America's forests to the 'natural' roots. *Science*, **287**, 573–575.

Koh, S. (1995). The responses of four species of spring flowering perennial herbs to grazing by white tailed deer in Southern Ontario. MS dissertation, York University, Ontario, Canada.

Laurance, W. F. (2000). Do edge effects occur over large spatial scales? *Trends in Evolution and Ecology*, **15**, 134–135.

Laurance, W. F. & Yensen, E. (1991). Predicting the impacts of edge effects in fragmented habitats. *Biological Conservation*, **55**, 77–92.

Li, C. (2000). Fire regimes and their simulation with reference to Ontario. In *Ecology of a Managed Terrestrial Landscape*, eds. A. H. Perera, D. L. Euler & I. D. Thompson, pp. 115–140. Toronto, Ontario: University of British Columbia Press and Ministry of Natural Resources, Ontario.

Lienkaemper, G. W. & Swanson, F. J. (1987). Dynamics of large woody debris in streams in old-growth Douglas-fir forests. *Canadian Journal of Forest Research*, **17**, 150–156.

Lutz, H. J. (1940). *Disturbance of Forest Soil Resulting from the Uprooting of Trees*, School of Forestry Bulletin no. 45. New Haven, CT: Yale University.

Macmillan, D. C., Towers, W., MacLeay, S. & Kupiec, J. (1997). Modelling the potential distribution of the native woodland resource in the Cairngorms. *Scottish Forestry*, **51**, 70–75.

Mannan, R. W. & Meslow, E. C. (1984). Bird populations and vegetation characteristics in managed and old-growth forests, northeastern Oregon. *Journal of Wildlife Management*, **48**, 1219–1238.

Martin, W. H. (1992). Characteristics of old-growth mixed mesophytic forests. *Natural Areas Journal*, **12**, 127–135.

Matlack, G. R. (1993). Microenvironment variation within and among forest edge sites in the eastern United States. *Biological Conservation*, **66**, 185–194.

Matsuda, A., Ito, S., Sato, S. & Nogami, K. (1998). Effects of topography on stand structure of an evergreen broadleaved forest. *Bulletin of the Kyushu Branch, Japanese Forestry Society*, **51**, 59–60.

Matthews, J. D. (1989). *Silvicultural Systems.* Oxford: Clarendon Press.

McCollin, D. (1998). Forest edges and habitat selection in birds: a functional approach. *Ecography*, **21**, 247–260.

McLachlan, S. M. (1997). Multiple-scale approaches to the restoration of deciduous forest in southwestern Ontario, Canada. PhD dissertation, York University, Ontario, Canada.

McLachlan, S. M. & Bazely, D. R. (2001). Recovery patterns of understory herbs and their use as indicators of deciduous forest regeneration. *Conservation Biology*, **15**, 98–110.

McVean, D. N. & Ratcliffe, D. A. (1962). *Plant Communities of the Scottish Highlands.* London: HMSO.

Meier, A. J., Bratton, S. P. & Duffy, D. C. (1995). Possible ecological mechanisms for loss of vernal-herb diversity in logged eastern deciduous forests. *Ecological Applications*, **5**, 935–946.

Mitchell, P. L. & Kirby, K. J. (1989). *Ecological Effects of Forestry Practices in Long-Established Woodland, and their Implications for Nature Conservation*, Oxford Forestry Institute Occasional Paper no. 39. Oxford: Forestry Institute.

Murcia, C. (1995). Edge effects in fragmented forests: implications for conservation. *Trends in Ecology and Evolution*, **10**, 58–62.

Nakagawa, M. & Ito, S. (1997). Species richness in sugi (*Cryptomeria japonica* D. Don) plantation forests: effect of stand age and site conditions in the Miyazaki University Forest. *Bulletin of the Kyushu Branch, Japanese Forestry Society*, **50**, 87–88.

O'Sullivan, P. E. (1977). Vegetation history and the native pinewoods. In *Native Pinewoods of Scotland,* eds. R. G. H. Bunce & J. N. Jeffers, pp. 60–69. Cambridge: Institute of Terrestrial Ecology.

Parker, V. T. & Pickett, S. T. A. (1997). Restoration as an ecological process: implications of the modern ecological paradigm. In *Restoration Ecology and Sustainable Development*, eds. K. M. Urbanska, N. R. Webb & P. J. Edwards, pp. 17–32. Cambridge: Cambridge University Press.

Parks Canada (2000). *Unimpaired for Future Generations? Protecting Ecological Integrity with Canada's National Parks*, Report of the Panel on Ecological Integrity of Canada's National Parks. Ottawa, Ontario: Parks Canada.

Parviainen J., Little, D., Doyle, M., O'Sullivan, A., Kettunen, M. & Korhonen, M. (eds.) (1999). *Research in Forest Reserves and Natural Forests in European Countries*, European Forest Institute Proceedings no. 16. Joansuu, Finland: European Forest Institute.

Patton, D. R. (1992). *Wildlife Habitat Relationships in Forested Ecosystems*. Portland, OR: Timber Press.

Pearce, C. M. (1996). *Identification and Spatial Analysis of Land Cover within and adjacent to Point Peleé National Park using LANDSAT TM Imagery*. Cornwall, Ontario: Department of Canadian Heritage, Parks Canada.

Peterken, G. F. (1974). A method of assessing woodland flora for conservation using indicator species. *Biological Conservation*, **6**, 239–245.

Peterken, G. F. (1996). *Natural Woodland: Ecology and Conservation in Northern Temperate regions*. Cambridge: Cambridge University Press.

Peterken, G. F. (1998). Woodland composition and structure. In *Native Woodland Restoration in Southern Scotland: Principles and Practice*, eds. A. C. Newton & P. Ashmole, pp. 22–26. Jedburgh, UK: Borders Forest Trust.

Peterken, G. F. (2000). *Natural Reserves in English Woodlands*, English Nature Research Report no. 384. Peterborough, UK: English Nature.

Peterken, G. F. & Game, M. (1984). Historical factors affecting the number and distribution of vascular plant species in the woodlands of central Lincolnshire. *Ecology*, **72**, 155–182.

Peterken, G. F., Auscherman D., Buchenau M. & Formann, R. T. T. (1992). Old-growth conservation within British upland conifers. *Forestry*, **65**, 127–144.

Petts, G. E. (1990). Forested river corridors: a lost resource. In *Water, Engineering and Landscape*, eds. D. Cosgrove & G. E. Petts, pp. 12–34. London: Belhaven Press.

Pickett, S. T. A. & Thompson, J. N. (1978). Patch dynamics and the design of nature reserves. *Biological Conservation*, **13**, 27–37.

Pollard, E. & Yates, T. J. (1993). *Monitoring Butterflies for Ecology and Conservation*. London: Chapman & Hall.

Pyatt, D. G. & Suarez, J. C. (1997). *An Ecological Site Classification for Forestry in Great Britain, with Special Reference to Grampian, Scotland*, Forestry Commission Technical Paper no. 20. Edinburgh, UK: Forestry Commission.

Quine, C. P., Coutts, M. P., Gardiner, B. A. & Pyatt, D. G. (1995). *Forests and Wind: Management to Minimise Damage*, Forestry Commission Bulletin no. 114. London: HMSO.

Rackham, O. (1980). *Ancient Woodland: Its History, Vegetation and Uses in England*. London: Edward Arnold.

Ratcliffe, D. A. (ed.) (1977). *A Nature Conservation Review*, vol. 1. Cambridge: Cambridge University Press.

Reive, D., Sharp, M. & Stephenson, W. R. (1992). Integrating ecological restoration with ecosystem conservation objectives: Point Peleé National Park. In: *Proceedings of the 4th Congress of the Society for Ecological Restoration*, pp. 12–14. Waterloo, Ontario: Society for Ecological Restoration.

Riley, J. L. & Mohr, P. (1994). *The Natural Heritage of Southern Ontario's Landscapes: A Review of Conservation and Restoration Ecology for Land-Use and Landscape Planning*. Chatham, Ontario: Ontario Ministry of Natural Resources, Southern Region.

Rodwell, J. S. (ed.) (1991). *British Plant Communities*, vol. 1,

Woodlands and Scrub. Cambridge: Cambridge University Press.

Rodwell, J. S. & Patterson, G. S. (1994) *Creating New Native Woodlands*, Forestry Commission Bulletin no. 112. London: HMSO.

Romme, W. H. & Despain, D. G. (1989). The Yellowstone fires. *Scientific American*, **61**, 21–29.

Rothermel, R. C. (1983). *How to Predict the Spread and Intensity of Wildland Fires* General Technical Report no. INT-143, Washington, DC: US Department of Agriculture Forest Service.

Royal Society for the Protection of Birds (1993). *Time for Pine: A Future for Caledonian Pinewoods*. Edinburgh, UK: RSPB.

Runkle, J. R. (1982). Patterns of disturbance in some old-growth mesic forests of eastern North America. *Ecology*, **63**, 1041–1051.

Runkle, J. R. (1991). Gap dynamics in old-growth eastern forests: management implications. *Natural Areas Journal*, **11**, 19–25.

Saunders, D. A., Hobbs, R. J. & Margules, C. R. (1991). Biological consequences of ecosystem fragmentation: a review. *Conservation Biology*, **5**, 18–32.

Schoenau, J. J., Stewart, J. W. B. & Bettany, J. R. 1989. Cycling of phosphorus in prairie and boreal forest soils. *Biogeochemistry*, **8**, 223–237.

Seymour, R. S. & Hunter, M. L. (1999). Principles of ecological forestry. In *Maintaining Biodiversity in Forest Ecosystems*, ed. M. L. Hunter, pp. 22–61. Cambridge: Cambridge University Press.

Simberloff, D. (1997). Flagships, umbrellas and keystones: is single-species conservation passé in the landscape era? *Biological Conservation*, **83**, 247–257.

Sinclair, W. T., Morman, J. D. & Ennos, R. A. (1998). Multiple origins for Scots pine (*Pinus sylvestris* L.) in Scotland: evidence from mitochondrial DNA variation. *Heredity*, **80**, 233–240.

Smout, T. C. (ed.) (1997). *Scottish Woodland History*. Edinburgh, UK: Scottish Cultural Press.

Sparks, T. H. (2000). The long-term phenology of woodland species in Britain. In *Long-Term Studies in British Woodland*, eds. K. J. Kirby & M. D. Morecroft, pp. 98–105. Peterborough, UK: English Nature.

Sparks, T. H. & Carey, P. D. (1995). The responses of

species to climate over two centuries: an analysis of the Marsham phenological record, 1736–1947. *Journal of Ecology*, **83**, 321–329.

Spencer, J. W. & Kirby, K. J. (1992). An inventory of ancient woodland for England and Wales. *Biological Conservation*, **62**, 77–93.

Spies, T. A. & Franklin, J. F. (1988). Old-growth and forest dynamics in the Douglas-fir forests of western Oregon and Washington. *Natural Areas Journal*, **8**, 190–201.

Steven, H. M. & Carlisle, A. (1959). *The Native Pinewoods of Scotland*. Edinburgh, UK: Oliver & Boyd.

Tanouchi, H. & Yamamoto, S. (1995). Structure and regeneration of canopy species in an old-growth evergreen forest in Aya district, southwestern Japan. *Vegetatio*, **117**, 51–60.

Thompson, I. D. & Anglestam, P. (1999). Special species. In *Maintaining Biodiversity in Forest Ecosystems*, ed. M. L. Hunter, pp. 434–459. Cambridge: Cambridge University Press.

van Devender, T. (1986). Climatic cadences and the composition of Chichuahuan desert communities: the late Pleistocene packrat midden record. In *Community Ecology*, eds. J. Diamond & T. J. Case, pp. 285–299. New York: Harper & Row.

van Wagner, C. E. (1987a). *Development and Structure of the Canadian Forest Weather Index System*, Forestry Technical Report no. 35. Ottawa, Ontario: Canadian Forest Service.

van Wagner, C. E. (1987b). Age-class distribution and the forest fire cycle. *Canadian Journal of Forest Research*, **8**, 220–227.

Whitbread, A. & Jenman, W. (1995). A natural method of conserving biodiversity in Britain. *British Wildlife*, **7**, 84–93.

Whitney, G. G. (1986). Relation of Michigan's presettlement pine forests to substrate and disturbance history. *Ecology*, **67**, 1548–1559.

Whitney, G. G. & Foster, D. R. (1988). Overstorey composition and age as determinants of the understorey flora of woods of Central New England. *Journal of Ecology*, **76**, 867–876.

Whittaker, R. H. (1975). *Community and Ecosystem*, 2nd edn. New York: Macmillan.

Wilcove, D. S. (1993). Turning conservation goals into tangible results: the case of the spotted owl and old-growth forests. In *Large-Scale Ecology and Conservation Biology*, eds. P. J. Edwards, R. M. May & N. R. Webb, pp. 313–329. Oxford: British Ecological Society and Blackwell.

Wilcox, B. A. & Murphy, D. A. (1985). Conservation strategy: the effects of fragmentation on extinction. *American Naturalist*, **125**, 879–887.

Wilson, B. R., Moffat, A. J. & Nortcliff, S. (1997). The nature of three ancient woodland soils in southern England. *Journal of Biogeography*, **24**, 633–646.

23 • Tropical moist forest

KAREN D. HOLL

INTRODUCTION

Tropical forests are the most diverse ecosystems on earth. Tropical forests are home to >50% of all plant and animal species, despite occupying less than 7% of the terrestrial surface of the planet (Wilson, 1988). Tropical moist forest (TMF) is generally defined by high rainfall (>1700 mm annually), even distribution of solar radiation throughout the year, constant high temperatures (mean monthly temperature >24 °C) and lack of frost (Grainger, 1993). Some TMFs, those referred to strictly as rainforests, receive relatively equal rainfall throughout the year, while others experience a distinct dry season. TMFs are found on a wide range of soil types, although they are primarily found on highly weathered soils that have high clay content and are low in phosphorus. Tropical soils also often have high concentrations of aluminium and iron. While much of TMF is found near sea level, tropical forests vary along elevation gradients and numerous unique species are found in montane and cloud forests.

Tropical moist forests are located in Latin America, Africa and Southeast Asia within 23.5° N or S of the equator, coincident with high rainfall and temperature conditions. More than half of TMF are found in Latin America (Fig. 23.1). The largest remaining area is in the northern half of South America with smaller areas in Central America and the Caribbean. In Asia, much of Papua New Guinea, the Philippines, Indonesia, Malaysia and Singapore were covered by forests, although much has already been cleared by humans. While much of Africa is savanna, significant portions of central Africa and Madagascar were once covered with forest. The geographic distribution of tropical forests has expanded over the past 10 000 years since the last ice age. For example, much of Central America was covered by savanna during the late Pleistocene. Many of the areas with the highest diversity of tropical forest, such as the Amazon Basin, coincide with areas that were not glaciated during the last ice age (Whitmore, 1998).

RATIONALE FOR RESTORATION

Tropical moist forests are being cleared at an alarming rate. While rates of tropical deforestation are difficult to calculate, they are unquestionably high. The most recent estimates of tropical deforestation from the Food and Agriculture Organisation (1999) indicate that less than one-third of rainforest remains in Southeast Asia and Africa. Nearly all the forest in western Africa has been cleared; a large section remains in central Africa, but 3.5% of this forest was lost between 1990 and 1995 alone. In Central America less than half of TMF remains. In South America an estimated 14% of the Amazon Basin has been deforested, but with infrastructure developments planned by the Brazilian government, it is estimated that 40% will be cleared within the next two decades (Laurance et al., 2001). These figures are likely to underestimate the amount of forest highly impacted by humans, as they often do not include areas where forests are not directly cleared, but impoverished due to hunting, fragmentation, selective logging, fires and extraction of mineral resources (Bawa & Dayanandan, 1998; Nepstad et al., 1999).

The causes of deforestation are complex and interrelated, and their relative importance varies by region. Causes of deforestation include clearing forest for pasture for cattle grazing, logging for wood exports, collection of firewood, commercial and

Fig. 23.1. Primary, tropical, montane moist forest in southern Costa Rica near La Amistad National Park. Common tree families include Lauraceae, Moraceae and Palmae.

the atmosphere, which are affecting the global climate; an estimated 23% of anthropogenic CO_2 emissions are from tropical forest clearing (Houghton, 1997). Tropical forests also provide important erosion control that maintains water quality. Because the vast majority of nutrients in tropical forest are retained in the living biomass and about 50% of the rain that falls is transpired by plants, clearing forest substantially changes both hydrological and nutrient cycling. Modelling studies of the Amazon predict a 20% reduction in rainfall with complete conversion of forest to pasture (Lean & Warrilow, 1989; Shukla et al., 1990). It is impossible to estimate the numbers of species that are being lost, given that tropical forests are so poorly inventoried. It is estimated that extinction rates are currently 100 to 1000 times pre-human levels with a large proportion in tropical forests (Pimm et al., 1995). Many of these species have important agricultural and medicinal uses. Tropical forests are also home to numerous indigenous peoples that are increasingly being displaced and exposed to diseases.

Interest in restoring TMF is increasing for many reasons, and the number of studies on TMF has increased exponentially in the past five years. First, in much of the tropics agricultural land rapidly loses productivity due to poor soils, and is soon abandoned (Uhl et al., 1990; Nepstad et al., 1991). Only about half of tropical land cleared for agriculture is used for this purpose for more than a few years (Houghton, 1995). Second, in some countries government subsidies that encourage forest conversion to agricultural land are being changed (Butterfield & Fisher, 1994). Third, the drastic effects of the loss of forests on ecosystem processes, species conservation and human well-being are rapidly becoming apparent. Finally, tropical forests have become an increasing focus of strategies to reduce net fluxes of CO_2, as early-successional tropical forest may uptake substantial quantities (Silver et al., 2000). While it is clear that tropical reforestation will not alone compensate for fossil fuel emissions, paying landowners in tropical countries to restore degraded forests to serve as carbon sinks may provide substantial funds for tropical reforestation efforts (Kremen et al., 2000).

subsistence agriculture, population growth and re-settlement programmes, and fire. Feedback loops between these causes result in increased deforestation. For example, Cochrane et al. (1999) have shown that fires to clear cattle pasture in the Amazon are increasingly likely to spread because of decreases in rainfall due to regional deforestation. These fires are likely to be exacerbated in the future by increasing alterations in precipitation due to anthropogenically elevated levels of greenhouse gases.

Tropical deforestation has profound effects on carbon cycling, soil stability, the hydrological cycle and biodiversity. Forest clearing results in the release of large quantities of CO_2 and trace gases to

This chapter reviews both factors that influence the natural recovery process and strategies to restore TMF. Given the high diversity of species and complex interactions in TMF, reintroducing the myriad species simply is not possible. Therefore, it is necessary to identify stages where the recovery process is most limited. Then, this information can be used to design strategies to accelerate natural recovery. It is important to note a few limitations in the scope of this chapter. First, the majority of research reviewed here is from the neotropics (Latin America and the Caribbean), which reflects the large amount of research on forest restoration in this region in the past few years. Second, most efforts to restore TMF have only been initiated in the past ten years (Holl & Kappelle, 1999); therefore, conclusions must be evaluated with this in mind. It is impossible to say what the long-term effect of restoration strategies will be, but we do not have time to wait for final results given the critical conservation questions at hand. Third, as anyone working in restoration knows, the term 'restoration' is variably defined. In this chapter the focus is on efforts to restore some approximation of pre-disturbance forest. There is a great deal of important research on agro-forestry systems that provide resources to people living in these regions while minimising alterations to habitat quality and ecosystem processes. Such strategies are essential to forest conservation efforts, but are not the focus of this chapter.

FACTORS LIMITING RECOVERY

One management option for degraded ecosystems is to remove existing stresses, such as grazing, and to allow the ecosystem to recover naturally. While slash-and-burn agricultural systems have long resulted in abandoned agricultural lands (Lugo, 1988), in the past these lands were subjected to less intensive and shorter-term disturbance compared to current practices. A number of studies on abandoned agricultural and logged lands in the tropics suggest that over a period of 15–60 years a number of forest species establish and accumulate biomass levels that approach those of intact forests (Reiners et al., 1994; Guariguata et al., 1997; Aide et al., 2000; Finegan & Delgado, 2000; Silver et al.,

2000) (Box 23.1). Research in various regions suggests that forest recovery is slower in sites that have been more intensively disturbed (Buschbacher et al., 1988; Uhl et al., 1988; Reiners et al., 1994; Pinard et al., 2000). Finegan & Delgado (2000) found an overrepresentation of wind-dispersed species in one 30-year-old secondary forest in Costa Rica. Chapman & Chapman (1997) argue that succession in abandoned lands in Africa where they have done research is slower than rates reported from studies in Latin America, largely due to the lack of rapidly colonising tree species. These studies suggest that in some cases simply leaving areas to recover naturally may be a viable strategy for the recovery of many species, but that recovery will depend a great deal on the ecology and disturbance history of the area. In most cases, it is simply impossible to judge whether these forests will recover without human intervention, given the short time that recovery has been monitored at most sites.

As recovery is slow on many disturbed tropical lands, there is increasing interest in restoring these areas for conservation of both species and ecosystem processes. In order for forests to recover in abandoned pasture a number of processes must occur, including dispersal of seeds, avoidance of predation, and germination of seeds, as well as survival and growth of seedlings (Fig. 23.2). Each of these processes is, in turn, influenced by a number of factors (Fig. 23.2); for example, seedling predation is affected by the abundance of herbivores, which is in turn influenced by vegetative cover. If such barriers are identified and attempts are made to overcome such barriers then recovery may be accelerated.

Seed availability

Almost all studies of TMF recovery world-wide concur that lack of seeds, principally due to lack of seed dispersal, is a primary factor limiting recovery (Holl & Kappelle, 1999). Most tropical forest seeds have an extremely short duration of viability (Garwood, 1989). Accordingly, several studies have shown that seeds of forest species are commonly absent from soil seed banks in pasture land (Uhl, 1987; Nepstad et al., 1996; Toh et al., 1999; Zimmerman et al., 2000). In agricultural land that has been used

Box 23.1   Tropical forest regeneration: is restoration necessary?

Tropical forests have long been cleared for human settlement and agriculture, and then have been abandoned as the land becomes less fertile. But the scale of human clearing of rain forests in the past 50 years is unprecedented with less than 50% remaining. Most studies of tropical forest restoration have documented trends for less than ten years, which begs the question of whether restoration efforts are required to accelerate forest recovery over the long term. Puerto Rico is unique in that forest covered <10% of the island at the beginning of the twentieth century, but forest cover has increased during the second half of the twentieth century due to pasture abandonment, as the island transitioned to a more industrial economy. Aide and colleagues at the University of Puerto Rico have been studying vegetation recovery on chronosequences of land abandoned for periods of 5–75 years at four different locations on the island (Aide *et al.*, 2000). They have surveyed forest composition, basal area, and biomass in 71 abandoned pastures. Their results show that basal area and biomass increase rapidly in the first 25 years after abandonment and by 40 years after abandonment the structure is similar to that in much older sites (Aide *et al.*, 2000). The number of plant species increases with time since abandonment and then plateaus after about 40 years. The composition of the oldest abandoned pastures, however, is quite different from old forest sites. These results suggest that either abiotic conditions are not appropriate for the establishment of certain species or that interactions with other species, such as dispersal and pollination have been lost. Other studies of natural recovery in Brazil (Uhl *et al.*, 1988; Parrotta & Knowles, 1999) and Costa Rica (Guariguata *et al.*, 1997; Finegan & Delgado, 2000) concur that forest structure recovery occurs fairly quickly in areas where seed sources exist, soils are not highly compacted, fires are not prevalent, and invasive exotic species are not abundant, but that certain forest species are absent from abandoned pastures. Therefore, enrichment planting or reintroduction of certain mutualists may be necessary to restore tropical forests. Also, as humans alter these other factors, recovery may not be as likely in the future.

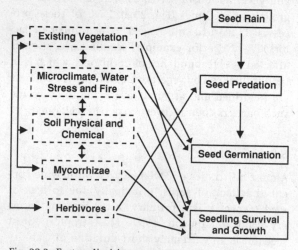

Fig. 23.2. Factors limiting pasture recovery. Processes that influence seedling establishment are in boxes with solid lines. Factors affecting the magnitude of these processes are in boxes with dashed lines. Modified from Holl *et al.* (2000).

for any length of time, roots of prior vegetation have been destroyed so there is no opportunity for resprouting. Therefore, all seeds of forest plant species must be recently dispersed into agricultural land.

The vast majority of TMF plant species are animal-dispersed (Howe, 1984) and many birds, bats and other mammals will not venture into disturbed areas (Wunderle, 1997). As a result, the numerous studies of seed rain in TMF world-wide suggest that seed rain declines rapidly within a few metres of the forest edge, and that most seeds falling in pasture land are from species already present in the pasture or from a few small-seeded pioneer species (Nepstad *et al.*, 1990; Aide & Cavelier, 1994; Hardwick *et al.*, 1997; Chapman & Chapman, 1999; Holl, 1999; Medellín & Gaona, 1999; Aide *et al.*, 2000; Zimmerman *et al.*, 2000). Bird activity and seed rain are generally concentrated under remnant trees or shrub patches (Guevara *et al.*, 1986; Kolb,

1993; Cardoso da Silva *et al.*, 1996; Nepstad *et al.*, 1996; Holl *et al.*, 2000). In summary, few forest seeds arrive in the pasture and those that do are highly patchily distributed.

Seeds of forest species that do arrive in the pasture are often subjected to high rates of predation (Uhl, 1987; Nepstad *et al.*, 1990; Aide & Cavelier, 1994; Osunkoya, 1994; Holl & Lulow, 1997; Chapman & Chapman, 1999). For example, in a seed predation study in Brazil, Nepstad *et al.* (1990) reported that all seeds of one tree species placed in the pasture were removed within 24 hours, while more than 80% of eight species were removed within 50 days. Similarly, a study in Costa Rica found that approximately two-thirds of all seeds were predated within 30 days (Holl & Lulow, 1997). Common seed predators include small mammals, ants and bruchid beetles. Rates of seed predation are quite variable between species, which affects patterns of forest recovery (Nepstad *et al.*, 1990; Aide & Cavelier, 1994; Holl & Lulow, 1997).

Seed germination

For seedlings of forest species to establish, seeds dispersed into the pasture must be able to germinate. Most studies of tropical forest recovery in abandoned pastures have shown a wide range of seed germination rates, with very high germination of some species and no germination of others (Aide & Cavelier, 1994; González Montagut, 1996; Hardwick *et al.*, 1997; Holl, 1999). A number of studies have also indicated that germination of some species is lower in areas cleared of pasture grasses where microclimatic conditions are more stressful (Aide & Cavelier, 1994; González Montagut, 1996; Hardwick *et al.*, 1997; Holl, 1999; Zimmerman *et al.*, 2000). Given the high variability in germination rates it is difficult to draw general conclusions about the importance of rates of seed germination in limiting recovery. Studies comparing germination rates in pasture and intact forest have shown that germination can be much lower in pasture for some species, but that lack of germination is probably not the primary factor limiting recovery for most species (González Montagut, 1996; Holl, 1999; Zimmerman *et al.*, 2000).

Competition with existing vegetation

Clearly, seedling establishment in abandoned pastures is limited foremost by lack of dispersal, and secondarily by high seed predation and low germination rates in some species. If seedlings do become established, a suite of abiotic and biotic factors including competition with aggressive existing vegetation, stressful microclimatic conditions, fire, lack of soil nutrients, reduced mycorrhizal inoculum and herbivory may reduce seedling survival and growth.

A major factor that has been repeatedly shown to limit survival and growth of forest species, particularly in agricultural land, is aggressive existing vegetation that was either planted or rapidly colonised after the land was abandoned. Exotic pasture grasses (such as *Axonopus scoparius*, *Brachiaria* spp., *Melinus minutiflora*, *Panicum* spp. and *Paspalum* spp.) often form a monoculture in previously grazed areas (Nepstad *et al.*, 1990; Guariguata *et al.*, 1995; Chapman & Chapman, 1999; Holl, 1999) (Fig. 23.3). In other cases dense ferns compete with forest seedlings (Aide *et al.*, 1995; Ashton *et al.*, 1997; Rivera *et al.*, 2000). This residual vegetation may limit recovery in a number of ways, including providing shelter for seed and seedling predators, such as rodents and leaf-cutter ants; competing for soil moisture, nutrients and light; increasing the probability of fire which kills tree seedlings; and emitting allelopathic chemicals which limit seedling growth (Nepstad *et al.*, 1990; Chapman & Chapman, 1999; Holl, 1999). For example, Holl *et al.* (2000) found that species richness and cover of broadleaved species were five times higher in areas cleared of pasture grasses compared to areas with dense pasture grasses. Sun & Dickinson (1996) found higher survival and growth of tree seedlings when pasture grasses were cleared in tropical forest in Australia.

In contrast, Aide & Cavelier (1994) suggest that grasses may actually facilitate recovery on severely degraded sites by reducing loss of soil moisture. Their research in Columbia showed higher seed germination in plots with grass than those that had been burned or cleared, while seedling growth was similar in all plots. Chapman *et al.* (in press) found that during a severe dry-season seedling

Fig. 23.3. Author standing in front of *Axonopus scoparius* to demonstrate height of African pasture grasses that compete with forest seedlings.

air and soil temperatures are commonly elevated, and humidity and soil moisture levels are reduced in agricultural clearings compared to forests (Chazdon & Fetcher, 1984; Kapos, 1989; Williams-Linera, 1990; Williams-Linera *et al.*, 1998). While exotic pasture grasses are often adapted to high light and low soil moisture levels, stressful microclimatic conditions may negatively affect seed germination, and seedling growth and survival of colonising woody species. Nepstad *et al.* (1996) found xylem pressure potential, a measure of water stress, was two to five times lower for seedlings planted in pasture compared to tree-fall gaps during the dry season in the Brazilian Amazon. Loik & Holl (1999) demonstrated that light intensity in open pasture is in excess of that needed for maximal photosynthesis, which can be stressful for seedlings of certain species once they reach a sufficient height to overtop the grasses.

In areas with distinct dry seasons, fire can also be critical in limiting seedling survival. Fires are becoming increasingly common, as rainfall decreases due to regional deforestation, and increased solar radiation in early second-growth areas reduces vegetation moisture content (Cochrane *et al.*, 1999; Nepstad *et al.*, 1999). Most tropical forest tree seedlings are not adapted to fire, so fire can set back recovery a number of years. In contrast, most pasture grasses quickly regrow from surviving roots.

Soil nutrients and microbial communities

In some cases, lack of nutrients appears to limit seedling growth in abandoned lands in the tropics with phosphorus the most commonly limiting (Vitousek & Sanford, 1986). Conclusions about whether or not nutrients limit recovery are variable among studies. For example, results of some reciprocal transplant and fertilisation studies suggest that seedlings are nutrient limited in agricultural land (Aide & Cavelier, 1994; Vieira *et al.*, 1994; Holl *et al.*, 2000), whereas other research suggests that nutrients do not limit seedling growth (Nepstad *et al.*, 1990; Nepstad *et al.*, 1996; Holl, 1999).

These varied conclusions probably reflect differences in soil types and plant adaptations. Much of the tropics are covered by oxisols and ultisols, which

death was higher in weeded compared to unweeded second-growth areas in Uganda. As mentioned previously, the higher humidity and soil moisture below grasses, compared to pasture without grass, may be more favourable for seed germination. In summary, existing vegetation may have both positive and negative effects on forest recovery depending on the aggressiveness of the species, the severity of disturbance, and the seasonality of the ecosystem.

Microclimate and fire

Stressful microclimatic conditions may also limit seedling survival and growth. Light levels and have low nutrient levels and high acidity. However, some areas have more fertile, volcanic soils, such as andisols and inceptisols. In many cases nutrient availability in mature forests is limited so plants have adaptations to low nutrient conditions. Conflicting results regarding nutrient limitation also reflect differences in disturbance histories and nutrient measurement techniques.

Janos (1980) suggested that tropical forest recovery may be limited by lack of mycorrhizae, as many colonising plants are facultatively mycorrhizal while mature forest species are often obligately mycorrhizal. Some studies have reported similar or higher levels of mycorrhizal spores or mycorrhizal infection of roots in tropical pastures compared to forests (Fischer et al., 1994; Johnson & Wedin, 1997; Allen et al., 1998). Other studies, however, have found different species composition of mycorrhizal spores in highly degraded pastures compared to primary or secondary forest (Allen et al., 1998; F. L. Carpenter, personal communication). However, there have been few field studies testing this conclusion and there is an immense need for research on belowground processes and their effect on forest recovery (Holl & Kappelle, 1999).

Herbivores

A final factor that may reduce seedling survival and growth is herbivory. While reports of extensive herbivory are limited, at some sites herbivory may be a major factor limiting recovery. Nepstad et al. (1990) reported that 30%–80% of seedlings of four tree species planted in abandoned pasture were defoliated by leaf-cutter ants (Atta sexdens) after 16 days. Holl & Quiros-Nietzen (1999) found, in a trial using native tree species for reforestation, that 64% of seedlings (0.1–0.5 m tall) were cut by rabbits (Sylvilagus spp.) within two years. There is no question that herbivory plays a major role in tropical forests, as demonstrated by the complicated antiherbivore mechanisms that have evolved in many tropical plants (Coley, 1996), but it is difficult to generalise the importance of herbivory in tropical forest restoration given the limited number of studies.

Social factors

As discussed in the previous sections, there are numerous ecological obstacles to TMF recovery. Any discussion of TMF restoration would not be complete, however, without mentioning a number of other obstacles to recovery. A detailed discussion of social and political factors is beyond the scope of this chapter, but two deserve special attention. First, the predominant obstacles to most restoration projects are cost and competing land uses (Holl & Howarth, 2000). While restored lands may provide long-term income from logging or other extractive activities, short-term income from cash crops, cattle or subsistence agriculture are lost. For this reason, much research has focused on agroforestry systems that integrate multiple land uses in an effort to balance immediate human needs and the maintenance of ecosystem services. A shift to more sustainable land uses will require a change in the current subsidy system which favours large landowners in many countries (Grainger, 1993). For example, the Costa Rican government has passed legislation that reduces logging taxes for landowners who subsequently reforest with native tree species (Butterfield & Fisher, 1994).

Second, while population growth rates have decreased in most countries in recent years, they still remain well above replacement rate in many countries. It is estimated that an additional 1 billion people will be added to the global population in the next 14 years alone, most of whom will reside in the developing world where most tropical forest is located (United Nations, 1999). As long as the population continues to grow rapidly, restoration efforts will increasingly compete with the need for additional land for human settlements and agriculture. I hasten to note that changes in land cover are caused by a complex set of factors including not only population size, but also individual consumption levels and patterns and technology (Grainger, 1993). Many of the resources extracted from tropical forest are consumed in other countries. For instance, most of the wood from tropical forests is exported to Japan, Europe and the United States (Barbier, 1998). Population control, changes in consumption patterns and strategies to meet basic needs of tropical peoples are essential components of tropical forest conservation and restoration efforts.

STRATEGIES FOR ACCELERATING RECOVERY

Because tropical forest recovery may be slow in highly degraded areas and tropical forests are of critical conservation concern, humans are increasingly intervening in an effort to accelerate forest recovery. Few results of restoration projects are available in the scientific literature, however, for a number of reasons. First, most tropical forest restoration efforts have been initiated in the past five to ten years, so long-term data are not available. Second, restoration tends to be a trial-and-error process and negative results are less likely to be published. Third, many efforts have been initiated by small, grass-roots groups that often do not widely disseminate their results. This discussion of TMF restoration draws on the small, though growing, number of published studies, as well as the author's research in southern Costa Rica.

Tropical forests host such a large number of species that reintroducing even a small proportion of the total species would be time and cost prohibitive. Therefore, efforts to restore TMF must necessarily focus on strategies for facilitating succession rather than trying to either seed or plant a wide variety of species, as is often attempted in temperate systems. Clearly, reintroduction of certain species of conservation concern, such as mahogany (*Swietenia macrophylla*) may be necessary. As discussed previously multiple factors retard recovery in abandoned tropical pastures. Some problems seem to be widespread (e.g. lack of seed dispersal and high seed predation), whereas other tend to be more site-specific (e.g. water stress and herbivory). Given this fact, the most effective restoration strategies will simultaneously overcome a range of obstacles to regeneration.

Planting native tree seedlings

The most commonly used strategy for accelerating tropical forest succession is planting seedlings of a few native forest tree species that are fast-growing, drought-resistant, and able to grow in low soil nutrient levels. Planting seedlings can result in higher understorey diversity (Parrotta, 1992) and improve soil structure and soil nutrient availability (Prinsley, 1991; Parrotta, 1992; Montagnini et al., 1995). Trees also serve to ameliorate extremes in temperature and soil moisture in the understorey. Ultimately, the planted saplings will provide the canopy architecture to encourage perching and seed dispersal by birds. Research by Parrotta (1995) suggests that overstorey composition resulting from planted seedlings has a large influence on understorey colonisation rates.

Until recently most tropical areas were planted with just three genera: *Pinus*, *Eucalyptus* and *Tectona* (Evans, 1992). Increasingly, projects are being initiated to test a wide variety of native tree species for their ability to grow rapidly in cleared areas (e.g., Alfaro Bonilla & Barrantes Arias, 1995; Butterfield, 1995; Bruenig, 1996; Meguro & Miyawaki, 1997; Haggar et al., 1998). Most often, seeds are collected and seedlings are raised in a nursery for three months to one year prior to planting. Some authors have suggested that direct seeding may be a viable option (Nepstad et al., 1990; Parrotta & Knowles, 1999), but most research suggests that predation rates are sufficiently high to preclude direct seeding of forest species (Holl & Lulow, 1997; Chapman & Chapman, 1999).

Starting and maintaining seedlings in the tropics can be challenging. First, seed collection can be difficult, as many TMF trees do not set seed every year (Janzen, 1978) and some species must be collected directly from the tree to ensure high germination rates. Second, little is known about the germination requirements for most tropical species, although this is improving with an increasing number of multi-species trials (Box 23.2). Third, many tropical forest seeds rapidly lose viability when dried, making storage impossible. Fourth, in areas where pasture grasses are dense, it is necessary to clear grasses from around planted seedlings with a machete every few months for a year or two, to prevent the grasses from shading out the seedlings. Fifth, as discussed previously, seedling herbivory can be high.

Despite these obstacles, a number of studies have demonstrated that some native species show

Box 23.2   Screening tropical forest trees

Replanting with native tree species is one of the most promising strategies for accelerating tropical moist forest recovery. These trees serve as perching structures to enhance seed dispersal, increase nutrient inputs from leaf litter, shade out pasture grasses, and provide shade for young tree seedlings. Until recently the vast majority of reforestation in the tropics was done using non-native tree species, primarily the genera *Pinus*, *Eucalyptus* and *Tectona*. Why are these species so widely used when there are thousands of species of rainforest trees? First, propagation methods are well documented for these commonly used exotic species. Second, they grow extremely quickly, whereas rainforest trees show a range of growth rates. In the past five to ten years there has been an increasing number of studies screening many native tree species for reforestation to select those that are easiest to propagate and that produce marketable wood the most rapidly. For example in Costa Rica, a country with more than 1400 native tree species, The Organization for Tropical Studies and the Costa Rican forest service with funding from the MacArthur Foundation and the US Agency for International Development have been screening over 100 native tree species for reforestation in different regions of Costa Rica (Butterfield, 1995; Haggar *et al.*, 1998; J. Calvo, personal communication). Seedling growth rates were measured for a large number of species and those that appeared to be most

promising were planted at multiple sites in a region to test their utility over a wide area (Haggar *et al.*, 1998; J. Calvo, personal communication). Different combinations of native species were planted in combination with a couple of exotic species to compare growth rates with each mixture planted at a minimum of four sites. The programme is also working to identify seed trees with high productive potential. Most of the research sites are on privately owned lands and project staff do outreach with farmers through newsletters and workshops. Results in the northern half of Costa Rica, after three to six years of large-scale screening tests, indicate that the species with the fastest growth rates were the exotic species *Gmelina arborea* and *Acacia magnum*, but both were susceptible to fungal and insect outbreaks; other commonly planted exotics, such as *Eucalyptus* spp. and *Pinus* spp. showed poor growth and survival rates (Haggar *et al.*, 1998). Some native species, such as *Goethalsia meintha*, *Ochroma pyramidale* and *Vochysia* spp., had growth rates only slightly below the fastest growing exotics and show good potential for forestry. While the focus of this study was identifying species for reforestation for plantations, information regarding propagation methods, appropriate habitat, and growth rates of a range of native species is critical for selecting species for restoring logged areas and abandoned agricultural land. It is also essential to work with small farmers in the process to facilitate information transfer.

growth rates in disturbed areas in the tropics similar to those of more commonly used exotic species (Kartawinata, 1994; Alfaro Bonilla & Barantes Arias, 1995; Butterfield, 1995; Ashton *et al.*, 1997; Stanley & Montagnini, 1999), and may serve to facilitate forest regeneration (Fig. 23.4). Many species have survival rates in excess of 80% and grow as much as 2.5 m per year. However, other studies have reported less favourable results (e.g. Uhl, 1987; Nepstad *et al.*, 1996; Holl *et al.*, 2000). Survival rates may be largely affected by the specific site conditions, such as slope, aspect and soil nutrients (Butterfield, 1995) and by the specific mother tree from which seeds were collected (F. L. Carpenter, personal communication). In seasonal TMF forest,

seedlings may not be able to survive extreme microclimatic conditions. Uhl (1987) reported that all seedlings planted in a recently abandoned pasture died with one year, while 90% of seedlings planted under the vegetation of a five-year-old abandoned farm survived for 4.5 years. In some cases efforts may be taken to overcome site-specific obstacles, such as fertilising where nutrients are low, fencing to protect seedlings from mammalian herbivory (Holl *et al.*, 2000), or using fungicides on leaf-cutter ant colonies to control ant herbivory (Nepstad *et al.*, 1990). Such efforts necessarily increase costs and can also have negative side-effects. For example, widespread use of fertilisers may inhibit succession by encouraging the establishment

Fig. 23.4. Ten-year-old stand of native tree seedlings in montane forest in Xalapa, Mexico. G. Williams-Linera (foreground) is monitoring their growth rates.

and growth of weedy species (Harcombe, 1977; Uhl, 1987).

One strategy to bypass problems with sporadic seed set, low seed germination, and high herbivory on young seedlings is to plant pieces of tree branches or trunks, referred to as 'stakes'. Stakes have been used as 'living fence posts' for many years, as they often resprout, providing a more permanent structure to which barbed wire is attached. Increasingly, the potential of this practice for restoration is being recognised and the ability of a wide variety of native tree species to resprout from stems is being tested (Aide *et al.*, 1995; Granzow de la Cerda & Garth, 1999*a*). For example, in northeast Nicaragua more than 20 species have been tested with mixed success (Granzow de la Cerda & Garth, 1999*a*); initial trials showed that sapling trunks, but not branches resprouted when cut and planted. In subsequent experiments, eight native species sprouted from trunk stakes of saplings. A second technique, air rooting, has proven successful for the establishment of 15 species (Granzow de la Cerda & Garth, 1999*b*). For air rooting, the cortex of a sapling is cut, treated with growth hormones, and wrapped with soil in plastic. After the sapling has developed roots at the site of the cut, the section is separated and planted.

More research is needed to determine whether these techniques are widely applicable. These techniques remove costs of seed collection, nursery maintenance, and grass clearing, and minimise seedling losses due to herbivory. However, they require the presence of a nearby source forest, which is often not available. Furthermore, the effect of extensive removal of saplings from existing forest must be considered.

In summary, a range of propagation methods is available for native tree seedlings. In most regions, planting of a carefully selected group of native tree species seems to be an efficient strategy for facilitating forest recovery as, once established, these seedlings overcome a number of obstacles to recovery.

Planting non-native tree seedlings as nurse trees

Clearly, planting native tree seedlings to accelerate tropical forest recovery is a desirable strategy. But, in some regions, the ability to reforest with native species may be limited by lack of nurseries

Box 23.3 Tropical moist forest recovery and restoration in Uganda

Much of our knowledge of tropical forest recovery and restoration stems from studies in the neotropics, despite high rates of deforestation in both Asia and Africa. For example, <10% of original tropical forest remains in western Africa and Madagascar. The remaining rainforest in Central Africa is being logged at a rapid rate, and past research shows that forest recovery in some logged and agricultural lands in this region is slow (Chapman & Chapman, 1997; Duncan & Chapman, 1999). Therefore, understanding natural recovery and developing restoration strategies in this ecosystem is particularly critical. Chapman and colleagues have been studying forest recovery in abandoned agricultural lands and felled exotic tree plantations in Kibale National Park in western Uganda (Chapman et al., 1997; Duncan & Chapman, 1999; Zanne & Chapman, 2001; Chapman et al., in press; R. S. Duncan, unpublished data). The forest is categorised as moist evergreen forest with an average annual rainfall of 1700 mm, and is located at an elevation of 1500 m. The diversity of trees is relatively low by tropical forest standards (68 species in 4.8 ha of vegetation sampling: Chapman et al., 1997), but hosts a wide variety of mammal and bird species, such as chimpanzees, monkeys and hornbills.

Research suggests that a number of factors limit recovery, primarily lack of seed dispersal, but also high seed predation, competition with pasture grass (Pennisetum purpureum) and herbs (Acanthus pubescens), periodic fire, and seedling damage by elephants (Chapman & Chapman, 1997; Duncan & Chapman, 1999). As has been found in a number of studies in the neotropics, isolated trees and shrubs serve as foci for seed dispersal by bats and birds in abandoned agricultural lands (Duncan & Chapman, 1999). Two sets of strategies have been tested for facilitating recovery: (1) seeding and planting in abandoned pasture and (2) felling exotic tree plantations and weeding. Direct seeding did not enhance seedling establishment, presumably due to high rodent predation of seeds. Cuttings of early-successional tree species were planted to serve as foci for seed dispersal, but were overtopped by grass after the third year. A more promising strategy appears to be felling exotic tree plantations. Pines (Pinus spp.) and cypress (Cupressus lusitanica) were planted in the early 1900s and have recently been felled to provide wood. Establishment of native forest tree seedlings is higher in areas where exotic tree species plantations were felled than in areas where the plantations were left intact. In particular, recolonisation of forest tree species was higher when snags were left in plantations to encourage seed dispersal by birds (S. R. Duncan, unpublished data). In contrast to most previous studies, weeding vines and herbaceous species over a three-year period in recently felled plantations did not increase the number of seedlings present or average seedling size (Chapman et al., in press). Results from these studies in Uganda indicate that responses to different treatments are highly species-specific, a trend reported at many tropical sites. This means that tropical forest restoration will be challenging, particularly given the large number of species and high level of endemism.

producing native tree seedlings, and slow growth and low survival of native tree species. Therefore, non-native tree plantations have been one strategy suggested to facilitate tropical forest recovery (Ashton et al., 1997; Lugo, 1997; Keenan et al., 1997). Tree plantations may help to shade out aggressive pasture grasses, increase nutrient levels and enhance seed dispersal, while also providing a source of income to landowners. For example, Keenan et al. (1997) found that forest plantations in Australian rainforest facilitate establishment of native species.

Ashton et al. (1997) demonstrated high seedling growth of some native tree species in plantations of Pinus caribea in Sri Lanka that had been thinned. Parrotta & Knowles (1999) compared vegetation composition in non-native commercial tree plantings and mixed native species plantings; not surprisingly, they reported higher biomass in the commerical plantations, but higher species diversity in mixed native species plantings. It is important that the use of non-native trees as nurse species be considered carefully with regard to their

aggressiveness, potential to spread and potential to alter soil chemistry. But, given the critical need for economically viable tropical forest restoration strategies, there are times when planting exotic trees as nurse crops could be a useful strategy.

Remnant trees and planting patches of trees

Much previous work has focused on the critical role remnant pasture trees play in natural forest recovery by increasing seed dispersal, ameliorating microclimatic conditions and increasing soil nutrients (Janzen, 1988; Guevara & Laborde, 1993; Rhoades *et al.*, 1998; Otero-Arnaiz *et al.*, 1999; Duncan & Chapman, 1999; Toh *et al.*, 1999). For example, Loik & Holl (1999) found much higher growth rates of forest tree seedlings below remnant pasture trees (Fig. 23.5), which was probably due to a combination of intermediate light levels and increased nutrients. Likewise, strips of trees along riparian corridors (González Montagut, 1996; Samuels, 1998) and planted as windbreaks (Harvey, 2000) can serve as important corridors for animal movement and seed dispersal, as well as increase seedling establishment. Therefore, leaving some seed trees in areas that are logged, and planting or maintaining trees in agricultural lands should be encouraged both to

Fig. 23.5. One-year old seedlings of *Ocotea glaucosericea* planted in open pasture (above) and below a remnant pasture tree, *Inga* sp. (below). Note rubber boot for scale.

improve the quality of the habitat while the land is used for agriculture and to facilitate recovery if the land is abandoned.

The important role of isolated trees and patches of trees for facilitating seed dispersal and seedling establishment suggests that planting patches of trees may be a cost-efficient method for facilitating recovery. In restoration efforts, large areas are commonly planted with seedlings, but increasingly large areas of tropical forest will need to be restored with limited funds. A number of researchers have documented the pattern of nucleation (Yarranton & Morrison, 1974) in tropical pastures (Kolb, 1993; Vieira *et al.*, 1994; Holl *et al.*, 2000); small patches of trees and shrubs rapidly spread in abandoned pastures. Therefore, planting in patches may not only be more cost-efficient, but may also provide a level of spatial diversity characteristic of the ecosystem. It may also be helpful to plant in the vicinity of remnant patches of trees to facilitate animal movement. More research is needed on the effect of landscape pattern on forest recovery.

Seeding shrubs

Increasingly, it is being recognised that naturally colonising shrubs may play a critical role in ameliorating adverse conditions and facilitating succession in abandoned tropical pastures (Vieira *et al.*, 1994; Aide *et al.*, 1995; Holl *et al.*, 2000). Many shrub species attract birds, thereby increasing dispersal of forest seeds (Willson & Crome, 1989; Saab & Petit, 1992; Vieira *et al.*, 1994; Cardoso da Silva *et al.*, 1996; Nepstad *et al.*, 1996). As with tree seedlings, shrubs may help to overcome other obstacles to recovery, for example by ameliorating stressful microclimatic conditions, increasing soil nutrients and shading out pasture grasses (Vieira *et al.*, 1994; Holl, 1998a, unpublished data). In some cases, however, aggressive shrubs may stall recovery (Zahawi & Augspurger, 1999).

Seeding early-successional shrubs may be an inexpensive strategy to accelerate recovery in regions where shrubs facilitate tree seedling establishment, as many shrubs, unlike most tree species, produce copious seeds year round and are easily collected. Therefore, sufficient seeds could be direct seeded to overcome high predation rates. This strategy

has received little testing, probably because shrubs have little economic value. One short-term study has shown that seeding shrubs along with clearing grasses at the time of seeding can increase establishment of shrubs (Holl *et al.*, 2000), but it is impossible to draw any conclusions without more research on the short- and long-term effects of seeding shrubs.

Artificial perching structures, slash piles and logs

Artificial bird perching structures and slash piles have been suggested as a means for facilitating recovery, particularly in pastures lacking remnant trees. Artificial structures in open areas would appear to have particular potential for use in tropical forests where the seeds of 50%–90% of canopy trees and nearly 100% of shrubs and subcanopy trees have adaptations for animal dispersal (Howe, 1984). Increased vertical structure may serve to attract birds farther into the pasture, thereby enhancing seed dispersal.

A number of recent studies have investigated the use of bird perching structures in tropical forests and have shown that their use in accelerating recovery is limited (Aide & Cavelier, 1994; Ferguson, 1995; Press, 1995; Holl, 1998b; Miriti, 1998). All these studies have used posts or branches ranging from 2 to 5 m height with a perching structure (crossbar or platform) on the top (Fig. 23.6). In all studies, a variety of fruit-eating birds were observed on perches and seed rain under perches was elevated over that in open pasture, but well below that in forest and under remnant trees. Most seeds falling below perches were of pasture or pioneer species, which is probably due to the fact that bats have been observed to be more important dispersers of larger-seeded species in open areas in the tropics (Medellín & Gaona, 1999). The few studies that have measured seedling establishment below perches, however, have not found higher seedling establishment unless pasture grasses were cleared (Ferguson, 1995; Miriti, 1998; Holl, 1998b). Combining the use of bird perching structures with efforts to reduce pasture grasses through clipping or herbicides may help to facilitate recovery (Ferguson, 1995; Miriti, 1998), but seems unlikely to be a practical tropical forest restoration strategy over a large scale. It would

Fig. 23.6. Crossbar perch with seed traps, to collect seeds for seed rain measurements, below at forest–pasture edge.

Clearing existing vegetation

As discussed previously, dense cover of exotic pasture grasses and other early successional vegetation may impede forest recovery and, therefore, reducing their cover often enhances forest recovery. Nepstad *et al.* (1990) recommend burning grasses to facilitate forest recovery, as burning serves to reduce grass competition and increase soil nutrient availability. While burning may facilitate initial establishment of woody plant species, it comprises a large risk to nearby ecosystems and human communities. Even with firebreaks, fires may rapidly spread beyond the desired area. Grasses may also be killed using herbicides, but hardy grasses often require repeated treatments with pesticides to eradicate all grasses. In many areas herbicides may not be available. The use of herbicides is cost-prohibitive on a large scale and may cause extensive soil and water pollution. Posada *et al.* (2000) found that introducing grazing animals at low density after reseeding reduced grass biomass and increased seedling growth. Reintroduction of animals comprises less risk than fire or herbicides but must be carefully timed.

Clearing all vegetative cover may have negative side-effects. For example, it may greatly increase erosion. As discussed earlier, grasses may also moderate the microclimate, so removing them without establishing other vegetation may provide stressful conditions for seedling establishment. Therefore, establishing less-aggressive vegetation that shades out grasses has generally proven to be the most economically and ecologically effective strategy. Initial clearing of vegetation to facilitate establishment or during the first year of seedling growth may be important. Reducing cover of these aggressive ground-cover species is critical to forest seedling establishment in many abandoned tropical lands.

necessarily be labour-intensive as most grasses regrow rapidly after clearing.

A number of studies have demonstrated that piles of branches and logs remaining after forest felling in agricultural areas reduce light levels and temperatures at the soil surface and provide safe sites for woody seedling establishment (Uhl *et al.*, 1981; Holl *et al.*, 2000; Peterson & Haines, 2000; Slocum, 2000). They also serve as perching structures and shelter for a number of bird species. Artificially created piles of logs resulted in higher woody seedling establishment during the first year following pasture abandonment in Venezuela (Uhl *et al.*, 1982). Although this strategy for facilitating recovery has not been widely tested, it seems more promising than perching structures, as the logs and slash piles enhance both seed dispersal and seedling establishment.

Fire prevention

While burning an area at the time of abandonment serves to facilitate recovery, fire after abandonment kills most woody seedlings, thereby setting back the process of succession (Uhl *et al.*, 1988). Human activities have increased both the frequency and intensity of fires in tropical forests in Latin America, and many tropical forest plant species are not adapted to fire (Uhl *et al.*, 1990; Nepstad *et al.*, 1991). Therefore, an essential component of any tropical forest restoration effort in areas with extended dry seasons, is fire prevention, by patrolling areas susceptible to burns and educating landowners of the risk of burning at dry times of the year. For example, significant resources are devoted to prevention of the outbreak and spread of fires in dry forest restoration in northern Costa Rica (Janzen, 1988, this volume).

CONCLUDING REMARKS

It is clear that a number of factors may limit forest recovery in disturbed tropical land. The relative importance of these factors depends on the original ecosystem, the history of disturbance and the landscape pattern. Lack of forest seeds appears to be the overriding factor in most cases. If few forest seeds are available, then grass competition, water stress and lack of soil nutrients limiting vegetation survival and growth become moot questions.

Past research has demonstrated a number of potential methods for facilitating succession. The most promising strategy appears to be planting native tree seedlings, as they overcome a number of different obstacles to recovery simultaneously and may also provide income to the landowner through selective logging. As discussed previously, research is desperately needed on the suitability of specific native tree species and other strategies for facilitating recovery. Bird perching structures, while an interesting strategy, seem to be a less effective strategy, as they do not have the potential to ameliorate microclimatic conditions or increase soil nutrients. Slash piles may serve as a relatively cheap strategy to facilitate recovery, though more long-term research is necessary. Reducing existing vegetation cover is an important component of recovery at many sites. Regardless of the strategy best suited to an area, this information must be made widely available. This research is often published in English for scientific purposes, but should also be made available in the language of the country and distributed through local information channels.

Although the number of studies on tropical forest restoration is increasing, our knowledge in this field is still minuscule. Most of the research thus far has been focused on establishment of plant species. As noted earlier, there has been little study of belowground processes. There has also been little research on how restoration efforts affect faunal communities, which is critical given the high diversity of insect, bird, mammals and amphibian communities in these regions and their close co-evolutionary relationships. There is also limited information on how successful tropical forest restoration will be in restoring ecosystem processes such as nutrient and hydrological cycling.

One obstacle to developing tropical forest restoration strategies is the high variability of the system. Because endemism is extremely high in most tropical forests and soil, altitudinal and climate gradients vary over small distances, the tree species suitable for reforestation may vary considerably over a small spatial scale. Factors limiting recovery may vary over similarly small distances. Therefore, results of studies on restoration strategies may have very localised applications. It is important to compare studies to determine what results can be generalised and which are likely to be more site-specific (Holl & Kappelle, 1999). This site-specificity highlights the importance of undertaking small-scale experiments to test strategies at particular sites.

Restoring TMFs will clearly be difficult. It is too early to know whether we can truly restore tropical forests to any semblance of pre-disturbance condition. It is unlikely that the numerous species present before disturbance will all recolonise. Given the complexity of these systems and our lack of knowledge of their resilience, we should view tropical forest restoration as a necessary activity given existing land uses, and must simultaneously focus efforts on conserving existing forest.

The rapid rate of forest destruction combined with the incredible complexity of TMF make the task of restoring these ecosystems appear overwhelming. But some changes give reason to be optimistic. First, a number of countries are expanding their efforts to protect and inventory minimally disturbed forest (Janzen, 1995). Second, with more environmental education efforts worldwide, people are becoming increasingly aware of the critical role tropical forests play in maintaining human wellbeing. Finally, as a result of this increased awareness, more projects are being initiated on both small and large scales to try to restore the structure and function of these ecosystems and, more often than previously, efforts are being made to disseminate the results widely. In some countries, government incentives are being offered to encourage TMF conservation and restoration efforts.

REFERENCES

Aide, T. M. & Cavelier, J. (1994). Barriers to lowland tropical forest restoration in the Sierra Nevada de Santa Marta, Colombia. *Restoration Ecology*, **2**, 219–229.

Aide, T. M., Zimmerman, J. K., Herrera, L., Rosario, M. & Serrano, M. (1995). Forest recovery in abandoned tropical pasture in Puerto Rico. *Forest Ecology and Management*, **77**, 77–86.

Aide, T. M., Zimmerman, J. K., Pascarella, J. B., Rivera, L. & Marcano-Vega, H. (2000). Forest regeneration in a chronosequence of tropical abandoned pastures: implications for restoration ecology. *Restoration Ecology*, **8**, 328–338.

Alfaro Bonilla, C. M. & Barrantes Arias, P. (1995). Estudio de adaptabilidad preliminar de especies de altura en la zona sur de Costa Rica. BSc dissertation, Instituto Tecnológico de Costa Rica.

Allen, E. B., Rincón, E., Allen, M. F., Pérez-Jimenez, A. & Huante, P. (1998). Disturbance and seasonal dynamics of mycorrhizae in a tropical deciduous forest in Mexico. *Biotropica*, **30**, 261–274.

Ashton, P. M. S., Gamage, S., Gunatilleke, I. A. U. N. & Gunatilleke, C. V. S. (1997). Restoration of Sri Lankan rainforest: using Caribbean pine *Pinus caribaea* as a nurse for establishing late-successional tree species. *Journal of Applied Ecology*, **34**, 915–925.

Bawa, K. S. & Dayanandan, S. (1998). Causes of tropical deforestation and institutional constraints to conservation. In *Tropical Rain Forest: A Wider Perspective*, ed. F. B. Goldsmith, pp. 175–198. London: Chapman & Hall.

Barbier, E. B. (1998). The economics of the tropical timber trade and sustainable forest management. In *Tropical Rain Forest: A Wider Perspective*, ed. F. B. Goldsmith, pp. 199–254. London: Chapman & Hall.

Bruenig, E. F. (1996). *Conservation and Management of Tropical Rainforests*. Wallingford, UK: CAB International.

Buschbacher, R., Uhl, C. & Serrao, E. A. S. (1988). Abandoned pastures in eastern Amazonia. 2: Nutrient stocks in the soil and vegetation. *Journal of Ecology*, **76**, 682–699.

Butterfield, R. P. (1995). Promoting diversity: advances in evaluating native species for reforestation. *Forest Ecology and Management*, **75**, 111–121.

Butterfield, R. P. & Fisher, R. F. (1994). Untapped potential for native species reforestation. *Journal of Forestry*, **92**(6), 37–40.

Cardoso da Silva, J. M., Uhl, C. & Murray, G. (1996). Plant succession, landscape management, and the ecology of frugivorous birds in abandoned Amazonian pastures. *Conservation Biology*, **10**, 491–503.

Chapman, C. A. & Chapman, L. J. (1997). Forest regeneration in logged and unlogged forests of Kibale National Park, Uganda. *Biotropica*, **29**, 396–412.

Chapman, C. A. & Chapman, L. J. (1999). Forest restoration in abandoned agricultural land: a case study from east Africa. *Conservation Biology*, **13**, 1301–1311.

Chapman, C. A., Chapman, L. J., Wrangham, R. W., Isabirye-Basuta, G. & Ben-David, K. (1997). Spatial and temporal variability in the structure of a tropical forest. *African Journal of Ecology*, **35**, 287–302.

Chapman, C. A., Chapman, L. J., Zanne, A. & Burgess, M. A. (in press). Does weeding promote regeneration of an indigenous tree community in felled pine plantations in Uganda? *Restoration Ecology*.

Chazdon, R. L. & Fetcher, N. (1984). Light environments of tropical forest. In *Physiological Ecology of Plants of the Wet Tropics*, eds. E. Medina, H. A. Mooney & C. Vázquez-Yánez, pp. 27–36. The Hague: Dr W. Junk.

Cochrane, M. A., Alencar, A., Schulze, M.D., Souza, C. M., Nepstad, D.C., Lefebvre, P. & Davidson, E. A. (1999). Positive feedbacks in the fire dynamic of closed canopy tropical forests. *Science*, **284**, 1832–1835.

Coley, P. D. (1996). Herbivory and plant defenses in tropical forests. *Annual Review of Ecology and Systematics*, **27**, 305–335.

Duncan, S. R. & Chapman, C. A. (1999). Seed dispersal and potential forest succession in abandoned agriculture in tropical Africa. *Ecological Applications*, **9**, 998–1008.

Evans, J. (1992). *Plantation Forestry in the Tropics*, 2nd edn. Oxford: Clarendon Press.

Ferguson, B. G. (1995). Overcoming barriers to forest regeneration in a degraded tropical pasture: an evaluation of restoration techniques. MSc dissertation, University of Michigan.

Finegan, B. & Delgado, D. (2000). Structural and floristic heterogeneity in a 30-year-old Costa Rican rainforest restored on pasture through natural secondary succession. *Restoration Ecology*, **8**, 380–393.

Fischer, C. R., Janos, D. P., Perry, D. A., Linderman, R. G. & Sollins, P. (1994). Mycorrhiza inoculum potentials in tropical secondary forest succession. *Biotropica*, **26**, 369–377.

Food and Agriculture Organisation of the United Nations (FAO) (1999). *State of the World's Forests.* Rome: FAO.

Garwood, N. C. (1989). Tropical soil seed banks: a review. In *Ecology of Soil Seed Banks*, eds. M. A. Leck, V. T. Parker & R. L. Simpson, pp. 149–208. San Diego, CA: Academic Press.

González Montagut, R. (1996). Establishment of three rainforest species along the riparian corridor–pasture gradient in Los Tuxtlas, México. PhD dissertation, Harvard University.

Grainger, A. (1993). *Controlling Tropical Deforestation.* London: Earthscan Publications.

Granzow de la Cerda, I. & Garth, R. (1999a). Tropical rainforest trees propagated using large cuttings (Nicaragua). *Ecological Restoration*, **17**, 84–85.

Granzow de la Cerda, I. & Garth, R. (1999b). Air-layering shows promise in propagating tropical trees (Nicaragua). *Ecological Restoration*, **17**, 85–86.

Guariguata, M. R., Rheingans, R. & Montagnini, F. (1995). Early woody invasion under tree plantations in Costa Rica: implications for forest restoration. *Restoration Ecology*, **3**, 252–260.

Guariguata, M. R., Chazdon, R. L., Denslow, J. S., Dupuy, J. M. & Anderson, L. (1997). Structure and floristics of secondary and old-growth forest stands in lowland Costa Rica. *Plant Ecology*, **132**, 107–120.

Guevara, S. & Laborde J. (1993). Monitoring seed dispersal at isolated standing trees in tropical pastures: consequences for local species availability. *Vegetatio*, **107/108**, 319–338.

Guevara, S., Purata, S. E. & Van der Maarel, E. (1986). The role of remnant forest trees in tropical secondary succession. *Vegetatio*, **66**, 77–84.

Haggar, J. P., Briscoe, B. C. & Butterfield, R. P. (1998). Native species: a resource for the diversification of forestry production in the lowland humid tropics. *Forest Ecology and Management*, **106**, 195–203.

Harcombe, P. A. (1977). The influence of fertilisation on some aspects of succession in a humid tropical forest. *Ecology*, **58**, 1375–1383.

Hardwick, K., Healey, J., Elliott, S., Garwood, N. & Anusarnunthorn, V. (1997). Understanding and assisting natural regeneration processes in degraded seasonal evergreen forests in northern Thailand. *Forest Ecology and Management*, **99**, 203–214.

Harvey, C. A. (2000). Windbreaks enhance seed dispersal into agricultural landscapes in Monteverde, Costa Rica. *Ecological Applications*, **10**, 155–173.

Holl, K. D. (1998a). Effects of above- and belowground competition of shrubs and grass on *Calophyllum brasiliense* (Camb.) seedling growth in abandoned tropical pasture. *Forest Ecology and Management*, **109**, 187–195.

Holl, K. D. (1998b). The role of bird perching structures in accelerating tropical forest recovery. *Restoration Ecology*, **6**, 253–261.

Holl, K. D. (1999). Factors limiting tropical moist forest regeneration in agricultural land: seed rain, seed germination, microclimate and soil. *Biotropica*, **31**, 229–242.

Holl, K. D. & Howarth, R. D. (2000). Paying for restoration. *Restoration Ecology*, **8**, 260–267.

Holl, K. D. & Kappelle, M. (1999). Tropical forest recovery and restoration. *Trends in Ecology and Evolution*, **14**, 378–379.

Holl, K. D. & Lulow, M. E. (1997). Effects of species, habitat, and distance from edge on post-dispersal seed predation in a tropical rainforest. *Biotropica*, **29**, 459–468.

Holl, K. D. & Quiros-Nietzen, E. (1999). The effect of rabbit herbivory on reforestation of an abandoned pasture in southern Costa Rica. *Biological Conservation*, **87**, 391–395.

Holl, K. D., Loik, M. E., Lin, E. H. V. & Samuels, I. A. (2000). Restoration of tropical rainforest in abandoned pastures in Costa Rica: overcoming barriers to dispersal and establishment. *Restoration Ecology*, **8**, 339–349.

Houghton, J. T. (1997). *Global Warming: The Complete Briefing.* Cambridge: Cambridge University Press.

Houghton, R. A. (1995). Global effects of deforestation. In *Handbook of Ecotoxicology*, eds. D. J. Hoffman, B. A. Rattner, G. A. Burton & J. Cairns, Jr, pp. 492–508. Boca Raton, FL: Lewis Publishers.

Howe, H. F. (1984). Implications of seed dispersal by animals for tropical reserve management. *Biological Conservation*, **30**, 261–281.

Janos, D. P. (1980). Mycorrhizae influence tropical succession. *Biotropica*, **12**, 56–64.

Janzen, D. H. (1978). Seeding patterns of tropical trees. In *Tropical Trees as Living Systems*, eds. P. B. Tomlinson & M. H. Zimmerman, pp. 83–127. Cambridge: Cambridge University Press.

Janzen, D. H. (1988). Guanacaste National Park: tropical ecological and biocultural restoration. In *Rehabilitating Damaged Ecosystems*, ed. J. Cairns, Jr, pp. 143–192. Boca Raton, FL: CRC Press.

Janzen, D. H. (1995). Neotropical restoration biology. *Vida Silvestre Neotropical*, **4**, 3–9.

Johnson, N. C. & Wedin, D. A. (1997). Soil carbon, nutrients, and mycorrhizae during conversion of dry tropical forest to grassland. *Ecological Applications*, **7**, 171–182.

Kapos, V. (1989). Effects of isolation on the water status of forest patches in the Brazilian Amazon. *Journal of Tropical Ecology*, **5**, 173–185.

Kartawinata, K. (1994). The use of secondary forest species in rehabilitation of degraded forest lands. *Journal of Tropical Forest Science*, **7**, 76–86.

Keenan, R., Lamb, D., Woldring, O., Irvine, T. & Jensen, R. (1997). Restoration of plant biodiversity beneath tropical tree plantations in northern Australia. *Forest Ecology and Management*, **99**, 117–131.

Kolb, S. R. (1993). Islands of secondary vegetation in degraded pasture of Brazil: their role in reestablishing Atlantic coastal forest. PhD dissertation, University of Georgia.

Kremen, C., Niles, J. O., Dalton, M. G., Daily, G. C., Ehrlich, P. R., Fay, J. P., Grewal, D. & Guillery, R. P. (2000). Economic incentives for rainforest conservation across scales. *Science*, **288**, 1828–1832.

Laurance, W. F., Cochrane, M. A., Bergen, S., Fearnside, P. M., Delamônica, P., Barber, C., D'Angelo, S. & Fernandes, T. (2001). The future of the Brazilian Amazon. *Science*, **291**, 438–439.

Lean, J. & Warrilow, D. A. (1989). Simulation of the regional climatic impact of Amazon deforestation. *Nature*, **342**, 411–413.

Loik, M. E. & Holl, K. D. (1999). Photosynthetic responses to light for rainforest seedlings planted to restore abandoned pasture, Costa Rica. *Restoration Ecology*, **7**, 253–261.

Lugo, A. E. (1988). The future of the forest: ecosystem rehabilitation in the tropics. *Environment*, **30**, 17–20, 41–45.

Lugo, A. E. (1997). The apparent paradox of reestablishing species richness on degraded lands with tree monocultures. *Forest Ecology and Management*, **99**, 9–19.

Medellín, R. A. & Gaona, O. (1999). Seed dispersal by bats and birds in forest and disturbed habitats of Chiapas, México. *Biotropica*, **31**, 478–485.

Meguro, S. I. & Miyawaki, A. (1997). A study of initial growth behaviour of planted Dipterocarpaceae trees for restoration of tropical rain forest in Borneo/Malaysia. *Tropical Ecology*, **38**, 237–245.

Miriti, M. N. (1998). Regeneracão florestal em pastagens abandonas na Amazônia Central: competicão, predacão, e disperso de sementes. In *Floresta Amazônica: Dimica, Regeneracâo e Manejo*, eds. C. Gascon & P. Moutinho, pp. 179–191. Manaus, Brazil: Instituto Nacional de Pesquisa da Amazônia.

Montagnini, F., Fanzeres, A. & Guimaraes da Vinha, S. (1995). The potentials of 20 indigenous tree species for soil rehabilitation in the Atlantic forest region of Bahia, Brazil. *Journal of Ecology*, **32**, 841–856.

Nepstad, D. C., Uhl, C. & Serrao E. A. (1990). Surmounting barriers to forest regeneration in abandoned, highly degraded pastures: a case study from Paragominas, Pará, Brazil. In *Alternatives to Deforestation: Steps Toward Sustainable Use of the Amazon RainForest*, ed. A. B. Anderson, pp. 215–229. New York: Columbia University Press.

Nepstad, D. C., Uhl, C. & Serrao, E. A. S. (1991). Recuperation of a degraded Amazonian landscape: forest recovery and agricultural restoration. *Ambio*, **20**, 248–255.

Nepstad, D. C., Uhl, C., Pereira, C. A. & Cardoso da Silva, J. M. (1996). A comparative study of tree establishment in abandoned pasture and mature forest of eastern Amazonia. *Oikos*, **76**, 25–39.

Nepstad, D. C., Verissimo, A., Alencar, A., Nobre, C., Lima, E., Lefebvre, P., Schlesinger, P., Potter, C., Moutinho, P., Mendoza, E., Cochrane, M. & Brooks, V. (1999). Large-scale impoverishment of Amazonian forests by logging and fire. *Nature*, **398**, 505–508.

Osunkoya, O. O. (1994). Postdispersal survivorship of north Queensland rainforest seeds and fruits: effects of forest, habitat and species. *Australian Journal of Ecology*, **19**, 52–64.

Otero-Arnaiz, A., Castillo, S., Meave, J. & Ibarra-Manríquez, G. (1999). Isolated pasture trees and the vegetation under their canopies in the Chiapas coastal plain, México. *Biotropica* **31**, 243–254.

Parrotta, J. A. (1992). The role of plantation forests in rehabilitating degraded tropical ecosystems. *Agriculture, Ecosystems and Environment*, **41**, 115–133.

Parrotta, J. A. (1995). Influence of overstory composition on understorey colonisation by native species in plantations on a degraded tropical site. *Journal of Vegetation Science*, **6**, 627–636.

Parrotta, J. A. & Knowles, O. H. (1999). Restoration of tropical moist forests on bauxite-mined lands in the Brazilian Amazon. *Restoration Ecology*, **7**, 103–116.

Peterson, C. J. & Haines, B. L. (2000). Early successional patterns and potential facilitation of woody plant colonisation by rotting logs in premontane Costa Rican pastures. *Restoration Ecology*, **8**, 361–369.

Pimm, S. L., Russell, G. J., Gittleman, J. L. & Brooks, T. M. (1995). The future of biodiversity. *Science*, **269**, 347–350.

Pinard, M. A., Barker, M. G. & Tay, J. (2000). Soil disturbance and post-logging forest recovery on bulldozer paths in Sabah, Malaysia. *Forest Ecology and Management*, **130**, 213–225.

Posada, J. M., Aide, T. M. & Cavelier, J. (2000). Cattle and weedy shrubs as restoration tools of tropical montane rainforest. *Restoration Ecology*, **8**, 370–379.

Press, D. T. (1995). The use of artificial perches to increase seed dispersal by birds in a pasture in southern Costa Rica. BSc dissertation, University of California, Santa Cruz.

Prinsley, R. T. (1991). The role of trees in sustainable agriculture: an overview. In *The Role of Trees in Sustainable Agriculture*, eds. R. T. Prinsley & J. Allnutt, pp. 87–117. Dordrecht: Kluwer.

Reiners, W. A., Bouwman, A. F., Parsons, W. F. J. & Keller, M. (1994). Tropical rainforest conversion to pasture: changes in vegetation and soil properties. *Ecological Applications*, **4**, 363–377.

Rhoades, C. C., Eckert, G. E. & Coleman, D. C. (1998). Effect of pasture trees on soil nitrogen and organic matter: implications for tropical montane forest restoration. *Restoration Ecology*, **6**, 262–270.

Rivera, L. W., Zimmerman, J. K. & Aide, T. M. (2000). Forest recovery in abandoned agricultural lands in a karst region of the Dominican Republic. *Plant Ecology*, **148**, 115–125.

Saab, V. A. & Petit, D. R. (1992). Impact of pasture development on winter bird communities in Belize, Central America. *Condor*, **94**, 66–71.

Samuels, I. A. (1998). The role of riparian corridors and isolated pasture trees in facilitating bird movement in tropical pasture. BSc dissertation, University of California, Santa Cruz.

Shukla, J., Nobre, C. A. & Sellers, P. (1990). Amazon deforestation and climate change. *Science*, **247**, 1322–1325.

Silver, W. L., Ostertag, R. & Lugo, A. E. (2000). The potential for carbon sequestration through reforestation of abandoned tropical agricultural and pasture lands. *Restoration Ecology*, **8**, 394–407.

Slocum, M. G. (2000). Logs and fern patches as recruitment sites in a tropical pasture. *Restoration Ecology*, **8**, 408–413.

Stanley, W. G. & Montagnini, F. (1999). Biomass and nutrient accumulation in pure and mixed plantations of indigenous tree species grown on poor soils in the humid tropics of Costa Rica. *Forest Ecology and Management*, **113**, 91–103.

Sun, D. & Dickinson, G. R. (1996). The competition effect of *Brachiaria decumbens* on the early growth of direct-seeded trees of *Alphitonia petriei* in tropical north Australia. *Biotropica*, **28**, 272–276.

Toh, I., Gillespie, M. & Lamb, D. (1999). The role of isolated trees in facilitating tree seedling recruitment at a degraded sub-tropical rainforest site. *Restoration Ecology*, **7**, 288–297.

Uhl, C. (1987). Factors controlling succession following slash-and-burn agriculture. *Journal of Ecology*, **75**, 377–407.

Uhl, C., Clark, K., Clark, H. & Murphy, P. (1981). Early plant succession after cutting and burning in the Upper Rio Negro Region of the Amazon Basin. *Journal of Ecology*, **69**, 632–649.

Uhl, C., Clark, H., Clark, K. & Marquirino, P. (1982). Successional patterns associated with slash-and-burn agriculture in the Upper Rio Negro Region of the Amazon Basin. *Biotropica*, **14**, 249–254.

Uhl, C., Buschbacher, R. & Serrao, E. A. S. (1988). Abandoned pastures in eastern Amazonia. 1: Patterns of plant succession. *Journal of Ecology*, **76**, 663–681.

Uhl, C., Nepstad, D., Buschbacher, R., Clark, K., Kauffman, B. & Subler, S. (1990). Studies of ecosystem response to natural and anthropogenic disturbances provide guidelines for designing sustainable land-use systems in Amazonia. In *Alternatives to Deforestation: Steps Toward Sustainable Use of the Amazon RainForest*, ed. A. B. Anderson, pp. 24–42. New York: Columbia University Press.

United Nations, Department of Economic and Social Affairs (1999). *World Population Prospects: The 1998 Revision.* New York: United Nations.

Vieira, I. C. G., Uhl, C. & Nepstad, D. (1994). The role of the shrub *Cordia multispicata* Cham. as a 'succession facilitator' in an abandoned pasture, Paragominas, Amazonia. *Vegetatio*, **115**, 91–99.

Vitousek, P. M. & Sanford, J., Jr (1986). Nutrient cycling in moist tropical forest. *Annual Review of Ecology and Systematics*, **17**, 137–167.

Whitmore, T. C. (1998). *An Introduction to Rain Forests*, 2nd edn. Oxford: Oxford University Press.

Williams-Linera, G. (1990). Origin and early development of forest edge vegetation in Panama. *Biotropica*, **22**, 235–241.

Williams-Linera, G., Dominguez-Gastelu, V. & Garcia-Zurita, M. E. (1998). Microenvironment and floristics of different edges in a fragmented tropical rainforest. *Conservation Biology*, **12**, 1091–1102.

Willson, M. F. & Crome, F. H. J. (1989). Patterns of seed rain at the edge of a tropical Queensland rainforest. *Journal of Tropical Ecology*, **5**, 301–308.

Wilson, E. O. (1988). The current state of biological diversity. In *Biodiversity*, ed. E. O. Wilson, pp. 3–18. Washington, DC: National Academy Press.

Wunderle, J. M. (1997). The role of animal seed dispersal in accelerating native forest regeneration on degraded tropical lands. *Forest Ecology and Management*, **99**, 223–235.

Yarranton, G. A. & Morrison, R. G. (1974). Spatial dynamics of a primary succession: nucleation. *Journal of Ecology*, **62**, 417–428.

Zahawi, R. A. & Augspurger, C. K. (1999). Early plant succession in abandoned pastures in Ecuador. *Biotropica*, **31**, 540–552.

Zanne, A. & Chapman, C. A. (2001). Expediting indigenous regeneration in African grasslands: plantations and the effects of distance and isolation from seed sources. *Ecological Applications*, **11**, 1610–1621.

Zimmerman, J. K., Pascarella, J. B. & Aide, T. M. (2000). Barriers to forest regeneration in an abandoned pasture in Puerto Rico. *Restoration Ecology*, **8**, 350–360.

Fig. 24.1 Aerial view of pasture–secondary successional forest mosaic during the dry season (March 1987, half-way between the words 'Santa Rosa' and 'Nancite' in Fig. 24.2). The ungrazed grass is 1–2 m tall jaragua (*Hyparrhenia rufa*), an introduction from East Africa that replaced the native grass pastures in the 1940s. The forest fragments, including the large one to the left, have been burned, logged and hunted for centuries.

Fig. 24.3 Jaragua–forest edge that was characteristic of tens of thousands of hectares of the ACG at the beginning of the restoration process (30 December 1980), half-way between the words 'Santa Rosa' and 'Pocosol' in Fig 24.2. This pasture is at least 200 years old, was formerly occupied by native grasses, and had been burned every one to three years. The forest to the left and background is old secondary succession oak (*Quercus oleoides*) with more than 100 other species of trees mixed in.

Fig. 24.4 Same view as in Fig. 24.3 (4 November 2000), following 17 years of elimination of anthropogenic fire. The canopy of the oak forest is still visible as horizon, and Winnie Hallwach's hand is 2 m above the ground. The isolated *Crescentia alata* tree in Fig. 24.3 is now totally hidden by the young forest. The bulk of the young forest is wind-dispersed *Rehdera trinervis* (Verbenaceae), intermixed with another 70 woody species. Such rapid natural forest invasion of pasture in the absence of fire is characteristic of tens of thousands of hectares of the ACG.

24 • Tropical dry forest: Area de Conservación Guanacaste, northwestern Costa Rica

DANIEL H. JANZEN

INTRODUCTION

For several decades prior to 1985, it was widely felt by conservationists that once tropical forest had been removed, it was gone forever. However, I had spent 25 years studying esoteric animal–plant interactions in secondary regrowth forest throughout the tropics, a habitat that by its very existence called this viewpoint into doubt. I often found myself doing research in 'old-growth' forest that was obviously regrown on fields abandoned centuries to millennia earlier. It is no mystery that when a patch of mainland wildland forest is cleared by a landslide, volcano, or hurricane/tornado, it quickly acquires a young forest that after a few hundred years is quite difficult to distinguish from the neighbouring 'pristine' forest (which will have its own history of perturbations as well). Naturally occurring forest restoration was, and is, widespread throughout the forested tropics. It is also no stranger to the tropical rural dweller. Shifting agriculture has long been omnipresent throughout the forested tropics and is nothing more than applied restoration ecology. Stop the assault and the forest will return – at different rates in different places – provided there are sources of inoculum. The question is not whether a tropical forest can be restored, but rather whether society will allow it to occur (Janzen, 1987a, b, 1988a, 1992a, 1995; Holl, 1999; Zahawi & Augspurger, 1999; Aide et al., 2000; Escofet, 2000; Silver et al., 2000).

In 1985, my wife Winnie Hallwachs and I found ourselves realising that if active steps were not taken to stop the assault on the secondary successional remnants of tropical dry forest in the 10 800 ha Parque Nacional Santa Rosa in northwestern Costa Rica (Fig. 24.1), the world would probably lose the last chance to have a conserved entire tropical dry forest ecosystem on the Pacific lowlands between Mazatlan, Mexico and the Panama Canal (Janzen, 1986a, 1988a; Allen, 2001). The Santa Rosa region was once totally clothed in semi-deciduous to deciduous dry forest (see Killeen et al., 1998 and Pennington et al., 2000 for recent descriptions of 'tropical dry forest') from the Pacific beaches to 400–600 m up the western slopes of the north–south Cordillera Guanacaste. However, its broken ecoscape had been assaulted since about 1600 by ranching, subsistence agriculture, irrigation, heterogeneous hunting, sporadic logging and anthropogenic fires. Throughout this region, dry-season fire was the standard tool for prevention of forest reinvasion of the hundreds of thousands of hectares of pasturelands intermingled with patches of all ages of woody secondary succession (Fig. 24.1). The technological recipe for the restoration of this large dry forest ecosystem was obvious: purchase the large tracts of marginal ranch and farmland adjacent to Santa Rosa and connect it with the wetter forests to the east, stop the fires, farming, and the occasional hunting and logging, and let nature take back its original terrain (Janzen 1986a, 1988a).

Santa Rosa became the nucleus for what has become the ten-times-larger Area de Conservación Guanacaste (ACG, 2001) (Fig. 24.2). The 110 000 terrestrial hectares of the ACG began as a huge experiment to see if a severely damaged dry forest ecosystem would restore itself to what will be an extraordinarily complex old-growth forest 500+ years from now. To date, the process of this experiment is only slightly documented (e.g. Janzen, 1983b, 1986a, b, 1987a, b, 1988b, c, d, 1992a, b, c; Janzen & Hallwachs, 1992; Gerhardt, 1993, 1996; Gerhardt &

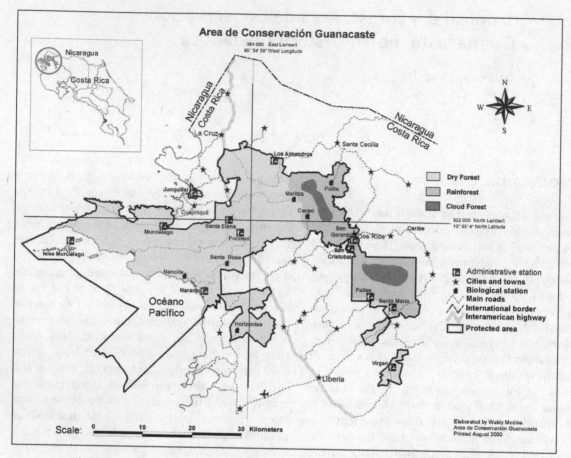

Fig. 24.2. Area de Conservación Guanacaste (ACG) and the approximate distribution of its four major ecosystems – 43 000 marine hectares and 110 000 hectares of dry forest, cloud forest and rainforest. The original dry forest portion of Parque Nacional Santa Rosa, today Sector Santa Rosa, is the area lying between the Interamerican Highway and the Pacific Ocean in the lower left of the ACG around the word 'Santa Rosa'.

Fredriksson, 1995). This is because the applied, and ongoing, sociology of initiating and sustaining the restoration process has taken precedence over the scientific and management desire to document and understand the technology of the restoration process in detail. The ACG is no longer a technological experiment, but quite simply, ecoscape management as nature runs its course (Janzen, 1999*a*, *b*, 2000*a*, *b*, *c*, *d*).

It is not too late to begin the detailed study of the restoration of the ACG biophysical unit. Indeed, it would be the study of a restoration process that has been ongoing since at least 1580 (and long before that under the influence of its pre-Colombian occupants). However, the driver for such an expenditure of resources will need to be scientific and managerial curiosity, and the search for information to be used elsewhere, rather than be an imperative for the restoration of the ACG itself. From a conservation standpoint, the ACG has already demonstrated unambiguously that a large and long-damaged tropical dry forest pasture/secondary forest mosaic can be put into full natural restoration mode by stopping the anthropogenic fires, farming, irrigation, hunting and logging, and letting nature take its course (see Figs. 24.3 and 24.4). No detailed study is required for the biological parts of this process to continue for centuries. Much further effort is required, however, to ensure a sociology that permits it to continue.

The ACG is landscape-level restoration of a large wildland island in an ocean of agroscape. It was far too expensive, and biologically dubious, to invest in processes to attempt to return any given hectare to some particular species composition. Furthermore, there is no remaining large-scale old-growth Mesoamerican dry forest standard to attempt to emulate. The only option was to generate a staff and administrative process that minimised the human assault as much as practicable.

The 235 000-plus species (Janzen, 1996) occupying the ACG forest remnants are simply being allowed to compete, mutualise and commensalise themselves into whatever patterns and habitats they arrive at given the abiotic and biotic heterogeneity intrinsic to the ACG. Roads, trails, buildings, observers and users (and some eventual local extinctions) add a bit more heterogeneity. The wild organisms are invading just as they must have when the entire area was levelled into a moonscape by a volcanic blast 1.5 million years ago. They will recreate/restore an old-growth dry forest and its adjacent marine, cloud forest and rainforest extensions, to whatever it was when humans arrived and began their assault, albeit lacking the exterminated megafauna (Janzen & Martin, 1982; Barlow, 2001), albeit as an ecological island in the agroscape, albeit suffering the contemporary assaults of global warming (e.g. Pounds et al., 1999) and pesticides (e.g. Standley & Sweeney, 1995).

The time-scale to 'completion' is centuries to millennia, varying from site to site. The ecosystem restoration ranges from that starting on bare ground created only a few days ago to continuing the largely natural restoration that began when a pre-Colombian village was abandoned more than four centuries ago. The latter, the Ojochal inland from Playa Naranjo, is occupied by what could fool you into thinking it is original forest except that all the huge trees have human-edible fruits and/or seeds and there are pottery shards, mollusc shells and large mammal bones in the soil below.

Here I summarise some of the restoration results of 15 years of severely reducing direct human interference in the ACG, and reflect lightly on the processes that these results reveal or suggest. My goal is to be useful to others elsewhere. I focus primarily on the ACG dry forest (and its interactions with its adjacent cloud forest and rainforest), though the restoration of the ACG cloud forests and wet forests is equally deserving of attention (see Holl, this volume, for a fuller treatment of tropical moist forest). The restoration process has yet another five centuries to run, at a minimum. For at least a millennium there will still be traces of the original assault visible in the age structures of populations of the longer-lived and slower-dispersing tree species. I only lightly cross-reference to restoration efforts in other parts of the tropics (but see Holl, this volume), and I leave detailed data-rich descriptions for later years when the conservation status of this biophysical unit is fully secure, and therefore we can endulge in such esoteria.

The overall mission of the ACG is to allow its maximum and naturally occurring ecosystem restoration. This is to allow the maximum possible chance of survival of its hundreds of thousands of species into perpetuity, at whatever intra- and interspecific relationships they attain through the forces of nature operating with minimal and unavoidable interference by humans.

SEASONALITY

The very strong wet-season–dry-season annual fluctuation in ACG rainfall (e.g. Janzen, 1986a, 1988b, 1993) has an enormous impact on the restoration process. First, the fact that there is a six-month dry season is the reason why dry forest is easily cleared through even light initial logging followed by dry season fires that block woody succession and favour grasses (and see Uhl & Buchsbacher, 1985). Second, organisms evolutionarily moulded by within- and between-year seasonally variable circumstances are generally more competent at dealing with microgeographic variation than are those evolved in more homogeneous environments (e.g. Janzen, 1967). In other words, dry-forest organisms have long been intensely selected to be resistant to the very challenges that are generated by fire, farming, logging, hunting and grazing. The dry forest biota is much more capable of rapidly invading the extreme and variable conditions of an abandoned pasture or field than is the rainforest bioto. And, thousands of species seasonally migrate out of the dry forest to pass the long dry season, and then return with the wet season (Janzen, 2002). This highly mobile portion of the biota is especially aggressive in the restoration process.

At the start of this landscape-level restoration of the ACG forest in 1985, the long dry season (January–May) and short dry season (late July to August) was creating maximal heterogeneity in physical and biotic conditions. The dry season ecoscape was a fine-scale habitat mosaic ranging from bare burned ground that had virtually no litter and was fully insolated (due to no vegetation or purely deciduous vegetation), dried by winds, and chilled at night, to deeply shaded moist springs and attendant mud in the middle of shady semi-deciduous (seemingly evergreen) old-growth forest. Even the intense dry season trade winds ranged from ground level in pastures to 30 m above the ground in the forest canopy. At sea level below the Santa Rosa plateau (300 m) there were valley bottoms ranging from old bean fields and banana plantings to 40 m tall forest, all baking in wind-free air.

As the terrain subject to this microclimatological heterogeneity has become clothed by the restoring forest, its dry season climate has ameliorated at ground level, and its microgeographic heterogeneity blurred. Dry sites are not as absolutely dry as before, and the comparatively moist sites are now less comparatively moist. The dry season now seems to be less dry to the human observer. When the entire ACG dry forest is once again old-growth forest, it will be biologically and meteorologically less dry.

Eight of the past ten years of ACG dry forest have had 'below average' rainy season rainfall (see ACG, 2001 for weather records). This has slowed the reforestation process in both pastures/fields and the young forest. During drier rainy seasons, saplings do not add as much vertical height and lateral branch foliage (and therefore they shade the pasture grasses less). Second, tree seedlings-of-the-year suffer more juvenile death in the subsequent dry season (presumably due to less well-developed root systems and lesser stored reserves). It may even be that seed crops are smaller following less wet rainy seasons, though this effect depends very much on whether a given species' seed output is light-limited (understorey species especially), water-limited, or based on supra-annual storage of reserves before seeding. There is also noticeably less wind in the first half of the dry seasons that follow drier rainy seasons. This shortens and reduces the wind-borne seed shadow downwind into old pastures.

A drier rainy season also generates a shorter and sparser grass cover (and may reduce grass seed crops as well). However, from a management standpoint this is probably less important than is the reduced rate of litter decomposition in a drier rainy season. This reduced rate leaves more fuel, which then fuels more intense dry season fires. These more intense fires kill yet larger trees, vines and shrubs, leading to more ground-level insolation in the following rainy season (and therefore more herbaceous vegetation, i.e. fuel, for the next fire). Dry rainy seasons also lead to a forest litter that dries more rapidly and more thoroughly at the beginning of the following dry season, leading to deeper penetration of the forest by fires started in open sites in the early dry season.

FIRE

The ACG restoration process has not struggled with the question of whether to let its fires burn. This is because other than that which might occur following a volcanic blast, there are no natural fires in the ACG. Lightning fires do not occur during the six-month rain-free dry season. A 'natural' fire may be set by a rainy season lightning strike into an ungrazed pasture with a massive amount of accumulated fuel, but such a fire dies out when it hits the shady (and moist) understorey of the adjacent forest. There is no evidence that the ACG had grasslands maintained by natural fires (or savannas as they are often called; and see Vieira & Marinho-Filho, 1998; Hoffman, 2000; Pennington et al., 2000), though pre-Colombian cultures probably burned off some hilltops (e.g., Cerro El Hacha, Pedregal Orosí) as ceremonial sites (e.g. Janzen 1986a, 1988a) and some forested areas cleared by volcanic blasts are extremely slow to reforest. While natural fire could have been part of the forest biology of the ACG area at some time in the distant past (see Ginsberg, 1998), now it is not.

The ACG does, however, face the question of whether to continue to occasionally burn designated grassy sites so as to maintain 'controls' for contrast with the adjacent regenerating dry forest decades and centuries from now. Four such 'fire plots', totalling about 9 ha, are currently burned annually for this purpose (Guacimo, Cruz de Piedra, Pitahaya, Principe). The 'final' long-term ACG management plan will contain several hundred hectares of fire plots in at least three more areas (Cerro El Hacha, Santa Elena, Horizontes). If cleverly sited, such burns can also be an integral part of the overall ACG fire elimination programme by serving as wildfire barriers. All of these grassy areas range from being dense near monocultures of jaragua (*Hyparrhenia rufa*, introduced from Africa in the 1940s) on 'good' soils, to complex mixes of introduced and native grasses, to near monocultures of native *Trachypogon plumosus* on 'bad' soils. It is tempting to think of also using cattle and fire to maintain a pasture–forest mix (as in Fig. 24.1) as yet another kind of comparison/control, though the motivation for this is currently minimal owing to the existence of tens of thousands of hectares of this habitat in other parts of Guanacaste Province, and owing to the management stress of maintaining a cattle herd for this purpose.

The annual to bi-annual fires in the ACG area prior to 1985 were most commonly set during the second half of the five- to six-month dry season. This timing of burning generates the most thorough exclusion of forest invasion because the longer the dry season has been drying the habitat, the higher the proportion of the habitat that is dry enough to carry a fire (e.g. Janzen 1988b, d). Early dry season fires (which also serve as very effective fire breaks later in the dry season) do much less damage to the restoration process than do late dry season fires.

There really are no forest fires in the ACG. Instead, there are grass, herb and litter fires. The ACG living woody plants do not generally carry a fire nor burn without additional fuel sources. However, if the fire is hot enough to kill or wound living trees, these in turn become prostrate or standing litter for the next year's fire. The more years since the last fire, the greater the fuel load and the larger the living woody stems that are killed or even consumed

by an ACG grass or litter fire. Early in the restoration process, when the intensity of ACG fires was still high because of large amounts of accumulating fuel, and the fire control system was still under development, the occasional fire in a forest–grass mix was more effective at removing adult trees (seed sources) than were the annual fires in the previously heavily grazed mosaic of grazing swards and forest patches. Once succession had proceeded for a few years, annual decomposition rates began to match annual production rates of readily combustible plant litter, and shade blocked herbaceous growth. The risk of restoration setback by escaped fires has declined.

Prior to fire control in Sector Santa Rosa, and on the many old extensive pasturelands on volcanic soils that were purchased and added to the ACG in 1986–90, the pasturelands were a mosaic of closely cropped (often jaragua) grass swards intersplotched with patches of regenerating secondary successional forest 100–400 years of age. Some of these patches were in moist ravines and others on rocky hillsides/cliffs (Janzen, 1988b, d; and see Larson et al., 2000). Others were on flatlands that for a variety of socio-economic reasons were not cleared to pasture. Much of the original ACG dry forest clearing followed forest invasion by herbs and grasses following high-grade logging and then repeated spread of free-ranging fires set to clear pastures (e.g. Uhl & Buschbacher, 1985). In the 1800s, and especially after the introduction of jaragua pasture grass in the 1940s, the cowboys were also instructed to set a dry season fire in the herbaceous material in any small natural clearing in the forest, which resulted in a gradually spreading 'pasture'.

Cattle kept the grass fuel load down to minimal levels in the pastures. The light fires nibbled into the forest edges in dry years (and in the late dry season) and the forest invaded the pasture in wet years (or following early dry season fires). When the cattle were removed from Sector Santa Rosa in 1976–7, the pastures quickly accumulated massive grass fuel loads, the heat from which was sufficient to destroy regenerating forest that had coexisted with the pastures for decades to centuries. The same occurred on neighbouring ranchlands purchased for addition to the growing ACG. Elimination of fire has left the forest restoration to move as rapidly as nature will deliver it. There was a need to leave cattle on some newly purchased pastures as biotic mowing machines until thorough fire control could be achieved. Once this was achieved, the cattle were removed because they were no longer needed to minimise fire impact, because the seeds they move into a site through their dung were a minority compared with those brought by other dispersal processes, and because they impacted the aquatic system dramatically through drinking, trampling and eating riparian vegetation during the dry season (but see below re Pleistocene megafauna).

Wind and vertebrates generate a massive seed rain from the intermingled patches of secondary successional forest into the ACG pastures (e.g. Janzen 1988c, 1989, 1990). However, the first four years following fire elimination were depressing and unrewarding to the fire control programme. During these first four years, the view across the sea of 50–150 cm tall grass contained no young forest. The millions of seedlings and saplings were invisible below the grass. However, from the fifth year onward, the pastures became rapidly more 'dirty' as the multi-species stands of young trees extended above the grass and created a closed and shade-generating canopy (Figs. 24.3 and 24.4).

In 1986–8 the ACG grew and planted thousands of native trees in large pastures in Sector Santa Rosa and Sector Pocosol. However, the natural forest invasion of these and other fire-free pastures quickly swamped all such efforts. Tree-planting was ceased in order to use the funds for the far more economical and effective process of fire control and other more important management costs such as the biological education programme, infrastructure, and development of ecosystem services, and purchase of more degraded ranchland. While dry forest rapidly fills an adjacent dry forest pasture or field if fire is excluded, re-establishing a 'normal' tree species composition can be hastened by human dispersal (or seedling planting) of the larger-seeded species, and of the species for which the only remaining breeding stock may be many kilometres distant.

Fire control in 1986 onward was achieved by hiring a small crew of former farmers and ranch hands

from an agroscape that has used fire for centuries to stop forest invasion of pasture. The ACG instruction was simply 'no fires'. They were also given the freedom to design the fire breaks (blackburns) and fire lookouts, and work schedules as they wished. They purchased the trucks, pumps, brooms, binoculars, radios and other simple equipment they needed. They extinguish 30–75 anthropogenic fires per year in or near the ACG through vigilance from high points and the fire lookout station at 1000 m elevation on the western flank of Volcan Cacao, radio communication, and six months of 24-hour anti-fire behaviour. When a large fire gets out of hand, government and private sector volunteers are brought in and supported for several days until it has been beaten and watered into submission. The cost of this programme (20% of the annual budget of the ACG) is a vastly cheaper method for generating tens of thousands of hectares of young restoring dry forest than could be any form of tree-planting. Further, if restoration had been attempted through tree-planting, there would still have been the need for a fire control programme.

While the 'pastures' of the ACG were best developed on good soils, the ranching practice was to burn indiscriminately. This created pastures of largely edible introduced grasses as the fire burned through the valleys and flats on volcanic and alluvial soils, but the drier ridges and uplands (serpentine as well as volcanic) were covered with largely inedible native grasses such as Trachypogon plumosus. When the fires are removed from such situations, the forest invades the valley bottoms much faster than it does the rocky ridges above. This is apparently not because the native grasses are superior competitors, but because of the harsher conditions there, because the vertebrates and wind do not direct seed rain at the ridges, and because the woody plant seedlings grow best on the more moist and nutrient-rich valley bottom and flatland soils.

SEED BANK

The true 'seed bank' for the ACG dry forest restoration is the annual and supra-annual reproduction by populations of plants and animals in the adjacent surviving forest fragments, such as those visible in Fig. 24.1. The vertebrate gut is a short-term ambulating wet anaerobic seed bank (e.g. Janzen, 1981a, b, 1983b, 1984; Janzen et. al., 1985).

There is very little multi-annual seed bank in ACG dry forest soils, be they in fields, pastures or forest. While many dry forest seed species can remain dormant for years to decades in a dry herbarium (Janzen & Vázquez-Yanes, 1991), in nature the great majority of ACG seeds and species germinate in the first or second rainy season after production (their great tolerance of desiccation was largely evolutionarily designed to pass a single dry season, and/or survive dispersal through the gut of a vertebrate, rather than to remain dormant waiting for a rainy year). An exception is Sesbania emerus. The seeds of this several-metres-tall annual herbaceous legume may remain dormant for many years in depressions while awaiting the occasional very wet year that converts their habitat into a swamp (Janzen, 1983c). However, once the ACG swampy areas have reacquired their forest cover, this light-loving herb may be one of the many species that will be locally extinguished or severely reduced by the restoration process.

The result is that when fire, farming and logging are ceased, the new seedlings are largely from seeds that dispersed into the site that year or subsequently. The gradual accumulation of growing species in an ACG restoration site over decades is largely due to incoming propagules.

The absence of a soil seed bank accumulated over many years is also produced by the intense year-round seed predation from Liomys salvini mice (Janzen, 1986c), larger mammals such as agoutis (Dasyprocta punctata) and peccaries (e.g. Hallwachs, 1986), doves, tinamous, bruchid beetles (Janzen, 1980), seed-predator ants, seed bugs, fungi, etc.

There is a second major source of plants on a site entering forest restoration. The old ACG extensive pastures were rich in root sprouts from burned-off stumps (or their roots at some distance from the stump) and inconspicuous dwarfed and pruned individuals (and see Harvey & Haber, 1999). It is commonplace for these species to occur at less than one individual per hectare. While they may require decades to become reproductive in the restoration process, these sucker shoots and inconspicuous individuals often carry the genomes of what were once common plants with omnipresent distributions. They often 'return' a species to a site much faster than do incoming seeds from forest fragments many kilometres away. Such plants are a kind of 'living dead' (Janzen, 2001) and are important contributors to the early biodiversity accumulation curve as forest succession begins, especially if the 'plot' is many hectares in size. The presence of such species is a major trait of extensive low-grade cattle ranches of the kind in the ACG, while pastures and farms on better soils are generally 'cleaned' of most of these species by ploughing and reseeding of yet better forage or crop plants.

SEED DISPERSAL

As mentioned above, the populations in the forest fragments are the primary source of the animals, plants and other organisms that invade and begin the ACG forest restoration process. Their distance and wind direction from any given site is paramount, but distance is not necessarily measured in metres. A seed-rich dung pile only 50 metres into the pasture may be as ecologically far from forest-bound seed predator mice as is a dung pile 500 m from forest (e.g. Janzen, 1982a, 1986c; and see Duncan & Duncan, 2000). The heterogeneity in a seed shadow generated by vertebrate perching preferences can be much greater than that generated by distances of perches from the parent tree (e.g. Janzen et al., 1976). Equally complicating is that a 'forest fragment' may have lost many of the populations that were once in it. Only by accumulating the propagule output from many such fragments does the seed rain at a point begin to take on the total species composition that the whole area will have when restored (cf. Janzen 1986d).

A plant species has a seed dispersal process and resultant seed shadow that is the accumulation of millions of years of anachronisms and contemporary functionalities that are/were operational for the challenges, opportunities and associated dispersal agents in some to many natural habitats/ecosystems (Janzen & Martin, 1982; Hallwachs, 1986). The abandoned pasture, field, logging site, trailside, etc. that is now the subject of dry forest restoration only crudely matches the detail of any of those circumstances. It grossly exaggerates some and shrinks others to nothing. Seed shadows and the intensity of seed rain will be created by the serendipity of parental proximity to the site, vagaries of animal survivals in this highly 'artificial' circumstance, ethnocentricities of directed animal and plant harvest by residents for previous centuries, serendipitous susceptibility to fire, herbicides, and soil contaminants, etc. The outcome is that two equal-aged adjacent regenerating pastures of the same size on the same soil can have grossly different plant species compositions and attendant different animal communities. These differences can persist for centuries in an ecoscape like that of the ACG with more than 500 species of woody plants with century or more life spans. It is presumed that the forests appearing on two adjacent pastures will eventually come to resemble each other, but it will take centuries to make up for the serendipity, for example, of having a single wind-dispersed *Ateleia herbert-smithii* tree (Janzen, 1989) upwind of one pasture and not of the other.

In the ACG dry forest, the winds of December to March (first half of the dry season) are major dispersers of hundreds of species of plants (from grasses and herbs to ancient vines (e.g. Bignoniaceae, Hippocrateaceae, Fabaceae, Polygalaceae) and centuries-old trees such as *Swietenia* mahogany (Janzen, 1983c), *Gliricidia* (Janzen, 1983c), *Ateleia* (Janzen, 1989) and *Lonchocarpus* (Janzen et al., 1990). The wind is not subject to hunting and habitat fragmentation and it is a major force in generating the early wave of woody invasion of ACG pastures. One closely watched pasture contains 19 of the 23 species of ACG wind-dispersed dry forest timber trees, all of which invaded during the first ten years of succession.

Wind-dispersed plants display heterogeneous invasion patterns. First, the forest invasion is from the upwind (generally eastward) side of the pasture. Second, wind almost never moves seeds more than several hundred metres downwind. Most seeds fall within a few tens of metres of the parent tree. Wind-dispersed seeds also demonstrate a temporal heterogeneity. As the crowns of early-successional trees grow rapidly upward, their dry-season leafless branches form a filter/screen that knocks wind-dispersed seeds out of the air. The result is that after five to ten years of woody secondary succession, the seed shadow from a given parent that initially extended far downwind into a pasture is severely shortened in the young forest. There may be a long hiatus before more wind-dispersed seeds arrive at the site from the new first-generation adults. These later seeds also fall in the shade of the vigorous parents. Since most of them are shade-intolerant, they are destined to sustain the seed and seedling predator community, and feed the litter community as the seedlings die of light deprivation.

Vertebrate-dispersed seeds initially move into the ACG dry forest pastures in small numbers through being defaecated or dropped by vertebrates passing through or over the pasture, and in much larger numbers at a few points by being defaecated under and near woody perches. The latter phenomenon generates enlarging patches of animal-dispersed species around isolated pasture trees at any cardinal direction from parent trees and at highly variable distances from forest patches (Janzen, 1988c, 1990; and see Nepstad et al., 1996; Toh et al., 1999; Slocum & Horvitz, 2000).

As these patches of vertebrate-dispersed species grow and coalesce with each other and with the more amorphous areas of wind-dispersed young forest, they create a forest species compositional structure that is visible for centuries. Most conspicuous are the individuals of huge 100–300-year-old trees of species that nearly always start in open sites and whose seeds are dispersed by large mammals, e.g. guanacaste (*Enterolobium cyclocarpum*: Janzen, 1981a, b), cenizaro (*Pithecellobium saman*: Janzen, 1982b), jicaro (*Crescentia alata*: Janzen, 1982c) and jobo (*Spondias mombin*: Janzen, 1985a). These are not to be confused with other mammal-dispersed dry forest trees, such as nispero (*Manilkara chicle*), guapinol (*Hymenaea courbaril*) and oak (*Quercus oleoides*) that may start in shade or sun in the first year of succession, and continue to maintain a viable population as the site converts to unperturbed old-growth forest. There are also the species with intermediate biology, such as nance (*Byrsonima crassifolia*) and hojaraspa (*Curatella americana*).

Heliophilic and serendipitously fire-tolerant, these two vertebrate-dispersed species may persist in low-quality pastures even with an annual burning regime. However, when the pastures are eliminated through forest restoration, they persist as low-density species in the forest through sapling survival in tree falls, cliff edges, ridge tops and other areas with a higher-than-usual frequency of canopy rupture.

The large expanses of young forests of tens of species of wind-dispersed species occupying hundreds of hectares of old ACG pastures are being gradually invaded by hundreds of species of vertebrate-dispersed species that are defaecated or cached (e.g. by agoutis: see Hallwachs, 1986). These species are slow to invade and become reproductive populations largely because the successional stage dominated by (the first invading) wind-dispersed species rarely has major tree falls until the trees reach senescence 40–100+ years hence. However, many of the wind-dispersed species seem to die at a somewhat programmed age in the 30–50-year age bracket (e.g. *Cochlospermum vitifolium*, *Lonchocarpus acuminatus* and *L. minimiflorus*). At times, nearly pure stands of one of these species may die within a five- to ten-year period, leaving many insolated opportunities for growth and survival for shade-supressed vertebrate-dispersed species that have been accumulating in the understorey.

In 1985, the plant species composition and relative abundance on any given hectare of the ACG bore little or no resemblance to the community structure under which most, if not all, of these species evolved. Likewise, the seed-dispersing vertebrates occurred in patterns and abundances unrelated to what they will again some day display when the entire ACG is old-growth forest (albeit an ecological island). The 'match' is always somewhat sloppy between the size and timing of the fruit crop of any given portion of a species of plant and the congruent frugivores that contribute the coterie of seed dispersers. When the Pleistocene megafaunal extinctions were layered onto this process, a major long-term and variously permanent mismatch was set in place (Janzen & Martin, 1982). When the chainsaw, shotgun, pesticides and habitat fragmentation/rarification are applied to this same set of

vertebrates, any match becomes even more serendipitous. We expect a multitude of shifting degrees and qualities of mismatches to be characteristic of the ACG plant–seed-disperser interaction as restoration moves ahead. It is commonplace to find a tree in one part of the ACG having all of its fruits consumed (and seeds dispersed to some degree) while a conspecific in some other habitat fragment has almost no fruit removed by dispersers. Equally, trees that produce very large fruit crops at supra-annual intervals commonly satiate all available fruit-eaters, with a very high proportion of the fruits remaining on or below the parent tree and having their seeds gradually killed by seed predators.

There is one trait of old-growth landscapes that may take an exceptionally long time to restore in the ACG. Each of the habitats had its own distinctive sets of species, and these sets were maintained (and at some particular density) not only by reproduction of the residents, but also by incoming seed flow from other habitats (Janzen, 1986d, e). This means that some of the extant forested ravines, for example, do not today have the plant species composition that they once had when subjected to the seed shadows of the trees growing on the adjacent flats. Also, as the forest puts itself back together on the reforesting flats, it will not attain its 'true' old-growth composition until the ravines have also obtained theirs, a composition that will depend in part on seed flow from an old-growth forest on the flat. How much of each complex species-rich tropical forest in a given habitat is made up of resident species that self-maintain their populations, and how much is made up of 'lost plants' remains to be determined (Janzen 1986e). This old-growth trait cannot be well studied until there are again some large intact old-growth landscapes with their interdependent plant associations.

The above processes apply to much more than just the interaction between plants and their dispersers (and seed predators). Society and the biologist have come to perceive and study the pre-restoration wobbly *status quo* as 'natural' and therefore 'normal' in some sense. Then as successional changes occur through forest restoration, the changes can be both startling and distressing. This is especially true if the restoration is part of a conservation movement intended to save familiar nature. In the oceans of ACG secondary successional forest, many species of ACG butterflies and birds are much less visible today than they were in 1985. The white-tailed deer (*Odocoileus virginianus*), much appreciated by the public, will become rare almost to extinction in the ACG dry forests as the decades pass (and see below). We can envision conservation area staff being sued by an angry public for not having 'protected' the deer. The guanacaste tree, the namesake of the province and the ACG, and the national tree of Costa Rica, will go locally extinct in the ACG unless at least one Pleistocene megafauna, the horse, is put back into the ACG. Ironically, the guanacaste tree is also going extinct on the Costa Rican agroscape as agro-industry cleans up the field edges and the motorised horse replaces the seed-dispersing horse.

EARLY STAGES OF SUCCESSION

An ACG pasture, or any other open area, receives a microgeographically heterogeneous seed rain containing almost all possible species. Only some of the really large seeds that are today carried primarily by forest-loving agoutis or tree-dwelling primates are largely missing from this rain, and they would have been in the Pleistocene seed rain. Seedlings and saplings in the initial stages of ACG primary and secondary succession therefore range from extremely short-lived 'early succession' species to species with successional life spans measured in centuries. All of these species also exist in old-growth forest, albeit in totally different patterns and abundances than are encountered in the initial stages of succession.

In addition to the above life span heterogeneity, the act of irregular burning, clearing, logging, farming, irrigating, grazing etc. this large area of dry forest landscape has created hundreds of species distributions that are very heterogeneously distributed over that landscape – common, rare and absent here and there. Even after ACG fire control was instituted in 1985–7, there have been 30–100 small fires per year. These have maintained some habitat and distributional heterogeneity. Such anthropogenerated species distributions are poor indicators of where any specific organism's population will end up common, rare and absent in the eventual old-growth situation. For example, *Casearia corymbosa* is a common forest-edge shrub and medium-age secondary dry forest understorey shrub, with most members of the population today in these two habitats. However, once the ACG dry forest has returned to natural old-growth status, *C. corymbosa* will be a very rare shrub restricted to particular kinds of cliff edges and ravine banks. It will take a minimum of two to five centuries for each of the ACG species to settle out into what might be called an 'old-growth population and microgeographic structure' (and see Kellman *et al.*, 1998; Williams-Linera *et al.*, 1998).

Once an ACG abandoned pasture has filled in with a dense stand of wind-dispersed species, interwoven with growing islands of animal-dispersed species, there is a distinctive process that will need decades to centuries to sort itself out (Finegan & Delgado, 2000). These large stands of relatively even-aged plants cast a deep shade that is not broken until decades to centuries later as large individuals die. The insolation-demanding smaller plants, and the animals associated with them, rapidly disappear (or seem to disappear) from the site.

Seeds disperse into a restoring forest site on a species-by-species basis. However, they do not survive and establish as isolated units but rather as members of obligatory to facultative interactions with pollinators, mycorrhizal associates, seed-dispersers, microclimate generators, herbivores, etc. For example, for ant-acacias to arrive and survive, acacia-ants have to arrive and survive (Janzen, 1983*c*). However, for acacia-ants to arrive and survive, ant-acacias have to arrive and survive. Many ACG populations are creeping only slowly into the large expanses of restoring forests, as a spreading of extant remnant populations, rather than rapidly as they would be if each species invaded as an independent unit.

SOILS

The ACG dry forest soils range from weathered serpentine 85 million years above the sea (much of the Santa Elena Peninsula: cf. Janzen, 1998*b*) to recent marine sediments to weathered volcanic tuff and other debris a few decades to 1.5 million years of age to latosols of older volcanic origin. Alluvial deposits of material eroded from all of these origins (pure and mixed) are currently under restoring forest of all ages. All have been subjected to long and complex burning regimes. Natural forest restoration is occurring most rapidly on the recent marine sediments both because the soils are better in texture, nutrients and water storage, and because these areas generally were not converted to grass pastures. Dry forest is slowest to invade the grasslands created by anthropogenic fires on the serpentines of the Santa Elena Peninsula, the white tuff soils of the lower western slopes of Volcan Rincón de la Vieja, and the ancient latosols of Cerro El Hacha (the 6-million-year-old core of a volcano).

Clearing ACG dry forest for pastures creates a litter-poor thin soil that is rock-hard in the dry season and has high rates of rainwater runoff in the rainy season. It appears as though it will not support forest again. However, the experience has been that during the 5–15 years of natural forest invasion once fire is eliminated, these grassy hardpans on volcanic substrates revert to a deep surface loam that is rich in organic material. On serpentine soils, the forest and soil restoration process is much slower, but it occurs. The slowness is the combined effect of poor soils (and hence low plant productivity) and low seed-disperser density (as well as reduced seed crops) as compared with forest invading on volcanic soils.

A particularly dramatic example of ACG soil restoration through management has occurred through the deposition of a 10–30-cm-deep layer of processed (essential-oil-extracted) orange peels on centuries-old pastures filled with jaragua grasses and broadleaved herbs (Livernash, 1998; Janzen, 1999a, b, 2000c; Escofet, 2000). This treatment has created a 5–10-cm-deep black porous loam where there was a hard and nutrient-poor degraded pasture soil. This loam supports a 50–100-cm-tall dense stand of 50–100 species of herbs (including sedges and grasses) and young woody plants. The orange-peel layer asphyxiated the jaragua roots.

The initial invasion of the 3-ha site was from the broadleaf plant seeds in the soil that survived the anoxia and violence of the one to two years of orange-peel biodegradation by native fly larvae and micro-organisms. This is overlain by a wave of bird- and wind-dispersed species. This biodegradation has been used very effectively as a environmental service paid for by the neighbouring agro-industry, and will be used by the ACG to create jaragua-free firebreaks and hasten forest invasion in old jaragua-rich pastures in future years (Janzen, 2000c).

EVERGREEN TREES

The ACG old-growth dry forest extended from the middle–lower slopes (400–800 m) of the western Cordillera Guanacaste (Volcán Orosí, Volcán Cacao, Volcán Rincón de la Vieja/Volcán Santa Maria) westward and down to the coastal dunes, mangroves and freshwater swamps (Fig. 24.2). Along this gradient, there was a conspicuous but gradual increase in dry season deciduousness. This was most pronounced on dry ridges (and low-grade agricultural soils) and slowest to be expressed in valleys.

A representative dry forest on moderately flat and good soil (e.g. the small remnant old-growth forest in the Bosque Humedo in Sector Santa Rosa 3 km north of the administration area), tended to have a canopy with 5–15 species of evergreens making up 30%–60% of the crowns, and the remaining canopy space occupied by 100–200 seasonally deciduous species (each at low density). These were leafless throughout much, to all, of the five- to six-month dry season. Plants in the understorey, both juveniles of the overstorey species and those that reached reproductive maturity, followed the same pattern, but tended to be more evergreen (the 'evergreen' overstorey trees, such as Manilkara chicle, Hymenaea courbaril, Swietenia macrophylla, Sloanea terniflora and Quercus oleoides, are not really evergreen; each drops the entire leaf crop and makes another within one to three weeks at a particular time of year, while understorey evergreens drop old leaves below while making new leaves above and are truly evergreen).

When such a 'semi-deciduous' forest is cleared or heavily logged, it is the deciduous trees whose crowns quickly form the canopy for the site. They are fast-growing and annually have large crops of relatively small seeds. Old-growth dry forest is now so rare in Mesoamerica that these highly deciduous secondary successional forests are widely viewed as 'original' (and often studied as natural exemplars of tropical dry forest), which they are not. With time (various centuries, depending on the distance to seed sources and the presence of seed-dispersing vertebrates), the evergreens gradually regain their proportional abundance in the canopy. They are slow-growing, supra-annually seeding, and produce comparatively few (large) seeds. This slow invasion process is one of the principal reasons why the ACG dry forests will require many centuries of restoration before they will fool the alert biologist into being viewed as 'undisturbed' by European-style agriculture. For example, *Hymenaea courbaril* trees in forest do not even begin to produce significant seed crops until they are 100+ years old. This process of reinvasion of the evergreens (and attendant readjustment downward of deciduous species proportional abundances) can be speeded up by expenditures on planting seedlings of these large evergreens into the early-successional seres. However, restoration funds are probably better spent by buying more hectares of degraded pasture containing forest fragments, thereby creating larger final areas of old growth many centuries hence.

The restorational shift from highly deciduous to much more evergreen renders very dramatic changes in everything biological. For example, in the early stages in the ACG, when there were few such evergreen patches, they were sites of concentration of insects seeking shade and humidity to pass the dry season as reproductively dormant individuals. However, when the rains begin in May, these shady sites were quickly abandoned for the more insolated (more productive, more edible) deciduous forest (Janzen, 1983c). As evergreen-ness re-establishes itself across the landscape, this phenomenon is becoming less visible. Increased evergreenness also causes insects to appear to be less abundant because during the dry season they cannot be found as readily as before. A forest becoming richer in evergreen (and longer-lived) crowns

becomes more inhospitable to the juveniles of fast-growing deciduous trees, and offers less light resource for sexual reproduction by the understorey plants that reproduce there.

In the early stages of landscape-level restoration, the small patches of remaining semi-deciduous old-growth forest are also sites of concentration for vertebrates in the rainy season. They offer shade, moist soil in dry spells, nesting trees and tree holes, and distinctive species of seeds and fruits. This in turn generates massive animal-dispersed seed flow from deciduous species in the ocean of surrounding secondary succession into the more evergreen forest. This in turn alters the species composition and competition regime of natural succession in tree falls (Janzen, 1983a), which in its turn alters the nature of the old-growth forest. While these old-growth patches are very important seed sources for the restoration process outside of them, this animal behaviour also leads to their alteration. This has led to the suggestion that they may do better surrounded by rice fields than by oceans of secondary succession (as in so-called 'buffer zones'), if they are to retain their old growth structure (Janzen, 1983a).

The ACG herbivores are being dramatically impacted by the successional shift from strongly deciduous to semi-deciduous forest. First, as the evergreen species come to occupy progressively more of the overall photosynthetic terrain, herbivore species richness and biomass carrying capacity is declining because there are relatively few species that can, or do, eat the leaves of evergreens. This also means that seeds and saplings of deciduous species become less abundant in the understorey, with the same consequence for the herbivores. Second, the evergreen species not only store more of their photosynthate in denser and longer-lived (and chemically more toxic) wood, they also parcel out their seed and fruit crops at multi-annual intervals rather than annually. This creates a more erratic food source for seed and seedling predators/parasites, and their predators/parasites in turn. For example, the evergreen *Hymenaea courbaril* in the Bosque Humedo old-growth forest have not had a mast seed crop since 1983. Third, the annually discarded leaf crop of the evergreens is much more resistant to biodegredation than are the leaves of the deciduous species (and the latter fall as caterpillar frass much more than is the case with the evergreens). This means that the litter dynamics of an old-growth forest is qualitatively (as well as quantitatively) quite different from that of the secondary successional more deciduous forest that it replaces. The latter, in turn, has its forest-floor litter far more heavily exposed to wind, heat, drying, and UV than does the forest floor below the old-growth semi-deciduous forest.

The broken terrain of the ACG dry forest, coupled with the highly heterogenous way in which it was logged and cleared, resulted in many cases of a single or few ancient evergreens left isolated on a cliff, ravine bank, or even in a pasture. These were mostly 'living dead' until the ACG restoration process started in 1985. Their large seed crops (often made in greater abundance than by conspecifics in old-growth forest) were falling below the parent and dying there as seeds or seedlings owing to an absence of seed-dispersers, or at least seed-dispersers that would move the seedlings to where they would be free of fire. However, as the reforestation process gets under way, these isolated individuals are now finding themselves, and their seed crops, to be once again a member of a forest with its attendant traits.

Watercourses through ACG dry forest pastures are displaying among the most dramatic of the changes through forest restoration. As the streams and rivers become once again subject to forest-ameliorated surface runoff of both water and organic debris, they also become deeply shaded. This shade is particularly intense due to individuals in riparian sites remaining leafy longer than their conspecifics on dry hillsides, and due to the riparian habitat supporting evergreen species. This in turn alters the use of the site by both terrestrial and aquatic biotas. It also leads to the streams flowing for more days into the dry season than they did when the hillsides were deforested.

ACG ECOLOGICAL INSULARITY

The ACG is not simply a monstrously complex case of heterogenous secondary succession. It is simul-taneously undergoing the shifts and shrinkages associated with settling into the large ecological island in the agroscape that is its permanent destiny. The ACG is no exception to the general rule that all tropical conserved wildlands are destined to be ecological islands in an ocean of agroscape.

As the tiny Parque Nacional Santa Rosa expanded to the ten-times-larger restoring ACG, simultaneously, the ACG was beginning to settle into the traits it will have as an ecological island. This island is a kaleidoscope of overlapping habitat islands, each of which will eventually come to have some (lower) new equilibrium density of species (and their microgeographic distributions). As we watch the shifting fates of the multi-specific and multi-habitat successional process, we are simultaneously watching a site suffer the losses attributable to ecological insularity, much as happens when a peninsula tip is cut off from the mainland to become an island (e.g. Riddle *et al.*, 2000). The demographic expressions of these dynamics will intersect and confound, and there is nothing we can do about it but to observe. As the mountain lion (*Felis concolor*) density declines in the ACG with restoration (owing to the decline in deer density), it may also decline through insularity (owing to the decline in area of habitats). As the jaguar (*Panthera onca*) density increases in the ACG with restoration, it may also increase with relaxed insularity as the society of the neighbouring agroscape becomes more forgiving of jaguars and hence less predatory on them.

ANIMALS AS SEED PREDATORS AND PLANT PARASITES

When the ACG dry forest lost its Pleistocene megafaunal seed-dispersers and plant predators and parasites (i.e. herbivores) (e.g. Janzen & Martin, 1982; Janzen, 1984), the consequent shifts in vegetation structure, plant demographics, seed shadows, and primary and secondary consumers must have been spectacular. What we contemporarily call old-growth is this dry forest without its megafauna. The last four centuries of direct vegetation and faunal assault by humans produced a much more dramatic alteration. When we initiated the restoration process in the ACG in 1985, we were setting the site back on to multiple trajectories of recovery (varying with soil, proximity to seed and animal sources, thoroughness of pasture and field clearing, etc.). Within this heterogeneity the fauna is also being restored. Some seed-dispersers and foliage herbivores are becoming much more common (e.g. agoutis, folivorous monkeys) while others are conspicuously declining in density (e.g. white-tailed deer, collared peccaries [Tayassu tajacu], many species of caterpillars). Again, as with the impact of insularity layered onto this multi-species process, the impact of phenomena such as the amphibian decline (Houlahan et al., 2000) and the rising elevation of the cloud forest through global warming (Pounds et al., 1999) are layered on to, and possible causes for, the changes that appear to be caused by succession.

A telling example is the following. In two decades of watching the ACG dry forest succession over a very large area, it is obvious that the elimination of fire, and with it the hundreds of kilometres of edge between grazing sward and forest, the density of white-tailed deer is gradually declining. The deer maintained (and still maintains) a large and heathy population of a specialised deer predator, the mountain lion. What does such a long-lived smart predator do as its usual prey decline? It shifts to other prey, such as peccaries and juvenile tapirs (Tapirus bairdi), among others. Both of these species are displaying a conspicuous decline in numbers as the dry forest fills in the pastures with dense stands of juvenile trees, which neither have low browse for the deer nor are (yet) dropping large quantities of edible fruits and seeds. The older stages of succession in which the peccaries and tapirs have always found a large proportion of their food have not diminished in area, nor are these vertebrates being poached.

The ACG community of small herbivores is clearly changing with succession. The overall abundance of caterpillars and leaf-cutter ant nests (Atta spp.), which together perform easily 90% of the green foliage consumption in the ACG dry forest, appears to have been on a steady decline since 1977–8 when I first began to pay attention to herbivory in general (Janzen, 1993). Many species of moths were very common when the ACG was a mosaic of secondary forest and grazing swards. Their density has plummeted during 15 years of restoration. Severe defoliations of a particular species of food plant by its particular host-specific caterpillar (Janzen 1985b, 1988f) have also become much rarer to almost non-existent in the 1990s. The extreme population fluctuations of Liomys salvini, seed- and pupae-predating mice, have dampened substantially since the first four-year cycle documented in 1983–7.

INTRODUCED/INVASIVE SPECIES

Essentially all of the biota of the ACG is 'introduced' in the sense that it has come into the site following a massive volcanic cleaning of the slate 1.5 million years ago (and a variety of smaller volcanic perturbations since) (e.g. Janzen, 1986f). Essentially all of the ACG biota has a distribution ranging from many degrees of latitude to the north to many to the south. There is no way to know in what ecological island each of the ACG species evolved to its current form. What is clear, however, is that the species arrived and have 'ecologically fit' (Janzen, 1985c) together into the 'communities' that we see today. The concept of an 'introduced' or 'invasive' species is very arbitrary.

The terrestrial ACG, like many Mesoamerican continental semi-wild (extensively ranched and subsistence farmed) habitats, is not plagued by any introduced species of animals (other than those actively or passively maintained by humans and their dwellings). We have been, to date, spared feral populations of the mongoose (Herpestes ichneumon) and Asiatic buffalo (Syncerus caffer). The horse, an example of Pleistocene megafauna, is perhaps the closest approximation. Both the horse and cow are especially noticed and reviled as destroyers of waterholes and streams during the dry season, and are unambiguously removed by the classical national park process. However, the true 'natural' state of these waterholes, under which they have evolved for tens of millions of years, is as one today encounters dry-season watering sites in east African national parks – muddy pits churned by frequent bathing, defaecating, trampling and drinking by large herbivorous mammals.

The aquatic ACG is a different matter. Tilapia, an African food fish now widely introduced (and 'escaped') throughout the tropics, is a serious predaceous weed that is increasingly common in Costa Rican rivers. It is in the lower reaches of the Rio Tempisque, whose upper tributaries all flow from the ACG and many of which have sufficient water volume to sustain *Tilapia*. Other ACG stream systems are under constant threat of someone introducing tilapia downstream. Full restoration of the ACG 'terrestrial' ecosystem may well have to simply live with a riverine ecosystem thoroughly altered by the introduction of this predator. It is as if the mongoose had been liberated in the ACG. It is hoped that the threat of trout will be avoided by the fact that all of the cold-water upstream portions of ACG rivers and streams are within the ACG.

The African honeybee (*Apis mellifera*) arrived in the ACG in the early 1980s. It occupied all ages of successional forest and was extremely abundant (as was the usual case on its advancing front from south to north through Central America). Within five years, however, its density fell to where today it is just one more very low-density member of the local bee fauna. However, over the same course of two decades, the overall solitary and social bee density has also declined very noticeably throughout the ACG dry forest. It is possible that this is the consequence of elimination of the pastures and forest edge (which dramatically lowers flower abundance, at least while the pastures are filling with non-reproductive young trees). The general decline in wild native bee biodiversity and abundance in the remainder of Guanacaste Province, probably attributed correctly to agro-industrialisation (Frankie *et al.*, 1998), is not counterbalanced by a bee increase in the large ACG dry forests being restored. However, the massive numbers of bees and bee species encountered in the 1960–80 period in Guanacaste Province were present under circumstances that were very favourable to these insects. There were large expanses of broken and patchy forest with highly favourable nest sites and huge flower crops from insolated herbs and shrubs on edges.

The ACG ctenosaur or spiny-tailed iguana (*Ctenosaurus similis*) is another conspicuous vertebrate whose 'ideal' habitat is severely reduced by forest restoration and whose density is declining.

The once omnipresent bare insolated ground for egg nest tunnel construction (Janzen, 1983c) and the masses of herbaceous vegetation that generate insects for the new (insectivorous) hatchlings are being drastically reduced. Equally, prime succulent new vegetation, for the mostly ground-dwelling and ground-foraging, largely herbivorous adults, is reduced in amount. Whether the new predator regime being restored favours or hinders the ctenosaur population, it certainly is different from that when the area was clothed in extensive cattle ranches. And the ctenosaur population is now generating much less small to large prey biomass for the predator (and parasite) regime.

Mesoamerica is densely clothed with more than 150 species of African grasses introduced during the past century (Parsons, 1972). Their populations are largely maintained by the pasturing process and anthropogenic direct disturbance (roadsides, construction sites, anthropogenerated flooding, etc.). The only species that is sufficiently common in the ACG dry forest restoration arena to be of habitat-level concern is jaragua (a major native ingredient in Serengeti grazing swards). Jaragua was widely and deliberately introduced into all ACG dry forest pastures to improve cattle yield. When the cattle are removed, it creates 1–2-m-tall dense stands that carry a high wall of fire in the dry season. When the ACG cattle were removed in the late 1970s, year after year this fire was progressively eliminating the patches of forest that had coexisted with the grazed swards for decades (and for centuries with other grasses). However, jaragua is very shade-sensitive. The straightforward management act of stopping the fire and allowing the forest with its shade to invade eliminates the large stands of jaragua. It remains to be seen whether it will persist on forest edges and as a naturalised ruderal.

The removal of jaragua by the forest restoration process unambiguously lowers the amount of food available to those few native species that ate jaragua, and subsequently the species that are higher on the food chain that ate them. For example the rice rat (*Sigmodon hispidus*) was very common in the 1970s and 1980s in the ACG jaragua pastures and adjacent old fields (e.g. Bonoff & Janzen, 1980). It now appears to be nearly extinct. The subsequent impact of this reduction in coyote (*Canis latrans*), owl and diurnal snake food may be in part responsible for the conspicuous decline in numbers of these top predators in the ACG restoring forest. Equally, the large numbers of somewhat generalist acridid and tettigoniid grasshoppers that were common in the old pastures are now reduced to the occasional individual encountered on a roadside.

The jaragua pasture (and the pastures of other species before it) was not a plant monoculture, but rather one where 90%+ of the vegetation was a single introduced species. Interwoven with the jaragua and its place-by-place variation in fire pattern and intensity, there were at least 100 species of other native (and a few introduced) grasses, herbs (Cyperaceae, Liliaceae, Fabaceae, Orchidaceae, Turneraceae, Melanthiaceae, Scrophulariaceae, Gesneriaceae, etc.), and small shrubs, both annual and perennial (the majority). The thousands of hectares of *Trachypogon plumosus* 'pastures' (on the Santa Elena Peninsula, on Cerro El Hacha) contain yet more (and somewhat different) such species. How many of these will survive within the ACG once the pastures are gone? Most will survive in the other kinds of (micro-) habitats that have frequent or continuous ground-level insolation, and at densities and relationships much more akin to the 'natural' situation than at present. However, an effort to favour their continued existence in the ACG is represented by the grass pasture sites within the ACG that will be burned at annual or multi-annual intervals. These plants (and associated other organisms) will experience the usual consequences of fragmentation of a once large and semi-continuous distribution.

The same process will occur in the few ACG seasonal marshes as they are restored back to forested swamps, though these habitats have always been highly insular. On the other hand, there may be some species that today are barely surviving in these insolated marshes but that will once again become common when their rainy-season waters and dry-season mud are heavily shaded by a forest canopy.

Pasture elimination is dramatically shrinking the peculiar habitat called 'pasture edge' down to its much more limited and fragmented extension as cliff and landslide edges, rocky ridges, ravine and river banks, and tree falls (each being in itself a special kind of natural restoration habitat). Many populations of animals and plants existed in a state of high density, high reproduction, lush vegetative growth, etc. on the thousands of kilometres of ACG pasture edge. With the restoration of the pastures to forest, these species are moving toward grossly different population and microgeographic characteristics, and some may go locally extinct. As emphasised earlier, the hundreds of species of caterpillars that were once very abundant on ACG forest edges are now substantially less abundant than they were in the 1960–80 period.

While jaragua (and other pasture grasses) offers spectacular examples of isolated terrain being clothed by a new arrival, in the year 2000 the ACG discovered an apparently self-introduced terrestrial understorey African orchid, *Oeceoclades maculata* (J. Ackerman, personal communication; B. Hammel, personal communication) in the dry-forest–wet-forest intergrade (Estación Maritza, 600–700 m). It is quite possible that this plant will spread to thoroughly occupy the forest understorey as it has in Puerto Rico (J. Ackerman, personal communication) but on the other hand it may be found that on the mainland it only flourishes in secondary succession and becomes just one more low density member of the flora in 'old-growth' forest.

MIGRATION

ACG seasonal migration is both local (to and from moist places), and more distant (to and from the rainforests and cloud forests: e.g. Hunt *et al.*, 1999). It is an integral part of the natural history of ACG dry forest species (e.g. Janzen, 1987*c*, 1988*g*, in press). This propensity to move in search of seasonal resources (moisture, food, sun, carnivore-free space) is part of the reason why ACG pastures are quickly colonised by the dry forest biota once allowed to do so. Equally, the seasonal movements of seed-dispersers (and the microbes contained within migrants of all types) are a major factor in colonisation of the old pastures, fields and logged sites.

Forest restoration unambiguously changes the overall migration phenomenon. As species of mobile animals change their density with the loss of edge and pasture, and with the increased biomass of forest, so will change the annual output of migrating propagules and the annual absorptive power for incoming migrants. This applies to both the resident Costa Rican species, and the migrant birds that arrive to pass the northern winter in the ACG.

The detailed impact of restoration on migrant biology will vary case by case, and understanding will require a large investment in natural history observations and data. For example, reforestation of the ACG dry forest means tens of thousands more hectares covered with the understory rubiaceous shrubs, *Psychotria nervosa*, *P. horizontalis* and *P. microdon*. These are the primary dry forest food plants for the large caterpillars of the highly migratory sphinx moths *Xylophanes chiron*, *X. anubus*, *X. maculator*, *X. ceratomioides*, *X. pluto*, *X. porcus* and *X. libya*. These moths arrive from the rainforest side of the ACG (and the remainder of Costa Rica) with the first rains in May. It is unknown if the size of the next generation (the outgoing generation) is determined by the number to arrive or the amount of food plant available. Since it is more likely to be the former than the latter in this particular case (the caterpillars are thinly spread among their food plants), restoration of the ACG dry forest will probably have little impact. It is possible that restoration of rainforest habitat in the eastern wetter ACG will, however, generate more of these moths since their food plant density is much lower in the ACG rainforest. This may in turn increase the numbers of incoming migrants to the dry forest, and perhaps these increased numbers will reach the level where dry forest restoration would have an impact.

The large white-lipped peccary (*Tayassu pecari*) is particularly mobile in the ACG. The herd(s) normally circulates over many tens of square kilometres during the year. These animals currently visit the ACG dry forest in January–March, feeding on the newly fallen acorns and other dry season fruits/seeds, and then apparently return to wetter forest. As the ACG dry forests return to an old-growth state, it is very likely that the amount of habitat attractive to white-lipped peccaries will enormously increase. The outcome is likely to be that this 'migratory' animal will actually spend more time in the ACG dry forest than

it does at present (with consequences for understorey plants, seed shadows, seed predation regimes, etc.).

DRY FOREST IN CONTRAST WITH CLOUD FOREST AND WITH RAINFOREST

The original intent in 1985 of the ACG was the full restoration of the its dry forest ecosystem (Janzen, 1988a). However, it quickly became evident that the adjacent middle-elevation to cloud forests of the Cordillera Guanacaste to the east, and even the rainforests further to the east, were essential as seasonal refuges for the dry forest biota (and vice versa). They have subsequently also become important as wetter ecosystems into which dry forest organisms can move with the drying and heating of the ACG dry forest that is occurring with global warming. These realisations have led to both closer examination of cloud forest and rainforest pastures and the restoration process, and led to buying rainforest pasture, fields and logged areas for restoration. This in turn has allowed direct comparison of the restoration process in three very different ecosystems that both intergrade and lie only a few kilometres apart.

The most glaring contrast is in the slow rate of forest invasion of rainforest and cloud forest pastures clothed with various species of African grasses. As mentioned earlier, the dry forest invasion process begins as soon as the anthropogenic fires are eliminated, and within a decade on average-quality soils, there is a closed canopy and the bulk of the grasses are rapidly being shaded out. In the wetter pastures, however, a clean, ungrazed and unburned pasture may exist for decades before becoming covered with naturally colonising woody plants (see also Aide et al., 1995; Slocum, 2000). It can take as many as five decades before an abandoned rainforest pasture has disappeared beneath a naturally occurring young regenerating rainforest, and even longer at elevations above 600–800 m (Zahawi & Augspurger, 1999; Duncan & Duncan, 2000; Holl, 2000; Holl et al., 2000, Wijdeven & Kuzee, 2000).

There are at least four major obstacles to colonisation of pastures by the ACG wetter and cooler forests (increasing elevation exacerbates all four of these restoration-delaying processes).

First, there are almost no wind-dispersed species in the forests adjacent to these pastures. The restoration process is almost entirely dependent on vertebrates for seed movement.

Second, the wet forest seed-dispersing vertebrates are much less inclined to move through and forage in the wet forest pastures than are dry forest vertebrates in dry forest pastures. This results in conspicuously less vertebrate-generated seed rain into the wet forest pastures than into dry forest pastures. This process is complicated, however, by the fact that a hectare of ACG wet forest adjacent to a wet forest pasture has substantially less biomass of seed-dispersing vertebrates circulating through it than does its dry forest equivalent.

Third, when a rainforest seedling germinates in a rainforest pasture, it has been moved into substantially more sunny, dry, hot and windy circumstances than is the 'normal' situation in the adjacent rainforest understorey and tree falls. In contrast, dry forest seedlings in a dry forest pasture are in circumstances not as different from what they have to survive in the dry season in the forest.

Fourth, admittedly more speculative, it may be that tree seedlings in rainforest pastures have more difficulty finding their mycorrhizal associates than is the case in dry forest (Janos, 1996; Allen *et al.*, 1998; Holl *et al.*, 2000). In the dry forest, spores of mycorrhizae, formed in abundance before or during the dry season, are likely to be blown or wafted throughout the habitat by the dry and windy weather of the dry season(s). The forest seedling that germinates in the pasture in the next rainy season has a better chance of being able to establish the appropriate mycorrhizal associations than is the case in the wet forest pastures where mycorrhizae creep into the pasture from its margins as tree roots do.

Whatever the causes, the lethargic return to forest secondary succession in ACG wet forest pastures has led to several moderately costly attempts to establish forest beginnings with native species by various combinations of ploughing, burning, seedling-planting and mycorrhizal inoculum. In all cases, a few trees survive but nothing approximates the closed canopy needed to shade out the dense stands of pasture grasses. However, these wet forest pastures will fill through forest invasion from their margins over many decades (e.g. Aide *et al.*, 2000; Finegan & Delgado, 2000; Slocum, 2000). This means that available funds would best be put into buying more pasture–forest mosaic for restoration rather than into trying to hasten the forest restoration process artificially.

However, there is one unambiguous way to eliminate these ACG wet forest pastures in three to five years (Janzen, 2000*a*). Fill them with a commercial plantation of gmelina (*Gmelina arborea*). Grown throughout the tropics for fibre and cheap wood, this tree has a dense canopy which closes in about a year. The dense shade kills the grass. The understorey rapidly fills with a very diverse community of rainforest trees, shrubs and vines. They are brought by the small vertebrate community, drawn from the species pool of the neighbouring rainforest, and shade-tolerant as juveniles. After six to eight years of this management, the gmelina will either be killed in place and left to rot, or harvested and sold to bolster the ACG endowment. The understorey plant mortality from gmelina poisoning or harvest can be tolerated by the millions of juveniles now left to continue the forest succession. This method has been used with a variety of other species of trees in commercial plantings as a device to restore forest ecosystems on degraded tropical lands as part of agroscape management (Parrotta & Turnbull, 1997).

The interface of dry forest with rainforest and cloud forest interacts strongly with the forest restoration process. As every frontier farmer and rancher knows, clearing forest dries and warms the site. This means that the dry forest biota invades the rainforest and cloud forest through the pasture and field habitats (and associated roads and trails). In the ACG, examples are the movement of white-tailed deer, coyotes, ctenosaurs, rice rats, rattlesnakes (*Crotalus durissus*) and a multitude of plant and insect species into mid- to upper elevation wet/cloud forests, and into lower elevation rainforests. Restoration of these fields and pastures back into the wetter forest that once occupied their soils and rainfall regimes gradually expels these dry forest organisms. This does, however, raise the management question as to whether some of these rainforest pastures should be maintained in a 'disclimax' state both for the diversity of dry-forest—wet-forest interactions they display, and for the living convenience of staff and visitors.

It is not clear whether a restoring forest on an abandoned ACG dry forest pasture takes longer to return to its original total standing biomass of sequestered carbon (above and below ground) than is the case with a wet forest ACG pasture, once there is a young forest in place (see Silver *et al.*, 2000). There are many growth processes that appear to accelerate per year with year-round rainfall, but decomposition processes do as well. I have also been impressed by how many centuries it takes to replace a canopy of fast-growing dry forest deciduous trees with the much slower-growing evergreen species (with much denser wood).

RECOLONIZATION AND POPULATION AUGMENTATION

The ACG, owing to its size and the very heterogeneously distributed assault on its ecosystems, has lost very few species during its 400 years of European-style agro-ranching. The giant anteater (*Myrmecophaga tridactyla*) is now extinct in the ACG, as well as in the remainder of Costa Rica. The restoration of its population could be attempted from South America, but will a reforested insular ACG have enough of the right kind of habitat to sustain a breeding population? As recently as 1975 the scarlet macaw (*Ara macaw*) still occurred in the ACG, and it has been seen again in 1999. It will probably reintroduce itself from points south (Palo Verde, Carara, Osa Peninsula) once the ACG has enough old-growth forest to support it. Alternatively, it could be reintroduced as with the giant anteater. The green macaw (*Ara ambigua*), once common in the wetter forests of the volcanic moister east end of the ACG, is presumably now absent because it does not find the habitat to be suitable. While a pair has been seen in the western slopes of Volcan Rincón de la Vieja, it appears to be near extinction throughout the country. Whether there is enough prey-rich habitat to sustain reintroduction or self-reintroduction by the three Costa Rican eagle-sized raptors remains to be seen (a crested eagle [*Morphnus guianensis*] was seen in 1997 in the northeastern ACG rainforest). As indicated earlier, white-lipped peccaries are generally absent from the ACG dry forest but will presumably self-reintroduce themselves by expansion from their current wet forest habitat as the old-growth dry forest with its larger-seeded trees comes to occupy more terrain.

There is no basis for introduction/addition of individuals of any species to "augment" the ACG's naturally present populations. Each of the species will expand (or contract) on its own to match the (still expanding) carrying capacity of the restoring ACG. Their genetic diversities and parasite loads will become what they become. Introducing other genomes from other parts of Costa Rica would simply modify whatever unique dry forest (or dry-cloud–rainforest) genome the ACG animals are carrying (though it is arguable that selection will take the altered genome back to whatever it was before). The restored ACG will be an ecological island (except for those species that range freely over the agroscape) and its populations will be genetically whatever they can be on an island of this size.

Free-ranging horses are a Spanish gift from the Pleistocene (Janzen, 1981*a*, *b*, 1982*b*, *c*; Janzen & Martin, 1982). They were standard herbivores and seed-dispersal agents in Mesoamerican Pleistocene dry forest, and can arguably have a place in the ACG 'restored' dry forest. They are also essential for survival of the guanacaste tree population in the ACG (e.g. Janzen, 1981*a*, *b*). The ACG will eventually establish such a 'feral' horse population on some sector of the ACG dry forest, complete with rifle or castration as a surrogate for sabretooth cats. At present, they have been removed from the restoration process because of their nuisance value for researchers and tourists. Furthermore, since horses are in no imminent danger of extinction, there is less impetus to worry about them. I note that were a ground sloth population to be discovered, the world would be happy to have this Pleistocene mammal survive in the ACG.

It is tempting to view the few large conserved Mesoamerican wildlands, all undergoing restoration, as potential Noah's Arks for species threatened with extinction elsewhere. Species occurring in other Mesoamerican dry forests, for example, could be introduced into the ACG and probably maintain 'normal' populations there. Since the bulk of the entire ACG biota came from elsewhere (e.g. Janzen, 1986e), it is very hard to argue against such a rescue mission. Whether it would result in the extermination of resident species is unknowable. This proposes the yet more subjective question of which species was indeed more 'important' to keep.

GENOMES AND EVOLUTION

Many (but certainly not all) of the ACG dry forest populations were both severely reduced in numbers and area that they occupied for decades to centuries. Are they now threatened by having genomes with reduced variation? While the dry forest alteration in the ACG was undoubtedly massive, on-site genetic bottlenecks or founder effects may have occurred only in a very few cases. Populations reduced to fewer than 10–100 individuals in well less than 1% of the species. Given the extreme heterogeneity of ACG habitats (dry forest to cloud forest spread over 2000 m elevation) a species population probably contained about as much genetic heterogeneity as it could. For example, while the ACG monkey populations were reduced to a few tens of troops by habitat reduction, those few tens of troops were scattered across almost all habitat types that the three species originally occupied in Costa Rica. Now as they expand back into restoring forests in all of these habitat types (e.g. Fedigan et al., 1996, 1998), their genetic landscape will not resume its original dimensions.

What is genomically more problematic is the role of restoration biology in the omnipresent speciation event that is being launched by fragmentation of once-gigantic and continuous geographic distributions throughout the tropics. If one were to set out to create a swarm of sibling species, the way to do it would be to fragment one of the hundreds of thousands of species that are distributed from Mexico/Mesoamerica down into South America, and occupy many ecosystems. By erecting oceans of agroscape isolating fragments of these ecosystems, we are creating hundreds of small to medium-sized populations. Each will occupy a somewhat different set of ecological circumstances, each of which is more uniform than the area occupied by the once-widespread population. Speciation will occur. This is the destiny of the species in most small tropical conserved wildlands, even though some of the larger ones, like the ACG, will have substantial ecosystem and habitat diversity within them. Restoration biology may recreate wildland islands, or allow some fragments to coalesce, but it will not eliminate the massive species-generation process set in motion by fragmentation of once-widespread populations.

What is the role of landscape-level restoration in such a scenario? By buying up a very large and but only moderately damaged area, such as the ACG, and allowing all possible ecosystems and habitats (and species in them) to restore themselves and their interactions, a maximally diverse evolutionary template is re-created on which species may reside (and evolve) into perpetuity. In general terms, bigger is better for such a scenario and process – especially if the 'bigger' is aimed at acquiring habitat and ecosystem diversity.

In today's world of thorough human occupation of the globe, conservation through restoration of large expanses is far more effective at conserving biodiversity than is searching across the landscape for small fragments of intact forest/biota. Fighting with society to save them as jewels is illusory. Each is fated to undergo the rapid extinction events and eventually, the evolutionary events, characteristic of small oceanic islands. While it is still practical to save large blocks of biodiverse tropical old growth in some wet forest mountains and major river basins, for nearly all of the tropical dry forest and its adjacent wetter areas this option is long gone (Janzen 1986a, 1988a). Furthermore, buying up large parcels of seemingly exploitable old-growth dry forest for conservation and biodiversity development is socially unwelcome. On the other hand, restoring areas of low-grade real estate and marginal agro/ranching land into dry forest meets minimal resistance.

Conservation through restoration has an additional pragmatic advantage. The restoration process itself is a kind of biotic engineering that can incorporate the roads, buildings, biodegredation sites, animal control, tourist facilities and impacts, and other activities essential for wildland biodiversity conservation through biodiversity development (Janzen 1988*a*, *e*, 2000*a*, *b*, *c*, *d*). Once the restoration process has run its centuries to become old-growth forest, these human footprints will seem as normal as are rivers, swamps, beaches and volcanoes in the few remaining old-growth tropical jewels.

ACKNOWLEDGMENTS

The experiences that have led to these comments on the ACG have been supported generously for 37 years by the US National Science Foundation (NSF), by the international scientific and donor community, and by the government and people of Costa Rica. More specifically, the personnel of the Area de Conservación Guanacaste (ACG), the Instituto Nacional de Biodiversidad (INBio), the Fundación de Parques Nacionales (FPN) and the Ministerio del Ambiente y Energia (MINAE) have provoked and facilitated the initiation and existence of the ACG. I particularly thank Alvaro Umaña, Rodrigo Gámez, Alvaro Ugalde, Mario Boza, Alfio Piva, Pedro Leon, Luis Diego Gomez, Rene Castro, Randall Garcia, Johnny Rosales, Luis Daniel Gonzales, Karla Ceciliano, Jose Maria Figueres, Maria Marta Chavarria, Roger Blanco, Angel Solìs, Isidro Chacon, Nelson Zamora, Jorge Corrales, Manuel Zumbado, Eugenia Phillips, Jesus Ugalde, Carlos Mario Rodriguez, Alonso Matamoros, Jorge Jimenez, Alejandro Masis, Ana Sittenfeld, Felipe Chavarria, Julio Quiros, Julio Diaz, Jose Jaramillo, Guisselle Mendez, Xiomara Driggs, Jorge Baltodano, Luz Maria Romero and Sigifredo Marin, and all of the parataxonomists of the ACG and INBio, for their specially insightful and inspirational input over the last 15 years. While it is clear that the international cast of contributors to a project of this nature is enormous, I particularly thank Winnie Hallwachs, Kenton Miller, Randy Curtis, Geoff Barnard, Walt Reid, Peter Raven, Tom Eisner, Jerry Meinwald, Ed Wilson, Don Stone, Paul Ehrlich, Hal Mooney, Kris Krishtalka, Jim Edwards, Gordon Orians, Monte Lloyd, Mike Robinson, Steve Young, Preston Scott, Leif Christoffersen, Odd Sandlund, Mats Segnestam, Eha Kern, Bern Kern, Hiroshi Kidono, Frank Joyce, Ian Gauld, Jon Jensen, Murray Gell-Mann, Steve Viederman, Staffan Ulfstrand, Carlos Herrera, Steve Blackmore, Meredith Lane, Jim Beach, John Pickering, Scott Miller, Amy Rossman, Bob Anderson, Terry Erwin, Don Wilson, Diana Wall, Keith Langdon, Neal Smith, Craig Guyer, Gary Hartshorn, Don Stone, Chris Thompson, Marilyn Roossnick, Dan Brooks, Charles Michener, Bob Sokal, John Vandermeer, Jack Longino, Rob Colwell, Chris Vaughan and Tom Lovejoy for their investment in this process.

REFERENCES

Aide, T. M., Zimmerman, J. K., Herrera, L., Rosario, M. & Serrano, M. (1995). Forest recovery in abandoned tropical pastures in Puerto Rico. *Forest Ecology and Management*, **77**, 77–86.

Aide, T. M., Zimmerman, J. K., Pascarella, J. B., Rivera, L. & Marcano-Vega, H. (2000). Forest regeneration in a chronosequence of tropical abandoned pastures: implications for restoration ecology. *Restoration Ecology*, **8**, 328–338.

Allen, W. (2001). *Green Phoenix: Restoring the Tropical Forests of Guanacaste, Costa Rica.* New York: Oxford University Press.

Allen, E. B., Rincon, E., Allen, M. F., Perez-Jimenez, A. & Huante, P. (1998). Disturbance and seasonal dynamics of mycorrhizae in a tropical deciduous forest in Mexico. *Biotropica*, **30**, 261–274.

Area de Conservación Guanacaste (ACG) (2001). www.acguanacaste.ac.cr

Barlow, C. (2001). *The Ghosts of Evolution: Nonsensical Fruits, Missing Partners, and other Ecological Anachronisms.* New York: Basic Books.

Bonoff, M. B. & Janzen, D. H. (1980). Small terrestrial rodents in 11 habitats in Santa Rosa National Park, Costa Rica. *Brenesia*, **17**, 163–174.

Duncan, R. S. & Duncan, V. E. (2000). Forest succession and distance from forest edge in an afro-tropical grassland. *Biotropica*, **32**, 33–41.

Escofet, G. (2000). Costa Rican orange-peel project turns sour. *EcoAmericas*, **2**, 6–8.

Fedigan, L. M., Rose, L. M. & Morera Avila, R. (1996). Tracking capuchin monkey (*Cebus capucinus*) populations in a regenerating Costa Rican dry forest. In *Adaptive Radiations of Neotropical Primates*, eds. M. A.

Norconk, A. L. Rosenberger & P. A. Garber, pp. 289–307. New York: Plenum Press.

Fedigan, L. M., Rose, L. M. & Morera Avila, R. (1998). Growth of mantled howler groups in a regenerating Costa Rican dry forest. *International Journal of Primatology*, **19**, 405–432.

Finegan, B. & Delgado, D. (2000). Structural and floristic heterogeneity in a 30-year-old Costa Rican rainforest restored on pasture through natural secondary succession. *Restoration Ecology*, **8**, 380–391.

Frankie, G. W., Vinson, S. B., Rizzardi, M. A., Griswold, T. L., O'Keefe, S. & Snelling, R. R. (1998). Diversity and abundance of bees visiting a mass flowering tree species in disturbed seasonal dry forest, Costa Rica. *Journal of the Kansas Entomological Society*, **70**, 281–296.

Gerhardt, K. (1993). Tree seedling development in tropical dry abandoned pasture and secondary forest in Costa Rica. *Journal of Vegetation Science*, **4**, 95–102.

Gerhardt, K. (1996). Germination and development of sown mahogany (*Swietenia macrophylla* King) in secondary tropical dry forest habitats in Costa Rica. *Journal of Tropical Ecology*, **12**, 275–289.

Gerhardt, K. & Fredriksson, D. (1995). Biomass allocation by broad-leaf mahogany seedlings, *Swietenia macrophylla* (King), in abandoned pasture and secondary dry forest in Guanacaste, Costa Rica. *Biotropica*, **27**, 174–182.

Ginsberg, J. R. (1998). Perspectives on wildfire in the humid tropics. *Conservation Biology*, **12**, 942–943.

Hallwachs, W. (1986). Agoutis (*Dasyprocta punctata*): the inheritors of guapinol (*Hymenaea courbaril*: Leguminosae). In *Frugivores and Seed Dispersal*, eds. A. Estrada & T. Fleming, pp. 285–304. Dordrecht, Dr W. Junk.

Harvey, C. A. & Haber, W. A. (1999). Remnant trees and the conservation of biodiversity in Costa Rican pastures. *Agroforestry Systems*, **44**, 37–68.

Hoffmann, W. A. (2000). Post-establishment seedling success in the Brazilian Cerrado: a comparison of savanna and forest species. *Biotropica*, **32**, 62–69.

Holl, K. D. (1999). Tropical forest recovery and restoration. *TREE*, **14**, 378–379.

Holl, K. D. (2000). Factors limiting tropical rainforest regeneration in abandoned pasture: seed rain, seed germination, microclimate, and soil. *Biotropica*, **31**, 229–242.

Holl, K. D., Loik, M. E., Lin, E. H. V. & Samuels, I. A. (2000). Tropical montane forest restoration in Costa Rica: overcoming barriers to dispersal and establishment. *Restoration Ecology*, **8**, 339–349.

Houlahan, J. E., Findlay, C. S., Schmidt, B. R., Meyers, A. H. & Kuzmin, S. L. (2000). Quantitative evidence for global amphibian population declines. *Nature*, **404**, 752–755.

Hunt, J. H., Brodie, R. J., Carithers, T. P., Goldstein, P. Z. & Janzen, D. H. (1999). Dry season migration by Costa Rican lowland paper wasps to high elevation cold dormancy sites. *Biotropica*, **31**, 192–196.

Janos, D. P. (1996). Mycorrhizas, succession, and the rehabilitation of deforested lands in the humid tropics. In *Fungi and Environmental Change*, eds. J. C. Frankland, N. Magan & G. M. Gadd, pp. 129–162. Cambridge: Cambridge University Press.

Janzen, D. H. (1967). Why mountain passes are higher in the tropics. *American Naturalist*, **101**, 233–249.

Janzen, D. H. (1980). Specificity of seed-attacking beetles in a Costa Rican deciduous forest. *Journal of Ecology*, **68**, 929–952.

Janzen, D. H. (1981a). *Enterolobium cyclocarpum* seed passage rate and survival in horses, Costa Rican Pleistocene seed dispersal agents. *Ecology*, **62**, 593–601.

Janzen, D. H. (1981b). Guanacaste tree seed-swallowing by Costa Rican range horses. *Ecology*, **62**, 587–592.

Janzen, D. H. (1982a). Removal of seeds from horse dung by tropical rodents: influence of habitat and amount of dung. *Ecology*, **63**, 1887–1900.

Janzen, D. H. (1982b). Cenízero tree (Leguminosae: *Pithecellobium saman*) delayed fruit development in Costa Rican deciduous forests. *American Journal of Botany*, **69**, 1269–1276.

Janzen, D. H. (1982c). How and why horses open *Crescentia alata* fruits. *Biotropica*, **14**, 149–152.

Janzen, D. H. (1983a). No park is an island: increase in interference from outside as park size decreases. *Oikos*, **41**, 402–410.

Janzen, D. H. (1983b). Dispersal of seeds by vertebrate guts. In *Coevolution*, eds. D. J. Futuyma & M. Slatkin, pp. 232–262. Sunderland, MA: Sinauer Associates.

Janzen, D. H. (ed). (1983c). *Costa Rican Natural History*. Chicago, IL: University of Chicago Press.

Janzen, D. H. (1984). Dispersal of small seeds by big herbivores: foliage is the fruit. *American Naturalist*, **123**, 338–353.

Janzen, D. H. (1985a). *Spondias mombin* is culturally deprived in megafauna-free forest. *Journal of Tropical Ecology*, **1**, 131–155.

Janzen, D. H. (1985b). A host plant is more than its chemistry. *Illinois Natural History Bulletin*, **33**, 141–174.

Janzen, D. H. (1985c). On ecological fitting. *Oikos*, **45**, 308–310.

Janzen, D. H. (1986a). *Guanacaste National Park: Tropical Ecological and Cultural Restoration*. San José, Costa Rica: Editorial Universidad Estatal a Distancia.

Janzen, D. H. (1986b). The eternal external threat. In *Conservation Biology: The Science of Scarcity and Diversity*, ed. M. E. Soulé, pp. 286–303. Sunderland, MA: Sinauer Associates.

Janzen, D. H. (1986c). Mice, big mammals, and seeds: it matters who defecates what where. In *Frugivores and Seed Dispersal*, eds. A. Estrada & T. H. Fleming, pp. 251–271. Dordrecht: Dr W. Junk.

Janzen, D. H. (1986d). Blurry catastrophes. *Oikos*, **47**, 1–2.

Janzen, D. H. (1986e). Lost plants. *Oikos*, **46**, 129–131.

Janzen, D. H. (1986f). Biogeography of an unexceptional place: what determines the saturniid and sphingid moth fauna of Santa Rosa National Park, Costa Rica, and what does it mean to conservation biology? *Brenesia*, **25/26**, 51–87.

Janzen, D. H. (1987a). Forest restoration in Costa Rica. *Science*, **235**, 15.

Janzen, D. H. (1987b). How to grow a tropical national park: basic philosophy for Guanacaste National Park, northwestern Costa Rica. *Experientia*, **43**, 1037–1038.

Janzen, D. H. (1987c). When, and when not to leave. *Oikos*, **49**, 241–243.

Janzen, D. H. (1988a). Guanacaste National Park: tropical ecological and biocultural restoration. In *Rehabilitating Damaged Ecosystems*, vol. 2, ed. J. J. Cairns Jr, pp. 143–192. Boca Raton, FL: CRC Press.

Janzen, D. H. (1988b). Tropical dry forests: the most endangered major tropical ecosystem. In *Biodiversity*, ed. E. O. Wilson, pp. 130–137. Washington, DC: National Academy Press.

Janzen, D. H. (1988c). Management of habitat fragments in a tropical dry forest: growth. *Annals of the Missouri Botanical Garden*, **75**, 105–116.

Janzen, D. H. (1988d). Complexity is in the eye of the beholder. In *Tropical Rainforests: Diversity and Conservation*, eds. F. Almeda & C. M. Pringle, pp. 29–51,

San Francisco, CA: California Academy of Science and AAAS.

Janzen, D. H. (1988e). Tropical ecological and biocultural restoration. *Science*, **239**, 243–244.

Janzen, D. H. (1988f). Ecological characterisation of a Costa Rican dry forest caterpillar fauna. *Biotropica*, **20**, 120–135.

Janzen, D. H. (1988g). The migrant moths of Guanacaste. *Orion Nature Quarterly*, **7**, 38–41.

Janzen, D. H. (1989). Natural history of a wind-pollinated Central American dry forest legume tree (*Ateleia herbert-smithii* Pittier). *Monographs in Systematic Botany from the Missouri Botanical Garden*, **29**, 293–376.

Janzen, D. H. (1990). An abandoned field is not a tree fall gap. *Vida Silvestre Neotropical*, **2**, 64–67.

Janzen, D. H. (1992a). The neotropics: a broad look at prospects for restoration in Central and South America raises some basic questions about methods, about goals, and about the restorationist's role in evolution. *Restoration and Management Notes*, **10**, 8–13.

Janzen, D. H. (1992b). A dry tropical forest ecosystem restored. *Earth Summit Times, New York Daily News*, Issue 28, 1 April 1992, p. 1.

Janzen, D. H. (1992c). Tropical reforestation: the Guanacaste Project. In *Proceedings of the First Pan American Furniture Manufacturers' Symposium on Tropical Hardwoods*, 1991, ed. R. Sullivan, pp. 25–36. Grand Rapids, MI: Center for Environmental Study.

Janzen, D. H. (1993). Caterpillar seasonality in a Costa Rican dry forest. In *Caterpillars: Ecological and Evolutionary Constraints on Foraging*, eds. N. E. Stamp & T. M. Casey, pp. 448–477. New York: Chapman & Hall.

Janzen, D. H. (1995). Neotropical restoration biology. *Vida Silvestre Neotropical*, **4**, 3–9.

Janzen, D. H. (1996). Prioritisation of major groups of taxa for the All Taxa Biodiversity Inventory (ATBI) of the Guanacaste Conservation Area in northwestern Costa Rica, a biodiversity development project. *ASC Newsletter*, **26**, 45, 49–56.

Janzen, D. H. (1998a). Gardenification of wildland nature and the human footprint. *Science*, **279**, 1312–1313.

Janzen, D. H. (1998b). How to grow a wildland: the gardenification of nature. *Insect Science and Application*, **17**, 269–276.

Janzen, D. H. (1998c). *Conservation Analysis of the Santa Elena Property, Peninsula Santa Elena, Northwestern Costa*

Rica, Report to the Government of Costa Rica, Area de Conservacion Guanacaste. Liberia, Costa Rica: ACG.

Janzen, D. H. (1999*a*). Gardenification of tropical conserved wildlands: multitasking, multicropping, and multiusers. *Proceedings of the National Academy of Sciences, USA*, **96**, 5987–5994.

Janzen, D. H. (1999*b*). La sobrevivencia de las areas silvestres de Costa Rica por medio de su jardinificación. *Ciencias Ambientales*, **16**, 8–18.

Janzen, D. H. (2000*a*). Costa Rica's Area de Conservación Guanacaste: a long march to survival through non-damaging biodiversity and ecosystem development. In *Norway/UN Conference on the Ecosystem Approach for Sustainable Use of Biological Diversity*, pp. 122–132. Trondheim, Norway: Norwegian Directorate for Nature Research and Norwegian Institute for Nature Research.

Janzen, D. H. (2000*b*). How to grow a wildland: the gardenification of nature. In *Nature and Human Society*, eds. P. H. Raven & T. Williams, pp. 521–529. Washington, DC: National Academy Press.

Janzen, D. H. (2000*c*). Costa Rica's Area de Conservación Guanacaste: a long march to survival through non-damaging biodevelopment. *Biodiversity*, **1**, 7–20.

Janzen, D. H. (2000*d*). Wildlands as gardens. *National Parks Magazine*, **74**, 50–51.

Janzen, D. H. (2001). Latent extinctions: the living dead. In *Encyclopedia of Biodiversity*, vol. 3, ed. S. A. Levin, pp. 689–699. San Diego, CA: Academic Press.

Janzen, D. H. (2002). Ecology of dry forest wildland insects in the Area de Conservación Guanacaste, northwestern Costa Rica. In *Biodiversity Conservation in Costa Rica: Learning the Lesson in Seasonal Dry Forest*, eds. G. Frankie & B. Vinson. Berkeley, CA: University of California Press.

Janzen, D. H. & Hallwachs, W. (1992). La restauración de la biodiversidad tropical: experiencias del Area de Conservación Guanacaste y posibles aplicaciones en México. In *México ante los retos de la biodiversidad*, eds. J. Sarukhan, R. Dirzo, & compilers, pp. 243–250. Mexico City: Comisión Nacional para el Conocimiento y Uso de la Biodiversidad.

Janzen, D. H. & Martin, P. S. (1982). Neotropical anachronisms: the fruits the gomphotheres ate. *Science*, **215**, 19–27.

Janzen, D. H. & Vázquez-Yanes, C. (1991). Aspects of tropical seed ecology of relevance to management of tropical forest wildlands. In *Rainforest Regeneration and Management*, eds. A. Gomez-Pompa, T. C. Whitmore & M. Hadley, pp. 137–157. Paris: UNESCO.

Janzen, D. H., Miller, G. A., Hackforth-Jones, J., Pond, C. M., Hooper, K. & Janos, D. P. (1976). Two Costa Rican bat-generated seed shadows of *Andira inermis* (Leguminosae). *Ecology*, **57**, 1068–1075.

Janzen, D. H., Demment, M. W. & Robertson, J. B. (1985). How fast and why do germinating guanacaste seeds (*Enterolobium cyclocarpum*) die inside cows and horses? *Biotropica*, **17**, 322–325.

Janzen, D. H., Fellows, L. E. & Waterman, P. G. (1990). What protects *Lonchocarpus* (Leguminosae) seeds in a Costa Rican dry forest? *Biotropica*, **22**, 272–285.

Kellman, M., Tackaberry, R. & Rigg, L. (1998). Structure and function in two tropical gallery forest communities: implications for forest conservation in fragmented systems. *Journal of Applied Ecology*, **35**, 195–206.

Killeen, T. J., Jardim, A., Mamani, F. & Rojas, N. (1998). Diversity, composition and structure of a tropical semideciduous forest in the Chiquitanía region of Santa Cruz, Bolivia. *Journal of Tropical Ecology*, **14**, 803–827.

Larson, D. W., Matthes, U., Gerrath, J. A., Larson, N. W. K., Gerrath, J. M., Nekola, J. C., Walker, G. L., Porembski, S. & Charlton, A. (2000). Evidence for the widespread occurrence of ancient forests on cliffs. *Journal of Biogeography*, **27**, 319–331.

Livernash, R. (1998). The biodiversity agenda: beyond central government. In *Valuing the Global Environment: Actions and Investments for a 21st Century*, ed. R. Livernash, pp. 87–88. Washington, DC: Global Environmental Facility.

Nepstad, D. C., Uhl, C., Pereira, C. A. & Cardoso da Silva, J. M. (1996). A comparative study of tree establishment in abandoned pasture and mature forest of Eastern Amazonia. *Oikos*, **76**, 25–39.

Parrotta, J. A. & Turnbull, J. W. (eds.) (1997). Catalysing native forest regeneration on degraded tropical lands. *Forest Ecology and Management*, **99**, 1–290.

Parsons, J. J. (1972). The spread of African pasture grasses into the New World tropics. *Journal of Range Management*, **20**, 13–17.

Pennington, R. T., Prado, D. E. & Pendry, C. A. (2000). Neotropical seasonally dry forests and Quarternary vegetation changes. *Journal of Biogeography*, **27**, 261–273.

Pounds, J. A., Fogden, M. P. L. & Campbell, J. H. (1999). Biological response to climate change on a tropical mountain. *Nature*, **398**, 611–615.

Riddle, B. R., Hafner, D. J., Alexander, L. F. & Jaeger, J. R. (2000). Cryptic vicariance in the historical assembly of a Baja California peninsular desert biota. *Proceedings of the National Academy of Sciences, USA*, **97**, 14438–14443.

Silver, W. L., Ostertag, R. & Lugo, A. E. (2000). The potential for carbon sequestration through reforestation of abandoned tropical agricultural and pasture lands. *Restoration Ecology*, **8**, 394–403.

Slocum, M. G. (2000). Logs and fern patches as recruitment sites in a tropical pasture. *Restoration Ecology*, **8**, 408–413.

Slocum, M. G. & Horvitz, C. C. (2000). Seed arrival under different genera of trees in a neotropical pasture. *Plant Ecology*, **149**, 51–62.

Standley, L. J. & Sweeney, B. W. (1995). Organochlorine pesticides in stream mayflies and terrestrial vegetation of undisturbed tropical catchments exposed to long-range atmospheric transport. *Journal of the North American Benthological Society*, **14**, 38–49.

Toh, I., Gillespie, M. & Lamb, D. (1999). The role of isolated trees in facilitating tree seedling recruitment at a degraded sub-tropical rainforest site. *Restoration Ecology*, **7**, 288–297.

Vieira, E. M. & Marinho-Filho, J. (1998). Pre- and post-fire habitat utilisation by rodents of Cerrado from Central Brazil. *Biotropica*, **30**, 491–496.

Uhl, C. & Buschbacher, R. (1985). A disturbing synergism between cattle ranch burning practices and selective tree harvesting in the eastern Amazon. *Biotropica*, **17**, 265–268.

Wijdeven, S. M. J. & Kuzee, M. E. (2000). Seed availability as a limiting factor in forest recovery processes in Costa Rica. *Restoration Ecology*, **8**, 414–424.

Williams-Linera, G., Dominguez-Gastelu, V. & Garcia-Zurita, M. E. (1998). Microenvironment and floristics of different edges in a fragmented tropical rainforest. *Conservation Biology*, **12**, 1091–1102.

Zahawi, R. A. & Augspurger, C. K. (1999). Early plant succession in abandoned pastures in Ecuador. *Biotropica*, **31**, 540–552.

Index

Page numbers in italics refer to information in a box, figure or table